Trigonometric Identities

$\sin^2 A = 0.5[1 - \cos 2A]$

$\cos^2 A = 0.5[1 + \cos 2A]$

$\sin^2 A + \cos^2 A = 1$

$\sin(A \pm B) = \sin A \cos B \pm \cos A \sin B$

$\cos(A \pm B) = \cos A \cos B \mp \sin A \sin B$

$\sin A \sin B = 0.5[\cos(A - B) - \cos(A + B)]$

$\cos A \cos B = 0.5[\cos(A + B) + \cos(A - B)]$

$\sin A \cos B = 0.5[\sin(A + B) + \sin(A - B)]$

$e^{j\theta} = \cos\theta + j\sin\theta$

$$\cos\theta = \frac{e^{j\theta} + e^{-j\theta}}{2}$$

$$\sin\theta = \frac{e^{j\theta} - e^{-j\theta}}{2j}$$

Communication Systems

Analysis and Design

COMMUNICATION SYSTEMS
Analysis and Design

Harold P. E. Stern
University of Alabama, Tuscaloosa
Samy A. Mahmoud
Carleton University, Ottawa
Lee Elliott Stern
Graphic Developer

Upper Saddle River, NJ 07458

Library of Congress Cataloging-in-Publication Data

Stern, Harold P. E
Communication systems: analysis and design / Harold P. E. Stern, Samy A. Mahmoud.
p. cm
ISBN 0-13-040268-0
1. Telecommunication systems--Design and construction. I. Mahmoud, Samy A. II. Title

TK5103.S72 2004
621.382--dc22

2003062300

Vice President and Editorial Director, ECS: *Marcia J. Horton*
Publisher: *Tom Robbins*
Associate Editor: *Alice Dworkin*
Vice President and Director of Production and Manufacturing, ESM: *David W. Riccardi*
Executive Managing Editor: *Vince O'Brien*
Managing Editor: *David A. George*
Production Editor: *Kevin Bradley*
Director of Creative Services: *Paul Belfanti*
Art Director: *Jayne Conte*
Cover Designer: *Bruce Kenselaar*
Art Editor: *Greg Dulles*
Manufacturing Manager: *Trudy Pisciotti*
Manufacturing Buyer: *Lynda Castillo*
Marketing Manager: *Holly Stark*

About the Cover: John Lund/Corbis.

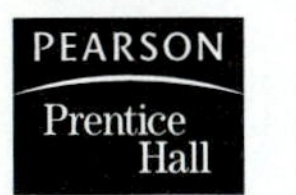

Pearson Prentice Hall
Pearson Education, Inc.
Upper Saddle River, NJ 07458

Printed in the United States of America

10 9 8 7 6 5 4 3 2 1

ISBN 0-13-040268-0

Pearson Education Ltd., *London*
Pearson Education Australia Pty. Ltd., *Sydney*
Pearson Education Singapore, Pte. Ltd.
Pearson Education North Asia Ltd., *Hong Kong*
Pearson Education Canada, Inc., *Toronto*
Pearson Educación de Mexico, S.A. de C.V.
Pearson Education—Japan, *Tokyo*
Pearson Education Malaysia, Pte. Ltd.
Pearson Education, Inc., *Upper Saddle River, New Jersey*

To Lee—my soulmate and the love of my life;
to Mom—for your unconditional love and support;
and to Dylan—my son, you bring me so much pride.

HPES

To my wife Maha and my son Omar, for their unwavering support
and encouragement, and for helping me to appreciate better every
day the value of keeping the communication channels open.

SAM

Contents

Preface xiii
Acknowledgments xvii

Chapter 1 Introduction 1

1.1 Components of a Communication System 1
1.2 An Overview of Trade-Offs in Communication System Design 4
Problems 5

Chapter 2 Frequency Domain Analysis 6

2.1 Why Study Frequency Domain Analysis? 6
2.2 The Fourier Series 13
2.2.1 Trigonometric Form of the Fourier Series 13
2.2.2 Other Forms of the Fourier Series 18
2.3 Representing Power in the Frequency Domain 38
2.3.1 The One-Sided Average Normalized Power Spectrum 41
2.3.2 Formally Defining the Term "Bandwidth" 48
2.3.3 The Two-Sided Average Normalized Power Spectrum 48
2.4 The Fourier Transform 52
2.5 Normalized Energy Spectral Density 60
2.6 Properties of the Fourier Transform 65
2.7 Using the Unit Impulse Function to Represent Discrete Frequency Components as Densities 68
Problems 69

Chapter 3 Digital Baseband Modulation Techniques 75

3.1 Goals in Communication System Design 75
3.2 Baseband Modulation Using Rectangular Pulses and Binary Pulse Amplitude Modulation 76

3.3 Pulse Shaping to Improve Spectral Efficiency 89
3.3.1 The Sinc-Shaped Pulse 89
3.3.2 The Raised Cosine Pulse (Damped Sinc-Shaped Pulse) 101
3.4 Building a Baseband Transmitter 111
Problems 115

Chapter 4 Receiver Design (and Stochastic Mathematics, Part I) 123

4.1 Developing a Simple Pulse Amplitude Modulation Receiver 123
4.1.1 Establishing an Expression for Probability of Bit Error 124
4.1.2 Stochastic Mathematics—Part I (Random Variables) 127
4.1.3 Examining Thermal Noise 132
4.1.4 The Gaussian Probability Density Function 137
4.1.5 Simplifying the Expression for Probability of Bit Error 144
4.2 Building the Optimal Receiver (The Matched Filter or Correlation Receiver) 149
4.2.1 Basic Structure for the Optimal Receiver 149
4.2.2 Implications of Employing Optimum Processing 153
4.2.3 A Graphical Interpretation of Probability of Bit Error for the Optimal Receiver 156
4.2.4 Designing the Correlation Receiver for More General Signals 162
4.3 Synchronization 172
4.3.1 Basic Structure of Continuous-Time Phase Locked Loops 173
4.3.2 Analysis of the PLL with Linearized Dynamics 173
4.3.3 Frequency Synthesizers 178
4.3.4 Timing Recovery 181
4.3.5 Further Reading on Synchronization 189
4.4 Equalization 190
4.4.1 Intersymbol Interference 192
4.4.2 Linear Transversal Equalizers 196
4.4.3 Least-Mean-Square Equalizers 198
4.4.4 Other Types of Equalizers 199
4.4.5 Further Reading on Equalization 199
4.5 Multi-Level PAM (M-ary PAM) 200
Problems 207

Chapter 5 Digital Bandpass Modulation and Demodulation Techniques (and Stochastic Mathematics, Part II) 211

5.1 Binary Amplitude Shift Keying 212
5.2 Other Binary Bandpass Modulation Techniques 219
5.2.1 Binary Frequency Shift Keying 219
5.2.2 Binary Phase Shift Keying 221
5.2.3 Calculating Average Normalized Power Spectral Density for Binary FSK and Binary PSK 222

5.3 Coherent Demodulation of Bandpass Signals 225
5.3.1 Developing a Coherent PSK Receiver 227
5.3.2 Developing a Coherent ASK Receiver 230
5.3.3 Developing a Coherent FSK Receiver 232
5.3.4 Comparing Coherent PSK, FSK, and ASK 235
5.4 Stochastic Mathematics—Part II (Random Processes) 236
5.4.1 Random Processes 238
5.4.2 The Wiener–Khintchine Theorem 245
5.4.3 Ergodicity 252
5.5 Noncoherent Receivers for ASK and FSK 253
5.5.1 The Envelope Detector 254
5.5.2 Noncoherent Demodulation of ASK 255
5.5.3 Noncoherent Demodulation of FSK 255
5.5.4 Performance of Noncoherent ASK and FSK Receivers 256
5.6 Differential (Noncoherent) PSK 267
5.6.1 Demodulation of Binary DPSK 268
5.6.2 Probability of Bit Error for a DPSK Receiver 270
5.7 A Comparison of Binary Bandpass Systems 271
5.8 M-ary Bandpass Techniques 274
5.8.1 Quaternary Phase Shift Keying 274
5.8.2 Differential Quaternary Phase Shift Keying 284
5.8.3 M-ary Phase Shift Keying 286
5.8.4 M-ary Frequency Shift Keying 292
5.8.5 Multiparameter M-ary Bandpass Signaling 298
Problems 301

Chapter 6 Analog Modulation and Demodulation 306

6.1 Transmitting an Amplitude Modulated Signal 306
6.2 Coherent Demodulation of AM Signals 309
6.3 Noncoherent Demodulation of AM Signals 315
6.4 Single Sideband and Vestigial Sideband AM systems 326
6.5 Frequency Modulation and Phase Modulation 334
6.6 Generating and Demodulating FM and PM Signals 343
6.6.1 FM and PM Modulators 343
6.6.2 FM and PM Demodulators 345
6.6.3 Noise in FM and PM Systems 347
6.7 A Comparison of Analog Modulation Techniques 355
Problems 357

Chaper 7 Multiplexing Techniques 362

7.1 Time Division Multiplexing 364
7.2 Frequency Division Multiplexing 368
7.3 Code Division Multiplexing and Spread Spectram 370
7.3.1 Direct Sequence Spread Spectrum 371
7.3.2 Frequency-Hopping Spread Spectrum 381
Problems 385

Chapter 8 Analog-to-Digital and Digital-to-Analog Conversion 388

8.1 Sampling and Quantizing 390
8.1.1 Sampling Baseband Analog Signals 392
8.1.2 Practical Considerations in Baseband Sampling 397
8.1.3 Sampling Bandpass Analog Signals 399
8.1.4 The Quantizing Process 400
8.2 Differential Pulse Coded Modulation 407
8.3 Delta Modulation and Continuously Variable Slope Delta Modulation 411
8.3.1 Delta Modulation 411
8.3.2 Continuously Variable Slope Delta Modulation 415
8.4 Further Reading on Analog-to-Digital and Digital-to-Analog Conversion 417
Problems 417

Chapter 9 Fundamentals of Information Theory, Data Compression, and Image Compression 421

9.1 Information Content, Entropy, and Information Rate of Independent Sources 421
9.2 Variable-Length, Self-Punctuating Coding for Data Compression 424
9.2.1 Prefix Coding and the Tree Diagram 426
9.2.2 Huffman Coding 431
9.3 Sources with Dependent Messages 439
9.3.1 Static Dictionary Encoding 439
9.3.2 LZW Compression—an Example of Dynamic Dictionary Encoding 442
9.4 Still-Image Compression 444
9.4.1 Facsimile 444
9.4.2 Monochromatic Gray Scale Images 447
9.4.3 Color Images (the DCT and JPEG) 447
9.5 Moving-Image Compression 450
Problems 454

Chapter 10 Basics of Error Control Coding 458

10.1 Channel Capacity 460
10.2 Field Theory and Modulo-2 Operators 461
10.2.1 Galois Field of Order 2 461
10.2.2 Matrix Representation and Manipulation 463
10.3 Hamming Codes 464
10.4 A Geometric Interpretation of Error Control Coding 472
10.5 Cyclic Codes 481
10.5.1 Cyclic Redundancy Check Codes 485
10.5.2 Bose Chaudhuri Hocquenghem Codes 487

10.6 Hybrid FEC/ARQ Codes 488
10.7 Correcting Burst Errors 489
10.7.1 Interleaving 489
10.7.2 Reed–Solomon Codes 490
10.8 Convolutional Codes and Viterbi Decoding 490
10.8.1 Convolutional Encoding 490
10.8.2 Creating a Trellis 496
10.8.3 Decoding and the Viterbi Algorithm 497
Problems 506

References 509

Answers to Selected Problems 515

Index 519

Preface

Communication systems analysis and design is not only complex and exciting, but also often daunting. We're very pleased to offer a new approach to learning the subject: We present the material mathematically, graphically, and intuitively. In our combined 43 years of classroom instruction, we've found that any dedicated engineering student can learn the subject, but that everyone learns differently. *Communication Systems: Analysis and Design* maintains mathematic rigor while reinforcing concepts through extensive graphic illustration and by bolstering a student's intuitive grasp of the subject.

Combining mathematical, graphical, and intuitive techniques allows students to learn using the technique they understand best, then reinforces their learning using the other two techniques. This strategy helps to provide a more complete understanding of the material and strengthens a student's abilities in all three learning techniques. Homework problems in each chapter emphasize concepts as well as standard quantitative analysis skills.

Communication Systems: Analysis and Design emphasizes what we feel is the heart of communication system design: performance-versus-cost trade-offs in the context of the parameters important for communications systems: transmission speed, accuracy, reliability, equipment complexity (hardware and software), bandwidth, and power. Trade-offs, emphasized in examples, are used to motivate students to understand all the different approaches and techniques involved in communication system design.

By the time students are ready to take a course in communication systems (typically in their junior or senior year), they are generally quite proficient at mathematical manipulation, but many do not have a solid understanding of the concepts behind the mathematics or an understanding of when or why particular mathematical techniques are used. This is also true among many first-year graduate students. Of course, we want students to be mathematically strong (there is no substitute for rigorous mathematics), but we also need them to develop a solid understanding of the concepts behind the math and an intuitive feel for engineering.

Frequency domain analysis and stochastic processes are the two main hurdles in a typical communications course, where many students give up their urge to understand

and succumb to a "plug in the math and grind" approach. To avoid this mindset, we devote considerable time to discussing frequency domain analysis (one entire chapter), even though the typical student has already been exposed to most of this material in a prerequisite signals and systems course. By dedicating Chapter 2 to the fundamentals of frequency domain analysis, we hope to impart more insight and understanding of this important subject by capitalizing on the fact that most students have had some previous exposure to it. We first provide extensive practical applications and convincing reasons for mastering this material, then we provide graphical and intuitive teaching approaches to reinforce the mathematics.

Stochastic processes is taught in two parts: First, random variables are introduced to deal with the effects of noise (Chapter 4); then random processes are introduced in Chapter 5 to deal with the uncertainty associated with the transmitted message itself. In each case, the mathematics are coupled with practical applications to reinforce the concepts. Separating random variables and random processes allows us to build insight gradually in the difficult subject of stochastic mathematics. The simpler concept of random variables is introduced first and its applications to communications systems analysis are discussed at length before introducing the more abstract concept of random processes.

The basics of multiplexing, information theory, data compression, and error control coding are provided in later chapters, so that students can realize that "transmitting smarter" is as important to communications systems engineering as "transmitting faster." Chapter 8 provides the basics of analog-to-digital conversion. Information theory, data compression, and error control coding cannot be discussed in depth in an introductory communications textbook (indeed, entire textbooks have been written about each subject), but enough material has been included to help students understand why each topic is important, as well as to prepare them to perform in-depth research on each topic, either on their own or in more advanced courses.

The included CD-ROM contains PowerPoint®-based animated presentations designed to reinforce certain examples within the book. The chosen examples are labeled in the text with a mouse symbol . The CD-ROM also contains pdf files (Adobe Acrobat®) with full-color versions of selected figures from the book. The chosen figures are mostly plots with multiple curves, where color differentiates the curves at a glance, so students can concentrate on the meaning of the illustration, rather than unraveling dotted and dashed lines. The chosen figures are labeled in the text with a CD-ROM symbol .

This book has been written to accommodate many different paths through the material (see the flow chart on page xv). A sequential approach through the chapters provides a discussion of digital systems first, then analog systems, then an introduction to the more advanced topics. If the instructor desires, analog systems can be studied first by assigning Chapter 6 immediately after Chapter 2. If such an approach is used, analog-to-digital conversion (Chapter 8) can be assigned immediately after Chapter 6 to motivate the subsequent study of digital systems (Chapters 3–5). The material in Chapter 6 has been written to accommodate an "analog first" approach by providing direct references to information on stochastic processes covered in Chapters 4 and 5. The stochastic processes material has been written so that it can be referenced directly from Chapter 6 without the need to read the digital material in Chapters 3–5 first. The accompanying flow chart shows suggested paths through the textbook.

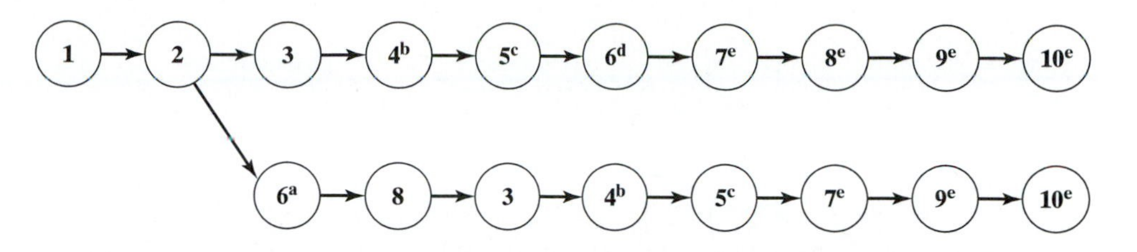

Suggested Paths Through *Communication Systems: Analysis and Design*

[a]Portions of Chapter 4 (Gaussian noise and random variables) and Chapter 5 (representation of narrowband noise) will need to be studied alongside Chapter 6. These portions are referenced within Chapter 6 and have been written to allow easy incorporation into the Chapter 6 material.

[b]Portions concerning synchronization, equalization, and M-ary baseband techniques may each be selected for coverage or skipped as desired—these portions are at the end of the chapter.

[c]Portions concerning various M-ary bandpass modulation techniques may be selected for coverage or skipped as desired—these portions are at the end of the chapter.

[d]For a purely digital course, Chapter 6 can be skipped, but basic AM principles will need brief coverage when FDM is studied in Chapter 7.

[e]Order of study for Chapters 7–10 is independent. Note, however, that there are a few concepts in Chapter 10 (information content and entropy) that will require brief references back to the first section of Chapter 9.

The conversational tone we have used throughout this book makes students feel comfortable with the material while they learn about this complex subject. We've learned where to anticipate questions from the fastest learners. We've also learned when to encourage the rest of the class in order to let them know that everything will soon fall into place.

HAROLD P. E. STERN
SAMY A. MAHMOUD

Acknowledgments

We are extremely grateful for all the help we received in writing this book. Our thanks especially to Lee E. Stern, graphic developer, artist, and a leading creative force in the development of this book. Lee's graphics were often our first attempt to communicate a particular engineering concept. Her innovations and designs determined the approach, and in many cases, the specific language used in the book. The innovations that this book offers, both in terms of engineering (i.e., determining how engineers approach specific tasks and what tools they apply) and in terms of education, are due as much to Lee as they are due to us.

We also thank an extraordinary group of people at Pearson Education/Prentice Hall. Alice Dworkin has been outstandingly patient, wise, and resourceful, providing expert assistance and guidance through the various steps involved in writing and producing this book. Tom Robbins and David George have provided outstanding guidance and editorial counsel. We are deeply indebted to Kevin Bradley at Sunflower Publishing Services for his skill in guiding the book through production, and to Paula Grant, our copy editor, for her exceptional work. Kevin has shown great patience with novice authors, and Paula has provided clarity and consistency throughout the text (no small feat, considering our writing skills) while allowing us to maintain our own styles and voices. We also thank the skilled personnel at Laserwords, especially their graphic artists.

Thanks also to our reviewers: Dale Callahan, University of Alabama at Birmingham; Jon Bredeson, Texas Tech University; Derong Liu, University of Illinois at Chicago; Hong Man, Stevens Institute of Technology; and Christos Douligeris, University of Miami. Their many suggestions have greatly improved this book. Finally, thanks to our colleagues and key members of our respective institutes for their support and for providing an environment conducive to learning and research.

Harold Stern would like to thank his Mom (Myrna Longenecker), Robin Stern, Michael Pickett, Elizabeth Book, Charles Roth, Francis Bostick, and Roberto Garcia-Munoz, who, through their love, patience, and example, have all taught him how to teach. He also thanks his Dad (Morris Stern), Hal Sobol, Jack Fitzer, John McElroy, and Jack Mueller, who taught him how to learn. Thanks to his students through the years, especially Michael Casey, Jayoung Koo, Jody Adams, Jason Schock, Ick Don Lee,

Blair Fonville, Jeffrey Parker, Christopher and Michael Blaylock, and Jeff McClure, for their many valuable insights and suggestions. Thanks to his colleagues, especially Raghvendra Pandey, L.A. ("Pete") Morley, and Russell Pimmel for their support, advice, and encouragement. Thanks also to Rachel Falk of Prentice Hall for her encouragement in the initial stages of the project, and to Robert Wells, interim vice president for research at the University of Alabama, for his support. A final word of thanks, again, to Lee, for her inspiring encouragement, love, and support.

Samy Mahmoud would like to acknowledge the contributions of Len MacEachern, Samy Ghoniemy, and Nagui Michael for many fruitful suggestions and for the experimental setup that generated the eye diagrams in Section 4.4. Special thanks go to David Falconer for providing the set of problems on equalization. He would also like to express his gratitude for the support of the Staff of the Office of the Dean of Engineering at Carleton University, Anne Waddell, Yvonne Clevers, and Jim Simpson. Their competence and efficiency made it possible for Professor Mahmoud to participate in this exciting project while serving as Faculty Dean.

Chapter 1

Introduction

1.1 Components of a Communication System

Welcome to the most exciting topic in electrical engineering today—communications. Communication systems impact our lives in many ways—through televisions, telephones, radios, CDs, fax machines, the Internet, DVDs, and automated teller machines. It is almost impossible to imagine our society without these systems and the services they provide. Communication system analysis and design is an important, dynamic, "leading edge" field in electrical engineering. We hope that you will enjoy learning how to analyze and design communication systems as much as we enjoy sharing our knowledge with you.

Perhaps the best way to begin is by defining the term "communication."[1] *Communication* is the transmission of information from a source to a user. This definition explains one term and introduces three others (*information, source*, and *user*). Before we define these terms, let's observe that our definition of communication suggests the block diagram of a system shown in Figure 1-1.

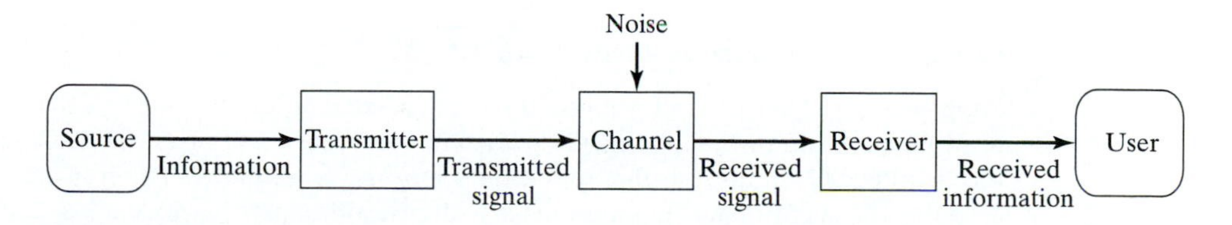

Figure 1-1 General block diagram of a communication system.

[1]The following approach is the work of Claude Shannon, whose ground-breaking efforts (see [1.1] in the References) were motivated by the challenge of breaking German spy codes during World War II. Shannon formalized the study of coding and data compression, a subject known as *information theory*.

Now, let's define the components and signals in the block diagram, including the terms "information," "source," and "user." The block diagram and the following definitions will be referred to throughout this book.

Information: Information is data that the user did not possess prior to communicating with the source. This definition is both important and deceptively simple. The key concept is that the data the user receives were *unknown* to the user before the source began communicating.

Source: The source is the originator of the information (a human voice, music, digital data from a computer).

Channel: The channel is the physical transmission medium over which the communication is sent (wires, air, fiber optics). All channels have physical limitations that distort, attenuate, and add noise to the transmitted signal. Thus, the received signal will not be an exact duplicate of the transmitted signal.

Transmitter: We could try to send the source output directly through the channel, but, generally speaking, the channel will add unacceptable amounts of noise and distortion to the signal. The transmitter *reformats* and *reshapes* the signal to reduce distortion from the channel and corruption from the noise. After reformatting and reshaping, the transmitter may also boost the power of the signal.

Receiver: The receiver attempts to translate the received signal back into the original information sent by the source. This involves basically two steps:

1. Compensating, as much as possible, for the noise and distortion added by the channel, and
2. "Undoing" the reformatting and reshaping performed by the transmitter.

User: The user is the recipient of the information.

The information from the source may be either an analog signal (such as voice or music) or a digital signal (such as the output from a computer). This book examines both systems that transmit analog signals and systems that transmit digital signals.

Consider the following example to illustrate the components and signals in a communication system.

Example 1.1 An automated teller machine[2]

Suppose I'm in Seattle and I want to use an automated teller machine (ATM) to withdraw money from my bank, which is located in San Francisco. I insert my bank card in the Seattle ATM and push the appropriate buttons to withdraw $50 from my bank account. The ATM then communicates my withdrawal request to my bank's computer in San Francisco.

[2]Some of the examples in this book have accompanying animated demonstrations. These examples are marked with a mouse symbol. The demonstrations, written in PowerPoint, are included in the CD-ROM provided with this book.

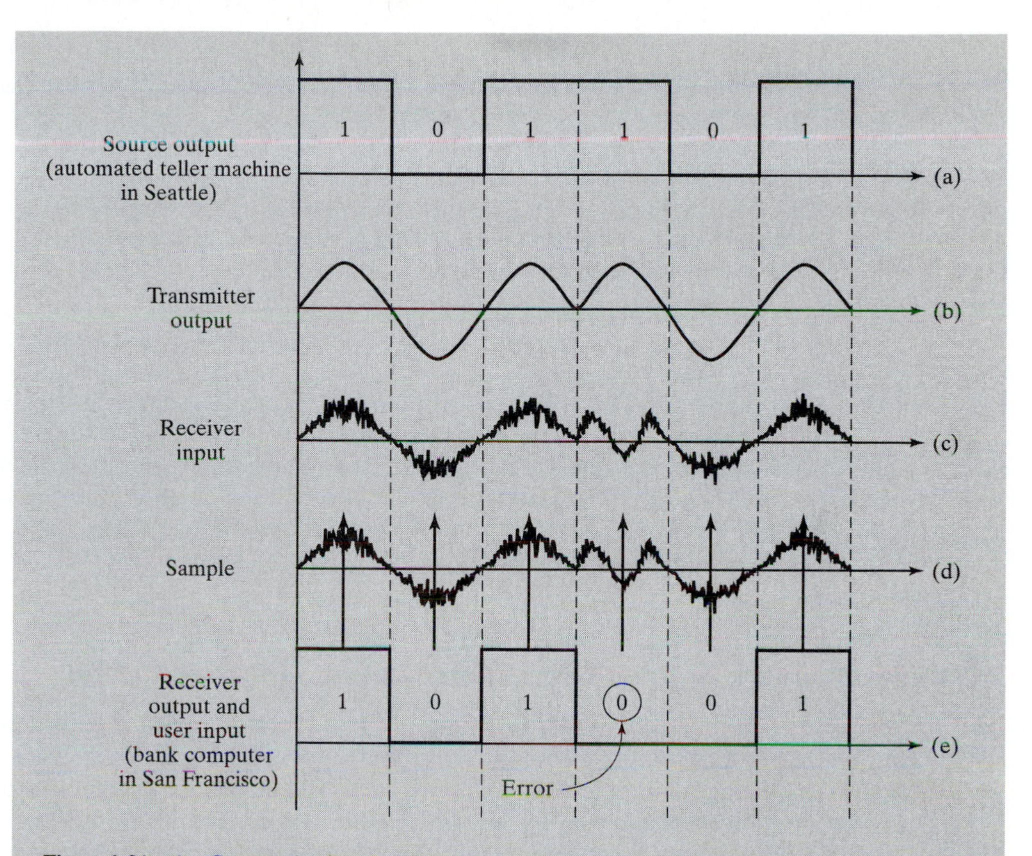

Figure 1-2(a–e) Communication signals in an automated teller machine transaction.

1. The source (the ATM) outputs digital information corresponding to the request—say that "101101" represents my request—using a high voltage for a "1" and a low voltage for a "0." This is illustrated in Figure 1-2a.
2. This information needs to be sent across a channel from Seattle to San Francisco. In this case, the channel will be a telephone line that passes only low-frequency signals, say between 300 and 3400 Hz.

   If we sent the source output signal directly through the telephone lines, the signal would be severely distorted. This happens because the source output signal has a dc component that the telephone line will not pass, as well as some high-frequency components (evidenced by the sharp corners of the signal) that the telephone line will filter out.

   The transmitter reshapes the source output signal into a waveform that passes through the channel with much less distortion. This reshaping is called *modulation* (which means "to change," even to non-engineers). The transmitter outputs a smooth, positive pulse when the source data is "1" and a smooth negative pulse when the data is "0." The transmitter output, shown in Figure 1-2b, passes through the channel with much less distortion and requires much less power to transmit than the original signal from the source.

3. The signal is then sent across the channel to the receiver, but the channel adds *noise*, corrupting the signal. It also attenuates the signal and adds some distortion. The signal at the receiver input might look like Figure 1-2c.
4. The receiver in San Francisco must translate the distorted and corrupted signal back into binary data to feed to the bank computer. This process is called *demodulation*. One way to do this is to sample the received signal at the center of each bit period, as shown in Figure 1-2d. If the sample is positive, the receiver outputs a "1"; if the sample is negative, the receiver outputs a "0". This is shown in Figure 1-2e.

Note that channel noise and distortion caused an error in the fourth received bit. How can we prevent such errors? Maybe there's a better pulse shape to transmit than the one that was chosen (in other words, maybe there's a better *modulation technique*). We will study modulation techniques in Chapter 3. Or, maybe there's a better way for the receiver to translate the received signal back into its binary form (a better *demodulation technique*). Possible improvements include increasing the number of samples the receiver takes per bit period or integrating the received signal to find out whether its total area during a bit period is positive or negative. We will study demodulation techniques in Chapter 4.

1.2 An Overview of Trade-Offs in Communication System Design

Example 1.1 is intuitively understandable, but more formal tools are needed to determine good modulation and demodulation techniques. Let's start by considering the factors involved. Communication system design—and indeed *all* engineering design—involves trade-offs of *performance versus cost*.

In communication systems, performance parameters include speed of transmission, accuracy of reception, and reliability of equipment. Cost parameters include complexity of equipment (types and number of parts, computational power required, amount of effort involved in development), channel bandwidth required by equipment, and power required to transmit the signal. All six of these performance and cost parameters are interrelated.

What do we need to design a communication system? Which components are under our control? Let's refer back to the block diagram in Figure 1-1 on page 1.

We can't design the source, nor do we want to put a lot of restrictions on how the source can behave. In fact, we want the system to be as transparent to the source as possible. Similarly, we can't design the user nor restrict how the user can behave.

We also don't have much control over the channel or its characteristics. Sometimes we can choose among different types of channels (for example, we could use a radio or a telephone line in the earlier ATM example), but sometimes we have no choice (for instance, we can't use wire telephone lines for a cellular phone).

Thus, our design effort will focus on building a transmitter and a receiver to optimize the performance versus cost trade-offs for our particular application. (Note that different applications require different trade-offs. The system that sends back pictures from the space shuttle, for instance, needs to be more reliable than the system that transmits MTV. Space shuttle systems are also less likely to be constrained, or chosen solely, by budget.

In summary, communication system design can be described as follows:

1. First, characterize the channel or choice of channels available for the communication system. Describe the distortion, attenuation, and noise that the channel will impose on the transmitted signal.
2. Second, design a transmitter and a receiver to modulate the information signal prior to transmission, and demodulate the received signal after reception. The purpose of the modulation and demodulation is to counteract, as much as possible, the effects of noise and distortion caused by the channel. The transmitter and receiver are designed, with knowledge of the channel's characteristics, to optimize the performance versus cost trade-offs for a particular application.

How do we characterize the channel? How do we optimize the trade-offs? We can use our intuition and experience to *qualitatively* characterize a channel (for example, we know that coaxial cable produces less distortion than telephone wires) or to *qualitatively* evaluate trade-offs (for example, we know that increasing the power of a transmitted signal will improve the accuracy of the received signal). Unfortunately, intuition alone is insufficient to *quantitatively* describe the channel or to *quantitatively* evaluate trade-offs (for example, if we double the power of the transmitted signal, how much do we improve accuracy?). Although intuition and qualitative analysis are important in guiding our design, quantitative analysis is necessary to ensure that we optimize performance versus cost trade-offs. To evaluate our trade-offs quantitatively, we need to develop some mathematical tools. The first group of these tools, frequency domain analysis, is discussed in Chapter 2.[3]

PROBLEMS

1.1 In one or two paragraphs, state why you want to study the analysis and design of communication systems.

1.2 What specific information would you like to learn in this class?

For Problems 1.3–1.5, provide enough explanation to justify each answer. Use your imagination.

1.3 Give an example of a communication system where *accuracy of reception* is the most important parameter.

1.4 Give an example of a communication system where *reliability* is the most important parameter.

1.5 Give an example of a communication system where *equipment simplicity* (lack of complexity in hardware and software) is the most important parameter.

[3]This discussion is not meant to belittle the importance of intuition. In fact, one mission of this book is to sharpen and improve your intuition. Throughout the text, whenever we perform a mathematical operation, prove a theorem, or obtain a result, we will ask, "In plain English, what does this mean?" and "Does this result make sense?"

Chapter 2

Frequency Domain Analysis

2.1 Why Study Frequency Domain Analysis?

As we discussed in Chapter 1, communication system design involves characterizing the channel and then designing the transmitter and receiver to optimize performance versus cost trade-offs (transmission speed, accuracy, reliability, complexity, bandwidth, and power) for a particular application. As we also mentioned, our intuition and experience provide good guidelines, but we need to develop some mathematical tools in order to quantitatively evaluate our trade-offs. In this chapter, we will develop some of those tools.

Our mathematical tools must provide us with the capability to analyze the electrical characteristics of the transmitter, channel, and receiver. We can perform these analyses in two ways:

1. Time domain analysis. In *time domain analysis*, we examine the amplitude-versus-time characteristics of the waveform.
2. Frequency domain analysis. In *frequency domain analysis*, we replace the waveform with a summation of sinusoids that produce an equivalent waveform. We then examine the relative amplitudes, phases, and frequencies of the sinusoids.

The case for using frequency domain analysis in communication systems Frequency domain analysis looks harder than time domain analysis because we have to keep track of amplitudes, phases, and frequencies for a whole series of sinusoids. Furthermore, all of our intuition and years of experience involve time domain analysis. So why use frequency domain analysis? There are six reasons frequency domain analysis is preferred for communication systems:

1. If the channel is linear, the concepts of impedance (transfer function) and superposition can be used to reduce channel analysis from problems that require differential equations in the time domain to problems that require only simple algebra in the frequency domain.

Let's start with a formal definition of linearity:

If $f(x)$ is a system's (or a channel's) response to an input x and if $f(y)$ is a system's (or a channel's) response to an input y, then the system is *linear* if and only if for any constants A and B, $f(Ax + By) = Af(x) + Bf(y)$.

Pictorially, this linearity can be shown as:

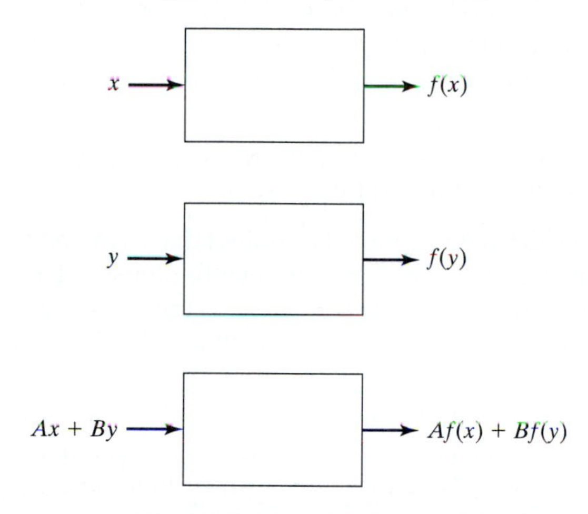

Figure 2-1 A pictorial representation of linearity.

Why is this definition useful? Because, if the channel is linear, it is relatively easy to determine the channel's response to a sinusoid (we can use the concept of *impedance*, just as in network theory), but it is hard to determine the channel's response to a more general waveform (such as a square wave). A linear channel's response to a summation of sinusoids is the same as the summation of the channel's response to each of the individual sinusoids, and the response to each of the individual sinusoids can be easily calculated. Thus, instead of performing one difficult task (determining the channel's response to a general waveform), we can perform a series of simple tasks by

a. Expressing the general waveform as a summation of sinusoids (i.e., expressing the general waveform in the frequency domain),

b. Using impedance to calculate the channel's response to each sinusoid in the summation, and then

c. Adding together all of the channel's responses to each sinusoid. By superposition, this sum is the same as the channel's response to the general waveform.

Graphically, we can represent these tasks as shown in Figure 2-1. Example 2.1 will further illustrate the concept.

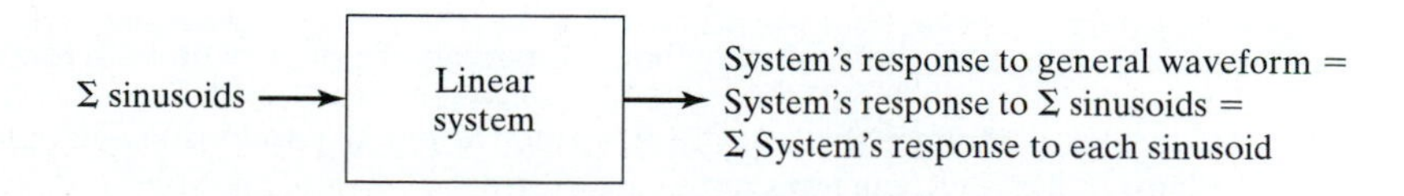

Figure 2-2 Exploiting linearity to simplify system analysis.

The calculation of a linear system's response to a general waveform requires convolution by the impulse response in the time domain but requires only multiplication by the *transfer function* in the frequency domain.

2. The same concepts of impedance, transfer function, linear systems, and multiplication versus convolution can be applied to the analysis of how the transmitter creates the transmitted waveform and how the receiver responds to the received waveform.
3. Certain characteristics of a signal are easier to identify and measure using frequency domain analysis rather than time domain analysis. (See Example 2.2 on page 12.)
4. Channel bandwidth can be easily evaluated in the frequency domain but not in the time domain. Bandwidth is important because:
   a. Sometimes the bandwidth of the channel is limited due to the physical characteristics of the channel. For instance, the bandwidth of a standard audio telephone line is approximately 3400 Hz. If you try to use the telephone lines to transmit a signal requiring a bandwidth greater than 3400 Hz, significant distortion will occur.
   b. Other times, channel bandwidth is limited due to regulations. For example, the federal government limits the bandwidth of channels for commercial FM radio communications to 200 kHz. When the government limits the channel bandwidth, it often does so to prevent one source from interfering with another source that is employing an adjacent frequency band; for instance, without bandwidth restrictions television's Channel 4 broadcast could cause interference with Channel 5's broadcast. (The federal government auctioned off frequency channels for new-generation cellular telephones. The auction netted millions of dollars, providing a graphic example of how valuable bandwidth can be.) Frequency domain analysis allows us to easily determine the frequency content of the transmitted signal and see if it conforms to our bandwidth limitations.
5. Power calculations are often easier to perform in the frequency domain.
6. Analyzing the effects of noise is easier in the frequency domain.

For all these reasons, we will use frequency domain analysis for most of our communication system evaluations.

Examples 2.1 and 2.2 show that it's easier to analyze communication systems using the frequency domain. Please make sure that you understand these examples before going on to Section 2.2.

Example 2.1 Channel analysis using time domain and frequency domain tools

Consider a channel that can be described using the following RC model shown in Figure 2-3a. Given the transmitted waveform

$$s(t) = A_c[1 + \cos(\omega_m t)]\cos(\omega_c t)$$

describe the received waveform $r(t)$.[1]

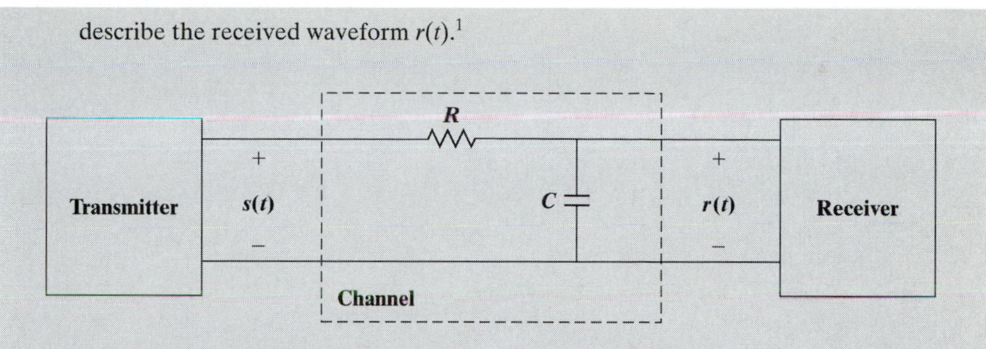

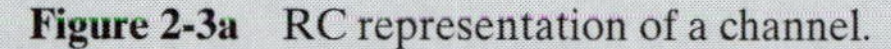

Figure 2-3a RC representation of a channel.

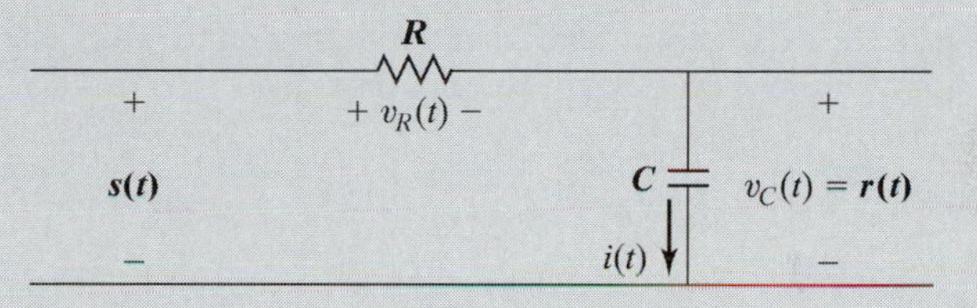

Figure 2-3b Analyzing the channel in the time domain.

Solution using time domain

Using Figure 2-3b and applying Kirchoff's voltage law,

$$s(t) = v_R(t) + v_C(t)$$

Substituting and assuming zero volts across the capacitor at time $t = 0$,

$$A[1 + \cos(\omega_m t)]\cos(\omega_c t) = i(t)R + \frac{1}{C}\int_0^t i(t)\,dt$$

We must now solve this differential equation for $i(t)$, then find $v_c(t)$, which is tedious and difficult. Let's try another way.

Solution using frequency domain

Step 1: Express s(t) as a summation of sinusoids.

$$\begin{aligned} A[1 + \cos(\omega_m t)]\cos(\omega_c t) &= A\cos(\omega_c t) + A\cos(\omega_m t)\cos(\omega_c t) \\ &= A\cos(\omega_c t) + \frac{A}{2}\{\cos[(\omega_c - \omega_m)t] + \cos[(\omega_c + \omega_m)t]\} \\ &= A\cos(\omega_c t) + \frac{A}{2}\cos[(\omega_c - \omega_m)t] + \frac{A}{2}\cos[(\omega_c + \omega_m)t] \end{aligned}$$

[1]The RC model in this example is realistic for many low-frequency wireline channels, and the transmitted signal $s(t)$ is an amplitude modulated (AM) signal typical of the type transmitted in many radio systems. In practical systems, $\omega_c \gg \omega_m$.

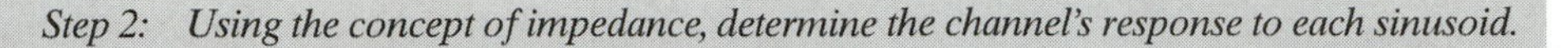

Step 2: Using the concept of impedance, determine the channel's response to each sinusoid.

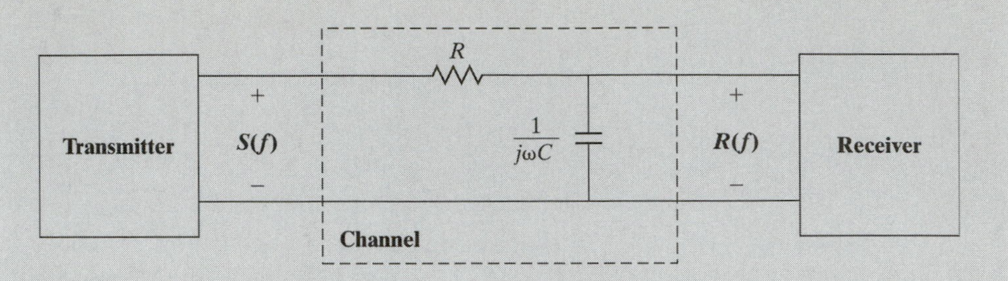

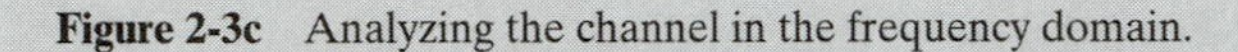

Figure 2-3c Analyzing the channel in the frequency domain.

Figure 2-3c is a representation of the channel in the frequency domain. The resistor and capacitor are characterized by their impedances. By voltage division,

$$R(f) = \frac{S(f)}{1 + j\omega RC}$$

so, for $S_1(f) = A\cos(\omega_c t)$, the channel's response is $R_1(f) = \dfrac{A\cos(\omega_c t)}{1 + j\omega RC}$

for $S_2(f) = \dfrac{A}{2}\cos[(\omega_c - \omega_m)t]$, the channel's response is $R_2(f) = \dfrac{0.5A\cos[(\omega_c - \omega_m)t]}{1 + j(\omega_c - \omega_m)RC}$

for $S_3(f) = \dfrac{A}{2}\cos[(\omega_c + \omega_m)t]$, the channel's response is $R_3(f) = \dfrac{0.5A\cos[(\omega_c + \omega_m)t]}{1 + j(\omega_c + \omega_m)RC}$

Step 3: Using superposition, the channel's total response is the sum of the individual responses.

$$R(f) = R_1(f) + R_2(f) + R_3(f)$$

$$= \frac{A\cos(\omega_c t)}{1 + j\omega_c RC} + \frac{0.5A\cos[(\omega_c - \omega_m)t]}{1 + j(\omega_c - \omega_m)RC} + \frac{0.5A\cos[(\omega_c + \omega_m)t]}{1 + j(\omega_c + \omega_m)RC}$$

Magnitude and phase spectra

Another advantage of frequency domain analysis is that graphical techniques (*spectra*) are available to illustrate the relationships among relative amplitudes, phases, and frequencies of the sinusoids.

Magnitude spectrum

The *magnitude spectrum* is the plot of the magnitude of each sinusoid versus the sinusoid's frequency. Figures 2-4a and b compare the magnitude spectra of the transmitted and received signals in Example 2.1, showing how the channel distorts the magnitude of the transmitted waveform. We can see that the channel attenuates higher-frequency components more than lower-frequency components.

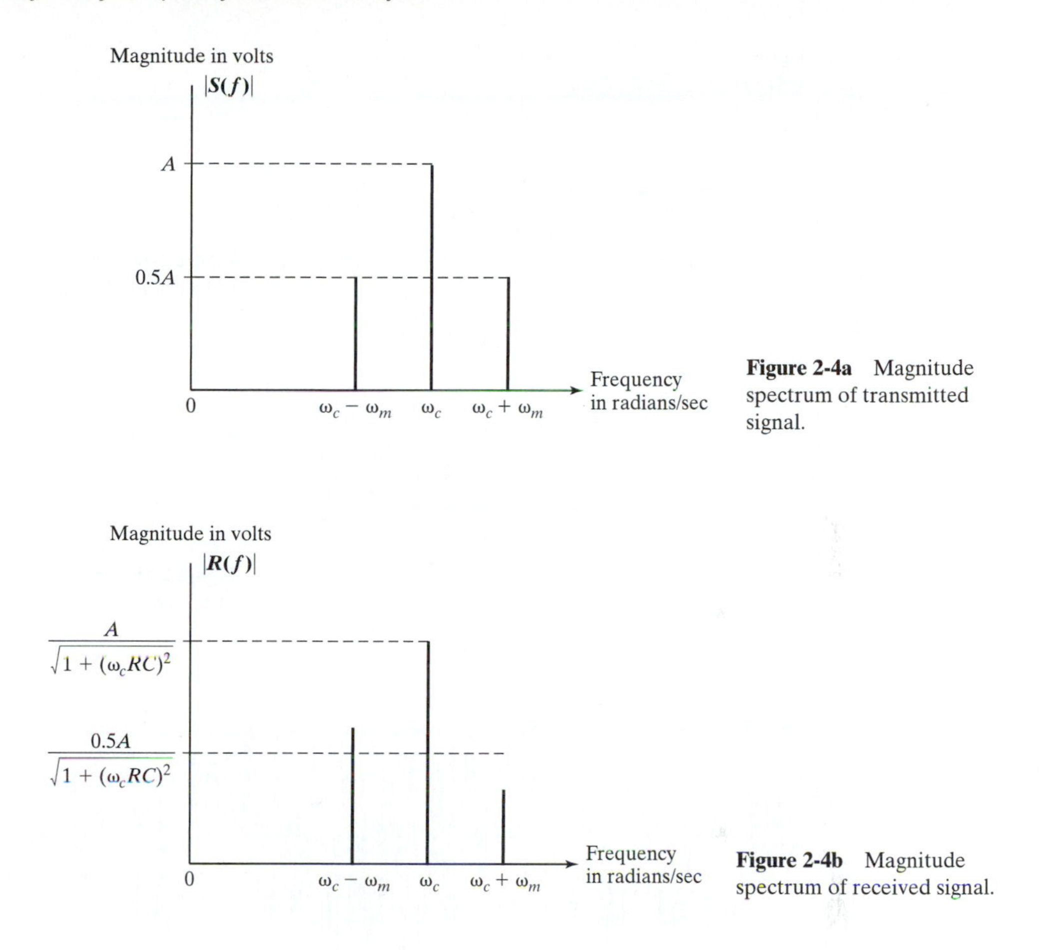

Figure 2-4a Magnitude spectrum of transmitted signal.

Figure 2-4b Magnitude spectrum of received signal.

Phase spectrum

The *phase spectrum* is the plot of the relative phase of each sinusoid versus the sinusoid's frequency. Figures 2-5a and 2-5b compare the phase spectra of the transmitted and received signals in Example 2.1, showing how the channel distorts the phase of the transmitted waveform.

Example 2.2 Extracting signal from noise in the frequency domain

This example illustrates how certain characteristics of a signal may be easier to identify and measure using frequency domain analysis rather than time domain analysis. Figure 2-6a is the time domain representation of a 1 volt (peak) 10 MHz sinusoid buried in noise. (The signal could represent a transmitted message from a distant satellite and the noise could be from the atmosphere.) The effects of the noise cause the sinusoid to be virtually unnoticeable. Figure 2-6b is the magnitude spectrum of the same signal. Note that the effects of the noise are spread out over all frequencies, making the presence of the 10 MHz signal very noticeable.

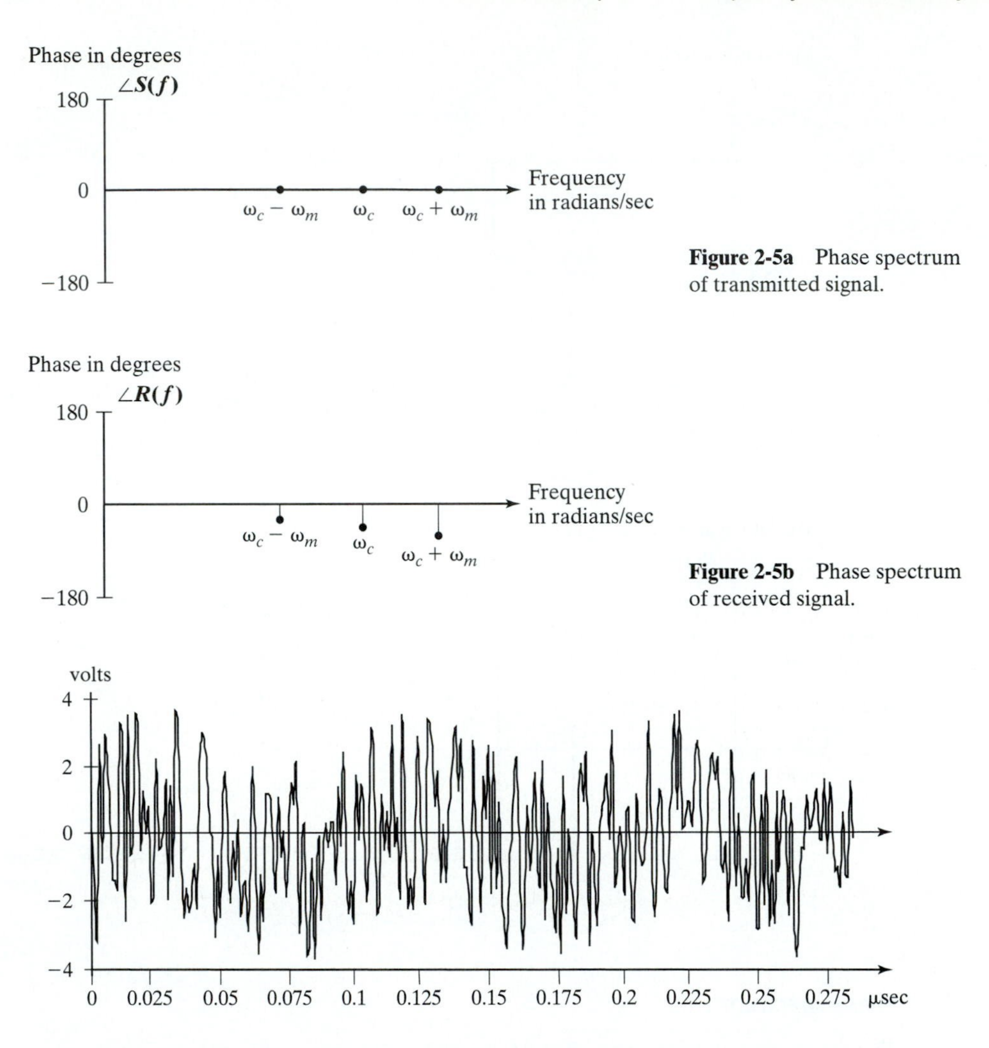

Figure 2-5a Phase spectrum of transmitted signal.

Figure 2-5b Phase spectrum of received signal.

Figure 2-6a Time domain representation of the received signal. (Note that the effects of the noise cause the sinusoid to be virtually unnoticeable.)

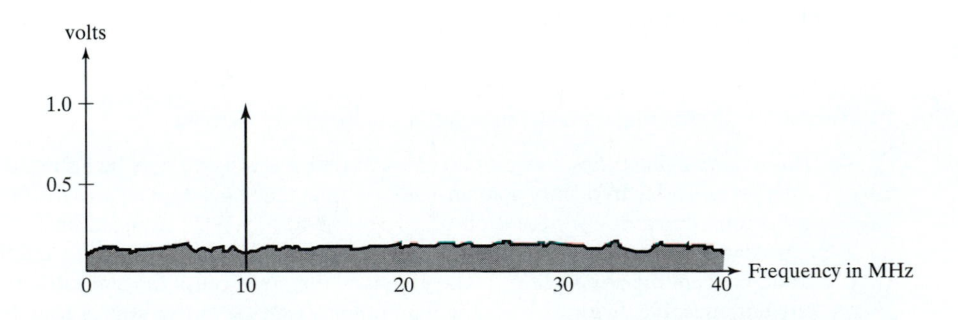

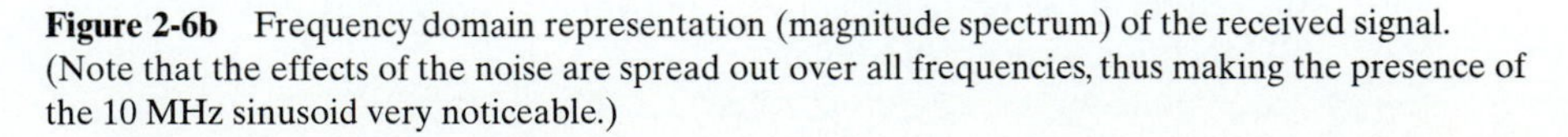

Figure 2-6b Frequency domain representation (magnitude spectrum) of the received signal. (Note that the effects of the noise are spread out over all frequencies, thus making the presence of the 10 MHz sinusoid very noticeable.)

These justifications for frequency domain analysis will become clearer and stronger beginning in Chapter 3. As mentioned earlier, we have all developed extensive intuition based on years of working with time domain analysis. It is important that we begin to develop similar intuition for frequency domain analysis. As we study communication systems, it is *extremely* important not to view frequency domain analysis as merely a mathematical exercise and to regard our work as more significant than simply grinding out mathematical "turn the crank" problems. At each step in communication system analysis and design, it is important to understand exactly *what* we are doing and *why* we are doing it. It is important to understand *how* each signal and each component in the communication system interacts. Throughout our studies, we need to constantly ask questions such as: "In plain English, what are we doing?" "Why are we doing this?" "What does this result mean?" and "Does this result make sense?"

In Section 2.2, we will begin to develop techniques for representing signals in the frequency domain.

2.2 The Fourier Series

Now that we understand the usefulness of frequency domain analysis, we must determine ways to convert signals from a time domain representation to a frequency domain representation. In other words, we must determine how to take a signal and create a series of sinusoids that, when added together, produce a signal equivalent to the original signal. How can we do this?

1. Sometimes we can use trigonometric identities to convert a signal from the time domain to the frequency domain (as we did in Example 2.1). *This method, however, can only be applied in certain rare cases.*
2. If the signal is periodic and has finite energy per period, we can use the *Fourier series.*

2.2.1 Trigonometric Form of the Fourier Series

Let $s(t)$ represent a signal in the time domain. If $s(t)$ is a periodic function with period T and if $s(t)$ has finite energy per period (i.e., $\int_{t_o}^{t_o+T} |s(t)^2|dt < \infty$), then

$$s(t) = a_0 + \sum_{n=1}^{\infty}(a_n \cos 2\pi n f_o t + b_n \sin 2\pi n f_o t) \tag{2.1}$$

where t_o represents an arbitrary time,

$$f_o = \frac{1}{T} \tag{2.2}$$

and

$$a_0 = \frac{1}{T}\int_{t_o}^{t_o+T} s(t)dt \tag{2.3}$$

$$a_n = \frac{2}{T}\int_{t_o}^{t_o+T} s(t)\cos(2\pi n f_o t)dt \tag{2.4}$$

$$b_n = \frac{2}{T}\int_{t_o}^{t_o+T} s(t)\sin(2\pi n f_o t)dt \tag{2.5}$$

What do these equations say? If $s(t)$ is periodic and has finite energy per period, we can replace $s(t)$ with an equivalent waveform that is the sum of a constant plus a series of sinusoids whose frequencies are all integral multiples of $1/T$. This is exactly what we wanted for frequency domain analysis. In fact, we can even think of the constant as a 0 Hz cosine.

Terminology

- We call the constant term, a_0, the *dc term* (or *average value*). We produce a_0 by averaging the signal over one period (see Equation (2.3)).
- We call the terms $a_1 \cos 2\pi f_o t$ and $b_1 \sin 2\pi f_o t$ the *fundamental frequency components* because they have the same period as the original waveform $s(t)$.
- We call all the other terms (which are all at higher frequencies) the *harmonic frequency components*.

Equations (2.1)–(2.5) can be derived from the orthogonality of harmonic sine and cosine terms. This orthogonality is discussed and proven in Problem 2.4. Derivation of Equations (2.1)–(2.5) is then performed in Problem 2.5.

Let's work an example of a Fourier series converting a periodic waveform into a constant plus a series of sinusoids.

Example 2.3 Part A: Trigonometric Fourier series representation of a "stairstep" signal

Express $s(t)$ in Figure 2-7 in the frequency domain (i.e., as a constant plus a series of sinusoids).

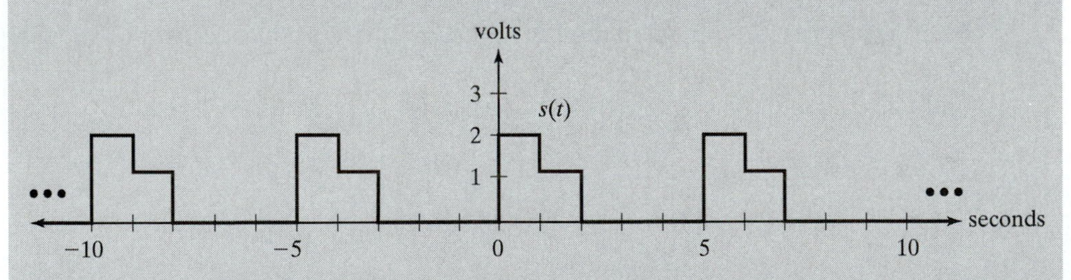

Figure 2-7 Time domain representation of a signal $s(t)$.

Solution

Can we use the Fourier series? Let's check our constraints.

1. Is $s(t)$ periodic? Yes, with a period of five seconds.
2. Does $s(t)$ contain finite energy within a period?

$$\int_{t_o}^{t_o+T} |s(t)|^2 dt = \int_0^1 2^2 dt + \int_1^2 1^2 dt + \int_2^5 0^2 dt = 4 + 1 = 5 < \infty$$

Since the constraints are met, we can use the Fourier series to represent $s(t)$ as a constant plus a series of sines and cosines. Let's start with the basic form of the Fourier series, solve for f_o, and expand.

$$\begin{aligned}
s(t) &= a_0 + \sum_{n=1}^{\infty}(a_n \cos 2\pi n f_o t + b_n \sin 2\pi n f_o t) \\
&= a_0 + \sum_{n=1}^{\infty}\left(a_n \cos\frac{2\pi nt}{5} + b_n \sin\frac{2\pi nt}{5}\right) \\
&= a_0 + a_1\cos(0.4\pi t) + b_1\sin(0.4\pi t) + a_2\cos(0.8\pi t) + b_2\sin(0.8\pi t) + \ldots
\end{aligned} \tag{2.2A}$$

Let's now solve for the constants a_0, a_1, b_1, a_2, b_2, etc.

$$a_0 = \frac{1}{T}\int_{t_o}^{t_o+T} s(t)\,dt = \frac{1}{5}\left[\int_0^1 2\,dt + \int_1^2 1\,dt + \int_2^5 0\,dt\right] = \frac{3}{5}\text{ volt, our average or dc value}$$

$$\begin{aligned}
a_1 &= \frac{2}{T}\int_{t_o}^{t_o+T} s(t)\cos 2\pi f_o t\,dt = \frac{2}{5}\left[\int_0^1 2\cos(0.4\pi t)\,dt + \int_1^2 \cos(0.4\pi t)\,dt + \int_2^5 0\cos(0.4\pi t)\,dt\right] \\
&= \frac{2}{5}\left[\frac{2}{0.4\pi}\sin(0.4\pi t)\Big|_0^1 + \frac{1}{0.4\pi}\sin(0.4\pi t)\Big|_1^2\right] \\
&= \frac{2}{2\pi}[2\sin(0.4\pi) - 2\sin(0) + \sin(0.8\pi) - \sin(0.4\pi)] \\
&= \frac{1}{\pi}[\sin(0.4\pi) + \sin(0.8\pi)] = 0.4898 \text{ volts}
\end{aligned}$$

Generally,

$$\begin{aligned}
a_n &= \frac{2}{T}\int_{t_0}^{t_o+T} s(t)\cos 2\pi n f_o t\,dt \\
&= \frac{2}{5}\left[\int_0^1 2\cos(0.4n\pi t)\,dt + \int_1^2 \cos(0.4n\pi t)\,dt + \int_2^5 0\cos(0.4n\pi t)\,dt\right] \\
&= \frac{2}{5}\left[\frac{2}{0.4n\pi}\sin(0.4n\pi t)\Big|_0^1 + \frac{1}{0.4n\pi}\sin(0.4n\pi t)\Big|_1^2\right] \\
&= \frac{1}{n\pi}[\sin(0.4n\pi) + \sin(0.8n\pi)] \text{ volts}
\end{aligned}$$

So

$$a_2 = \frac{1}{2\pi}[\sin(0.8\pi) + \sin(1.6\pi)] = -0.05782 \text{ volts}$$

$$a_3 = \frac{1}{3\pi}[\sin(1.2\pi) + \sin(2.4\pi)] = 0.03854 \text{ volts}$$

etc. We can, of course, also find the value of a_1, from the general expression.

Solving for b_n,

$$b_n = \frac{2}{T}\int_{t_o}^{t_o+T} s(t)\sin 2\pi n f_o t\,dt = \ldots = \frac{1}{n\pi}[2 - \cos(0.4n\pi) - \cos(0.8n\pi)] \text{ volts}$$

So

$$b_1 = \frac{1}{\pi}[2 - \cos(0.4\pi) - \cos(0.8\pi)] = 0.7958 \text{ volts}$$

$$b_2 = \frac{1}{2\pi}[2 - \cos(0.8\pi) - \cos(1.6\pi)] = 0.3979 \text{ volts}$$

$$b_3 = \frac{1}{3\pi}[2 - \cos(1.2\pi) - \cos(2.4\pi)] = 0.2653 \text{ volts}$$

etc.

We can now substitute all the constants back into Expansion (2.2A) on page 15:

$$\begin{aligned} s(t) &= a_0 + a_1\cos(0.4\pi t) + b_1\sin(0.4\pi t) + a_2\cos(0.8\pi t) + b_2\sin(0.8\pi t) \\ &\quad + a_3\cos(1.2\pi t) + b_3\sin(1.2\pi t) + \ldots \\ &= 0.6 + 0.4898\cos(0.4\pi t) + 0.7958\sin(0.4\pi t) - 0.05782\cos(0.8\pi t) \\ &\quad + 0.3979\sin(0.8\pi t) + 0.03854\cos(1.2\pi t) + 0.2653\sin(1.2\pi t) + \ldots \text{ volts} \end{aligned}$$

We have thus replaced $s(t)$ with a constant plus a summation of sinusoids.

Example 2.3 Part B: Convergence

Now, let's plot the results of Part A to show that the summation of sines and cosines we developed really does produce the "stairstep" signal $s(t)$ shown in Part A. Figure 2-8 shows five plots. The first plot, Figure 2-8a, represents the sum of the dc term plus the first five harmonics (i.e., a_0 plus the sines and cosines with frequencies up to and including $5f_o$). The second plot, Figure 2-8b, represents the sum of the dc term plus the first ten harmonics. Figures 2-8c and 2-8d represent the sum of the dc term plus the first fifteen harmonics and the dc term plus the first twenty-five harmonics, respectively. Figure 2-8e, the fifth plot, represents the sum of the dc term plus the summation of all harmonics (remember that the sum goes out to infinity). You'll perform these reconstructions in Problem 2.6.

Note that as we add more harmonics, the reconstructed signal more closely represents the original stairstep signal. This observation is important because the channel in a communication system may attenuate or eliminate the higher order harmonics. The number of harmonics that the channel accurately passes is related to the bandwidth of the channel (we'll be defining bandwidth formally later in this chapter). The greater the channel bandwidth, the higher the number of harmonics passed by the channel. If more harmonics arrive at the receiver, the transmitted signal is more accurately reconstructed.

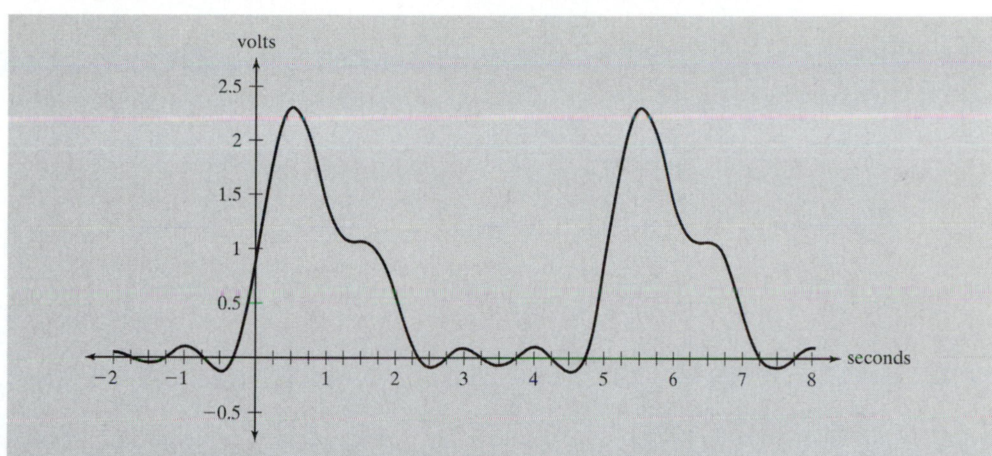

Figure 2-8a Summation of dc term plus the first five harmonics.

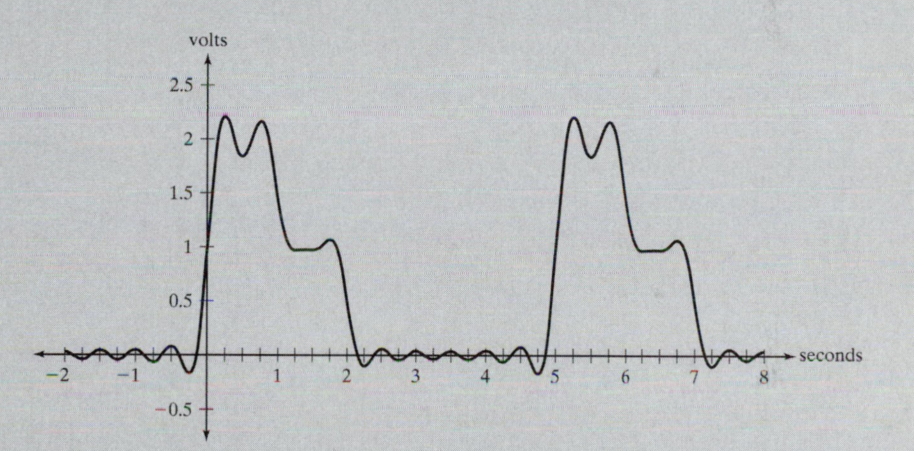

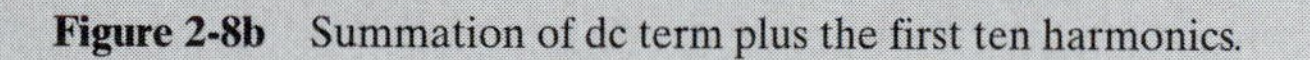

Figure 2-8b Summation of dc term plus the first ten harmonics.

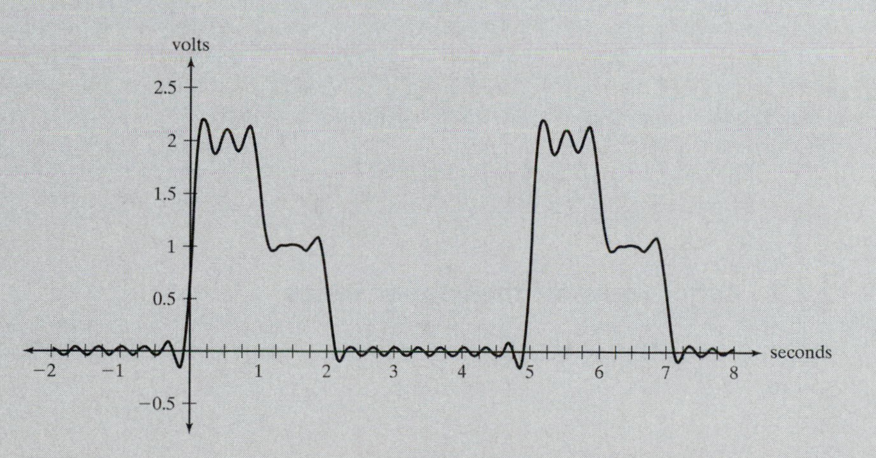

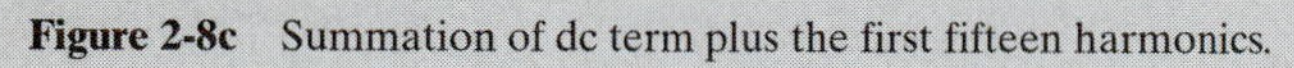

Figure 2-8c Summation of dc term plus the first fifteen harmonics.

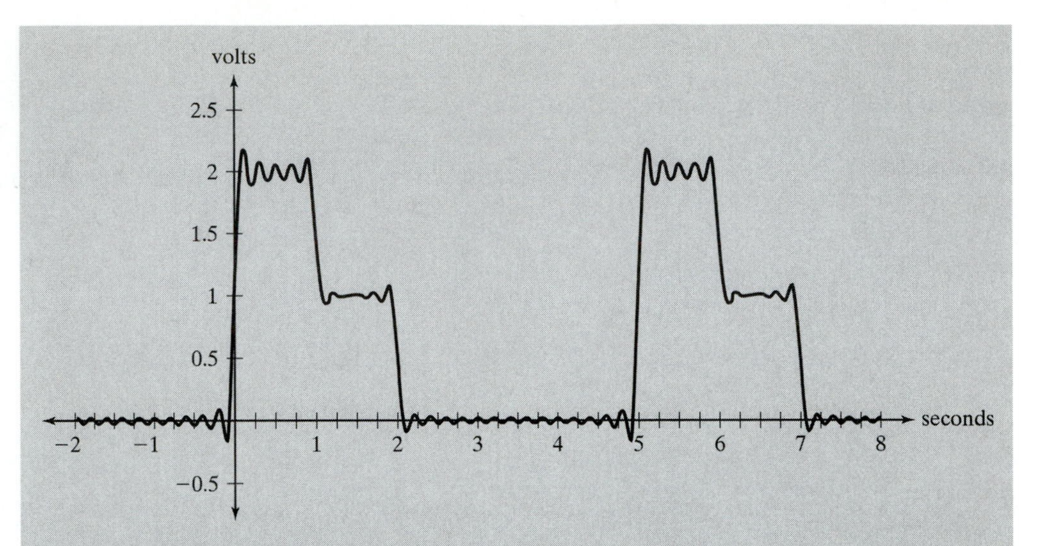

Figure 2-8d Summation of dc term plus the first twenty-five harmonics.

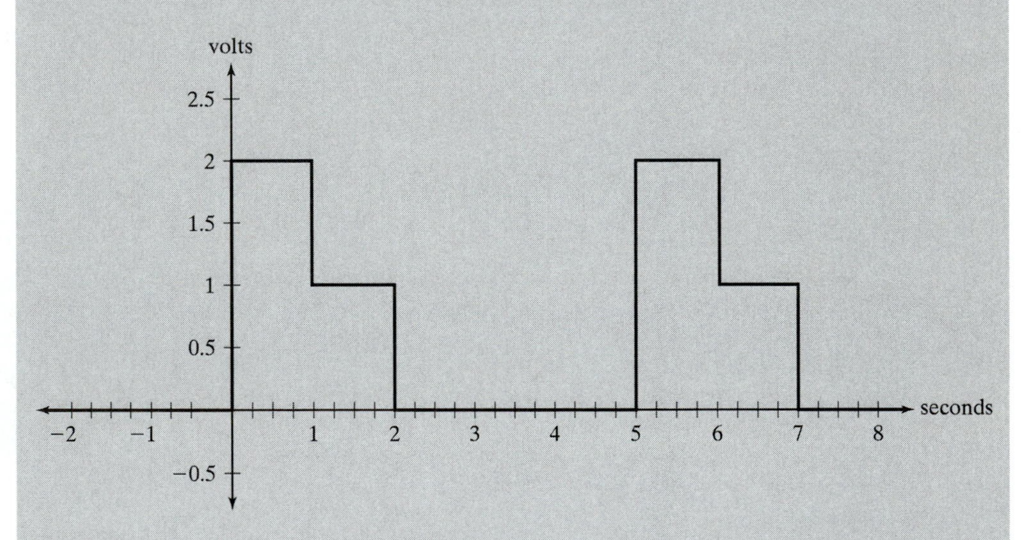

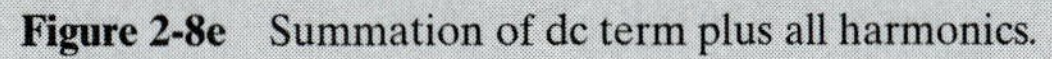

Figure 2-8e Summation of dc term plus all harmonics.

Thus, we can see from this example that bandwidth and accuracy are related: greater bandwidth produces greater accuracy, an illustration of a performance versus cost trade-off.

2.2.2 Other Forms of the Fourier Series

The Fourier series can be represented in many equivalent forms. The form we've used so far,

$$s(t) = a_0 + \sum_{n=1}^{\infty}(a_n \cos 2\pi n f_o t + b_n \sin 2\pi n f_o t) \qquad (2.1R)$$

is called the *trigonometric form of the Fourier series*.[2] We can also express the Fourier series in other forms, some of which may be more convenient than the trigonometric form for certain applications.

2.2.2.1 *The one-sided form* How can we express the Fourier series in another form? Note that

$$a_n \cos 2\pi n f_o t + b_n \sin 2\pi n f_o t = \sqrt{a_n^2 + b_n^2} \cos(2\pi n f_o t - \gamma_n) \tag{2.6}$$

where

$$\gamma_n = \arctan\left(\frac{b_n}{a_n}\right) \tag{2.7}$$

This is simply rectangular-to-polar conversion, as shown in Figure 2-9a, representing a_n and b_n as phasors (you'll do this conversion in Problem 2.9). Note that the b_n phasor *lags* the a_n phasor by 90 degrees because sine lags cosine. The two phasors a_n and b_n are summed to form one equivalent phasor.

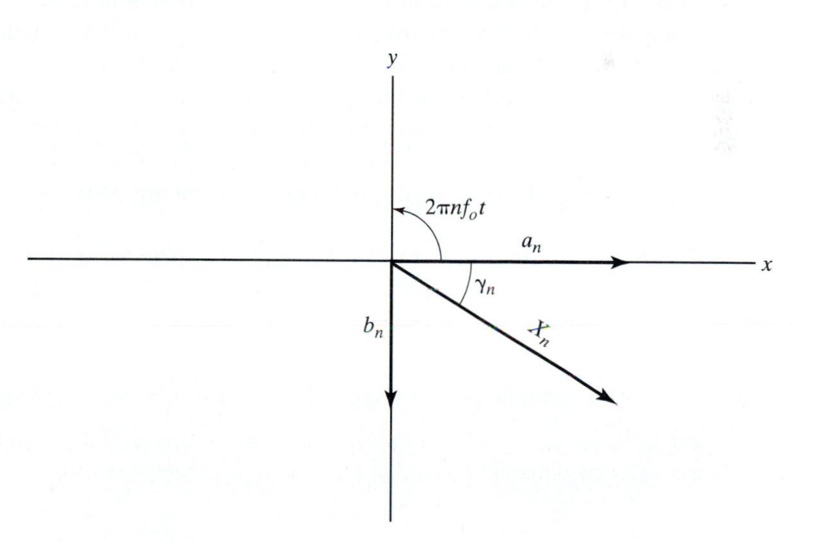

Figure 2-9 Relationship of one-sided and trigonometric Fourier series components.

[2]Please take a moment to examine the equation number. Any time we repeat a previously-seen equation, we will use its original number followed by an *R*. This will make it easier for you if you need to go back to study the origins of the equation.

Thus, we can write

$$\begin{aligned} s(t) &= a_0 + \sum_{n=1}^{\infty}\left(a_n \cos 2\pi n f_o t + b_n \sin 2\pi n f_o t\right) \\ &= a_0 + \sum_{n=1}^{\infty}\sqrt{a_n^2 + b_n^2}\cos(2\pi n f_o t - \gamma_n) \\ &= X_0 + \sum_{n=1}^{\infty} X_n \cos(2\pi n f_o t + \phi_n) \end{aligned} \tag{2.8}$$

where

$$X_0 = a_0 \tag{2.9}$$

$$X_n = \sqrt{a_n^2 + b_n^2} \tag{2.10}$$

$$\text{and } \phi_n = -\gamma_n = \arctan\left(\frac{-b_n}{a_n}\right) \tag{2.11}$$

We call X_n the *magnitude* and ϕ_n the *phase*, which is the angle by which the single equivalent phasor leads the a_n phasor. Note the relationship between γ_n and ϕ_n and the need for the minus sign in the *numerator* of Equation (2.11). Use of ϕ_n other than γ_n may look more complicated than necessary, but it will be needed for consistency when we establish the next form of the Fourier series. Also remember to keep track of the sign in the numerator and denominator of the arctangent function, since there are two solutions (180 degrees apart) for the arctangent and only one will be correct.

This form, called the *one-sided form of the Fourier series*, is useful because it easily shows the phase and magnitude of each frequency component of $s(t)$. In other words, the one-sided form allows us to easily plot magnitude and phase spectra (as we did for Example 2.1 on page 11). These spectra are graphic tools that can provide a - better understanding of how a signal behaves by allowing us to "see" the signal in the frequency domain.

Suppose we want to convert from the trigonometric form of the Fourier series to the one-sided form. Let's try this by extending Example 2.3 (from page 14).

Example 2.4 One-sided phase and magnitude spectra of the "stairstep" signal

Give the trigonometric form of the Fourier series for the waveform below, then draw the single-sided magnitude and phase spectra for the waveform.

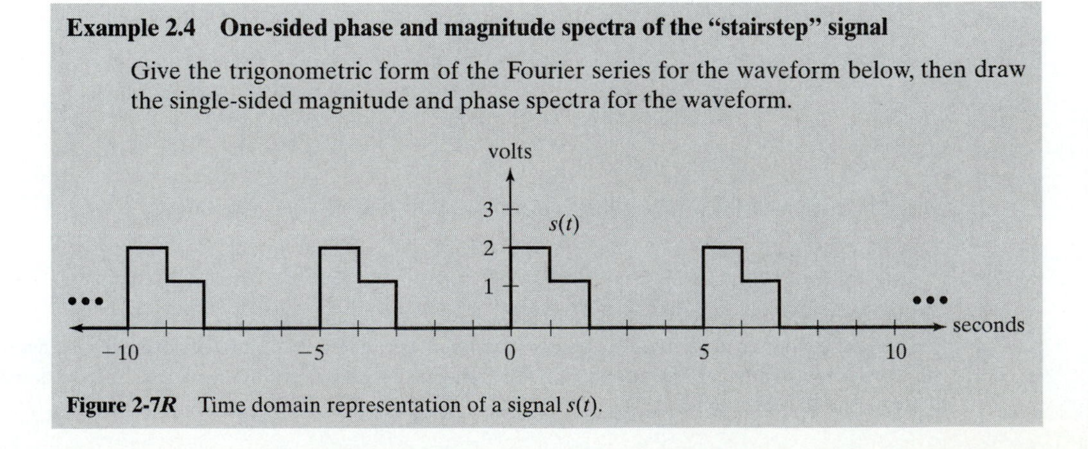

Figure 2-7R Time domain representation of a signal $s(t)$.

Solution

As we determined in Example 2.3, the trigonometric form is

$$s(t) = a_0 + \sum_{n=1}^{\infty}(a_n \cos 2\pi n f_o t + b_n \sin 2\pi n f_o t)$$

where $f_o = 0.2$ Hz

$$a_0 = 0.6 \text{ volts}$$

$$a_n = \frac{[\sin(0.4n\pi) + \sin(0.8n\pi)]}{n\pi} \text{ volts}$$

$$b_n = \frac{[2 - \cos(0.4n\pi) - \cos(0.8n\pi)]}{n\pi} \text{ volts}$$

To plot the single-sided magnitude and phase spectra, we must use the single-sided form for the Fourier series, where, as we've previously determined, the magnitude and phase of the *n*th term are expressed as:

$$X_0 = a_0 \tag{2.9R}$$

$$X_n = \sqrt{a_n^2 + b_n^2} \tag{2.10R}$$

$$\text{and } \phi_n = \arctan\left(\frac{-b_n}{a_n}\right) \tag{2.11R}$$

Values for a_0, a_n, and b_n, the magnitude (X_n), and the phase (ϕ_n) are given in Table 2-1 for the first 15 harmonics. Magnitude and phase spectra are also plotted in Figures 2-10a and 2-10b.

Table 2-1 Trigonometric and One-Sided Fourier Series Coefficients for $s(t)$

Harmonic (n)	a_n	b_n	Magnitude (X_n)	Phase (ϕ_n) in Degrees
0 (dc term)	0.6	—	0.6	0
1	0.490	0.796	0.934	−58.4
2	−0.058	0.398	0.402	−98.3
3	0.039	0.265	0.268	−81.7
4	−0.122	0.199	0.234	−121.6
5	0	0	0	*
6	0.082	0.133	0.156	−58.4
7	−0.017	0.114	0.115	−98.3
8	0.014	0.099	0.101	−81.7
9	−0.054	0.088	0.104	−121.6
10	0	0	0	*
11	0.045	0.072	0.085	−58.4
12	−0.010	0.066	0.067	−98.3
13	0.009	0.061	0.062	−81.7
14	−0.035	0.057	0.067	−121.6
15	0	0	0	*

*Technically, phase is irrelevant if magnitude = 0. These terms will, however, be plotted as having 0 degrees phase on the phase spectrum.

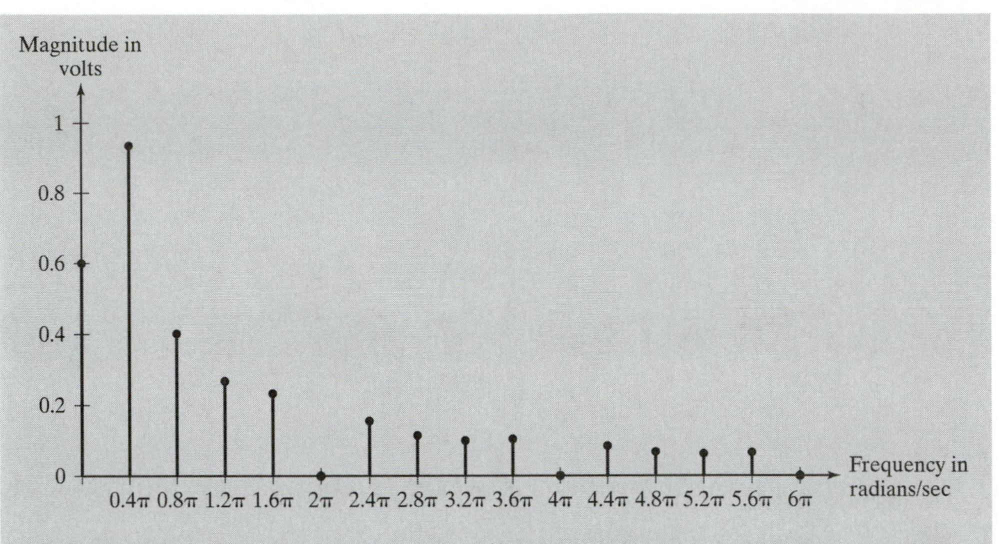

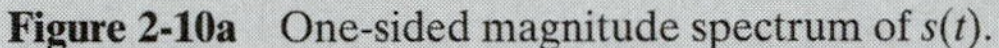
Figure 2-10a One-sided magnitude spectrum of $s(t)$.

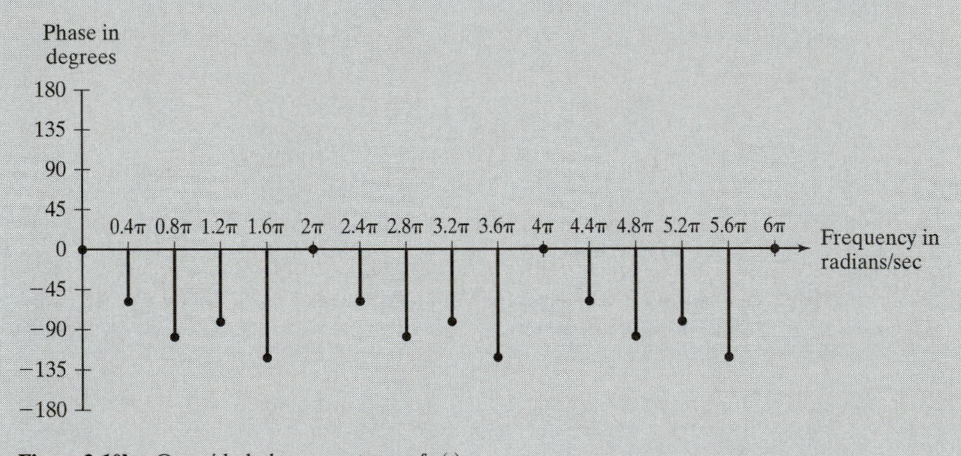

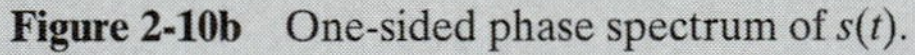
Figure 2-10b One-sided phase spectrum of $s(t)$.

Thus,

$$
\begin{aligned}
s(t) &= 0.6 + 0.490\cos(0.4\pi t) + 0.796\sin(0.4\pi t) - 0.058\cos(0.8\pi t) \\
&\qquad + 0.398\sin(0.8\pi t) + 0.039\cos(1.2\pi t) + 0.265\sin(1.2\pi t) + \ldots \text{volts} \\
&= 0.6 + 0.934\cos(0.4\pi t - 58.4^\circ) + 0.402\cos(0.8\pi t - 98.3^\circ) \\
&\qquad + 0.268\cos(1.2\pi t - 81.7^\circ) + \ldots \text{volts}
\end{aligned}
$$

Note the phase angles in Table 2-1. If you are not careful in keeping track of the signs of the numerator and denominator when you evaluate the arctangent, the phase angles you calculate may be in error by 180 degrees. Some calculators and software programs keep track of these signs, others will not.

2.2.2.2 The complex exponential (or two-sided) form Now consider the following form of the Fourier series:

$$s(t) = \sum_{n=-\infty}^{\infty} c_n e^{j2\pi n f_o t} \quad \text{where } c_n = \frac{1}{T}\int_{t_o}^{t_o+T} s(t)e^{-j2\pi n f_o t}dt \tag{2.12}$$

This form, called the *two-sided Fourier series* or *complex exponential Fourier series*, is useful for three reasons. First, it is often easier to mathematically manipulate exponentials than sines and cosines. Second, the dc term of the complex exponential form is calculated in exactly the same way as all the other harmonics (this isn't true for the trigonometric or one-sided forms). Third, and most importantly, the complex exponential form can be extended into the Fourier transform, which will allow us to also represent nonperiodic signals in the frequency domain.

Let's try to gain some insight into the complex exponential form through Euler's identity:

$$e^{j\theta} = \cos\theta + j\sin\theta \tag{2.13}$$

Euler's identity says, "Even when you see complex exponentials, think *sines* and *cosines*."

The equivalence of the trigonometric and complex exponential forms of the Fourier series

As another step in gaining insight into the complex exponential form and in understanding how components in the three forms of the Fourier series are related, let's prove that the trigonometric form and the exponential (or two-sided) form of the Fourier series are equivalent. In other words, let's prove

$$a_0 + \sum_{n=1}^{\infty}(a_n \cos 2\pi n f_o t + b_n \sin 2\pi n f_o t) = \sum_{n=-\infty}^{\infty} c_n e^{j2\pi n f_o t} \tag{2.14}$$

where, as we've established,

$$a_0 = \frac{1}{T}\int_{t_o}^{t_o+T} s(t)dt \tag{2.3R}$$

$$a_n = \frac{2}{T}\int_{t_o}^{t_o+T} s(t)\cos(2\pi n f_o t)dt \tag{2.4R}$$

$$b_n = \frac{2}{T}\int_{t_o}^{t_o+T} s(t)\sin(2\pi n f_o t)dt \tag{2.5R}$$

and

$$c_n = \frac{1}{T}\int_{t_o}^{t_o+T} s(t)e^{-j2\pi n f_o t}dt \qquad (2.12R)$$

Proof:

Our strategy will be to show an equivalence between the terms on the left-hand side of Equation (2.14) and a corresponding group of terms on the right-hand side. We will do this in the following way:

Part 1. Show that the $n = 0$ term on the right-hand side is equal to a_0 on the left-hand side.

Part 2. Show that for $i > 0$, the $n = i$ terms on the left-hand side are equal to the sum of the $n = i$ and $n = -i$ terms on the right-hand side.

The key to this proof is Euler's identity,

$$e^{j\theta} = \cos\theta + j\sin\theta \qquad (2.13R)$$

Part 1

$$c_0 = \frac{1}{T}\int_{t_o}^{t_o+T} s(t)e^{-j0}dt = \frac{1}{T}\int_{t_o}^{t_o+T} s(t)[\cos(0) - j\sin(0)]dt = \frac{1}{T}\int_{t_o}^{t_o+T} s(t)dt = a_0 \qquad (2.15)$$

Part 2

For $n = i$ $(i > 0)$, the terms of the trigonometric Fourier series are

$$a_i\cos(2\pi i f_o t) + b_i\sin(2\pi i f_o t)$$

The sum of the $n = i$ and $n = -i$ terms of the complex exponential Fourier series are

$$c_i e^{j2\pi i f_o t} + c_{-i}e^{-j2\pi i f_o t}$$

For the trigonometric and complex exponential forms to be equivalent, we need to prove that the two sets of terms are equal. Starting with the equation for the c_i coefficient and applying Euler's identity,

$$\begin{aligned} c_i &= \frac{1}{T}\int_{t_o}^{t_o+T} s(t)e^{-j2\pi i f_o t}dt = \frac{1}{T}\int_{t_o}^{t_o+T} s(t)[\cos 2\pi i f_o t - j\sin 2\pi i f_o t]dt \\ &= \frac{1}{T}\int_{t_o}^{t_o+T} s(t)\cos(2\pi i f_o t)\,dt - j\frac{1}{T}\int_{t_o}^{t_o+T} s(t)\sin(2\pi i f_o t)\,dt = \frac{1}{2}(a_i - jb_i) \end{aligned} \qquad (2.16)$$

Similarly,

$$c_{-i} = \frac{1}{T}\int_{t_o}^{t_o+T} s(t)e^{-j2\pi(-i)f_ot}dt = \frac{1}{T}\int_{t_o}^{t_o+T} s(t)[\cos 2\pi i f_o t + j\sin 2\pi i f_o t]\,dt$$
$$= \frac{1}{T}\int_{t_o}^{t_o+T} s(t)\cos(2\pi i f_o t)\,dt + j\frac{1}{T}\int_{t_o}^{t_o+T} s(t)\sin(2\pi i f_o t)\,dt = \frac{1}{2}(a_i + jb_i) \quad (2.17)$$

Thus,

$$c_i e^{j2\pi i f_o t} + c_{-i}e^{-j2\pi i f_o t} = \frac{1}{2}(a_i - jb_i)e^{j2\pi i f_o t} + \frac{1}{2}(a_i + jb_i)e^{-j2\pi i f_o t}$$
$$= \frac{a_i}{2}(e^{j2\pi i f_o t} + e^{-j2\pi i f_o t}) - j\frac{b_i}{2}(e^{j2\pi i f_o t} - e^{-j2\pi i f_o t}) \quad (2.18)$$
$$= a_i\left(\frac{e^{j2\pi i f_o t} + e^{-j2\pi i f_o t}}{2}\right) + b_i\left(\frac{e^{j2\pi i f_o t} - e^{-j2\pi i f_o t}}{2j}\right)$$
$$= a_i \cos 2\pi i f_o t + b_i \sin 2\pi i f_o t$$

since, based on Euler's identity, we can express cosine and sine as:

$$\cos 2\pi n f_o t = \frac{e^{j2\pi n f_o t} + e^{-j2\pi n f_o t}}{2} \quad (2.19)$$

$$\sin 2\pi n f_o t = \frac{e^{j2\pi n f_o t} - e^{-j2\pi n f_o t}}{2j} \quad (2.20)$$

Comments:

In addition to proving that the trigonometric and complex exponential forms of the Fourier series are equivalent, we have also derived the expressions for conversion of the Fourier coefficients from one form to the other. Equation (2.15) shows that $c_0 = a_0$, and Equations (2.16) and (2.17) show that for $n > 0$,

$$c_n = \frac{1}{2}(a_n - jb_n) \text{ and } c_{-n} = \frac{1}{2}(a_n + jb_n) \quad (2.21)$$

Equation (2.21) shows us that

$$c_{-n} = c_n^* \quad (2.22)$$

where * symbolizes the complex conjugate. Equations (2.15) and (2.21) can also be easily extrapolated to show the relationship of the one-sided and two-sided Fourier coefficients.

For $n = 0$,

$$c_0 = X_0 \tag{2.23}$$

and for $n > 0$,

$$|c_n| = |c_{-n}| = \frac{X_n}{2} \tag{2.24}$$

$$\angle c_n = \phi_n \tag{2.25}$$

$$\angle c_{-n} = -\phi_n \tag{2.26}$$

Now we're ready to consider an example.

Example 2.5 Part A: Two-sided or complex exponential Fourier series representation of the "stairstep" signal

Determine the complex exponential form of the Fourier series for the waveform below. (Note that this is the same waveform used in Examples 2.3 and 2.4.)

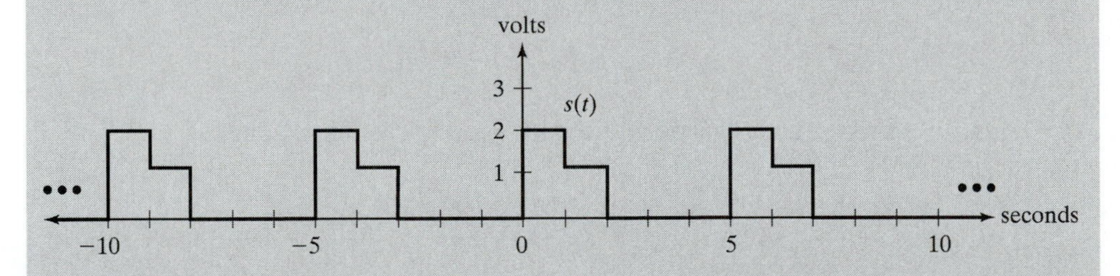

Figure 2-7*R* Time domain representation of a signal $s(t)$.

Solution

We have three options for obtaining the complex exponential Fourier series coefficients.

Option 1:

Calculate the c_n coefficients of the complex exponential Fourier series form by using Equation (2.12),

$$c_n = \frac{1}{T}\int_{t_o}^{t_o+T} s(t)e^{-j2\pi n f_o t}dt$$

Option 2:

In Example 2.3, we determined the trigonometric form of the Fourier series of $s(t)$. We thus know the a_n and b_n coefficients. Using Equations (2.15) and (2.21),

$$\text{for } n = 0,\ c_0 = a_0 \tag{2.15R}$$

$$\text{for } n > 0,\ c_n = \frac{1}{2}(a_n - jb_n) \text{ and } c_{-n} = \frac{1}{2}(a_n + jb_n) \tag{2.21R}$$

We can thus determine the complex exponential Fourier series coefficients (i.e., the c_n values) as shown in Table 2-2.

Table 2-2 Complex Exponential Fourier Series Coefficients for $s(t)$

n	a_n	b_n	c_n	Phase (ϕ_n) in Degrees
−15	—	—	$0 + j0$	*
−14	—	—	$-0.017 + j0.028$	121.6
−13	—	—	$0.004 + j0.031$	81.7
−12	—	—	$-0.005 + j0.033$	98.3
−11	—	—	$0.022 + j0.036$	58.4
−10	—	—	$0 + j0$	*
−9	—	—	$-0.027 + j0.044$	121.6
−8	—	—	$0.007 + j0.050$	81.7
−7	—	—	$-0.008 + j0.057$	98.3
−6	—	—	$0.041 + j0.066$	58.4
−5	—	—	$0 + j0$	*
−4	—	—	$-0.061 + j0.099$	121.6
−3	—	—	$0.019 + j0.133$	81.7
−2	—	—	$-0.029 + j0.199$	98.3
−1	—	—	$0.245 + j0.398$	58.4
0 (dc term)	0.6	—	$0.600 + j0$	0
1	0.490	0.796	$0.245 - j0.398$	−58.4
2	−0.058	0.398	$-0.029 - j0.199$	−98.3
3	0.039	0.265	$0.019 - j0.133$	−81.7
4	−0.122	0.199	$-0.061 - j0.099$	−121.6
5	0	0	$0 + j0$	*
6	0.082	0.133	$0.041 - j0.066$	−58.4
7	−0.017	0.114	$-0.008 - j0.057$	−98.3
8	0.014	0.099	$0.007 - j0.050$	−81.7
9	−0.054	0.088	$-0.027 - j0.044$	−121.6
10	0	0	$0 + j0$	*
11	0.045	0.072	$0.022 - j0.036$	−58.4
12	−0.010	0.066	$-0.005 - j0.033$	−98.3
13	0.009	0.061	$0.004 - j0.031$	−81.7
14	−0.035	0.057	$-0.017 - j0.028$	−121.6
15	0	0	$0 + j0$	*

* Technically, phase is irrelevant if magnitude = 0. These terms will, however, be plotted as having 0 degrees phase on the phase spectrum.

Option 3:

In Example 2.4 we determined the magnitudes and phases for the one-sided form of the Fourier series of $s(t)$. Using Equations (2.23)–(2.26) we can determine the complex exponential Fourier series coefficients. (Stop now and perform these calculations. Your c_n values and c_{-n} values should match those in Table 2-2.)

Example 2.5 Part B: Two-sided phase and magnitude spectra

In Example 2.4 we drew the magnitude and phase spectra for $s(t)$ based on the one-sided form of the Fourier series. We stated that the spectra are graphic tools that help us better visualize the behavior of a signal. Are there similar graphic tools based on the complex exponential form? If so, how do we develop them? How should they relate to the spectra based on the one-sided form?

Let's answer all of these questions together. First, note that the one-sided and complex exponential forms of the Fourier series are equivalent. Thus, spectra developed using the one-sided form and spectra developed using the complex exponential form should be equivalent; that is, they should contain the same information (although they may present that information in different ways). With this in mind, consider the structure of the one-sided magnitude spectrum. The one-sided magnitude spectrum is a plot of the magnitude of each term in the one-sided Fourier series (X_n) versus its corresponding frequency (nf_o). Similarly, the one-sided phase spectrum is a plot of the phase of each term in the one-sided Fourier series (ϕ_n) versus its corresponding frequency (nf_o). Refer back to Figures 2-10a and 2-10b on page 22. Note that the one-sided magnitude and phase spectra contain all the information necessary to reconstruct the original signal. The original signal is just the summation of cosines of the frequencies, magnitudes, and phases shown in the spectra.

Can we develop complex exponential magnitude and phase spectra using the same concepts as we used with the one-sided spectra? Let's consider creating a complex exponential magnitude spectrum by plotting the magnitude of each term of the complex exponential Fourier series (i.e., $|c_n|$) versus nf_o. In a similar way, we can create the complex exponential phase spectrum by plotting ϕ_n versus nf_o for all terms of the complex exponential Fourier series. Figures 2-11a and 2-11b show the complex exponential magnitude and phase spectra for $s(t)$, developed as we've just described. Note the following:

a. In comparing the complex exponential spectra with the one-sided spectra (i.e., comparing Figures 2-11a and 2-11b with Figures 2-10a and 2-10b), note that the complex exponentials have terms to the left and to the right of the y axis while the one-sided spectra have terms only to the right of the y axis. For this reason, the complex exponential spectra are often called two-sided spectra and the complex exponential form of the Fourier series is often called the two-sided form. (We can now also see where the one-sided form got its name.) The terms to the left-hand side of the y axis in the two-sided spectra correspond to the negative values of n in the summation in Equation (2.12).

b. In comparing the two-sided and one-sided magnitude spectra (i.e., comparing Figure 2-11a with Figure 2-10a), note that the dc term is the same in both spectra, that the terms to the right of the dc term in the two-sided magnitude spectrum are only half as large as the corresponding terms in the one-sided magnitude spectrum, and that the two-sided magnitude spectrum is symmetrical about the y axis. These observations are consistent with Equations (2.22)–(2.24) (be sure you understand why).

c. In comparing the two-sided and one-sided phase spectra (Figures 2-10b and 2-11b), note that all terms to the right of the y axis are the same for both spectra and that the two-sided phase spectrum exhibits odd symmetry about the y axis. These observations are consistent with Equations (2.22), (2.25), and (2.26) (be sure you understand why).

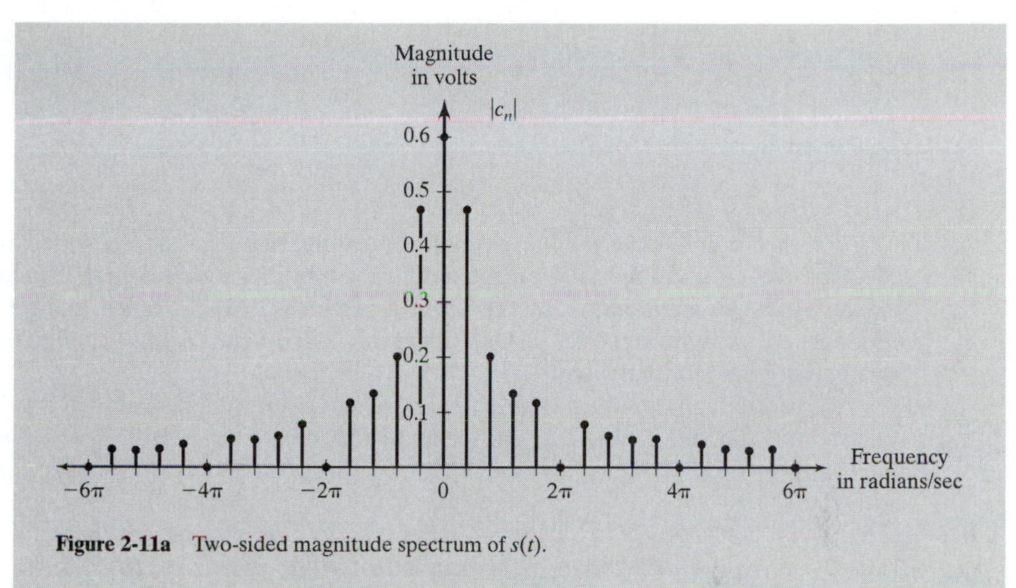

Figure 2-11a Two-sided magnitude spectrum of $s(t)$.

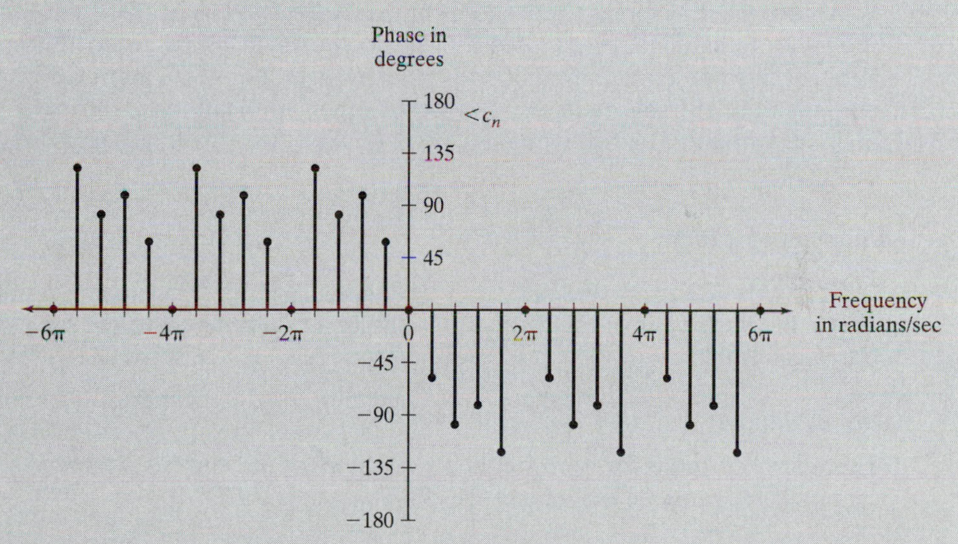

Figure 2-11b Two-sided phase spectrum of $s(t)$.

d. In examining the two-sided magnitude and phase spectra, be sure you understand what the x axis represents. Since we are plotting $|c_n|$ (magnitude spectrum) or ϕ_n (phase spectrum) versus nf_o, the x axis represents nf_o for various values of n. The units for the x axis thus involve frequency and are usually either given in Hz or radians/sec. Does this mean that the portion of the x axis to the left of zero represents negative frequency? The answer is yes in a *mathematically abstract* sense, but no in a physical sense. In a physical sense, a component in the left half of the two-sided magnitude spectrum (i.e., a component at a "negative" frequency) merely represents an *additional contribution* to the corresponding positive frequency. Consider, for instance, the two-sided magnitude spectrum of $s(t)$ in Figure 2-11a. The physical magnitude of the sinusoid at, say, 1.2π radians/sec is the *sum* of the components at 1.2π and -1.2π (i.e., 0.13 volts +0.13 volts = 0.26 volts). This result is, of course, consistent with the one-sided magnitude spectrum of $s(t)$ given in Figure 2-10a.

This last point—consistency in physical meaning between the one-sided and two-sided spectra—is extremely important and philosophically fundamental. Changing the way in which we mathematically represent a waveform (for instance, using the two-sided form of the Fourier series instead of the one-sided form) cannot change the physical characteristics of the waveform.

Be sure you understand the relationships between the one-sided and two-sided spectra, and the significance of the terms to the left of the y axis in the two-sided spectra. Also be sure you understand that all three forms of the Fourier series are equivalent.

2.2.2.3 *Why we use three different, equivalent forms of the Fourier series*

Why have we developed three different, equivalent forms of the Fourier series? As shown in Examples 2.3–2.5, and as will be shown in more detail throughout this book, each form has advantages and disadvantages. For some applications, one form may be more suitable than the others; for other applications, a different form may be more suitable. The following is a partial list of the advantages and disadvantages of each form:

Trigonometric form:

Advantages:

This is the most basic of the three forms and is very intuitive (we see $s(t)$ as a summation of sinusoids).

Disadvantages:

There are two terms for each frequency component (a_n and b_n), which means that we cannot easily generate the magnitude and phase spectra.

One-sided form:

Advantages:

1. This form has only one term at each frequency component, so it is very easy to generate magnitude and phase spectra. These spectra are powerful graphic tools that can help us visualize the behavior of a signal.
2. This form is still intuitive (we still see $s(t)$ as a summation of sinusoids).

Disadvantages:

The dc component of the magnitude spectrum is calculated in a manner different from the other components.

Complex exponential, or two-sided, form:

Advantages:

1. It is often easier to mathematically manipulate exponentials than sines and cosines.
2. The dc component is calculated in the same way as all other components, which is not the case for either the one-sided or the trigonometric form.
3. The complex exponential or two-sided form can be conceptually extended to include nonperiodic waveforms (the Fourier transform). We will do this in Section 2.4.

Disadvantages:

This form is nonintuitive for two reasons:

1. The form uses complex exponentials; it is now harder to see $s(t)$ as a summation of sinusoids. The form also uses complex coefficients.
2. The form introduces the abstract mathematical concept of "negative" frequency. (As discussed previously on page 29, for the magnitude spectrum we should consider the "negative" frequency component as physically representing an additional contribution at the corresponding positive frequency.)

Example 2.6 Fourier series for a rectangular pulse train

The rectangular pulse train $v(t)$ shown in Figure 2-12 is useful for many communication systems applications. Determine its frequency domain representation.

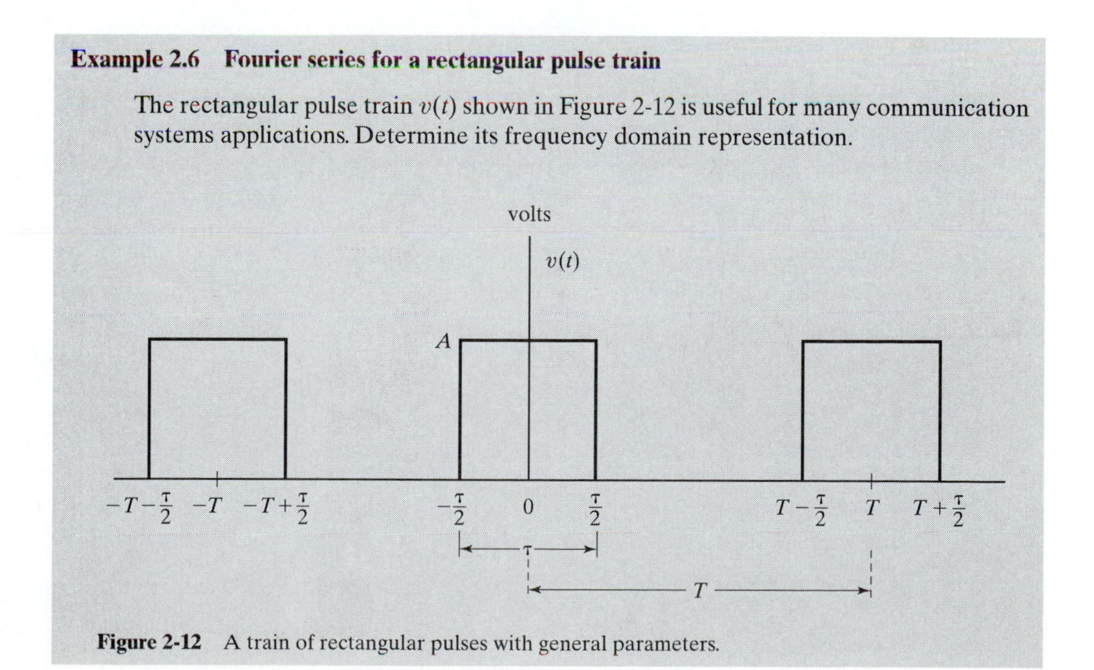

Figure 2-12 A train of rectangular pulses with general parameters.

Solution

Can we use the Fourier series?

1. Is $v(t)$ periodic? Yes, period $= T$.
2. Does $v(t)$ contain finite energy within a period?

$$\int_{t_o}^{t_o+T} |v(t)|^2 dt = \int_{-\frac{\tau}{2}}^{\frac{\tau}{2}} A^2 dt = A^2\tau < \infty$$

The constraints are thus met and we can use the Fourier series. Let's try the Fourier series expansion using the complex exponential or two-sided form:

$$v(t) = \sum_{n=-\infty}^{\infty} c_n e^{j2\pi n f_o t} \text{ where } c_n = \frac{1}{T}\int_{t_o}^{t_o+T} v(t)e^{-j2\pi n f_o t}dt \tag{2.12R}$$

Solving for the complex exponential or two-sided Fourier coefficients:

$$\begin{aligned} c_n &= \frac{1}{T}\int_{-\frac{T}{2}}^{\frac{T}{2}} v(t)e^{-j2\pi n f_o t}dt = \frac{1}{T}\int_{-\frac{\tau}{2}}^{\frac{\tau}{2}} Ae^{-j2\pi n f_o t}dt = \frac{1}{T}\left(\frac{A}{-j2\pi n f_o}\right)e^{-j2\pi n f_o t}\Bigg|_{-\frac{\tau}{2}}^{\frac{\tau}{2}} \\ &= \frac{A}{-j2\pi n}\left(e^{-j\pi n f_o \tau} - e^{j\pi n f_o \tau}\right) = \frac{A}{\pi n}\left(\frac{e^{j\pi n f_o \tau} - e^{-j\pi n f_o \tau}}{2j}\right) = \frac{A\sin(\pi n f_o \tau)}{\pi n} \\ &= \frac{A\sin(\frac{\pi n \tau}{T})}{\pi n} \end{aligned}$$

Let's manipulate this expression a little further (you'll see why shortly). Multiplying and dividing by $\frac{\tau}{T}$,

$$c_n = \frac{A\sin(\frac{\pi n \tau}{T})}{\pi n} = \left(\frac{\frac{\tau}{T}}{\frac{\tau}{T}}\right)\left(\frac{A\sin(\frac{\pi n \tau}{T})}{\pi n}\right) = \left(\frac{A\tau}{T}\right)\left(\frac{\sin(\frac{\pi n \tau}{T})}{\frac{\pi n \tau}{T}}\right)$$

The function $\frac{\sin(x)}{x}$ will occur many times in communication systems analysis. We will call this the *sinc* function

$$\text{sinc}(x) \equiv \frac{\sin(x)}{x} \tag{2.27}$$

Thus,

$$c_n = \frac{A\tau}{T}\,\text{sinc}\left(\frac{\pi n \tau}{T}\right)$$

and we can express the rectangular pulse train $v(t)$ in the frequency domain as:

$$v(t) = \sum_{n=-\infty}^{\infty} c_n e^{j2\pi n f_o t} \text{ where } c_n = \left(\frac{A\tau}{T}\right)\text{sinc}\left(\frac{\pi n \tau}{T}\right)$$

Example 2.7 Magnitude and phase spectra of a rectangular pulse train

Let's further our understanding of the results of Example 2.6 by plotting two-sided magnitude and phase spectra for $v(t)$ given particular values of A (amplitude), τ (pulse width), and T (period). Let's say $A = 3$ volts, $\tau = 2$ milliseconds, and $T = 10$ milliseconds, as shown in Figure 2-13.

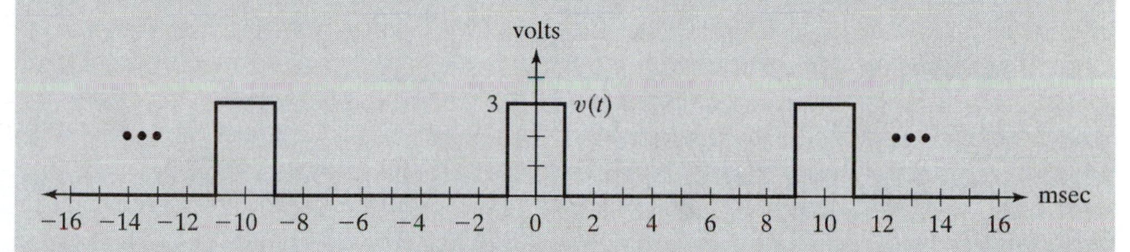

Figure 2-13 Rectangular pulse train with 3-volt amplitude, 2-millisecond pulse width, and 10-millisecond period.

Calculating the Fourier coefficients,

$$c_1 = \frac{A\tau}{T}\operatorname{sinc}\left(\frac{\pi n\tau}{T}\right) = \frac{(3)(0.002)}{(0.01)}\left[\frac{\sin\left(\frac{0.002}{0.01}\pi\right)}{\left(\frac{0.002}{0.01}\pi\right)}\right] = 0.6\left(\frac{\sin(0.2\pi)}{0.2\pi}\right) = 0.561 \text{ volts}$$

$$c_2 = \frac{A\tau}{T}\operatorname{sinc}\left(\frac{\pi n\tau}{T}\right) = 0.6\left(\frac{\sin(0.4\pi)}{0.4\pi}\right) = 0.454 \text{ volts}$$

$$c_3 = 0.6\left(\frac{\sin(0.6\pi)}{0.6\pi}\right) = 0.303 \text{ volts}$$

$$c_4 = 0.6\left(\frac{\sin(0.8\pi)}{0.8\pi}\right) = 0.140 \text{ volts}$$

$$c_5 = 0.6\left(\frac{\sin(\pi)}{\pi}\right) = 0 \text{ volts}$$

$$c_6 = 0.6\left(\frac{\sin(1.2\pi)}{1.2\pi}\right) = -0.094 \text{ volts}$$

$$c_7 = 0.6\left(\frac{\sin(1.4\pi)}{1.4\pi}\right) = -0.130 \text{ volts}$$

$$\vdots$$

Also note that

$$c_{-1} = \frac{A\tau}{T}\operatorname{sinc}\left(\frac{-\pi\tau}{T}\right) = 0.6\left(\frac{\sin(-0.2\pi)}{-0.2\pi}\right) = c_1$$

Similarly, $c_{-2} = c_2$, $c_{-3} = c_3$, etc. (Generally $c_{-n} = c_n^*$, as shown in Equation (2.22), but since all coefficients are real in this example, $c_n = c_{-n}$.)

What about c_0?

$$c_0 = \frac{A\tau}{T}\text{sinc}(0) = 0.6\left(\frac{\sin(0)}{0}\right) = 0.6\left(\frac{0}{0}\right) \text{ Oops!}$$

Let's try another approach. Considering the limiting case and then applying L'Hôpital's rule:

$$c_0 = \frac{A\tau}{T}\lim_{\epsilon\to 0}\frac{\sin\epsilon}{\epsilon} = \frac{A\tau}{T}\lim_{\epsilon\to 0}\frac{\frac{d}{d\epsilon}\sin(\epsilon)}{\frac{d}{d\epsilon}\epsilon} = \frac{A\tau}{T}\lim_{\epsilon\to 0}\frac{\cos(\epsilon)}{1} = \frac{A\tau}{T} = 0.6 \tag{2.28}$$

Observing that $f_o = 1/T = 100$ Hz or 0.1 kHz, we can now plot the two-sided magnitude and phase spectra for $v(t)$ in Figures 2-14a and 2-14b.

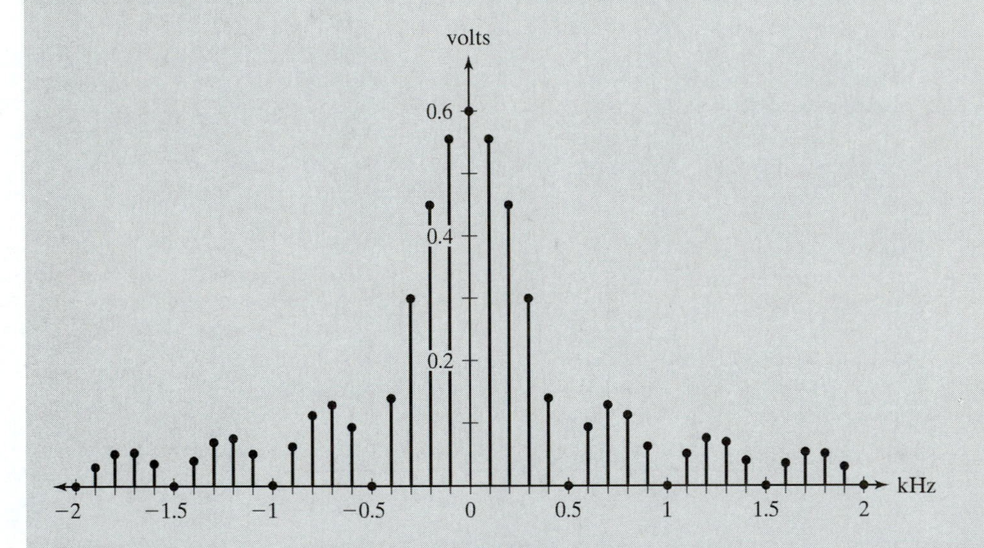

Figure 2-14a Two-sided magnitude spectrum of $v(t)$.

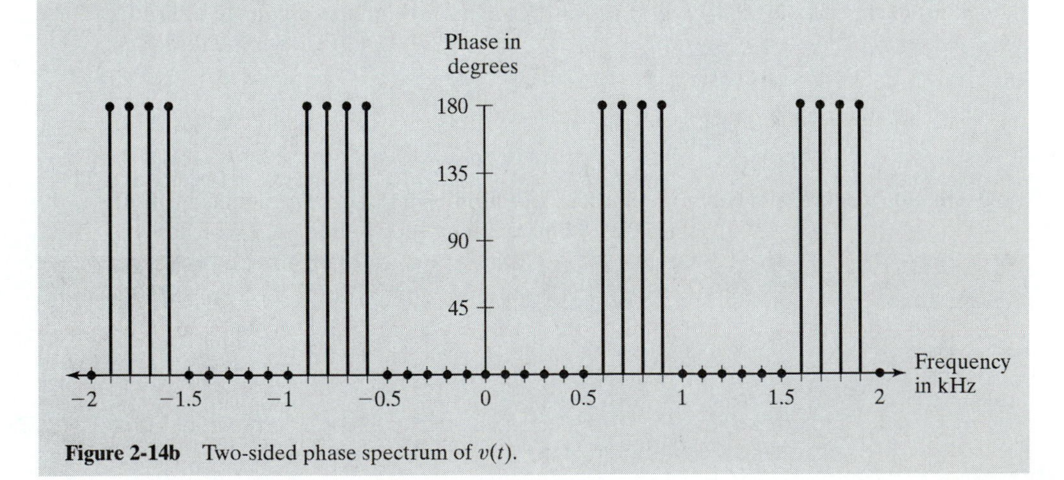

Figure 2-14b Two-sided phase spectrum of $v(t)$.

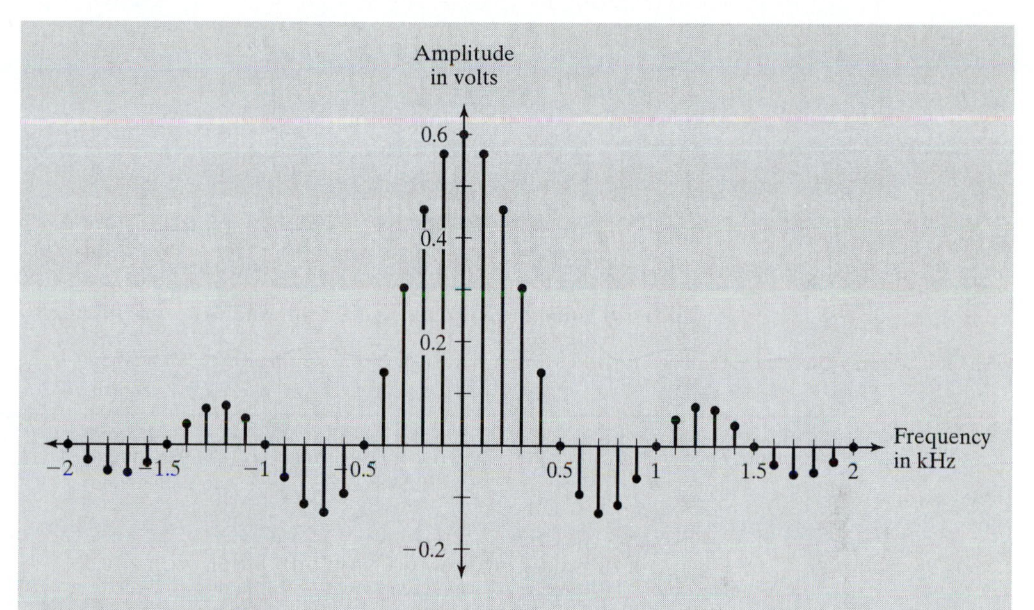

Figure 2-14c Two-sided amplitude spectrum of $v(t)$.

Let's consider another type of spectrum we can plot when all the c_n values are real—an amplitude spectrum. As we've just seen, when c_n is real, phase will be either 0 degrees (when c_n is positive) or 180 degrees (when c_n is negative). Thus, for waveforms where all Fourier coefficients are real, rather than plot two spectra (phase and magnitude), we can plot just an *amplitude* spectrum, consolidating all the information of the magnitude and phase spectra into a single plot as shown in Figure 2-14c.

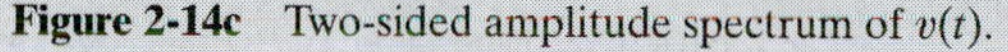

We can make the following observations about the amplitude spectrum of $v(t)$ in Example 2.7:

1. The amplitude coefficients follow the envelope $\frac{A\tau}{T}\operatorname{sinc}(\pi f \tau)$.
2. The value of the dc term is $\frac{A\tau}{T}$, since $\operatorname{sinc}(0) = 1$ (by applying L'Hôpital's rule).
3. The frequency spacing between adjacent coefficients is $\frac{1}{T}$ Hz.
4. The zero crossings of the envelope occur at integral multiples of $\frac{1}{\tau}$ Hz.

None of these observations is restricted by our selection of $A = 3$ volts, $\tau = 2$ milliseconds, and $T = 10$ milliseconds in Example 2.7. We can use these observations to draw a generalized amplitude spectrum for a rectangular pulse train of amplitude A, pulse width τ, and pulse period T (i.e., the pulse train in Figure 2-12), as shown in Figure 2-15.

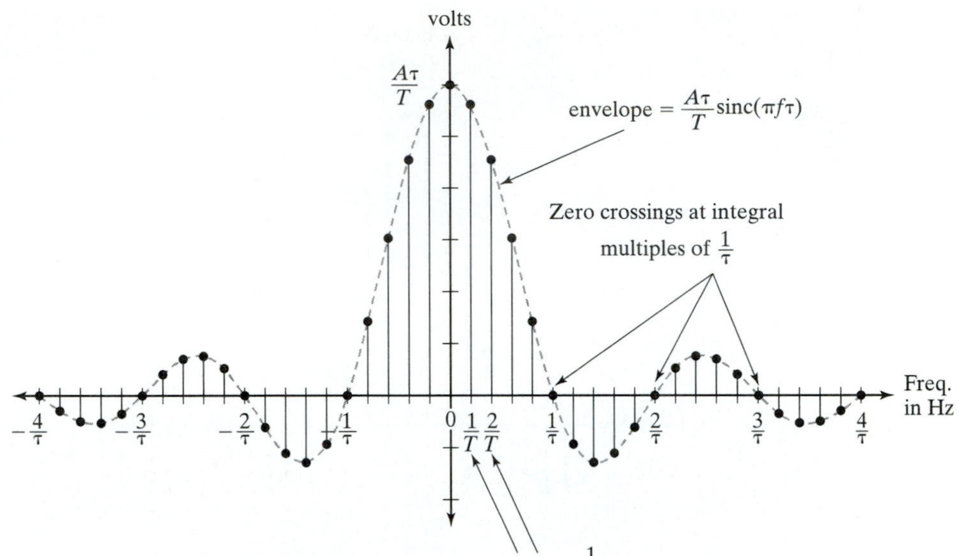

Figure 2-15 Amplitude spectrum for a rectangular pulse train of generalized amplitude A, pulse width τ, and pulse period T.

Let's now examine the effect that pulse width (τ) has on the shape of the amplitude spectrum.

Example 2.8 Effect of pulse width on the amplitude spectrum of a rectangular pulse train

Consider the rectangular pulse train shown in Figure 2-16. Draw the two-sided amplitude spectra for the following values of τ:

a. $\tau = 0.0625$ sec
b. $\tau = 0.125$ sec
c. $\tau = 0.25$ sec

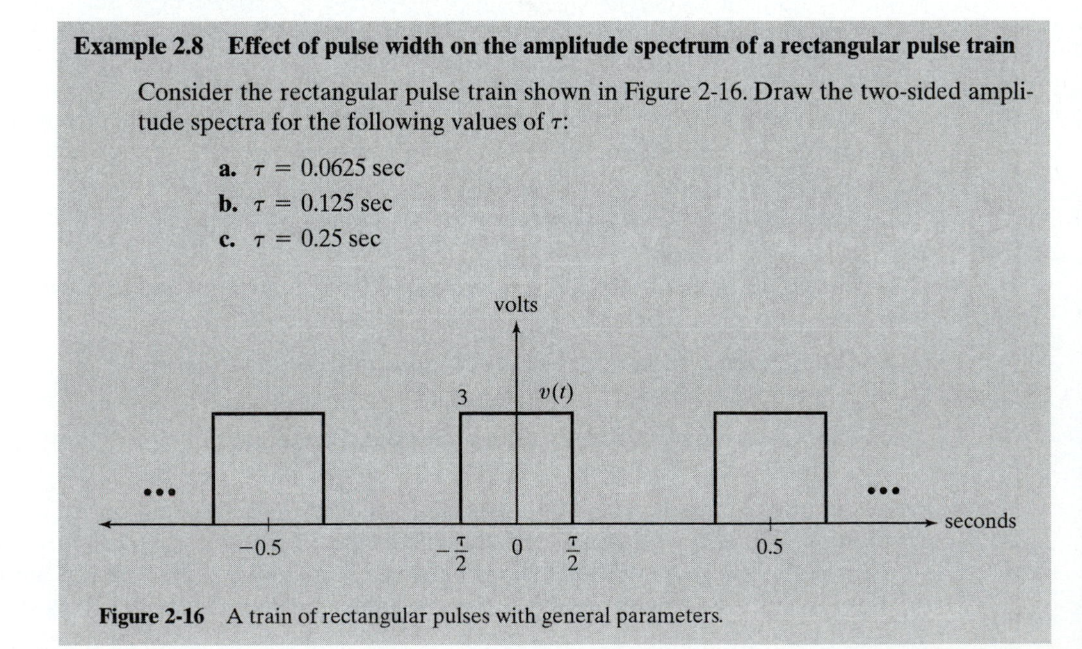

Figure 2-16 A train of rectangular pulses with general parameters.

Solution

In Example 2.6 we determined that

$$v(t) = \sum_{n=-\infty}^{\infty} c_n e^{j2\pi n f_o t} \quad where\ c_n = \left(\frac{3\tau}{0.5}\right)\text{sinc}\left(\frac{\pi n \tau}{0.5}\right) = 6\tau\text{sinc}(2\pi n\tau)$$

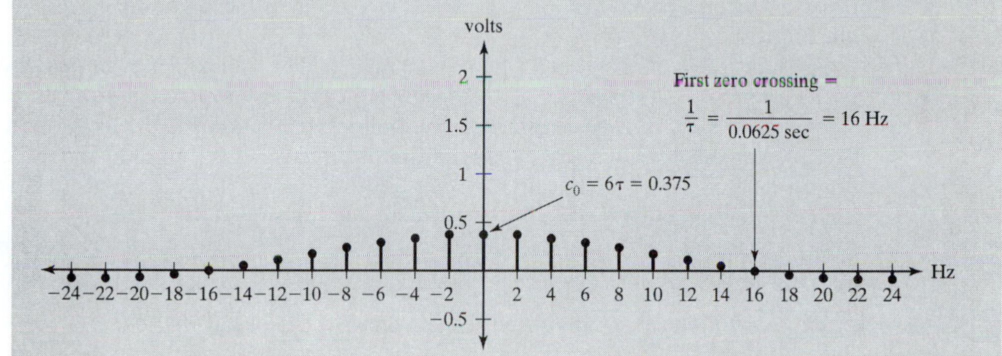

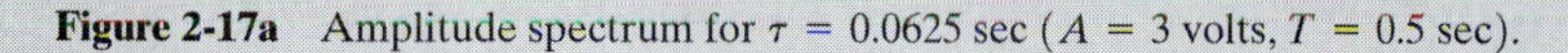

Figure 2-17a Amplitude spectrum for $\tau = 0.0625$ sec ($A = 3$ volts, $T = 0.5$ sec).

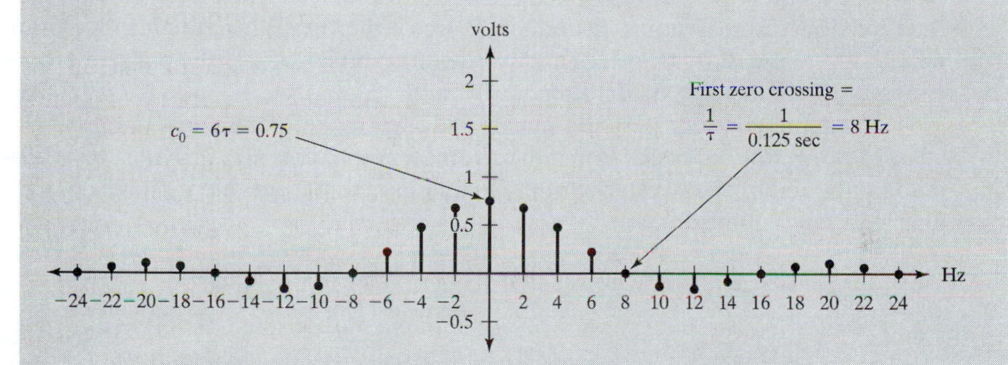

Figure 2-17b Amplitude spectrum for $\tau = 0.125$ sec ($A = 3$ volts, $T = 0.5$ sec).

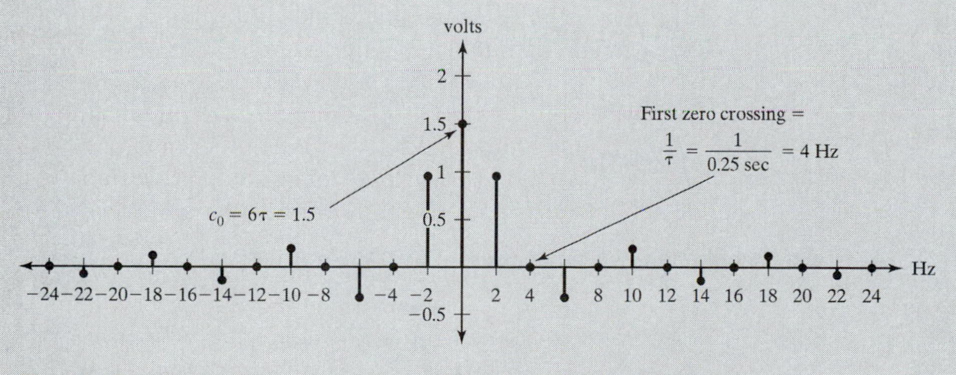

Figure 2-17c Amplitude spectrum for $\tau = 0.25$ sec ($A = 3$ volts, $T = 0.5$ sec).

For Parts a, b, and c above we know that the amplitude of the 0 Hz component of the spectrum is $c_0 = 6\tau$. We also know that the distance between adjacent components in the spectrum is $1/0.5 = 2$ Hz for Parts a, b, and c. Additionally, we know that zero crossings occur when $f\tau$ is a nonzero integer and that the envelope for the amplitude spectrum is a sinc function.

Amplitude plots for Parts a, b, and c are shown in Figures 2-17a, b, and c. Be sure to observe and understand how the differences in τ affect the shape of the corresponding amplitude spectrum.

We can see from Figures 2-17a, b, and c that smaller values of τ cause an increase in the frequency of the first zero crossing and larger values of τ cause a decrease in the frequency at the first zero crossing. As we shall see in Section 2.3, the location of the first zero crossing in the magnitude spectrum is directly related to the bandwidth required by a digital communication system transmitting rectangular impulses. We will further show that pulse width τ is directly related to transmission speed. Section 2.3 will thus provide the techniques and insight necessary to begin quantitatively examining communication system trade-offs involving bandwidth versus transmission speed.

2.3 Representing Power in the Frequency Domain

Now let's talk about how we represent the power of a transmitted or received signal in the time domain and frequency domain. Power is important because it's the parameter we will use to measure distortion and bandwidth, and it's one of the parameters we will use to measure the effects of noise.

Let's first consider periodic signals and start by defining a new parameter called *average normalized power* (P_s) as the average power that $s(t)$ provides to a resistive load multiplied by the resistance of the load. Let R represent the resistance of a particular load. In the time domain,

$$\begin{aligned} P_s &\equiv R(\text{average power that } s(t) \text{ provides to the load}) \\ &= R\left[\frac{1}{T}\int_{t_o}^{t_o+T} (\text{voltage across load})(\text{current through load})\,dt\right] \\ &= R\left[\frac{1}{T}\int_{t_o}^{t_o+T} s(t)\left(\frac{s(t)}{R}\right)dt\right] \\ &= \frac{1}{T}\int_{t_o}^{t_o+T} s^2(t)\,dt \end{aligned} \tag{2.29a}$$

Thus, in the time domain we can represent average normalized power as:

$$P_s = \frac{1}{T}\int_{t_o}^{t_o+T} s^2(t)\,dt \tag{2.29b}$$

Units for P_s are watt-ohms or volts^2. Be sure you understand why these two units are equivalent and why each is appropriate. Conceptually, we can think of the value of average normalized power (although not the units) as the average power provided by $s(t)$ if $s(t)$ were to appear across the terminals of a 1Ω resistor.

How can we represent average normalized power in the frequency domain? Let's start by considering the one-sided Fourier series representation for $s(t)$:

$$\underbrace{s(t)}_{\text{time domain representation}} = \underbrace{X_0 + \sum_{n=1}^{\infty} X_n \cos(2\pi n f_o t + \phi_n)}_{\text{frequency domain representation}} \tag{2.8R}$$

The left-hand side of the equation is the time domain representation of $s(t)$, and the right-hand side is a frequency domain representation of $s(t)$. Since the two representations are equivalent, the average normalized power in the time domain representation must equal the average normalized power in the frequency domain representation. As we discussed earlier, how we represent something mathematically cannot change its physical properties.

Let's start with the time domain definition of average normalized power:

$$P_s = \frac{1}{T}\int_{t_o}^{t_o+T} s^2(t)\,dt \tag{2.29R}$$

Next, substitute in the frequency domain representation of $s(t)$.

$$P_s = \frac{1}{T}\int_{t_o}^{t_o+T} s^2(t)\,dt = \frac{1}{T}\int_{t_o}^{t_o+T} \left\{X_0 + \sum_{n=1}^{\infty} X_n \cos(2\pi n f_o t + \phi_n)\right\}^2 dt$$

$$= \frac{1}{T}\int_{t_o}^{t_o+T} \left\{X_0 + \sum_{n=1}^{\infty} X_n \cos(2\pi n f_o t + \phi_n)\right\}\left\{X_0 + \sum_{m=1}^{\infty} X_m \cos(2\pi m f_o t + \phi_m)\right\} dt$$

(Note that the second index has been changed from n to m in order to preserve the cross terms.)

$$= \frac{1}{T}\int_{t_o}^{t_o+T} X_0^2\,dt + \frac{1}{T}\int_{t_o}^{t_o+T} 2X_0 \sum_{n=1}^{\infty} X_n \cos(2\pi n f_o t + \phi_n)\,dt$$

$$+ \frac{1}{T}\int_{t_o}^{t_o+T} \left\{\sum_{n=1}^{\infty} X_n \cos(2\pi n f_o t + \phi_n)\right\}\left\{\sum_{m=1}^{\infty} X_m \cos(2\pi m f_o t + \phi_m)\right\} dt \tag{2.30}$$

Solving the three terms one at a time,

$$\textit{Term 1:}\ \frac{1}{T}\int_{t_o}^{t_o+T} X_0^2\,dt = X_0^2 \tag{2.31}$$

$$\textit{Term 2:}\ \frac{1}{T}\int_{t_o}^{t_o+T} 2X_0\sum_{n=1}^{\infty} X_n\cos(2\pi n f_o t + \phi_n)\,dt$$

$$= \frac{2X_0}{T}\sum_{n=1}^{\infty} X_n\left\{\int_{t_o}^{t_o+T}\cos(2\pi n f_o t + \phi_n)\,dt\right\} = \frac{2X_0}{T}\sum_{n=1}^{\infty}\{0\} = 0 \tag{2.32}$$

since each integral involves a whole number of cycles of cosine.

$$\textit{Term 3:}\ \frac{1}{T}\int_{t_o}^{t_o+T}\left\{\sum_{n=1}^{\infty} X_n\cos(2\pi n f_o t + \phi_n)\right\}\left\{\sum_{m=1}^{\infty} X_m\cos(2\pi m f_o t + \phi_m)\right\}dt$$

$$= \frac{1}{T}\sum_{n=1}^{\infty}\sum_{m=1}^{\infty} X_n X_m\left\{\int_{t_o}^{t_o+T}\cos(2\pi n f_o t + \phi_n)\cos(2\pi m f_o t + \phi_m)\,dt\right\} \tag{2.33}$$

Due to the orthogonality of the one-sided Fourier series terms (which you can prove in the same way you prove the orthogonality of the trigonometric Fourier series components in Problem 2.4),

$$\int_{t_o}^{t_o+T}\cos(2\pi n f_o t + \phi_n)\cos(2\pi m f_o t + \phi_m)\,dt = \begin{cases} 0 & \text{if } m \neq n \\ 0.5T & \text{if } m = n \neq 0 \end{cases} \tag{2.34}$$

Thus,

$$\frac{1}{T}\sum_{n=1}^{\infty}\sum_{m=1}^{\infty} X_n X_m\left\{\int_{t_o}^{t_o+T}\cos(2\pi n f_o t + \phi_n)\cos(2\pi m f_o t + \phi_m)\,dt\right\}$$

$$= \frac{1}{T}\sum_{n=1}^{\infty} X_n^2(0.5T) = \sum_{n=1}^{\infty}\frac{X_n^2}{2} \tag{2.35}$$

Adding the effects of all three terms,

$$P_s = \underbrace{\frac{1}{T}\int_{t_o}^{t_o+T} s^2(t)\,dt}_{\text{time domain representation}} = \underbrace{X_0^2 + \sum_{n=1}^{\infty}\frac{X_n^2}{2}}_{\text{frequency domain representation}} \tag{2.36}$$

Equation (2.36), known as *Parseval's theorem*, makes good intuitive sense. The time domain integral on the left-hand side of Equation (2.36) is summing all of the signal's average normalized power as represented in the time domain, and the frequency domain summation on the right-hand side of Equation (2.36) is summing all of the signal's average normalized power as represented in the frequency domain. In both cases we are representing the same physical parameter (average normalized power)—we're just using different mathematical representations. As we know, changing a parameter's mathematical representation cannot change its physical properties. Section 2.3.3 will show that we can also represent Parseval's theorem using the two-sided form of the Fourier series.

2.3.1 The One-Sided Average Normalized Power Spectrum

The work we have just done allows us to translate a magnitude spectrum into an average normalized power spectrum, which shows us graphically how a signal distributes its power throughout a frequency band. As we will see shortly, such a spectrum is an extremely valuable tool for calculating and visualizing a signal's bandwidth.

Figures 2-18a and b show a magnitude spectrum and the corresponding average normalized power spectrum for a typical signal. In examining Figures 2-18a and b, note the following:

a. The orthogonality of the terms in the Fourier series means that each component in the average normalized power spectrum is related only to the corresponding term in the magnitude spectrum (i.e., the average normalized power at frequency f_o is related only to X_1—the magnitude of the Fourier series term at f_o; the average normalized power at frequency $2f_o$ is related only to X_2—the magnitude of the Fourier series term at $2f_o$; etc.). In other words, when we square the magnitude spectrum to produce an average normalized power spectrum, *there are no cross terms*. We see this graphically in Figures 2-18a and b, and mathematically in Equations (2.33)–(2.35).

b. Observe that the dc term for the average normalized power spectrum is calculated in a manner slightly different from the other terms: the dc term is $X_0{}^2$, and the other terms are $\frac{1}{2}X_n{}^2$. We can see the difference by comparing Equations (2.31) and (2.35). We can also relate this difference to our experiences in calculating dc power and average ac power in electric circuits.

c. Observe the units in the average normalized power spectrum (Figure 2-18b). Be sure you understand why these units are consistent and appropriate.

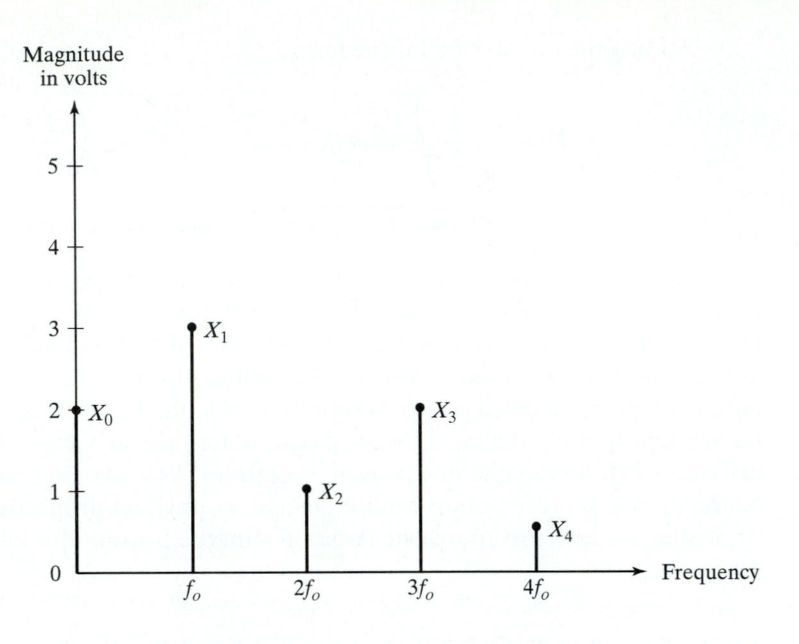

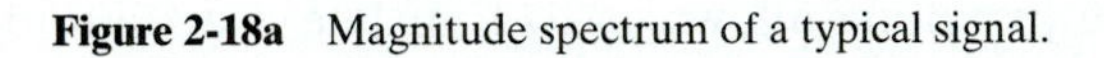
Figure 2-18a Magnitude spectrum of a typical signal.

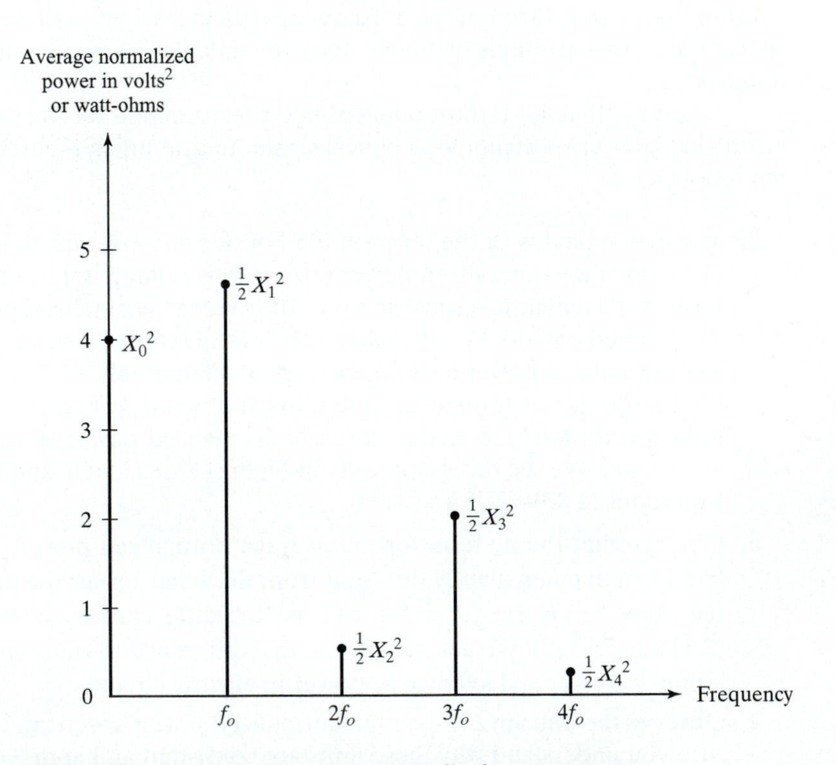

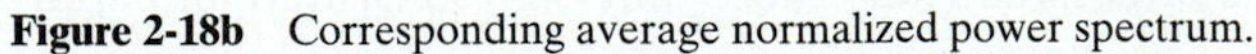
Figure 2-18b Corresponding average normalized power spectrum.

Example 2.9 Drawing an average normalized power spectrum

The average normalized power spectrum shows graphically how a signal distributes its power through a frequency band. Draw the average normalized power spectrum for the rectangular pulse train $v(t)$ in Example 2.7 on p. 33.

Solution

In Example 2.7 we determined the complex exponential Fourier series coefficients for $v(t)$. Converting these coefficients to magnitudes of the one-sided form (see Equations (2.23) and (2.24)),

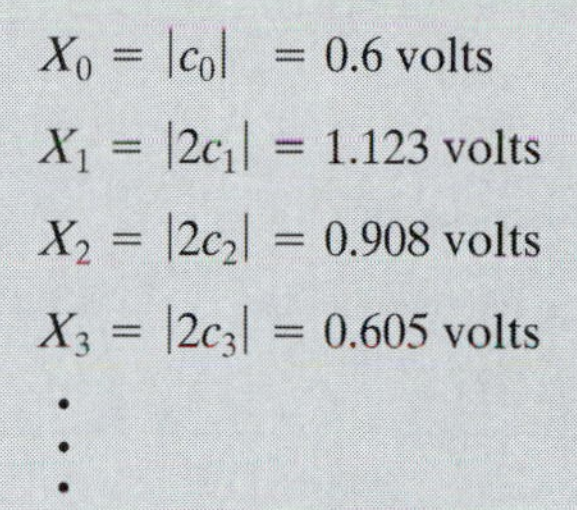

$$X_0 = |c_0| = 0.6 \text{ volts}$$

$$X_1 = |2c_1| = 1.123 \text{ volts}$$

$$X_2 = |2c_2| = 0.908 \text{ volts}$$

$$X_3 = |2c_3| = 0.605 \text{ volts}$$

$$\vdots$$

The one-sided magnitude spectrum of $v(t)$ is thus plotted in Figure 2-19a.

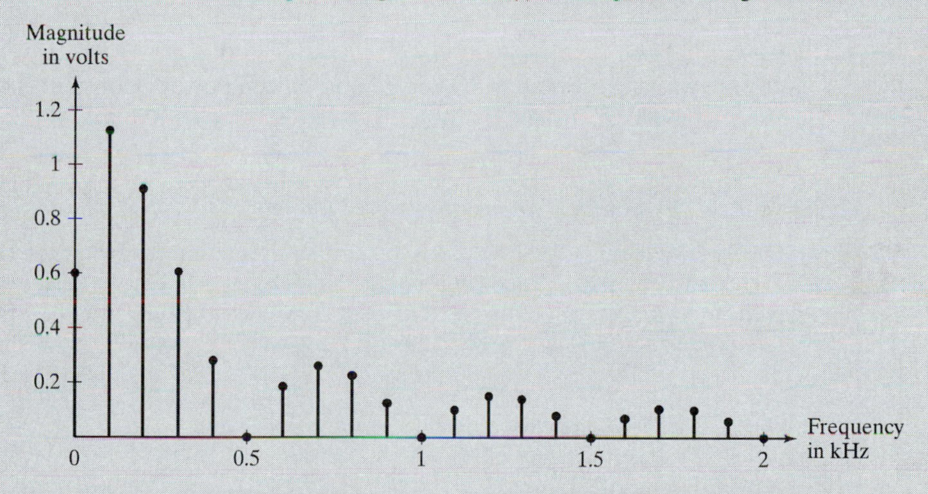

Figure 2-19a One-sided magnitude spectrum for $v(t)$. (Compare to Figure 2-14a.)

For the average normalized power spectrum, the dc term is

$$X_0^2 = (0.6 \text{ volts})^2 = 0.36 \text{ volts}^2$$

The term at the first harmonic is

$$\frac{X_1^2}{2} = \frac{(1.123 \text{ volts})^2}{2} = 0.630 \text{ volts}^2$$

The term at the second harmonic is

$$\frac{X_2^2}{2} = \frac{(0.908 \text{ volts})^2}{2} = 0.412 \text{ volts}^2$$

$$\vdots$$

We can thus plot the average normalized power spectrum as shown in Figure 2-19b.

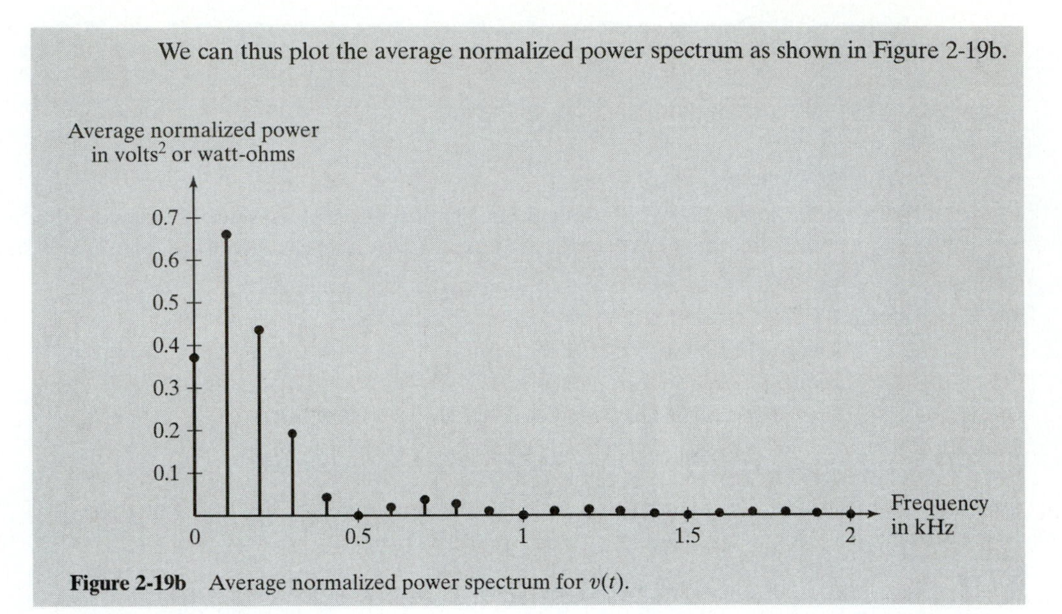

Figure 2-19b Average normalized power spectrum for $v(t)$.

Why are we so concerned with average normalized power? Consider Example 2-10:

Example 2.10 Relevance of average normalized power

Suppose a communication system is attempting to transmit the rectangular pulse train $v(\text{t})$ from Examples 2.7 and 2.9 across a channel that acts like an ideal lowpass filter with cutoff frequency $f_{\text{cutoff}} = 300$ Hz. (Note that such a channel is called an *ideal baseband channel* with a bandwidth of 300 Hz.)

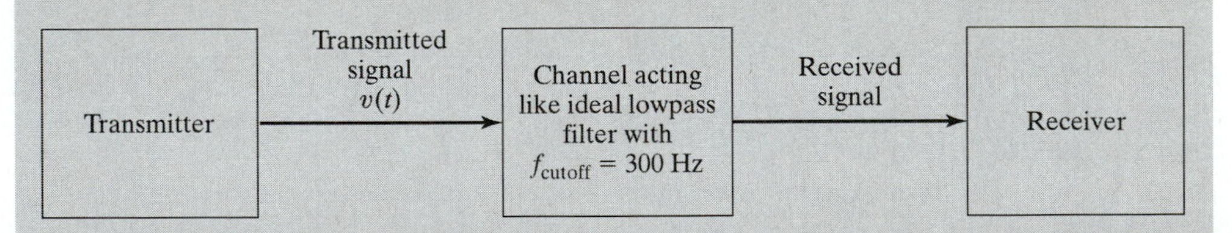

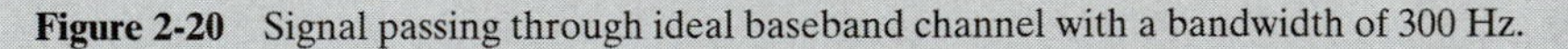

Figure 2-20 Signal passing through ideal baseband channel with a bandwidth of 300 Hz.

a. How much average normalized power passes through the channel?

b. What percentage of the average normalized power of $v(t)$ lies in the frequency band passed by the channel?

Solutions

a. This type of calculation is difficult in the time domain but is much simpler in the frequency domain. We know that the channel passes all frequency components from dc up to 300 Hz, but not any frequency components above 300 Hz. Using the one-sided average normalized power spectrum of $v(t)$ (previously plotted in Figure 2-19b), we can see that the channel passes only the dc component and the first three harmonics of $v(t)$. The average normalized power of $v(t)$ lying in the frequency band from dc to 300 Hz is

$$X_0^2 + \frac{X_1^2}{2} + \frac{X_2^2}{2} + \frac{X_3^2}{2} = 0.36 + 0.630 + 0.412 + 0.183 = 1.586 \text{ volts}^2$$

b. The percentage of the average normalized power lying within the bandwidth of the channel is

$$\frac{1.586 \text{ volts}^2}{P_v}$$

where P_v represents all of the average normalized power of $v(t)$.

We can calculate P_v as

$$P_v = \frac{1}{T}\int_{t_o}^{t_o+T} v^2(t)\,dt \quad \text{or as} \quad P_v = X_0^2 + \sum_{n=1}^{\infty} \frac{X_n^2}{2}$$

whichever is simpler for the particular signal. For the rectangular pulse train, it is simpler to calculate P_v in the time domain.

$$P_v = \frac{1}{T}\int_{t_o}^{t_o+T} v^2(t)\,dt = \frac{1}{0.01}\int_{-0.001}^{0.009} v^2(t)\,dt = \frac{1}{0.01}\int_{-0.001}^{0.001} 3^2\,dt = 1.8 \text{ volts}^2$$

Thus, the percentage of the average power of $v(t)$ lying within the bandwidth of the channel is

$$\frac{1.586 \text{ volts}^2}{1.8 \text{ volts}^2} = 88\%$$

As we discussed in Example 2.3, Part B on page 16, the amount of distortion in the received signal depends on which harmonics of the transmitted signal pass through the channel and which harmonics do not pass through and are therefore not present at the receiver. Specifically, the higher the percentage of average normalized power within the harmonics that pass through the channel, the smaller the distortion of the received signal. Thus, for an ideal baseband channel, we can quantify the channel's distorting effects as the percentage of the transmitted signal's power that lies outside the band of frequencies passed by the channel.

Figure 2-21a shows the received signal if $v(t)$ is transmitted through an ideal baseband channel with a bandwidth of 300 Hz. Note how the corners of the rectangular pulses are rounded (this is why we didn't transmit rectangular pulses in Example 1.2). Figures 2-21b–d show the received signal if the channel bandwidth is 500, 1000, and 2000 Hz, respectively.

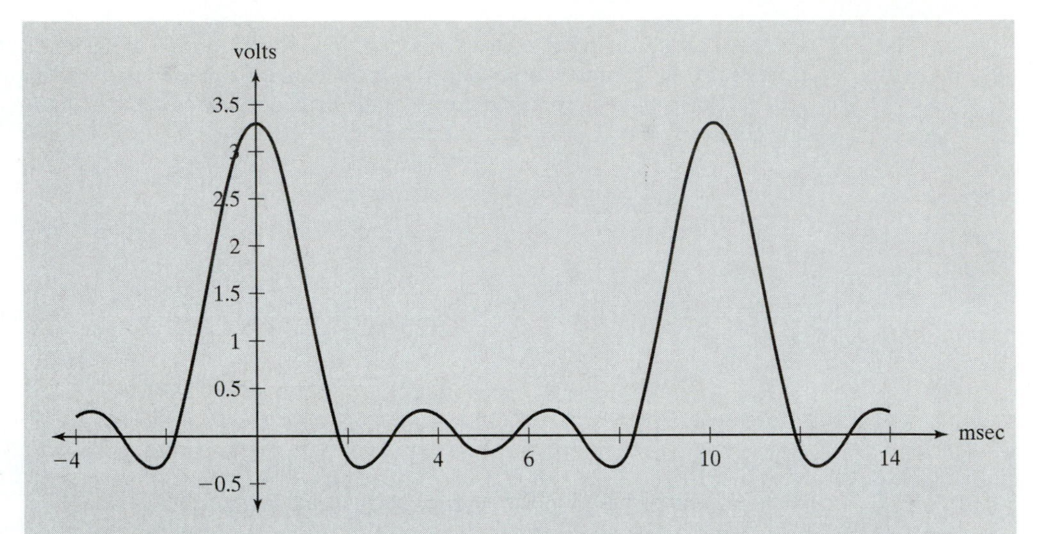

Figure 2-21a Received signal if channel bandwidth = 300 Hz. 88% of transmitted signal's average normalized power lies within the frequency band passed by the channel.

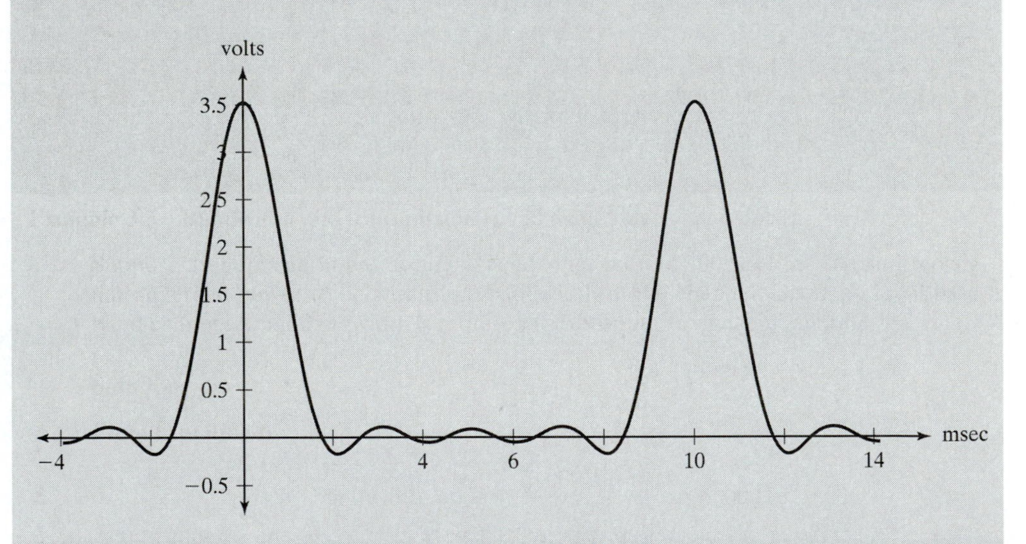

Figure 2-21b Received signal if channel bandwidth = 500 Hz. 90% of transmitted signal's average normalized power is "in band."

How do the calculations concerning percentage of in-band average normalized power in Example 2.10 relate to average in-band power for a general load of R ohms rather than a normalized 1Ω load? Looking at Example 2.10 the average power if $v(t)$ appears across an $R\Omega$ load is

$$P_{R\Omega\,\text{load}} = \frac{1}{T}\int_{t_o}^{t_o+T} v_v(t)i_v(t)\,dt = \frac{1}{T}\int_{t_o}^{t_o+T} v(t)\left(\frac{v(t)}{R}\right)dt = \frac{1}{R}\left[\frac{1}{T}\int_{t_o}^{t_o+T} s^2(t)\,dt\right] = \frac{P_v}{R} \quad (2.37)$$

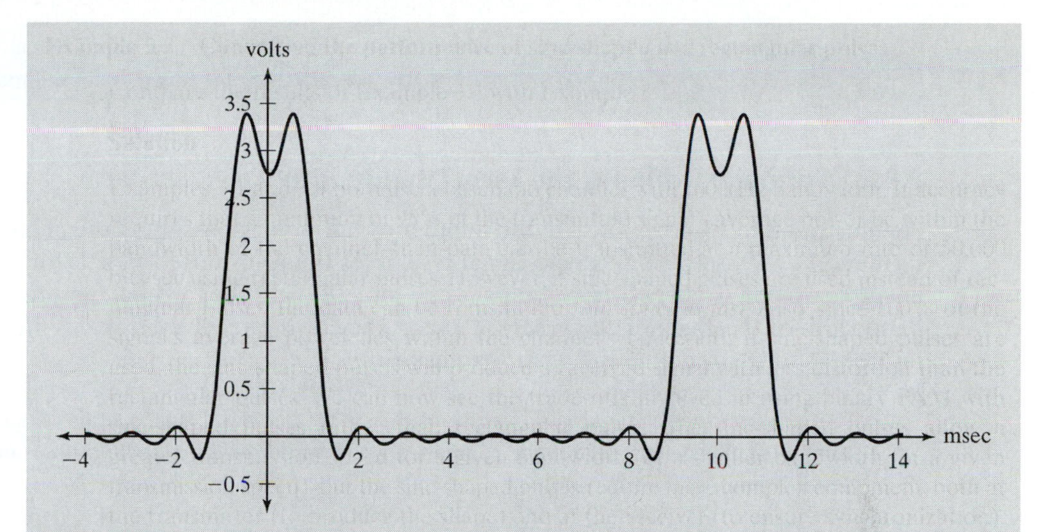

Figure 2-21c Received signal if channel bandwidth = 1000 Hz. 95% of transmitted signal's average normalized power is "in band."

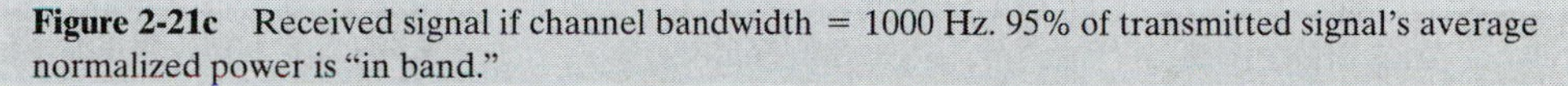

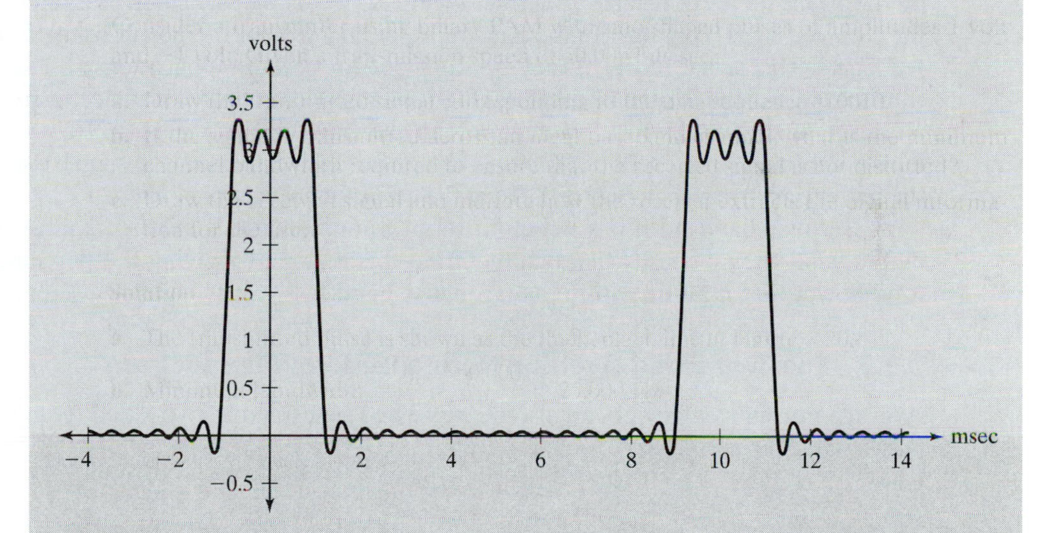

Figure 2-21d Received signal if channel bandwidth = 2000 Hz. 97.5% of transmitted signal's average normalized power is "in band."

Similarly, the power spectrum for $v(t)$ appearing across an $R\Omega$ load has a dc term of $X_0^2/(2R)$, a term at the first harmonic of $X_1^2/(2R)$, a term at the second harmonic of $X_2^2/(2R)$, etc. The calculation for percentage of in-band power will have both the numerator and denominator divided by R. Since these divisions cancel out, the result will be the same as in Example 2.10 unchanged. Thus, the result in Example 2.10 is the percentage of average power lying within the bandwidth of the channel *regardless of the resistance of the load.* For ratios and percentages, the adjective "normalized" is therefore not needed and we are not concerned with the actual value of the load; it doesn't matter.[3]

2.3.2 Formally Defining the Term "Bandwidth"

As illustrated in Figures 2-21a–d, the number of harmonics passed by the channel determines how closely the received signal resembles the transmitted signal (for another example, review Figures 2-8a–e on pages 17 and 18). In other words, the greater the bandwidth of the channel, the greater the accuracy of the received signal. With this in mind, let's formally define the term "bandwidth."

> The *bandwidth* of a signal is the width of the frequency band that contains a sufficient number of the signal's frequency components to reproduce the signal without an unacceptable amount of distortion.

The percentage of the signal's power within a frequency band is a good indication of the amount of distortion that will be caused by not including the frequency components that are outside the frequency band. What percentage of power is sufficient? It depends on the application. For instance, consider Figure 2-21a. As we calculated in Example 2.10, the received signal contains components representing 88% of the power of the transmitted signal. This signal received may be sufficient for certain applications (for instance, transmitting rap music), but is insufficient for other applications (for instance, transmitting an image of an X-ray) where we might prefer the waveform in Figure 2-21d, which contains components representing 97.5% of the power of the transmitted signal. We obtain the better performance in Figure 2-21d at the cost of increased bandwidth.

Because bandwidth is a nebulous term, communications engineers must always define what they mean by bandwidth. Often this is done by giving the percentage of total power that must be within the frequency band.[4]

2.3.3 The Two-Sided Average Normalized Power Spectrum

Average normalized power can also be expressed by using the two-sided (complex exponential) Fourier series:

$$P_s = \frac{1}{T}\int_{t_o}^{t_o+T} s^2(t)\,dt = \sum_{n=-\infty}^{\infty} c_n c_{-n} = \sum_{n=-\infty}^{\infty} c_n c_n^* = \sum_{n=-\infty}^{\infty} |c_n|^2 \qquad (2.38)$$

[3]Warning: Because the actual value of the load is irrelevant for calculations involving ratios or percentages, communication systems engineers often drop the adjective "normalized" and just call the results of Equations (2.29) and (2.36) "signal power." Similarly, they call a spectral plot such as Figure 2-18b, a "power spectrum." This practice is obviously incorrect but is nonetheless quite common.

[4]For an excellent listing of many possible quantitative definitions of bandwidth and a discussion of why no one definition can be universally relevant, see Section 1.7.2 of Sklar [2.1].

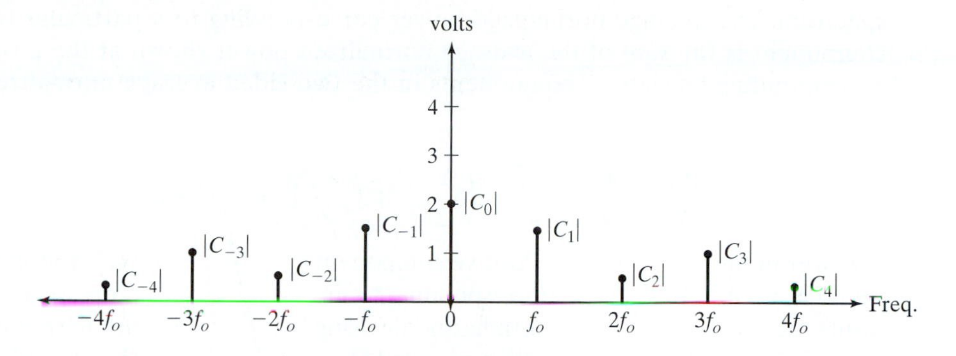

Figure 2-22a Two-sided magnitude spectrum of a typical signal. Note consistency with Figure 2-18a.

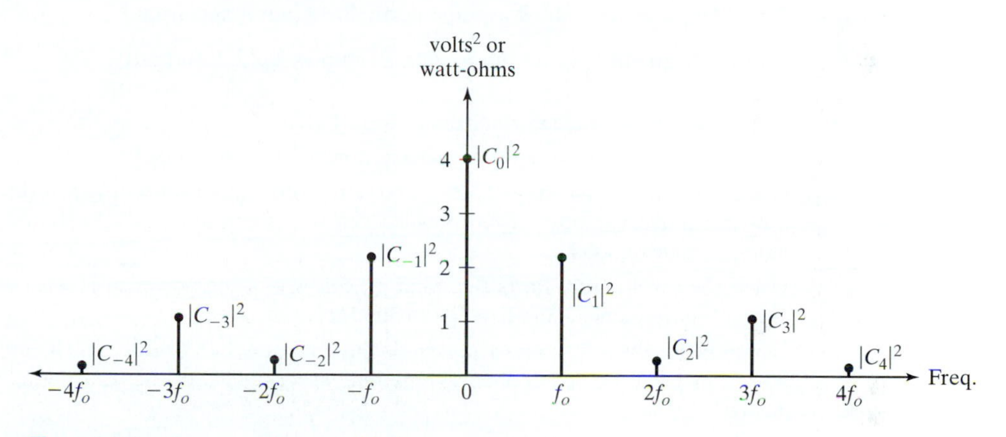

Figure 2-22b Corresponding two-sided average normalized power spectrum.

You'll derive Equation (2.38) in Problem 2.17. Note that Equation (2.38) is Parseval's theorem using the two-sided Fourier series.

As with the one-sided form, you can easily translate from a two-sided magnitude spectrum to a two-sided average normalized power spectrum. Figures 2-22a and b show the two-sided magnitude spectrum and the two-sided average normalized power spectrum of a typical signal. In comparing Figures 2-18a and b on page 42 (one-sided magnitude and average normalized power spectra) with Figures 2-22a and b, note that the two-sided average normalized power spectrum is actually easier to derive, since the dc term isn't treated differently from the other terms.

What is the physical significance of the terms to the left of the y axis in a two-sided average normalized power spectrum? When this question was asked in Section 2.2.2 concerning the two-sided magnitude spectrum, we observed that a component in the left half of the two-sided magnitude spectrum merely represented an additional contribution at the corresponding positive frequency. As Example 2.11 will illustrate, the same concept is true for the two-sided average normalized power spectrum. The average normalized power corresponding to a particular frequency component is the *sum* of the average normalized power shown at the positive and corresponding "negative" components in the two-sided average normalized power spectrum.

$$\frac{X_n^2}{2} = \frac{X_n^2}{4} + \frac{X_n^2}{4} = \left(\frac{X_n}{2}\right)^2 + \left(\frac{X_n}{2}\right)^2 = |c_n|^2 + |c_{-n}|^2$$

Power in the nth harmonic using the one-sided form

Positive component corresponding to the nth harmonic using the two-sided form

Negative component corresponding to the nth harmonic using the two-sided form

Example 2.11 Using the two-sided average normalized power spectrum

Given the rectangular pulse train used in Examples 2.7, 2.9, and 2.10,

a. Determine the average normalized power of $v(t)$.

b. Draw the two-sided average normalized power spectrum of $v(t)$.

c. Using the two-sided average normalized power spectrum drawn in Part b, determine how much average normalized power from $v(t)$ is contained within the frequency band from dc to 300 Hz.

d. Using the results from Parts a–c, what percentage of the power of $v(t)$ is contained within the frequency band from dc to 300 Hz?

e. Using the results from Parts a–d, what is the minimum bandwidth of a channel which must pass frequency components containing at least 88% of the transmitted signal's power?

Solutions

a. From Example 2.10 on page 44,

$$P_v = \frac{1}{T}\int_{t_o}^{t_o+T} v^2(t)\,dt = \frac{1}{0.01}\int_{-0.001}^{0.009} v^2(t)\,dt = \frac{1}{0.01}\int_{-0.001}^{0.001} 3^2\,dt = 1.8 \text{ volts}^2$$

b. In Example 2.7 on page 33, we determined the complex exponential (or two-sided) Fourier series coefficients for $v(t)$

$$c_n = \frac{A\tau}{T}\,\text{sinc}\left(\frac{\pi n \tau}{T}\right) = 0.6 \text{ sinc } (0.2\, n\pi)$$

and we plotted the two-sided magnitude spectrum (see Figure 2-14a, which is plotted again below). We can thus easily plot the two-sided average normalized power spectrum in Figure 2-23.

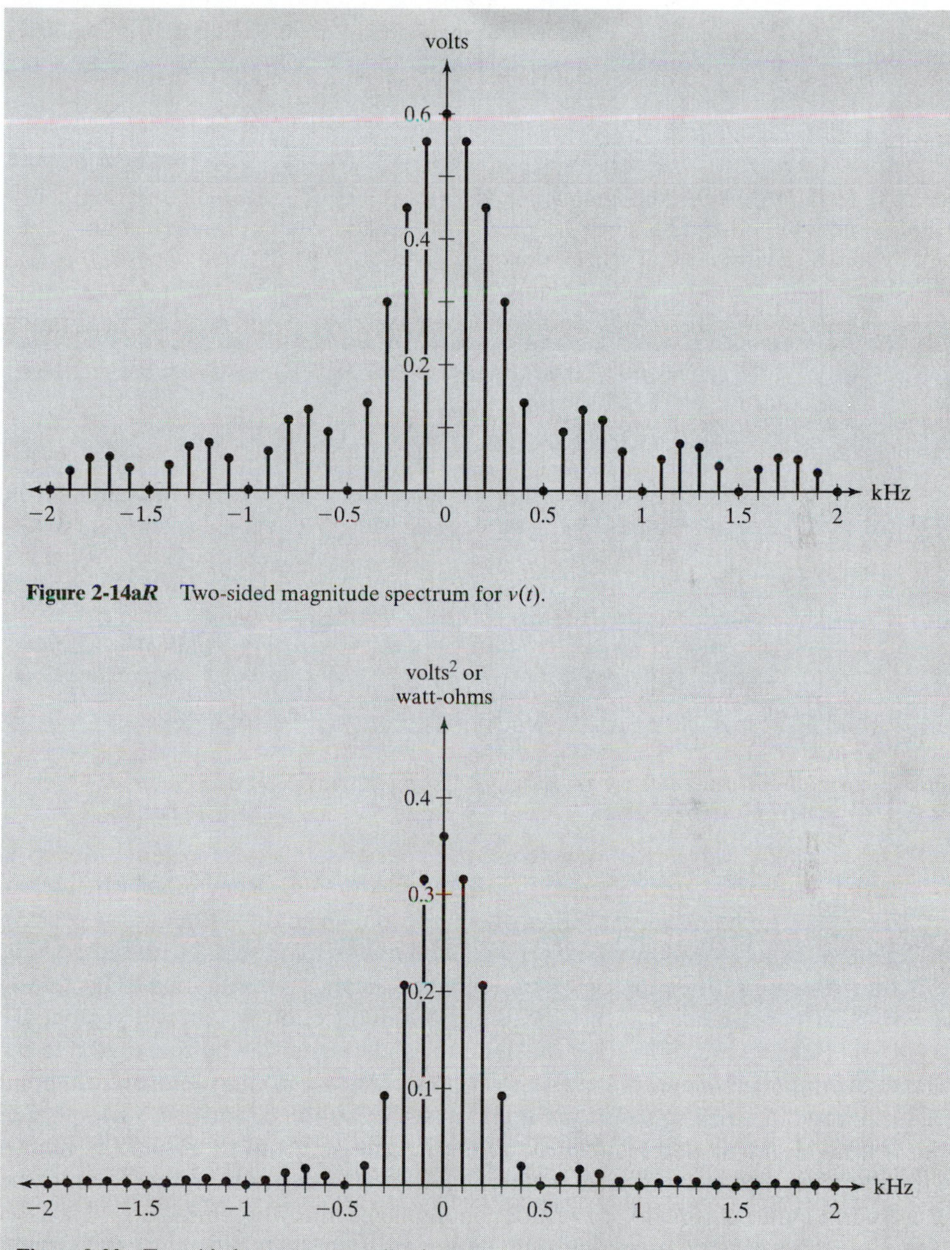

Figure 2-14a*R* Two-sided magnitude spectrum for $v(t)$.

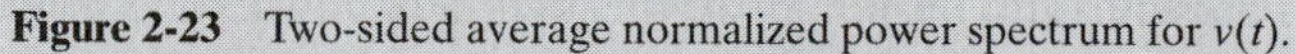

Figure 2-23 Two-sided average normalized power spectrum for $v(t)$.

c. Using Figure 2-23 and remembering that the terms to the left of the y axis represent contributions to the corresponding positive frequency terms, the average normalized power lying in the frequency band from dc to 300 Hz can be calculated as:

$$|c_{-3}|^2 + |c_{-2}|^2 + |c_{-1}|^2 + |c_0|^2 + |c_1|^2 + |c_2|^2 + |c_3|^2$$
$$= 0.092 + 0.206 + 0.3165 + 0.36 + 0.3165 + 0.206 + 0.092 = 1.586 \text{ volts}^2$$

This is the same result we obtained in Example 2.10, as it must be. How we mathematically represent a signal (using the one-sided form or the two-sided form) cannot change the physical characteristics of that signal.

d. From Part a, $P_v = 1.8 \text{ volts}^2$ and from Part c, average normalized power within the frequency band from dc to 300 Hz = 1.586 volts^2. Thus, the percentage of power within the frequency band from dc to 300 Hz =

$$\frac{1.586 \text{ volts}^2}{1.8 \text{ volts}^2} = 88\%$$

Does this value agree with Example 2.10 Part b, which was calculated using a one-sided average normalized power spectrum? Yes—it has to. Does the answer change if the system has a load resistance other than 1 Ω? No. Refer back to the comments after Example 2.10 on page 44.

e. The frequency band from dc to 300 Hz contains 88% of the signal's transmitted power, thus the minimum bandwidth of the channel is 300 Hz. Looking back at Figure 2-23 and Parts b and c on pages 50 and 51, you may be tempted to think that the bandwidth should be 600 Hz (the width from the c_{-3} component to the c_3 component), but we need to remember that components to the left of the y axis (i.e., "negative frequency" components) actually represent contributions at the corresponding positive frequency. *Thus, both the positive and "negative" frequency components in the two-sided average normalized power spectrum must be summed when determining normalized power within a frequency band, but bandwidth is only the width of the positive frequencies.*

2.4 The Fourier Transform

In Section 2.2 we learned how to use the Fourier series to represent any periodic signal with finite energy per period in the frequency domain. (Remember why we want to use frequency domain analysis rather than time domain analysis.) Let's now consider a more general issue. How do we convert a general *nonperiodic* signal with finite energy from a time domain representation to a frequency domain representation? This question is important because many of the signals in a communication system will be nonperiodic.

Let's start by observing that a nonperiodic signal can be considered to be periodic *with infinite period* ($T = \infty$). This is very interesting from a philosophical point of view, but how does it help us? If the signal is periodic (albeit, periodic with infinite period), intuitively we should be able to use the Fourier series to represent the function in the frequency domain. Let's see what happens mathematically to the Fourier series as we stretch period (T) to ∞.

Let's use the two-sided form of the Fourier series, since it is the easiest form to work with mathematically.

$$s(t) = \sum_{n=-\infty}^{\infty} c_n e^{j2\pi n f_o t}$$

$$\text{where } c_n = \frac{1}{T}\int_{t_o}^{t_o+T} s(t)e^{-j2\pi n f_o t}\,dt \qquad (2.12R)$$

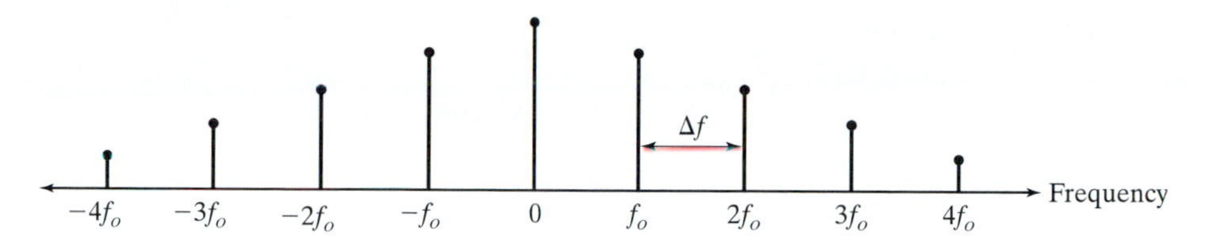

Figure 2-24 An illustration of Δf, the spacing between adjacent harmonics.

Before we stretch T to ∞, let's make a few observations:

a. We know that f_o represents the fundamental frequency in the Fourier series and that $f_o = \dfrac{1}{T}$.

b. Let's define another symbol, Δf, as the spacing, in Hz, between adjacent harmonics in the spectrum. Δf is illustrated in Figure 2-24. What is the value of Δf? The spacing between any two adjacent harmonics (say, the nth harmonic and the n-1st harmonic) is

$$\Delta f = nf_o - (n - 1)f_o = f_o \tag{2.39}$$

Thus f_o and Δf have the same value, but they physically represent very different things.

c. As $T \rightarrow \infty$, $f_o \rightarrow 0$ and $\Delta f \rightarrow 0$.

Let's now see what happens to the Fourier series as we let T approach infinity.

$$\begin{aligned}
\lim_{T\rightarrow\infty} s(t) &= \lim_{T\rightarrow\infty}\left[\sum_{n=-\infty}^{\infty} c_n e^{j2\pi n f_o t}\right] \\
&= \lim_{f_o\rightarrow 0}\left[\sum_{n=-\infty}^{\infty} c_n e^{j2\pi n f_o t}\right] \quad \text{since } f_o = \frac{1}{T} \\
&= \lim_{\Delta f\rightarrow 0}\left[\sum_{n=-\infty}^{\infty} c_n e^{j2\pi n \Delta f t}\right] \quad \text{by substitution, since } \Delta f = f_o \\
&= \lim_{\Delta f\rightarrow 0}\left[\sum_{n=-\infty}^{\infty} \frac{c_n}{\Delta f} e^{j2\pi n \Delta f t}\Delta f\right] \quad \text{multiplication and division by } \Delta f \\
&= \sum_{n=-\infty}^{\infty}\left[\lim_{\Delta f\rightarrow 0} \frac{c_n}{\Delta f} e^{j2\pi n \Delta f t}\Delta f\right] \quad \text{interchanging } lim \text{ and } \Sigma \\
&= \int_{-\infty}^{\infty}\left(\lim_{\Delta f\rightarrow 0}\frac{c_n}{\Delta f}\right)e^{j2\pi f t}df \quad \text{since } \Delta f \text{ approaches 0 and } n \text{ spans from } -\infty \text{ to } \infty, \text{ we can express } n\Delta f \text{ as a continuous variable and the summation as an integral} \\
&= \int_{-\infty}^{\infty} S(f)e^{j2\pi f t}df
\end{aligned} \tag{2.40}$$

where we define

$$S(f) \equiv \lim_{\Delta f \to 0} \frac{c_n}{\Delta f} \tag{2.41}$$

Thus,

$$\lim_{T \to \infty} s(t) = \int_{-\infty}^{\infty} S(f) e^{j2\pi ft} df \tag{2.40R}$$

But what is $\lim_{T \to \infty} s(t)$? If $s(t)$ is nonperiodic, then $\lim_{T \to \infty} s(t) = s(t)$.

Thus,

$$s(t) = \int_{-\infty}^{\infty} S(f) e^{j2\pi ft}\, df \tag{2.42}$$

As we shall see shortly, Equation (2.42) is called the *inverse Fourier transform*.

This is all well and good, but we should feel a little bit cheated since we established $S(f)$ above merely by definition. Let's now solve for $S(f)$.

$$S(f) \equiv \lim_{\Delta f \to 0} \frac{c_n}{\Delta f}$$

$$= \lim_{\Delta f \to 0} \left[\frac{\frac{1}{T} \int_{t_o}^{t_o+T} s(t) e^{-j2\pi n f_o t}\, dt}{\Delta f} \right]$$ by substitution using Equation (2.12) for c_n

$$= \lim_{\Delta f \to 0} \left[\frac{\Delta f \int_{t_o}^{t_o+T} s(t) e^{-j2\pi n f_o t}\, dt}{\Delta f} \right]$$ since $\frac{1}{T} = f_o = \Delta f$

$$= \lim_{\Delta f \to 0} \int_{t_o}^{t_o+T} s(t) e^{-j2\pi n \Delta f t}\, dt$$

$$= \lim_{\Delta f \to 0} \int_{-\frac{T}{2}}^{\frac{T}{2}} s(t) e^{-j2\pi n \Delta f t}\, dt$$ Choosing to integrate over the period from $-\frac{T}{2}$ to $\frac{T}{2}$

$$= \int_{-\infty}^{\infty} s(t) e^{-j2\pi ft}\, dt$$ In the limiting case, $n\Delta f$ can be expressed as a continuous variable f, since n is summed from $-\infty$ to ∞ and $\Delta f \to 0$.

Thus, $$S(f) = \int_{-\infty}^{\infty} s(t) e^{-j2\pi ft}\, dt \tag{2.43}$$

Equation (2.43) is called the *Fourier transform*.

Let's look closer at Equations (2.42) and (2.43):

$$s(t) = \int_{-\infty}^{\infty} S(f)e^{j2\pi ft}df \tag{2.42R}$$

$$S(f) = \int_{-\infty}^{\infty} s(t)e^{-j2\pi ft}\,dt \tag{2.43R}$$

What do these equations say? Equation (2.42) says, "$s(t)$ can be replaced by an infinite summation (*the integral*) of sinusoids (*the complex exponential*) with the frequency differences between adjacent sinusoids infinitesimally small (*f is continuous*). Coefficients of the sinusoids ($S(f)$) are no longer scalar values (*volts*) but rather are densities (*volts/Hz*)." We can also understand the units for $S(f)$ from the way we defined $S(f)$ in establishing Equation (2.41):

$$S(f) \equiv \lim_{\Delta f \to 0} \frac{c_n}{\Delta f} \tag{2.41R}$$

Equation (2.43) says, "We can calculate $S(f)$, the function that gives us the continuous spectrum of $s(t)$."

Equation (2.43) is called the Fourier transform because it *transforms* the time domain representation of a signal ($s(t)$) into the frequency domain representation of the signal ($S(f)$). Equation (2.42) is called the inverse Fourier transform because it transforms the frequency domain representation of the signal ($S(f)$) back into the time domain representation of the signal ($s(t)$). Note that the two equations don't change the physical signal, just its mathematical representation.

Let's see if we can correlate our findings for nonperiodic signals with our previous findings for periodic signals. As you review Table 2.3, compare both the mathematical descriptions and the English descriptions.

Table 2-3 Frequency Domain Expressions for Periodic and Nonperiodic Signals

Periodic		Nonperiodic	
The spectrum of a periodic signal has discrete components, and the magnitude of the components is scalar (c_n is expressed in volts).	$s(t) = \sum_{n=-\infty}^{\infty} c_n e^{j2\pi n f_o t}$ $c_n = \frac{1}{T}\int_{t_o}^{t_o+T} s(t)e^{-j2\pi n f_o t}\,dt$	$s(t) = \int_{-\infty}^{\infty} S(f)e^{j2\pi ft}\,df$ $S(f) = \int_{-\infty}^{\infty} s(t)e^{-j2\pi ft}\,dt$	The spectrum of a nonperiodic signal is continuous, and the magnitude of the components is expressed as a density. ($S(f)$ is expressed in volts/Hz rather than as a scalar value—remember how we defined $S(f)$.)

Let's now work an example expressing a nonperiodic waveform in the frequency domain.

Example 2.12 Frequency domain representation of a single rectangular pulse

Consider a single rectangular pulse of amplitude A and width τ. Find the frequency domain representation of this pulse.

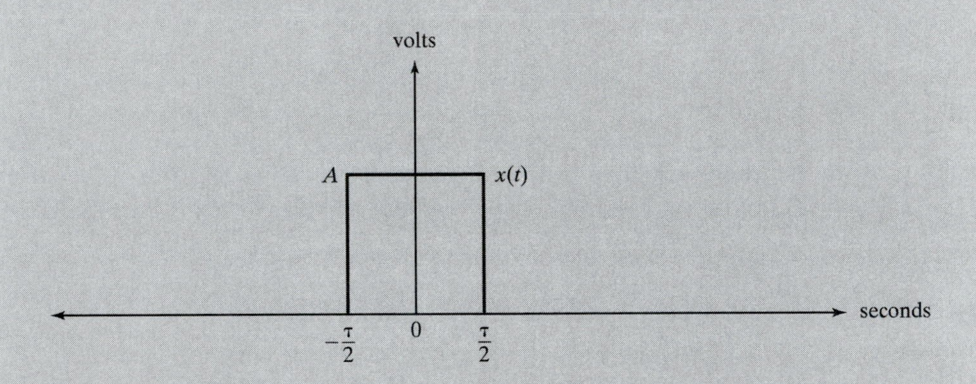

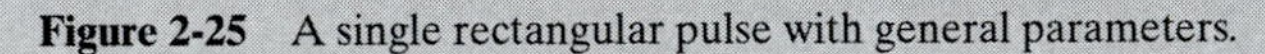

Figure 2-25 A single rectangular pulse with general parameters.

Solution

The pulse is nonperiodic and has finite energy. We can thus take its Fourier transform

$$\underbrace{\mathcal{F}\{x(t)\}}_{\substack{\text{This representation} \\ \text{means "the Fourier} \\ \text{transform of } x(t)\text{"}}} = X(f) = \int_{-\infty}^{\infty} x(t)e^{-j2\pi ft}\,dt$$

$$= \int_{-\frac{\tau}{2}}^{\frac{\tau}{2}} Ae^{-j2\pi ft}\,dt = \frac{-A}{j2\pi f}e^{-j2\pi ft}\Bigg|_{-\frac{\tau}{2}}^{\frac{\tau}{2}}$$

$$= \frac{-A\left[e^{-j2\pi f\frac{\tau}{2}} - e^{j2\pi f\frac{\tau}{2}}\right]}{j2\pi f} \tag{2.44}$$

$$= \frac{A}{\pi f}\left[\frac{e^{j\pi f\tau} - e^{-j\pi f\tau}}{2j}\right] = \left[\frac{A}{\pi f}\right]\sin(\pi f\tau)$$

$$= (A\tau)\left(\frac{\sin(\pi f\tau)}{\pi f\tau}\right) = A\tau\,\mathrm{sinc}(\pi f\tau)\text{ volts/Hz}$$

Drawing the magnitude and phase spectra in Figures 2-26a and b:

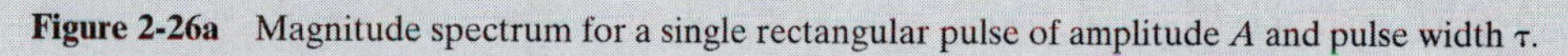

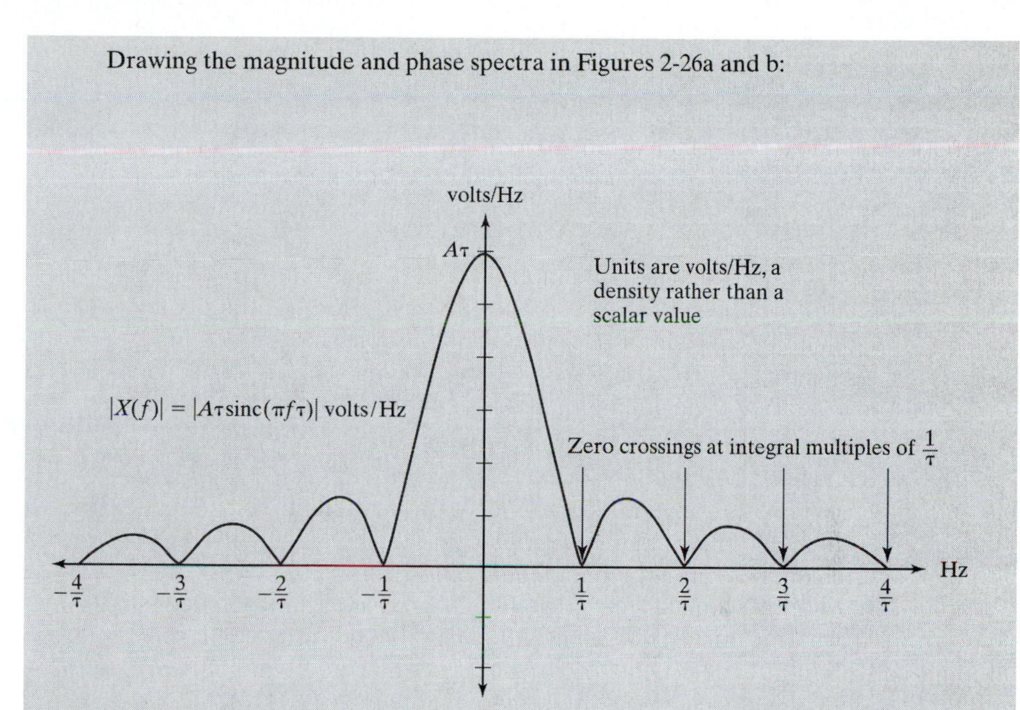

Figure 2-26a Magnitude spectrum for a single rectangular pulse of amplitude A and pulse width τ.

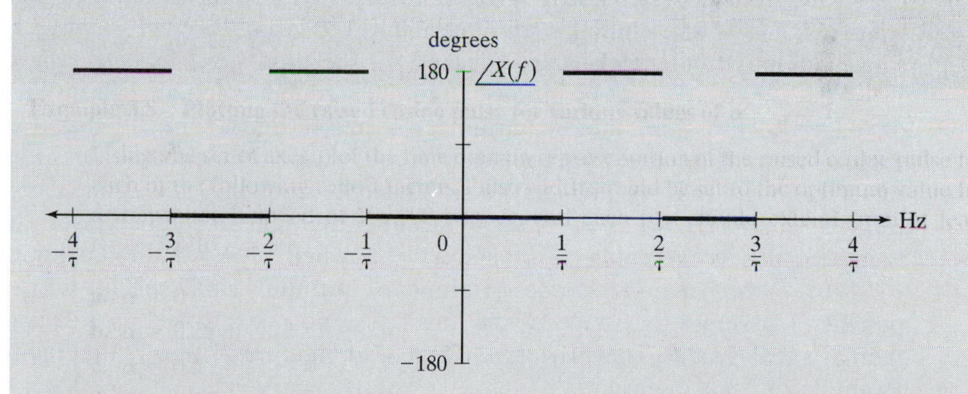

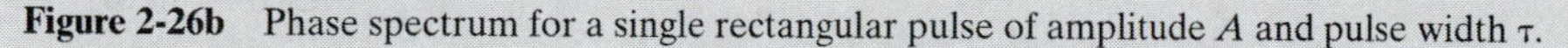

Figure 2-26b Phase spectrum for a single rectangular pulse of amplitude A and pulse width τ.

Since, for this example, $S(f)$ has no imaginary component, the phase is always 0° or 180° and we can combine the magnitude and phase spectra to produce an amplitude spectrum as shown in Figure 2-26c.

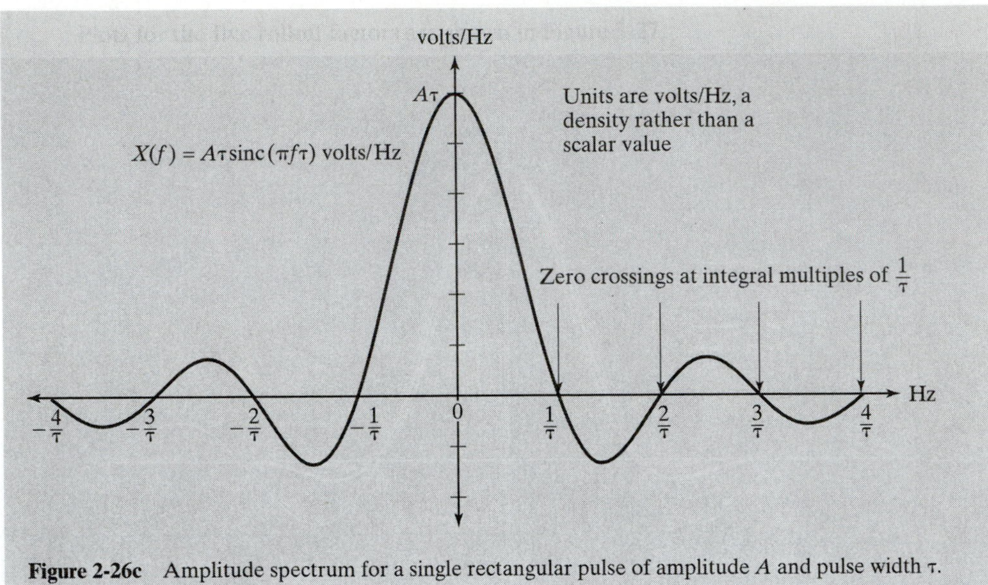

Figure 2-26c Amplitude spectrum for a single rectangular pulse of amplitude A and pulse width τ. (Compare to Figure 2-15.)

Do the results of Example 2.12 make sense relative to our earlier examples with periodic rectangular pulse trains? To find out, let's again examine the correlation between the Fourier series and Fourier transform. What happens to a periodic pulse train as $T \rightarrow \infty$? Stretching the pulse period (T) to infinity converts a pulse train to a single pulse. All other pulses are "pushed off the time axis," leaving a single pulse. What does pushing T out to infinity do in the frequency domain? From Figure 2-15 on page 36 (the amplitude spectrum for a pulse train), remember that changing T changes the spacing between adjacent frequency components but does *not* change the envelope. The spacing between adjacent frequency components = $(2\pi/T)$ rad/sec = $(1/T)$ Hz, so as $T \rightarrow \infty$, adjacent components become closer and closer together (this is the same as saying $\Delta f \rightarrow 0$ when $T \rightarrow \infty$). In the limiting case ("$T = \infty$") the function becomes continuous ("infinitely small spacing between components"), assumes the shape of the envelope, and is expressed as a density rather than as a scalar value. This process, illustrated in Figure 2-27, shows that the results of Example 2.12 make sense.

The relationship between the Fourier series and Fourier transform has been established mathematically in Equations (2.40)–(2.43), has been summarized in Table 2-3, and has been illustrated for a particular waveform in Figure 2-27. Be sure you understand this relationship.

Let's make one last observation about the time and frequency domain representations of a rectangular pulse. Zero crossings occur at $f = k/\tau$ for all nonzero integer values of k. Thus, as pulse width τ decreases, the spacing between zero crossings increases. Figure 2-28 illustrates this process. Thus, just as in the periodic case, a narrower pulse width requires more bandwidth.

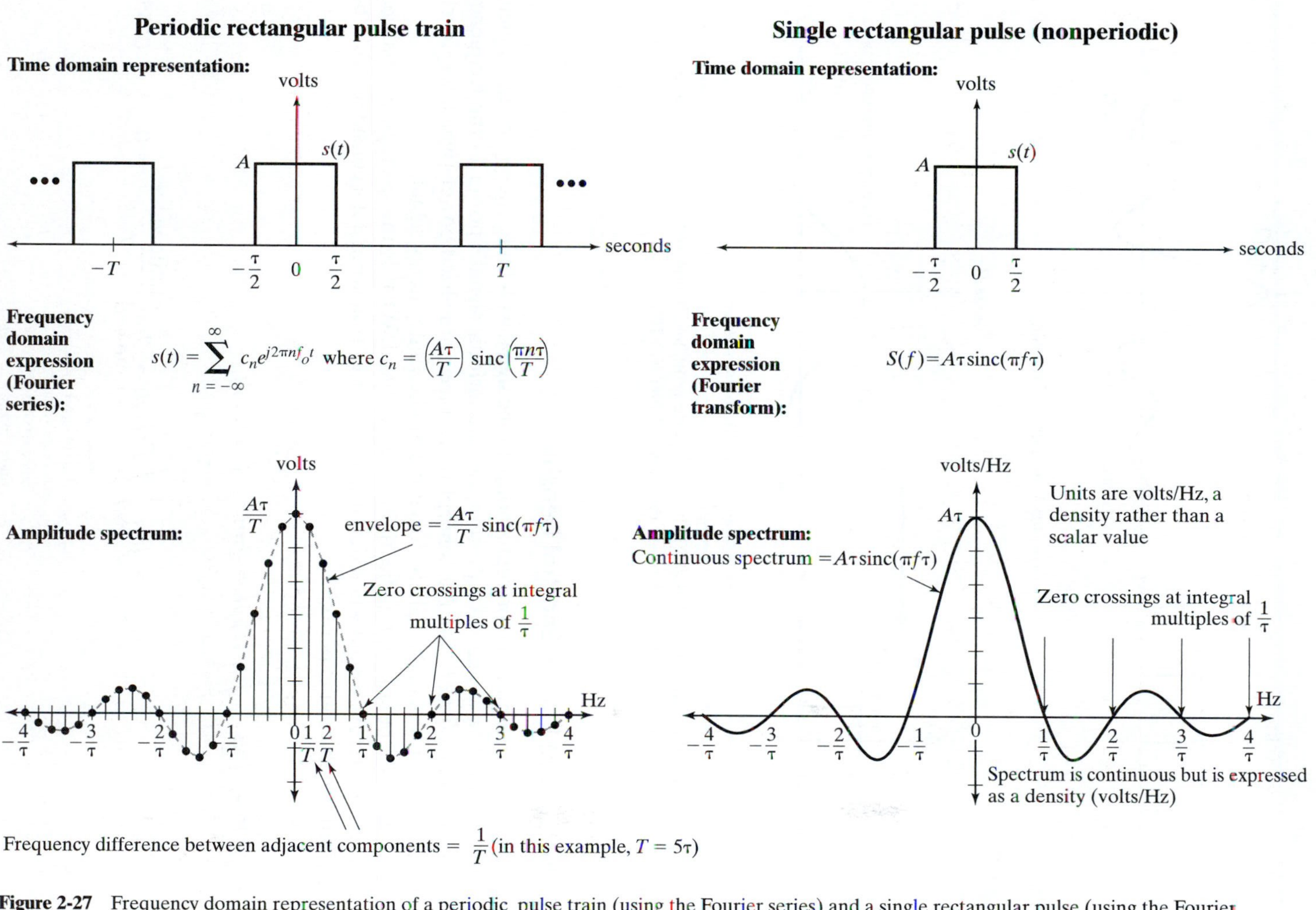

Figure 2-27 Frequency domain representation of a periodic pulse train (using the Fourier series) and a single rectangular pulse (using the Fourier transform).

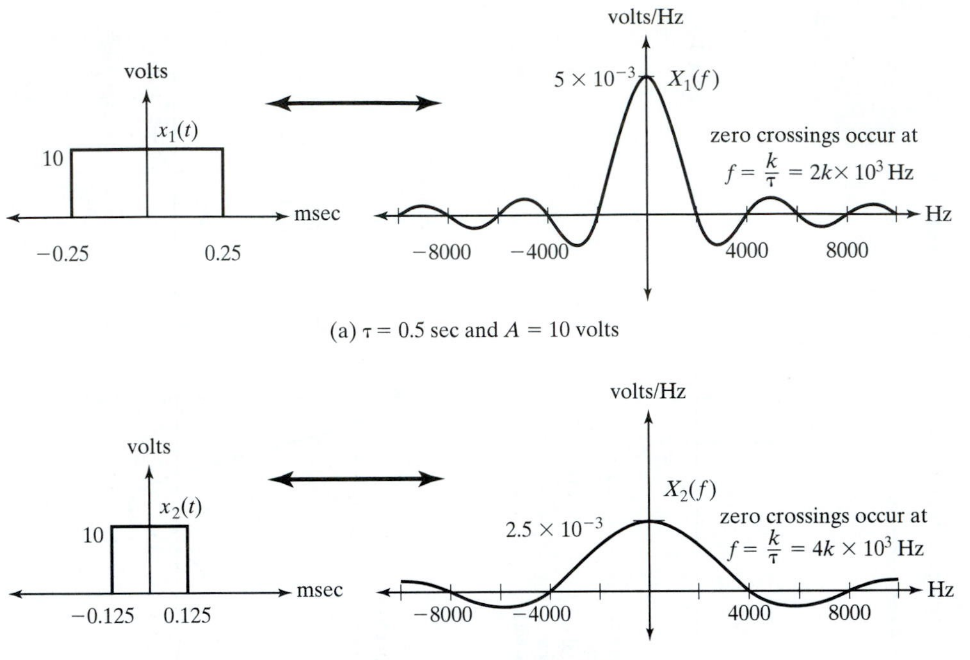

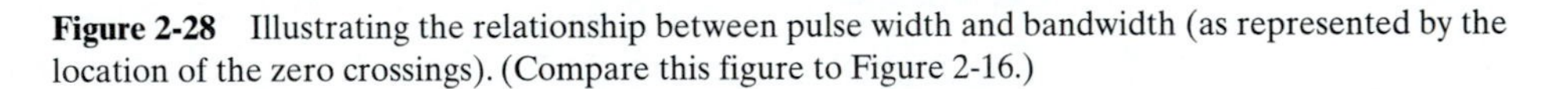

Figure 2-28 Illustrating the relationship between pulse width and bandwidth (as represented by the location of the zero crossings). (Compare this figure to Figure 2-16.)

2.5 Normalized Energy Spectral Density

We've now seen how to represent a nonperiodic signal with finite energy in the frequency domain. How would we represent the nonperiodic signal's power in the frequency domain? Let's extrapolate the concept of average normalized power (which we developed earlier for periodic signals) to a nonperiodic, finite energy signal.

Let $s(t)$ be a nonperiodic, finite energy signal. What is the average normalized power of $s(t)$? As we did when deriving the Fourier transform, let's consider $s(t)$ to be periodic with infinite period.

$$P_s = \lim_{T\to\infty} \frac{1}{T} \int_{t_o}^{t_o+T} s^2(t)\,dt = \lim_{T\to\infty} \frac{1}{T} \int_{-T/2}^{T/2} s^2(t)\,dt$$

$$= \lim_{T\to\infty} \frac{1}{T} \underbrace{\int_{-\infty}^{\infty} s^2(t)\,dt}_{\substack{\text{This integral} \\ \text{has finite value,} \\ \text{since } s(t) \text{ has} \\ \text{finite energy.}}} = \lim_{T\to\infty} \frac{\text{finite value}}{T} = 0 \tag{2.45}$$

Does our result make sense? Yes, since

$$\text{Power} = \frac{\text{Energy} \leftarrow \text{finite}}{\text{Time} \leftarrow \text{infinite}}$$

Our result means that normalized power is not a good parameter for nonperiodic, finite energy signals. What might be a useful parameter? How about *normalized energy*? Let's define the normalized energy (E_s) of a finite-energy, nonperiodic waveform $s(t)$ as the energy that $s(t)$ provides to a resistive load multiplied by the resistance of the load. Let R represent the resistance of a particular load.

$$\begin{aligned} E_s &\equiv R\,(\text{Total energy that } s(t) \text{ provides to the load}) \\ &= R\left[\int_{-\infty}^{\infty} (\text{voltage across load})(\text{current through load})\, dt\right] \\ &= R\left[\int_{-\infty}^{\infty} s(t)\left(\frac{s(t)}{R}\right) dt\right] = \int_{-\infty}^{\infty} s^2(t)\, dt \end{aligned} \quad (2.46)$$

Units for E_s are joule-ohms or volts^2-sec. Be sure you understand why these two units are equivalent and why each is appropriate. Conceptually, the value of normalized energy (although not the units) can be thought of as the energy provided by $s(t)$ if $s(t)$ were to appear across the terminals of a 1Ω resistor.

What does normalized energy (E_s) look like in the frequency domain? Using the same approach we followed in Section 2.3 for periodic signals and average normalized power, we can show that the components in the Fourier transform are orthogonal, and therefore the normalized energy of a nonperiodic signal $s(t)$ can be calculated in the frequency domain as:

$$E_s = \int_{-\infty}^{\infty} |S(f)|^2\, df \quad (2.47)$$

Compare Equation (2.47) for the nonperiodic signal with Equation (2.38) for the periodic signal. Combining Equations (2.46) and (2.47),

$$E_s = \int_{-\infty}^{\infty} s^2(t)\, dt = \int_{-\infty}^{\infty} |S(f)|^2\, df \quad (2.48)$$

Equation (2.48), known as *Parseval's energy theorem*, makes good intuitive sense. The time domain integral on the left-hand side of Equation (2.48) is summing all of the signal's normalized energy as represented in the time domain, and the frequency domain integral on the right-hand side of Equation (2.48) is summing all of the signal's normalized energy as represented in the frequency domain. In both cases we are representing the same physical parameter (normalized energy)—we're just using different mathematical expressions. As we know, changing a parameter's mathematical representation cannot change its physical properties.

As we did concerning average normalized power of a periodic signal, we can use spectral plots to illustrate how a nonperiodic signal's normalized energy is distributed throughout the frequency band. Figure 2-29a shows a magnitude spectrum and Figure 2-29b shows the corresponding normalized energy spectrum for a typical nonperiodic, finite energy signal.

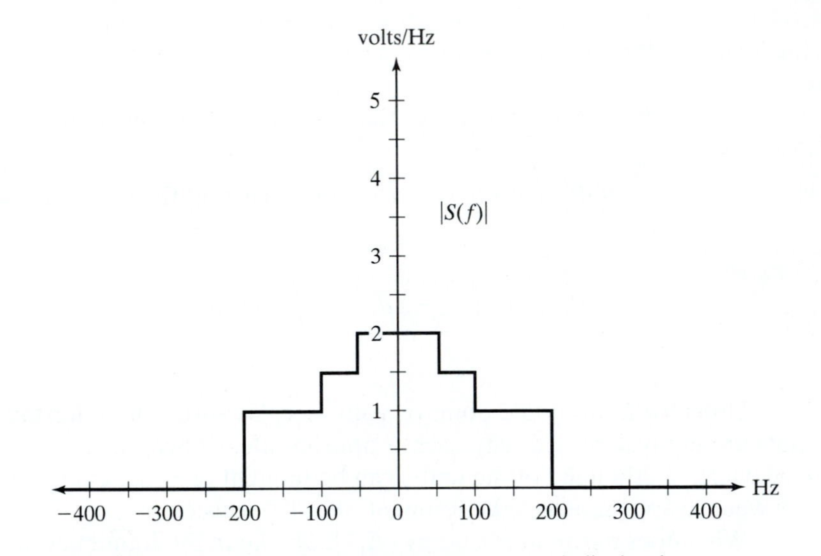

Figure 2-29a Magnitude spectrum of a typical nonperiodic signal.

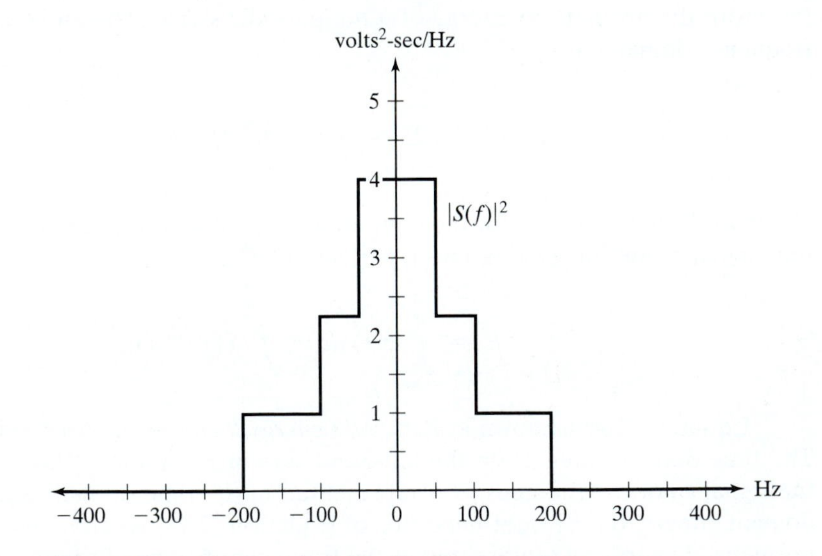

Figure 2-29b Corresponding normalized energy spectrum.

The quantity $|S(f)|^2$ in Figure 2-29b is a normalized energy spectral density (it shows how normalized energy is distributed through the frequency band). Note that this quantity is a density and its units are volts^2/Hz^2, which can also be expressed as volts^2-sec/Hz (this latter form has more of a physical meaning, since volts^2-sec is the unit for normalized energy). Be sure that you understand how the units were obtained. $|S(f)|^2$ is such an important quantity that it has its own symbol: $\Psi(f)$.

$$|S(f)|^2 \equiv \Psi(f) = \text{Normalized energy spectral density} \tag{2.49}$$

To find the normalized energy within a given frequency band, we merely integrate the energy spectral density over that frequency band (very similar to what we did for average normalized power with periodic signals in Section 2.3). We must remember to include the "negative frequency" band in our calculations.

Example 2.13 Using normalized energy spectral density

Given a nonperiodic, finite energy waveform $s(t)$ with the magnitude spectrum $S(f)$ shown in Figure 2-29a,

a. Find the normalized energy of the waveform.
b. If $s(t)$ is passed through an ideal lowpass filter with a cutoff frequency of 75 Hz, find the normalized energy of the waveform at the output of the filter.
c. What percentage of the signal's energy lies in the frequency band passed by the filter?

Solution

a. We can calculate E_s in the time domain as:

$$E_s = \int_{-\infty}^{\infty} s^2(t)\, dt$$

or we can calculate E_s in the frequency domain as

$$E_s = \int_{-\infty}^{\infty} \Psi(f)\, df = \int_{-\infty}^{\infty} |S(f)|^2\, df$$

Using Figure 2-29b and calculating E_s in the frequency domain,

$$E_s = \int_{-200}^{-100} 1\, df + \int_{-100}^{-50} 2.25\, df + \int_{-50}^{50} 4\, df + \int_{50}^{100} 2.25\, df + \int_{100}^{200} 1\, df = 825 \text{ volts}^2\text{-sec}$$

b. Normalized energy passed by filter =

$$\int_{-75}^{75} \Psi(f)\, df = \int_{-75}^{-50} 2.25\, df + \int_{-50}^{50} 4\, df + \int_{50}^{75} 2.25\, df = 512.5 \text{ volts}^2\text{-sec}$$

c. Percentage of energy passed by filter $= \dfrac{512.5}{825} = 62.1\%$

Look carefully at how Example 2.13 resembles Example 2.10 on page 44 (which concerned average normalized power of a periodic signal). In particular, note the comments in Example 2.10 concerning relevance; these comments also apply to Example 2.13.

Now that we've seen how to evaluate the effects of an ideal filter on a signal's energy, what about a general (nonideal) filter? We can represent the filter as a linear system with a transfer function of $H(f)$.[5] As discussed earlier in this chapter, if we use frequency domain analysis, we can express the output of the filter as

$$S_{\text{out}}(f) = S_{\text{in}}(f)H(f) \tag{2.50}$$

Equation (2.50) is illustrated in Figure 2-30a. The implications concerning the signal's energy are shown in Theorem 2.1 and the corresponding proof.

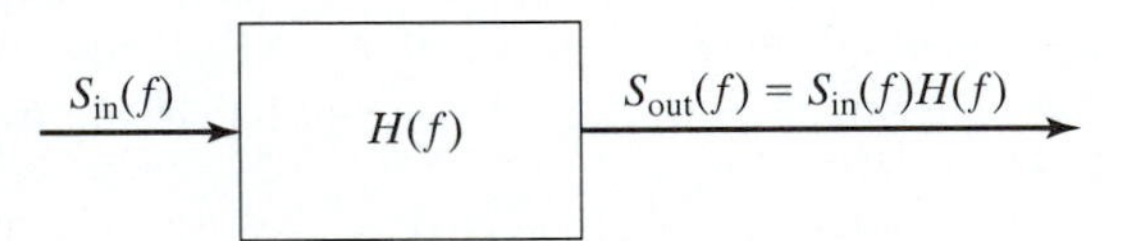

Figure 2-30a Tracking the flow of a signal through a linear system (e.g., a filter).

Theorem 2.1:

If a nonperiodic, finite energy signal $S_{\text{in}}(f)$ has a normalized energy spectral density $\Psi_{\text{in}}(f)$ and is passed through a filter (or any linear system) having a frequency response (or transfer function) of $H(f)$, then the signal $S_{\text{out}}(f)$ coming out of the filter (or linear system) has a normalized energy spectral density

$$\Psi_{\text{out}}(f) = \Psi_{\text{in}}(f)|H(f)|^2 \tag{2.51}$$

Proof:

$$\begin{aligned} S_{\text{out}}(f) &= S_{\text{in}}(f)H(f) \\ \Psi_{\text{out}}(f) &= |S_{\text{out}}(f)|^2 \text{ by definition} \\ &= |S_{\text{in}}(f)H(f)|^2 \\ &= [S_{\text{in}}(f)H(f)][S_{\text{in}}(f)H(f)]^* \end{aligned}$$

[5] $H(f)$ is the Fourier transform of the system's impulse response $h(t)$. $H(f)$ is also often called the *frequency response* of the system.

$$= S_{in}(f)H(f)S_{in}^*(f)H^*(f)$$
$$= S_{in}(f)S_{in}^*(f)H(f)H^*(f)$$
$$= |S_{in}(f)|^2|H(f)|^2$$
$$= \Psi_{in}(f)|H(f)|^2$$

We can now directly track the flow of energy through a linear system, as shown in Figure 2-30b.

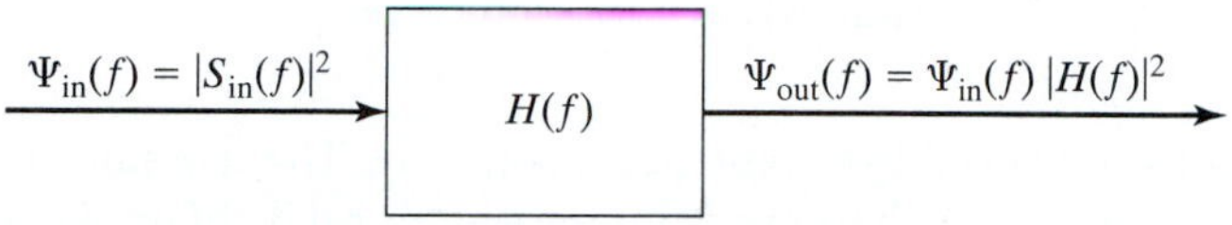

$\Psi_{in}(f) = |S_{in}(f)|^2$ → $H(f)$ → $\Psi_{out}(f) = \Psi_{in}(f)\,|H(f)|^2$

Figure 2-30b Tracking the flow of a signal's energy (and energy spectral density) through a linear system.

2.6 Properties of the Fourier Transform

Let's now determine some properties of the Fourier transform that will be especially helpful in communication system analysis and design. We've already established one of these properties

$$s(t)*h(t) \leftrightarrow S(f)H(f) \tag{2.52}$$

The double-sided arrow means that the left-hand side and right-hand side of Equation (2.52) represent a *transform pair*. One side represents signals and operations in the time domain (in this case, the left-hand side of Equation (2.52)), while the other side represents signals and equivalent operations in the frequency domain.

As we just established, the usefulness of Equation (2.52) is twofold:

a. As shown in Figure 2-31, in the frequency domain we can easily trace signals through a linear system.

Time domain:

$s_{in}(t)$ → Linear system with impulse response $h(t)$ → $s_{out}(t) = s_{in}(t)*h(t)$

Frequency domain:

$S_{in}(f)$ → Linear system with impulse response $h(t)$ → $S_{out}(f) = S_{in}(f)H(f)$

$H(f) \equiv \mathcal{F}\{h(t)\}$
We call $H(f)$ the *transfer function.*

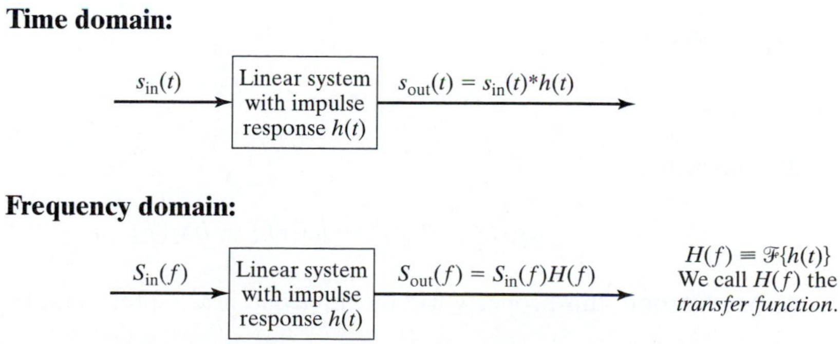

Figure 2-31 Tracking the flow of a signal through a linear system (e.g., a filter).

b. As shown in Figure 2-30b, in the frequency domain we can easily trace energy and normalized energy spectral density through a linear system:

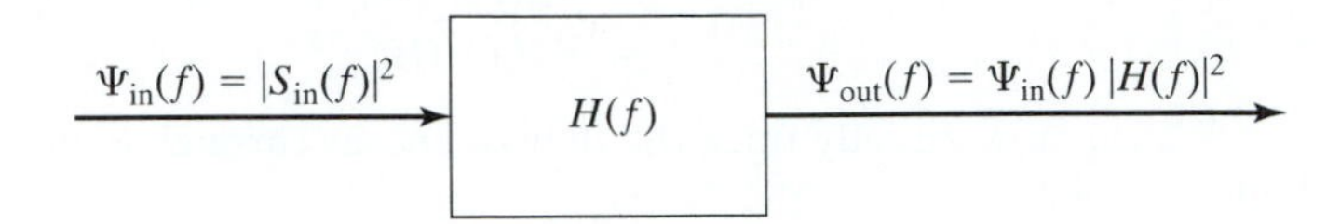

Figure 2-30b*R* Tracking the flow of a signal's energy (and energy spectral density) through a linear system.

Let's now establish some new properties. The usefulness of these properties will be discussed in later chapters as we encounter relevant applications; for now, it is sufficient that you just familiarize yourself with these properties. Also note that while only the first property below is proven, all subsequent properties can be proven in the same manner—by starting with the definition of the Fourier transform and then performing appropriate mathematical manipulations.

1. Time shift

$$s(t - t_o) \leftrightarrow S(f)e^{-j2\pi f t_o} \tag{2.53}$$

Proof:

$$\mathcal{F}\{s(t - t_o)\} = \int_{-\infty}^{\infty} s(t - t_o)e^{-j2\pi f t}\, dt$$

$$= \int_{-\infty}^{\infty} s(\tau)e^{-j2\pi f(\tau + t_o)}d\tau \quad \text{by change of variable, let } \tau = t - t_o;\ \ d\tau = dt$$

$$= e^{-j2\pi f t_o}\int_{-\infty}^{\infty} s(\tau)e^{-j2\pi f \tau}d\tau$$

$$= e^{-j2\pi f t_o}S(f)$$

2. Frequency shift

$$S(f + f_o) \leftrightarrow s(t)e^{-j2\pi f_o t} \tag{2.54}$$

3. Linearity

$$as_1(t) + bs_2(t) \leftrightarrow aS_1(f) + bS_2(f) \tag{2.55}$$

4. Modulation (this property will be important in Chapters 5 and 6)

$$s(t)\cos(2\pi f_o t) \leftrightarrow \tfrac{1}{2}[S(f - f_o) + S(f + f_o)] \tag{2.56}$$

5. Time scaling

$$s(at) \leftrightarrow \frac{1}{|a|} S\left(\frac{f}{a}\right) \tag{2.57}$$

be careful to observe the absolute value sign

6. Frequency scaling

$$S(af) \leftrightarrow \frac{1}{|a|} s\left(\frac{t}{a}\right) \tag{2.58}$$

7. Differentiation

$$\frac{d^n s(t)}{dt^n} \leftrightarrow (j2\pi f)^n S(f) \tag{2.59}$$

8. Integration

$$\int_{-\infty}^{t} s(\lambda)d\lambda = (j2\pi f)^{-1} S(f) + \tfrac{1}{2} S(0)\delta(f) \tag{2.60}$$

9. Multiplication

$$s_1(t)s_2(t) \leftrightarrow S_1(f) * S_2(f) \tag{2.61}$$

10. Duality

$$S(t) \leftrightarrow s\{(-f)\} \tag{2.62}$$

Other properties concerning even/odd symmetry and real versus imaginary parts of $s(t)$ and $S(f)$ are shown in Table 2-4.

Table 2-4 Recognizing Patterns in $s(t)$ and $S(f)$.

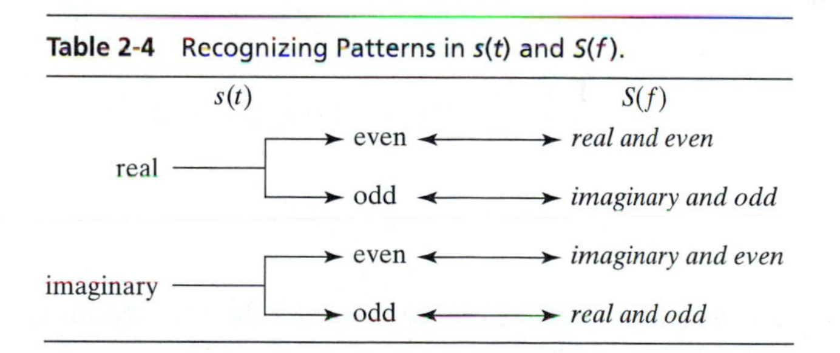

s(t)
S(f)
real
even
real and even
odd
imaginary and odd
imaginary
even
imaginary and even
odd
real and odd

2.6 Using the Unit Impulse Function to Represent Discrete Frequency Components as Densities

Consider the signal

$$y(t) = [1 + s(t)] \cos 2\pi f_c t \text{ volts} \tag{2.63}$$

where $s(t)$ is nonperiodic with finite energy. Signals of this type are often seen in communication systems such as commercial AM broadcasting. How can we represent this signal in the frequency domain?

Let's deconstruct the signal into two parts:

$$\begin{aligned} y(t) &= [1 + s(t)] \cos 2\pi f_c t \\ &= \cos 2\pi f_c t + s(t) \cos 2\pi f_c t \text{ volts} \end{aligned} \tag{2.64}$$

We see that the first part of the signal is periodic with finite power and the second part is nonperiodic with finite energy. Using the two-sided form of the Fourier series, we can express the first term as:

$$\cos 2\pi f_c t = \frac{1}{2} e^{-j2\pi f_c t} + \frac{1}{2} e^{j2\pi f_c t} \tag{2.65}$$

Using the Fourier transform and Equation (2.56), we can express the second part of the signal as:

$$\mathcal{F}\{s(t) \cos 2\pi f_c t\} = \frac{1}{2} S(f - f_c) + \frac{1}{2} S(f + f_c) \tag{2.66}$$

The frequency domain representation of $y(t)$ is thus:

$$Y(f) = \frac{1}{2} e^{-j2\pi f_c t} + \frac{1}{2} e^{j2\pi f_c t} \text{ volts} + \frac{1}{2} S(f + f_c) + \frac{1}{2} S(f - f_c) \text{ volts/Hz} \tag{2.67}$$

Equation (2.67) is peculiar—conceptually we understand that $Y(f)$ has both discrete components and a density, but having two different, inconsistent units in the equation limits its usefulness. For example, how can we draw a consistent magnitude spectrum of $Y(f)$? To solve this problem, we need to find a way to express the discrete components in Equation (2.67) as a density.

Consider the unit impulse function $\delta(x)$, which is defined by the following two properties

$$\delta(x) = 0 \text{ for all } x \neq 0 \tag{2.68}$$

$$\int_{-\infty}^{\infty} \delta(x)\, dx = 1 \tag{2.69}$$

With these two properties we can represent a discrete frequency component as a density. Using Equation (2.65),

$$\cos 2\pi f_c t = \frac{1}{2} \delta(f - f_c) + \frac{1}{2} \delta(f + f_c) \text{ volts/Hz} \tag{2.70}$$

The frequency domain representation of $y(t)$ can now be written as:

$$Y(f) = \frac{1}{2} \delta(f + f_c) + \frac{1}{2} \delta(f - f_c) + \frac{1}{2} S(f + f_c) + \frac{1}{2} S(f - f_c) \text{ volts/Hz} \tag{2.71}$$

Suppose the magnitude spectrum of $S(f)$ is plotted in Figure 2-32a; the magnitude spectrum of $Y(f)$ can then be plotted in Figure 2-32b. Note that the impulse function is

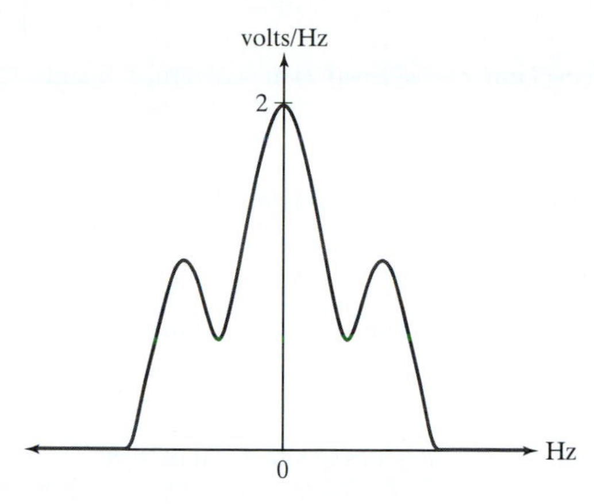

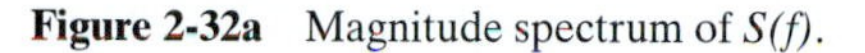
Figure 2-32a Magnitude spectrum of $S(f)$.

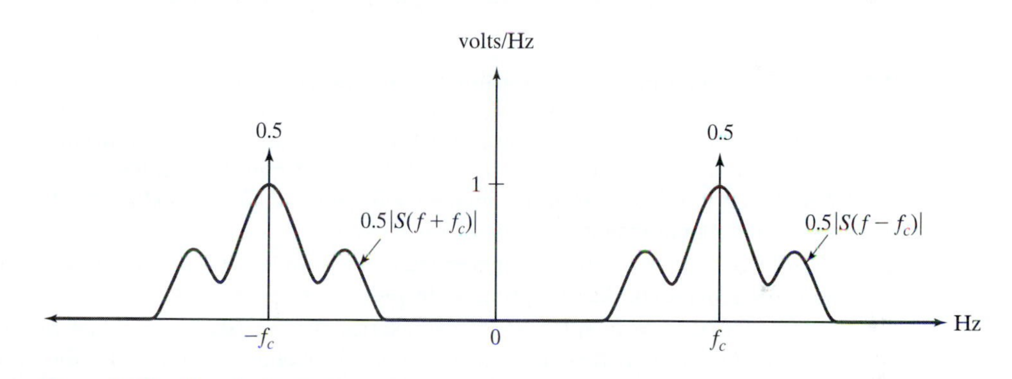

Figure 2-32b Magnitude spectrum of $Y(f)$.

represented as a vertical arrow with a number on top corresponding to the coefficient (or strength) of the impulse.

We've now established the basic frequency domain analysis tools necessary to begin communication system analysis and design.

PROBLEMS

2.1 In your own words, define the following terms:

- **a.** Time domain analysis
- **b.** Frequency domain analysis

2.2 Give an example of a communication system application where bandwidth is very important.

2.3 Why is it important to analyze the effects of the channel in a communication system?

2.4 Show that the set of harmonically-related sines and cosines is orthogonal. In other words, show that for any positive integer values of m and n

a. $\int_{t_o}^{t_o+T} \cos(2\pi n f_o t)\cos(2\pi m f_o t)\,dt = 0$ if $m \neq n$ and

$$\int_{t_o}^{t_o+T} \cos(2\pi n f_o t)\cos(2\pi m f_o t)\,dt \neq 0 \text{ if } m = n$$

b. $\int_{t_o}^{t_o+T} \sin(2\pi n f_o t)\sin(2\pi m f_o t)\,dt = 0$ if $m \neq n$ and

$$\int_{t_o}^{t_o+T} \sin(2\pi n f_o t)\sin(2\pi m f_o t)\,dt \neq 0 \text{ if } m = n$$

c. $\int_{t_o}^{t_o+T} \cos(2\pi n f_o t)\sin(2\pi m f_o t)\,dt = 0$ for all values of m and n

2.5 Using the concept of orthogonality, derive the equations for the coefficients of the trigonometric form of the Fourier series (i.e., derive the expressions for a_0, a_n, and b_n). (Hint: To find the expression for a_n, start with the expression for the trigonometric form of the Fourier series, multiply both the left-hand and right-hand sides by $\cos 2\pi m f_o t$, and integrate over one period of the fundamental frequency. The a_0 and b_n expressions can be found in a similar manner.)

2.6 Using the Fourier coefficients determined in Example 2.3, Part A, and an appropriate software graphics package, reproduce the plots from Figures 2-8a–d.

2.7 Express the waveform in Figure P2-7 as a series of sinusoids. Your expression should use the actual Fourier coefficient values (i.e., the values of a_0, a_n, and b_n) for the dc term and at least the first four harmonics. Your expression should also use the actual value of the fundamental frequency (f_o). You may wish to set up a spreadsheet or software program to calculate your Fourier coefficients, since you will need to calculate the values for higher-order harmonics in Problem 2.8.

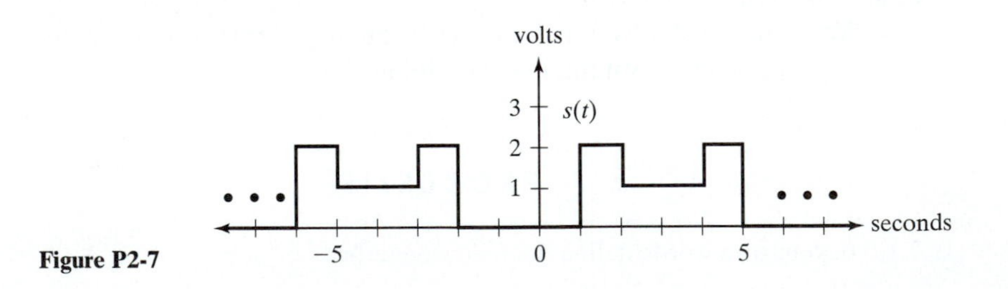

Figure P2-7

2.8 To see how many harmonics are needed for the Fourier series expression developed in Problem 2.7 to adequately replicate the original waveform,

a. Plot the waveform produced by summing the dc component and the sinusoids associated with the first five harmonics.

b. Plot the waveform produced by summing the dc component and the sinusoids associated with the first ten harmonics.

c. Plot the waveform produced by summing the dc component and the sinusoids associated with the first fifteen harmonics.

d. Plot the waveform produced by summing the dc component and the sinusoids associated with the first twenty-five harmonics.

e. Comment on how well the Fourier series is converging to produce the original waveform.

2.9 Prove that the trigonometric form of the Fourier series and one-sided form of the Fourier series are equivalent. Think about the relationship between rectangular and polar coordinates and see how that concept relates to this proof.

2.10 Express the waveform in Figure P2-7 as a series of cosines using the one-sided Fourier series. Draw one-sided magnitude and phase spectra for the waveform.

2.11 Prove Euler's identity. (Hint: Consider power series expansions of both the left-hand and right-hand sides of the identity.)

2.12 Prove that the trigonometric form of the Fourier series and complex exponential form of the Fourier series are equivalent.

2.13 Express the waveform in Figure P2-7 using the complex exponential Fourier series. Draw two-sided magnitude and phase spectra for the waveform.

2.14 Discuss the relationship between the one-sided spectra (magnitude and phase) drawn in Problem 2.10 and the two-sided spectra drawn in Problem 2.13.

2.15 Draw the two-sided amplitude spectrum for the waveform in Figure P2-15 for each of the following sets of parameters:

a. $T = 50$ milliseconds, $\tau = 10$ milliseconds
b. $T = 50$ milliseconds, $\tau = 25$ milliseconds
c. $T = 50$ milliseconds, $\tau = 5$ milliseconds
d. $T = 20$ milliseconds, $\tau = 10$ milliseconds
e. $T = 100$ milliseconds, $\tau = 10$ milliseconds

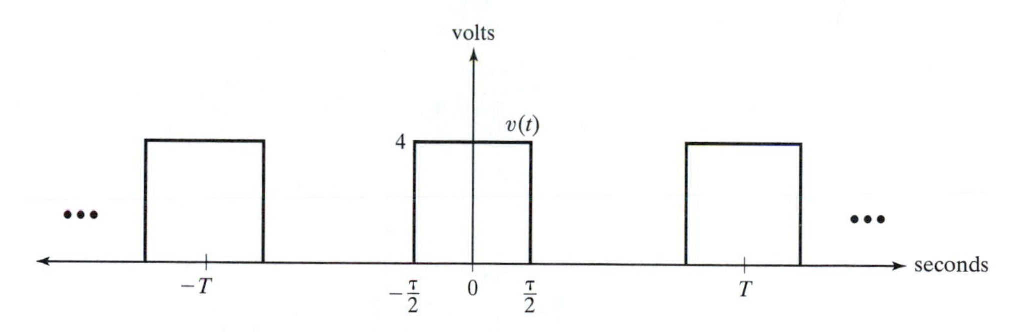

Figure P2-15

2.16 Discuss the advantages and disadvantages of each of the three forms of the Fourier series.

2.17 Prove that average normalized power can also be expressed in the frequency domain as:

$$P_s = \sum_{n=-\infty}^{\infty} |c_n|^2$$

(Hint: First show that $P_s = \sum_{n=-\infty}^{\infty} c_n c_{-n}$ and then show that for all values of n, $c_n c_{-n} = |c_n|^2$.)

2.18 In Problem 2.15, you drew the two-sided amplitude spectrum for the Figure P2-15 waveform for five sets of parameters:

a. $T = 50$ milliseconds, $\tau = 10$ milliseconds
b. $T = 50$ milliseconds, $\tau = 25$ milliseconds
c. $T = 50$ milliseconds, $\tau = 5$ milliseconds
d. $T = 20$ milliseconds, $\tau = 10$ milliseconds
e. $T = 100$ milliseconds, $\tau = 10$ milliseconds

For each of the waveforms in Problem 2.15 a–e,

1. Calculate the average normalized power using the time domain.
2. Draw the two-sided average normalized power spectrum.
3. Determine the bandwidth of the waveform (define bandwidth as the distance from dc (0 Hz) to the first zero crossing of the power spectrum).
4. Determine the percentage of each waveform's power that is contained within its bandwidth.

2.19 Explain why, when using the two-sided Fourier series, the "negative" frequency components must be included when performing power calculations but not when determining bandwidth.

2.20 Why is orthogonality of the Fourier series components important when representing power in the frequency domain?

2.21 Draw the time domain waveform that corresponds to the amplitude spectrum in Figure P2-21.

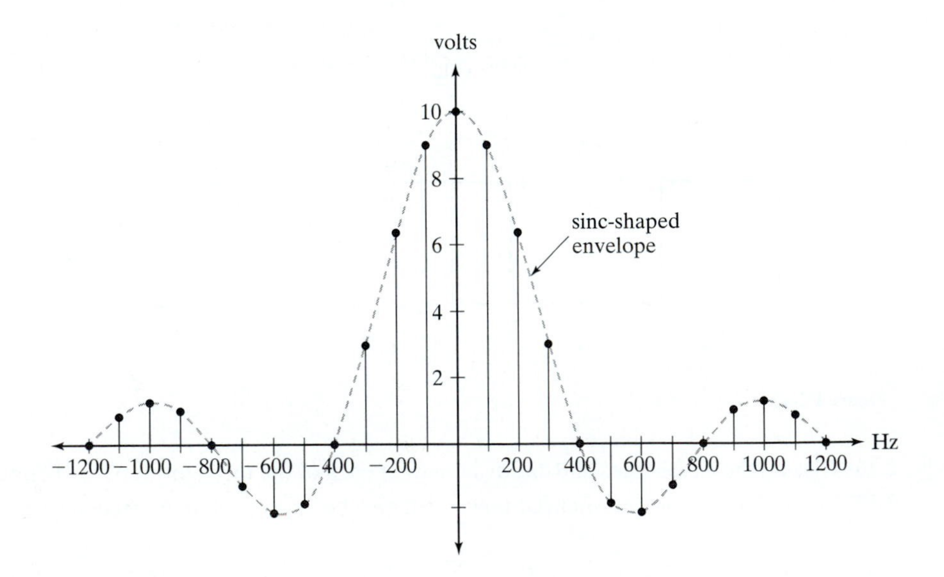

Figure P2-21

2.22 For the waveform in Figure P2-21, determine the bandwidth and the percentage of the waveform's power that lies within the bandwidth. Define the bandwidth of the waveform as the distance from dc (0 Hz) to the first zero crossing in the waveform's power spectrum.

2.23 Determine the frequency domain expression for each of the nonperiodic waveforms in Figure P2-23 and draw their magnitude and phase spectra.

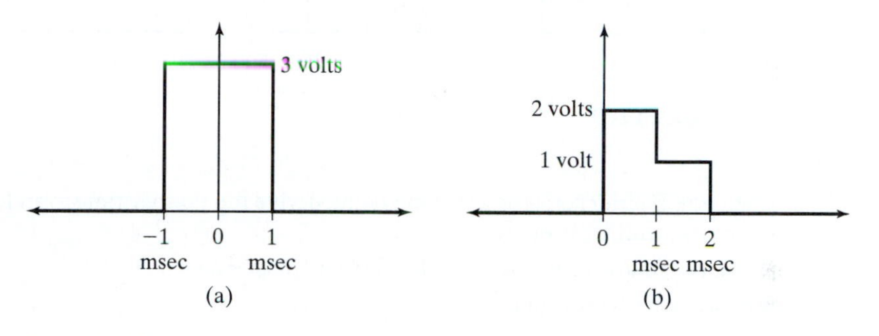

Figure P2-23

2.24 Draw an amplitude spectrum for the rectangular waveform in Figure P2-23a. Why is this spectrum possibly more useful than phase and magnitude spectra? Are there any restrictions on the type of waveform that can be represented using an amplitude spectrum?

2.25 Draw the normalized energy spectrum for the waveforms in Figures P2-23a and b.

2.26 Suppose the two waveforms in Figures P2-23a and b are passed through a filter with the transfer function shown in Figure P2-26. For each of the two waveforms, draw the magnitude spectrum and normalized energy spectrum at the filter's output. Also, calculate the normalized energy at the filter output for each of the two waveforms (you may leave your expression in integral form).

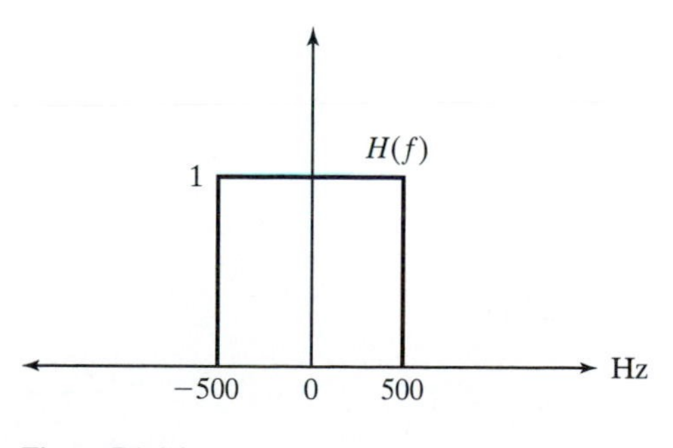

Figure P2-26

2.27 Repeat Problem 2.26 using a filter with the transfer function in Figure P2-27.

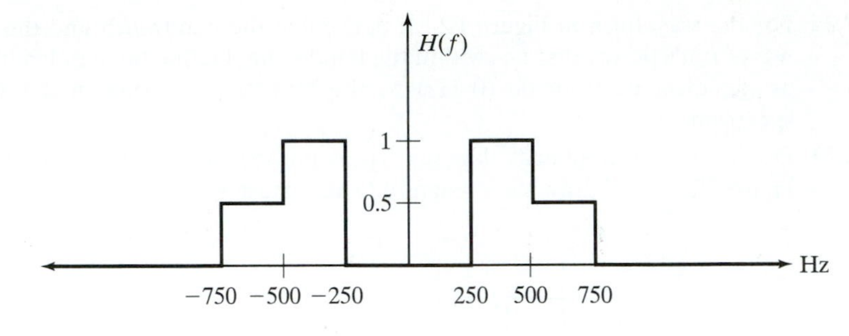

Figure P2-27

2.28 Using the Fourier series as a starting point, derive the Fourier transform for a nonperiodic waveform with finite energy.

2.29 Prove the following property of the Fourier transform:

For any nonzero constant a,

$$x(at) \leftrightarrow \frac{1}{|a|} X\left(\frac{f}{a}\right)$$

(Hint: Perform separate proofs for $a > 0$ and $a < 0$.)

2.30 Determine the Fourier transform of the signal $s(t) = 3\cos(2000\pi t) + 5e^{-|t|}$ volts and draw its magnitude spectrum.

Chapter 3

Digital Baseband Modulation Techniques

3.1 Goals in Communication System Design

With the mathematical tools we now have, we are ready to start analyzing and designing communication systems. The best place to begin is to review the block diagram of a communication system that we developed in Chapter 1:

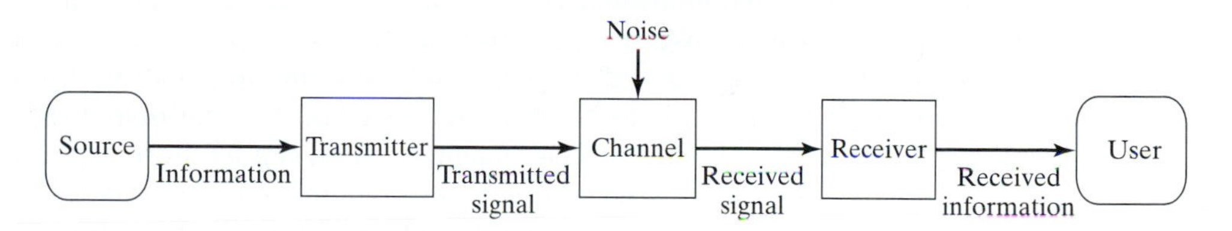

Figure 1-1*R* General block diagram of a communication system.

In Chapter 1, we also defined the various blocks and signals in the system:

Information: Information is data that the user did not possess prior to communicating with the source.

Source: The source is the originator of the information (e.g., a human voice, music, digital data from a computer). The information from the source may be either an analog signal (such as voice or music) or a digital signal (such as the output from the serial ports of a computer). In this chapter, we will examine digital communication systems that transmit information over a channel known as a baseband channel. In later chapters, we will examine other types of digital communication systems and analog communication systems.

Channel: The channel is the physical transmission medium over which the communication is sent (e.g., wires, air, fiber optics). As we discussed in Chapter 2, all channels have physical limitations that distort and attenuate the transmitted signal and add noise to it. Thus, the received signal will not be an exact duplicate of the transmitted signal.

Transmitter: Because the channel, generally speaking, adds unacceptable amounts of noise and distortion to the information signal, the transmitter is used to reformat and reshape the signal so that the channel will not distort it as much. This reformatting and reshaping is called *modulation*. After reformatting and reshaping, the transmitter may also boost the power of the signal.

Receiver: The receiver attempts to translate the received signal back into the original information sent by the source. This process, called *demodulation*, involves basically two steps:

1. Compensating, as much as possible, for the noise and distortion added by the channel, and
2. "Undoing" the reformatting and reshaping performed by the transmitter.

User: The user is the recipient of the information.

In Example 1.1 in Chapter 1 we examined a communication system designed for a specific application (an automated teller machine in Seattle, communicating over a telephone line with a bank computer in San Francisco) and we tracked the signals through the entire system. We discussed how communication system design involves trade-offs between performance and cost, and we noted that performance parameters include transmission speed, accuracy, and reliability, while cost parameters include hardware complexity, computational power, channel bandwidth, and the required power to transmit the signal. We've since observed that these parameters are all interrelated and that the proper trade-offs depend on the system's application (for instance, a medical imaging system needs to be more accurate and reliable than a system that transmits elevator music, but the medical imaging system can be more expensive). We concluded Chapter 1 by summarizing our goal in communication system design: Given an application, a source, a user, and a particular channel (or a limited choice of channels), our goal is to design a transmitter and a receiver to optimize the performance-versus-cost parameters for the specific application.

3.2 Baseband Modulation Using Rectangular Pulses and Binary Pulse Amplitude Modulation

A good place to start our analysis and design effort is by characterizing the channel. All channels can be classified as one of two types:

Baseband: A *baseband channel* efficiently passes frequency components from dc to f_c Hz, where f_c is called the *cutoff frequency*. Frequency components above f_c

are not efficiently passed by baseband channels. Wires and coaxial cables are examples of baseband channels.

Bandpass: A *bandpass channel* efficiently passes frequency components within a certain band, say, between f_l and f_h Hz. Components outside the frequency band ($<f_l$ or $>f_h$) are not efficiently passed by bandpass channels. Airwaves and fiber optic lines are examples of bandpass channels.

Given a particular channel—say, a baseband channel with a cutoff frequency of f_c—our job in communication system engineering is to design a transmitter and a receiver (i.e., develop modulation and demodulation techniques) that, for a specific application, optimize the performance-versus-cost trade-offs mentioned in Section 3.1. Let's begin by focusing on one specific performance parameter—accuracy—and let's suppose that our goal is to minimize error. Reviewing the general block diagram in Figure 1-1 and Example 1.1, we can conclude that there are two causes of error in communication systems: noise and distortion. First, let's simplify our problem:

1. Let's assume for now that the channel is noiseless.
2. Let's also assume that the channel can be modeled as an ideal lowpass filter. (As established in Chapter 2, we call such a channel an ideal baseband channel.)

If we start by assuming that the channel is noiseless, then the only thing that can cause errors is distortion, and distortion occurs in an ideal baseband channel only if the bandwidth of the transmitted signal exceeds the bandwidth of the channel (remember Example 2.10 and Figures 2-21a–d in Chapter 2). Thus, to ensure accuracy, we need to design the modulation technique to produce a signal that has all its significant frequency components within the bandwidth of the channel.

To find a signal suitable for transmission across an ideal baseband channel, we need to examine the bandwidth of individual signals. As established in Chapter 2, we do this by using frequency domain analysis and examining how the signal's energy is distributed throughout the frequency band.

Let's start by re-examining the rectangular pulse. From Chapter 2, we know the time and frequency domain representations of a single rectangular pulse of height A and width τ are as shown in Figure 3-1.

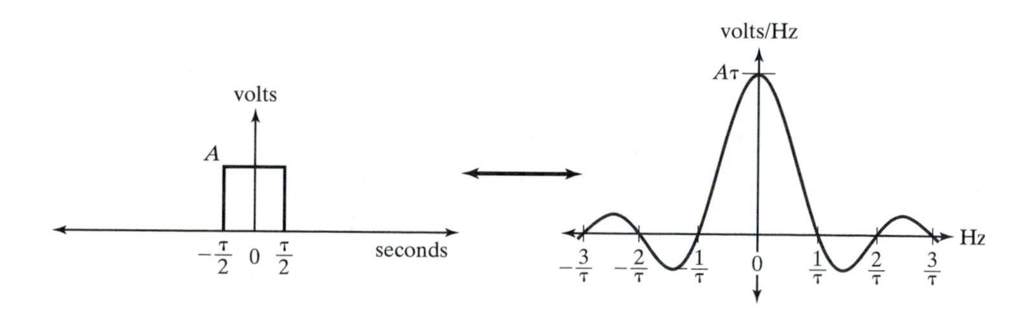

Figure 3-1 Time and frequency domain representations of a single rectangular pulse.

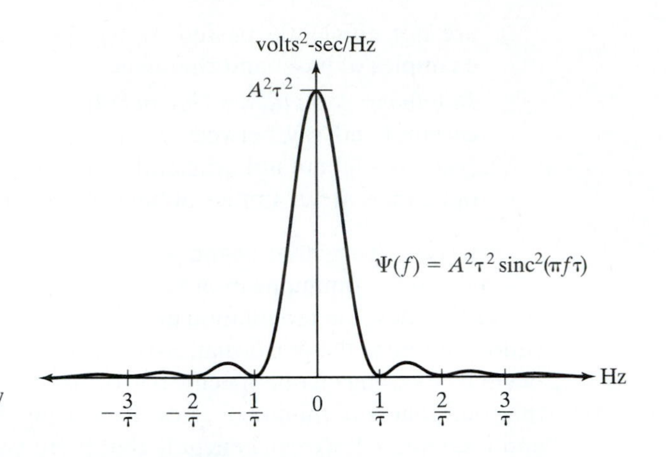

Figure 3-2 Normalized energy spectral density of a single rectangular pulse.

To determine bandwidth, we need to examine how the pulse's energy is distributed throughout the frequency band (i.e., we need to examine the pulse's energy spectral density). As we established in Section 2.5 in Chapter 2, normalized energy spectral density is obtained by squaring the magnitude spectrum. Figure 3-2 shows the normalized energy spectral density of a single, positive rectangular pulse.

Let's also examine the energy spectral density of a single negative rectangular pulse. As shown in Figures 3-3a–c, note that the normalized energy spectral density of a single negative rectangular pulse is the same as a single positive rectangular pulse. Be sure you understand why.

Now suppose that we're going to transmit digital information by sending a positive rectangular pulse to signify each "1" and a negative rectangular pulse to signify

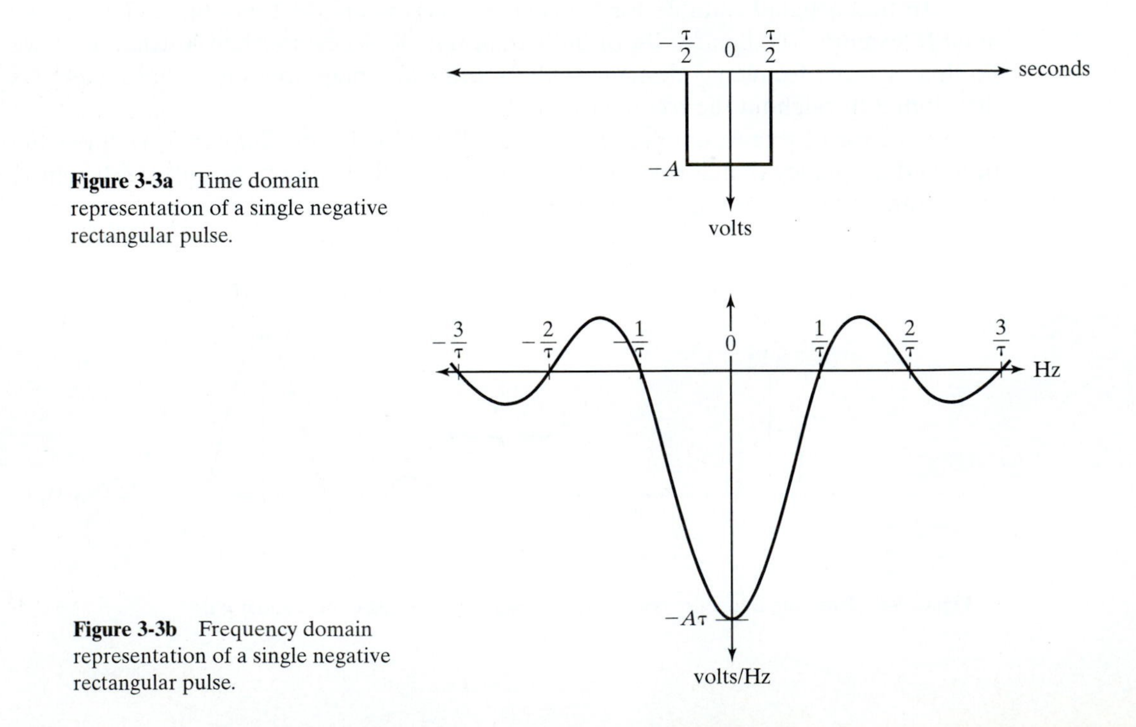

Figure 3-3a Time domain representation of a single negative rectangular pulse.

Figure 3-3b Frequency domain representation of a single negative rectangular pulse.

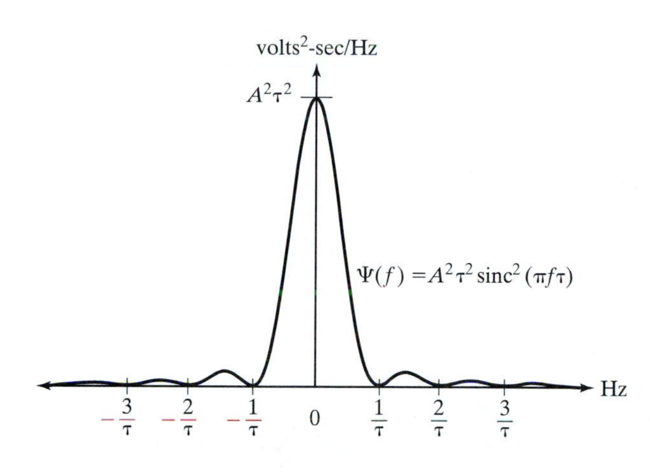

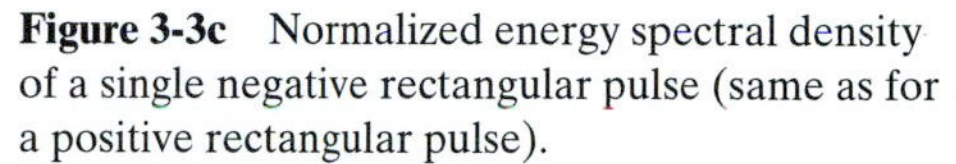
Figure 3-3c Normalized energy spectral density of a single negative rectangular pulse (same as for a positive rectangular pulse).

each "0." In other words, we're going to transmit a typical digital data sequence—say, "10110"—as shown in Figure 3-4.

How do we determine the normalized energy spectral density of a series of rectangular pulses? To do this formally, we will need another set of mathematical tools (stochastic mathematics) and a powerful theorem known as the Wiener-Khintchine theorem. These topics will be discussed in Chapters 4 and 5 and are essential to a complete understanding of communication systems. For now, however, let's take a less rigorous approach, which is valid provided that within the data stream of "1"s and "0"s, each bit has the same probability of being a "1" or a "0" and that the value of a particular bit has no influence on the values of any other bits (i.e., each bit is independent).

Let's start by considering a data sequence one bit long. The transmitted signal may be either a positive pulse (corresponding to a "1") or a negative pulse (corresponding to a "0"). Either way, the normalized energy spectral density of the transmitted signal is $\Psi(f) = A^2\tau^2\text{sinc}^2(\pi f\tau)$, as illustrated in Figures 3-2 and 3-3c.

Let's now consider a data sequence two bits long. The bits in this sequence could have four possible combinations (all equally likely): "00," "01," "10," and "11." As shown in Figures 3-5a–d, each bit sequence produces a unique waveform. Note that the

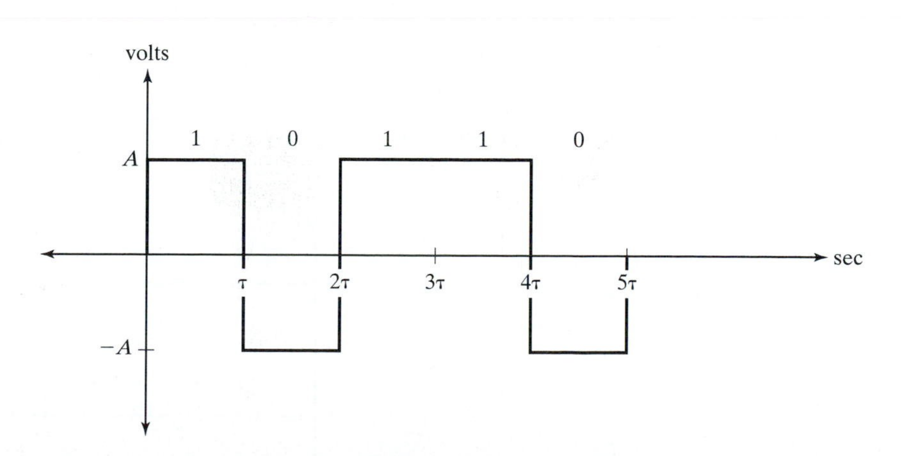

Figure 3-4 Transmitting digital data using rectangular pulses.

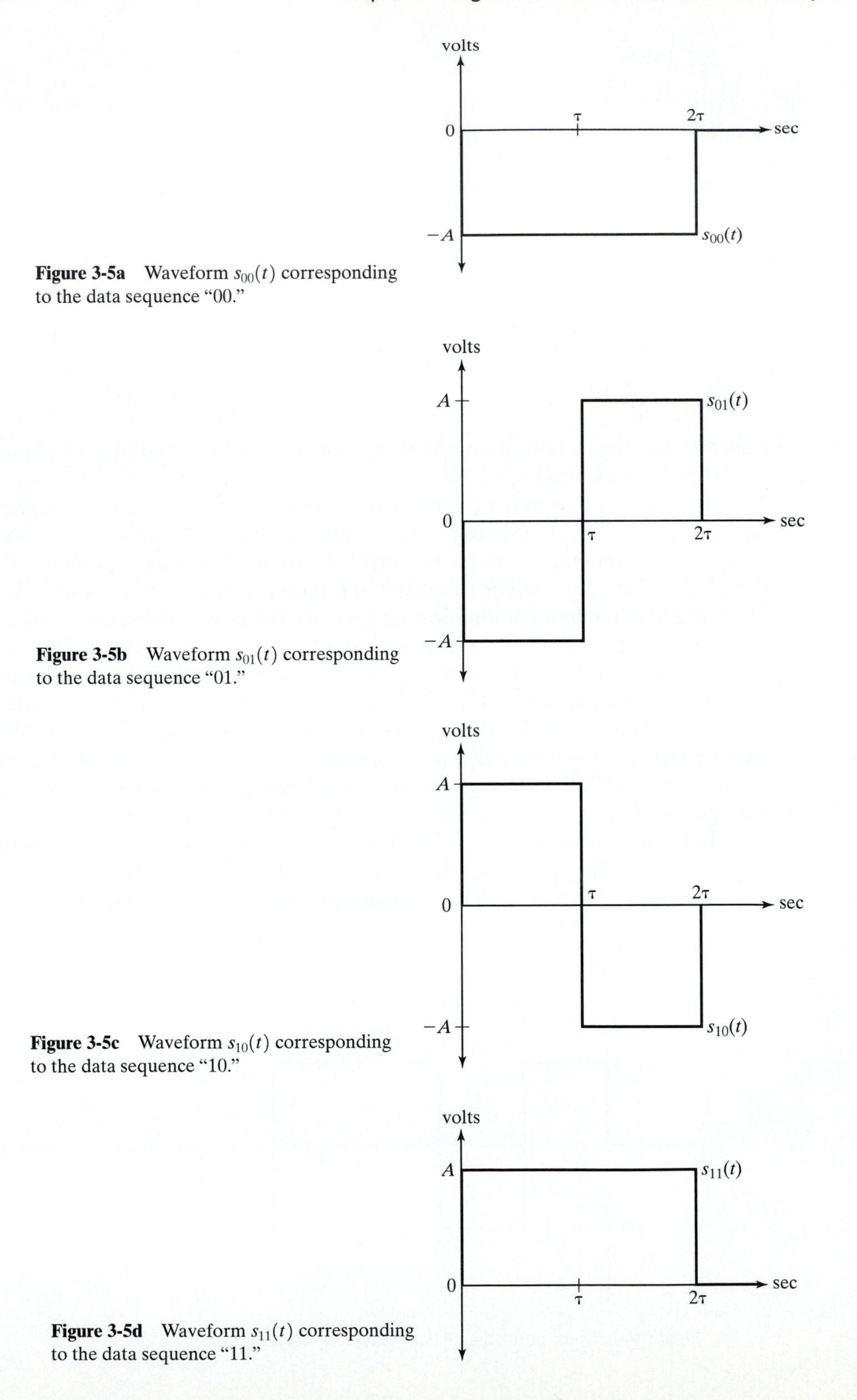

Figure 3-5a Waveform $s_{00}(t)$ corresponding to the data sequence "00."

Figure 3-5b Waveform $s_{01}(t)$ corresponding to the data sequence "01."

Figure 3-5c Waveform $s_{10}(t)$ corresponding to the data sequence "10."

Figure 3-5d Waveform $s_{11}(t)$ corresponding to the data sequence "11."

waveforms are denoted using the variable s and that the subscript denotes the bit pattern. Thus, $s_{01}(t)$ is the waveform corresponding to the data sequence "01." To determine the bandwidth of a two-bit data sequence transmitted using rectangular pulses, we need to examine the normalized energy spectral density of each of the four waveforms in Figures 3-5a–d. The normalized energy spectral densities of the four waveforms are shown in Figures 3-6a–d. Again, note that subscripts are used to denote the bit pattern.

We see from Figures 3-6a–d that the normalized energy spectral density of the transmitted two-bit waveform depends on the bit sequence from the source. As shown in Figure 3-6e, if we *average* the four possible normalized energy spectral densities, the result is a sinc-squared-shaped average normalized energy spectral density that corresponds to double the normalized energy spectral density of the single rectangular pulse shown in Figure 3-2. In a similar way (as you will see in Problem 3.1), we can show that for a three-bit sequence, if we average the normalized energy spectral densities corresponding to the eight possible waveforms, the result is a sinc-squared-shaped average normalized energy spectral density that corresponds to triple the energy spectral density of the single rectangular pulse. This observation can be extended to n bits: The average normalized energy spectral density of a series of rectangular pulses representing n data bits is a sinc-squared-shaped pulse corresponding to n times the normalized energy spectral density of a single rectangular pulse. This average normalized energy spectral density is shown in Figure 3-7.

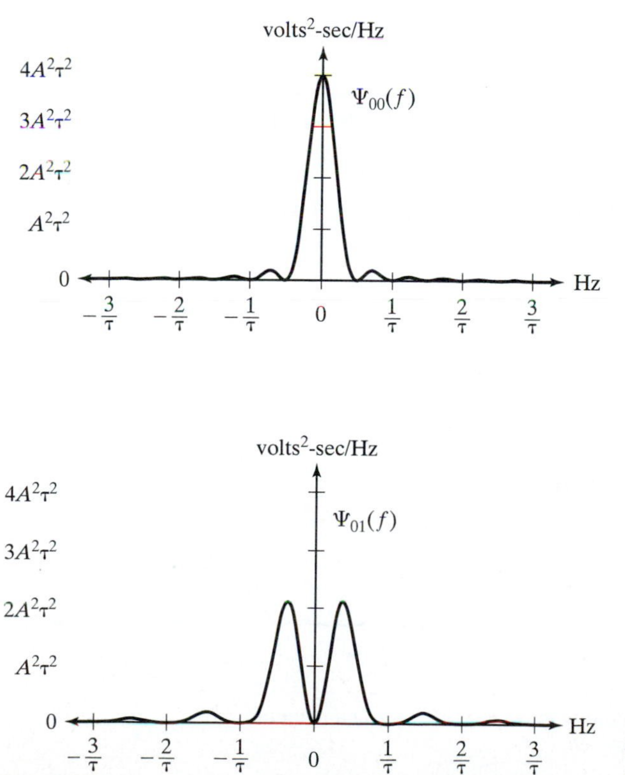

Figure 3-6a Normalized energy spectral density corresponding to the data sequence "00."

Figure 3-6b Normalized energy spectral density corresponding to the data sequence "01."

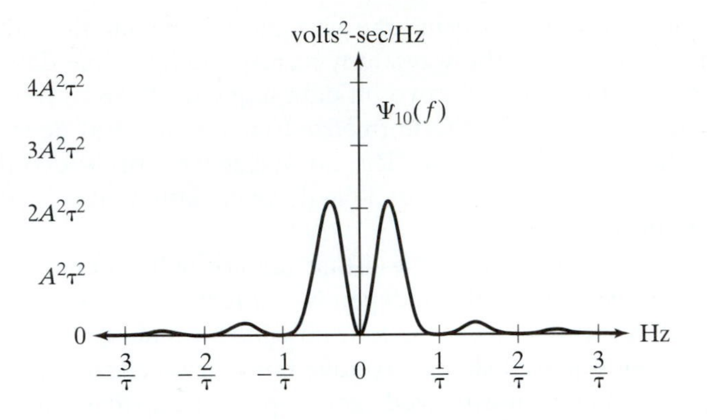

Figure 3-6c Normalized energy spectral density corresponding to the data sequence "10."

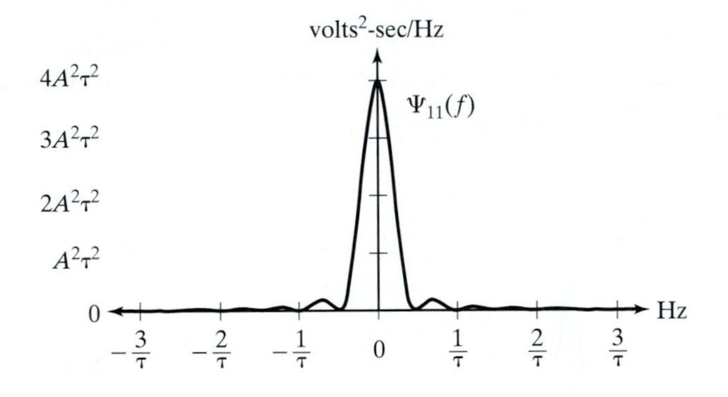

Figure 3-6d Normalized energy spectral density corresponding to the data sequence "11."

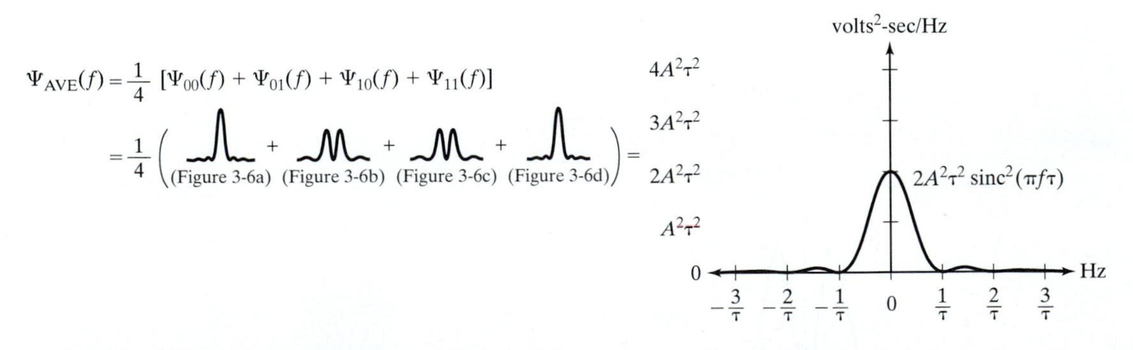

Figure 3-6e Average normalized energy spectral density for a series of two rectangular pulses.

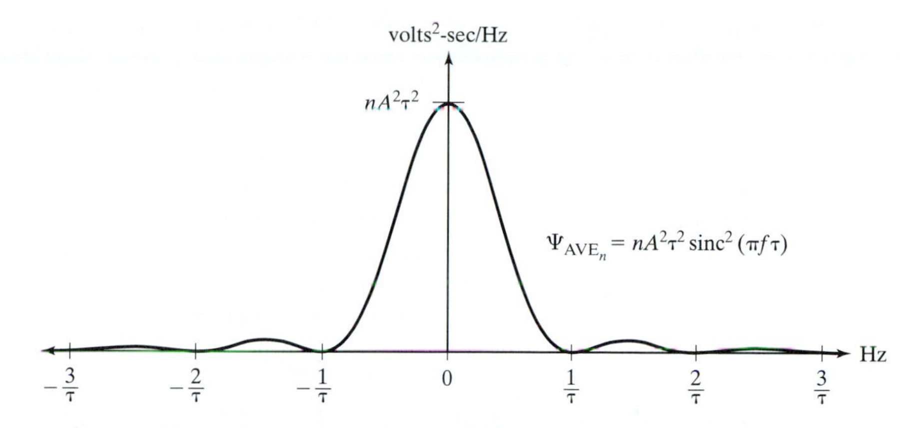

Figure 3-7 Average normalized energy spectral density for an n-bit sequence of rectangular pulses.

Let's now define the *power of a series of pulses* as

$$\text{Power of a series of pulses} = \frac{\text{Energy in a series of pulses}}{\text{Time to transmit the series of pulses}} \tag{3.1}$$

Extending this concept, we can plot an *average normalized power spectral density* of a series of pulses, symbolized as $G(f)$, using the average normalized energy spectral density and dividing it by the amount of time taken to transmit the series of pulses. The average normalized power spectral density of a series of rectangular pulses representing n data bits is thus,

$$G(f) = \frac{nA^2\tau^2\text{sinc}^2(\pi f\tau)}{n\tau} = A^2\tau\text{sinc}^2(\pi f\tau) \text{ volts}^2/\text{Hz} \tag{3.2}$$

As plotted in Figure 3-8, the average normalized power spectral density is sinc-squared-shaped with zero points at integral multiples of $1/\tau$. The average normalized power spectral density for a series of rectangular pulses has the same shape as the normalized energy

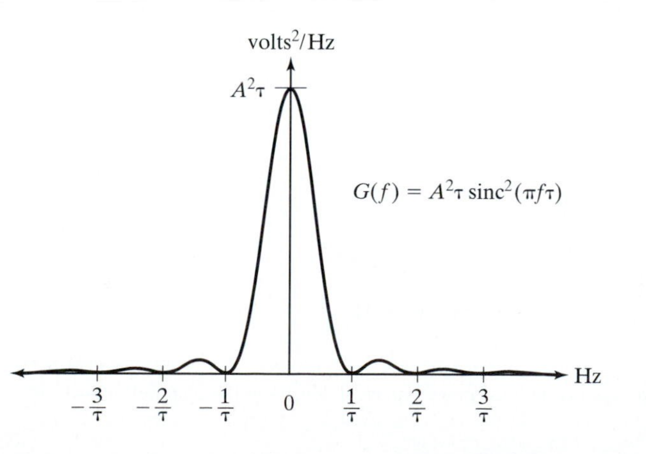

Figure 3-8 Average normalized power spectral density of a series of rectangular pulses.

spectral density of a single rectangular pulse. Although our approach for determining average normalized spectral density hasn't been rigorous (we'll get to the rigor later—see Chapter 5, Section 5.4.2), under the conditions we set earlier (bits are independent and equally likely to be "1"s and "0"s), the approach is valid.

Let's now go back to our original task: determining how much distortion is produced when we transmit various signals through an ideal baseband channel. In a manner similar to Theorem 2.1 in Chapter 2 on page 64, we can show that if a transmitted signal having an average normalized power spectral density $G_{\text{in}}(f)$ is passed through a channel (or filter) having a transfer function of $H(f)$, then the average normalized power spectral density of the signal at the output of the channel (or filter) is

$$G_{\text{out}}(f) = G_{\text{in}}(f)|H(f)|^2 \tag{3.3}$$

For an ideal baseband channel,

$$H(f) = \begin{cases} 1 & |f| \le f_c \\ 0 & |f| > f_c \end{cases} \tag{3.4}$$

Thus, as illustrated in Figure 3-9, if a signal consisting of a series of rectangular pulses of pulse width τ is transmitted through an ideal baseband filter with a cutoff frequency f_c, the average normalized power spectral density of the received signal will be

$$G_{\text{out}}(f) = G_{\text{in}}(f)|H(f)|^2 = \begin{cases} A^2\tau\text{sinc}^2(\pi f t) & |f| \le f_c \\ 0 & |f| > f_c \end{cases} \tag{3.5}$$

The ideal baseband filter thus truncates (chops off) the average normalized power spectral density for all frequencies above f_c.

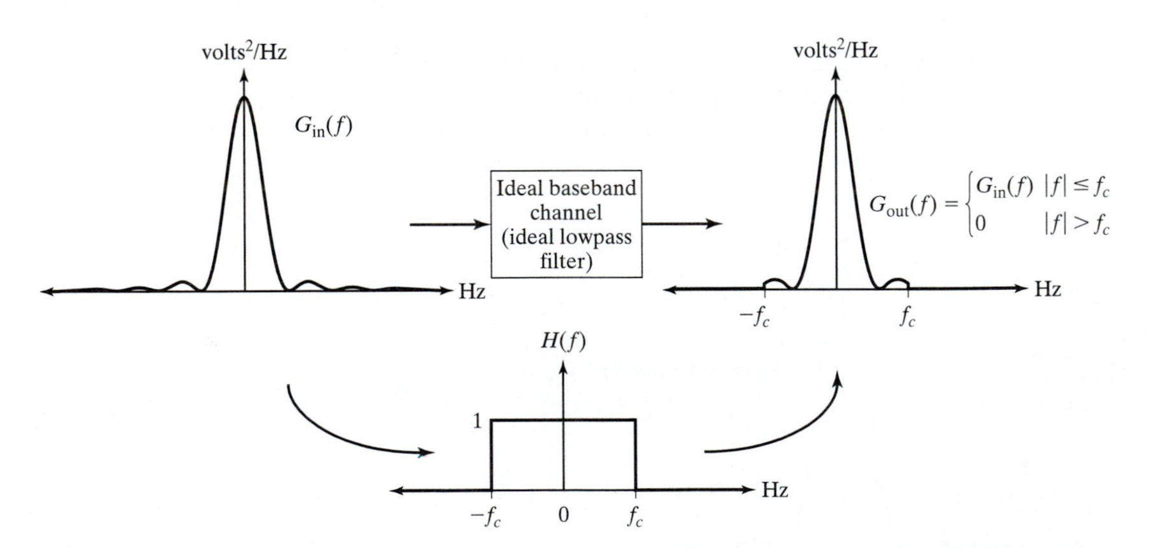

Figure 3-9 Ideal baseband channel truncates average normalized power spectral density at the channel's cutoff frequency.

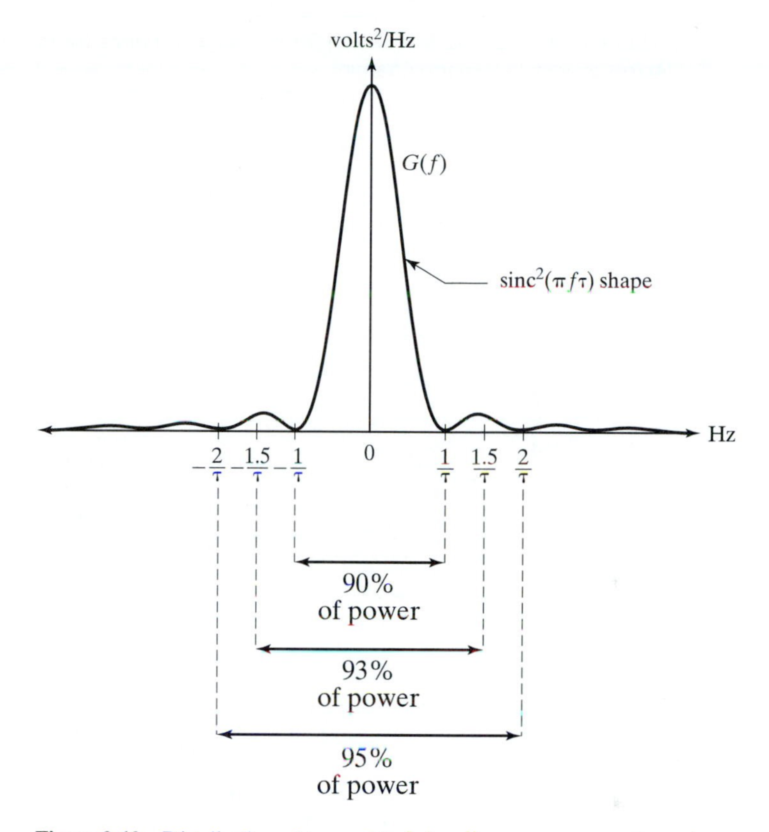

Figure 3-10 Distribution of transmitted signal's average power through the frequency band.

As we established in Example 2.10 in Chapter 2 on page 44, we can quantify the accuracy of the received signal for an ideal baseband channel by stating the percentage of the transmitted signal's power that lies within the frequency band passed by the channel. Figure 3-10 and Table 3-1 relate the accuracy of the received signal to the channel's bandwidth, stating the channel's bandwidth in terms of τ (the width of a rectangular pulse representing one bit in the transmitted bit stream).

We now see how accuracy, bandwidth, and pulse width (τ) are related. Let's go a step further and relate pulse width to transmission speed (another primary communication system parameter). Let's consider a source outputting r_b bits per second. The transmitter sends one rectangular pulse of width τ to represent each bit. A typical transmitted signal is shown in Figure 3-11. The amount of time between adjacent bits or pulses is called the *bit period* and is symbolized as T. Note that $T = 1/r_b$.

We know that bandwidth is related to pulse width τ, so what is the optimum pulse width? We want the pulses to be as wide as possible (this minimizes bandwidth), but we don't want the pulses to overlap. This is accomplished by selecting a pulse width of $\tau = T$, as shown in Figure 3-12. Note that if $\tau > T$, pulses overlap. If $\tau < T$, pulses do not overlap but bandwidth is wasted (since bandwidth is proportional to τ).

Table 3-1 Accuracy vs. Bandwidth for Rectangular Pulses Transmitted Across an Ideal Baseband Channel

Cutoff Frequency of the Ideal Baseband Channel	Average Power Lying within the Passband of the Channel
$f_c = \frac{1}{\tau}$	90%
$f_c = \frac{1.5}{\tau}$	93%
$f_c = \frac{2}{\tau}$	95%
$f_c = \frac{3}{\tau}$	96.5%
$f_c = \frac{4}{\tau}$	97.5%
$f_c = \frac{5}{\tau}$	98%

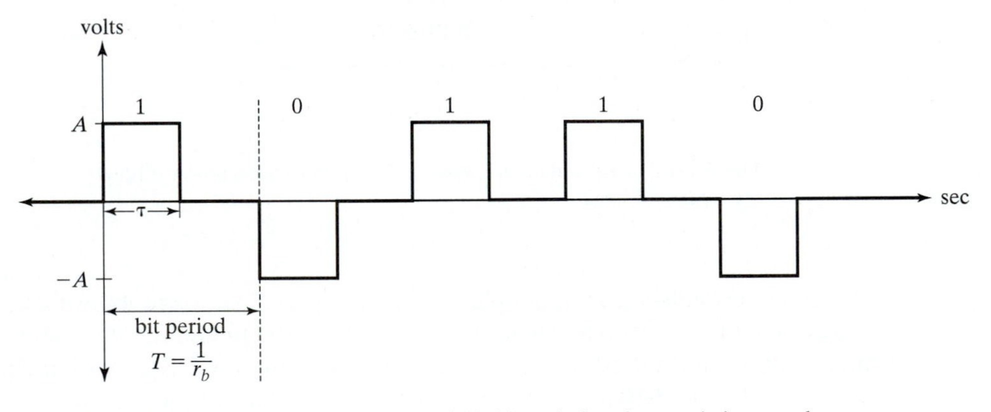

Figure 3-11 The relationships among pulse width, bit period, and transmission speed.

The optimum choice for pulse width is thus $\tau = T$, making each pulse as wide as its corresponding bit period. Note from Figure 3-11 that $T = 1/r_b$. Thus, we can relate the optimum pulse width to transmission speed

$$\tau_{optimum} = \frac{1}{r_b} \tag{3.6}$$

Selecting the optimum pulse width and using Equation (3.6), we can now relate the frequency axis of the transmitted signal's average normalized power spectral density (see Figure 3-10) to transmission speed. This relationship is shown in Figure 3-13.

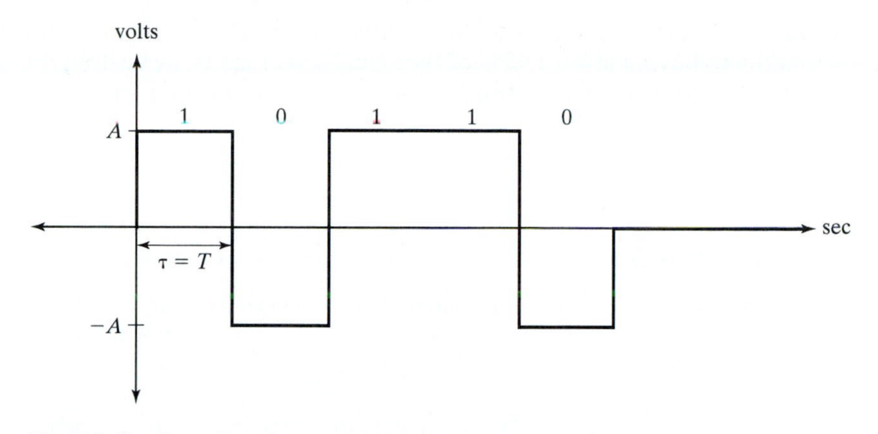

Figure 3-12 If $\tau = T$, pulses touch but don't overlap.

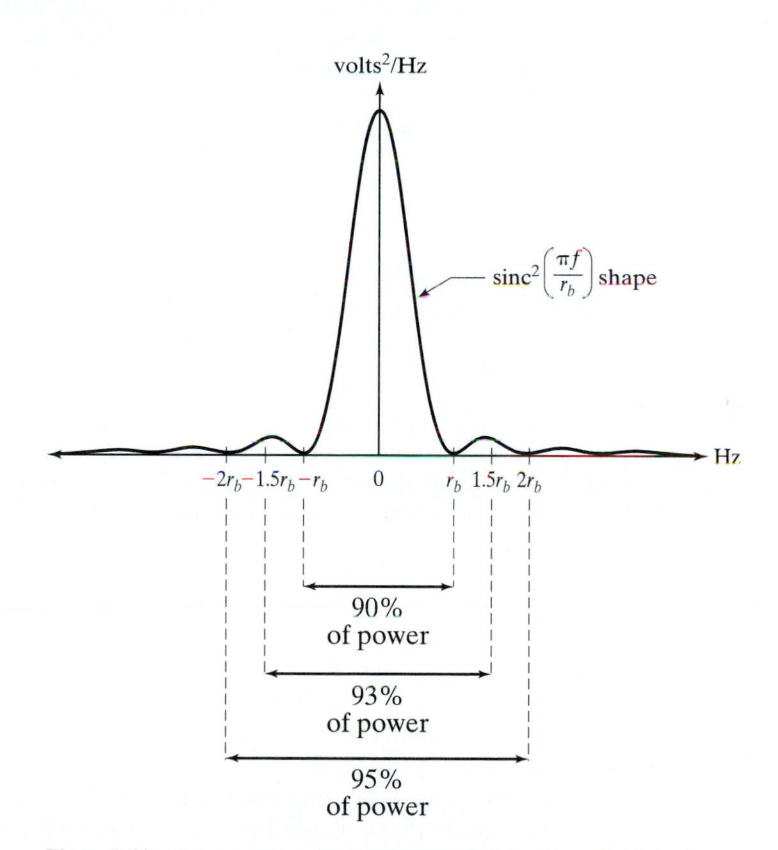

Figure 3-13 Average normalized power spectral density as it relates to transmission speed. (Compare this figure to Figure 3.10.)

We've reached a very important point in our analysis. If we want to transmit digital data across an ideal baseband channel by using rectangular pulses, we now know how to quantitatively relate accuracy, transmission speed, and bandwidth. For example, if a baseband channel has a bandwidth of B Hz (i.e., the channel passes frequencies from 0 to B Hz) and if a particular application requires a degree of accuracy corresponding to having at least 95% of the signal's average power within the channel bandwidth, then the maximum transmission speed of the system is $r_{b,\max} = \frac{B}{2}$ bits per second. Let's work on a few examples emphasizing the relationship among accuracy, transmission speed, and bandwidth.

Example 3.1 Relating transmission speed, accuracy, and bandwidth

A baseband channel has a bandwidth of 100 kHz. Accuracy requirements dictate that 95% of the signal's average power must be within the bandwidth of the channel. Determine the maximum transmission speed of the system.

Solution

From Figure 3-13, we know that for a given transmission speed of r_b bits/sec, 95% of the signal's average power lies within the frequency band between 0 and $2r_b$ Hz. Thus, to ensure that at least 95% of the average power is within the channel's bandwidth,

$$r_{b,\max} = \frac{B}{2} = \frac{100{,}000}{2} = 50{,}000 \text{ bits per second}$$

Example 3.2 Trading off accuracy for increased speed

Repeat Example 3.1 to reflect an application where accuracy requirements can be relaxed so that only 90% of the signal's power must be within the bandwidth of the channel.

Solution

From Figure 3-13, we know that for a given transmission speed r_b, 90% of the signal's average power lies in the frequency band between 0 and r_b Hz. So, to ensure that at least 90% of the signal's average power is within the channel's bandwidth,

$$r_{b,\max} = B = 100{,}000 \text{ bits per second}$$

In comparing Examples 3.1 and 3.2, we see how we can increase transmission speed by accepting reduced accuracy or how we can increase accuracy by reducing transmission speed (i.e., for a channel with a fixed bandwidth, we can trade off accuracy and transmission speed).

The transmission technique of sending a rectangular pulse of amplitude A to signify each "1" and a rectangular pulse of amplitude $-A$ to signify each "0" in the data pattern belongs to a family of modulation techniques known as *binary pulse amplitude modulation* (binary PAM). Since the term "modulate" means "to change," the term "pulse amplitude modulation" means "changing the amplitude of pulses." This is exactly what we are doing: We are using two different pulse amplitudes, A and $-A$, to represent the two different possible values of a data bit, "1" and "0." The term "binary" means that each pulse is being used to represent one of two possible data values.

3.3 Pulse Shaping to Improve Spectral Efficiency

Section 3.2 shows how accuracy, transmission speed, and bandwidth are related when transmitting digital data using rectangular pulses and binary PAM. Suppose we want to increase transmission speed without reducing accuracy or increasing bandwidth. (As an example, consider trying to develop a faster telephone modem for the Internet. The bandwidth of the telephone line is fixed, so we can't increase the system's bandwidth, and there are minimum requirements for accuracy.) How can we increase transmission speed without reducing accuracy or increasing bandwidth? Let's consider transmitting pulses that require less bandwidth than the rectangular pulse. What pulse shape should we choose? Consider the following requirements for a "good" pulse shape:

1. We want a "smooth" shape because we know that such a shape has an energy spectrum consisting of lower frequency components (thus minimizing bandwidth).
2. We need to be sure that the pulse transmitted to represent a particular bit (say, the nth bit) does not interfere at the receiver with the pulse transmitted previously (the pulse for the $n - 1$st bit) or the pulse to be transmitted next (the pulse for the $n + 1$st bit). The phenomenon of pulses interfering with each other is known as *intersymbol interference* (ISI).

3.3.1 The Sinc-Shaped Pulse

Keeping in mind our two requirements (smooth shape for spectral efficiency and no ISI), consider the pulse shape shown in Figure 3-14. This is the sinc function, which we used extensively in Chapter 2. Note, however, that in Chapter 2 the sinc function was

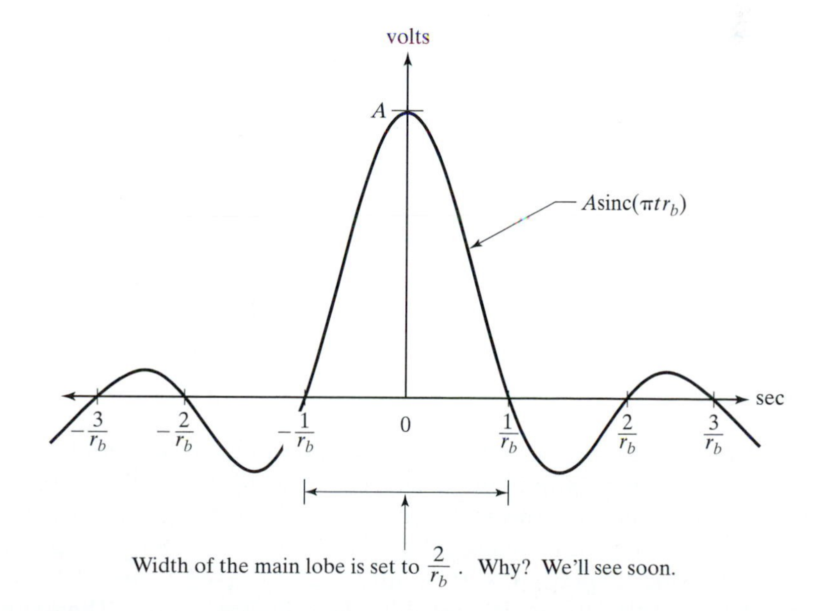

Figure 3-14 A pulse that is sinc-shaped in the time domain.

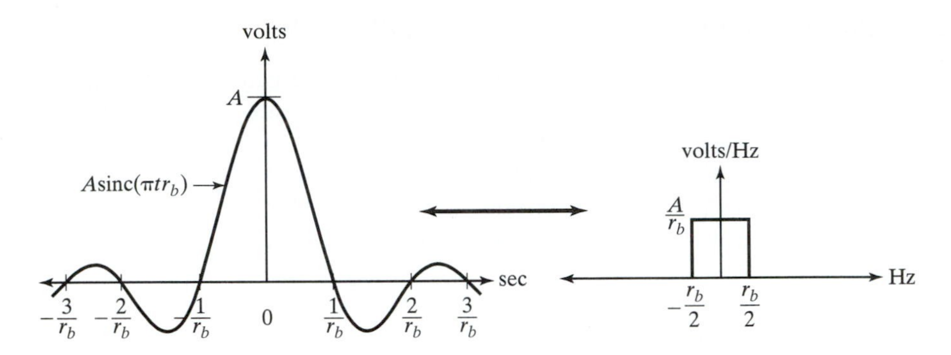

Figure 3-15 A sinc-shaped time domain pulse has a rectangular-shaped magnitude spectrum.

used in the frequency domain, while in Figure 3-14 the sinc function is used in the time domain.

What is the frequency domain representation of the sinc-shaped pulse in Figure 3-14? Remembering the duality property of the Fourier transform (Equation 2.62), a pulse that is sinc-shaped in the time domain has a rectangular shape magnitude spectrum, as illustrated in Figure 3-15.

Squaring the magnitude spectrum produces the normalized energy spectral density shown in Figure 3-16. All of the pulse's energy lies within the frequency band from 0 to $r_b/2$ Hz, meaning that the sinc-shaped pulse is extremely spectrally efficient. For a sinc shaped pulse,

$$\text{bandwidth} = \frac{r_b}{2}, \text{ or } r_b = 2(\text{bandwidth}) \tag{3.7}$$

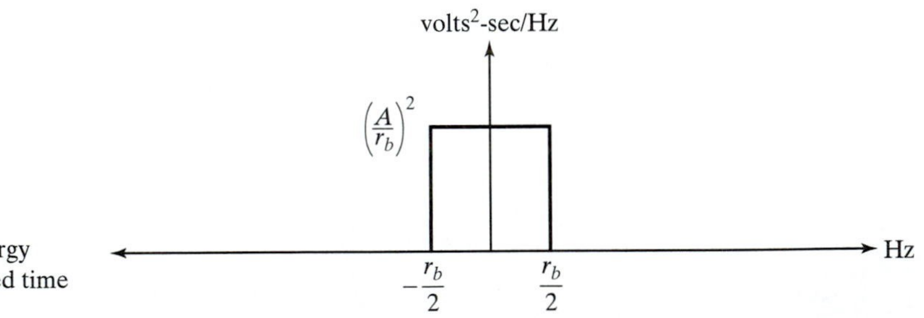

Figure 3-16 Normalized energy spectral density for sinc-shaped time domain pulse.

Figure 3-16 shows that the sinc-shaped pulse is spectrally compact (our first requirement for a "good" pulse shape). What about avoiding ISI (our second requirement)? Consider the transmitted waveform for a typical data sequence (say, 110100) at a speed of r_b bits/sec. As we did with the rectangular pulses, let's transmit a positive sinc-shaped pulse to signify each "1" (see Figure 3-17a) and a negative sinc-shaped pulse to signify each "0" (see Figure 3-17b). The sequence of pulses corresponding to the data sequence "110100" transmitted at r_b bits/sec is shown in Figure 3-17c.

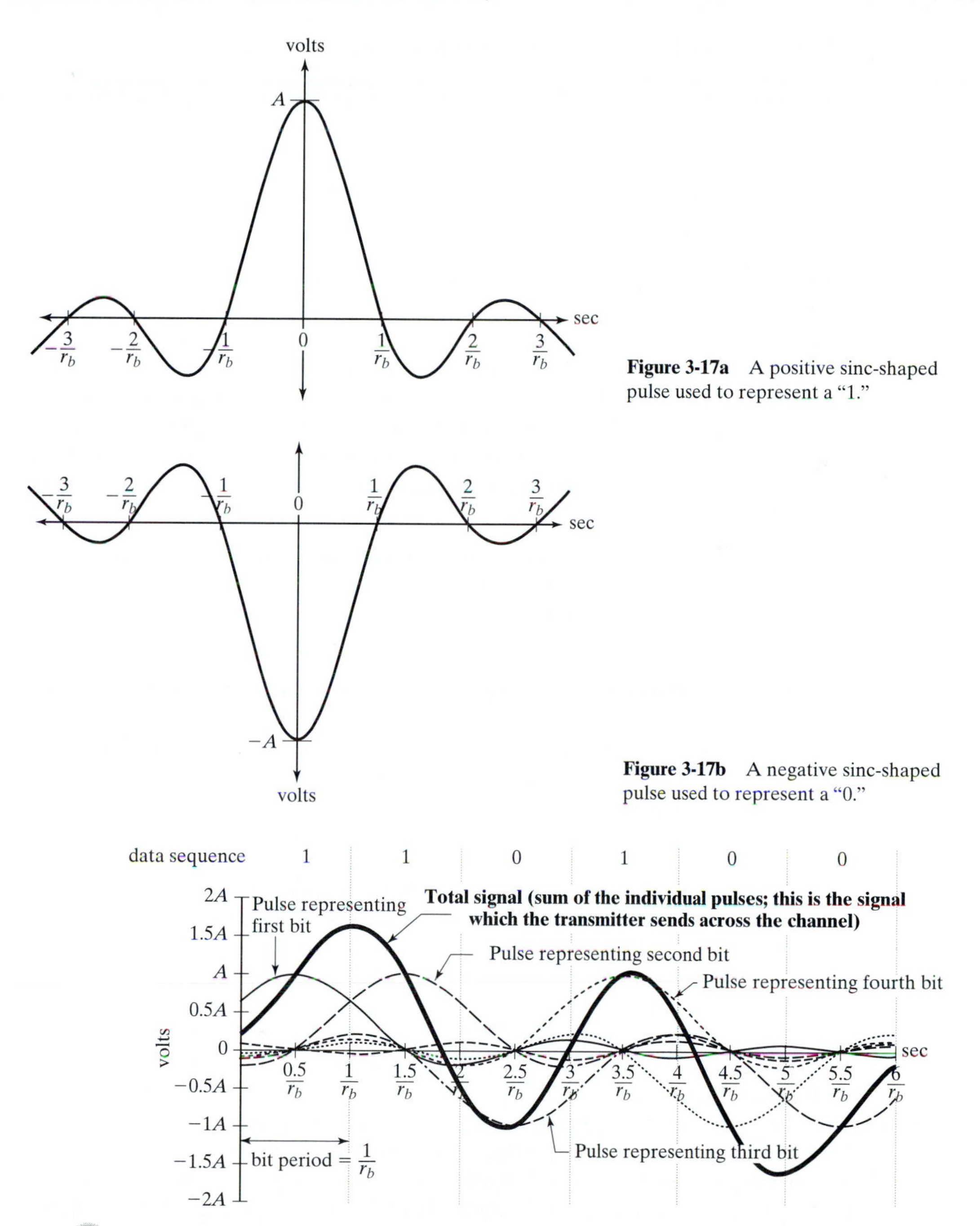

Figure 3-17a A positive sinc-shaped pulse used to represent a "1."

Figure 3-17b A negative sinc-shaped pulse used to represent a "0."

Figure 3-17c Sequence of sinc-shaped pulses corresponding to the data sequence "110100."[1]

[1]This symbol, which is used throughout this book, indicates that the companion CD-ROM has a color version of this figure, which may help you to better see certain features. Insert the CD-ROM into any computer with Adobe® Acrobat® Reader and follow the on-screen instructions. If you have a color printer, you will also be able to print out a color version of the figure.

Let's examine Figure 3-17c. Note the following:

1. Each pulse is centered within its respective bit period (i.e., the pulse representing the first bit is centered at $0.5/r_b$, the pulse representing the second bit is centered at $1.5/r_b$, etc.).
2. The actual signal that will be sent across the channel by the transmitter is the sum of the individual pulses. This signal is shown in Figure 3-17c by a thick, black line marked "total signal."
3. Most of the time the total signal is affected by more than one individual pulse (thus producing ISI), but note that *at exactly the center of each bit period, all of the pulses are at 0 volts except for the single pulse representing the corresponding data bit.* Thus, at the center of the first bit period, the transmitted signal has an amplitude of A, corresponding only to the amplitude of the first pulse; at the center of the second bit period, the transmitted signal also has an amplitude of A, corresponding only to the amplitude of the second pulse; at the center of the third bit period, the transmitted signal has an amplitude of $-A$, corresponding only to the amplitude of the third pulse, etc.

This last observation is extremely important: *There is no ISI at exactly the center of each bit period.* Using this observation and assuming a noiseless, ideal baseband channel with sufficient bandwidth (a bandwidth of at least $r_b/2$ Hz), the receiver can demodulate the transmitted signal by using the following process:

a. Sample the received signal exactly in the center of each bit period. If a particular sample has a value of A volts (or, in fact, any value greater than or equal to 0 volts), then the corresponding data from the source was a "1"; if the sample value is $-A$ volts (or, in fact, any value less than zero volts), then the corresponding data from the source was a "0."
b. Output the demodulated data (i.e., the series of "1"s and "0"s) to the user.

Figures 3-18a and 3-18b illustrate the complete modulation/demodulation process for the data sequence "110100" using sinc-shaped pulses of amplitudes +2 volts and −2 volts, and using a bit rate of 10,000 bits/sec.

Let's closely inspect Figures 3-18a and 3-18b. Note in Figure 3-18a that since the transmission speed is 10,000 bits/sec, the bit period is 100 μsec, or $1/r_b$ (the bit boundaries are shown using dashed vertical lines). The individual pulses are centered within each bit period, and the width of the main lobe of each individual pulse is 200 μsec or $2/r_b$. Observe also that there is no ISI in the exact center of each bit period: At 50 μsec only the first pulse is nonzero, at 150 μsec only the second pulse is nonzero, etc. Thus, at 50 μsec the transmitted signal is affected by only the value of the first bit from the source, at 150 μsec the transmitted signal is affected by only the value of the second bit from the source, etc. This concept is key to demodulating the received signal, shown in Figure 3-18b. The receiver has advance knowledge of the transmission speed (in this case, 10,000 bits/sec) and thus knows that the bit period is 100 μsec. By sampling the received waveform in the exact center of the first bit period (i.e., at 50 μsec), the receiver obtains a value of +2 volts and therefore knows that the first bit output by the source was a "1." Similarly, sampling at 150 μsec produces +2 volts, thereby indicating

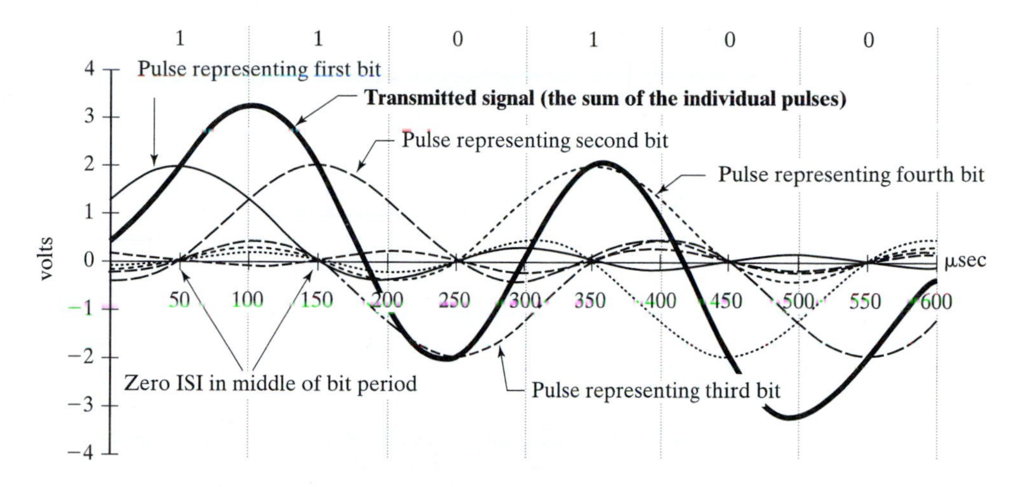

Figure 3-18a Modulation process for transmitting the data sequence "110100" at 10,000 bits/sec using sinc-shaped pulses. Note that the thick black waveform is the only signal that is transmitted.

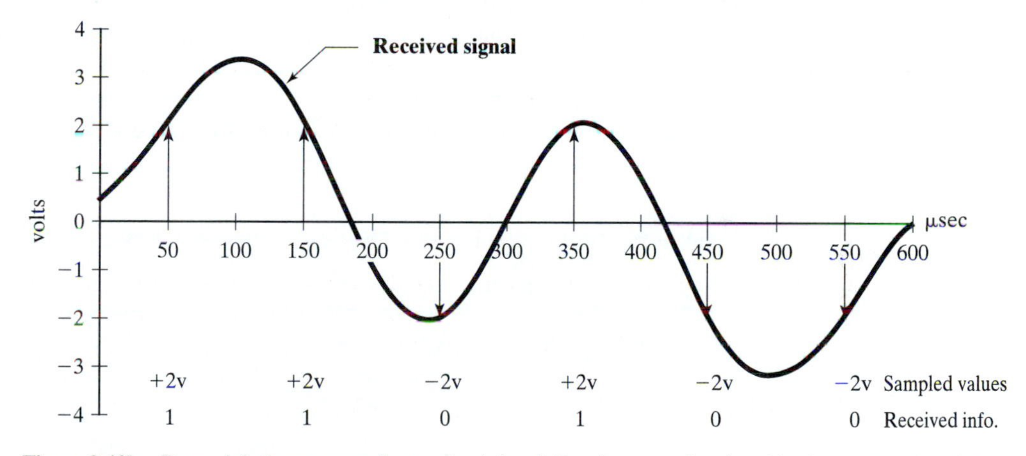

Figure 3-18b Demodulation process for received signal. Receiver samples signal in the center of each bit period (i.e., at 50 μsec, 150 μsec, 250 μsec, etc.). If sampled value is +2 volts, corresponding data is a "1." If sampled value is −2 volts, corresponding data is a "0."

that the second bit was also a "1," sampling at 250 μsec produces −2 volts, indicating that the third bit was a "0," etc. It is very important that the receiver be synchronized with the transmitter so that the received signal is sampled exactly in the center of each bit period. We will discuss synchronization in detail later in this chapter and in Chapter 4.

Let's make one more observation: In Figure 3-14, the width of the main lobe of the sinc-shaped pulse was set to $2/r_b$; this value is used in Figure 3-17 and in the above example. Is this the optimum width for the main lobe of the sinc-shaped pulse? We know that we want a pulse width that produces no ISI but that does produce the lowest bandwidth possible for a given bit rate. Figure 3-19 shows a general sinc-shaped pulse and its corresponding magnitude spectrum. If the main lobe of the pulse has a

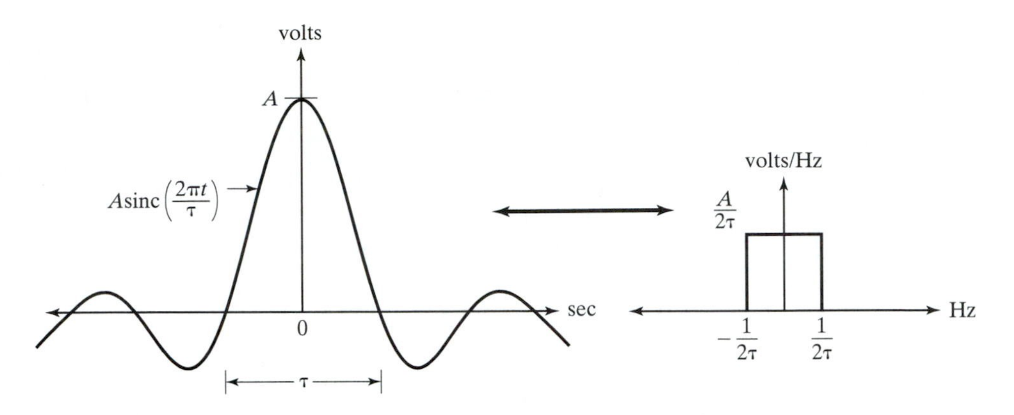

Figure 3-19 Relationship between bandwidth and width of main lobe of sinc-shaped pulse.

width of τ, then the magnitude spectrum shows a bandwidth of $1/(2\tau)$. We thus want to choose a τ value as large as possible without producing ISI. As shown in Figure 3-17, choosing $\tau = 2/r_b$ produces no ISI if the received signal is sampled in the center of each bit period. By making a few diagrams, you can see that while certain smaller values of τ (such as $\tau = 1/r_b$) also produce no ISI if the received signal is sampled in the center of the bit period, there are no larger values of τ for which this is true. Thus, $\tau = 2/r_b$ represents the optimum width for the main lobe of the sinc-shaped pulse.

Example 3.3 Bandwidth and transmission speed using sinc-shaped pulses

Suppose an application has accuracy needs that require 100% of the transmitted signal's average power to lie within the bandwidth of the channel. Given an ideal baseband channel, what bandwidth is required to transmit the signal in Figure 3-18?

Solution

From Equation (3.7),

$$\text{bandwidth} = \frac{r_b}{2} = \frac{10{,}000 \text{ bits/sec}}{2} = 5000 \text{ Hz}$$

Example 3.4 Maximum transmission speed for a fixed-bandwidth system

A baseband channel has a bandwidth of 100 kHz. Accuracy requirements dictate that 100% of the signal's average power must be within the bandwidth of the channel. Determine the maximum transmission speed of the system.

Solution

Again using Equation (3.7),

$$100{,}000 \text{ Hz} \geq \frac{r_b}{2}, \text{ so } r_{b,\max} = 2(100{,}000) = 200{,}000 \text{ bits/sec}$$

Example 3.5 Comparing the performance of sinc-shaped and rectangular pulses

Compare the results of Example 3.4 with Example 3.1.

Solution

Examples 3.1 and 3.4 both use a baseband channel with 100 kHz bandwidth. If accuracy requires that a minimum of 95% of the transmitted signal's average power be within the bandwidth of the channel, then data can be transmitted at a maximum rate of 50,000 bits/sec using rectangular pulses. However, if sinc-shaped pulses are used instead of rectangular pulses, the data can be transmitted *four times as fast*. Also, since 100% of the signal's average power lies within the channel's bandwidth if sinc-shaped pulses are used, the sinc-shaped pulses will produce a received signal with less distortion than the rectangular pulses. We can now see the trade-offs involved in using binary PAM with sinc-shaped pulses rather than rectangular pulses: The sinc-shaped pulses allow a greater transmission speed for a given bandwidth (or a smaller bandwidth for a given transmission speed), but the sinc-shaped pulses require more complex equipment, both at the transmitter (to produce the shape) and at the receiver (to ensure synchronization).

Example 3.6 Designing a binary PAM system using sinc-shaped pulses

Consider a transmitter using binary PAM with sinc-shaped pulses of amplitudes 1 volt and −1 volt. Given a transmission speed of 50,000 bits/sec.:

a. Draw the transmitted signal corresponding to the data sequence "10010."

b. If the signal is transmitted across an ideal baseband channel, what is the minimum channel bandwidth required to ensure that the received signal is not distorted?

c. Draw the received signal and indicate how the receiver extracts the digital information for the user.

Solution

a. The transmitted pulse is shown as the thick, black line in Figure 3-20a.

b. Minimum bandwidth $= \frac{r_b}{2} = \frac{50{,}000}{2} = 25{,}000$ Hz.

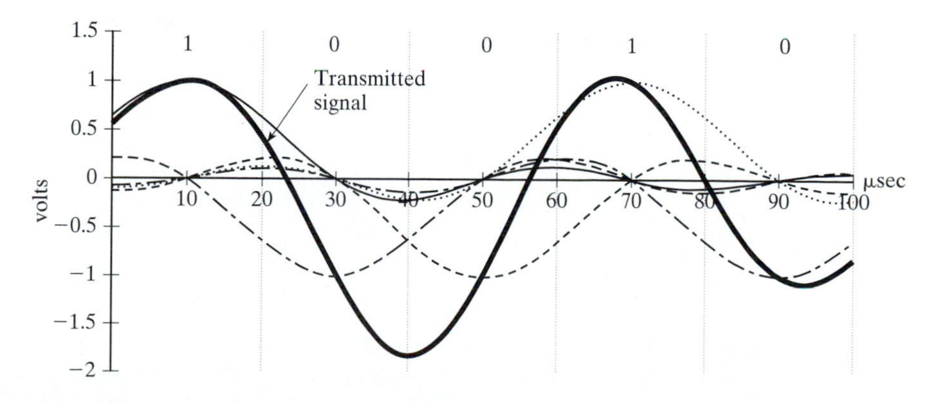

Figure 3-20a Transmitted binary PAM waveform for the data sequence "10010" using sinc-shaped pulses at a transmission speed of 50,000 bits/sec.

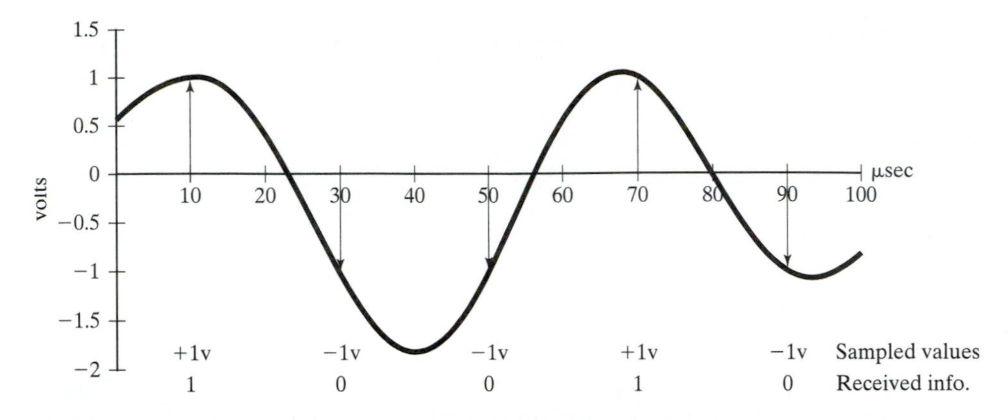

Figure 3-20b Demodulation of received binary PAM waveform.

c. The received signal and the demodulation process are shown in Figure 3-20b. The receiver samples the received signal in the center of each bit period (i.e., at 10 μsec, 30 μsec, 50 μsec, etc.). If a received sample has a value of 1 volt, then the corresponding data delivered to the user is a "1." If a received sample has a value of −1 volt, then the corresponding data delivered to the user is a "0."

Example 3.6 is important because it shows a problem that can occur with binary PAM when we use a sinc-shaped pulse. There is no ISI if the receiver samples the received waveform exactly in the center of each bit period (i.e., at 10 μsec, 30 μsec, 50 μsec, etc.). Let's look at what happens, however, if the receiver is not completely synchronized with the transmitter (this lack of complete synchronization is called *timing jitter*, or *jitter*).

Example 3.7 Examining the effects of losing synchronization

Consider the waveform transmitted in Example 3.6: binary PAM with sinc-shaped pulses of amplitudes 1 volt and −1 volt, a transmission speed of 50,000 bits/sec, and a data sequence of "10010." Let's examine the sample at the receiver corresponding to the third bit. If the receiver is perfectly synchronized, it will sample the waveform at exactly 50 μsec (the exact middle of the bit period, where there is no ISI) and the sampled value will be −1 volt. Suppose, however, that the receiver is not perfectly synchronized. Determine the value of the receiver's sample for the following three cases:

a. The receiver is slightly out of synchronization, and sampling occurs at 52 μsec.

b. The receiver is a little farther out of synchronization, and sampling occurs at 54 μsec.

c. The receiver is even farther out of synchronization, and sampling occurs at 56 μsec.

Solution

a. Figure 3-20a shows the transmitted waveform, which is the sum of five individual pulses corresponding to the five data bits. The pulse representing the first data bit (shown as the thin solid line) is centered in the middle of the first bit period and has zero crossings in the centers of all other bit periods. Mathematically, we can express the first pulse as:

$$p_1(t) = \text{sinc}\left(\frac{\pi(t - 10\ \mu\text{sec})}{20\ \mu\text{sec}}\right)$$

Be sure you understand the reason for each portion of the expression for $p_1(t)$. Similarly, the pulse representing the second bit is:

$$p_2(t) = -\text{sinc}\left(\frac{\pi(t - 30\ \mu\text{sec})}{20\ \mu\text{sec}}\right)$$

The negative sign is necessary because the second bit is a "0." The expressions for the third, fourth, and fifth pulses are:

$$p_3(t) = -\text{sinc}\left(\frac{\pi(t - 50\ \mu\text{sec})}{20\ \mu\text{sec}}\right)$$

$$p_4(t) = \text{sinc}\left(\frac{\pi(t - 70\ \mu\text{sec})}{20\ \mu\text{sec}}\right)$$

$$p_5(t) = -\text{sinc}\left(\frac{\pi(t - 90\ \mu\text{sec})}{20\ \mu\text{sec}}\right)$$

The transmitted waveform is the sum of the individual pulses:

$$s(t) = p_1(t) + p_2(t) + p_3(t) + p_4(t) + p_5(t)$$

At 52 μsec,

$$\begin{aligned} s(52\ \mu\text{sec}) &= \text{sinc}(2.1\pi) - \text{sinc}(1.1\pi) - \text{sinc}(0.1\pi) + \text{sinc}(-0.9\pi) - \text{sinc}(-1.9\pi) \\ &= 0.0468 + 0.0894 - 0.9836 + 0.1093 + 0.0518 \\ &= -0.6863 \text{ volts} \end{aligned}$$

b. At 54 μsec,

$$\begin{aligned} s(54\ \mu\text{sec}) &= \text{sinc}(2.2\pi) - \text{sinc}(1.2\pi) - \text{sinc}(0.2\pi) + \text{sinc}(-0.8\pi) - \text{sinc}(-1.8\pi) \\ &= -0.3567 \text{ volts} \end{aligned}$$

c. At 56 μsec,

$$s(56\ \mu\text{sec}) = -0.0290 \text{ volts}$$

The sampled values at 50, 52, 54, and 56 μsec, which illustrate the received signal's susceptibility to jitter, are illustrated in Figure 3-21. We see that if the receiver is out of synchronization and samples the received signal later than it should (for instance, at 52 μsec instead of 50 μsec), the magnitude of the sampled voltage corresponding to the third bit is reduced. In fact, the farther the receiver is out of synchronization, the closer the third bit's sampled value is to the threshold of zero volts. This is problematic because if the sampled value for the third bit happens to be greater than

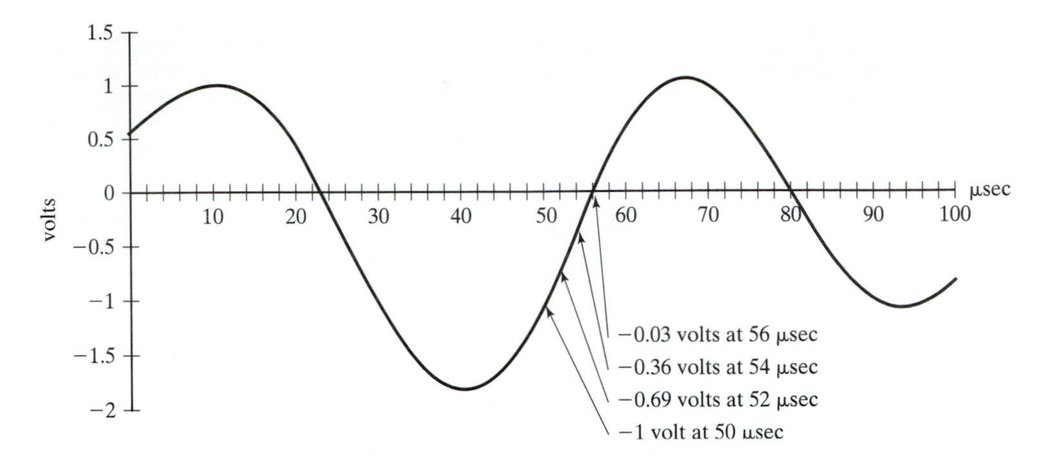

Figure 3-21 Received waveform's susceptibility to jitter.

zero volts, an error will occur: The receiver will demodulate the sample as a "1" instead of a "0." As will be discussed in Chapter 4, achieving and keeping synchronization in the receiver can be difficult and can involve significant equipment complexity.

In Example 3.7, it appears that the receiver would have to be out of synchronization by at least 6.5 μsec (or 32.5% of a bit period) in order for the third bit to be received in error. The problem becomes worse when we consider that the example assumed no noise and an ideal baseband channel. In practical situations, the received signal will be corrupted by noise and the channel will attenuate the transmitted signal (reduce the signal's amplitude) and possibly distort the transmitted signal. Figure 3-22a shows the same transmitted signal as in Example 3.7 ("10010" transmitted using binary PAM with sinc-shaped pulses at 50,000 bits/sec), but Figure 3-22b shows the received signal after passing through a channel that adds noise and produces losses that attenuate the transmitted signal by 50%. The received signal has been halved (from the attenuation) and "hashed" (from the noise). We are now no longer confident about correctly receiving the third bit if the receiver is out of synchronization by more than approximately 3 μsec.

We have seen that the farther the sampled value of a noiseless signal is from the threshhold (0 volts), the more likely it is that noise will not corrupt the signal badly enough to cause an error. Let's define *noise margin* as the distance between a noiseless received signal and the threshhold at the instant of sampling. For example, consider the signal shown in Figure 3-21. If this signal is transmitted across an ideal baseband channel (no attenuation occurs), then the noise margin corresponding to the third bit will be 1 volt if the receiver is perfectly synchronized, 0.69 volts if the receiver samples the signal 2 μsec late, 0.36 volts if the receiver samples the signal 4 μsec late, etc. On the other hand, if the signal in Figure 3-21 is transmitted across a channel that causes a 50% attenuation, the noise margin corresponding to the third bit will be 0.5 volts if the receiver is perfectly synchronized, 0.34 volts if the receiver samples the signal 2 μsec late, 0.18 volts if the receiver samples the signal 4 μsec late, etc.

How can we increase noise margin in the presence of jitter (and thus reduce susceptibility to noise when the receiver is not perfectly synchronized)? One way is to

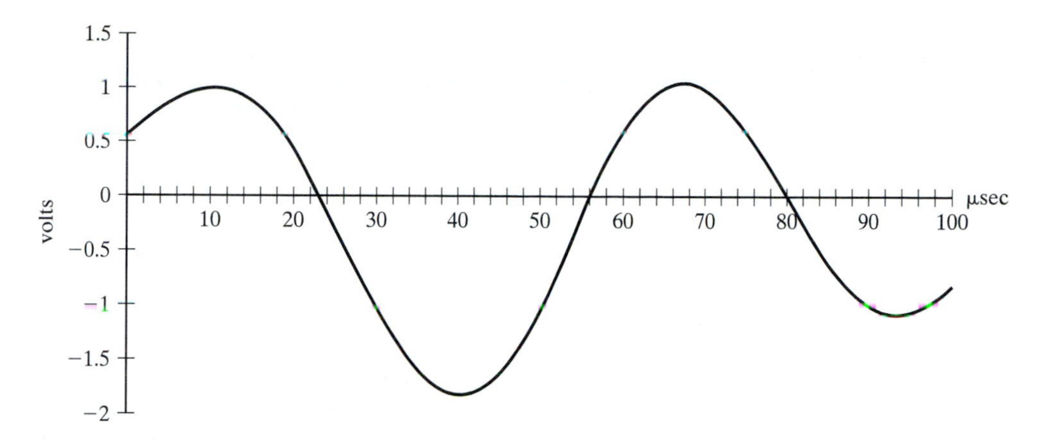

Figure 3-22a Transmitted signal for "10010" using sinc-shaped pulses.

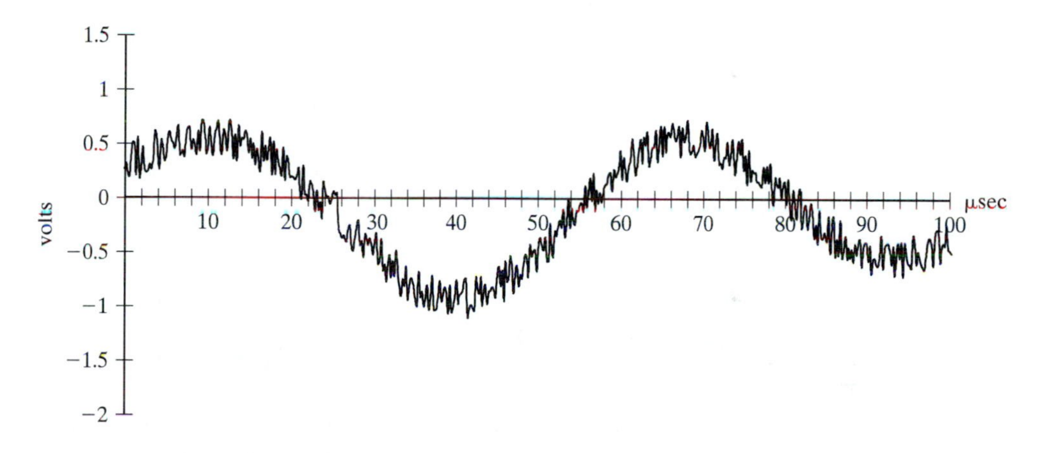

Figure 3-22b Received signal—the channel is attenuating the transmitted signal levels by 50% and corrupting the signal with noise.

increase the amplitude of the transmitted signal (which, of course, requires an increase in the power of the transmitted signal). Thus, if we can increase the power of the transmitted signal, we can increase the accuracy of the received signal. Such a trade-off, however, may not be desirable or even feasible in many applications. (For example, increasing transmitted power in a cell phone reduces the operating time until the battery needs recharging and also causes increased interference to other users.)

Another approach to reducing susceptibility to jitter is to use a different pulse shape. Figure 3-23a shows a binary PAM signal representing the same data pattern and transmitted at the same speed as the signal shown in Figure 3-22a, except that rectangular-shaped pulses are used instead of sinc-shaped pulses. In Figure 3-23b, the signal has been passed through a channel adding the same amount of noise and producing the same amount of attenuation as in the signal shown in Figure 3-22b. In comparing Figures 3-22a and b with Figures 3-23a and b, note that transmitting with rectangular

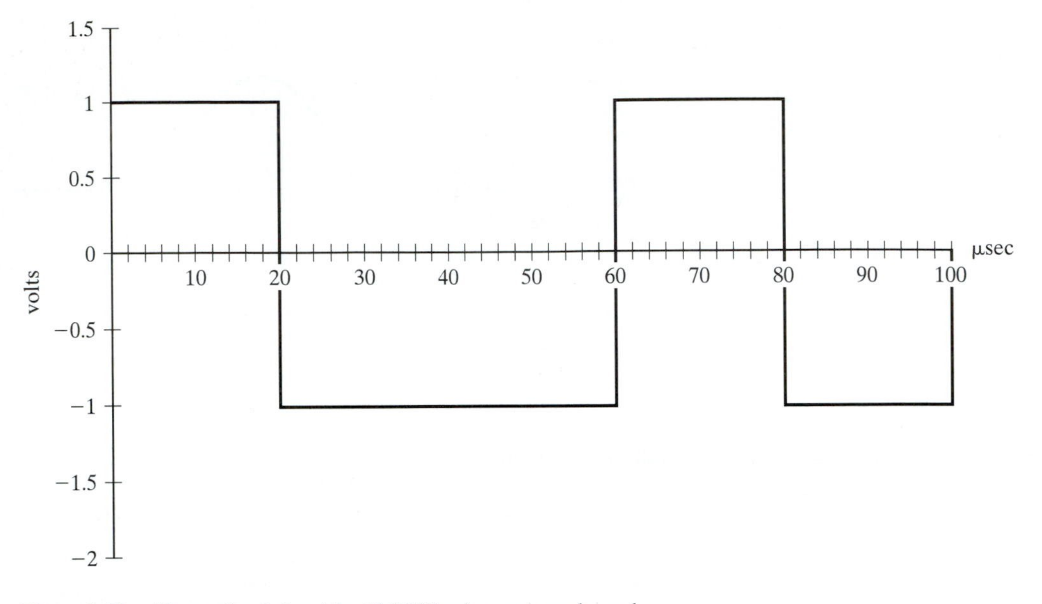

Figure 3-23a Transmitted signal for "10010" using rectangular pulses.

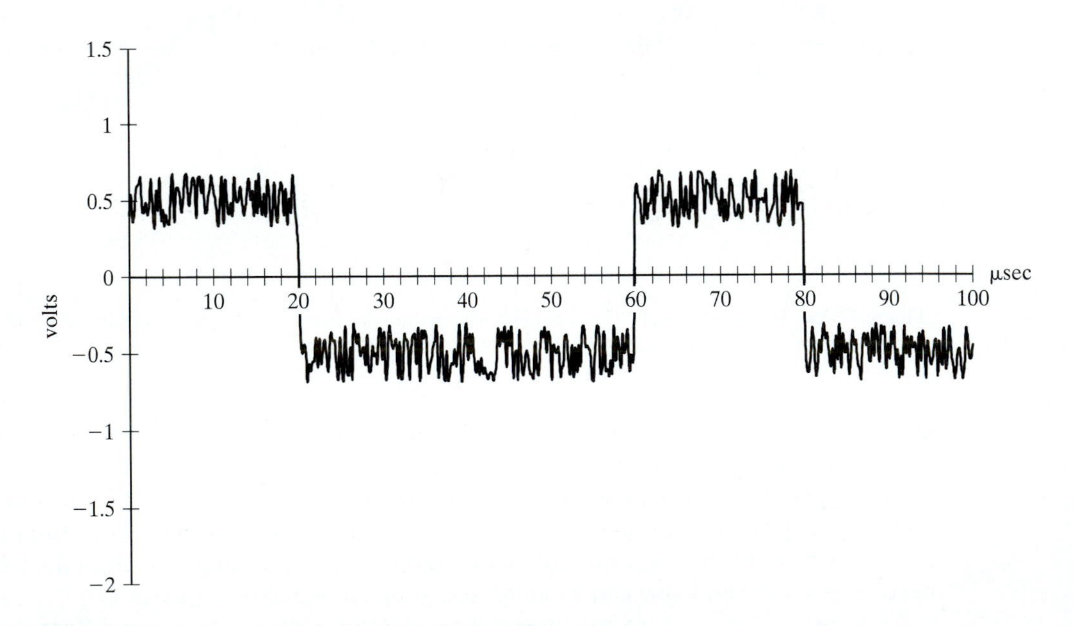

Figure 3-23b Received signal with channel attenuation and noise.

pulses rather than sinc-shaped pulses produces a signal with a constant noise margin, even if the receiver is out of synchronization by as much as ±49% of one bit period.

We've now seen that transmitting binary PAM with rectangular-shaped pulses requires much more bandwidth than does transmitting with sinc-shaped pulses, but

Table 3-2 A Comparison of Rectangular and Sinc-Shaped Pulses for Binary PAM

Pulse Shape	Bandwidth	Advantages	Disadvantages
Rectangular, pulse width $= T_b$	$2r_b$ (95% in-band power)	No ISI; minimal susceptibility to jitter; simple to build transmitter and receiver	High bandwidth
Sinc, main lobe $= 2T_b$	$0.5r_b$ (100% in-band power)	Low bandwidth; no ISI if receiver is perfectly synchronized to transmitter	Susceptible to timing jitter at receiver; equipment more complex

transmitting with rectangular pulses produces a signal that is much less susceptible to timing jitter at the receiver. The performance-versus-cost parameters of our observations are listed in Table 3-2.

Is there a compromise between the two pulse shapes? Can we find a pulse shape that produces low bandwidth (like the sinc pulse) while providing good protection against timing jitter (like the rectangular pulse)? To find out, let's begin by asking, "What are the physical characteristics of the sinc pulse that cause it to be so susceptible to jitter?" Let's revisit Figure 3-20a (in Example 3.6) and examine what is happening to the transmitted waveform between $t = 50\ \mu$sec and $t = 60\ \mu$sec.

As we discussed earlier in this section, at $t = 50\ \mu$sec all pulses are zero except for the third pulse, which has a value of -1 volt. Thus, at $t = 50\ \mu$sec, the transmitted waveform has a value of -1 volt and there is no ISI. At $t = 52\ \mu$sec, the third pulse has a value of -0.98 volts (still a large negative number), but the values of the first, second, fourth, and fifth pulses are no longer zero. In fact, as we established in Example 3.7, at $t = 52\ \mu$sec the value of the first pulse is 0.05 volts, the value of the second pulse is 0.09 volts, the value of the fourth pulse is 0.11 volts, and the value of the fifth pulse is 0.05 volts. Thus, the value of the transmitted signal at $t = 52\ \mu$sec is $0.05 + 0.09 - 0.98 + 0.11 + 0.05 = -0.68$ volts.

The ISI from the first, second, fourth, and fifth pulses erodes the noise margin by 0.3 volts, and the situation worsens as t approaches 60 μsec. How can we reduce the amount of ISI? In looking back at Figure 3-20a, we see that we can reduce the ISI from the fourth pulse by making its main lobe thinner, and we can reduce the ISI from the first, second, and fifth pulses by making their tails flatter. In other words, to produce low bandwidth but reduce susceptibility to timing jitter, we should use a pulse that is smooth like a sinc-shaped pulse but that has a narrower main lobe and flatter tails than a sinc pulse.

3.3.2 The Raised Cosine Pulse (Damped Sinc-Shaped Pulse)

How can we produce a pulse that is smooth like a sinc pulse but has a narrower main lobe and flatter tails? Consider the waveform in Figure 3-24, which is a sinc-shaped pulse multiplied by a damping factor.

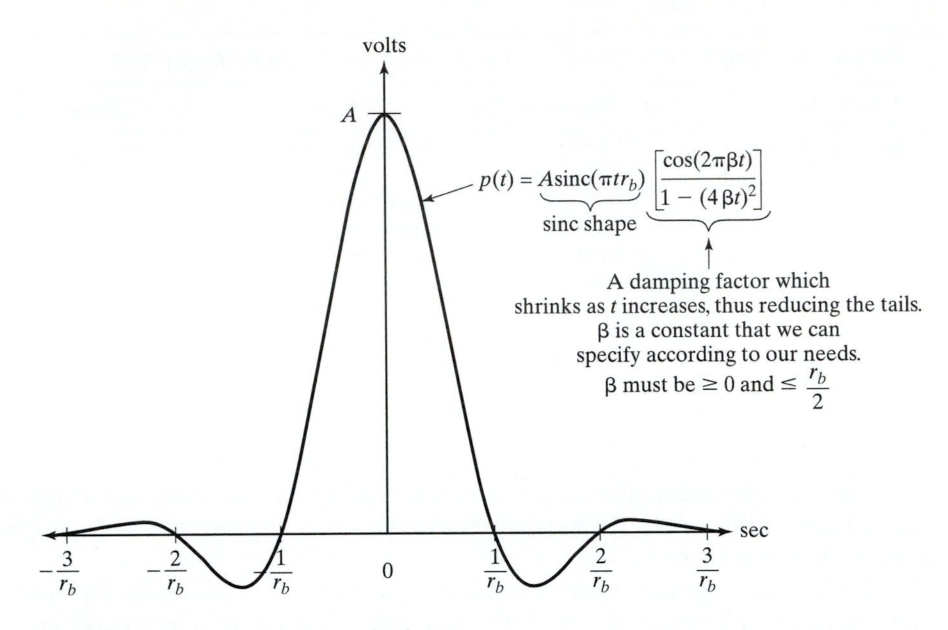

Figure 3-24 A damped sinc-shaped pulse.

Figure 3-25 shows the new pulse for various damping factors, from $\beta = 0$ (no damping at all) to $\beta = r_b/2$. Note that the larger the value of β, the narrower the main lobe and the flatter the tails of the pulse. Thus, the larger the value of β, the less the effects of ISI and thus the less susceptible the transmitted signal will be to timing jitter.

What is the trade-off for this reduced susceptibility to jitter? Figure 3-26 shows the magnitude spectrum of the damped sinc-shaped pulse. Note that the bandwidth of the pulse is

$$\text{bandwidth} = \frac{r_b}{2} + \beta \tag{3.8}$$

Thus, the greater the value of β, the greater the bandwidth of the transmitted signal but the less susceptible the transmitted signal is to timing jitter.

The Fourier transform of the damped sinc-shaped pulse is

$$P(f) = \begin{cases} \dfrac{A}{r_b} & \text{for } |f| \leq \dfrac{r_b}{2} - \beta \\[2ex] \dfrac{A}{2r_b}\left[1 + \cos\left\{\dfrac{\pi}{2\beta}\left(|f| - \dfrac{r_b}{2} + \beta\right)\right\}\right] & \text{for } \dfrac{r_b}{2} - \beta < |f| < \dfrac{r_b}{2} + \beta \\[2ex] 0 & \text{for } |f| \geq \dfrac{r_b}{2} + \beta \end{cases} \tag{3.9}$$

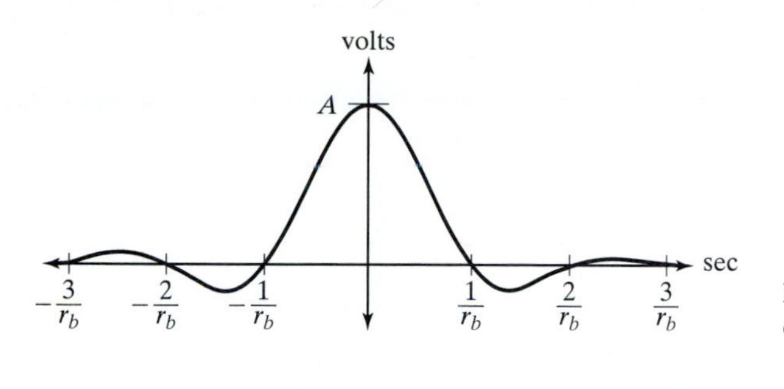

Figure 3-25a Damped sinc-shaped pulse with damping factor $\beta = 0$.

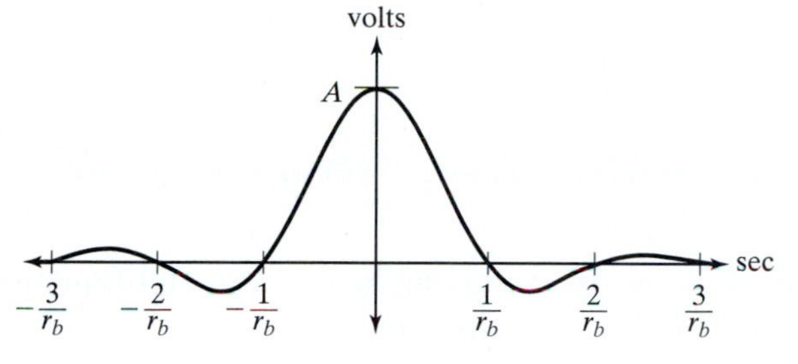

Figure 3-25b Damped sinc-shaped pulse with damping factor $\beta = \frac{r_b}{8}$.

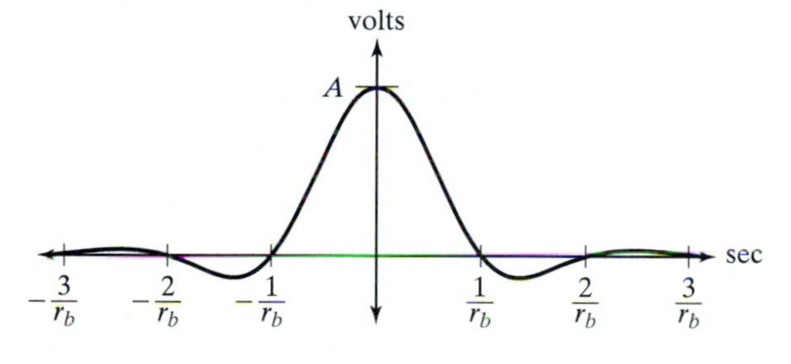

Figure 3-25c Damped sinc-shaped pulse with damping factor $\beta = \frac{r_b}{4}$.

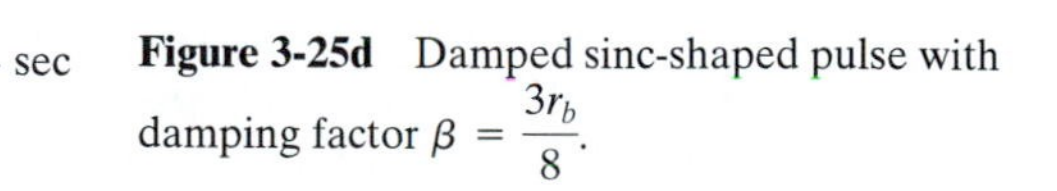

Figure 3-25d Damped sinc-shaped pulse with damping factor $\beta = \frac{3r_b}{8}$.

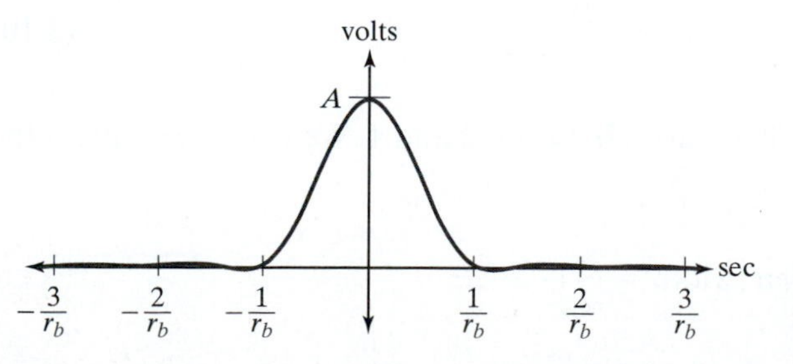

Figure 3-25e Damped sinc-shaped pulse with damping factor $\beta = \frac{r_b}{2}$.

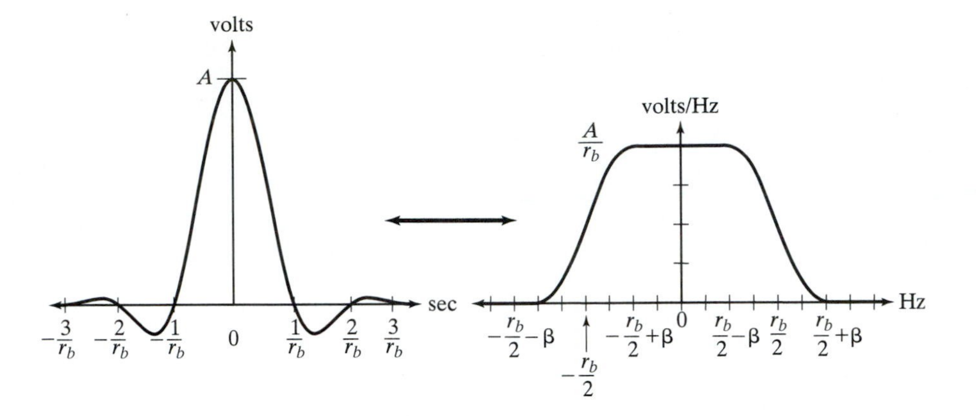

Figure 3-26 A damped sinc-shaped time domain pulse has a raised-cosine-shaped magnitude spectrum.

We call this particular shape a *raised cosine*. In examining Figure 3-26 and Equation (3.9), note that the raised cosine has the following properties:

1. The raised cosine has a constant value of A/r_b for $|f| \leq (r_b/2) - \beta$ and a constant value of zero for $|f| \geq (r_b/2) + \beta$.
2. In the region between these two constants (i.e., for $(r_b/2) - \beta < |f| < (r_b/2) + \beta$), the waveform is odd-symmetric about the points $|f| = r_b/2$.
3. $\beta = 0$ corresponds to a rectangular shape in the frequency domain and a sinc-shaped pulse in the time domain.
4. Bandwidth $= (r_b/2) + \beta$, so larger values of β require larger bandwidth. Note from Figure 3-25 that larger values of β provide more damping (i.e., flatter tails and a narrower main lobe) for the corresponding time domain waveform, thus reducing ISI. Choosing a value for β therefore involves trading off increased bandwidth for reduced susceptibility to timing jitter (we'll illustrate this point in Examples 3.8 and 3.9).

The fourth observation is especially important because it allows us to optimize system performance for particular applications. Often, we are interested in expressing the additional bandwidth involved in the trade-off not in terms of Hz, but rather as a percentage of the minimum possible bandwidth, $r_b/2$. With this in mind, let's define a quantity called *rolloff factor* (α):

$$\alpha \triangleq \frac{2\beta}{r_b} \tag{3.10}$$

Since β varies between 0 and $r_b/2$, α varies between 0 and 1. We can now express the bandwidth as

$$\text{bandwidth} = \frac{r_b}{2}(1 + \alpha) \tag{3.11}$$

Using rolloff factor (α) instead of β allows us to express the trade-off of additional bandwidth for less susceptibility to jitter in a manner that is independent of transmission speed.

So far, we've investigated binary PAM using three different pulse shapes: rectangular pulses, sinc-shaped pulses, and damped sinc-shaped pulses. General practice in industry and the profession is to describe the damped sinc-shaped pulse by the name of its *frequency domain shape* rather than its time domain shape. The shape is thus called a *raised cosine pulse*, and the term *raised cosine pulse shaping* refers to the pulse that looks like a raised cosine in the frequency domain but a damped sinc shape in the time domain. Note that this is not consistent with the terminology for the other pulse shapes (which we call rectangular and sinc shaped, based on their time domain shapes), but it has become standard practice.[2]

We can use the following notation to mathematically describe each of the pulse shapes:

$p(t)$ = time domain representation of a pulse

$P(f)$ = frequency domain representation of a pulse

τ = width of the pulse in the time domain. For the rectangular pulse, τ is the span of all nonzero values. For the sinc-shaped and raised cosine pulse shapes, τ is the width of the main lobe of the pulse (the portion of the pulse between the first zero crossing to the left of the y axis and the first zero crossing to the right of the y axis).

We know that the optimum pulse width for the rectangular pulse is $\tau = 1/r_b$, and the optimum pulse width for sinc-shaped and raised cosine pulses is $\tau = 2/r_b$. We're now ready to work some examples illustrating how different rolloff factors affect the characteristics of the raised cosine pulse.

Example 3.8 Plotting the raised cosine pulse for various values of α

Using one set of axes, plot the time domain representation of the raised cosine pulse for each of the following rolloff factors. Pulse width should be set to the optimum value for a transmission speed of 100,000 bits/sec, and each plot should extend from at least $t = -3/r_b$ to $t = 3/r_b$.

a. $\alpha = 0$

b. $\alpha = 0.25$

c. $\alpha = 0.5$

d. $\alpha = 0.75$

e. $\alpha = 1$

Solution

$$\tau_{\text{optimum}} = \frac{2}{r_b} = 20\ \mu\text{sec}$$

[2]See the discussion of Nyquist pulse shaping in Schwartz [3.1] for a different way to derive the raised cosine shape.

Plots for the five rolloff factors are shown in Figure 3-27.

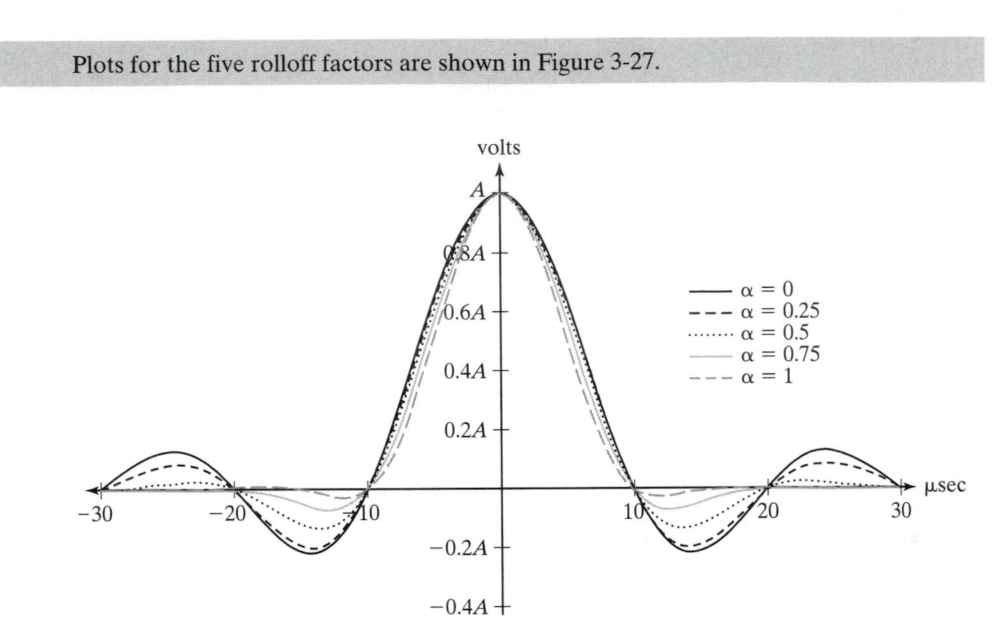

Figure 3-27 A comparison of the time domain representation of the raised cosine pulse for five different rolloff factors.

Example 3.9 Plotting PAM waveforms with raised cosine pulses

A source outputs data at the rate of 50,000 bits/sec. The transmitter uses binary PAM with raised cosine pulse shaping of optimum pulse width. Each pulse has a maximum amplitude of 1 volt. For each of the five rolloff factors in Example 3.8:

1. Draw the transmitted waveform for the data pattern "10010."
2. Determine the bandwidth of the transmitted waveform.

Solution

1. $\tau_{\text{optimum}} = \dfrac{2}{r_b} = 40\ \mu\text{sec}$

   Plots are shown in Figures 3-28–3-32.

2. Bandwidth $= (r_b/2)(1 + \alpha)$. The bandwidths of the waveforms in Part 1 of this example are as follows:

   a. Bandwidth $= 25{,}000(1 + 0) = 25$ kHz
   b. Bandwidth $= 25{,}000(1 + 0.25) = 31.25$ kHz
   c. Bandwidth $= 25{,}000(1 + 0.5) = 37.5$ kHz
   d. Bandwidth $= 25{,}000(1 + 0.75) = 43.75$ kHz
   e. Bandwidth $= 25{,}000(1 + 1) = 50$ kHz

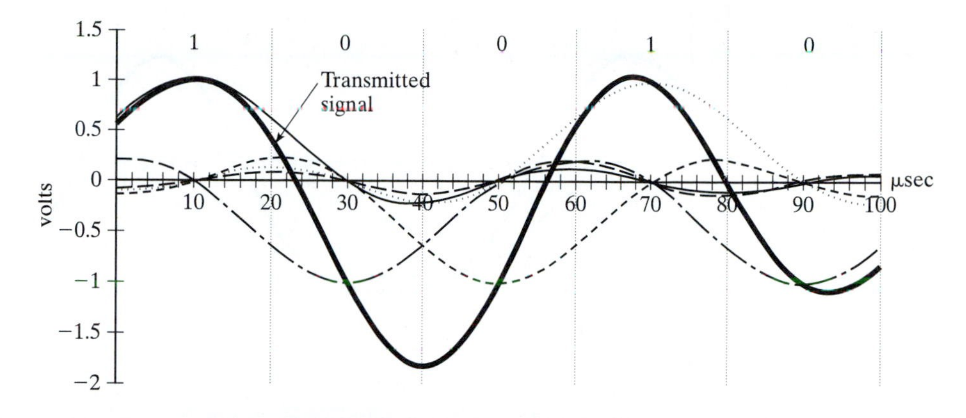

Figure 3-28 Transmitted binary PAM waveform for the data sequence "10010" using sinc-shaped pulses (raised cosine pulse shaping, $\alpha = 0$) at a transmission speed of 50,000 bits/sec. Note that this is the same plot as in Figure 3-20a.

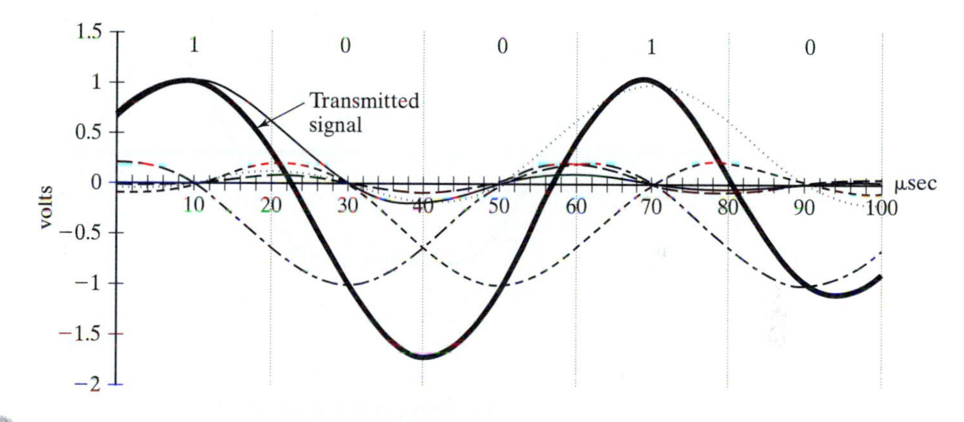

Figure 3-29 "10010" with raised cosine pulse shaping, $\alpha = 0.25$.

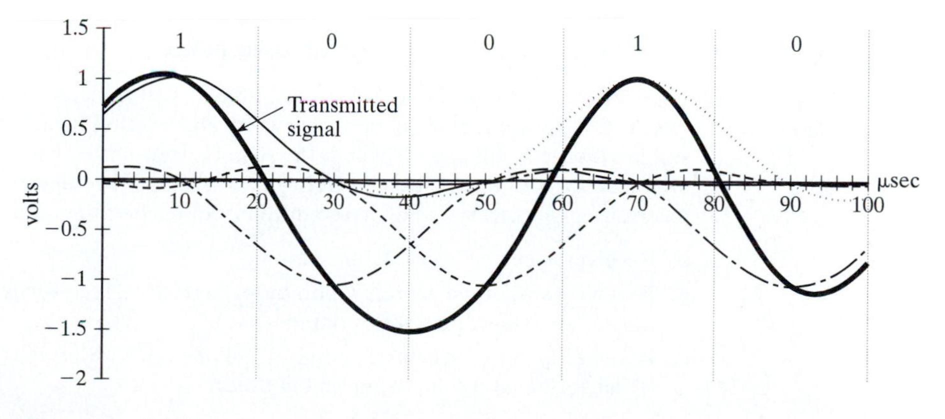

Figure 3-30 "10010" with raised cosine pulse shaping, $\alpha = 0.5$.

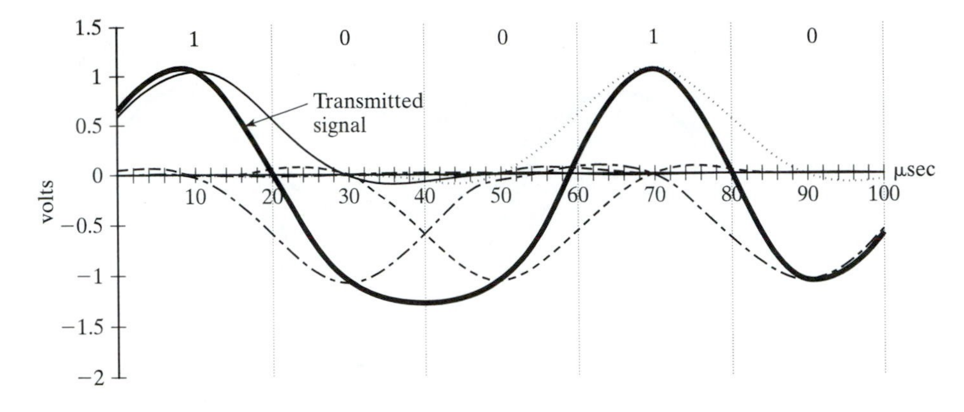

Figure 3-31 "10010" with raised cosine pulse shaping, $\alpha = 0.75$.

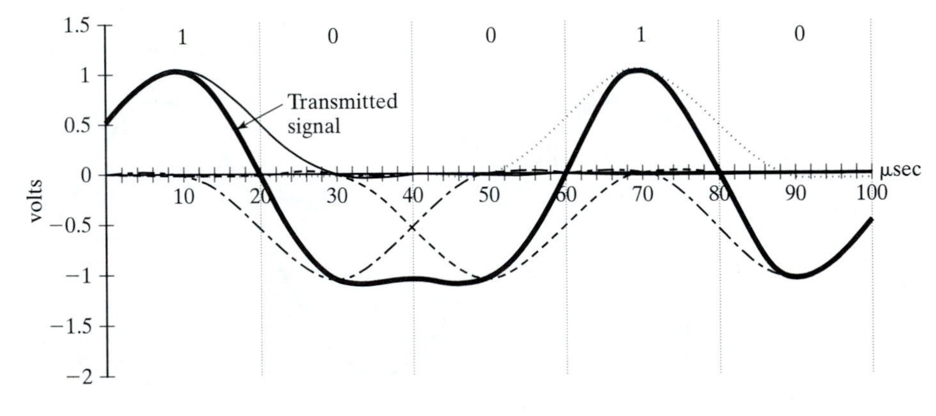

Figure 3-32 "10010" with raised cosine pulse shaping, $\alpha = 1$.

Example 3.10 Susceptibility to jitter using raised cosine pulses

1. Consider the data pattern "10010" transmitted at 50,000 bits/sec using PAM with raised cosine pulse shaping of optimum width and a rolloff factor of $\alpha = 0$. Pulse amplitudes are 1 volt and −1 volt. If the signal is transmitted over a channel that produces a 50% attenuation in voltage, determine the noise margin of the received signal corresponding to the third bit for each of the following cases:

   a. Receiver is perfectly synchronized.

   b. Receiver is out of synchronization by +2 μsec (i.e., receiver is sampling the signal 2 μsec later than the optimum time).

   c. Receiver is out of synchronization by +4 μsec (i.e., receiver is sampling the signal 4 μsec later than the optimum time).

d. Receiver is out of synchronization by +6 μsec (i.e., receiver is sampling the signal 6 μsec later than the optimum time).

e. Receiver is out of synchronization by more than +6 μsec; what happens to the demodulated bit?

2. Repeat Part 1, except use a rolloff factor of $\alpha = 0.5$.
3. Repeat Part 1, except use a rolloff factor of $\alpha = 1$.

Solution

Note that the data pattern and transmission speed in this example are the same as in Example 3.9. Therefore, the transmitted waveforms corresponding to $\alpha = 0$, $\alpha = 0.5$, and $\alpha = 1$ are the same as those shown in Figures 3-28, 3-30, and 3-32, respectively. Fifty percent attenuation means that the signal at the receiver will have half the amplitude of the transmitted signal, so the received signals for $\alpha = 0$, $\alpha = 0.5$, and $\alpha = 1$ are shown in Figures 3-33a, b, and c, respectively. Table 3-3 shows the noise margins at the receiver for the third bit if the receiver is perfectly synchronized, if the receiver is out of synchronization by +2 μsec, if the receiver is out of synchronization by +4 μsec, and if the receiver is out of synchronization by +6 μsec. Table 3-3 shows that increasing the rolloff factor significantly reduces the effects of the receiver being out of synchronization (but at the cost of additional bandwidth). This concept is also illustrated in Figure 3-33d.

Table 3-3 Noise Margin for Third Bit of Received Signal Using Raised Cosine Pulse Shaping with Various Rolloff Factors

Receiver Synchronization	Rolloff Factor (in Volts)		
	$\alpha = 0$	$\alpha = 0.5$	$\alpha = 1$
Perfect	0.50	0.50	0.50
+2 μsec out of sync	0.34	0.40	0.46
+4 μsec out of sync	0.18	0.28	0.38
+6 μsec out of sync	0.014	0.14	0.28

If the receiver is too far out of synchronization, the noise margin is completely eliminated and the probability that the bit will be demodulated in error is very high. In this example, if $\alpha = 0$ and the receiver is out of synchronization by more than 6 μsec, the noiseless sampled value for the third bit will be positive, and hence in error, even in the absence of noise. For $\alpha = 0.5$, the receiver can be out of synchronization by 8 μsec before a similar point is reached, and for $\alpha = 1$, the receiver can be out of synchronization by almost 10 μsec before a similar point is reached.

We now have the capability to quantify and analyze many of the performance-versus-cost trade-offs involved in designing a communication system for a baseband channel. Table 3-4 summarizes some comparisons concerning the three different baseband binary PAM techniques. Review Table 3-4, asking yourself how each of the performance-versus-cost parameters (equipment complexity, bandwidth, transmitted signal power, transmission speed, accuracy, and reliability) are quantitatively involved in determining the optimum pulse shape for a particular application.

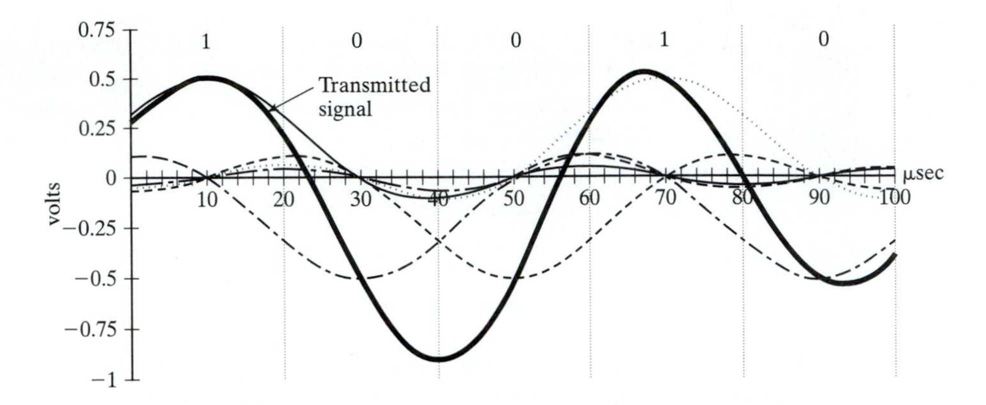

Figure 3-33a Received signal, $\alpha = 0$, bandwidth = 25 kHz.

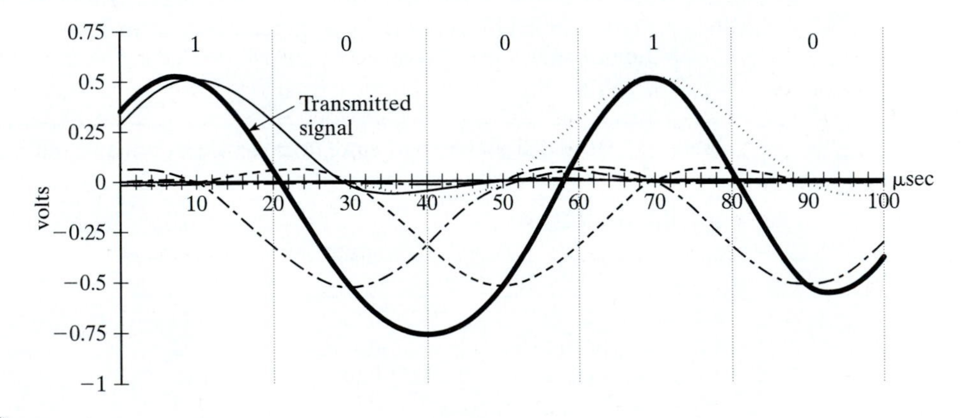

Figure 3-33b Received signal, $\alpha = 0.5$, bandwidth = 37.5 kHz.

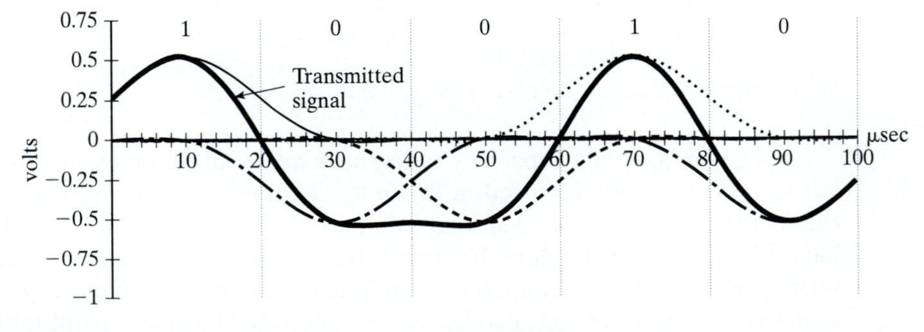

Figure 3-33c Received signal, $\alpha = 1$, bandwidth = 50 kHz.

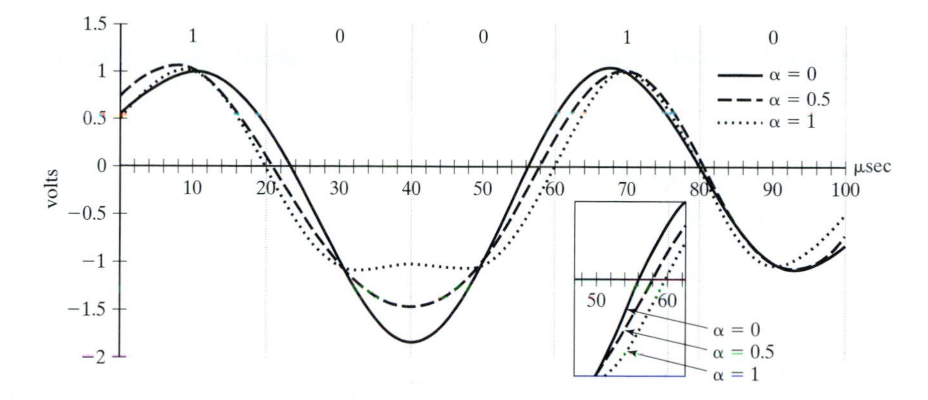

Figure 3-33d Illustrating the influence of rolloff factor on noise margin.

3.4 Building a Baseband Transmitter

Table 3-4 only partially addresses the issue of equipment complexity. We know that obtaining tighter synchronization requires more complex equipment within the receiver; thus, transmitting raised cosine pulses will reduce the receiver's complexity relative to transmitting sinc-shaped pulses. We'll discuss building receivers in detail in Chapter 4, but what about the complexity of the transmitter? Let's determine how to build a transmitter capable of generating sinc-shaped and raised-cosine-shaped pulses.

Generating raised-cosine-shaped pulses is difficult using only analog circuitry. A filter, composed of discrete components, is designed to produce an impulse response resembling a time-delayed version of the raised cosine pulse. A series of narrow pulses

Table 3-4 Trade-Offs—Selecting Pulse Shapes for Binary PAM

Pulse Shape	Bandwidth	Advantages	Disadvantages
Rectangular $\tau = \frac{1}{r_b}$	$2r_b$ (95% in-band power)	No ISI; minimal susceptibility to jitter	High bandwidth
Sinc $\tau = \frac{2}{r_b}$	$0.5r_b$ (100% in-band power)	Low bandwidth; no ISI if receiver timing is perfect	Susceptible to timing jitter at the receiver
Raised cosine* $\tau = \frac{2}{r_b}$	$\frac{r_b}{2}(1 + \alpha)$ (100% in-band power)	No ISI; less susceptibility to jitter than sinc	Requires more bandwidth than sinc, but still less bandwidth than rectangular

*Refers to *frequency domain* shape

For any particular application, choose the value of α that produces the best trade-off between bandwidth and susceptibility to jitter (larger α means more bandwidth but less susceptibility to jitter).

(approximating impulses) is then input into the filter, one pulse per bit period, with a positive narrow pulse representing each "1" and a negative narrow pulse representing each "0". The output of the filter will approximate the raised cosine waveform that corresponds to the appropriate data. The drawback of this analog approach is that the filter requires a large number of discrete components to produce an impulse response that resembles a time-delayed raised cosine pulse, especially for low rolloff factors. In fact, a filter that produces a time-delayed true sinc-shaped impulse response requires an infinite number of poles (its transfer function must be rectangular). Thus, before digital circuitry became readily available and inexpensive (i.e., before the 1980s), baseband transmitters using raised-cosine-shaped pulses were too impractical for most applications.

Generating raised cosine and sinc-shaped pulses becomes much simpler if digital circuitry is used. While the sinc-shaped and raised cosine pulses have nonzero values for $-\infty < t < \infty$, practically speaking, the values are so small outside the range $-3.5T_b < t < 3.5T_b$ that they can be neglected. Thus, the pulse for the ith bit has a practical effect on the transmitted waveform for only seven bit periods—from the $i - 3$rd bit period through the $i + 3$rd bit period. Similarly, from a practical standpoint, only seven bits affect the shape of the transmitted waveform during the ith bit period—the $i - 3$rd bit, the $i - 2$nd bit, the $i - 1$st bit, the ith bit, the $i + 1$st bit, the $i + 2$nd bit, and the $i + 3$rd bit. This concept is illustrated in Figure 3-34. Suppose, for example, we want to

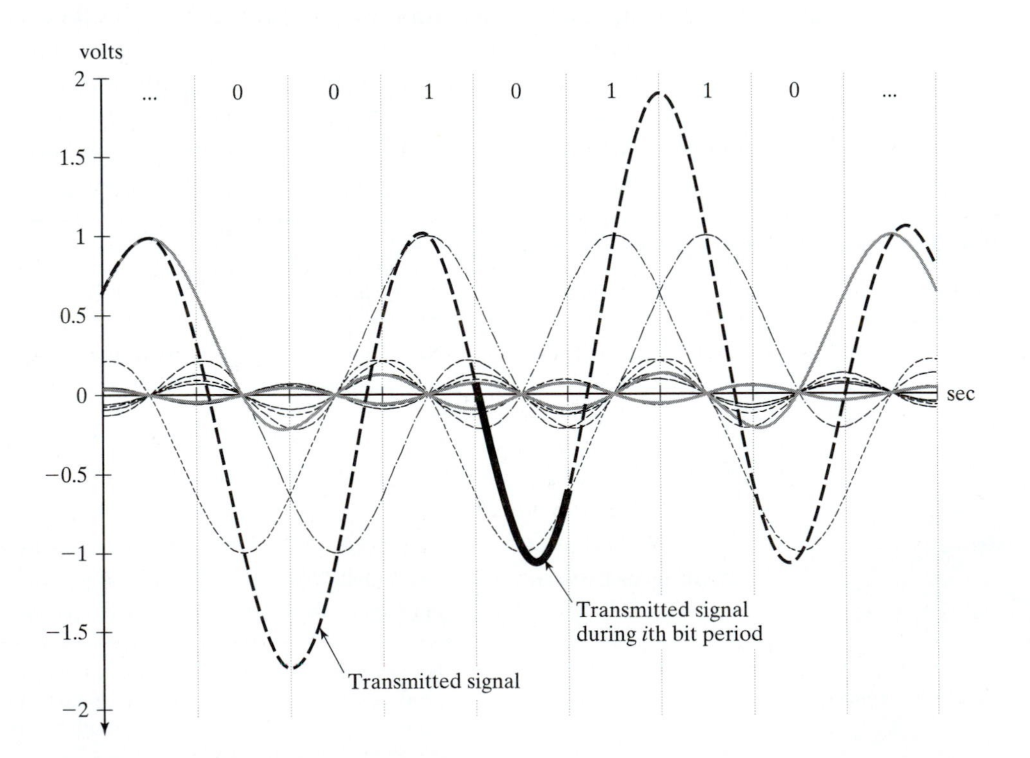

Figure 3-34 Practically speaking, the shape of the transmitted waveform during the ith bit period is influenced by only the seven closest pulses (from bits $i - 3$ to $i + 3$). For greater clarity, view the color version of this figure on CD-ROM.

generate the transmitted waveform corresponding to the fifth bit period of a particular data sequence. We could do so if we knew the values of bits 2 through 8.

Figure 3-35 shows a schematic for a raised-cosine- or sinc-shaped PAM modulator. The 74LS164A chip in the upper left corner is a shift register that holds seven bits. The value generated by these seven bits points to a 32-byte-wide location in the 27128A EPROM (this is accomplished by connecting the seven bits of the shift register to EPROM address lines A5–A11, as you can see by following the schematic). This 32-byte-wide block of memory in the EPROM contains the values of 32 sample points for one bit period of the transmitted waveform. As shown in Figure 3-36, the sample points are equally spaced in time, representing the waveform for the bit period corresponding to the center bit of the seven-bit sequence stored in the shift register. A clock, 32 times as fast as the bit rate, is input into the 74LS163A counters in the lower-left corner of the schematic, and the output from the counters causes the EPROM to sequence through the 32 memory locations containing the sample points for the appropriate waveform (follow the lines in the schematic from the counters to EPROM address lines A0–A4). The EPROM output is latched (the 74S412 latch), then converted from digital to analog, and then amplified and transmitted.

As an example, consider Figure 3-36. Note that the 32 sample points for the seven-bit sequence "0010110" are stored in memory locations 02C0–02DF (corresponding to the addresses generated when address lines A11, A10, A9, A8, A7, A6, and A5 are set to 0, 0, 1, 0, 1, 1, and 0, respectively). The 74LS163A counters cause the EPROM to sequence through the 32 sample points, generating the appropriate transmitted waveform. After the sequencing is completed, a new bit of data is input into the shift register in Figure 3-35, changing the values of the address lines A5–A11 and causing the EPROM to cycle through a new 32-sample sequence, corresponding to the transmitted waveform for the next data bit.

As a final feature of the Figure 3-35 schematic, note that if the EPROM is large enough, then different portions of the EPROM can be used to store waveforms corresponding to raised cosine pulses with different rolloff factors. This is shown in the schematic by the DIP switches (upper center) connected to the higher EPROM address lines A12 and A13.

Raised cosine transmitters were prohibitively costly when only analog circuitry was available, but using digital electronics and an EPROM-based lookup table, a raised cosine transmitter can now be easily built. This illustrates a very important point: *Performance-versus-cost trade-offs are not static.* Often the designer is required to try to predict what the next generation of technology (usually appearing every six to 18 months) will be able to provide. This is part of the fun of engineering.

We now have a more solid understanding of Table 3-4. Pulse shaping does not require overly complex transmission equipment, but synchronization requirements may dictate complex equipment at the receiver. Our analysis to this point has been very useful, but there are still issues that have not been completely addressed. For instance, we've determined that synchronization at the receiver is important, but we have yet to discuss how the receiver achieves synchronization. Also, we've seen that one way to demodulate a received signal is to sample the signal in the center of each bit period. Are there other ways to demodulate the signal and, if so, might these other methods provide superior performance for certain applications? Chapter 4 answers these questions and others.

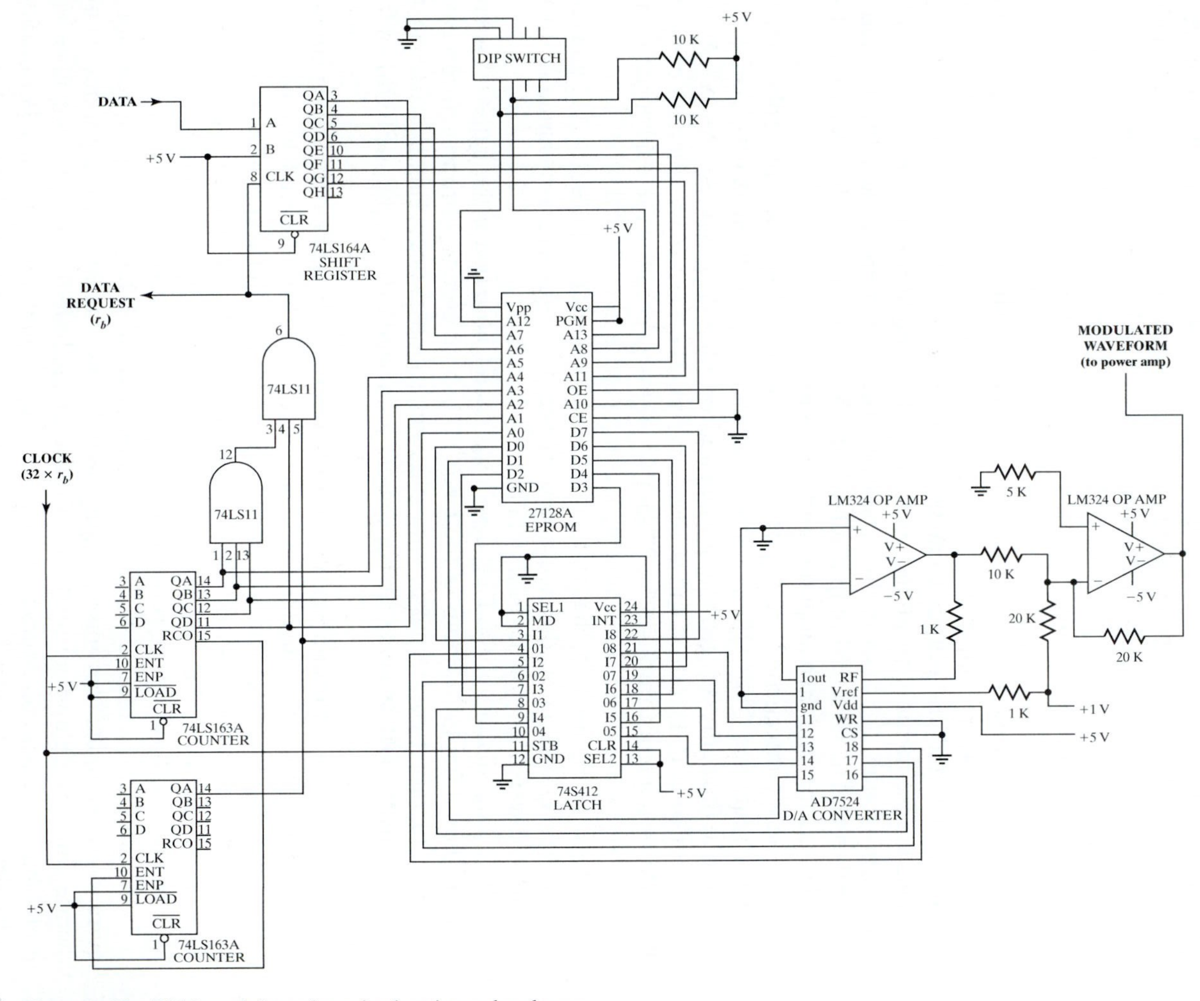

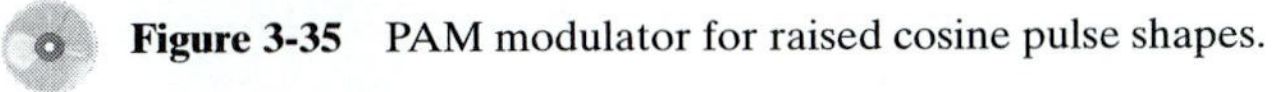

Figure 3-35 PAM modulator for raised cosine pulse shapes.

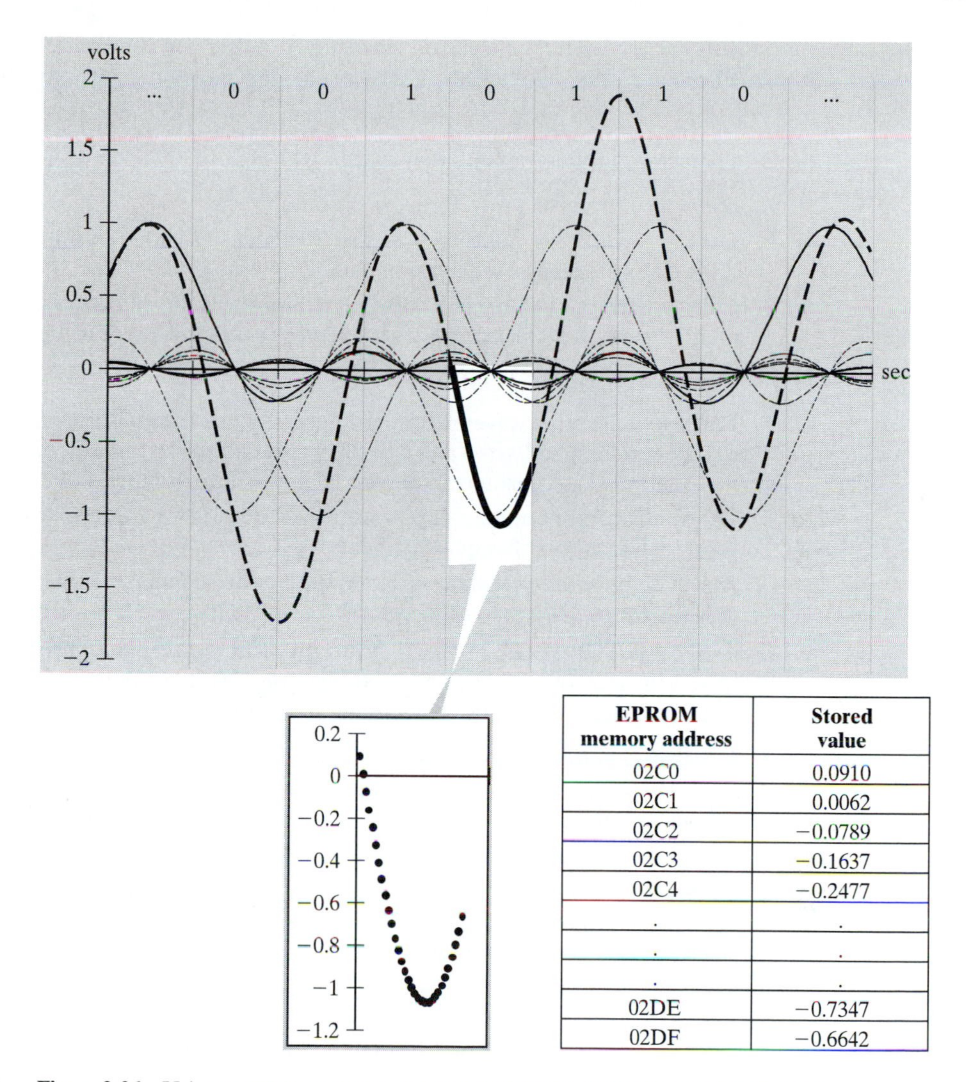

EPROM memory address	**Stored value**
02C0	0.0910
02C1	0.0062
02C2	−0.0789
02C3	−0.1637
02C4	−0.2477
.	.
.	.
.	.
02DE	−0.7347
02DF	−0.6642

Figure 3-36 Using an EPROM-based lookup table with a sequence of stored values to generate the appropriate raised cosine waveform ($\alpha = 0$).

PROBLEMS

3.1 Early in this chapter we calculated the average normalized energy spectral density for a binary PAM waveform representing two data bits using rectangular pulses. Using the same approach, calculate the average normalized energy spectral density for a binary PAM waveform representing three data bits using rectangular pulses of amplitude A and width τ.

3.2 Define the term "noise margin" and explain its significance in determining the accuracy of a received signal.

3.3 Investigate the use of rectangular and sinc-shaped pulses for data transmission over a baseband channel using binary PAM. Consider the data pattern "100011010" transmitted at a rate of 100,000 bits/sec. Let pulse amplitudes be +1 and −1 volts.

a. Using rectangular pulses of optimum width:

1. Draw the transmitted waveform for the data.
2. Draw the average normalized power spectral density for a transmitted waveform representing a general sequence of data.
3. Calculate the minimum bandwidth needed for the channel if accuracy requirements dictate that at least 95% of the signal's average power must be in-band.

b. Using sinc-shaped pulses of optimum width:

1. Plot the transmitted waveform for the data (if you want to use a computer to do this, be sure to read Parts c and d of this problem first, so that you develop your program with the capability to answer all parts of the problem).
2. Draw the average normalized power spectral density for a transmitted waveform representing a general sequence of data.
3. Calculate the minimum bandwidth required of the channel (allow 100% of the signal's average power to be in-band).

c. If the signal you developed in Part a of this problem (rectangular pulses) is transmitted over a channel that produces 50% attenuation in voltage, what is the noise margin of the received signal corresponding to the sixth bit in each of the following cases:

1. Receiver is perfectly synchronized.
2. Receiver is out of synchronization by +1 μsec (i.e., receiver is sampling the signal 1 μsec later than the optimum time).
3. Receiver is out of synchronization by +2 μsec (i.e., receiver is sampling the signal 2 μsec later than the optimum time).
4. Receiver is out of synchronization by +3 μsec (i.e., receiver is sampling the signal 3 μsec later than the optimum time).

d. If the signal you developed in Part b of this problem (sinc-shaped pulses) is transmitted over a channel that produces 50% attenuation in voltage, what is the noise margin of the received signal corresponding to the sixth bit in each of the following cases:

1. Receiver is perfectly synchronized.
2. Receiver is out of synchronization by +1 μsec (i.e., receiver is sampling the signal 1 μsec later than the optimum time).
3. Receiver is out of synchronization by +2 μsec (i.e., receiver is sampling the signal 2 μsec later than the optimum time).
4. Receiver is out of synchronization by +2.5 μsec (i.e., receiver is sampling the signal 2.5 μsec later than the optimum time).
5. Describe what happens to the demodulated bit if the receiver is out of synchronization by more than +2.5 μsec.

e. Using the results from Parts a–d of this problem, discuss the trade-offs involved in determining whether to use rectangular or sinc-shaped pulses for binary PAM.

3.4 Discuss the advantages and disadvantages of using raised cosine pulse shaping for PAM. Explain the significance of the rolloff factor.

3.5 A source outputs data at the rate of 50,000 bits/sec. The transmitter uses binary PAM with raised cosine shaping and transmits across an ideal baseband channel.

a. Suppose the ideal baseband channel has a cutoff frequency of 30,000 Hz. What is the maximum rolloff factor (α) that can be applied to the pulse shaping?

b. Suppose a rolloff factor higher than the value calculated in Part a is used. What effect will this have on the received waveform? Why might this be undesirable?

c. Suppose a rolloff factor lower than the value calculated in Part a is used. What effect will this have on the received waveform? Why is this undesirable?

3.6 Consider a binary PAM system using raised cosine pulse shaping (i.e., the frequency domain representation of each pulse is a raised cosine). The system transmits at the speed of 50,000 bits/sec and the transmitted waveform has a bandwidth of 35,000 Hz. Each pulse has a peak amplitude (time domain) of 4 volts.

a. Draw the magnitude spectrum of one pulse. Be sure to include sufficient detail in your drawing.

b. Write the time domain equation for one pulse (for simplicity, you may assume in your equation that the pulse is centered at $t = 0$).

3.7 Investigate the use of raised cosine pulses for data transmission over a baseband channel using binary PAM. Consider the same data pattern, transmission rate, and pulse amplitudes used in Problem 3.3. For rolloff factors of $\alpha = 0.25$, $\alpha = 0.5$, and $\alpha = 0.75$:

a. Plot the transmitted waveform for the data (if you want to use a computer to do this, be sure to read Part d of this problem first so that your program provides sufficient capability to also address that part).

b. Draw the average normalized power spectral density for a transmitted waveform representing a general sequence of data.

c. Calculate the minimum bandwidth required of the channel (allow 100% of the signal's average power to be in-band).

d. Suppose the signal is transmitted over a channel that produces 50% attenuation in voltage. For each of the three rolloff factors given above, establish the noise margin of the received signal corresponding to the sixth bit in each of the following cases:

1. Receiver is perfectly synchronized.
2. Receiver is out of synchronization by $+1\ \mu$sec (i.e., receiver is sampling the signal $1\ \mu$sec later than the optimum time).
3. Receiver is out of synchronization by $+2\ \mu$sec (i.e., receiver is sampling the signal $2\ \mu$sec later than the optimum time).
4. Receiver is out of synchronization by $+3\ \mu$sec (i.e., receiver is sampling the signal $3\ \mu$sec later than the optimum time).
5. Receiver is out of synchronization by $+4\ \mu$sec (i.e., receiver is sampling the signal $4\ \mu$sec later than the optimum time).

e. Using the results from Parts a–d in this problem, discuss the trade-offs involved in selecting a rolloff factor for the raised cosine pulse for a binary PAM system.

f. Using the results from Problem 3.3 and earlier parts of this problem, discuss the trade-offs involved in determining the pulse shape to use for binary PAM (when discussing the raised cosine pulse, be sure to include the trade-offs involved in selecting the rolloff factor).

3.8 **a.** For the sinc and raised cosine pulse shapes, what is the largest value of the main lobe width (τ) that can be chosen and still produce no ISI at the sampling points?

b. Justify your answers to Part a of this problem.

c. Discuss why the answer to Part a is significant with respect to the bandwidth of the transmitted signal.

3.9 A source outputs data at the rate of 1,000,000 bits/sec. The transmitter uses binary PAM with raised cosine pulse shaping. If the ideal baseband channel has a cutoff frequency of 850 kHz, what is the maximum rolloff factor that can be applied to the pulse shaping?

3.10 Determine the bandwidth of a typical transmitted signal with a transmission speed of 200,000 bits/sec for binary PAM with each of the following pulse shapes:

a. Rectangular pulse shape, optimum pulse width

b. Sinc-shaped pulse, optimum pulse width

c. Raised cosine pulse, $\alpha = 0$, optimum pulse width

d. Raised cosine pulse, $\alpha = 0.5$, optimum pulse width

e. Raised cosine pulse, $\alpha = 1.0$, optimum pulse width

3.11 Figure P3-11 shows the frequency domain representation of a raised cosine pulse. Determine the rolloff factor (α) of the pulse and the pulse width (τ) of the corresponding time domain representation.

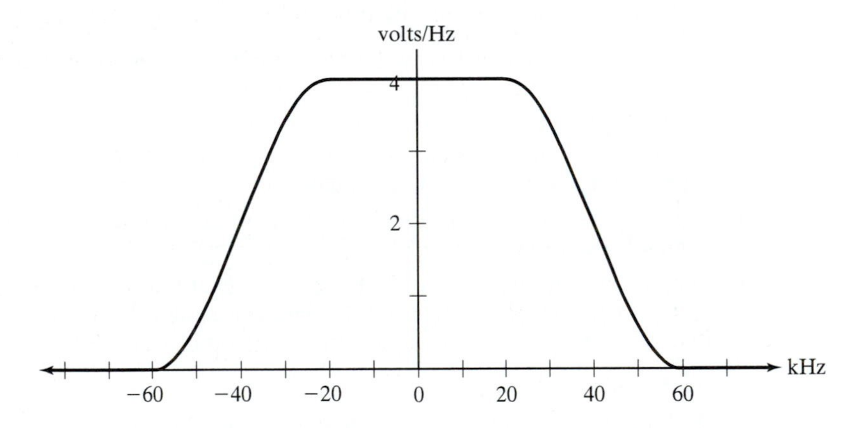

Figure P3-11 Frequency domain representation of a raised cosine pulse.

3.12 Consider a binary PAM system using a triangular pulse shape of width $2/r_b$, as shown in Figure P3-12. A pulse with a positive amplitude (1 volt) will be transmitted for a "1" and a pulse with a negative amplitude (-1 volt) will be transmitted for a "0."

a. Draw the transmitted waveform for a system sending the data pattern "10110" at a transmission speed of 100,000 bits per second. Be sure to label times and amplitudes on your sketch.

b. Assuming that the energy spectral density for a single triangular pulse is the same shape as the power spectral density for a typical series of triangular pulses, what is the bandwidth of a series of triangular pulses (assume at least 93% of power must be in-band)?

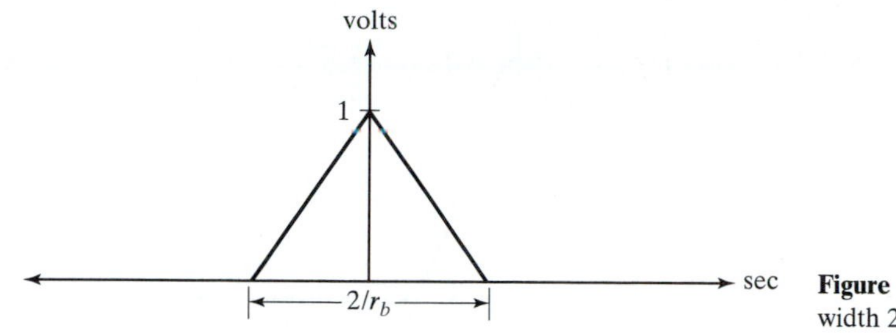

Figure P3-12 A triangular pulse of width $2/r_b$.

c. If the receiver is perfectly synchronized with the transmitter, how much ISI is present in the received waveform? (You may use the waveform in Part a of this problem to explain your answer if you want.)

d. Suppose the channel attenuates the waveform amplitude by 80%. What is the noise margin for the second bit (assuming perfect synchronization)?

e. Suppose the receiver is out of synchronization and that it samples the received waveform 2 μsec after the optimum time. Using the attenuation values from Part c of this problem, what is the noise margin for the second bit?

3.13 Figure P3-13 shows three different raised cosine pulses. One of the three pulses has a rolloff factor of 0, another has a rolloff factor of 0.5, and the remaining pulse has a rolloff factor of 1.

a. Identify which pulse has which rolloff factor.

b. Let $T_b = 1/r_b = 25\ \mu$sec. Draw the energy spectral density for each of the three pulses. Be sure to label your axes appropriately.

c. Explain the significance of the rolloff factor and discuss the trade-offs involved in increasing or decreasing the rolloff factor when raised cosine pulses are used for transmission of signals over baseband channels.

3.14 Examine Waveform A in Figure P3-14. This waveform represents the output of a transmitter using PAM with raised cosine pulse shaping at a data rate of 50,000 bits/sec. The transmitter sends pulses of amplitude 1 volt to signify a "1" and pulses of amplitude −1 volt to signify a "0."

a. What is the data pattern represented by Waveform A? Explain how you obtained your answer.

b. Examine Waveforms B and C in Figure P3-14. Waveforms A, B, and C all represent the output of transmitters using PAM with raised cosine pulse shaping. *All three waveforms represent the same data sequence, use pulses of the same amplitude, and operate at the same transmission speed. The only difference between the three waveforms is that they were produced using different rolloff factors.* One of the three waveforms uses $\alpha = 0$, another uses $\alpha = 0.5$, and the remaining waveform uses $\alpha = 1$. Identify which rolloff factor is used by each of the three waveforms. Explain how you obtained your answer.

c. What is the bandwidth of each of the three waveforms?

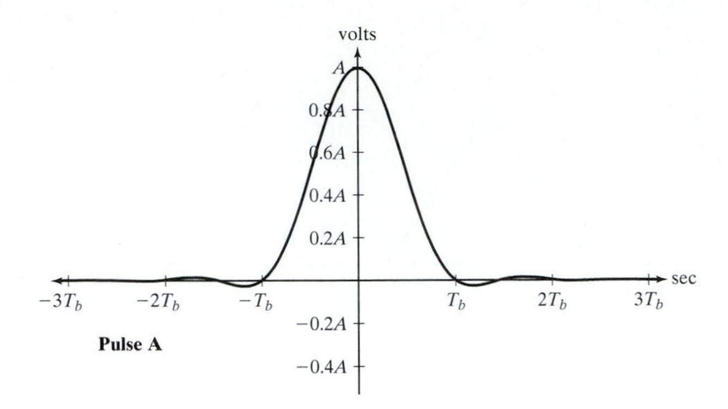

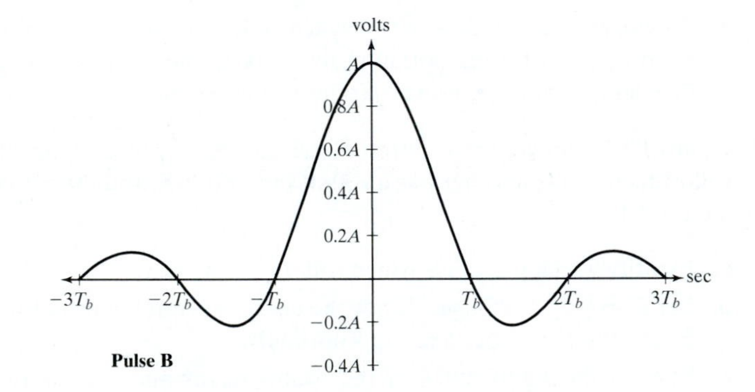

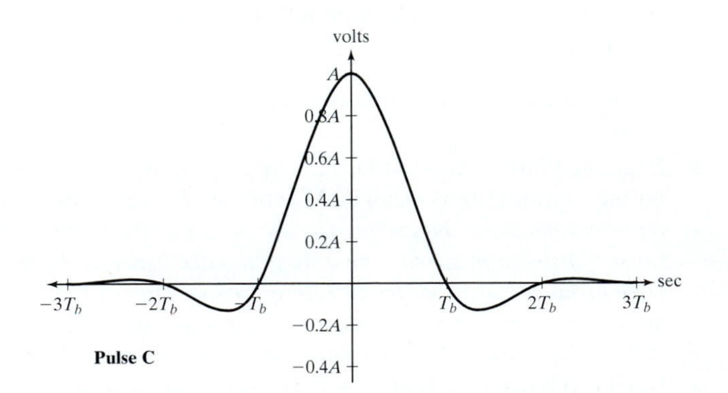

Figure P3-13 Three pulses with different rolloff factors.

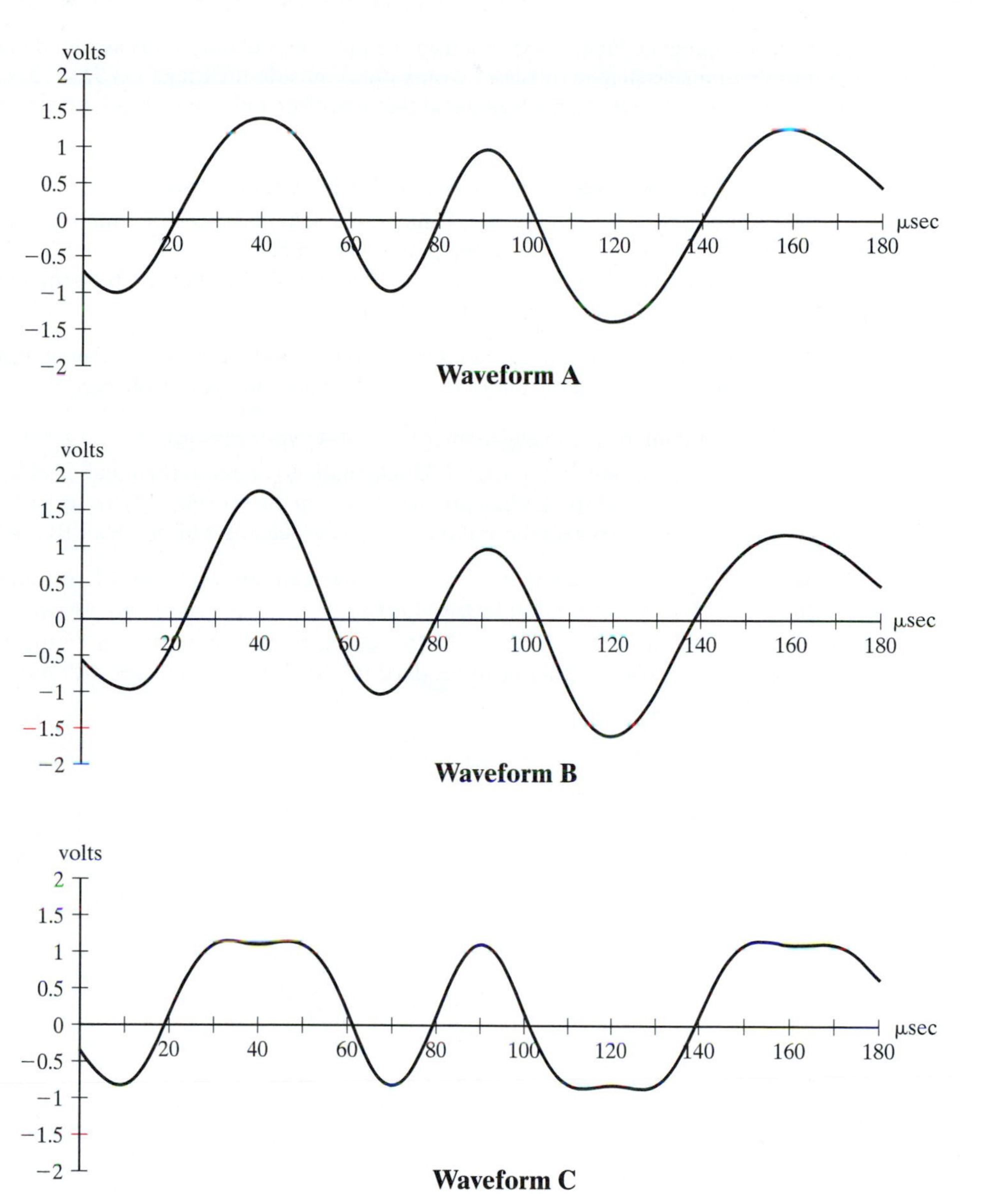

Figure P3-14 Three waveforms representing the same data transmitted at the same rate using different rolloff factors.

3.15 Explain, in detail, the purpose of the 74LS163A counters in the Figure 3-35 schematic of the raised cosine pulse modulator.

3.16 Explain, in detail, the purpose of the DIP switches in the Figure 3-35 schematic of the raised cosine pulse modulator.

3.17 Explain, in detail, the purpose of the 74LS164A shift register in the Figure 3-35 schematic of the raised cosine pulse modulator.

3.18 In developing the Figure 3-35 raised cosine pulse modulator, it was assumed that the magnitude of a sinc-shaped or raised cosine pulse outside the range $-3.5T_b < t < 3.5T_b$ was so small that it could be neglected and that therefore only seven bits had a practical effect on the waveform during the ith bit period.

a. Do you think this is a valid assumption? State your reasons.

b. Suppose that we modify the Figure 3-35 schematic to account for the values of the sinc-shaped or raised cosine pulse in the range $-4.5T_b < t < 4.5T_b$. What are the effects on the schematic? (Be sure to consider all the chips on the schematic and the addressing scheme within the EPROM.)

3.19 In developing the Figure 3-35 raised cosine pulse modulator, it was assumed that 32 sampled values would be sufficient to recreate the pulse during one bit period.

a. Do you think this is a valid assumption? State your reasons.

b. Suppose we modify the Figure 3-35 schematic to increase the number of sampled values per bit period to 64. What are the effects on the schematic? (Be sure to consider all the chips on the schematic and the addressing scheme within the EPROM.)

3.20 In the Figure 3-35 schematic, determine the contents of EPROM memory locations 02E0–02FF. Assume a rolloff factor of zero.

3.21 Explain how the memory in the EPROM can be segmented to allow the Figure 3-35 raised cosine pulse modulator to produce raised cosine waveforms with different rolloff factors.

Chapter 4

Receiver Design (and Stochastic Mathematics, Part I)

4.1 Developing a Simple Pulse Amplitude Modulation Receiver

To develop a deeper understanding of baseband communication system design, we will now take a more detailed look at receivers. In Chapter 3, we developed a very simple demodulation technique. The receiver sampled the received signal in the center of each bit period. If the sampled value was ≥ 0 volts, the receiver assumed that the corresponding data from the source was a "1" (since a "1" from the source would produce a positive pulse from the transmitter). If the sampled value was <0 volts, the receiver assumed that the corresponding data from the source was a "0" (since a "0" from the source would produce a negative pulse from the transmitter). A block diagram for this simple receiver is shown in Figure 4-1. Note that the receiver also contains a lowpass filter with the same bandwidth as the transmitted signal—as we will shortly see, this filter will help reduce the effects of noise.

In analyzing our simple receiver, we determined in Chapter 3 that under real-world conditions (a channel that attenuates the transmitted signal and also corrupts the signal by adding noise), errors could occur. In other words, the receiver could demodulate a sampled value as a "1" even though the corresponding data from the source was actually a "0," or the receiver could demodulate a sampled value as a "0" even though the corresponding data bit from the source was a "1." In Chapter 3, we quantified this inaccuracy by measuring noise margin. Let's now develop another, much more powerful, measure of accuracy—*probability of bit error*.

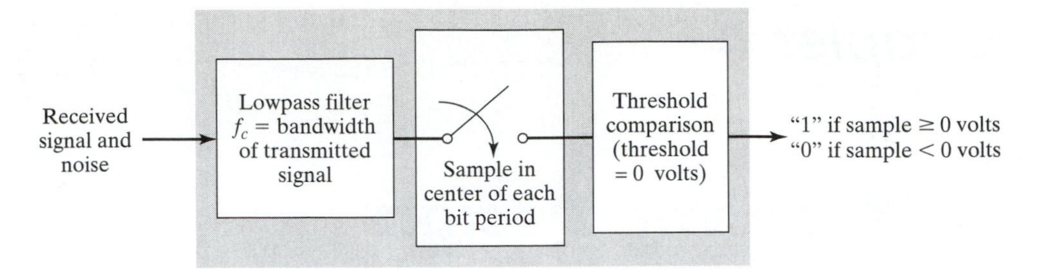

Figure 4-1 A simple PAM receiver.

4.1.1 Establishing an Expression for Probability of Bit Error

Let's quantify accuracy by calculating the probability that a particular bit—say, the ith bit—is demodulated in error. For the simple PAM receiver in Figure 4-1, we begin our analysis by developing the following notation:

Let b_i ≡ the ith transmitted bit ($b_i = 1$, or $b_i = 0$)
Let $s(t)$ ≡ the transmitted signal
Let $r(t)$ ≡ the received signal, which has been attenuated and corrupted by noise
Let $z(t)$ ≡ the voltage at the input of the sampler in the receiver
Let $n(t)$ ≡ noise at the input of the receiver
Let $n_o(t)$ ≡ noise at the output of the receiver's lowpass filter

As in Chapter 3, T_b = bit period. Note that $T_b = 1/r_b$. This looks like a lot of variables, but we'll redefine each one the first time we use it.

The sampling time for the ith bit is

$$t = (i - 1)T_b + \frac{T_b}{2}$$

This time corresponds to the center of the ith bit period. Thus, using our new notation, the voltage at the output of the sampler corresponding to the ith bit is

$$z\left[(i - 1)T_b + \frac{T_b}{2}\right]$$

We can derive the probability of bit error as follows:

The probability that the ith bit is demodulated in error =

$P\{i\text{th bit} = 1$, but the sampled voltage at the receiver during the ith bit period $< 0\}$ +

$P\{i\text{th bit} = 0$, but the sampled voltage at the receiver during the ith bit period $\geq 0\}$ =

$$P\left\{b_i = 1 \text{ and } z\left[(i - 1)T_b + \frac{T_b}{2}\right] < 0\right\} + P\left\{b_i = 0 \text{ and } z\left[(i - 1)T_b + \frac{T_b}{2}\right] \geq 0\right\} =$$

$$P\{b_i = 1\}P\left\{z\left[(i-1)T_b + \frac{T_b}{2}\right] < 0 \text{ when } b_i = 1\right\} + P\{b_i = 0\}P\left\{z\left[(i-1)T_b + \frac{T_b}{2}\right] \geq 0 \text{ when } b_i = 0\right\} \tag{4.1a}$$

These last expressions,

$$P\left\{z\left[(i-1)T_b + \frac{T_b}{2}\right] < 0 \text{ when } b_i = 1\right\} \text{ and } P\left\{z\left[(i-1)T_b + \frac{T_b}{2}\right] \geq 0 \text{ when } b_i = 0\right\},$$

are *conditional probabilities*. We can develop a shorthand notation, using the symbol "|" instead of the word "when." Equation (4.1a) can thus be expressed as

The probability that the ith bit is demodulated in error =

$$P\{b_i = 1\}P\left\{z\left[(i-1)T_b + \frac{T_b}{2}\right] < 0 \middle| b_i = 1\right\} + P\{b_i = 0\}P\left\{z\left[(i-1)T_b + \frac{T_b}{2}\right] \geq 0 \middle| b_i = 0\right\} \tag{4.1b}$$

Conditional probabilities will be discussed in more detail in Section 4.1.2.7.

Now assume $P\{b_i = 1\} = P\{b_i = 0\} = \frac{1}{2}$ (i.e., "1" and "0" are *equiprobable*) and that the value of each bit is independent of the value of all other bits in the data stream.[1] Substituting into Equation (4.1b),

$P\{i\text{th bit is demodulated in error}\}$ =

$$\frac{1}{2}P\left\{z\left[(i-1)T_b + \frac{T_b}{2}\right] < 0 \middle| b_i = 1\right\} + \frac{1}{2}P\left\{z\left[(i-1)T_b + \frac{T_b}{2}\right] \geq 0 \middle| b_i = 0\right\} \tag{4.2}$$

Let's take a closer look at $z(t)$, the input to the sampler. The input to the sampler is the received signal after it has been passed through the lowpass filter (see Figure 4-1 again). In other words,

$$z(t) = \text{received signal after filtering}$$

Before we investigate the effects of filtering, let's examine the received signal. Assuming that the channel has a sufficient bandwidth (i.e., the channel does not distort the transmitted signal, although it can attenuate the signal),

$$\text{received signal} = \text{attenuated transmitted signal} + \text{noise}$$

Expressing this symbolically,

A constant reflecting attenuation of the channel — Noise at receiver input

$$r(t) = \gamma s(t) + \overbrace{n(t)}$$

[1]The assumption of equiprobable, independent bits is realistic for a well-designed system. As we will see in Chapter 9, data which does not consist of equiprobable, independent bits contains redundancy that can be eliminated through compression. For the sake of completeness, we will evaluate probability of bit error for nonequiprobable bits in Section 4.2.4.2.

So

$$r\left[(i-1)T_b + \frac{T_b}{2}\right] = \gamma s\left[(i-1)T_b + \frac{T_b}{2}\right] + n\left[(i-1)T_b + \frac{T_b}{2}\right] \tag{4.3}$$

where γ is the portion of the transmitted signal that arrives at the receiver (the rest is lost due to attenuation in the channel).

Let's now look at the effects of filtering the received signal. As established in Equation (4.3), the received signal has two components: an attenuated version of the transmitted signal, $\gamma s(t)$, and the noise at the receiver input, $n(t)$. Since the filter has the same bandwidth as the transmitted signal, $\gamma s(t)$ passes through the filter unaffected (assuming an ideal lowpass filter). The noise, however, has a greater bandwidth than the filter. We will use the variable $n_o(t)$ to represent the noise at the filter output. Adding the components due to the transmitted signal and the noise, the voltage at the sampler input is thus

$$z(t) = \gamma s(t) + \overbrace{n_o(t)}^{\text{noise at receiver filter output}}$$

So

$$z\left[(i-1)T_b + \frac{T_b}{2}\right] = \gamma s\left[(i-1)T_b + \frac{T_b}{2}\right] + n_o\left[(i-1)T_b + \frac{T_b}{2}\right] \tag{4.4}$$

Let's say that the transmitted signal is binary PAM composed of rectangular, sinc, or raised cosine pulses with peak amplitude of A if $b_i = 1$ and $-A$ if $b_i = 0$. Since there is no ISI in the exact center of the bit period,

$$s\left[(i-1)T_b + \frac{T_b}{2}\right] = A \text{ when } b_i = 1, \text{ and } s\left[(i-1)T_b + \frac{T_b}{2}\right] = -A \text{ when } b_i = 0.$$

Substituting into Equation (4.4),

$$z\left[(i-1)T_b + \frac{T_b}{2}\right] = \begin{cases} \gamma A + n_o\left[(i-1)T_b + \frac{T_b}{2}\right] \text{ if } b_i = 1 \\ -\gamma A + n_o\left[(i-1)T_b + \frac{T_b}{2}\right] \text{ if } b_i = 0 \end{cases} \tag{4.5}$$

Thus, we can express Equation (4.2) as

$$P\{i\text{th bit is demodulated in error}\} =$$

$$\frac{1}{2}P\left\{\gamma A + n_o\left[(i-1)T_b + \frac{T_b}{2}\right] < 0\right\} + \frac{1}{2}P\left\{-\gamma A + n_o\left[(i-1)T_b + \frac{T_b}{2}\right] \geq 0\right\} \tag{4.6}$$

Mathematically manipulating Equation (4.6),

$$P\{i\text{th bit is demodulated in error}\} =$$

$$\frac{1}{2}P\left\{n_o\left[(i-1)T_b + \frac{T_b}{2}\right] < -\gamma A\right\} + \frac{1}{2}P\left\{n_o\left[(i-1)T_b + \frac{T_b}{2}\right] \geq \gamma A\right\} \tag{4.7}$$

For the case of independent bits and equiprobable "1"s and "0"s, Equation (4.7) means that the ith bit is demodulated in error if, at the time of the ith sampling at the receiver, the noise is stronger than the attenuated transmitted signal and is of the opposite sign. Be sure this makes sense to you.

Can we determine the two probabilities on the right side of Equation (4.7)? To do so, we need to know a little about the noise in the system.

4.1.2 Stochastic Mathematics—Part I (Random Variables)

Although we cannot say for certain how much noise will occur in the system at any particular time, we can determine the probability that the amount of noise at that time will be less than a certain value. In other words, we may not be able to determine in advance the value of $n[(i-1)T_b + T_b/2]$, but we can determine the probability that it will be $<-\gamma A$ volts. Note that this is a very different approach (*probabilistic*, or *stochastic, mathematics*) compared to what we're used to (*deterministic mathematics*). Deterministic mathematics says, "I can tell you with certainty what the value of a particular variable will be at any time t." Probabilistic or stochastic mathematics says, "I don't know enough about a particular variable (like $n(t)$) to be able to tell you with certainty what its value will be at a particular time in the future (like $t = (i-1)T_b + T_b/2$), but I can tell you something about the *tendencies* of $n(t)$ at time $t = (i-1)T_b + T_b/2$. For instance, I can tell you the probability that $n(t)$ will be less than a certain value at time $t = (i-1)T_b + T_b/2$." A probabilistic description isn't as good as a deterministic description, but often it's the best we can do and is better than nothing. A phenomenon that we cannot describe deterministically, but that we can describe probabilistically is called *random*. We don't know exactly how it will behave at any particular time, but, through experimentation and analysis of the physical occurrences that produce the random event, we do know its tendencies. Let's develop a set of mathematical parameters that we can use to describe the tendencies of a random phenomenon such as noise.

4.1.2.1 Probability distribution function As we saw in calculating probability of error, one parameter we are interested in concerning noise is the probability that $n(t)$ will be less than a certain value $(-\gamma A)$ at a certain time $[t = (i-1)T_b + T_b/2]$. Generally speaking, we are interested in knowing the probability that the result of a random phenomenon (e.g., noise voltage) is less than or equal to a certain value at a certain time. Let's use the variable **X** to describe the voltage of the noise at a particular time (**X** is called a *random variable*). The *probability distribution function* for **X** (also known as the *distribution function*, or *cumulative distribution function*) is symbolized as $F_{\mathbf{X}}(a)$ and is defined as

$$F_{\mathbf{X}}(a) \equiv P\{\mathbf{X} \leq a\} \tag{4.8}$$

In other words, $F_{\mathbf{X}}(a)$ is the probability that the random variable **X** is less than or equal to some specific value a.

The probability distribution function has the following properties:

$$F_{\mathbf{X}}(-\infty) = 0 \tag{4.9}$$

$$F_{\mathbf{X}}(\infty) = 1 \tag{4.10}$$

$$\text{If } b > a, \text{ then } F_{\mathbf{X}}(b) \geq F_{\mathbf{X}}(a) \tag{4.11}$$

Example 4.1 Probability distribution for a single die

A professor rolls a single die and announces the number. Determine the probability distribution function describing the number appearing on the die.

Solution

Let $\mathbf{X}$ represent the value on the die after it has been rolled. The following probability distribution function $F_{\mathbf{X}}(a)$ describes the probability that the number appearing on the die is less than or equal to the number a. (Note that a can be any rational number.)

$$F_{\mathbf{X}}(a) \equiv P\{\mathbf{X} \le a\} = \begin{cases} 0 & a < 1 \\ 1/6 & 1 \le a < 2 \\ 1/3 & 2 \le a < 3 \\ 1/2 & 3 \le a < 4 \\ 2/3 & 4 \le a < 5 \\ 5/6 & 5 \le a < 6 \\ 1 & a \ge 6 \end{cases}$$

This probability distribution function is plotted in Figure 4-2.

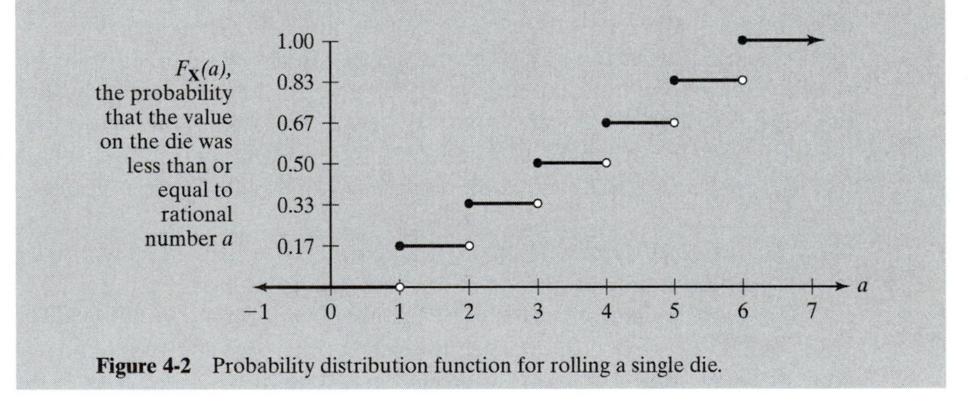

Figure 4-2 Probability distribution function for rolling a single die.

Returning to our original objective, calculating the probability that the noise at the receiver $n(t)$ will be less than or equal to a certain value $(-\gamma A)$ at a certain time $[t = (i - 1)T_b + T_b/2]$, we need to determine the probability distribution function for noise. The most prevalent type of noise in the channel (and actually in the front end of the receiver, too) is noise due to the random motion of electrons within electronic devices. This is called *thermal noise* and has the probability distribution function

$$F_{\mathbf{X}}(a) = \int_{-\infty}^{a} \frac{1}{\sqrt{2\pi}\sigma_n} e^{-\left(\frac{(x-\mu_n)^2}{2\sigma_n^2}\right)} dx \tag{4.12}$$

where μ_n (known as the *mean*) $= 0$ and σ_n^2 (known as the *variance*) $=$ average normalized noise power.

We now have enough information to characterize thermal noise sufficiently to determine probability of bit error, but first let's develop some additional parameters for stochastic mathematics. We will need them later.

4.1.2.2 Probability density function The *probability density function* is symbolized as $f_{\mathbf{X}}(x)$, is defined as

$$f_{\mathbf{X}}(x) \equiv \frac{dF_{\mathbf{X}}(x)}{dx} \tag{4.13}$$

and can be used to express the probability that the random variable $\mathbf{X}$ lies between two values, say, a and b ($b > a$).

$$\begin{aligned} P\{a < \mathbf{X} \leq b\} &= P\{\mathbf{X} \leq b\} - P\{\mathbf{X} \leq a\} \\ &= F_{\mathbf{X}}(b) - F_{\mathbf{X}}(a) \\ &= \int_{-\infty}^{b} f_{\mathbf{X}}(x)dx - \int_{-\infty}^{a} f_{\mathbf{X}}(x)dx \\ &= \int_{a}^{b} f_{\mathbf{X}}(x)dx \end{aligned} \tag{4.14}$$

Do you see why $f_{\mathbf{X}}(x)$ is called a probability density function?

The probability density function has the following properties:

$$f_{\mathbf{X}}(x) \geq 0 \text{ for } -\infty \leq x \leq \infty \tag{4.15}$$

$$\int_{-\infty}^{\infty} f_{\mathbf{X}}(x)dx = 1 \tag{4.16}$$

$$F_{\mathbf{X}}(a) = \int_{-\infty}^{a} f_{\mathbf{X}}(x)dx \tag{4.17}$$

Combining Equations (4.16) and (4.17), we can determine the probability that the value of $\mathbf{X}$ is greater than a given value, say, c

$$\begin{aligned} P\{\mathbf{X} > c\} &= 1 - P\{\mathbf{X} \leq c\} = 1 - F_{\mathbf{X}}(c) \\ &= 1 - \int_{-\infty}^{c} f_{\mathbf{X}}(x)dx = \int_{-\infty}^{\infty} f_{\mathbf{X}}(x)dx - \int_{-\infty}^{c} f_{\mathbf{X}}(x)dx = \int_{c}^{\infty} f_{\mathbf{X}}(x)dx \end{aligned} \tag{4.18}$$

Example 4.2 Probability density function for a single die

Consider Example 4.1, in which a professor rolls a single die. The probability density function for the value on the die is

$$f_{\mathbf{X}}(x) = \begin{cases} 1/6 & \mathbf{X} = 1, 2, 3, 4, 5, \text{ or } 6 \\ 0 & \text{all other values of } \mathbf{X} \end{cases}$$

4.1.2.3 *Relevance to thermal noise* As discussed previously, thermal noise has the probability distribution function

$$F_{\mathbf{X}}(a) = \int_{-\infty}^{a} \frac{1}{\sqrt{2\pi}\sigma_n} e^{-\left(\frac{(x-\mu_n)^2}{2\sigma_n^2}\right)} dx \qquad (4.12R)$$

where $\mu_n = 0$ and σ_n^2 = average normalized noise power. Thus, the probability density function for thermal noise is

$$f_{\mathbf{X}}(x) = \frac{1}{\sqrt{2\pi}\sigma_n} e^{-\left(\frac{(x-\mu_n)^2}{2\sigma_n^2}\right)} \qquad (4.19)$$

This function is known as the *Gaussian probability density function*, or *Gaussian function*, and it appears in many natural phenomena. The Gaussian probability density function will be discussed in detail in Section 4.1.3.

4.1.2.4 *Mean (or expected value)* The *mean* of a random variable **X** is symbolized as $\mu_{\mathbf{X}}$, or sometimes $m_{\mathbf{X}}$, and is defined as

$$\mu_{\mathbf{X}} \equiv \int_{-\infty}^{\infty} x f_{\mathbf{X}}(x) dx \qquad (4.20a)$$

In other words, $\mu_{\mathbf{X}}$ is the average value of the random variable **X**. In fact, the mean is often called the *expected value* of **X** and symbolized $E\{\mathbf{X}\}$. Using this new notation,

$$E\{\mathbf{X}\} \equiv \mu_{\mathbf{X}} \equiv \int_{-\infty}^{\infty} x f_{\mathbf{X}}(x) dx \qquad (4.20b)$$

Example 4.3 Mean for a single die

Consider again the example in which a professor rolls a single die. The mean (or average) value of the die is

$$\mu_{\mathbf{X}} = \int_{-\infty}^{\infty} x f_{\mathbf{X}}(x) dx = \frac{1}{6}(1) + \frac{1}{6}(2) + \frac{1}{6}(3) + \frac{1}{6}(4) + \frac{1}{6}(5) + \frac{1}{6}(6) = 3.5$$

In this example, we could have intuitively calculated the average value of the die. Do you see how Equation (4.20b) produces an average value?

4.1.2.5 *Variance* The *variance* of a random variable **X** is symbolized as $\sigma_{\mathbf{X}}^2$ and is defined as

$$\sigma_{\mathbf{X}}^2 = \int_{-\infty}^{\infty} (x - \mu_{\mathbf{X}})^2 f_{\mathbf{X}}(x) dx \qquad (4.21a)$$

Equation (4.21a) can also be expressed as

$$\sigma_{\mathbf{X}}^2 = \int_{-\infty}^{\infty} (x - \mu_{\mathbf{X}})^2 f_{\mathbf{X}}(x)dx = E\{(\mathbf{X} - \mu_{\mathbf{X}})^2\} \tag{4.21b}$$

Variance is a measure of how much the values of the random variable **X** fluctuate (or *vary*) around its average. Thus, variance is a measure of the unpredictability of **X**.

Example 4.4 Variance for a single die versus flipping two coins

Consider again the example in which a professor rolls a single die. The variance of the die is

$$\sigma_{\mathbf{X}}^2 = \int_{-\infty}^{\infty} (x - \mu_{\mathbf{X}})^2 f_{\mathbf{X}}(x)dx = \int_{-\infty}^{\infty} (x - 3.5)^2 f_{\mathbf{X}}(x)dx$$

$$= (1/6)(-2.5)^2 + (1/6)(-1.5)^2 + (1/6)(-0.5)^2 + (1/6)(0.5)^2 + (1/6)(1.5)^2 + (1/6)(2.5)^2$$

$$= 2.917$$

Now suppose that the professor flips two coins and counts the number of coins that land heads up (let's use the random variable **Y** to denote the number of coins that land heads up). The probability density function for the number of heads-up coins is

$$f_{\mathbf{Y}}(y) = \begin{cases} 0.25 & \mathbf{Y} = 0 \\ 0.50 & \mathbf{Y} = 1 \\ 0.25 & \mathbf{Y} = 2 \\ 0 & \text{All other values of } \mathbf{Y} \end{cases}$$

The mean of **Y** is

$$\mu_{\mathbf{Y}} = \int_{-\infty}^{\infty} y f_{\mathbf{Y}}(y)dy = \frac{1}{4}(0) + \frac{1}{2}(1) + \frac{1}{4}(2) = 1$$

The variance of **Y** is

$$\sigma_{\mathbf{Y}}^2 = \int_{-\infty}^{\infty} (y - \mu_{\mathbf{Y}})^2 f_{\mathbf{Y}}(y)dy = \int_{-\infty}^{\infty} (y - 1)^2 f_{\mathbf{Y}}(y)dy$$

$$= 0.25(-1)^2 + 0.50(0)^2 + 0.25(1)^2 = 0.50$$

Comparing the variance of the die roll (2.917) with the variance of the two-coin flip (0.50), we see that the outcome of the two-coin flip is much more predictable.

Note that if a random variable has zero mean, then its variance is equal to its average normalized power. This can be proven in a mathematically straightforward manner:

$$\sigma_{\mathbf{X}}^2 = \int_{-\infty}^{\infty} (x - \mu_{\mathbf{X}})^2 f_{\mathbf{X}}(x)dx = \int_{-\infty}^{\infty} (x - 0)^2 f_{\mathbf{X}}(x)dx = \int_{-\infty}^{\infty} x^2 f_{\mathbf{X}}(x)dx \tag{4.22}$$

which is the average value of $\mathbf{X}^2$, corresponding to the average normalized power of $\mathbf{X}$.

4.1.2.6 *Joint probabilities* Joint probabilities are useful when determining how two random variables may be related to each other.

Joint probability distribution function

Consider two random variables, $\mathbf{X}$ and $\mathbf{Y}$. The *joint probability distribution function* is defined as

$$F_{\mathbf{X},\mathbf{Y}}(a, b) \equiv P\{\mathbf{X} \leq a \text{ and } \mathbf{Y} \leq b\} \tag{4.23}$$

If the two random variables are not related to each other (for instance, if $\mathbf{X}$ represents the noise in a particular computer modem's receiver and $\mathbf{Y}$ represents the amount of data being transmitted), then the two variables are *independent*. If $\mathbf{X}$ and $\mathbf{Y}$ are independent, then

$$F_{\mathbf{X},\mathbf{Y}}(a,b) \equiv P\{\mathbf{X} \leq a \text{ and } \mathbf{Y} \leq b\} = P\{\mathbf{X} \leq a\}P\{\mathbf{Y} \leq b\} = F_{\mathbf{X}}(a)F_{\mathbf{Y}}(b) \tag{4.24}$$

Joint probability density function

The *joint probability density function* of two random variables, $\mathbf{X}$ and $\mathbf{Y}$, is defined as

$$f_{\mathbf{X},\mathbf{Y}}(x,y) \equiv \frac{\partial^2 F_{\mathbf{X},\mathbf{Y}}(x,y)}{\partial x \, \partial y} \tag{4.25}$$

This definition is consistent with the way the probability density function for a single random variable was defined in Equation (4.13).

4.1.2.7 *Conditional probabilities* Conditional probabilities can also be used to express relationships between random occurrences. As we established when developing Equations (4.1a and b), we use the symbol $|$ "given" to represent a conditional probability, and the notation $P(\mathbf{X} > a|\text{event Z})$ to express "the probability that random variable $\mathbf{X}$ is greater than a given that event Z has occurred." Note that

$$P\{\mathbf{X} > a \text{ and event Z}\} = P\{\text{event Z}\}P\{\mathbf{X} > a|\text{event Z}\} \tag{4.26}$$

4.1.2.8 *Autocorrelation and random processes* These more advanced parameters and mathematical concepts will be defined and examined in Chapter 5, Section 5.4.

4.1.3 Examining Thermal Noise

We are now ready to determine the characteristics of noise in the channel so that we can calculate probability of error for a PAM receiver. As mentioned earlier, the most prevalent type of noise in the channel (and actually in the front end of the receiver, too) is the noise caused by the random motion of electrons within electronic devices. This is called *thermal noise* and, through experimentation and analysis of the physical events that produce the noise (the random motion of electrons), we can determine the following tendencies, or stochastic parameters, for the noise voltage:

1. Its mean (average value) is 0. $\mu_n = 0$.
2. It has a Gaussian probability density function

$$f_{\mathbf{X}}(x) = \frac{1}{\sqrt{2\pi}\sigma_n} e^{-\left(\frac{(x-\mu_n)^2}{2\sigma_n^2}\right)} \tag{4.19R}$$

and thus the probability that the noise voltage $n(t)$ at time t_o will be less than or equal to a certain value, say $-\gamma A$, is

$$P\{n(t_o) \leq -\gamma A\} = F_{\mathbf{X}}(-\gamma A) = \int_{-\infty}^{-\gamma A} \frac{1}{\sqrt{2\pi}\sigma_n} e^{-\left(\frac{(x-\mu_n)^2}{2\sigma_n^2}\right)} dx \tag{4.27}$$

and the probability that $n(t_o)$ is greater than a certain value, say γA, is

$$P\{n(t_o) > \gamma A\} = 1 - F_{\mathbf{X}}(\gamma A) = \int_{\gamma A}^{\infty} \frac{1}{\sqrt{2\pi}\sigma_n} e^{-\left(\frac{(x-\mu_n)^2}{2\sigma_n^2}\right)} dx \tag{4.28}$$

Since $\mu_n = 0$, the variance σ_n^2 is equal to the average normalized noise power (see Equation (4.22)). A typical Gaussian probability density function (for $\mu_n = 0$ and $\sigma_n^2 = 1$) is shown in Figure 4-3.

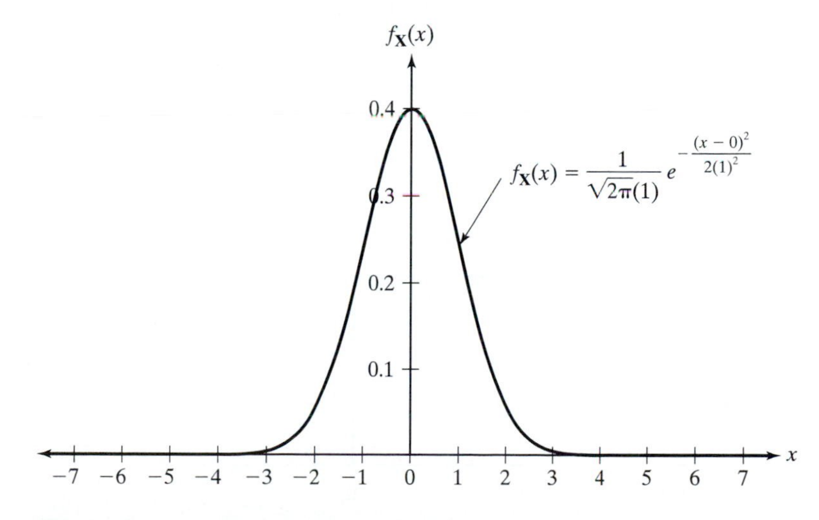

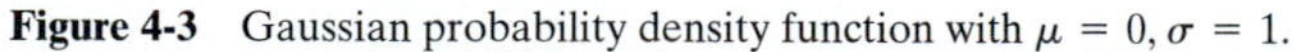

Figure 4-3 Gaussian probability density function with $\mu = 0, \sigma = 1$.

3. The probabilistic properties of $n(t)$ do not change with time ($n(t)$ is said to be *stationary*). Thus,

$$P\{n(t_o) \leq a\} = P\{n(t_1) \leq a\} = P\{n(t_2) \leq a\} \ldots$$

for any t_o, t_1, t_2. This is *not* to say that $n(t)$ is the same value at times t_o, t_1, and t_2, but merely that it has the same probabilistic tendencies.

4. Thermal noise has an average normalized power spectral density that is constant from dc to approximately 10^{12} Hz. This power spectral density, shown in Figure 4-4, means that thermal noise contains equal amounts of all frequency components in the band from dc to 10^{12} Hz. To draw an analogy, consider the color white. The color white is composed of the sum of equal amounts of all other colors; similarly, thermal noise is composed of the sum of equal strength components of all frequencies. In fact, thermal noise is often called *white noise* or, more specifically, *additive white Gaussian noise*.

Summarizing, we say that the noise at the input to the receiver's lowpass filter is thermal noise, which is white and can be modeled as an additive Gaussian random variable with zero mean and variance equal to average normalized noise power.

Now that we know the stochastic properties of $n(t)$—the thermal noise at the input of the receiver in Figure 4-1—we can determine the stochastic properties of $n_o(t)$, the thermal noise after it has passed through the receiver's lowpass filter. If the lowpass filter has an ideal frequency response and a cutoff frequency of f_c, then the average normalized power spectrum of the noise at the filter output is as shown in Figure 4-5.

If a random signal with a Gaussian probability density function is passed through a linear, time-invariant filter, then the output of the filter is a random signal that also has a Gaussian probability density function (Taub & Schilling [4.1]). Thus, $n_o(t)$ still has a Gaussian probability density function. The mean of $n_o(t)$ will be zero, like $n(t)$, but its variance will not be the same as the variance of $n(t)$. Adapting Equation (4.28), the probability that $n_o(t)$ is greater than some constant a at time $t = t_o$ can be expressed as

$$P\{n_o(t_o) > a\} = \int_a^{\infty} \frac{1}{\sqrt{2\pi}\sigma_o} e^{-\left(\frac{(x-\mu_o)^2}{2\sigma_o^2}\right)} dx \tag{4.29}$$

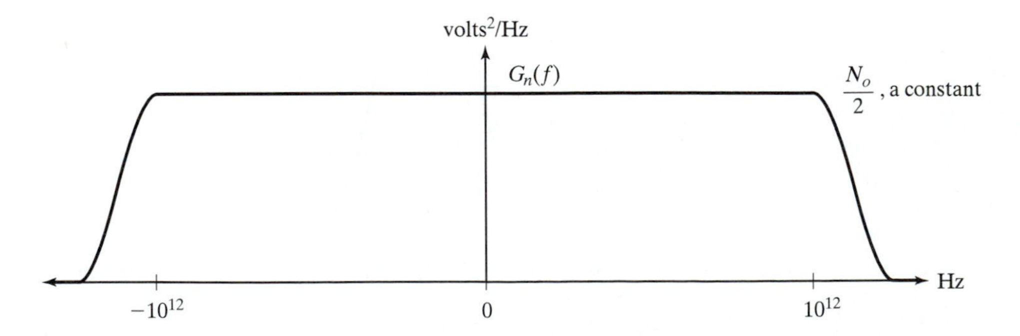

Figure 4-4 Average normalized power spectrum of thermal noise at input of the receiver.

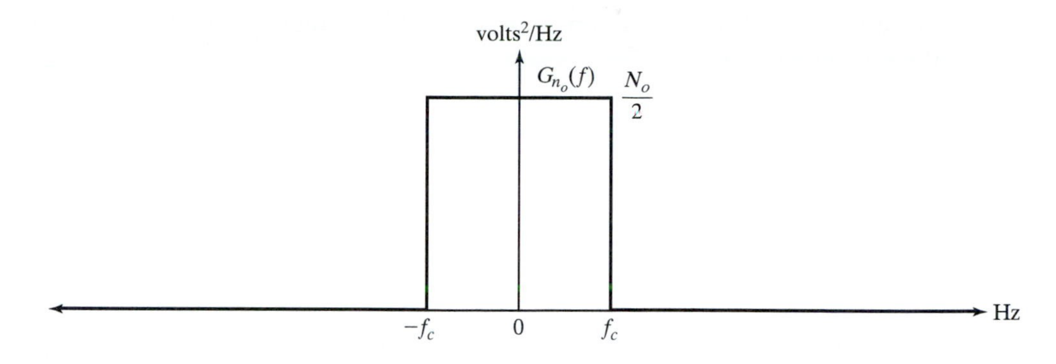

Figure 4-5 Average normalized power spectrum of thermal noise after lowpass filtering at the receiver.

where $\mu_o = 0$ and σ_o^2 = average normalized noise power at the filter's output. We can express σ_o^2 as

$$\sigma_o^2 = \int_{-\infty}^{\infty} |G_{n_o}(f)|^2\, df = \int_{-f_c}^{f_c} \frac{N_o}{2}\, df = N_o f_c \tag{4.30}$$

In examining Figures 4-4 and 4-5, note that σ_o^2 (the average normalized power of the noise at the filter output) is significantly smaller than σ_n^2 (the average normalized power of the thermal noise before filtering). (Remember that the average normalized power of a signal is the area under the curve of its power spectrum.) This concept is important because the smaller the value of σ_o^2, the smaller the value of the integral in Equation (4.29); thus the smaller the probability that the noise will exceed a given threshold a. Such a probability, as we have seen in Equation (4.7), is directly related to the probability that the transmitted signal is received in error.

Let's now continue from Equation (4.7), where we developed a mathematical expression for the probability that the ith transmitted bit is demodulated in error.

$$P\{i\text{th bit is demodulated in error}\}$$
$$= \frac{1}{2}P\left\{n_o\left[(i-1)T_b + \frac{T_b}{2}\right] < -\gamma A\right\} + \frac{1}{2}P\left\{n_o\left[(i-1)T_b + \frac{T_b}{2}\right] \geq \gamma A\right\}$$
$$= \frac{1}{2}P\{n_o(t_o) < -\gamma A\} + \frac{1}{2}P\{n_o(t_o) \geq \gamma A\} \tag{4.31}$$

for any time t_o since the noise is stationary.

From Equation (4.29), we know

$$P\{n_o(t_o) > \gamma A\} = \int_{\gamma A}^{\infty} \frac{1}{\sqrt{2\pi}\sigma_o} e^{-\left(\frac{(x-\mu_o)^2}{2\sigma_o^2}\right)} dx \tag{4.32a}$$

Since the probability density function of $n_o(t_o)$ is continuous and does not contain any discrete terms,

$$P\{n_o(t_o) \geq \gamma A\} = \int_{\gamma A}^{\infty} \frac{1}{\sqrt{2\pi}\sigma_o} e^{-\left(\frac{(x-\mu_o)^2}{2\sigma_o^2}\right)} dx \tag{4.32b}$$

Additionally, by inspecting Figure 4-3, we see that the Gaussian probability density function is symmetric about its mean, so

$$\begin{aligned} P\{n_o(t_o) < -\gamma A\} &= \int_{-\infty}^{-\gamma A} \frac{1}{\sqrt{2\pi}\sigma_o} e^{-\left(\frac{(x-\mu_o)^2}{2\sigma_o^2}\right)} dx = \int_{\gamma A}^{\infty} \frac{1}{\sqrt{2\pi}\sigma_o} e^{-\left(\frac{(x-\mu_o)^2}{2\sigma_o^2}\right)} dx \\ &= P(n_o(t_o) > \gamma A) \end{aligned} \tag{4.33}$$

Substituting into Equation (4.31),

$$\begin{aligned} &P\{i\text{th bit is demodulated in error}\} \\ &= \frac{1}{2}P\left\{n_o\left[(i-1)T_b + \frac{T_b}{2}\right] < -\gamma A\right\} + \frac{1}{2}P\left\{n_o\left[(i-1)T_b + \frac{T_b}{2}\right] \geq \gamma A\right\} \\ &= \frac{1}{2}P\{n_o(t_o) < -\gamma A\} + \frac{1}{2}P\{n_o(t_o) \geq \gamma A\} \\ &= \int_{\gamma A}^{\infty} \frac{1}{\sqrt{2\pi}\sigma_o} e^{-\left(\frac{(x-\mu_o)^2}{2\sigma_o^2}\right)} dx \end{aligned} \tag{4.34}$$

$P\{i\text{th bit is demodulated in error}\}$ is often called *probability of bit error* and is denoted as P_b.

Let's now work an example to calculate the probability of bit error.

Example 4.5 Probability of bit error for a simple PAM receiver

A transmitter uses raised cosine pulse shaping with pulse amplitudes of +3 volts and −3 volts. By the time the signal arrives at the receiver, the received signal voltage has been attenuated to half of the transmitted signal voltage and the signal has been corrupted with additive white Gaussian noise. The average normalized noise power at the output of the receiver's filter is 0.36 volts^2. Find P_b assuming perfect synchronization.

Solution

A = Magnitude of transmitted signal at the center of the bit period = 3 volts
γ = Fraction of the transmitted signal's voltage that arrives at the receiver = 0.5
σ_o^2 = Average normalized noise power = 0.36 volts^2

By Equation (4.34),

$$\begin{aligned} &P\{i\text{th bit is demodulated in error}\} \\ &= \int_{\gamma A}^{\infty} \frac{1}{\sqrt{2\pi}\sigma_o} e^{-\left(\frac{(x-\mu_o)^2}{2\sigma_o^2}\right)} dx = \int_{1.5}^{\infty} \frac{1}{\sqrt{2\pi}(0.6)} e^{-\left(\frac{(x-0)^2}{2(0.36)}\right)} dx \end{aligned}$$

We cannot solve this integral in closed form, but there are numerical techniques we can employ. Let's leave this example momentarily and focus on a general numerical way to solve the integral of the Gaussian probability density function.

4.1.4 The Gaussian Probability Density Function

As we have established, the Gaussian probability density function can be expressed as

$$f_{\mathbf{X}}(x) = \frac{1}{\sqrt{2\pi}\sigma} e^{\frac{-(x-\mu)^2}{2\sigma^2}} dx \qquad (4.19R)$$

Thus, if a random variable can be described using the Gaussian probability density function,

$$P\{\text{Value of Gaussian random variable} > a\} = \int_a^\infty \frac{1}{\sqrt{2\pi}\sigma} e^{\frac{-(x-\mu)^2}{2\sigma^2}} dx \qquad (4.28R)$$

Many random occurrences can be expressed using the Gaussian function, with the value of the mean and the value of the variance dependent on the particular phenomenon (the Gaussian function is the familiar "bell curve"). For example, thermal noise can be described using the Gaussian function with zero mean and a variance equal to the average normalized noise power, while IQ results can be described using the Gaussian function with a mean of 100 and a variance of 256. Often, σ (called *standard deviation*) is used to describe the function rather than variance (σ^2). Figures 4-6a–d show Gaussian probability densities for four different combinations of μ and σ. Notice that changing the value of μ corresponds to a shift along the x axis, and that changing the value of σ affects the "spread" of the function. Since σ is the square root of variance, it makes sense that larger values of σ should produce greater "spread," or variability.

We can see from Example 4.5 and the previous discussion that evaluating the integral in Equation (4.28R), which is related to the Gaussian probability distribution function, is extremely useful. As we discussed in Example 4.5, the integral cannot be solved in closed form and thus numerical solutions must be applied. Since we don't want to use a computer program every time we need to solve this integral, it will be valuable to create a table to store the value of the integral for various values of a, μ, and σ. Unfortunately, any such table with three independent variables would be extremely unwieldy (think of all the possible combinations of a, μ, and σ you would have to store). Let's consider instead the following approach:

1. Create a single table showing the values of the integral for various values of a with $\mu = 0$ and $\sigma = 1$. Such a table would correspond to the integral

$$\int_a^\infty \frac{1}{\sqrt{2\pi}(1)} e^{\frac{-(x-0)^2}{2(1)^2}} dx = \int_a^\infty \frac{1}{\sqrt{2\pi}} e^{\frac{-x^2}{2}} dx \qquad (4.35a)$$

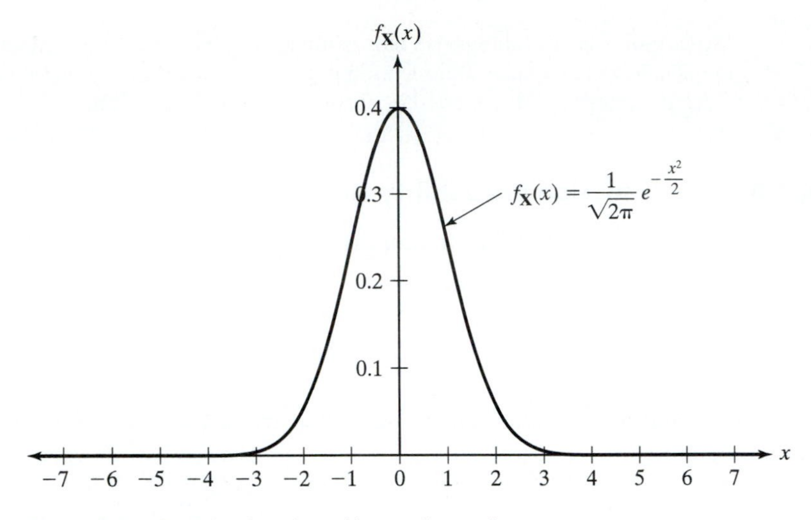

Figure 4-6a Gaussian function with $\mu = 0, \sigma = 1$.

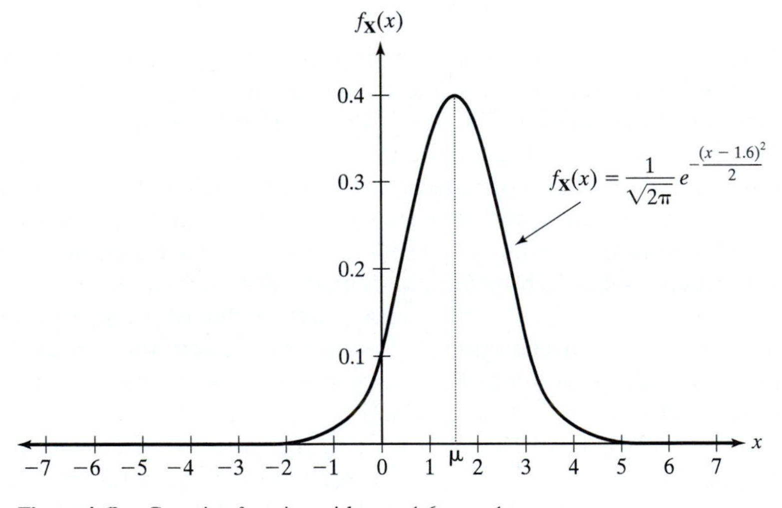

Figure 4-6b Gaussian function with $\mu = 1.6, \sigma = 1$.

which we will define as $Q(a)$. Thus,

$$Q(a) \equiv \int_a^\infty \frac{1}{\sqrt{2\pi}(1)} e^{\frac{-(x-0)^2}{2(1)^2}} dx = \int_a^\infty \frac{1}{\sqrt{2\pi}} e^{\frac{-x^2}{2}} dx \tag{4.35b}$$

The $Q(a)$ integral is shown graphically in Figure 4-7. Table 4-1 shows the values of $Q(a)$ for various values of a. The first two digits of a are found in the left column of the table. After selecting the row corresponding to the first two digits of a, the column corresponding to the third significant digit of a is found, thereby providing the value of $Q(a)$. For example, $Q(0.53) = 0.2981$. For $a \geq 3$ (which corresponds

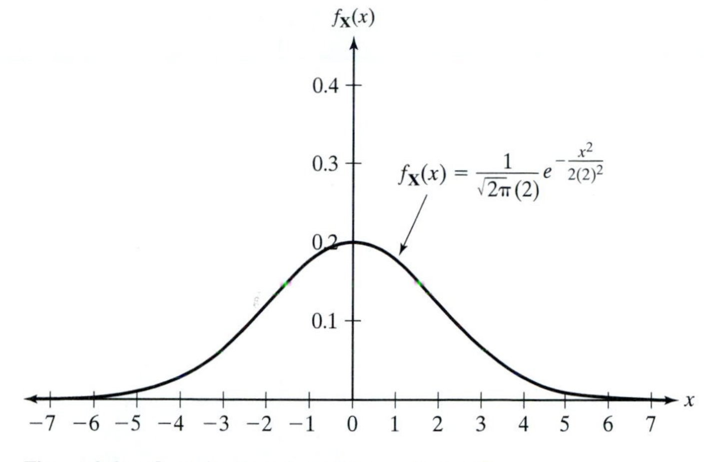

Figure 4-6c Gaussian function with $\mu = 0, \sigma = 2$.

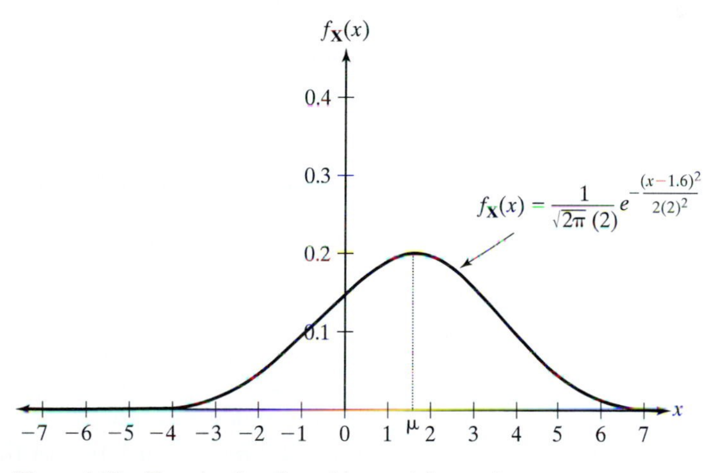

Figure 4-6d Gaussian function with $\mu = 1.6, \sigma = 2$.

to $Q(a) \leq .0014$), we can approximate $Q(a)$ as

$$Q(a) \cong \frac{1}{a\sqrt{2\pi}} e^{\frac{-a^2}{2}} \text{ for } a \geq 3 \tag{4.36}$$

With Table 4-1 and Equation (4.36), we can solve the integral in Equation (4.28*R*) for any value of a as long as $\mu = 0$ and $\sigma = 1$. The two small pairs of columns at the bottom of the table give the value of a for certain small values of $Q(a)$. These columns will be useful for system design. Calculators with statistical functions may also have the Q function.

2. What if μ is not equal to zero? Consider the integral

$$\int_a^\infty \frac{1}{\sqrt{2\pi}} e^{\frac{-(x-\mu)^2}{2}} dx \tag{4.37}$$

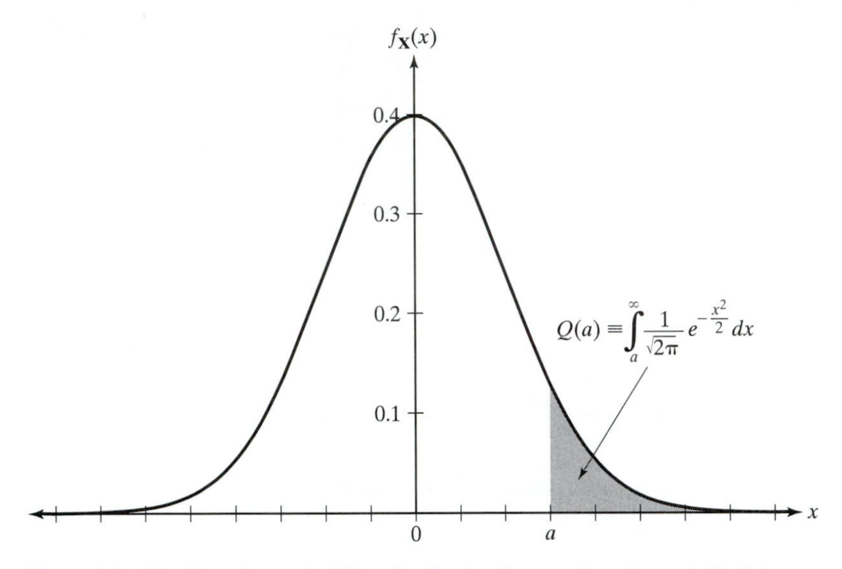

Figure 4-7 The Gaussian probability distribution for $\mu = 0$ and $\sigma = 1$ (defining the Q function).

By change of variable, let $w = x - \mu$. Performing this change of variable causes $dw = dx$ and changes the lower limit of the integral from a to $a - \mu$. Thus,

$$\int_a^\infty \frac{1}{\sqrt{2\pi}} e^{\frac{-(x-\mu)^2}{2}} dx = \int_{a-\mu}^\infty \frac{1}{\sqrt{2\pi}} e^{\frac{-w^2}{2}} dw \tag{4.38}$$

and the integral on the right side of Equation (4.38) is equivalent to $Q(a - \mu)$. This equivalency is shown geometrically in Figures 4-8a and b. The change in variable in Equation (4.38) corresponds geometrically to sliding the Gaussian probability density function to the left by μ. Our table for the Q function is thus useful for solving the integral of a Gaussian function for any value of a and any value of μ, as long as $\sigma = 1$.

3. What if σ is not equal to 1? Consider the integral

$$\int_a^\infty \frac{1}{\sqrt{2\pi}\sigma} e^{\frac{-(x-\mu)^2}{2\sigma^2}} dx \tag{4.39}$$

As in Step 2, we can perform a change of variable and let $w = x - \mu$. Performing this change of variable,

$$\int_a^\infty \frac{1}{\sqrt{2\pi}\sigma} e^{\frac{-(x-\mu)^2}{2\sigma^2}} dx = \int_{a-\mu}^\infty \frac{1}{\sqrt{2\pi}\sigma} e^{\frac{-w^2}{2\sigma^2}} dw \tag{4.40}$$

Now let's perform a second change of variable, letting $z = w/\sigma$. Performing this change of variable causes $dz = dw/\sigma$ and changes the lower limit of the integral from $a - \mu$ to $(a - \mu)/\sigma$.

Table 4-1 The Q Function (Gaussian Distribution with $\mu = 0$ and $\sigma = 1$)

$$Q(a) = \int_a^\infty \frac{1}{\sqrt{2\pi}} e^{\frac{-x^2}{2}} dx \qquad Q(a) \cong \frac{1}{a\sqrt{2\pi}} e^{\frac{-a^2}{2}} \text{ for } a \geq 3$$

	Third Significant Digit									
a	**0.00**	**0.01**	**0.02**	**0.03**	**0.04**	**0.05**	**0.06**	**0.07**	**0.08**	**0.09**
0.0	0.5000	0.4960	0.4920	0.4880	0.4840	0.4801	0.4761	0.4721	0.4681	0.4641
0.1	0.4602	0.4562	0.4522	0.4483	0.4443	0.4404	0.4364	0.4325	0.4286	0.4247
0.2	0.4207	0.4168	0.4129	0.4090	0.4052	0.4013	0.3974	0.3936	0.3897	0.3859
0.3	0.3821	0.3783	0.3745	0.3707	0.3669	0.3632	0.3594	0.3557	0.3520	0.3483
0.4	0.3446	0.3409	0.3372	0.3336	0.3300	0.3264	0.3228	0.3192	0.3156	0.3121
0.5	0.3085	0.3050	0.3015	0.2981	0.2946	0.2912	0.2877	0.2843	0.2810	0.2776
0.6	0.2743	0.2709	0.2676	0.2643	0.2611	0.2578	0.2546	0.2514	0.2483	0.2451
0.7	0.2420	0.2389	0.2358	0.2327	0.2297	0.2266	0.2236	0.2207	0.2177	0.2148
0.8	0.2119	0.2090	0.2061	0.2033	0.2005	0.1977	0.1949	0.1922	0.1894	0.1867
0.9	0.1841	0.1814	0.1788	0.1762	0.1736	0.1711	0.1685	0.1660	0.1635	0.1611
1.0	0.1587	0.1562	0.1539	0.1515	0.1492	0.1469	0.1446	0.1423	0.1401	0.1379
1.1	0.1357	0.1335	0.1314	0.1292	0.1271	0.1251	0.1230	0.1210	0.1190	0.1170
1.2	0.1151	0.1131	0.1112	0.1094	0.1075	0.1057	0.1038	0.1020	0.1003	0.0985
1.3	0.0968	0.0951	0.0934	0.0918	0.0901	0.0885	0.0869	0.0853	0.0838	0.0823
1.4	0.0808	0.0793	0.0778	0.0764	0.0749	0.0735	0.0721	0.0708	0.0694	0.0681
1.5	0.0668	0.0655	0.0643	0.0630	0.0618	0.0606	0.0594	0.0582	0.0571	0.0559
1.6	0.0548	0.0537	0.0526	0.0516	0.0505	0.0495	0.0485	0.0475	0.0465	0.0455
1.7	0.0446	0.0436	0.0427	0.0418	0.0409	0.0401	0.0392	0.0384	0.0375	0.0367
1.8	0.0359	0.0351	0.0344	0.0336	0.0329	0.0322	0.0314	0.0307	0.0301	0.0294
1.9	0.0287	0.0281	0.0274	0.0268	0.0262	0.0256	0.0250	0.0244	0.0239	0.0233
2.0	0.0228	0.0222	0.0217	0.0212	0.0207	0.0202	0.0197	0.0192	0.0188	0.0183
2.1	0.0179	0.0174	0.0170	0.0166	0.0162	0.0158	0.0154	0.0150	0.0146	0.0143
2.2	0.0139	0.0136	0.0132	0.0129	0.0125	0.0122	0.0119	0.0116	0.0113	0.0110
2.3	0.0107	0.0104	0.0102	0.0099	0.0096	0.0094	0.0091	0.0089	0.0087	0.0084
2.4	0.0082	0.0080	0.0078	0.0075	0.0073	0.0071	0.0069	0.0068	0.0066	0.0064
2.5	0.0062	0.0060	0.0059	0.0057	0.0055	0.0054	0.0052	0.0051	0.0049	0.0048
2.6	0.0047	0.0045	0.0044	0.0043	0.0041	0.0040	0.0039	0.0038	0.0037	0.0036
2.7	0.0035	0.0034	0.0033	0.0032	0.0031	0.0030	0.0029	0.0028	0.0027	0.0026
2.8	0.0026	0.0025	0.0024	0.0023	0.0023	0.0022	0.0021	0.0021	0.0020	0.0019
2.9	0.0019	0.0018	0.0018	0.0017	0.0016	0.0016	0.0015	0.0015	0.0014	0.0014
3.0	0.0014	0.0013	0.0013	0.0012	0.0012	0.0011	0.0011	0.0011	0.0010	0.0010
3.1	0.0010	0.0009	0.0009	0.0009	0.0008	0.0008	0.0008	0.0008	0.0007	0.0007
3.2	0.0007	0.0007	0.0006	0.0006	0.0006	0.0006	0.0006	0.0005	0.0005	0.0005
3.3	0.0005	0.0005	0.0005	0.0004	0.0004	0.0004	0.0004	0.0004	0.0004	0.0003
3.4	0.0003	0.0003	0.0003	0.0003	0.0003	0.0003	0.0003	0.0003	0.0003	0.0002

$Q(a)$	a
10^{-4}	3.73
5×10^{-5}	3.90
10^{-5}	4.27
5×10^{-6}	4.43

$Q(a)$	a
10^{-6}	4.76
10^{-7}	5.20
10^{-8}	5.61
10^{-9}	6.00

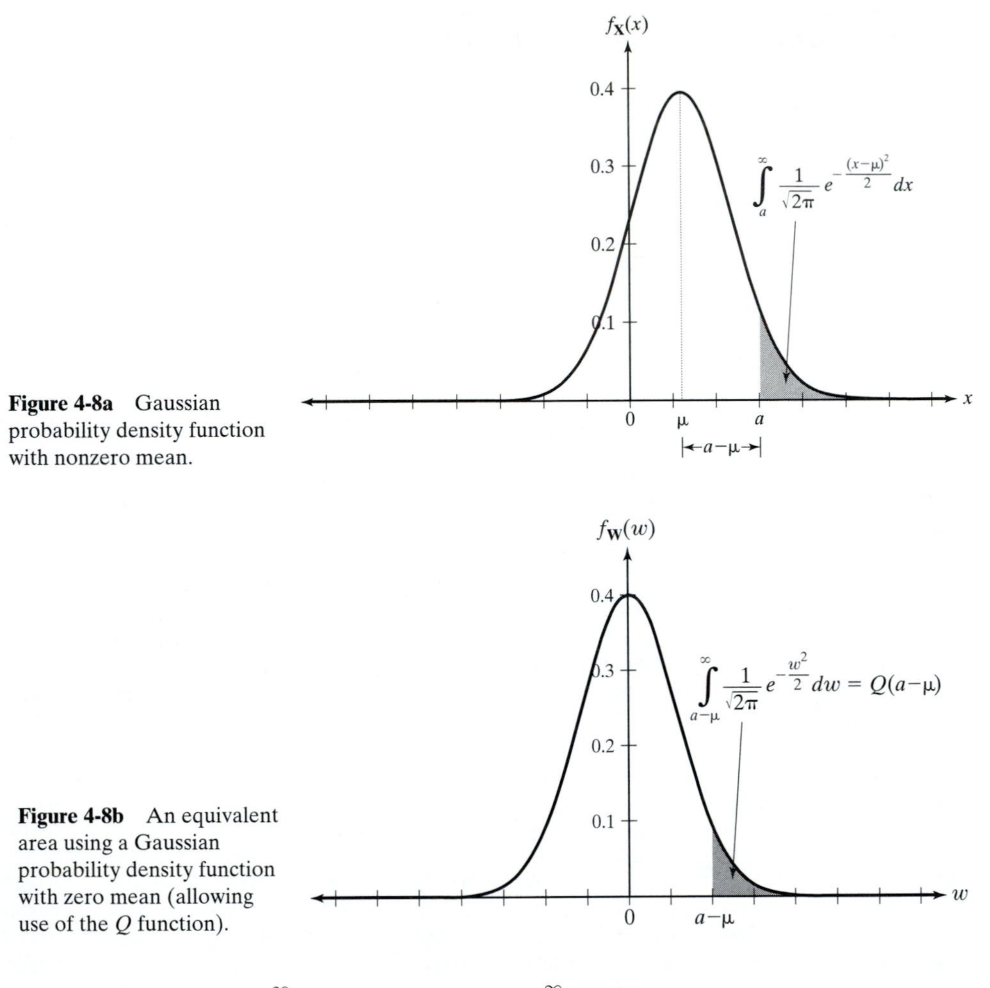

Figure 4-8a Gaussian probability density function with nonzero mean.

Figure 4-8b An equivalent area using a Gaussian probability density function with zero mean (allowing use of the Q function).

$$\int_{a-\mu}^{\infty} \frac{1}{\sqrt{2\pi}\sigma} e^{\frac{-w^2}{2\sigma^2}} dw = \int_{\frac{a-\mu}{\sigma}}^{\infty} \frac{1}{\sqrt{2\pi}} e^{\frac{-Z^2}{2}} dz \tag{4.41}$$

Thus, combining Equations (4.40) and (4.41):

$$\int_{a}^{\infty} \frac{1}{\sqrt{2\pi}\sigma} e^{\frac{-(x-\mu)^2}{2\sigma^2}} dx = \int_{\frac{a-\mu}{\sigma}}^{\infty} \frac{1}{\sqrt{2\pi}} e^{\frac{-z^2}{2}} dz = Q\left(\frac{a-\mu}{\sigma}\right) \tag{4.42}$$

and our table for the Q function can now be used to solve Equation (4.28) for any values of a, μ, and σ. The equivalency in Equation (4.40) is shown geometrically in Figures 4-9a and 4-9b; the equivalency in Equation (4.41) is shown geometrically in Figures 4-9b and 4-9c. Note that the combination of expanding the y axis by σ and compressing the x axis by σ in Figure 4-9c produces no net effect on the total shaded area under the curve.

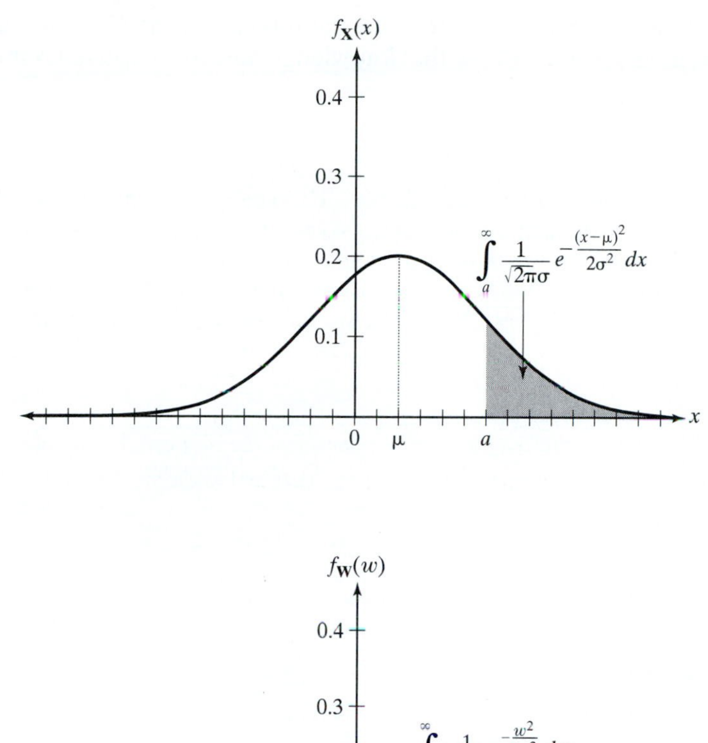

Figure 4-9a Gaussian probability density function with nonzero mean and $\sigma \neq 1$.

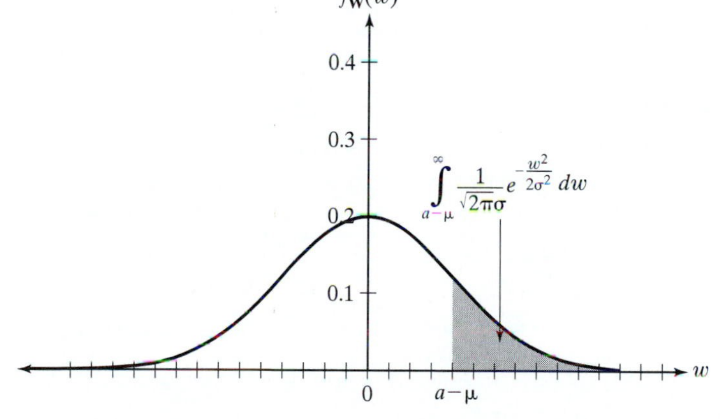

Figure 4-9b An equivalent area using a Gaussian probability density function with zero mean and $\sigma \neq 1$.

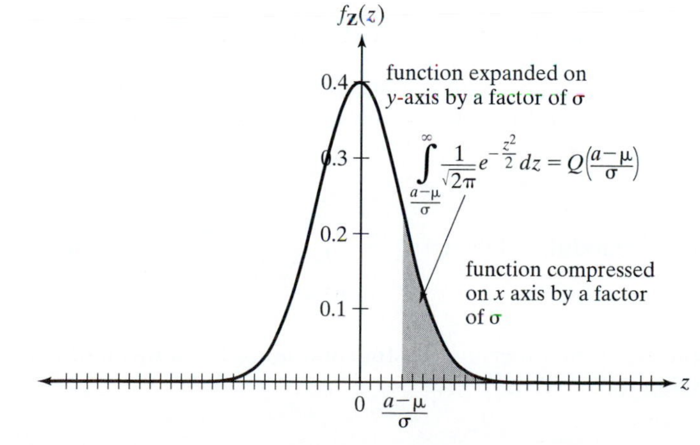

Figure 4-9c An equivalent area using a Gaussian probability density function with zero mean and $\sigma = 1$ (allowing use of the Q function). The combination of expanding the function along the y axis by σ and compressing the function along the x axis by σ produces no net effect on the shaded area.

We now see how to use Table 4-1 to solve the integral of the Gaussian function for any values of a, μ, and σ. We can use that knowledge now to complete Example 4.5.

Example 4.5 (revisited) Probability of bit error for a simple PAM receiver

A transmitter uses raised cosine pulse shaping with pulse amplitudes of +3 volts and −3 volts. By the time the signal arrives at the receiver, the received signal voltage has been attenuated to half of the transmitted signal voltage and the signal has been corrupted with additive white Gaussian noise. The average normalized noise power at the output of the receiver's filter is 0.36 volts². Find P_b assuming perfect synchronization.

Solution

A = Magnitude of transmitted signal at the center of the bit period = 3 volts
γ = Fraction of the transmitted signal's voltage that arrives at the receiver = 0.5
σ_o^2 = Average normalized noise power = 0.36 volts²

By Equations (4.34) and (4.42),

$$P\{i\text{th bit is demodulated in error}\} = \int_{\gamma A}^{\infty} \frac{1}{\sqrt{2\pi}\sigma_o} e^{-\left(\frac{(x-\mu_o)^2}{2\sigma_o^2}\right)} dx$$

$$= \int_{1.5}^{\infty} \frac{1}{\sqrt{2\pi}(0.6)} e^{-\left(\frac{(x-0)^2}{2(0.36)}\right)} dx$$

$$= Q\left(\frac{1.5 - 0}{0.6}\right) = Q(2.50) = .0062$$

Thus, using the Q table we can determine that the probability that the ith bit is demodulated in error is .0062, or 0.62%.

4.1.5 Simplifying the Expression for Probability of Bit Error

Let's now return to our original task: establishing the probability that the PAM receiver shown in Figure 4-1 would demodulate the ith bit of a stream of data in error. In Equation (4.34) we established that

$$P\{i\text{th bit is demodulated in error}\} = \int_{\gamma A}^{\infty} \frac{1}{\sqrt{2\pi}\sigma_o} e^{-\left(\frac{(x-\mu_o)^2}{2\sigma_o^2}\right)} dx \qquad (4.34R)$$

Using the Q function, and remembering that thermal noise has a mean of zero, we can simplify this expression to:

$$P\{i\text{th bit is demodulated in error}\} = \int_{\gamma A}^{\infty} \frac{1}{\sqrt{2\pi}\sigma_o} e^{-\left(\frac{(x-\mu_o)^2}{2\sigma_o^2}\right)} dx = Q\left(\frac{\gamma A}{\sigma_o}\right) \quad (4.43)$$

We can also provide a good, physical interpretation of Equation (4.43) as follows:

$$\begin{aligned} &P\{i\text{th bit is demodulated in error}\} \\ &= \int_{\gamma A}^{\infty} \frac{1}{\sqrt{2\pi}\sigma_o} e^{-\left(\frac{(x-\mu_o)^2}{2\sigma_o^2}\right)} dx = Q\left(\frac{\gamma A}{\sigma_o}\right) \\ &= Q\left(\frac{\text{noise margin of sampled value}}{\sigma_o}\right) \qquad (4.44) \\ &= Q\left(\frac{\text{noise margin of sampled value}}{\sqrt{\text{avg. normalized noise power at filter output}}}\right) \end{aligned}$$

Does Equation (4.44) make intuitive sense? Consider the following:

1. We know that the larger the value of a, the smaller the value of $Q(a)$. Thus, increasing the argument of the Q function in Equation (4.44) decreases the probability of bit error.
2. The argument of the Q function in Equation (4.44) is a *ratio* of the received signal strength (noise margin) versus noise (σ_o). The larger this *signal-to-noise ratio* (SNR), the lower the probability of error. This concept of probability of error being related to SNR makes inherent sense. It isn't just the strength of the received signal that's important, it's the strength of the received signal *relative to the noise*. For example, if you and I are at a football game and we're seated ten feet apart, you may not be able to understand what I'm saying even if I yell in your direction, whereas if you and I are in a classroom, we may be 30 feet apart and yet you can hear me when I speak in a normal tone.
3. We can now fully understand the purpose of the lowpass filter in the PAM receiver. The filter reduces the average normalized power of the noise (revisit Figures 4-1, 4-4, and 4-5) while not impacting the strength of the received signal (since the filter's bandwidth is compatible with the bandwidth of the signal). The filter thus increases the SNR, reducing the probability of bit error.
4. The derivation for Equation (4.44) can be performed in a similar manner even if the receiver is not perfectly synchronized (i.e., even if the receiver does not sample the signal exactly in the center of the bit period). The only difference is that the value of the noise margin will be different, as we observed in Chapter 3.

Be sure you understand Equation (4.44) and its derivation.

A few final comments concerning noise and probability of bit error. First, the value of the noise during each sample is unrelated to the value of the noise at any other sample (because thermal noise is uncorrelated). Therefore, P{*i*th bit is demodulated in error} = $P\{i + 1\text{st bit is demodulated in error}\}$, etc. This observation results in two different but equivalent interpretations for P_b:

1. P_b can be interpreted as the probability that a particular transmitted bit is demodulated in error.
2. P_b can also be interpreted as the percentage of all transmitted bits, which—on average—will be demodulated in error.

For example, if $P_b = 0.1$, then there is a 10% probability that a particular bit will be received in error (using interpretation 1) and, on average, 10 out of every 100 received bits will be demodulated in error (using interpretation 2).

Second, suppose that for a particular system $P_b = 10\%$. Even though this means that, on average, 90 bits out of every 100 will be demodulated correctly, we don't know *which* 90 bits out of the 100 are the correct bits. Thus, a system that exhibits a 10% probability of bit error is *not* 90% trustworthy. Suppose, for instance, that you have a friend who lies to you 10% of the time. Would you trust that friend, even though 90% of the time that friend is telling you the truth? Thus, even systems that exhibit seemingly small probabilities of error may be too unreliable for many applications.

Let's now work some more examples to calculate the probability of bit error.

Example 4.6 Accuracy of a PAM system using raised cosine pulses and a simple receiver

A transmitter uses raised cosine pulse shaping with pulse amplitudes of +1 volt and −1 volt. By the time the signal arrives at the receiver, the received signal voltage has been attenuated to half of the transmitted signal voltage and the signal has been corrupted with additive white Gaussian noise. The average normalized noise power at the output of the receiver filter is 0.035 volts^2. Find P_b assuming perfect synchronization.

Solution

A = Magnitude of transmitted signal at the center of the bit period = 1 volt

γ = Fraction of the transmitted signal's voltage that arrives at the receiver = 0.5

σ_o^2 = Average normalized noise power at output of the receiver filter = 0.035 volts^2

$$P_b = Q\left(\frac{\text{noise margin}}{\sigma_o}\right) = Q\left(\frac{\gamma A}{\sqrt{0.035}}\right) = Q\left(\frac{(0.5)(1)}{\sqrt{0.035}}\right) = Q(2.67) = 0.0038$$

Thus, on average, 0.38% of the transmitted bits will be received in error (or, using a different interpretation, on average, 3.8 out of every thousand bits will be received in error).

Example 4.7 Relationship of rolloff factor to accuracy

For raised cosine pulse shaping and a simple PAM receiver, does P_b depend on rolloff factor? Why or why not?

Solution

For raised cosine pulse shaping, P_b will not depend on rolloff factor if the receiver is perfectly synchronized, because there is no ISI in the exact center of a bit period. On the other hand, P_b will depend on rolloff factor if the receiver is not perfectly synchronized (we'll see this shortly in Example 4.10).

Example 4.8 Attenuation expressed in terms of signal power

Repeat Example 4.6 if the channel attenuates *power* such that the average normalized power of the received signal is half the average normalized power of the transmitted signal.

Solution

Normalized received power = 0.5 (normalized transmitted power). Since normalized power is proportional to the square of amplitude,

$$\text{received amplitude} = (\sqrt{0.5})\ (\text{transmitted amplitude})$$

$$\gamma = \sqrt{0.5} = 0.707$$

$$P_b = Q\left(\frac{\text{noise margin}}{\sigma_o}\right) = Q\left(\frac{\gamma A}{\sigma_o}\right) = Q\left(\frac{(0.707)(1)}{\sqrt{0.035}}\right) = Q(3.78)$$

The value 3.78 is too large for our Q table, so we must use the approximation in Equation (4.36)

$$Q(3.78) \cong \frac{1}{3.78\sqrt{2\pi}} e^{\frac{-3.78^2}{2}} = 8.33 \times 10^{-5}$$

Let's now try a more design-oriented problem.

Example 4.9 Designing a PAM receiver for a given level of accuracy

A transmitter uses raised cosine pulse shaping to transmit a signal. By the time the signal arrives at the receiver, its voltage is only 40% of the transmitted voltage. The average normalized noise power at the output of the receiver filter is 0.5 volts^2. Assuming that the receiver is perfectly synchronized, find the minimum pulse amplitude of the transmitted signal required to guarantee a probability of bit error less than or equal to 0.1%.

Solution

$$P_b \le 0.1\% = 0.001 \quad P_b = Q\left(\frac{\text{noise margin}}{\sigma_o}\right) = Q\left(\frac{\gamma A}{\sigma_o}\right) = Q\left(\frac{0.4A}{\sqrt{0.5}}\right) = Q(0.566A)$$

Thus, $P_b = Q(0.566A) \le 0.001$.

From Table 4-1, we see that $Q(3.08) = 0.001$, so we need $0.566A \ge 3.08$ (be sure you see why). Solving, $A \ge 3.08/0.566 = 5.44$ volts. Do you see how this is a design problem?

Let's now examine how badly accuracy is deteriorated by lack of synchronization.

Example 4.10 Impact of loss of synchronization

Reconsider Example 3.9. We transmitted the data pattern "10010" at 50,000 bits/sec using PAM with raised cosine pulse shaping of optimum width and a rolloff factor of $\alpha = 0$. Pulse amplitudes were +1 volt and −1 volt, and the signal was transmitted over a channel that produced a 50% attenuation in voltage. Let's now assume that the average normalized noise power at the output of the receiver filter is $\sigma_o^2 = 0.03$ volts^2 (additive white Gaussian noise).

1. Determine the probability of bit error for the third bit for each of the following cases:
   a. Receiver is perfectly synchronized
   b. Receiver is out of synchronization by +2 μsec
   c. Receiver is out of synchronization by +4 μsec
   d. Receiver is out of synchronization by +6 μsec
2. Repeat Part 1, except use a rolloff factor of $\alpha = 0.5$
3. Repeat Part 1, except use a rolloff factor of $\alpha = 1.0$

Solution:

In Example 3.9 in Chapter 3, we determined the noise margins for all of the above cases. The results are repeated in Table 3-3*R*.

Table 3-3*R* Noise Margin for Third Bit of Received Signal Using Raised Cosine Pulse Shaping with Various Rolloff Factors

Receiver Synchronization	Rolloff Factor (in Volts)		
	$\alpha = 0$	$\alpha = 0.5$	$\alpha = 1.0$
Perfect	0.5	0.5	0.5
+2 μsec out of sync.	0.34	0.40	0.46
+4 μsec out of sync.	0.18	0.28	0.38
+6 μsec out of sync.	0.014	0.14	0.28

Using Equation (4.44), we can determine the probability of error for the third bit in each of the above cases (see Table 4-2).

Table 4-2 Probability of Error for Third Bit of Received Signal Using Raised Cosine Pulse Shaping with Various Rolloff Factors

Receiver Synchronization	Rolloff Factor		
	$\alpha = 0$	$\alpha = 0.5$	$\alpha = 1.0$
Perfect	.0019	.0019	.0019
+2 μsec out of sync.	.0250	.0104	.0039
+4 μsec out of sync.	.1492	.0526	.0143
+6 μsec out of sync.	.4681	.2090	.0526

Be sure you understand how the values in Table 4-2 were determined.

Let's now interpret the values in Table 4-2. Looking down a particular column, we can see how loss of synchronization affects accuracy. For instance, suppose we are transmitting raised cosine pulses with a rolloff factor of $\alpha = 0.5$. Being 2 μsec out of synchronization increases the probability of bit error by a factor of 5; being 4 μsec out of synchronization increases probability of error by a factor of 27. Looking across a particular row, we can see how increasing the rolloff factor reduces the probability of error, given a certain lack of synchronization. For instance, suppose the receiver is 4 μsec out of synchronization. Using a rolloff factor of $\alpha = 1.0$ reduces the probability of error by more than a factor of 10 relative to using the sinc-shaped pulse ($\alpha = 0$). As we established in Chapter 3, the cost for this improved accuracy is a doubling of bandwidth.

Note that accuracy requirements for most systems are much more stringent than in the above examples. For instance, most digital voice systems (such as cellular telephones) require a probability of bit error of 10^{-3} or less, while most data systems require a probability of bit error of 10^{-5} to 10^{-7} or less (although much of this accuracy can be achieved using error control coding, which we will explore in Chapter 10). Thus, we often use Equation (4.36) rather than the Q table.

4.2 Building the Optimal Receiver (The Matched Filter or Correlation Receiver)

4.2.1 Basic Structure for the Optimal Receiver

In Figure 4-1, we established the following structure for our PAM receiver:

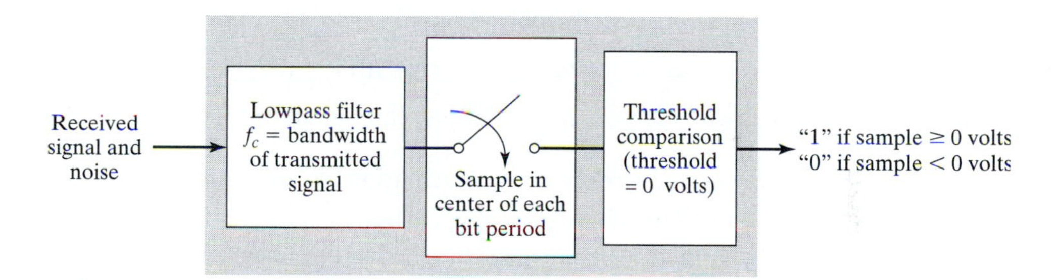

Figure 4-1R A simple PAM receiver.

Let's now investigate the possibility of building a receiver with even better performance. Intuitively we know that this may be possible. Remember in Example 1.1 and Figure 1-2 (the automated teller machine transmitting digital information over a telephone line) we discussed techniques for demodulating a signal that we felt might be more accurate than performing a single sampling at the center of the bit period.

Consider the general block diagram for a demodulation process shown in Figure 4-10.

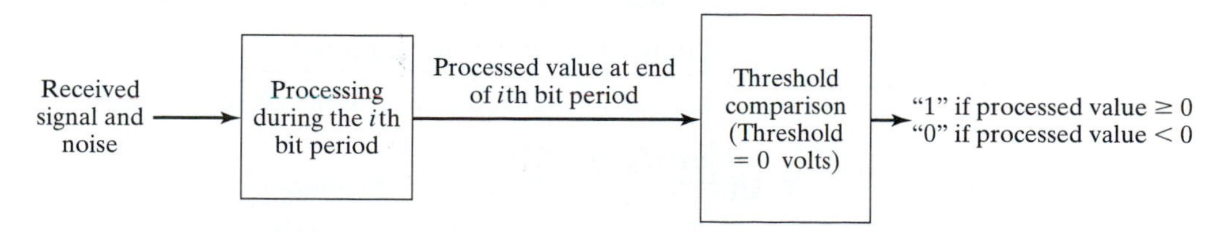

Figure 4-10 Block diagram for general demodulation process.

If we restrict our processing by requiring that it be linear, then the processing can be represented by a transfer function $H(f)$ and we can represent the demodulation process of Figure 4-10 with the receiver structure shown in Figure 4-11.

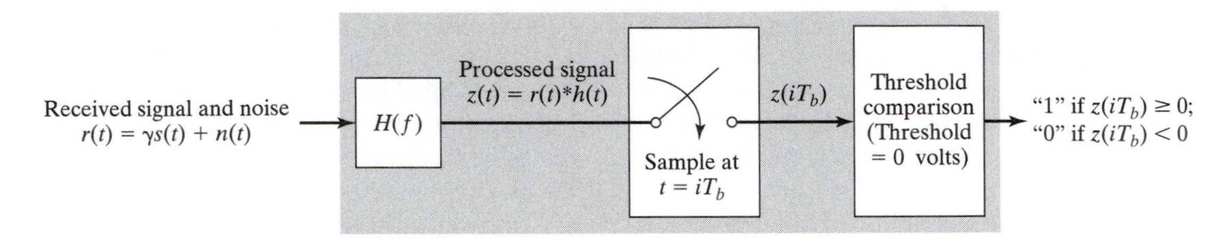

Figure 4-11 Receiver for implementing general demodulation process.

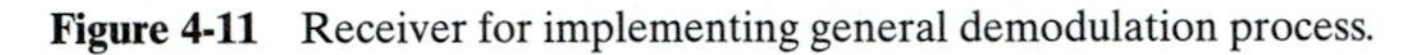

Reviewing the notation in Figure 4-11, which is similar to the notation in Section 4.1:

$r(t)$ is the received signal, which can be decomposed into a component due to the attenuated transmitted signal, $\gamma s(t)$, and a component due to noise, $n(t)$

$h(t)$ is the impulse response corresponding to the transfer function $H(f)$

$$h(t) = \mathcal{F}^{-1}[H(f)] \tag{4.45}$$

$z(t)$ is the processed received signal

$$z(t) = r(t)*h(t) = \gamma s(t)*h(t) + n(t)*h(t) \tag{4.46}$$

Let's also make two observations concerning the signals in Figure 4-11.

1. $n(t)$ has a Gaussian probability distribution, so $n(t)*h(t)$ also has a Gaussian probability distribution.
2. Processing of the received signal for the ith bit occurs during the time interval $(i - 1)T_b < t \leq iT_b$. At the end of this period, the processing produces a single value $z(iT_b)$, which is compared to the threshold to determine whether the ith bit is demodulated as a "1" or a "0."

For the simple PAM receiver shown in Figure 4-1, we determined

$$P_{b,\,simple\ receiver} = Q\left(\frac{\text{noise margin of sampled value}}{\sqrt{\text{avg. normalized noise power at filter output}}}\right) \tag{4.44R}$$

Remembering that $n(t)*h(t)$, the processed noise, exhibits a zero-mean Gaussian probability distribution, we can determine the probability of bit error for the receiver in Figure 4-11 by extending the derivation we used for Equation (4.44R).

$$P_{b,\,\text{Fig. 4-11}} = Q\left(\frac{\text{noise margin of sampled processed signal}}{\sqrt{\text{avg. normalized power of processed noise}}}\right) = Q\left(\frac{|\gamma s(iT_b)*h(iT_b)|}{\sigma_p}\right) \tag{4.47}$$

where σ_p^2 represents the average normalized power of the processed noise (and the variance of its probability distribution).

What type of processing will minimize the probability of bit error for the receiver in Figure 4-11? This is equivalent to asking what expression for $H(f)$ will optimize the receiver's accuracy. Minimizing $P_{b,\,\text{Fig. 4-11}}$ corresponds to maximizing the argument within the Q function in Equation (4.47)

$$\frac{|\gamma s(iT_b)*h(iT_b)|}{\sigma_p}$$

We know that σ_p is positive—it is related to the processed noise power. Thus, maximizing

$$\frac{|\gamma s(iT_b)*h(iT_b)|}{\sigma_p}$$

is equivalent to maximizing

$$\left(\frac{|\gamma s(iT_b)*h(iT_b)|}{\sigma_p}\right)^2$$

which can be expressed as

$$\frac{|\gamma s(iT_b)*h(iT_b)|^2}{\sigma_p^2}$$

Note that the quantities in the numerator and denominator have very basic intuitive meanings. The numerator represents the instantaneous normalized processed signal power at the end of the ith bit period, and the denominator represents the average normalized processed noise power. Thus, as we would intuitively believe, we minimize the probability of error when we maximize SNR.

So, how do we develop our process (i.e., specify $H(f)$) to maximize SNR? Let's start by expressing the processed received signal and processed noise in terms of $H(f)$:

$$\gamma s(t)*h(t) = \gamma \mathcal{F}^{-1}[S(f)H(f)] = \gamma \int_{-\infty}^{\infty} S(f)H(f)e^{j2\pi ft}\,df \tag{4.48}$$

Thus,

$$\gamma s(iT_b)*h(iT_b) = \gamma \int_{-\infty}^{\infty} S(f)H(f)e^{j2\pi fiT_b}\,df \tag{4.49}$$

and instantaneous processed signal power can be expressed as

$$|\gamma s(iT_b)*h(iT_b)|^2 = \gamma^2 \left| \int_{-\infty}^{\infty} S(f)H(f)e^{j2\pi fiT_b}\,df \right|^2 \tag{4.50}$$

For average processed noise power, we take a different approach. Let $G_{ni}(f)$ represent the average normalized noise power spectral density at the input to the receiver, and let $G_{np}(f)$ represent the power spectral density of the processed noise

$$\sigma_p^2 = \int_{-\infty}^{\infty} G_{np}(f)\,df = \int_{-\infty}^{\infty} G_{ni}(f)|H(f)|^2\,df = \int_{-\infty}^{\infty} \frac{N_o}{2}|H(f)|^2\,df \tag{4.51}$$

since the noise at the input to the receiver is additive white Gaussian noise. Using Equations (4.50) and (4.51),

$$\frac{|\gamma s(iT_b)*h(iT_b)|^2}{\sigma_p^2} = \frac{\gamma^2\left|\int_{-\infty}^{\infty} S(f)H(f)e^{j2\pi fiT_b}\,df\right|^2}{\int_{-\infty}^{\infty} \frac{N_o}{2}|H(f)|^2\,df} \tag{4.52}$$

As discussed earlier, we minimize the probability of error for the Figure 4-11 receiver when we maximize the argument of the Q function in Equation (4.47), which is equivalent to maximizing Equation (4.52). We thus want to find the transfer function $H(f)$ that maximizes Equation (4.52). Let's call this optimum transfer function $H_o(f)$.

How do we find $H_o(f)$? By Schwarz's inequality, we know

$$\left|\int_{-\infty}^{\infty} f_1(x)f_2(x)\,dx\right|^2 \le \int_{-\infty}^{\infty} |f_1(x)|^2\,dx \int_{-\infty}^{\infty} |f_2(x)|^2\,dx \tag{4.53}$$

with equality occurring only when $f_1(x) = kf_2{}^*(x)$ where * denotes complex conjugate and k is an arbitrary constant. Applying Schwarz's inequality to Equation (4.52),

$$\frac{\gamma^2\left|\int_{-\infty}^{\infty} S(f)H(f)e^{j2\pi fiT_b}\,df\right|^2}{\int_{-\infty}^{\infty} \frac{N_o}{2}|H(f)|^2\,df} \le \frac{\gamma^2\int_{-\infty}^{\infty} |S(f)e^{j2\pi fiT_b}|^2\,df \int_{-\infty}^{\infty} |H(f)|^2\,df}{\int_{-\infty}^{\infty} \frac{N_o}{2}|H(f)|^2\,df} \tag{4.54a}$$

$$= \frac{\gamma^2\int_{-\infty}^{\infty} |S(f)e^{j2\pi fiT_b}|^2\,df \int_{-\infty}^{\infty} |H(f)|^2\,df}{\frac{N_o}{2}\int_{-\infty}^{\infty} |H(f)|^2\,df} = \frac{2\gamma^2}{N_o}\int_{-\infty}^{\infty} |S(f)e^{j2\pi fiT_b}|^2\,df$$

Thus,

$$\frac{\gamma^2\left|\int_{-\infty}^{\infty} S(f)H(f)e^{j2\pi f iT_b}\,df\right|^2}{\int_{-\infty}^{\infty}\frac{N_o}{2}|H(f)|^2\,df} \le \frac{2\gamma^2}{N_o}\int_{-\infty}^{\infty}|S(f)e^{j2\pi f iT_b}|^2\,df \tag{4.54b}$$

with equality holding only when

$$H(f) = k[S(f)e^{j2\pi f iT_b}]^* \tag{4.55}$$

where * denotes complex conjugate and k is an arbitrary constant. The processed SNR

$$\frac{|\gamma s(iT_b)*h(iT_b)|^2}{\sigma_p^2}$$

is therefore maximized when

$$H(f) = k[S(f)e^{j2\pi f iT_b}]^*$$

a condition that can also be expressed as

$$H_o(f) = k[S(f)e^{j2\pi f iT_b}]^* = kS^*(f)e^{-j2\pi f iT_b} \tag{4.56}$$

where the o subscript denotes that the transfer function is "optimum."

Summarizing the above derivation, the receiver shown in Figure 4-11 is optimized (i.e., operates with the lowest probability of bit error) when the processing is as described by Equation (4.56).

4.2.2 Implications of Employing Optimum Processing

There are several interesting observations we can make if we use the receiver structure shown in Figure 4-11 and optimize the processing as described in Equation (4.56).

1. As discussed earlier, we minimize P_b when we maximize the argument inside the Q function in Equation (4.47). We maximize this argument by using the processing described in Equation (4.56), which makes Equation (4.54b) become an equality. Thus, when optimum processing is used,

$$\begin{aligned}\text{maximum}\,\frac{|\gamma s(iT_b)*h(iT_b)|^2}{\sigma_p^2} &= \frac{\gamma^2\left|\int_{-\infty}^{\infty} S(f)H_o(f)e^{j2\pi f iT_b}\,df\right|^2}{\int_{-\infty}^{\infty}\frac{N_o}{2}|H_o(f)|^2\,df} \\ &= \frac{2\gamma^2}{N_o}\int_{-\infty}^{\infty}\left|S(f)e^{j2\pi f iT_b}\right|^2 df = \frac{2E_b}{N_o}\end{aligned} \tag{4.57}$$

where $E_b = \gamma^2 \int_{-\infty}^{\infty} |S(f)e^{j2\pi fiT_b}|^2\, df = \gamma^2 \int_{-\infty}^{\infty} |S(f)|^2\, df$, the normalized energy over one bit period of the signal at the input to the receiver. Thus,

$$P_{b,\,minimum} = Q\left(\text{maximum}\,\frac{|\gamma s(iT_b)*h(iT_b)|}{\sigma_p}\right) \tag{4.58}$$
$$= Q\left(\sqrt{\text{maximum}\,\frac{|\gamma s(iT_b)*h(iT_b)|^2}{\sigma_p^2}}\right) = Q\left(\sqrt{\frac{2E_b}{N_o}}\right)$$

Note that when the receiver is optimized, P_b is dependent on only E_b (the normalized energy per bit of the signal at the input to the receiver) and N_o (the power spectral density of the noise at the input to the receiver), and not on the shape of the signal.

2. Let's now investigate the time domain representation of $H_o(f)$, since this representation may give us insight into how to build a receiver that implements optimum processing. From Equation (4.56),

$$H_o(f) = kS^*(f)e^{-j2\pi fiT_b} \tag{4.56R}$$

and thus,

$$h_o(t) = \mathcal{F}^{-1}\{H_o(f)\} = k\mathcal{F}^{-1}\{S^*(f)e^{-j2\pi fiT_b}\} \tag{4.59a}$$

If $s(t)$ is real, then $\mathcal{F}^{-1}\{S^*(f)\} = s(-t)$. Combining this property with the time shift property of the Fourier transform (see Equation (2.53)),

$$h_o(t) = \mathcal{F}^{-1}\{H_o(f)\} = k\mathcal{F}^{-1}\{S^*(f)e^{-j2\pi fiT_b}\} = ks(iT_b - t) \tag{4.59b}$$

Note that $ks(iT_b - t)$ is a scaled, time-delayed, time-inverted version of the transmitted signal $s(t)$ during the ith bit period. We thus call the optimum processing, $H_o(f)$ or $h_o(t)$, a *matched filter*, since the processing can be modeled as a filter at the input of the receiver with a shape matched to the shape of the transmitted signal, time-inverted and scaled by an arbitrary constant k.

If we know the shape of the transmitted signal, don't we also know whether the transmitter sent a signal corresponding to a "1" or a "0"? The answer is "not necessarily": If the signal transmitted to send a "1" and the signal transmitted to send a "0" are symmetric (as they are for the rectangular, sinc-shaped, and raised cosine pulses examined in Chapter 3), then the two signals have the same shape. The scaling factor k can be either positive or negative. Let's arbitrarily set k to 1, matching the filter to the positive pulse representing a "1." We will call this positive pulse $s_1(t)$.

3. To see how to build a matched filter, let's see what the filter's output $z(t)$ looks like during the ith bit period. As discussed above, without loss of generality, if the pulses are symmetric we can use the shape of a positive pulse and set $k = 1$. Remember that all processing for the ith bit is performed during the ith bit period: $(i - 1)T_b < t \leq iT_b$.

$$\begin{aligned} z(t)\big|_{i\text{th bit period}} &= r(t)^* h_o(t)\big|_{(i-1)T_b<t\le iT_b} \\ &= \int_{(i-1)T_b}^{t} r(\tau)h_o(t-\tau)\,d\tau \qquad (i-1)T_b < t \le iT_b \qquad (4.60) \\ &= \int_{(i-1)T_b}^{t} r(\tau)s_1[iT_b-(t-\tau)]\,d\tau \qquad (i-1)T_b < t \le iT_b \end{aligned}$$

Now let's examine the output of the sampler corresponding to the ith bit period. The output of the sampler is just $z(t)$ at time $t = iT_b$. Substituting into Equation (4.60),

$$z(iT_b) = \int_{(i-1)T_b}^{iT_b} r(\tau)s_1[iT_b-(iT_b-\tau)]\,d\tau = \int_{(i-1)T_b}^{iT_b} r(\tau)s_1(\tau)\,d\tau \qquad (4.61)$$

Equation (4.61) has significant implications concerning our receiver's design. The equation tells us that a receiver design with the optimum processing (i.e., the matched filter), as shown in Figure 4-12, is equivalent to the receiver design shown in Figure 4-13.

The particular receiver design in Figure 4-13 is called an *integrator*, or *correlator*. We are sampling only once per period, but the integration causes us to consider the received signal's behavior during the entire bit period. The integrator's value is reset to zero before processing begins on the $i + 1$st bit.

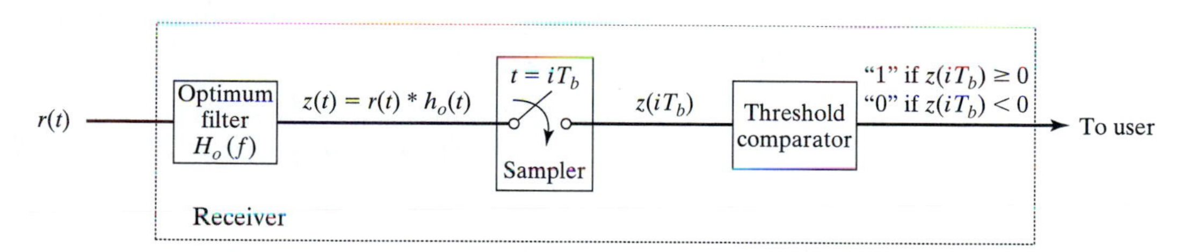

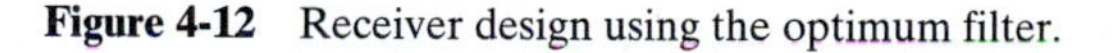
Figure 4-12 Receiver design using the optimum filter.

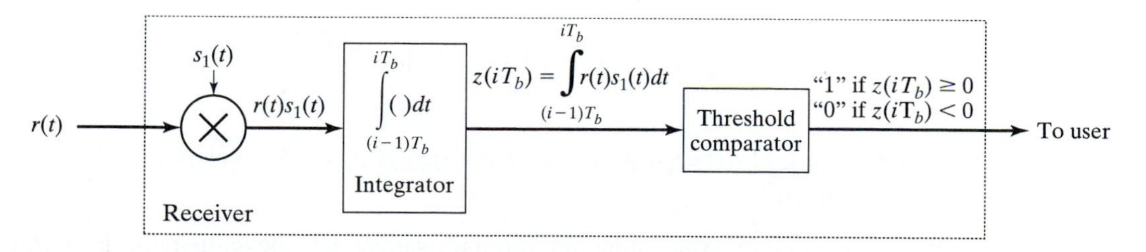

Figure 4-13 An equivalent design—the correlation receiver.

Let's make a few observations concerning the correlation receiver.

1. The mathematical operation of multiplying two signals together and integrating the product over a period of time is known as *correlation*. Consider two signals, $x(t)$ and $y(t)$, each having the same average magnitude. Over a given time period (say, from $t = 0$ to $t = t_o$), the larger the value of $\int_0^{t_o} x(t)y(t)\,dt$, the more the two signals resemble each other (i.e., the more the two signals are *co-related*, or *correlated*). Intuitively, we can support this claim in the same way we know that, given a rectangle with a fixed perimeter, the maximum area is produced by making the height and width equal. The receiver in Figure 4-13 calculates the correlation of the received signal and $s_1(t)$, the signal that would have been transmitted to signify a "1." If the correlation is high (i.e., if the received signal closely resembles $s_1(t)$), then a "1" is assumed to have been transmitted. If the correlation is low (i.e., if the received signal does not closely resemble $s_1(t)$), then a "0" is assumed to have been transmitted.
2. The correlation receiver is fairly simple to build, but note the continued importance of synchronization.
3. At the end of each bit period, the value of the integrator must be input to the threshold detector, and the integrator must then be immediately reset to zero so that it can start integrating for the next bit period. This procedure is known as *integrate-and-dump*.
4. As shown in Equation (4.58), probability of bit error for the optimum receiver (the matched filter or correlation receiver) is

$$P_b = Q\left(\sqrt{\frac{2E_b}{N_o}}\right) \tag{4.58R}$$

where E_b is the normalized energy per bit of the signal at the input to the receiver. E_b can be calculated in the frequency domain as shown earlier or it can be calculated in the time domain as

$$E_b = \int_{(i-1)T_b}^{iT_b} r^2(t)\,dt = \int_{(i-1)T_b}^{iT_b} \gamma^2 s^2(t)\,dt = \gamma^2 \int_0^{T_b} s_1^2(t)\,dt \tag{4.62}$$

since the signals used to transmit a "1" and a "0" are symmetric.

As a final step, Figure 4-14 plots probability of bit error versus E_b/N_o for the correlation receiver. Note that E_b/N_o is plotted in decibels, where

$$\frac{E_b}{N_o} \text{ in decibels (dB)} = 10\log_{10}\left(\frac{E_b}{N_o}\right) \tag{4.63}$$

4.2.3 A Graphical Interpretation of Probability of Bit Error for the Optimal Receiver

To improve our intuition concerning the optimal receiver, let's develop a graphical interpretation of probability of bit error for the matched filter or correlation receiver. We know

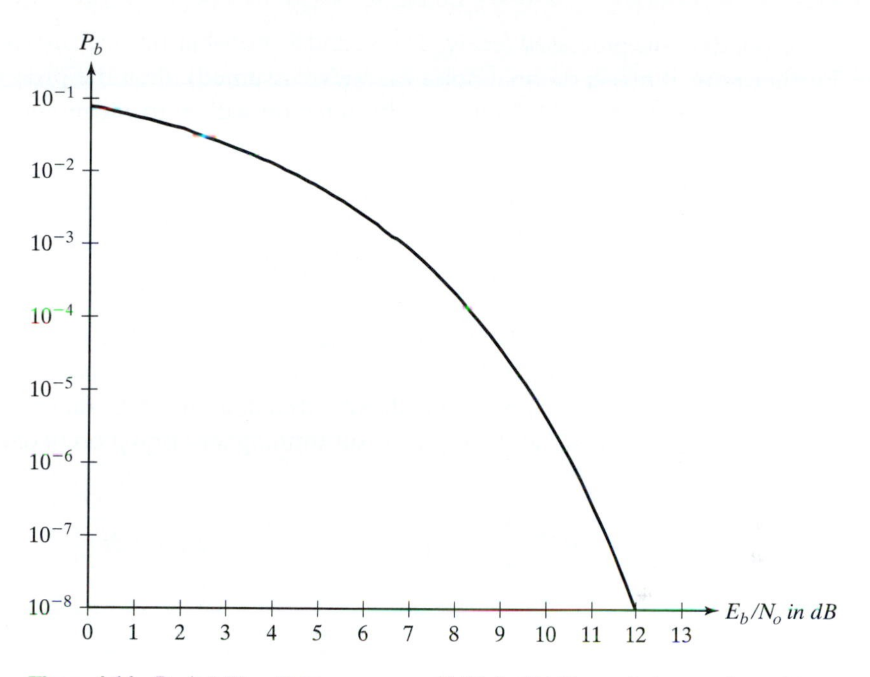

Figure 4-14 Probability of bit error versus E_b/N_o for PAM correlation receiver with symmetric signals.

$$\begin{aligned} P_{b,\,matched} &= P\{b_i = 1 \text{ and } z(iT_b) < 0\} + P\{b_i = 0 \text{ and } z(iT_b) \geq 0\} \\ &= P\{b_i = 1\}P\{z(iT_b) < 0 | b_i = 1\} + P\{b_i = 0\}P\{z(iT_b) \geq 0 | b_i = 0\} \\ &= \frac{1}{2}P\{z(iT_b) < 0 | b_i = 1\} + \frac{1}{2}P\{z(iT_b) \geq 0 | b_i = 0\} \end{aligned} \tag{4.64}$$

Furthermore, we know that if $b_i = 1$, then the signal $s_1(t)$ is transmitted, and from Figure 4-13 we can see

$$\begin{aligned} z(iT_b)\Big|_{b_i=1} &= \int_{(i-1)T_b}^{iT_b} r(t)s_1(t)\,dt\Big|_{b_i=1} = \int_{(i-1)T_b}^{iT_b} [\gamma s_1(t) + n(t)]s_1(t)\,dt \\ &= \underbrace{\int_{(i-1)T_b}^{iT_b} \gamma s_1^2(t)\,dt}_{\substack{\text{processed received} \\ \text{signal if } b_i=1}} + \underbrace{\int_{(i-1)T_b}^{iT_b} n(t)s_1(t)\,dt}_{\text{processed received noise}} \end{aligned} \tag{4.65}$$

The first term in Equation (4.65) corresponds to the processed received signal and is deterministic, since γ and $s_1(t)$ are deterministic. The second term in Equation (4.65) corresponds to the processed received noise and is probabilistic, since $n(t)$ is probabilistic. Furthermore, if $n(t)$ is thermal noise (as we've assumed), then the processed received noise has a Gaussian probability density function with zero mean and a variance of, say, σ_p^2. Thus, $z(iT_b)|_{b_{i=1}}$ is probabilistic with a Gaussian probability density function of variance σ_p^2 and mean

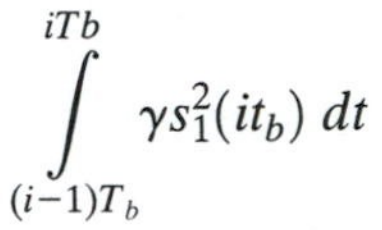

$$\int_{(i-1)T_b}^{iTb} \gamma s_1^2(it_b)\, dt$$

Figure 4-15a shows the probability density function of $z(iT_b)|_{b_{i=1}}$. Let's use the variable $a_1(iT_b)$ to represent the mean. We can see that $a_1(iT_b)$ has a physical interpretation—it represents the value of the input to the threshold comparator if the ith transmitted bit was a "1" and there is no noise at the receiver. Substituting into Equation (4.65),

$$z(iT_b)\Big|_{b_{i=1}} = a_1(iT_b) + \int_{(i-1)T_b}^{iT_b} n(t)s_1(t)\, dt \tag{4.66}$$

We've used $s_1(t)$ to symbolize the signal sent by the transmitter to represent the ith bit if the ith bit is a "1." Similarly, let $s_2(t)$ represent the signal transmitted if $b_i = 0$.

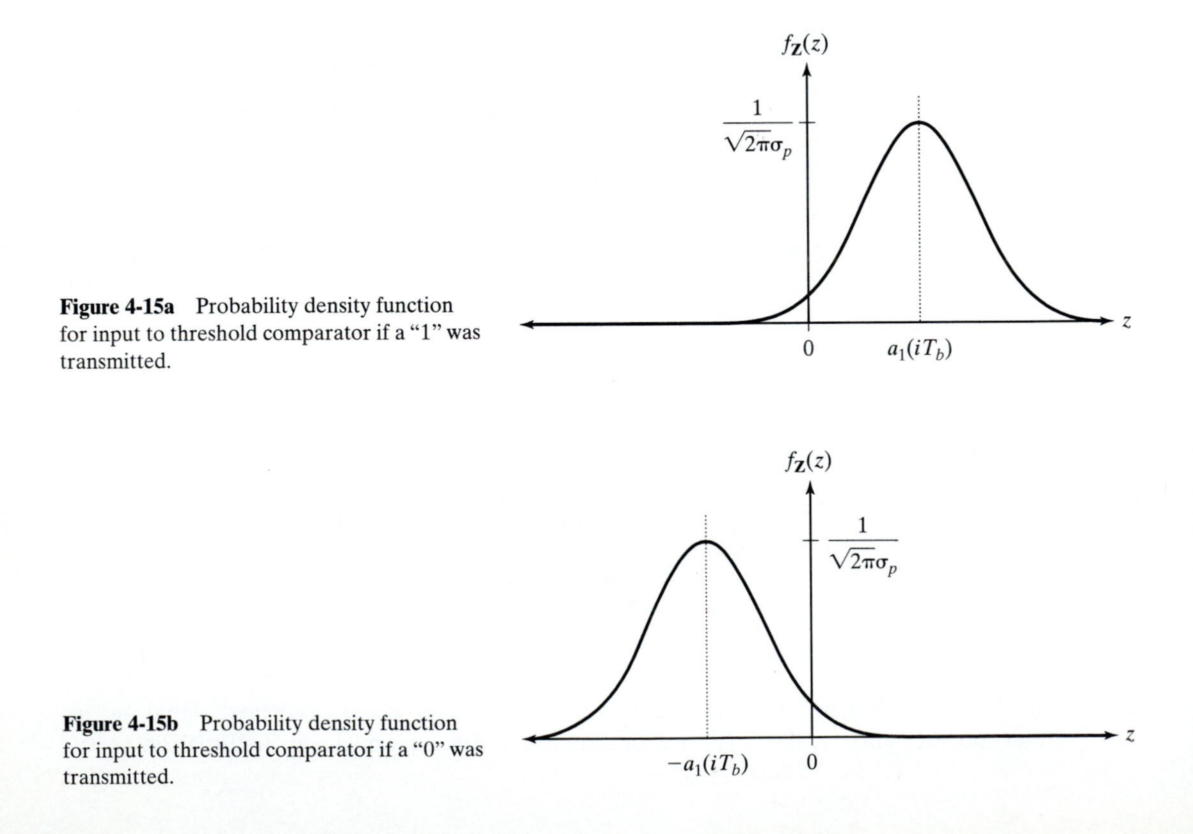

Figure 4-15a Probability density function for input to threshold comparator if a "1" was transmitted.

Figure 4-15b Probability density function for input to threshold comparator if a "0" was transmitted.

Since $s_1(t)$ and $s_2(t)$ are symmetric (i.e., $s_2(t) = -s_1(t)$),

$$z(iT_b)\Big|_{b_i=0} = \int_{(i-1)T_b}^{iT_b} r(t)s_1(t)\,dt\Big|_{b_i=0} = \int_{(i-1)T_b}^{iT_b} [\gamma s_2(t) + n(t)]s_1(t)\,dt$$

$$= \underbrace{\int_{(i-1)T_b}^{iT_b} \gamma s_2(t)s_1(t)\,dt}_{\substack{\text{processed received}\\ \text{signal if } b_i=0}} + \underbrace{\int_{(i-1)T_b}^{iT_b} n(t)s_1(t)\,dt}_{\text{processed received noise}}$$

$$= \int_{(i-1)T_b}^{iT_b} \gamma[-s_1(t)]s_1(t)\,dt + \int_{(i-1)T_b}^{iT_b} n(t)s_1(t)\,dt \tag{4.67}$$

$$= -a_1(iT_b) + \int_{(i-1)T_b}^{iT_b} n(t)s_1(t)\,dt$$

Analyzing Equation (4.67) in the same manner as Equation (4.65), we see that $z(iT_b)|_{b_i=0}$ is probabilistic with a Gaussian probability density function with a variance of σ_p^2 and a mean of $-a_1(iT_b)$. Figure 4-15b shows the probability density function of $z(iT_b)|_{b_i=0}$. Note that this is the same probability density function as $z(iT_b)|_{b_i=1}$ except that the mean is $-a_1(iT_b)$ rather than $a_1(iT_b)$. Such an observation makes sense because the transmitted signals are symmetric and the received signal has the same noise whether a "1" or "0" is transmitted.

We now have sufficient insight to graphically interpret Equation (4.64)—the probability of bit error for the matched filter or correlation receiver.

$$P_{b,\,matched} = \frac{1}{2}P\{z(iT_b) < 0|b_i = 1\} + \frac{1}{2}P\{z(iT_b) \geq 0|b_i = 0\} \tag{4.64R}$$

The first term, $0.5P\{z(iT_b) < 0|b_i = 1\}$, can be represented by the black shaded region in Figure 4-16a, the area under the curve to the left of zero for the probability density function of $0.5z(iT_b)|b_i = 1$. Similarly, the second term, $0.5P\{z(iT_b) \geq 0|b_i = 0\}$, can be represented by the gray shaded region in Figure 4-16b, the area under the curve to the right of zero for the probability density function $0.5z(iT_b)|b_i = 0$. Combining the two terms yields Figure 4-16c, where the probability of bit error for the correlation receiver is the black shaded area (corresponding to $0.5P\{z(iT_b) < 0|b_i = 1\}$) and the gray shaded area (corresponding to $0.5P\{z(iT_b) \geq 0|b_i = 0\}$). Be sure you see how Figure 4-16c corresponds to the probability of bit error expressed in Equation (4.64R).

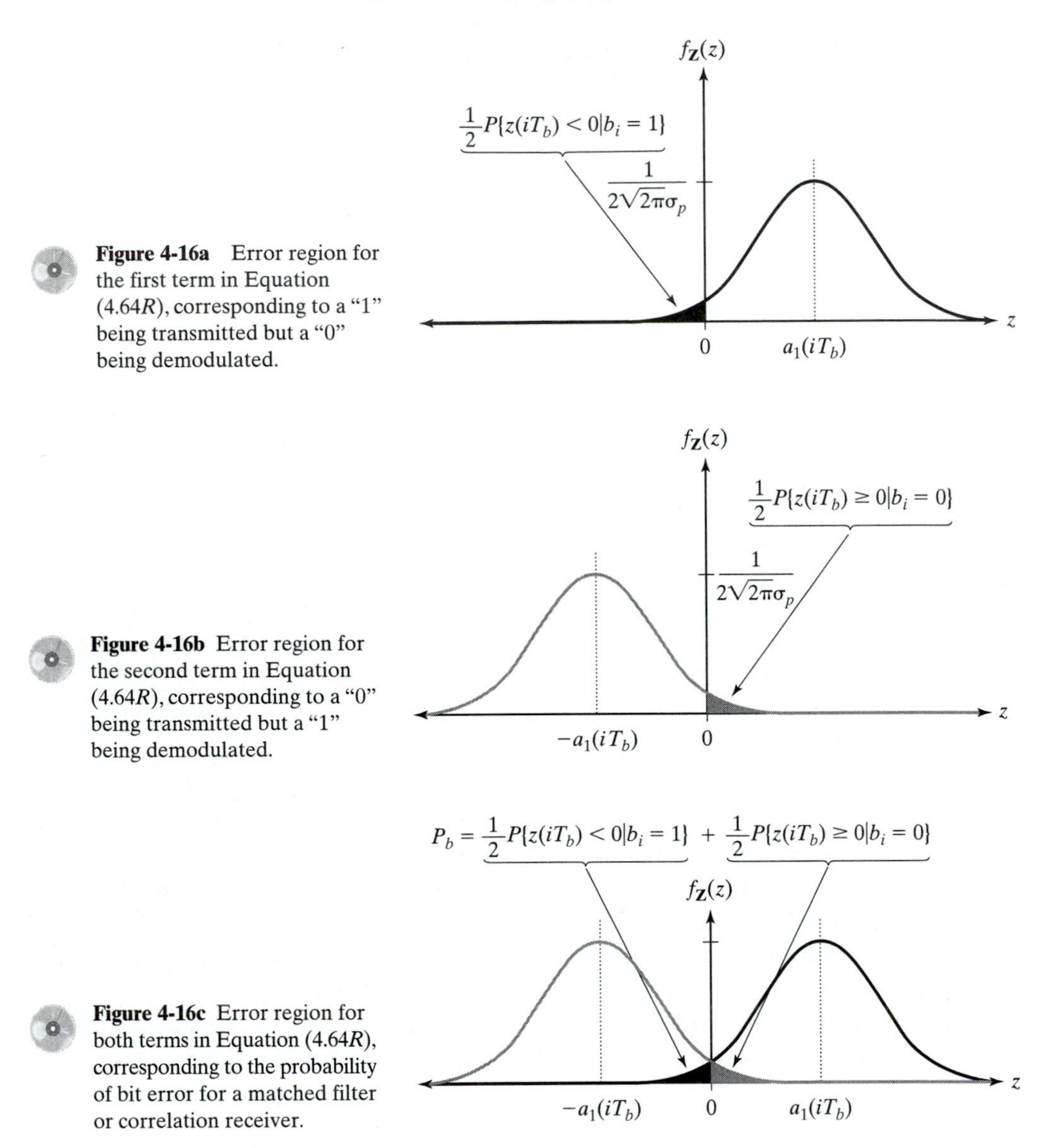

Figure 4-16a Error region for the first term in Equation (4.64*R*), corresponding to a "1" being transmitted but a "0" being demodulated.

Figure 4-16b Error region for the second term in Equation (4.64*R*), corresponding to a "0" being transmitted but a "1" being demodulated.

Figure 4-16c Error region for both terms in Equation (4.64*R*), corresponding to the probability of bit error for a matched filter or correlation receiver.

One use of the graphical interpretation of $P_{b,\,matched}$ is to determine the optimum value for the threshold comparator in the matched filter or correlation receiver (see Figure 4-13). Until now we've assumed (based mostly on intuition) that the optimum threshold value is zero volts: If $z(iT_b) \geq 0$ volts, then the ith bit is interpreted as a "1"; if $z(iT_b) < 0$ volts, then the ith bit is interpreted as a "0." Let's use τ to symbolize the threshold value. What value of τ will produce a lower value for $P_{b,\,matched}$? As established in Figure 4-16c, there are two cases of error, $0.5P\{z(iT_b) < \tau|b_i = 1\}$ (the black area in the figure) and $0.5P\{z(iT_b) \geq \tau|b_i = 0\}$ (the gray area in the figure), and the total probability of error is the sum of the two cases (i.e., the sum of the black and gray areas). If the threshold is set to $\tau = 0$ volts, then the probabilities for each of the two error cases are equal, as shown in Figure 4-17a. If, however, the threshold is positive, then the black area is increased by a value larger than the gray area is decreased, causing the sum of the two areas to increase, as shown in Figure 4-17b. Similarly, if the threshold is negative, then the gray area is increased by a value larger than the black area is decreased, again causing the sum of the two areas to increase, as shown in Figure 4-17c. Thus, we can see geometrically that the optimum value for the threshold is $\tau = 0$ volts. Note that we can also use calculus to find the optimum value mathematically, but calculus does not provide us with the same insight as does the geometric approach. We will, however, use a mathematical approach shortly for the case of nonequiprobable bits.

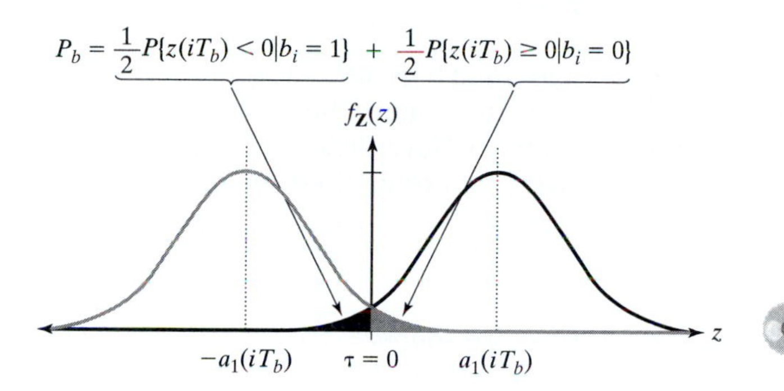

Figure 4-17a Probability of bit error (sum of shaded black and gray regions) if threshold $(\tau) = 0$ volts.

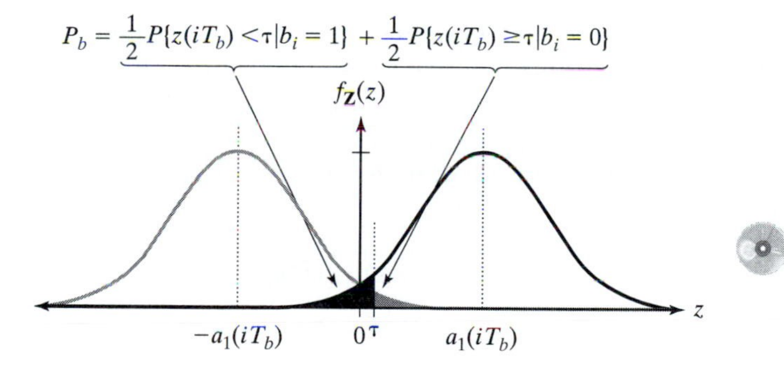

Figure 4-17b Probability of bit error (sum of shaded black and gray regions) if threshold (τ) is > 0 volts. Compare with Figure 4-17a and note that probability of bit error is lower when $\tau = 0$ volts.

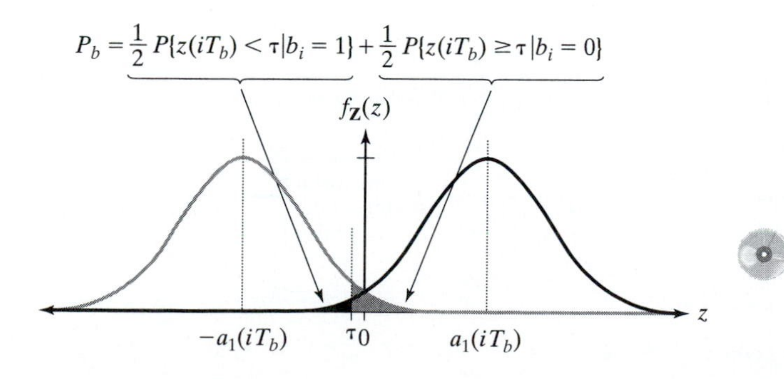

Figure 4-17c Probability of bit error (sum of shaded black and gray regions) if threshold (τ) is < 0 volts. Compare with Figure 4-17a and note that probability of bit error is lower when $\tau = 0$ volts.

4.2.4 Designing the Correlation Receiver for More General Signals

In developing the expressions for probability of bit error (Section 4.1) and in designing the correlation receiver (Sections 4.2.1 and 4.2.2), we've made the following assumptions concerning the transmitted signal:

a. The value of each bit output by the source has an equal probability of being a "1" or a "0."

b. The value of each bit output by the source is independent of the value of all other bits in the data stream.

c. The pulse shape transmitted to represent a "1" ($s_1(t)$) and the pulse shape transmitted to represent a "0" ($s_2(t)$) are symmetric.

As you will learn in Chapter 9, assumptions a and b are realistic for well-designed systems, which often employ data compression to achieve bit independence and equiprobability prior to transmission or data storage. Nevertheless, we can generalize our design and gain significant insight by removing the assumptions and observing how their removal affects our earlier analyses. Let's start by removing assumption c, that the pulse shapes are symmetric.

4.2.4.1 Correlation receiver for asymmetric pulse shapes We can transmit digital information using PAM with asymmetric pulse shapes: for instance, we can transmit a rectangular pulse of +3 volts to represent a "1" and −2 volts to represent a "0." Let's now establish how this asymmetry affects the design and performance of the optimum receiver. Optimum processing for the asymmetric case may be different than for the symmetric case, and similarly the optimum threshold for the asymmetric case may not be the same as for the symmetric case. Figure 4-18 shows the receiver structure for asymmetric PAM signals, where $h_{opt}(t)$ is the impulse response representing optimum processing for the asymmetric case and τ_{opt} represents the optimum threshold for the asymmetric case.

Calculating probability of bit error,

$$P_{b,\,asymmetric} = P\{b_i = 1 \text{ and } z(iT_b) < \tau\} + P\{b_i = 0 \text{ and } z(iT_b) \geq \tau\} =$$
$$P\{b_i = 1\}P\{z(iT_b) < \tau | b_i = 1\} + P\{b_i = 0\}P\{z(iT_b) \geq \tau | b_i = 0\} = \quad (4.68)$$
$$\frac{1}{2}P\{z(iT_b) < \tau | b_i = 1\} + \frac{1}{2}P\{z(iT_b) \geq \tau | b_i = 0\}$$

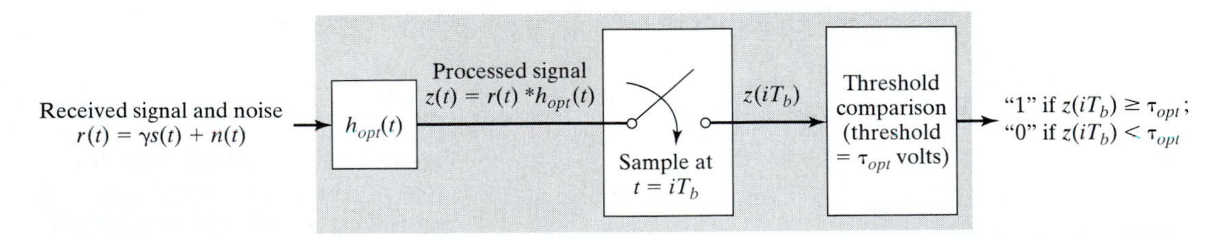

Figure 4-18 Optimal receiver structure for asymmetric PAM signals.

Using the same notation as in the previous sections, let $a_1(iT_b)$ represent the processed output for the ith bit if $b_i = 1$ and the receiver experiences no noise. This can be expressed mathematically as

$$a_1(iT_b) = \gamma s_1(t)*h_{opt}(t)|_{t=iT_b} \tag{4.69}$$

Similarly, let $a_2(iT_b)$ represent the processed output for the ith bit if $b_i = 0$ and the receiver experiences no noise. Mathematically,

$$a_2(iT_b) = \gamma s_2(t)*h_{opt}(t)|_{t=iT_b} \tag{4.70}$$

The noise at the input to the receiver is zero mean with a Gaussian probability density function, so the processed noise will be zero mean with a Gaussian probability density function. Let σ_o^2 symbolize the variance of the processed noise. Thus, $z(iT_b)|b_i = 1$ is probabilistic with a Gaussian probability density function with mean $a_1(iT_b)$ and variance σ_o^2. Figure 4-19a shows the probability density function of $0.5z(iT_b)|b_i = 1$, with the black area representing $0.5P\{z(iT_b) < \tau|b_i = 1\}$ for an arbitrary threshold τ. Similarly, $z(iT_b)|b_i = 0$ is probabilistic with a Gaussian probability density function with mean $a_2(iT_b)$ and variance σ_o^2. Figure 4-19b shows the probability density function of $0.5z(iT_b)|b_i = 0$, with the gray area representing $0.5P\{z(iT_b) \geq \tau|b_i = 0\}$ for an arbitrary threshold τ.

Finding the optimum threshold corresponds to finding the value of τ that minimizes Equation (4.68). As we established in Section 4.2.3, this corresponds to finding the value of τ that minimizes the sum of the black area in Figure 4-19a and the gray area in Figure 4-19b. Such a value,

$$\tau_{opt} = \frac{a_1(iT_b) + a_2(iT_b)}{2} \tag{4.71}$$

is shown in Figure 4-19c.

Now that we've determined the optimum threshold value for asymmetric, equiprobable pulses, let's calculate the corresponding P_b. We've determined the probability density functions for $z(iT_b)|b_i = 1$ and $z(iT_b)|b_i = 0$. Thus

$$\begin{aligned} &P_{b,\,\text{asymmetric, optimum threshold}} = \\ &\frac{1}{2}P\left\{z(iT_b) < \tau_{opt}\middle|b_i = 1\right\} + \frac{1}{2}P\left\{z(iT_b) \geq \tau_{opt}|b_i = 0\right\} \\ &= \frac{1}{2}\int_{-\infty}^{\tau_{opt}} \frac{1}{\sqrt{2\pi}\sigma_o} e^{\frac{-[x-a_1(iT_b)]^2}{2\sigma_o^2}} dx + \frac{1}{2}\int_{\tau_{opt}}^{\infty} \frac{1}{\sqrt{2\pi}\sigma_o} e^{\frac{-[x-a_2(iT_b)]^2}{2\sigma_o^2}} dx \end{aligned} \tag{4.72}$$

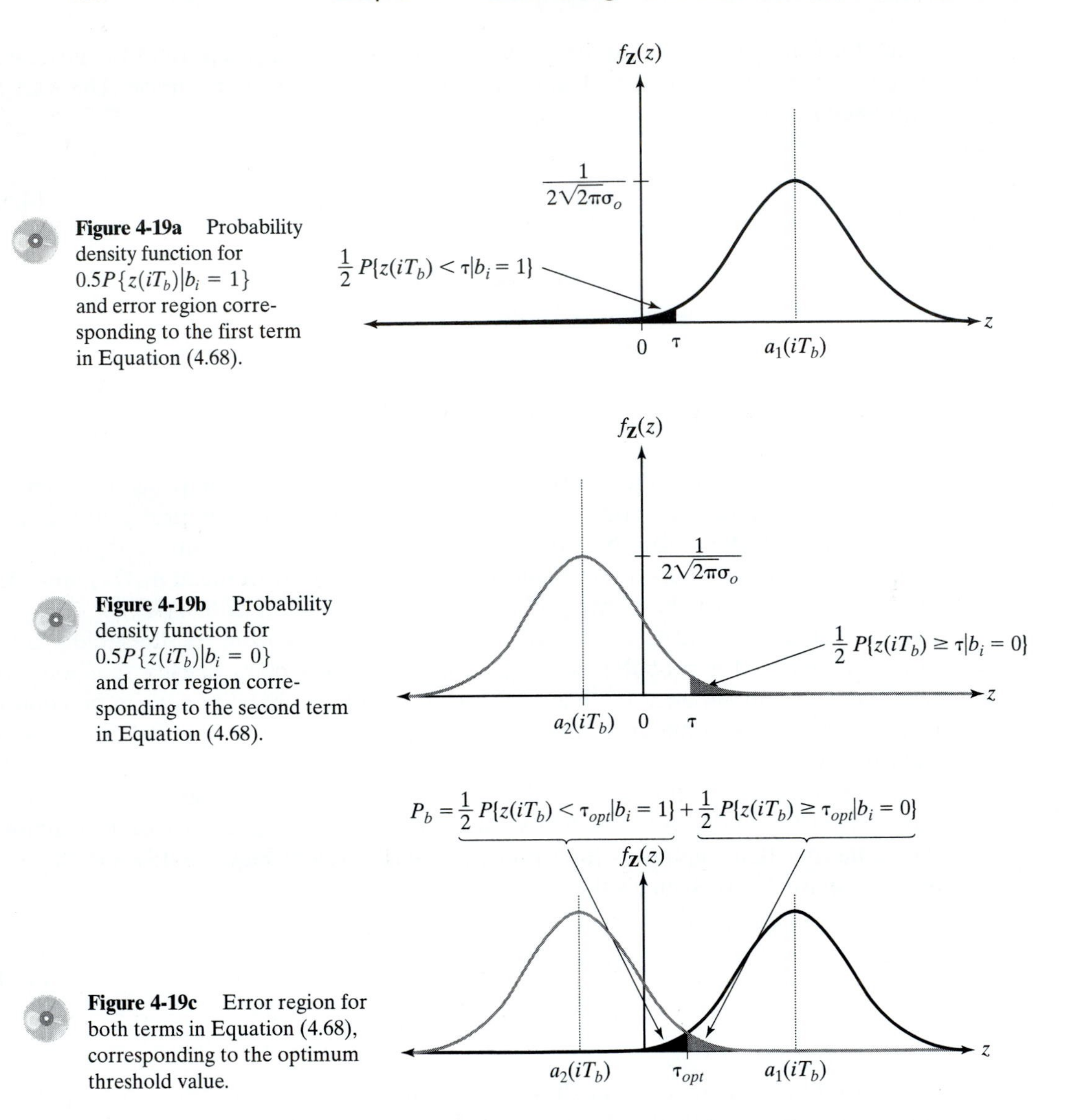

Figure 4-19a Probability density function for $0.5P\{z(iT_b)|b_i = 1\}$ and error region corresponding to the first term in Equation (4.68).

Figure 4-19b Probability density function for $0.5P\{z(iT_b)|b_i = 0\}$ and error region corresponding to the second term in Equation (4.68).

Figure 4-19c Error region for both terms in Equation (4.68), corresponding to the optimum threshold value.

We can express each of the two integrals in Equation (4.72) as Q functions. The second integral can be expressed as

$$\frac{1}{2}\int_{\tau_{opt}}^{\infty} \frac{1}{\sqrt{2\pi}\sigma_o} e^{\frac{-[x-a_2(iT_b)]^2}{2\sigma_o^2}}\, dx = \frac{1}{2}Q\left(\frac{\tau_{opt} - a_2(iT_b)}{\sigma_o}\right) \tag{4.73}$$

$$= \frac{1}{2}Q\left(\frac{\frac{a_1(iT_b) + a_2(iT_b)}{2} - a_2(iT_b)}{\sigma_o}\right) = \frac{1}{2}Q\left(\frac{a_1(iT_b) - a_2(iT_b)}{2\sigma_o}\right)$$

The first integral in Equation (4.72) can be calculated in either of two ways. The first way is to examine Figure 4-19c and note that the black area (corresponding to the first integral) is equal to the gray area (corresponding to the second integral), and thus

$$\frac{1}{2}\int_{-\infty}^{\tau_{opt}} \frac{1}{\sqrt{2\pi}\sigma_o} e^{\frac{-[x-a_1(iT_b)]^2}{2\sigma_o^2}} dx = \frac{1}{2}\int_{\tau_{opt}}^{\infty} \frac{1}{\sqrt{2\pi}\sigma_o} e^{\frac{-[x-a_2(iT_b)]^2}{2\sigma_o^2}} dx$$
$$= \frac{1}{2} Q\left(\frac{a_1(iT_b) - a_2(iT_b)}{2\sigma_o}\right) \tag{4.74}$$

A second approach exploits the symmetry of the Gaussian probability density function and calculates the first integral as follows:

$$\frac{1}{2}\int_{-\infty}^{\tau_{opt}} \frac{1}{\sqrt{2\pi}\sigma_o} e^{\frac{-[x-a_1(iT_b)]^2}{2\sigma_o^2}} dx = \frac{1}{2}\left(\int_{2a_1(iT_b)-\tau_{opt}}^{\infty} \frac{1}{\sqrt{2\pi}\sigma_o} e^{\frac{-[x-a_1(iT_b)]^2}{2\sigma_o^2}} dx\right)$$
$$= \frac{1}{2} Q\left(\frac{2a_1(iT_b) - \tau_{opt} - a_1(iT_b)}{\sigma_o}\right) = \frac{1}{2} Q\left(\frac{a_1(iT_b) - \tau_{opt}}{\sigma_o}\right) \tag{4.75}$$
$$= \frac{1}{2} Q\left(\frac{a_1(iT_b) - \dfrac{a_1(iT_b) + a_2(iT_b)}{2}}{\sigma_o}\right) = \frac{1}{2} Q\left(\frac{a_1(iT_b) - a_2(iT_b)}{2\sigma_o}\right)$$

Be sure you understand both approaches. Now substituting into Equation (4.72),

$$P_{b,\ asymmetric,\ optimum\ threshold}$$
$$= \frac{1}{2}\int_{-\infty}^{\tau_{opt}} \frac{1}{\sqrt{2\pi}\sigma_o} e^{\frac{-[x-a_1(iT_b)]^2}{2\sigma_o^2}} dx + \frac{1}{2}\int_{\tau_{opt}}^{\infty} \frac{1}{\sqrt{2\pi}\sigma_o} e^{\frac{-[x-a_2(iT_b)]^2}{2\sigma_o^2}} dx \tag{4.76}$$
$$= Q\left(\frac{a_1(iT_b) - a_2(iT_b)}{2\sigma_o}\right)$$

For the asymmetric, equiprobable pulse case, we've determined the optimum threshold (Equation (4.71)) and the probability of bit error (Equation (4.76)). Note, however, that these equations depend on $a_1(iT_b)$, $a_2(iT_b)$, and σ_o, and these values depend on the processing $h_{opt}(t)$, which we have yet to specify. Let's do that now. Using the same approach we used in Section 4.2.1 for the symmetric case, we want to determine the process $h_{opt}(t)$ that minimizes probability of bit error, as expressed in Equation (4.76). Minimizing P_b corresponds to maximizing the argument within the Q function, $[a_1(iT_b) - a_2(iT_b)]/(2\sigma_o)$, which is the same as maximizing $[a_1(iT_b) - a_2(iT_b)]/\sigma_o$ and the same as maximizing $[a_1(iT_b) - a_2(iT_b)]^2/\sigma_o^2$, since σ_o and $a_1(iT_b) - a_2(iT_b)$ are positive. Using the same type of derivation as used for the symmetric case, (i.e., employing Schwarz's inequality), $[a_1(iT_b) - a_2(iT_b)]^2/\sigma_o^2$ is maximized when we choose as our optimal receiver filter

$$H_o(f) = k[S_1(f) - S_2(f)]^* e^{-j2\pi f i T_b} \tag{4.77}$$

where k is an arbitrary constant.

The optimum filter is thus a filter matched to the *difference* between the two possible transmitted signals. It makes good intuitive sense that we're interested in the difference between the two signals, since this difference is what distinguishes one signal from the other.

Equation (4.77) has the following implications:

1. We can determine P_b for the optimum filter (using analysis similar to the symmetric case) as

$$P_b = Q\left(\frac{a_1(iT_b) - a_2(iT_b)}{2\sigma_o}\right) = Q\left(\sqrt{\frac{[a_1(iT_b) - a_2(iT_b)]^2}{4\sigma_o^2}}\right) = Q\left(\sqrt{\frac{E_d}{2N_o}}\right) \tag{4.78}$$

   where

$$E_d = \int_{(i-1)T_b}^{T_b} \{\gamma[s_1(t) - s_2(t)]\}^2 \, dt \tag{4.79}$$

   Note that E_d is the total energy over the bit period of the *difference* between the two possible received signals. We can interpret E_d as the energy expended to make a "1" and a "0" look different. The implications are very significant: If a correlation receiver is employed, any dc energy expended in transmitting a signal does not improve the signal's accuracy. (This will be illustrated in Example 4.11.)

2. In the time domain, the optimum filter is expressed as

$$h_o(t) = \mathcal{F}^{-1}\{H_o(f)\} = k[s_1(iT_b - t) - s_2(iT_b - t)] \tag{4.80}$$

   Letting $k = 1$, we can express the output of the filter during the ith bit period as

$$z(t) = r(t) * h_{opt}(t) = \int_{(i-1)T_b}^{t} r(\tau)[s_1(iT_b - t + \tau) - s_2(iT_b - t + \tau)] \, d\tau \tag{4.81}$$

   and the output of the sampler is thus

$$z(iT_b) = \int_{(i-1)T_b}^{T_b} r(\tau)[s_1(\tau) - s_2(\tau)] \, d\tau \tag{4.82}$$

Equation (4.82) suggests the receiver structure shown in Figure 4-20.

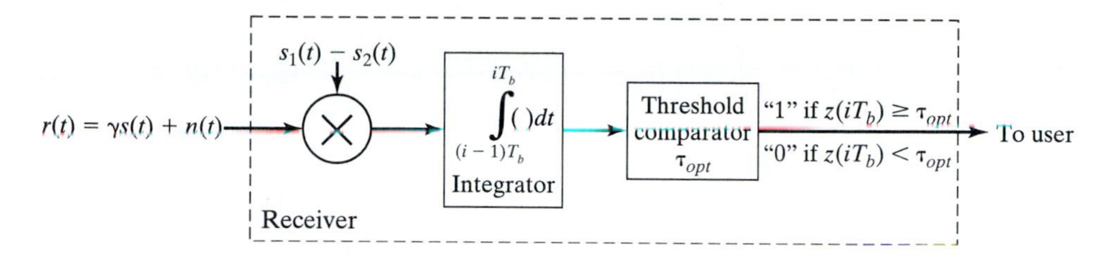

Figure 4-20 Optimal receiver structure for asymmetric, equiprobable PAM pulses.

We can also draw this receiver as shown in Figure 4-21.

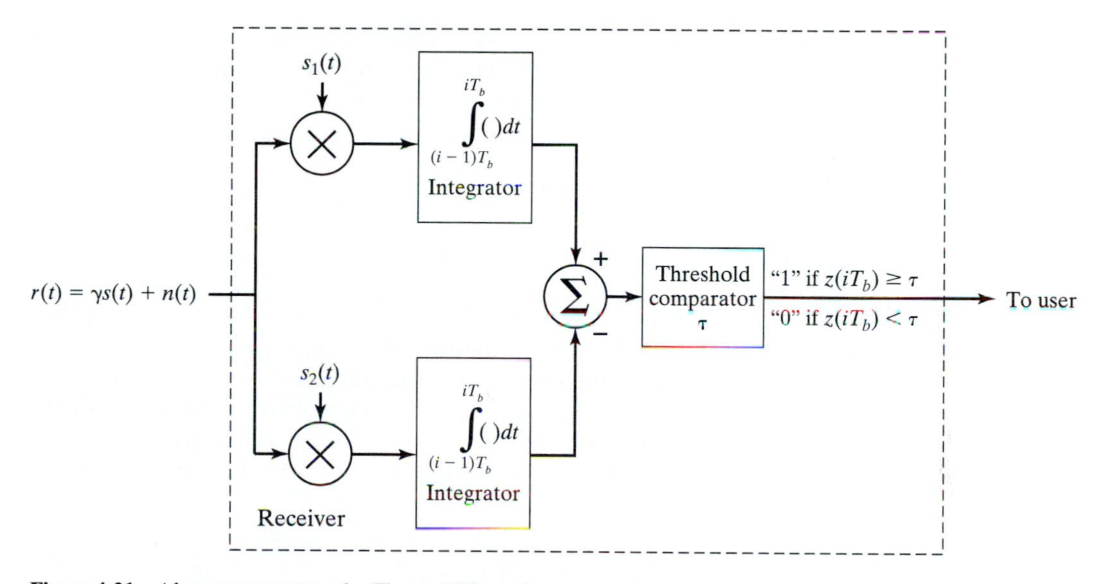

Figure 4-21 Alternate structure for Figure 4-20 receiver.

Figure 4-20 is a much more practical implementation of the receiver (it uses fewer parts), but Figure 4-21 shows us that, as with the symmetric case, the optimal receiver for the asymmetric PAM signals exploits correlation. The received signal is correlated with $s_1(t)$ and with $s_2(t)$, and the output is determined by which of the two correlations is stronger.

Finally, let's verify that the more general equations for optimum threshold and probability of bit error for asymmetric signals also hold for symmetric PAM signals. For symmetric signals, $a_2(iT_b) = -a_1(iT_b)$, and Equation (4.71) thus reduces to

$$\tau_{opt} = \frac{a_1(iT_b) + a_2(iT_b)}{2} = 0$$

which is consistent with our observations in Section 4.2.3. For symmetric signals, note that

$$E_d = \int_{(i-1)T_b}^{T_b} \{\gamma[s_1(t) - s_2(t)]\}^2\, dt = \int_{(i-1)T_b}^{T_b} [2\gamma s_1(t)]^2\, dt$$

$$= \int_{(i-1)T_b}^{T_b} 4[\gamma s_1(t)]^2\, dt = 4E_b$$

so, for symmetric signals Equation (4.78) reduces to

$$P_b = Q\left(\sqrt{\frac{E_d}{2N_o}}\right) = Q\left(\sqrt{\frac{4E_b}{2N_o}}\right) = Q\left(\sqrt{\frac{2E_b}{N_o}}\right)$$

which is consistent with Equation (4.58). The receiver structure in Figure 4-20 also reduces to Figure 4-13 when symmetry exists (the factor of two at the multiplier is insignificant, since the value of k in our analyses can be any nonzero constant).

Example 4.11 Bipolar versus unipolar signaling

Let's consider two types of signaling: bipolar and unipolar. Bipolar signaling uses symmetric pulse shapes to signify digital information (e.g., rectangular pulses of amplitude $+A$ to signify "1"s and amplitude $-A$ to signify "0"s). As its name implies, bipolar signaling uses both positive polarity and negative polarity signals. Unipolar signaling, on the other hand, uses signals of only a single polarity (so its signals are not symmetric). For example, unipolar signaling could use rectangular pulses of amplitude $+B$ to represent "1"s while "0"s could be represented by not transmitting any signal. An example of unipolar signaling is TTL logic, which uses rectangular pulses of +5 volts to signify a "1" and a pulse of 0 volts (no pulse at all) to signify a "0." Bipolar and unipolar signaling are illustrated in Figures 4-22a and b.

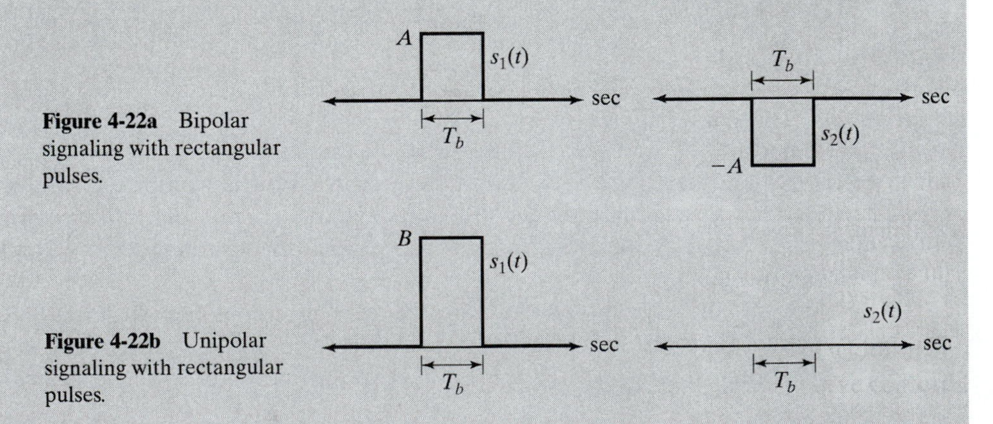

Figure 4-22a Bipolar signaling with rectangular pulses.

Figure 4-22b Unipolar signaling with rectangular pulses.

Less complex equipment can be used for unipolar signaling (bipolar logic and negative power supplies are not needed), but there is a price. For a given transmitted signal power, let's examine the relative accuracy of unipolar and bipolar signaling. Using the rectangular signals in Figures 4-22a and b, the average normalized transmitted signal power for bipolar signaling is

$$P_{ave,\ transmitted,\ bipolar} = \frac{1}{2}A^2 + \frac{1}{2}(-A)^2 = A^2$$

while the average normalized transmitted signal power for unipolar signaling is

$$P_{ave,\ transmitted,\ unipolar} = \frac{1}{2}B^2 + \frac{1}{2}(0)^2 = \frac{1}{2}B^2$$

Thus, for the bipolar and unipolar signaling to use the same transmitted signal power,

$$B = \sqrt{2}A$$

The transmitted energy difference between a "1" and a "0" for bipolar signaling is

$$E_{d,\ transmitted,\ bipolar} = \int_0^{T_b} [A - (-A)]^2\, dt = (2A)^2 T_b = 4A^2 T_b$$

The transmitted energy difference between a "1" and a "0" for unipolar signaling using the same average signal power is

$$E_{d,\ transmitted,\ unipolar} = \int_0^{T_b} [B - 0]^2\, dt = B^2 T_b = (\sqrt{2}A)^2 T_b = 2A^2 T_b$$

Thus, if unipolar and bipolar signals are transmitted with the same average power, the energy difference between a "1" and a "0" for unipolar signaling is only half as much as the energy difference for bipolar signaling. To achieve the same accuracy, unipolar signals must therefore be transmitted with twice the average power of bipolar signals.

What happens to the other 50% of the energy of the unipolar signal? Note that the bipolar signal has a dc value of 0 while the unipolar signal has a dc value of 1/2 B. The average normalized transmitted dc power of the unipolar signal is thus

$$P_{dc,\ transmitted,\ unipolar} = \left(\frac{1}{2}B\right)^2 = \frac{1}{4}B^2 = 0.5P_{ave,\ transmitted,\ unipolar}$$

Once again, we see that only energy expended to make a "1" look different than a "0" will have an effect on accuracy.

4.2.4.2 *Correlation receiver for nonequiprobable bits*

Now let's remove the assumption that the data bits have an equal probability of being a "1" or a "0." Let M represent the probability that $b_i = 1$, therefore 1-M represents the probability that $b_i = 0$. Let's use a receiver structure similar to the one used in Figure 4-18, except that the values of $h_{opt}(t)$ and τ_{opt} may be different than they were for the asymmetric, equiprobable case. Probability of bit error can be expressed as:

$$\begin{aligned} P_{b,\ nonequiprobable} &= P\{b_i = 1 \text{ and } z(iT_b) < \tau\} + P\{b_i = 0 \text{ and } z(iT_b) \geq \tau\} \\ &= P(b_i = 1)P\{z(iT_b) < \tau | b_i = 1\} + P(b_i = 0)P\{z(iT_b) \geq \tau | b_i = 0\} \\ &= MP\{z(iT_b) < \tau | b_i = 1\} + (1 - M)P\{z(iT_b) \geq \tau | b_i = 0\} \end{aligned} \tag{4.83}$$

Using the same notation as in the previous subsections, let $a_1(iT_b)$ represent the processed output for the ith bit if $b_i = 1$ and the receiver experiences no noise, and let $a_2(iT_b)$ represent the processed output for the ith bit if $b_i = 0$ and the receiver experiences no noise. The noise at the input to the receiver has a Gaussian probability density function with zero mean, and thus the processed noise will also have a Gaussian probability density function with zero mean. Let σ_o^2 represent the variance of the processed noise. Thus, $z(iT_b)|b_i = 1$ is probabilistic with a Gaussian probability density function with mean $a_1(iT_b)$ and variance σ_o^2. Similarly, $z(iT_b)|b_i = 0$ is probabilistic with a Gaussian probability density function with mean $a_2(iT_b)$ and variance σ_o^2. Figure 4-23a shows the probability density function $z(iT_b)|b_i = 1$ scaled by M and the probability density function $z(iT_b)|b_i = 0$ scaled by 1-M. (Note that in Figure 4-23a, M is approximately 1/3.) As with the symmetric and asymmetric equiprobable cases, probability of bit error is minimized by selecting a threshold at the point where the two scaled probability density functions intersect. This threshold, shown in Figure 4-23b as τ_{opt}, produces the lowest probability of bit error (i.e., the smallest sum of the black and gray areas).

The point where the two scaled probability density functions intersect, τ_{opt}, is determined by solving

$$M\left[\frac{1}{\sqrt{2\pi}\sigma_o}e^{\frac{-[\tau_{opt}-a_1(iT_b)]^2}{2\sigma_o^2}}\right] = (1 - M)\left[\frac{1}{\sqrt{2\pi}\sigma_o}e^{\frac{-[\tau_{opt}-a_2(iT_b)]^2}{2\sigma_o^2}}\right] \tag{4.84}$$

Figure 4-23a Probability density functions for $MP\{z(iT_b)|b_i = 1\}$ and $(1\text{-}M)P\{z(iT_b)|b_i = 0\}$. Optimum threshold occurs where the two functions intersect. (In this figure M is approximately 1/3.)

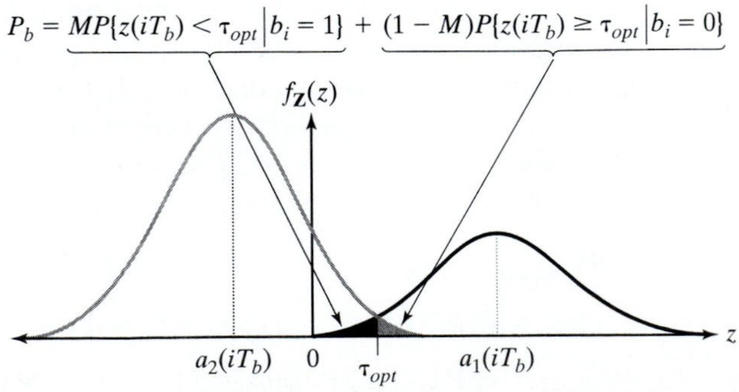

Figure 4-23b Error regions for non-equiprobable bits.

Dividing both sides by $\frac{1}{\sqrt{2\pi}\sigma_o}$ yields

$$Me^{\frac{-[\tau_{opt}-a_1(iT_b)]^2}{2\sigma_o^2}} = (1-M)e^{\frac{-[\tau_{opt}-a_2(iT_b)]^2}{2\sigma_o^2}}$$

Taking the base e logarithm of both sides,

$$\ln\left(Me^{\frac{-[\tau_{opt}-a_1(iT_b)]^2}{2\sigma_o^2}}\right) = \ln\left((1-M)e^{\frac{-[\tau_{opt}-a_2(iT_b)]^2}{2\sigma_o^2}}\right)$$

$$\ln(M) - \frac{[\tau_{opt}-a_1(iT_b)]^2}{2\sigma_o^2} = \ln(1-M) - \frac{[\tau_{opt}-a_2(iT_b)]^2}{2\sigma_o^2}$$

Solving for τ_{opt},

$$[\tau_{opt}-a_1(iT_b)]^2 - [\tau_{opt}-a_2(iT_b)]^2 = 2\sigma_o^2[\ln(M) - \ln(1-M)]$$

$$\tau_{opt}[-2a_1(iT_b) + 2a_2(iT_b)] = 2\sigma_o^2[\ln(M) - \ln(1-M)] - a_1(iT_b)^2 + a_2(iT_b)^2$$

and finally

$$\tau_{opt} = \frac{2\sigma_o^2 \ln\frac{M}{1-M} + a_2(iT_b)^2 - a_1(iT_b)^2}{2[a_2(iT_b) - a_1(iT_b)]} \tag{4.85}$$

Minimum probability of bit error is thus

$$\begin{aligned}
P_{b,\,minimum,\,nonequiprobable} &= P\{b_i = 1 \text{ and } z(iT_b) < \tau_{opt}\} + P\{b_i = 0 \text{ and } z(iT_b) \geq \tau_{opt}\} \\
&= P(b_i = 1)P\{z(iT_b) < \tau_{opt}|b_i = 1\} + P(b_i = 0)P\{z(iT_b) \geq \tau_{opt}|b_i = 0\} \\
&= MP\{z(iT_b) < \tau_{opt}|b_i = 1\} + (1-M)P\{z(iT_b) \geq \tau_{opt}|b_i = 0\} \qquad (4.86) \\
&= M\int_{-\infty}^{\tau_{opt}} \frac{1}{\sqrt{2\pi}\sigma_o} e^{\frac{-[x-a_1(iT_b)]^2}{2\sigma_o^2}}\,dx + (1-M)\int_{\tau_{opt}}^{\infty} \frac{1}{\sqrt{2\pi}\sigma_o} e^{\frac{-[x-a_2(iT_b)]^2}{2\sigma_o^2}}\,dx \\
&= MQ\left(\frac{a_1(iT_b) - \tau_{opt}}{\sigma_o}\right) + (1-M)Q\left(\frac{\tau_{opt} - a_2(iT_b)}{\sigma_o}\right)
\end{aligned}$$

Once again, let's verify that the more general equations for optimum threshold and probability of bit error for nonequiprobable signals also hold for equiprobable signals. For equiprobable signals $M = 0.5$, and Equation (4.85) thus reduces to

$$\begin{aligned}
\tau_{opt} &= \frac{2\sigma_o^2 \ln\frac{M}{1-M} + a_2(iT_b)^2 - a_1(iT_b)^2}{2[a_2(iT_b) - a_1(iT_b)]} = \frac{2\sigma_o^2 \ln(1) + a_2(iT_b)^2 - a_1(iT_b)^2}{2[a_2(iT_b) - a_1(iT_b)]} \\
&= \frac{a_2(iT_b)^2 - a_1(iT_b)^2}{2[a_2(iT_b) - a_1(iT_b)]} = \frac{a_2(iT_b) + a_1(iT_b)}{2}
\end{aligned}$$

which is consistent with Equation (4.71). Concerning P_b, Equation (4.86) easily reduces to Equation (4.76).

4.3 Synchronization

In everyday life we often see the need to establish a common time base when two or more people try to coordinate their activities from separate locations. The same need exists for communication systems that transmit and receive continuous signals at physically separated locations. Such systems typically use independent time bases that are often derived from devices called *crystal oscillators*. A typical system is shown in Figure 4-24. Crystal oscillators provide quite precise time references, but in practice their precision is not sufficient to ensure that the receiver can accurately demodulate the transmitted signal (remember how lack of synchronization affected the noise margin and hence the accuracy of the sinc- and raised-cosine-shaped pulses discussed in Chapter 3). To ensure correct recovery of the transmitted data, we must make the timing references identical on the basis of at least the long-term average. Making the timing references identical means that the source and user are *synchronized*.

In this section, we will examine the basic approaches for achieving synchronization between sources and users that have independent timing references. Underlying most synchronization techniques in practical communication systems is the *phase-locked loop* (PLL), which has many uses and applications. Three of the most important applications are

- Providing timing recovery. The main job of the PLL in this application is to generate a stable single-frequency waveform that is equal to the average bit rate of the received signal, while ignoring any fluctuations or transient variation of the bit rate that are caused by interference and/or noise.
- Providing carrier recovery. In this application, the main job is to track the phase of the carrier of the received signal as closely as possible (carrier signals will be discussed in detail in Chapter 5 when we examine bandpass modulation techniques). In practice, some channels tend to introduce significant fluctuations or impairments to the carrier phase and frequency. Accurate demodulation requires that all fluctuations in the carrier phase and frequency be replicated by the receiver. The PLL plays a vital role in carrier recovery techniques.

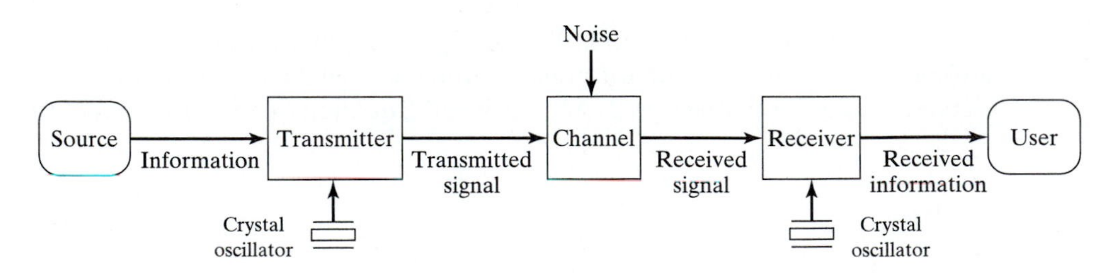

Figure 4-24 The transmitter and receiver, being at separate locations, use independent time bases (typically derived from crystal oscillators).

- Implementing frequency synthesizers. The crystal oscillator provides a frequency (timing) base for the source or user. In practice, we need to find ways to derive many frequencies from this base in order to perform channel switching or to generate different data rates. (These functions are important in multiplexing, which we will discuss in Chapter 7.) It is expensive to derive these different frequencies and rates using oscillators alone, since we would need a number of high-performance oscillators within each source or user. The use of a properly designed PLL in conjunction with a single crystal oscillator provides a cost-effective solution for building a programmable frequency synthesizer that has high resolution, fast switching time, and very-low-magnitude spurious signals and noise.

4.3.1 Basic Structure of Continuous-Time Phase Locked Loops

Basic PLL structure is shown in Figure 4-25. The input to the PLL is the signal $x(t)$. The function of the PLL is to track one of the attributes of the input signal: the bit clock timing, the carrier phase, or the carrier frequency. The output of the PLL is the signal $y(t)$ that tracks a specific attribute of the input signal $x(t)$. The PLL consists of three elements:

1. The *voltage-controlled oscillator* (VCO). The VCO is a device whose output has an instantaneous frequency that can be controlled by the input signal applied to the VCO.
2. The phase detector. The phase detector measures the phase error between the input signal $x(t)$ and the VCO output $y(t)$ and produces the difference in the form of an error signal $e(t)$ as shown in Figure 4-25. The phase detector can be either a simple device such as a multiplier or it can be very complicated. The degree of complexity will depend on the application, the impairments introduced by the channel, and the effects of the impairments on the transmitted signal.
3. The loop filter. The error signal $e(t)$ is filtered in order to generate the control signal $c(t)$ that drives the VCO.

4.3.2 Analysis of the PLL with Linearized Dynamics

PLL systems are inherently complex, and their analysis is further complicated by the nonlinear behavior of their VCOs and phase detectors. However, with the aid of some carefully selected simplifying assumptions, we can develop a useful analytical approach that will enable us to understand the fundamental behavior and characteristics of the PLL.

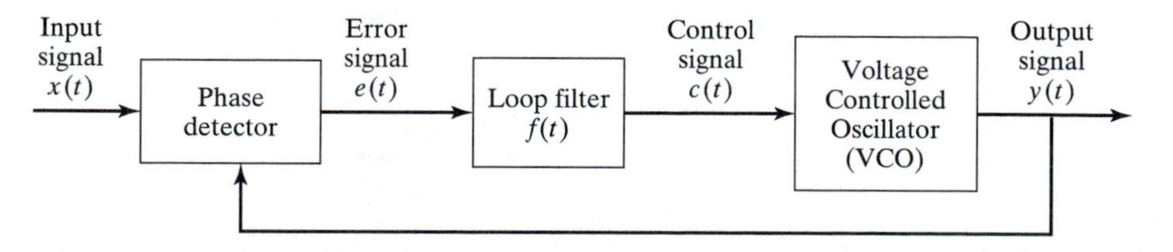

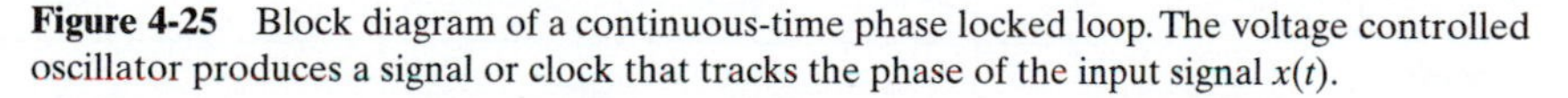

Figure 4-25 Block diagram of a continuous-time phase locked loop. The voltage controlled oscillator produces a signal or clock that tracks the phase of the input signal $x(t)$.

Our first assumption is related to the form of the input signal to the PLL:

$$x(t) = A_x \cos(\omega t + \theta(t)) \tag{4.87}$$

Let's assume that A_x and θ are constants. In practice, the amplitude, frequency, and phase of the input signal may have a complicated form, but for the purpose of our analysis the signal form given by Equation (4.87) is adequate as long as we select the appropriate phase detector. The output of the VCO, $y(t)$, has a similar form:

$$y(t) = A_y \cos(\omega t + \phi(t)) \tag{4.88}$$

When $\phi(t)$ is a constant, the frequency of the VCO output is ω and is called the *natural*, or *free-running*, frequency of the VCO.

The second assumption is related to the phase detector, whose main function is to detect the difference between the phase of the PLL input signal $x(t)$ and the VCO output signal $y(t)$. The phase detector can have many forms and structures. For the purpose of our analysis, we will consider a phase detector whose output is in the form:

$$e(t) = W(\theta(t) - \phi(t)) \tag{4.89}$$

Let's define the phase difference $\Psi(t)$ as

$$\Psi(t) = \theta(t) - \phi(t) \tag{4.90}$$

$\Psi(t)$ represents the phase error, which is the difference between the phase of the input signal to the PLL and the output signal of the VCO. If the error is equal to zero, we have a perfect tracking of the phase of the input signal. Figure 4-26 provides a representation of the phase error in the complex plane.

The function $W(\theta(t) - \phi(t))$ in Equation (4.89) is the output of the phase detector. Figure 4-27 shows one type of phase detector, called a *sawtooth phase detector*. This phase detector has a linear characteristic between $-\pi$ and π. Note, however, if the phase difference $\Psi(t) = \theta(t) - \phi(t)$ suddenly changes by 2π, this change will not be detected by the phase detector. For this reason, it is essential to state that the phase detector operation is

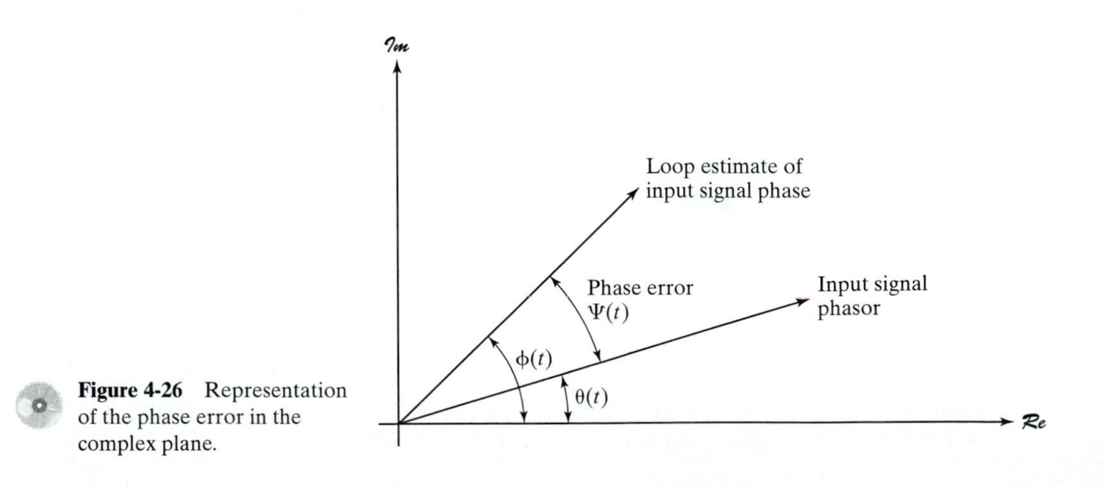

Figure 4-26 Representation of the phase error in the complex plane.

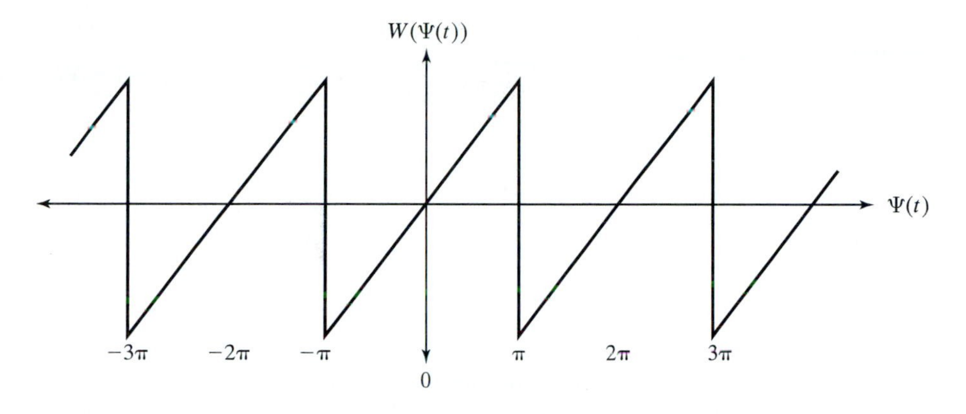

Figure 4-27 Output of the sawtooth phase detector as a function of phase error $\Psi(t)$.

valid only as long as phase difference is bounded by $\pm\pi$. For the sawtooth phase detector represented by Figure 4-27, the error $e(t)$ in Equation (4.89) will be

$$e(t) = k_d(\theta(t) - \phi(t)) \tag{4.91}$$

where k_d is the slope of the sawtooth line in Figure 4-27 between $\pm\pi$. It is possible to make this gain part of the loop filter and hence set $k_d = 1$ in Equation (4.91).

The third assumption is related to the VCO. As shown in Figure 4-25, a VCO has an input signal $c(t)$, called a *control signal*, and a sinusoidal output signal $y(t)$, where

$$y(t) = A_y \cos(\omega t + \phi(t)) \tag{4.88R}$$

The output signal $y(t)$ has the instantaneous frequency

$$\frac{d}{dt}[\omega t + \phi(t)] = \omega + \frac{d\phi(t)}{dt} \tag{4.92}$$

What we want to achieve in an ideal VCO is the ability to control the instantaneous frequency with the control signal $c(t)$. A practical VCO may have gain, k_v, but again we can make that gain part of the loop filter, and hence we set k_v equal to 1. The VCO should therefore be designed so that

$$c(t) = \frac{d\phi(t)}{dt} \tag{4.93a}$$

or, equivalently,

$$\phi(t) = \int c(t)\, dt \tag{4.93b}$$

As an example, consider an ideal VCO with a constant control signal $c(t) = K$. From Equations (4.92) and (4.93a), the output of the VCO will be a sinusoid of constant frequency $\omega + K$. Equation (4.93b) tells us that the ideal VCO functions as an integrator to produce instantaneous phase.

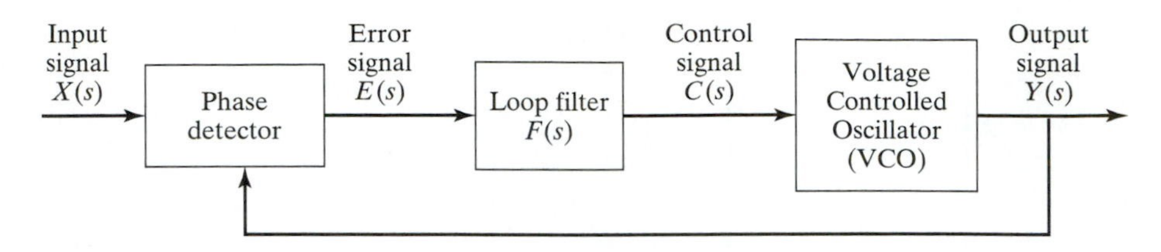

Figure 4-28 Block diagram of the PLL in the *s*-domain (Laplace transform). Compare with Figure 4-25.

With the assumptions related to the idealized input and output signals of the PLL, the idealized phase detector, and the VCO functions, it is possible now to derive the relationship between the phase $\theta(t)$ of the PLL input signal and the phase $\phi(t)$ of the output signal of the VCO. The PLL block diagrams shown in Figures 4-25 and 4-28 are useful in discerning this relationship. Figure 4-25 shows the time domain representation of the PLL. Figure 4-28 shows the *s*-domain (Laplace) representation (We hope you remember Laplace transforms and the *s*-domain from your electronics and mathematics classes.) Let $f(t)$ represent the impulse response of the loop filter, and let $F(s)$ represent the Laplace transform of $f(t)$. Similarly, let $E(s)$, $C(s)$, $\Theta(s)$, and $\Phi(s)$ represent the Laplace transforms of the error signal $e(t)$, the control signal $c(t)$, the phase of the input signal $\theta(t)$, and the phase of the output signal $\phi(t)$, respectively. We want to derive the transfer function that relates the phase of the output signal $\phi(t)$ to the phase of the input signal $\theta(t)$. From Equation (4.91), with $k_d = 1$ (as discussed previously), we can express the error signal in the *s*-domain as

$$E(s) = \Theta(s) - \Phi(s) \tag{4.94}$$

From Figure 4-28,

$$C(s) = E(s)F(s) \tag{4.95}$$

From Equation (4.93a), we note that

$$C(s) = s\Phi(s) \tag{4.96}$$

Combining Equations (4.94)–(4.96) and solving for the phase transfer function (see Problem 4.18),

$$\frac{\Phi(s)}{\Theta(s)} = \frac{F(s)}{[F(s) + s]} \tag{4.97}$$

Substituting Equations (4.96) and (4.97) into Equation (4.95), we can express the error signal $E(s)$ as

$$E(s) = \frac{s\Theta(s)}{[F(s) + s]} \tag{4.98}$$

A quick look at the right side of Equation (4.97) reveals the critical role that the loop filter plays in the behavior of the PLL. This is true for practical realizations as well as for the ideal PLL.

Our main interest is the steady-state behavior of the PLL—in particular, the steady-state error, which is the value of $e(t)$ after the PLL reaches steady-state condition. The steady-state error E_{ss} can be expressed as

$$E_{ss} = \lim_{t \to \infty} [e(t)] \tag{4.99}$$

Using the final value theorem of the Laplace transform, Equation (4.99) can also be written as

$$E_{ss} = \lim_{s \to 0} [sE(s)] \tag{4.100}$$

Substituting from Equation (4.98) into Equation (4.100), we obtain

$$E_{ss} = \lim_{s \to 0} \left[\frac{s^2 \Theta(s)}{[F(s) + s]} \right] \tag{4.101}$$

Let's now consider two examples, showing the critical role the loop filter plays in the behavior of the PLL.

Example 4.12 A first-order PLL

A first-order PLL has a simple loop filter $F(s)$ given by

$$F(s) = K$$

and the input to the PLL has the frequency offset

$$\theta(t) = \omega t + \theta$$

Determine the lock range and transfer function for the PLL. What restrictions are necessary on k to ensure stability?

Solution

Taking the Laplace transform of $\theta(t)$,

$$\Theta(s) = \frac{\omega}{s^2} + \frac{\theta}{s}$$

Substituting for $\Theta(s)$ and $F(s)$ in Equation (4.101), we obtain a steady-state error

$$E_{ss} = \frac{\omega}{K}$$

We know from earlier observations that this error must be bounded by $\pm\pi$, which is called the *lock range*. In other words,

$$\left| \frac{\omega}{K} \right| \leq \pi$$

which implies

$$|\omega| \leq \pi |K|$$

The transfer function is

$$\frac{\Phi(s)}{\Theta(s)} = \frac{F(s)}{[F(s) + s]} = \frac{K}{K + s}$$

which is stable as long as $K > 0$.

Example 4.13 A second-order PLL

Consider the case of a second-order PLL with a loop filter given by

$$F(s) = K\left[\frac{s + C_1}{s + C_2}\right]$$

where $K = 1, C_1 = 1$, and $C_2 = -0.6$. Let $\theta(t) = \omega t + \theta$ as in Example 4.12. Determine the lock range and transfer function for the PLL. Determine if the PLL is stable. Substituting into Equation (4.101), the steady-state error E_{ss} is given by

$$E_{ss} = \frac{\omega}{K[C_1/C_2]}$$

Solution

Again, we know that steady-state error must be bounded by $\pm\pi$. Thus,

$$\left|\frac{\omega}{K[C_1/C_2]}\right| \leq \pi$$

Substituting for K, C_1, and C_2, we obtain the lock range for proper PLL operation

$$|\omega| \leq \frac{\pi}{0.6}$$

We must also consider the loop stability, which can be determined from the transfer function. The transfer function is

$$\frac{\Phi(s)}{\Theta(s)} = \frac{F(s)}{[F(s) + s]} = \frac{Ks + KC_1}{s^2 + (K + C_2)s + KC_1}$$

which is stable if the poles have negative real values. This means that the conditions for stability are $C_2 > -K$ and $KC_1 > 0$. Both conditions are satisfied, since $K = C_1 = 1$ and $C_2 = -0.6$.

We see in Example 4.13 that the lock range is dependent on the filter parameters K, C_1, and C_2. Furthermore, two of these filter parameters, K and C_1, also have an effect on the bandwidth of the PLL. To appreciate this, examine Figure 4-29, which is a plot of the PLL transfer function $\Phi(j\omega)/\Theta(j\omega)$ versus frequency. The higher the value of the parameters K and C_1, the larger the bandwidth. In practice, our goal is to limit the PLL bandwidth in order to filter out higher distortion harmonics.

Example 4.13 also illustrates that we must be careful in selecting the filter parameters. The plot in Figure 4-29 shows that the gain for some range of frequencies may be greater than one. In fact, at frequency $\omega = \sqrt{KC_1}$, the gain is equal to $20 \log_{10} 12.5$, a value significantly greater than one. This means that some components of the phase of the input may be amplified. In practice, such a situation is often undesirable.

4.3.3 Frequency Synthesizers

In many communication systems, multiple sources and users share an entire frequency band, which is typically divided into sub-bands, or channels. This practice is called *frequency division multiplexing*, and we will learn more about this subject in Chapter 7.

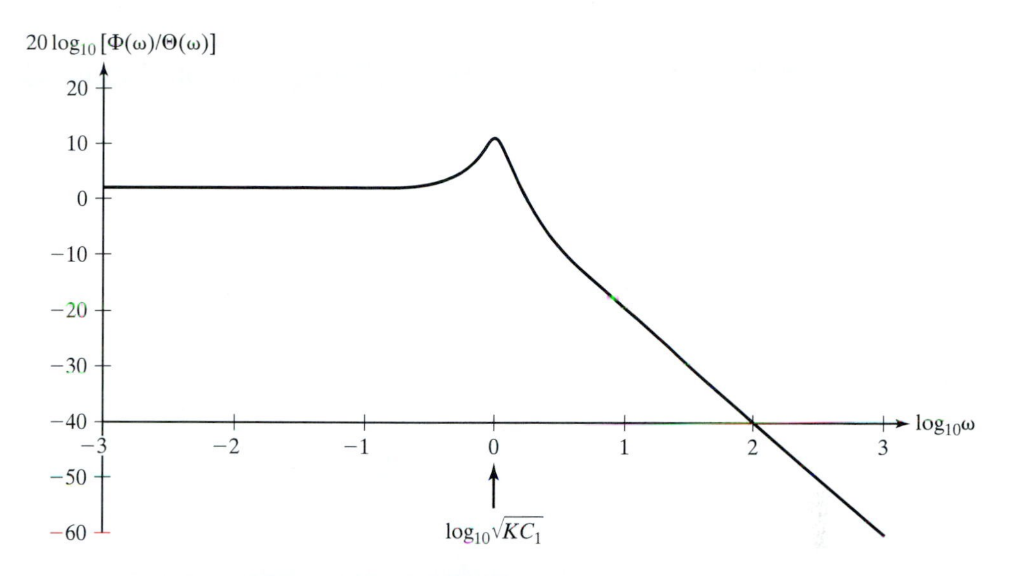

Figure 4-29 Frequency response of PLL in Example 4.13.

Frequency division multiplexing is employed, for instance, in many cellular radio systems. In North America, where cellular systems were first introduced, a system with a total available bandwidth of 60 MHz may divide that bandwidth into 30 kHz channels. In Europe, a cellular system called GSM (Groupe Speciale Mobile) is a common standard. In GSM, a typical system's bandwidth ranges from 5 MHz to 25 MHz and is divided into channels that are 200 kHz wide.

At the beginning of each communications session between two terminals,[2] a pair of channels is assigned by the system to the two terminals for two-way communications. At the end of the session, the channels are freed and may be assigned to other terminals. To operate effectively within a communications system in which a pool of channels is shared, it is essential for each terminal to be able to switch to any channel within the system's frequency band. This is accomplished using a *frequency synthesizer*. One effective design for frequency synthesizers is based on the PLL, as shown in Figure 4-30. The input frequency f_1 is produced by a crystal oscillator and is called the *reference frequency*. The signal with frequency f_1 is passed through a frequency divider, resulting in a signal with frequency f_1 divided by an integer m. The output of the frequency divider is then fed back to a PLL circuit that also has a frequency divider in its feedback loop. The VCO output frequency f_2 is thus divided by an integer n before it is fed to the phase detector of the PLL.

From the basic principle of the PLL, we know that when signal tracking is complete there is no phase error, meaning that

$$\frac{f_1}{m} = \frac{f_2}{n} \tag{4.102}$$

[2]In telephone communications and many other applications, each party may want to both send and receive information. Each party is thus both a source and a user. In such cases, each party is often called a *terminal*.

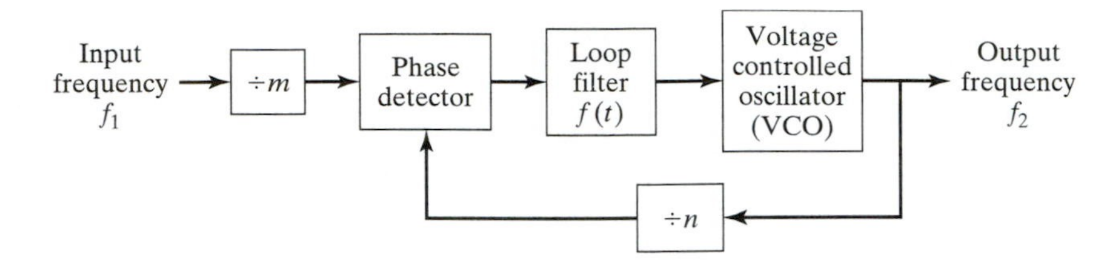

Figure 4-30 A frequency synthesizer employing a PLL. Note that output frequency $f_2 = nf_1/m$.

Solving for f_2,

$$f_2 = \frac{nf_1}{m} \tag{4.103}$$

Example 4.14 Determining the output frequency of a frequency synthesizer

Consider, for example, a PLL-based frequency synthesizer as shown in Figure 4-30, where $f_1 = 10$ MHz, $n = 4$, and $m = 2000$. In this case, f_2 will be 20 kHz.

It is possible to change the value of f_2 in steps by changing the coefficients m and n of the frequency dividers. The minimum frequency step size, Δf (which can be obtained through incrementing the value of the integer n by one) is given by

$$\Delta f = \frac{(n+1)f_1}{m} - \frac{nf_1}{m} = \frac{f_1}{m} \tag{4.104}$$

In Example 4.14, the minimum frequency increment Δf will be 5 kHz.

We must also consider the relationship between the minimum frequency increment and the bandwidth of the PLL. As we saw in Section 4.3.2, we can control the PLL's bandwidth by selecting the structure and the parameters of its loop filter. It is crucial to design the PLL so that its bandwidth is less than the minimum frequency increment Δf. This ensures that the ripples associated with Δf do not pass through the PLL. These ripples occur because in practice the output of the oscillator is not a pure sinusoid. This small bandwidth condition puts a significant constraint on the design of the PLL. It means that when small values of Δf are required, we must design the PLL to have an even narrower bandwidth. At some point, a very narrow loop bandwidth becomes undesirable, since it leads to more phase jitter and requires a longer time to reach a steady-state condition when the source wants to switch to another frequency.

To overcome this limitation, we can modify the frequency synthesizer system by adding another frequency divider at the output of the VCO, as shown in Figure 4-31. In this case the output frequency f_o is given by

$$f_o = \frac{nf_1}{mk} \tag{4.105}$$

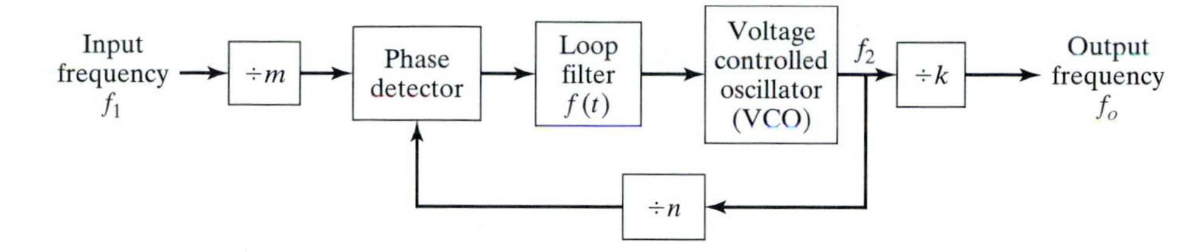

Figure 4-31 A frequency synthesizer employing a PLL with an extra divider to allow a larger loop bandwidth (output frequency $f_o = nf_1/km$).

where k is the integer of the frequency divider placed at the output of the VCO. The new minimum frequency increment associated with f_o will be

$$\Delta f = \frac{f_1}{mk} \tag{4.106}$$

This strategy of adding the extra divider allows us to make Δf as small as we want. We can do so by varying the value of the divider k while maintaining the loop bandwidth less than f_1/m. In other words, we are able to maintain the loop bandwidth at values that are a factor of k higher than the minimum frequency increment.

4.3.4 Timing Recovery

The data stream carried by the signal arriving at the receiver has a certain clock rate. Due to interference and noise, the timing phase of the received bits (or symbols) tends to encounter jitter in the form of fluctuations or transient variations of the bit rate. To recover the digital signal, it is important for the receiver to track the clock of the received signal and produce a frequency reference exactly representing the average bit (or symbol) rate of the received signal. A *clock recovery circuit* does this, and this circuit is an integral part of the receiver. Many clock recovery techniques exist, and some of these techniques are based on PLL circuits.

In this section, we will examine three aspects of clock recovery and acquisition of lock. First, we will study how we can examine the received signal and extract a component that carries information on the clock associated with the bit rate. This will allow us to design a clock recovery circuit to track and capture the exact frequency and phase of the clock. Second, we will study an example of a simple clock recovery circuit that uses a PLL. Third, we will use this example to obtain an intuitive understanding of how the PLL actually works and how it manages to acquire lock.

We will start by considering a class of timing recovery circuits in which a tone related to the clock frequency is extracted from the modulated waveform arriving at the receiver. In some communication systems, an explicit signal in the form of a tone with the clock frequency is transmitted along with the modulated waveform carrying the data. This explicit signal, called a *pilot tone*, is normally at a frequency outside the frequency band of the modulated signal. Clock recovery at the receiver starts by separating the pilot tone from the modulated signal and then using a PLL circuit to track the frequency of the pilot tone and hence recover the clock timing from it. Figure 4-32 shows such a system. The received signal consists of the pilot tone, $p(t)$, and the modulated waveform, $x(t)$. The pilot tone is separated and processed to obtain the clock frequency,

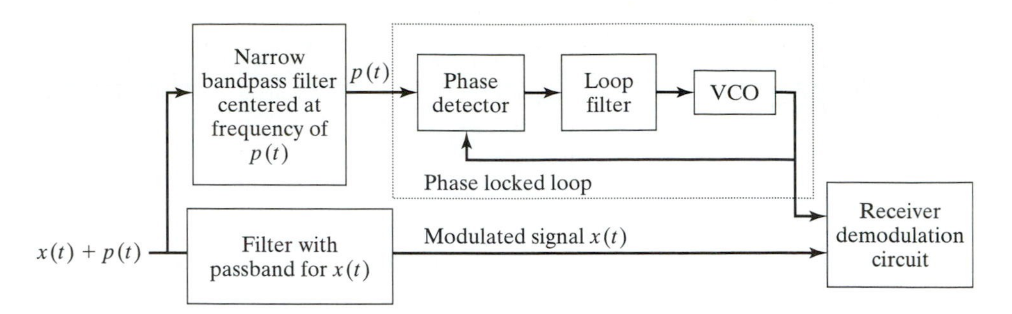

Figure 4-32 Timing recovery circuit using a pilot tone.

which is then used by the demodulator to sample the demodulated signal at the correct time instants. Using a pilot tone has the advantage of simplicity and the disadvantage of using more bandwidth and transmitter power.

In communication systems where the available bandwidth is limited and spectral efficiency is desirable, the pilot tone approach is not suitable. In these cases, only the modulated waveform $x(t)$ is transmitted, and the clock recovery circuit must use some form of signal processing on the received signal to extract information directly related to the clock frequency. This configuration is shown in Figure 4-33. The PLL circuit is once again used for tracking the clock signal extracted from $x(t)$. How we extract the clock timing tone from $x(t)$ will depend on the modulation format of $x(t)$ itself. We will consider two cases to illustrate this point.

Case 1: Timing recovery when demodulated signal has nonzero dc component

Consider a baseband PAM signal carrying digital information in the form of pulses. Let $p(t)$ represent the pulse shape (rectangular, sinc, or raised cosine, as discussed in Chapter 3) with a maximum amplitude of 1 volt, let s_1 be the amplitude of the pulse used to signify a "1," and let s_2 be the amplitude of the pulse used to signify a "0." Suppose we want to transmit n bits of data. We can represent the transmitted signal $x(t)$ as a series of pulses (remember that the pulses are centered in the middle of the bit periods)

$$x(t) = \sum_{i=1}^{n} d_k p\left\{t - \left[(i-1)T_b + \frac{T_b}{2}\right]\right\} \tag{4.107}$$

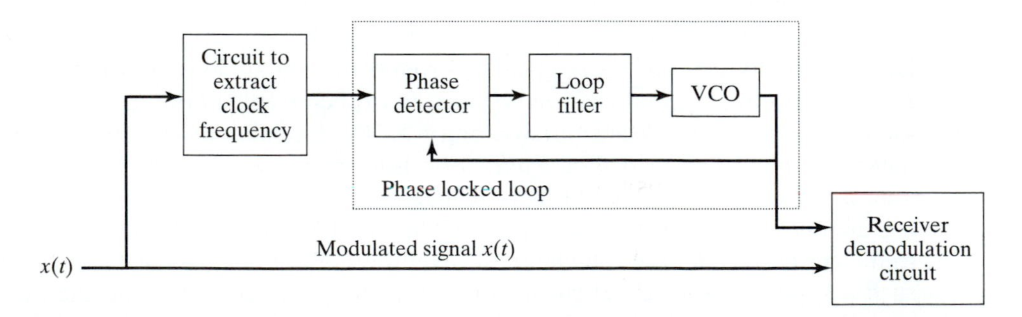

Figure 4-33 Timing recovery circuit using only the received modulated signal.

where

$$d_i = \begin{cases} s_1 & \text{if } b_i = 1 \\ s_2 & \text{if } b_i = 0 \end{cases} \tag{4.108}$$

and b_i represents the ith data bit. Let's now consider the mean value of the pulse amplitudes

$$\mu = \frac{s_1 + s_2}{2} \tag{4.109}$$

If μ is nonzero (i.e., if the pulses used to represent a "1" and a "0" are the same shape but are asymmetric), it is possible to rewrite Equation (4.107) as follows:

$$x(t) = \mu \sum_{i=1}^{n} p\left\{t - \left[(i-1)T_b + \frac{T_b}{2}\right]\right\} + \sum_{i=1}^{n} (d_k - \mu) p\left\{t - \left[(i-1)T_b + \frac{T_b}{2}\right]\right\} \tag{4.110}$$

The first term of Equation (4.110) is independent of the data and in fact is a periodic signal with period T_b. It can be considered a deterministic timing signal. The second component is data-dependent and is zero mean and random.

Example 4.15 Timing recovery for asymmetric rectangular pulses

The first signal in Figure 4-34 is a waveform $x(t)$, which can be expressed by Equation (4.107). A rectangular pulse with a 50% duty cycle is used to transmit the data: An amplitude of 2 volts is used to signify a "1" and an amplitude of -1 volt is used to signify a "0." The transmitted data is "100101." The waveform has nonzero mean, and so it can be decomposed into 2 components, as established in Equation (4.110). The first component, represented by the first term in Equation (4.110) and by the second signal in Figure 4-34, is independent of the data and is periodic with period T_b. We will use this as our timing signal. The second component, represented by the second term in Equation (4.110) and by the third signal in Figure 4-34, contains the data.

The fundamental frequency of the periodic signal (the second signal in Figure 4-34) is $f_o = 1/T_b = r_b$, the bit rate of the modulated signal. We can capture this fundamental frequency component using a narrow bandpass filter centered at r_b Hz. We can then input this fundamental frequency component to the PLL shown in Figure 4-33, and the steady-state output of the PLL will be the recovered clock signal. Examining Figure 4-33, we now know that the block marked "Circuit to extract clock frequency" is merely a bandpass filter.

It is important to remember that Example 4.15 is relevant only when the mean value of the pulse amplitudes is nonzero (i.e., when the pulse amplitudes are asymmetric). The pulses of the PAM waveform can be any shape—the 50% duty cycle rectangular pulses used in Figure 4-34 make an easy-to-see example, but they require excessive bandwidth (95% in-band power will require a bandwidth of $8r_b$). We can use sinc-shaped or raised-cosine-shaped pulses with maximum pulse width and achieve better spectral efficiency while still extracting the timing information using a narrow bandpass filter followed by a PLL. However, as we established in Section 4.2.4.1 and Example 4.11, asymmetric amplitude pulses require more transmitted power to achieve the same accuracy as symmetric pulses.

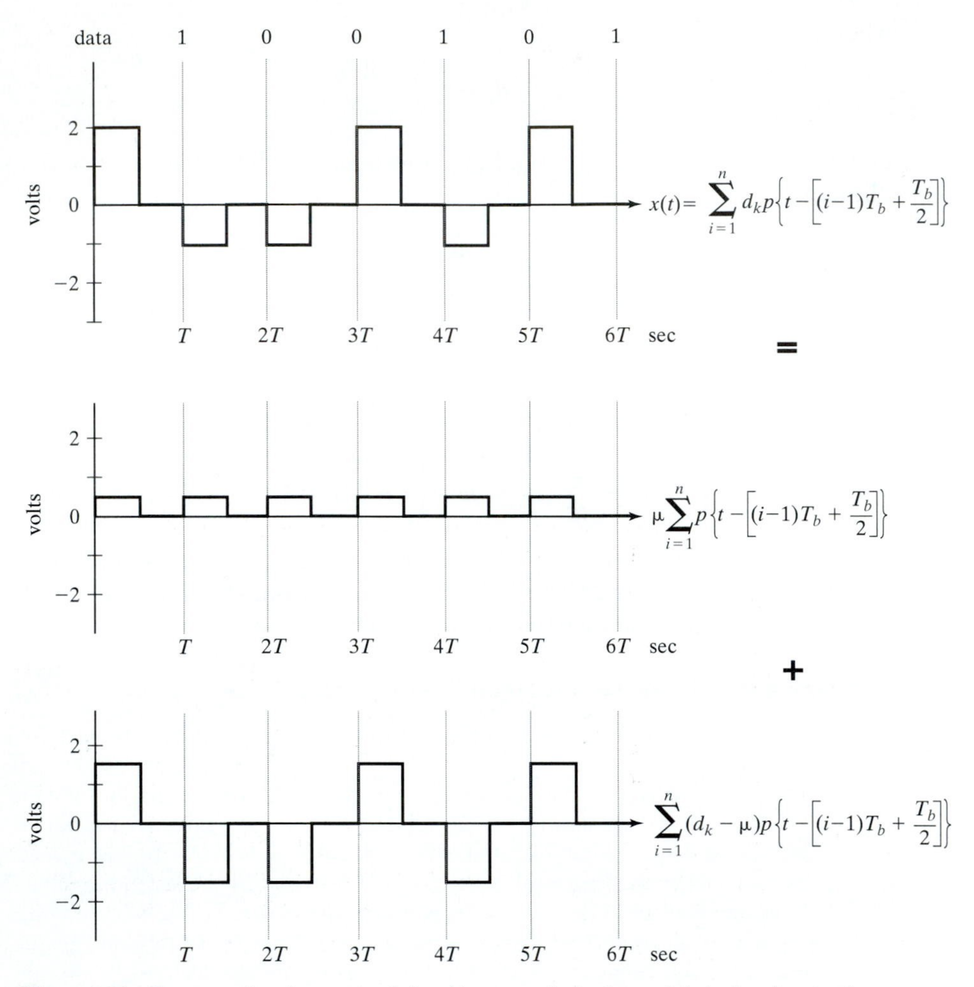

Figure 4-34 Decomposing the received signal into a periodic deterministic signal and a data-dependent signal.

Case 2: Timing recovery when demodulated signal has no dc component

We've seen that asymmetric pulses allow us to use simple circuitry to extract timing information, but at the cost of increased transmitted power. If we use symmetric pulses to reduce the required transmitted power, then the demodulated signal has no dc component and $\mu = 0$ in Equation (4.110), meaning that we cannot extract a signal from $x(t)$ that contains a harmonic component directly related to the clock frequency.

Can we develop an approach for extracting timing information in systems that require bandwidth efficiency and low transmitted power? We've established that such systems cannot use an out-of-band pilot tone or asymmetric pulses. One approach used in practice for symmetric pulses is to pass the received signal through a circuit with a memoryless nonlinearity. The main idea behind this approach is that the signal at the output of the nonlinear circuit will contain several harmonic components, and one of these components will be at the clock frequency of the signal $x(t)$ or at a multiple of the clock frequency. Such a component can be extracted using a suitable bandpass filter

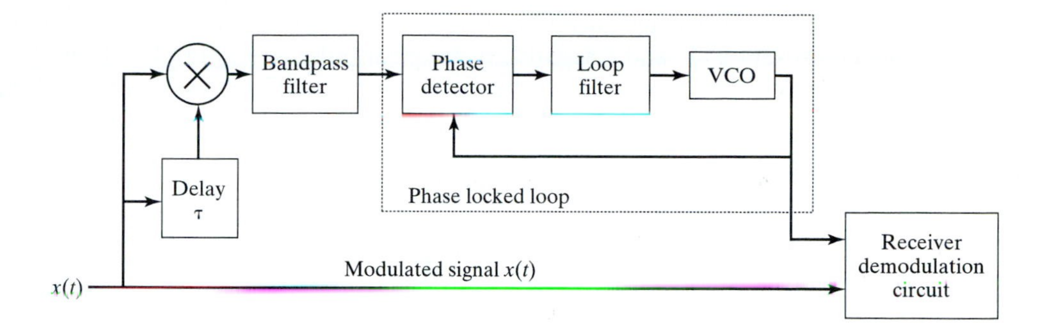

Figure 4-35 Timing recovery circuit for a modulated signal consisting of symmetric pulses.

and then applied as input to the PLL of the clock recovery circuit. We can, of course, expect that the harmonic component at the clock frequency will be mixed with noise as a result of passing the signal $x(t)$ through a nonlinear circuit.

An example of memoryless nonlinear processing of the input signal is shown in Figure 4-35. The signal $x(t)$ is passed through a filter with an ideal delay of τ seconds, where τ is a fraction of the bit period ($\tau < T_b$). The signal $x(t)$ is then multiplied by its delayed version $x(t - \tau)$ to produce the output signal $x(t)x(t - \tau)$. If the modulated signal $x(t)$ can be described as in Equation (4.107), then the output signal $x(t)x(t - \tau)$ will contain several spectral components, one of which will be at the clock frequency. As before, this component can be filtered using a bandpass filter and then applied as the input to the PLL circuit.

The analysis used to determine the exact value of the spectral or harmonic component at the clock frequency is fairly complicated. Instead, we will use a graphical approach to illustrate how this delay-and-multiply circuit works. In Figure 4-36, we start

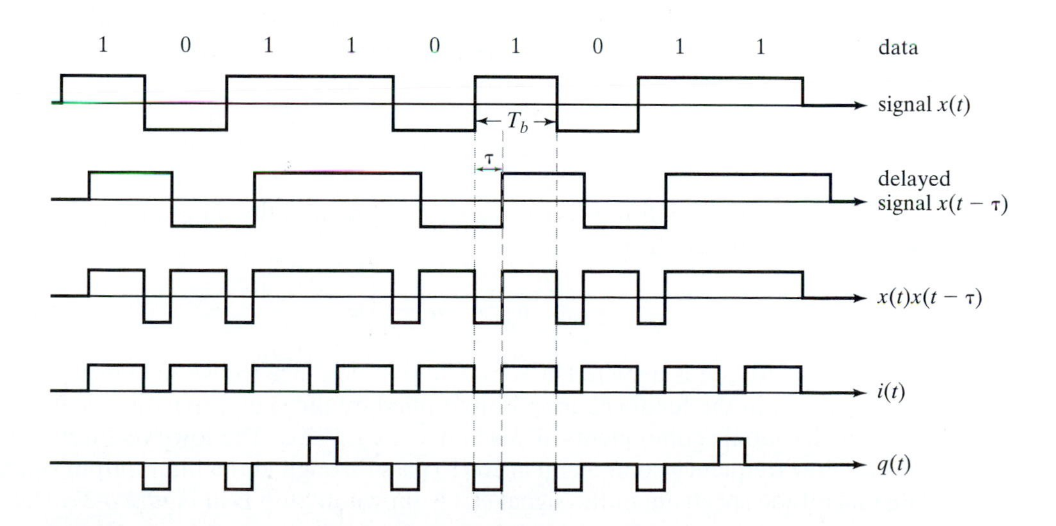

Figure 4-36 Timing diagram of delay circuit in Figure 4-35.

with a PAM signal $x(t)$ produced by using a positive rectangular pulse of amplitude 1 volt to represent a "1" and a negative rectangular pulse of amplitude -1 volt to represent a "0." Note that this signal can be represented using Equation (4.107) and that the mean of its pulse amplitudes is 0. Now multiply $x(t)$ by a delayed version of itself, $x(t - \tau)$. The delayed signal and the product are both shown in Figure 4-36 (in Figure 4-36, $\tau \approx 0.33T_b$ but, as stated earlier in this section, the only requirement on the delay τ is that it be less than the bit period).

As is also shown in Figure 4-36, the product $x(t)x(t - \tau)$ can be decomposed into the sum of two signals:

$$x(t)x(t - \tau) = i(t) + q(t) \tag{4.111}$$

where the signal $i(t)$ is periodic and independent of the data. Inspecting Figure 4-36 shows that $i(t) = x^2(t)$ during the portion of the bit period where $x(t)$ and $x(t + \tau)$ represent the same bit, and $i(t) = 0$ otherwise. As long as $\tau < T_b$, $i(t)$ will be periodic with period T_b. The signal $q(t)$ is pseudo-random and contains the data from the received, modulated signal $x(t)$. Since $i(t)$ is periodic with period T_b, its fundamental frequency is $f_o = 1/T_b = r_b$. This fundamental frequency can be extracted using a narrow bandpass filter, and the output of the bandpass filter can then be input to a PLL. The steady-state output of the PLL will be the recovered clock signal. Note that the component $q(t)$ is not used in the demodulation process; the demodulator extracts the data directly from the received signal $x(t)$ using the clock timing recovered from the output of the bandpass filter and the PLL circuit.

Let's now study the structure of a simple PLL circuit and see if we can form an intuitive understanding of how the PLL actually works and how it manages to acquire lock. The diagram at the top of Figure 4-37 is for a simple PLL circuit. The forward loop has a lowpass filter, which feeds a control signal to the VCO. The VCO produces an output signal $y(t)$, which represents the recovered clock timing. The phase detector of the PLL circuit is implemented using a multiplier. The input signal $r(t)$ has a spectral component at the clock frequency ω_{in}. The objective of the PLL circuit is to track this frequency, meaning that it should produce an output signal $y(t)$, which ideally is a pure harmonic at the clock frequency ω_{in} embedded in the signal $r(t)$. In practice, and as analyses in Equation (4.101) and Example 4.12 have shown, a steady-state error may exist in the tracking process. The designer of a PLL circuit aims to make this error very small and predictable.

At the initial stage of operation of the PLL circuit, the frequency of the input signal $r(t)$ is ω_{in} and the input to the VCO is zero. Thus, the VCO is initially running at the free frequency ω_F. The difference between the input and the output frequencies of the PLL is denoted by $\Delta\omega$. Thus,

$$\omega_{in} = \omega_F + \Delta\omega \tag{4.112}$$

The initial condition is depicted in Step A of the sequence shown in Figure 4-37. When the signal $y(t)$ in the feedback loop is multiplied by the signal $r(t)$, the resultant signal will have harmonic components at $\Delta\omega$ and at $2\omega_F + \Delta\omega$. The lowpass filter will allow only the low frequency component at $\Delta\omega$ to pass through in the filter output signal $c(t)$. The magnitude spectrum of the signal $c(t)$ is shown in Step B in Figure 4-37. The signal

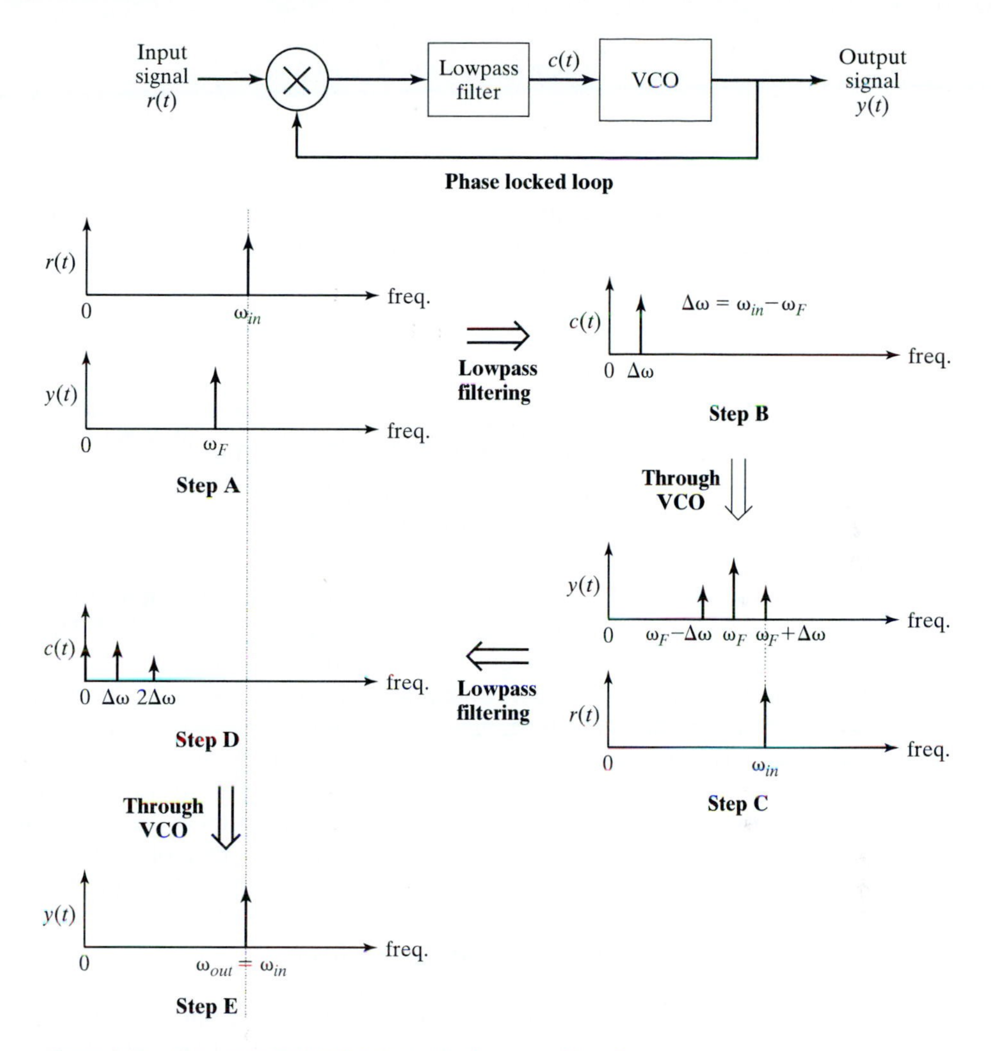

Figure 4-37 PLL acquisition behavior in the frequency domain.

$c(t)$ can thus be expressed as:

$$c(t) = K_c \cos(\Delta\omega t) \tag{4.113}$$

where K_c is a constant that takes into account the lowpass filter gain and the VCO constant. Recall from Equations (4.88) and (4.93b) that the VCO output is expressed as

$$y(t) = A_y \cos(\omega t + \int c(t)dt) \tag{4.114}$$

where $c(t)$ is the control signal. Substituting from Equation (4.113) into Equation (4.114), we obtain

$$\begin{aligned} y(t) &= A_y \cos\left(\omega_F t + K_c \int \cos(\Delta\omega t)dt\right) \\ &= A_y \cos\left(\omega_F t + \frac{K_c}{\Delta\omega}\sin(\Delta\omega t)\right) \end{aligned} \tag{4.115}$$

If we choose the PLL circuit parameters such that $(K_c/\Delta\omega)$ is very small, then $\cos[(K_c/\Delta\omega)\sin(\Delta\omega t)] \approx 1$ and $\sin[(K_c/\Delta\omega)\sin(\Delta\omega t)] \approx (K_c/\Delta\omega)\sin(\Delta\omega t)$, which allows us to simplify Equation (4.115) to

$$\begin{aligned} y(t) &= A_y \cos\left(\omega_F t + \frac{K_c}{\Delta\omega}\sin(\Delta\omega t)\right) \\ &= A_y\left\{\cos(\omega_F t)\cos\left[\frac{K_c}{\Delta\omega}\sin(\Delta\omega t)\right] - \sin(\omega_F t)\sin\left[\frac{K_c}{\Delta\omega}\sin(\Delta\omega t)\right]\right\} \\ &\approx A_y\left(\cos(\omega_F t) - \frac{K_c}{\Delta\omega}\sin(\omega_F t)\sin(\Delta\omega t)\right) \end{aligned} \tag{4.116}$$

Equation (4.116) shows that the signal $y(t)$ contains harmonic components at ω_F and at $\omega_F \pm \Delta\omega$. This is depicted in Step C of the sequence shown in Figure 4-37. The output signal $y(t)$ is then multiplied by the input signal $r(t)$; the resultant signal will have a dc component, two harmonic components at $\Delta\omega$ and $2\Delta\omega$, and higher frequency components close to $2\omega_F$. The high-frequency signal components will be filtered out, and the control signal $c(t)$ will have the spectrum shown in Step D. The signal $c(t)$ will thus drive the VCO from the free-running frequency ω_F to a frequency closer to ω_{in}. As the dc component of the control signal $c(t)$ becomes dominant, the harmonics in the signal will continue to diminish. Assuming that the magnitude of the dc component of the VCO control signal is K_d, then from Equation (4.114) the output of the VCO can be expressed as

$$y(t) \approx A_y \cos(\omega_F t + K_d t) \tag{4.117}$$

Thus, the frequency of the VCO output will be $\omega_F + K_d$. After one or more cycles, the dc control signal K_d will approach $\Delta\omega$ and the frequency of the PLL output $y(t)$ will be approximately equal to the frequency of the input signal $r(t)$, as shown in Step E. Steps A through D in Figure 4-37 will be repeated every time a shift $\Delta\omega$ emerges between the frequency of the input and output signals. In practice, the frequency tracking process is dynamic, meaning that the PLL monitors and tracks the clock frequency on a continuous basis.

In this analysis, we assumed that the lowpass filter is designed such that it passes the signal component with frequency $\Delta\omega$ (going from Step A to Step B). If this assumption does not hold true, then the control signal $c(t)$ in Step B will be too small to drive the VCO away from its free-running frequency and locking will never occur. In practice, we need to limit the bandwidth of the lowpass filter in order to minimize the noise at the output of the filter and to speed up the tracking process. Limiting the bandwidth of the lowpass filter implies that the PLL can operate successfully only when the frequency difference $\Delta\omega$ falls within a specific range. This range is called the *capture range*, or *acquisition range*. It is clear that the capture range is closely related to the bandwidth of the lowpass filter. This

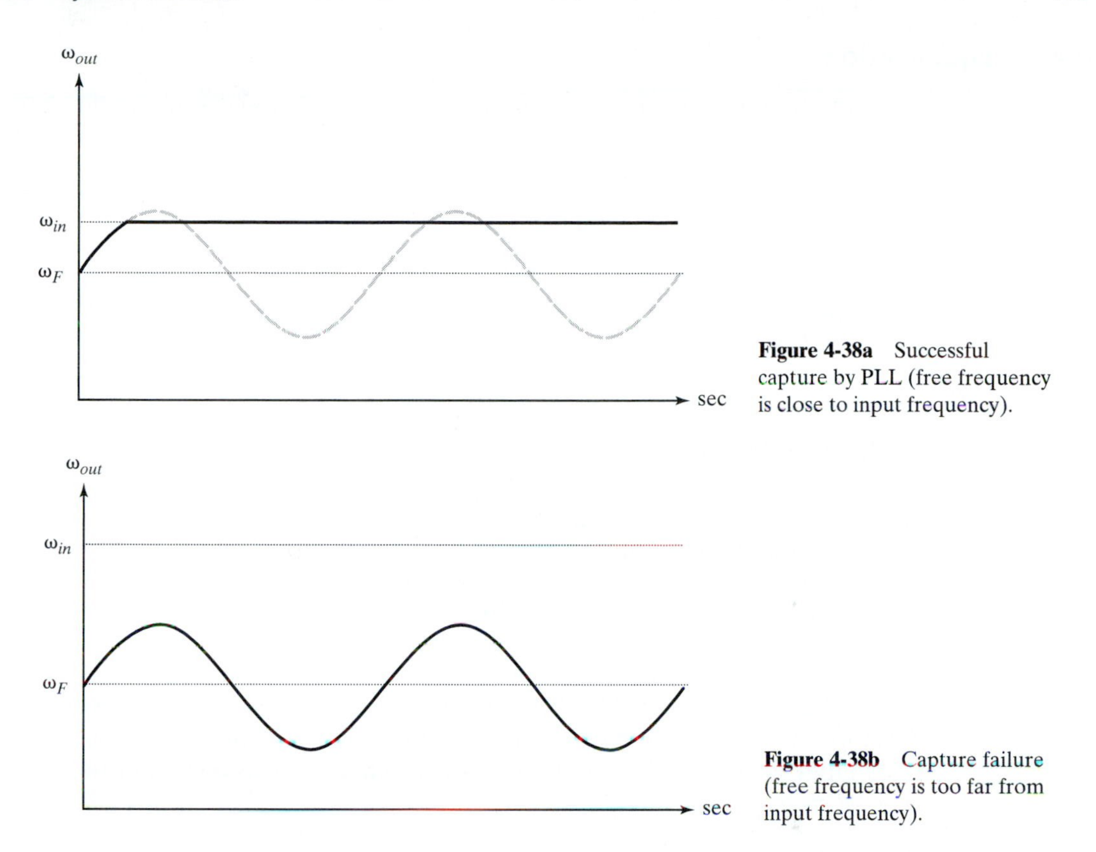

Figure 4-38a Successful capture by PLL (free frequency is close to input frequency).

Figure 4-38b Capture failure (free frequency is too far from input frequency).

point is illustrated in Figure 4-38. If ω_{in} is sufficiently close to the free-running frequency ω_F, frequency acquisition is achieved as the control signal drives the VCO output close to the input clock frequency. Figure 4-38a demonstrates such successful capture. If, however, ω_{in} is sufficiently distant from ω_F, the output of the lowpass filter will be negligible and hence not sufficient to drive the VCO toward lock. Figure 4-38b illustrates the situation where capture of clock frequency fails to materialize.

4.3.5 Further Reading on Synchronization

The topic of synchronization is vast, indeed. More information concerning the method and techniques for carrier and clock synchronization in communication systems can be found in a special issue of the *IEEE Transactions on Communications*, edited by Gardner and Lindsay [4.2]. Three articles (Franks [4.3], Scholtz [4.4], and Carter [4.5]) in this journal include excellent tutorials on clock and carrier recovery techniques and surveys of related publications. IEEE Press publishes a book that contains reprints of several important articles on the design, analysis, and implementation of a number of PLL structures in clock and carrier recovery circuits (Razavi [4.6]). In particular, the article by Roza [4.7] presents a general tutorial, and a related paper by Messerschmitt [4.8] presents detailed analysis of the underlying theory for clock and carrier synchronization. More material on synchronization can be found in several books (Stiffler [4.9], Lindsay and Simon [4.10], Lee and Messerschmitt [4.11], and Mehrotra [4.12]).

4.4 Equalization

Many communication channels, including telephone channels and some radio channels, can be characterized as band-limited linear filters. We can describe such channels by their frequency response $H(f)$, which can be expressed as

$$H(f) = A(f)e^{j\theta(f)} \tag{4.118}$$

where $A(f)$ is the amplitude response and $\theta(f)$ is the phase response. Let's consider an input signal, $S_{in}(f)$, transmitted over the channel. As established in Chapter 2, the output signal from this channel, $S_{out}(f)$, is a function of $S_{in}(f)$ and $H(f)$, and is given by

$$S_{out}(f) = S_{in}(f)H(f) \tag{4.119}$$

Substituting for $H(f)$ using Equation (4.118),

$$S_{out}(f) = A(f)e^{j\theta(f)}S_{in}(f) \tag{4.120}$$

In an ideal communication system, we would like the signal at the receiver to be a replica of the transmitted signal, even if there is a change in amplitude and some time delay. If we closely examine Equation (4.120), we find that such an ideal condition will exist only if, within the bandwidth of the input signal, the amplitude response $A(f)$ is constant and the phase response $\theta(f)$ is a linear function of the frequency f. If $A(f)$ is not constant, the input signal will be distorted after it passes through the channel. This type of distortion is called *amplitude distortion*. If $\theta(f)$ is not a linear function of frequency, the input signal will also be distorted by the channel. This type of distortion is called *delay distortion*.

Example 4.17 Delay distortion

As an example of the effect of delay distortion, consider the transmitted pulse $s_{in}(t)$ shown in Figure 4-39. This band-limited pulse has zeroes periodically spaced in time at points labeled $\pm T$, $\pm 2T$, etc. Transmission of this pulse over a channel that has constant amplitude response $A(f) = 1$ but that has the nonlinear phase response $\theta(f) = f^2$ results in the received pulse $s_{out}(t)$, also shown in Figure 4-39. Note that the zero crossings of the received pulse are no longer periodically spaced. This distortion occurs because, in this example, $\theta(f)$ is not a linear function of f.

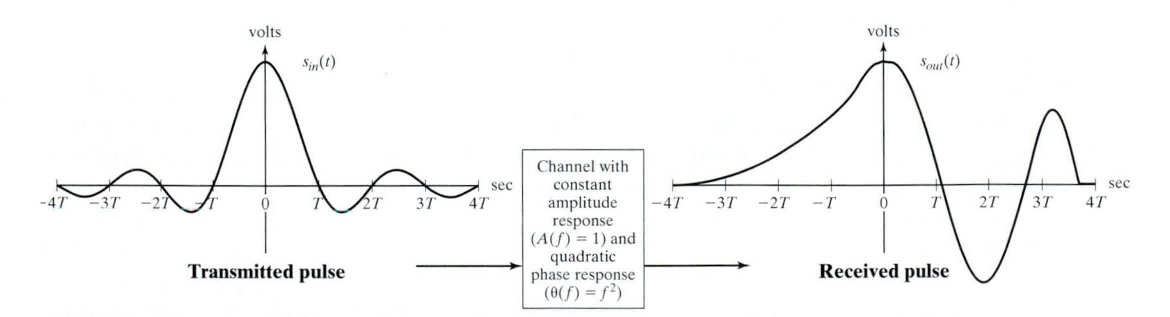

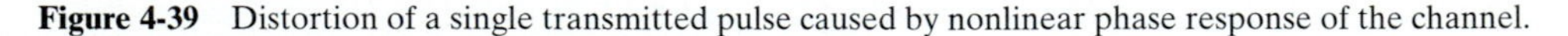

Figure 4-39 Distortion of a single transmitted pulse caused by nonlinear phase response of the channel.

Delay distortion is extremely detrimental. If we transmit a sequence of successive pulses over the nonideal channel in Example 4.17, the received signal is severely distorted. As shown in Figures 4-40a and b, successive pulses are smeared to the point that they are no longer distinguishable as well-defined pulses at the receiver. Thus, delay distortion can cause significant ISI. Delay distortion is one of the major obstacles to reliable high-speed data transmission over limited-bandwidth channels.

The idea behind equalization is simply to compensate for nonideal channel characteristics by additional filtering. In Section 4.4.1, we will develop a mathematical representation for the ISI caused by nonideal channel characteristics, and explain how we deal with ISI using channel equalization techniques.

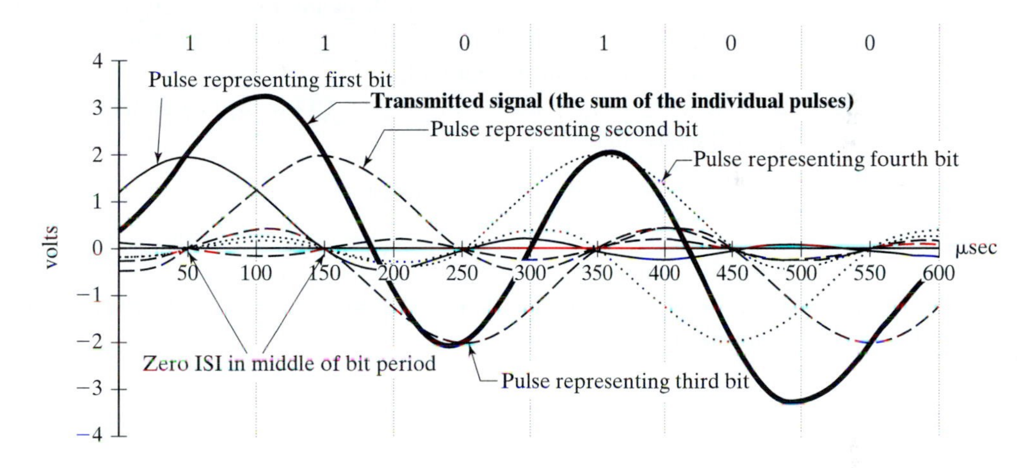

Figure 4-40a Transmitted data sequence using raised cosine pulses.

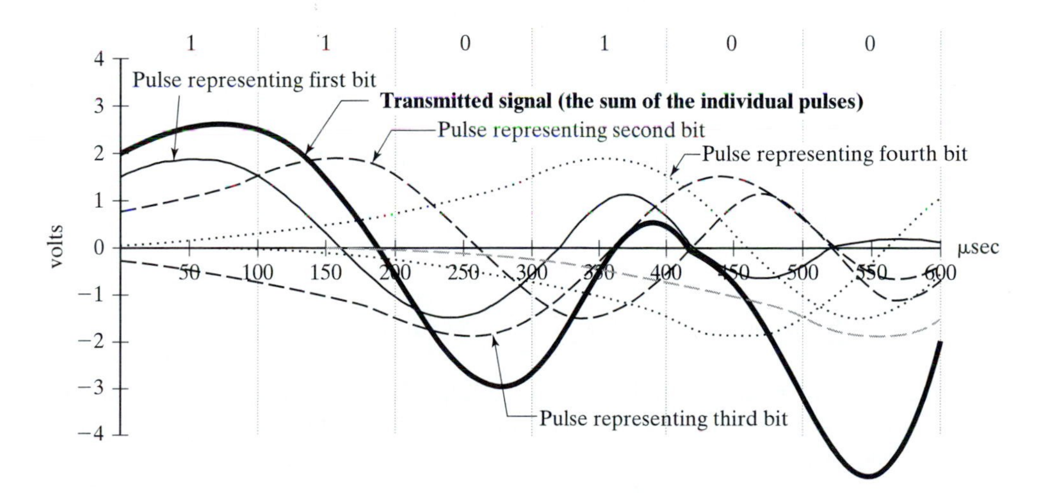

Figure 4-40b Received data sequence with distortion caused by nonlinear phase response of the channel.

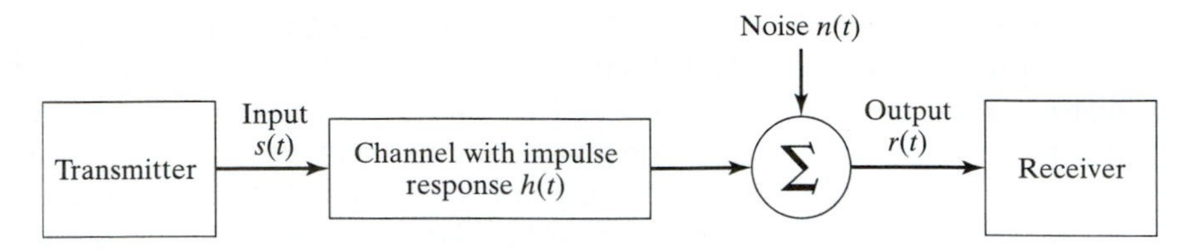

Figure 4-41 A baseband PAM system with nonideal channel.

4.4.1 Intersymbol Interference

Intersymbol interference occurs in all pulse-modulation systems, both baseband and bandpass, as long as the channel frequency response causes amplitude or delay distortion. Let's consider a baseband PAM system, as shown in Figure 4-41, transmitting a series of pulses to represent "1"s and "0"s. Let T_b represent the bit period, and let $h(t)$ represent the impulse response of the channel.

The output signal $r(t)$ of the channel is related to the input signal $s(t)$ and the impulse response of the channel by

$$r(t) = s(t)*h(t) + n(t) \tag{4.121}$$

as we established in Section 2.6. The term $n(t)$ represents the additive white Gaussian noise.

Let $p(t)$ represent the pulse shape (rectangular, sinc, or raised cosine, as discussed in Chapter 3), and define d_i as

$$d_i = \begin{cases} 1 & \text{if } b_i = 1 \\ -1 & \text{if } b_i = 0 \end{cases} \tag{4.122}$$

where b_i represents the ith data bit. Suppose we want to transmit n bits of data. We can now represent the transmitted signal $s(t)$ as a series of pulses (remember that the pulses are centered in the middle of the bit periods):

$$s(t) = \sum_{i=1}^{n} d_i p\left\{t - \left[(i-1)T_b + \frac{T_b}{2}\right]\right\} \tag{4.123}$$

Substituting into Equation (4.121), the received signal can be rewritten as

$$\begin{aligned} r(t) &= \left[\sum_{i=1}^{n} d_i p\left\{t - \left[(i-1)T_b + \frac{T_b}{2}\right]\right\}\right] * h(t) + n(t) \\ &= \sum_{i=1}^{n}\left[d_i p\left\{t - \left[(i-1)T_b + \frac{T_b}{2}\right]\right\} * h\left\{t - \left[(i-1)T_b + \frac{T_b}{2}\right]\right\}\right] + n(t) \end{aligned} \tag{4.124}$$

Suppose we want to sample the received signal at the time corresponding to the center of each bit period. If the channel introduces a delay of t_o, sampling for the jth bit occurs at time

$$t = (j-1)T_b + \frac{T_b}{2} + t_o \tag{4.125}$$

The sampled value of the received signal corresponding to the jth bit is

$$
\begin{aligned}
r\left\{(j-1)T_b + \frac{T_b}{2} + t_o\right\} &= \left[\sum_{i=1}^{n} d_i p\left\{(j-1)T_b + \frac{T_b}{2} - \left[(i-1)T_b + \frac{T_b}{2}\right]\right\}\right. \\
&\quad \left.\times h\left\{(j-1)T_b + \frac{T_b}{2} + t_o - \left[(i-1)T_b + \frac{T_b}{2}\right]\right\}\right] + n\left\{(j-1)T_b + \frac{T_b}{2} + t_o\right\} \\
&= \left[\sum_{i=1}^{n} d_i p\{(j-i)T_b\} h\{(j-i)T_b + t_o\}\right] \\
&\quad + n\left\{(j-1)T_b + \frac{T_b}{2} + t_o\right\}
\end{aligned} \tag{4.126}
$$

Note that t_o does not appear in the calculation of the pulses in the first term on the right side of Equation (4.126) because it represents a delay in the transmitted signal. Let's expand Equation (4.126) slightly:

$$
\begin{aligned}
& r\left\{(j-1)T_b + \frac{T_b}{2} + t_o\right\} \\
&\quad = \left[\sum_{i=1}^{n} d_i p\{(j-i)T_b\} h\{(j-i)T_b + t_o\}\right] + n\left\{(j-1)T_b + \frac{T_b}{2} + t_o\right\} \qquad (4.127) \\
&\quad = d_j p(0) h(t_o) + \left[\sum_{\substack{i=1 \\ i \neq j}}^{n} d_i p\{(j-i)T_b\} h\{(j-i)T_b + t_o\}\right] + n\left\{(j-1)T_b + \frac{T_b}{2} + t_o\right\}
\end{aligned}
$$

Let's now examine Equation (4.127) in detail. The first term on the right side is the most important and most useful for us, since it corresponds to the pulse representing the jth data bit. We know that $h(t_o)$ is a constant that we can obtain directly from the impulse response $h(t)$ at time t_o. Convolution by $h(t_o)$ can thus be replaced by multiplication by $h(t_o)$. We can see that the first term can be used to identify the jth data bit. The last term is the additive noise. The middle term contains components that are related to the neighboring pulses. Thus, the middle term represents the ISI. In other words, Equation (4.127) shows that the output signal at the time corresponding to the center of the jth bit period contains components contributed by the jth pulse, which is the desired signal, noise, and components contributed by the neighboring pulses, which we consider interference because they impede our ability to identify the data corresponding to the jth bit.

We can readily see from Equation (4.127) that if the middle term is zero, there will be no ISI; and the portion of Equation (4.127) related to the transmitted signal will be reduced to

$$
r\left\{(j-1)T_b + \frac{T_b}{2} + t_o\right\} = d_j p(0) h(t_o) \tag{4.128}
$$

This equation achieves our goal of being able to identify the jth transmitted pulse (and hence the jth data bit) once we receive the output signal $r(t)$ at the instant in the center

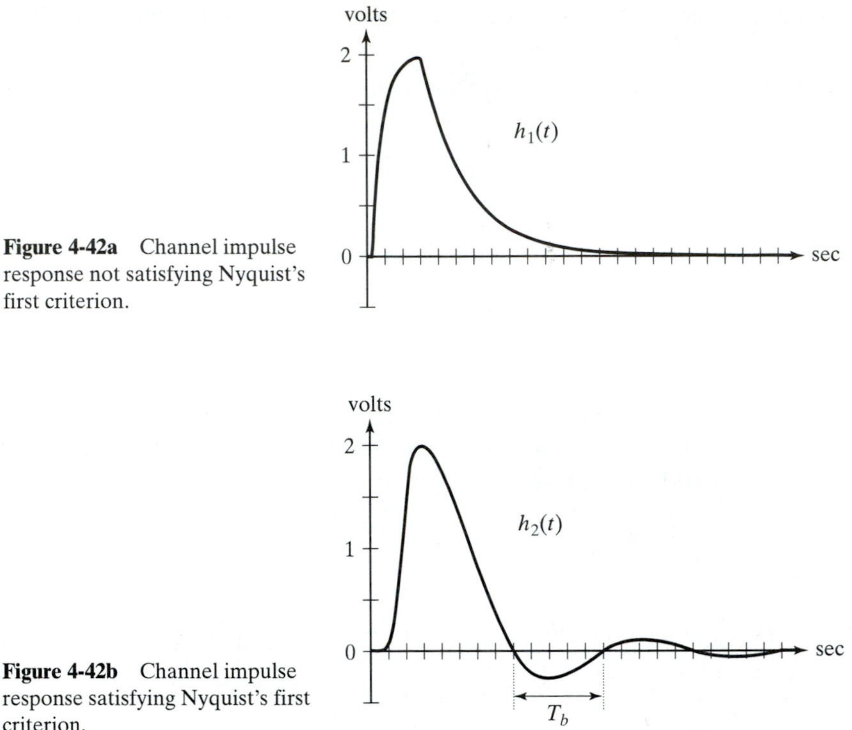

Figure 4-42a Channel impulse response not satisfying Nyquist's first criterion.

Figure 4-42b Channel impulse response satisfying Nyquist's first criterion.

of the jth bit period. The condition required for the middle term of Equation (4.127) to be equal to zero is that the impulse response of the channel, $h(t)$, must be equal to zero at each time instant $t = kT_b + t_o$ for every $k \neq 0$. In other words, the channel impulse response must have zero crossings at T_b-spaced intervals. When the impulse response has such uniformly spaced zero crossings, it is said to satisfy *Nyquist's first criterion*. The channel impulse response $h_1(t)$ shown in Figure 4-42a does not satisfy Nyquist's first criterion because it does not have zero crossings at uniformly T_b-spaced intervals. The channel impulse response $h_2(t)$ shown in Figure 4-42b does satisfy the criterion because it has zero crossings at uniformly T_b-spaced intervals.

The effect of ISI can be seen in practice when we trace the received signal on an oscilloscope with its time base synchronized to the bit rate. Figures 4-43a and b show such a trace for a binary PAM signal. We call this trace the *eye pattern*, or *eye diagram*. If the channel impulse response satisfies Nyquist's first criterion and no ISI exists, we will be able to see on the oscilloscope screen two distinct levels of $+1$ and -1 at the sampling instant t_o. The "eye" in Figure 4-43a is fully open and the distortion is zero. If ISI occurs, then we will observe the "eye" closing somewhat, as shown in Figure 4-43b and we will see distortion around the signal levels $+1$ and -1 (called *peak distortion*). The more severe the ISI, the larger the peak distortion, until we reach a situation where the eye pattern is totally smeared. Clearly, under such conditions, we will not be able to identify the signal that was transmitted.

We place an equalizer in the path of the received signal to reduce the ISI as much as possible in order to maximize the probability of correct demodulation. Now that we

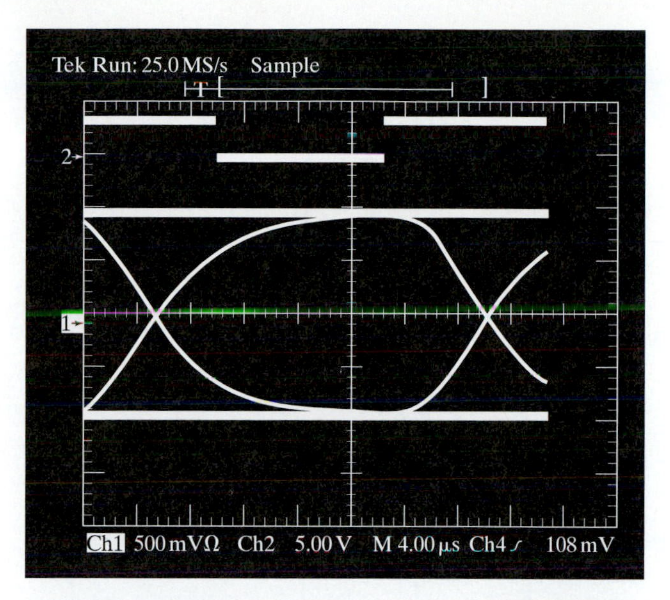

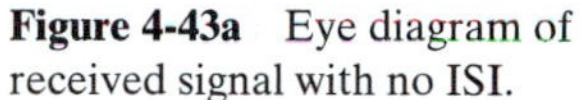
Figure 4-43a Eye diagram of received signal with no ISI.

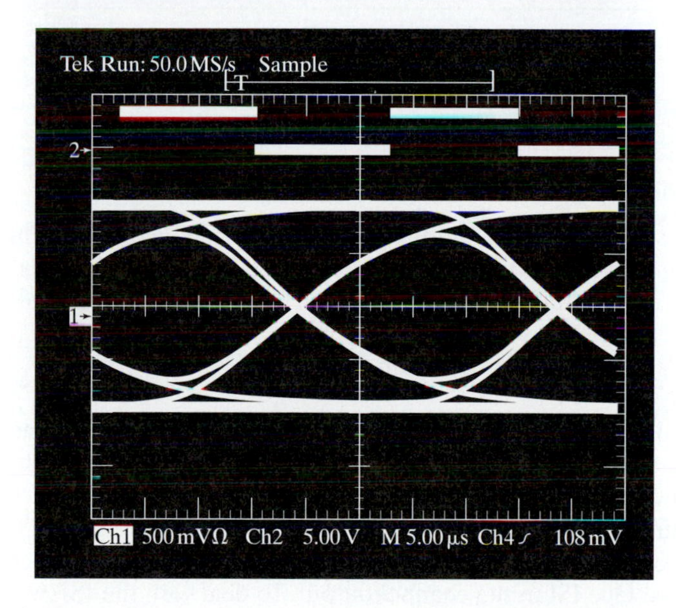

Figure 4-43b Eye diagram of received signal with moderate ISI.

understand how ISI occurs, we can focus on using channel equalizers to eliminate or reduce the presence of this interference in the signal at the output of the channel. As we know, different channels have different impulse response characteristics. For this reason, various equalizer structures have been developed to provide solutions that are suited for specific channel characteristics. Because it is impossible to fully cover such an extensive subject here, only the underlying principles of equalization techniques will be discussed in Sections 4.4.2–4.4.4 and references will be provided in Section 4.4.5 for readers with deeper interest in this subject.

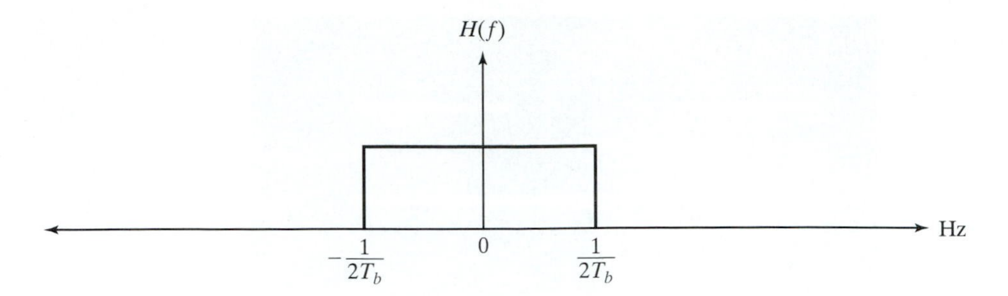

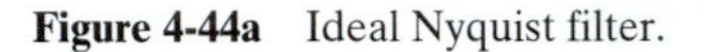
Figure 4-44a Ideal Nyquist filter.

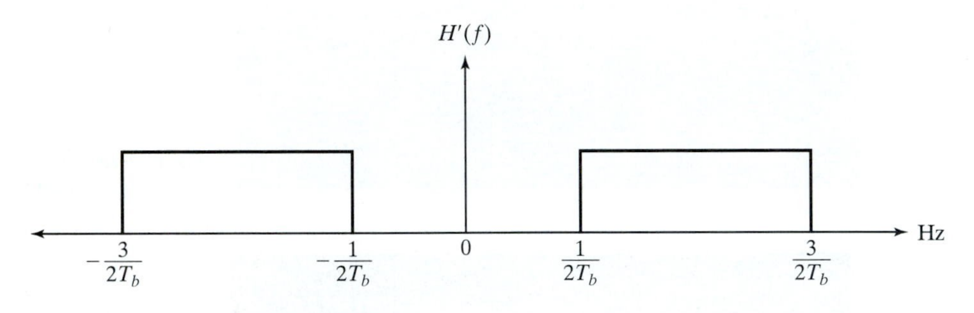

Figure 4-44b Folded ideal Nyquist filter.

4.4.2 Linear Transversal Equalizers

We have already established the condition under which no ISI exists: The impulse response of the channel, $h(t)$, must have zero crossings at T_b-spaced intervals, satisfying Nyquist's first criterion. In frequency domain terms, this condition is equivalent to:

$$H(f) = \text{constant for } |f| \leq \frac{1}{2T_b} \tag{4.129}$$

$H(f)$ is the channel frequency response and $H'(f)$ is the "folded" channel spectral response after bit rate sampling. The bandwidth $|f| \leq 1/(2T_b)$ is known as the Nyquist, or minimum, bandwidth. When $H(f) = 0$ for $|f| > 1/(2T_b)$, the folded response for $H(f)$ shown in Figure 4-44a will be $H'(f)$, shown in Figure 4-44b.

Let's now examine the case when $h(t)$ does not satisfy Nyquist's first criterion; in other words, when we face ISI in the channel output. To deal with the ISI, we insert the equalizer in tandem with the channel as shown in Figures 4-45a (time domain representation) and 4-45b (frequency domain representation). Through this approach, we hope that the channel plus the equalizer will have a combined frequency response that satisfies Nyquist's first criterion. In other words, we hope to achieve the following condition:

$$H'(f)C(f) = 1 \text{ for } |f| \leq \frac{1}{2T_b} \tag{4.130}$$

As we see from Equation (4.130), the equalizer is simply an inverse filter, which inverts the folded frequency response of the channel.

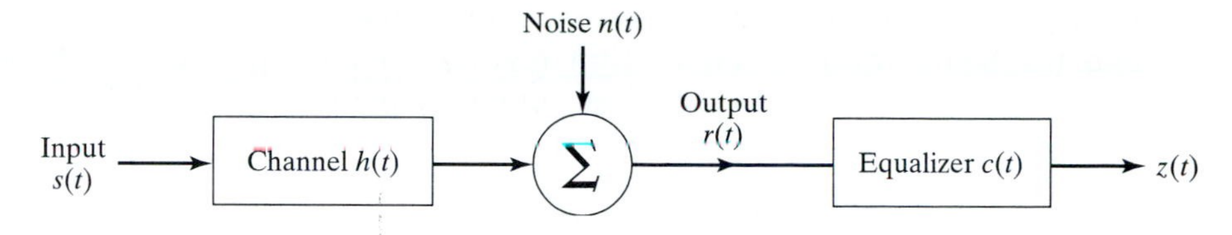

Figure 4-45a Time domain representation of channel and equalizer.

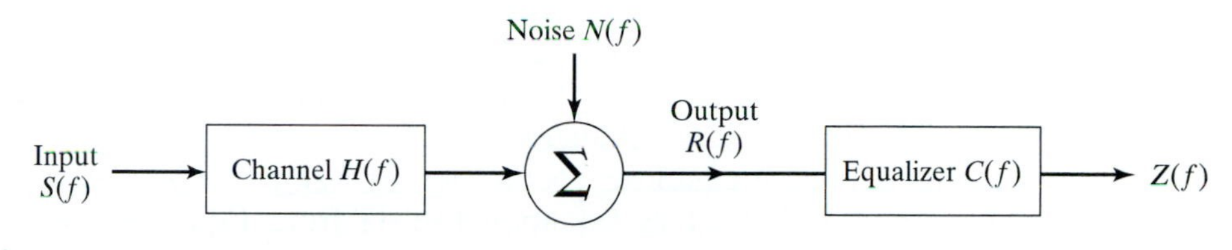

Figure 4-45b Frequency domain representation of channel and equalizer.

The next issue we need to address is how to implement the equalizer in practice so that we can achieve the result of Equation (4.130). One of the simplest techniques is to use a transversal equalizer in the form of a tapped-delay line, as shown in Figure 4-46. In such an equalizer, the current and past values of the received signal are linearly weighed by equalizer coefficients or tap gains, c_n, and summed to produce the output. If the delay and tap gains are analog, the continuous output of the equalizer, $z(t)$, is sampled at the rate $1/T_b$ and then sent to the decision stage. In practice, the equalizer is

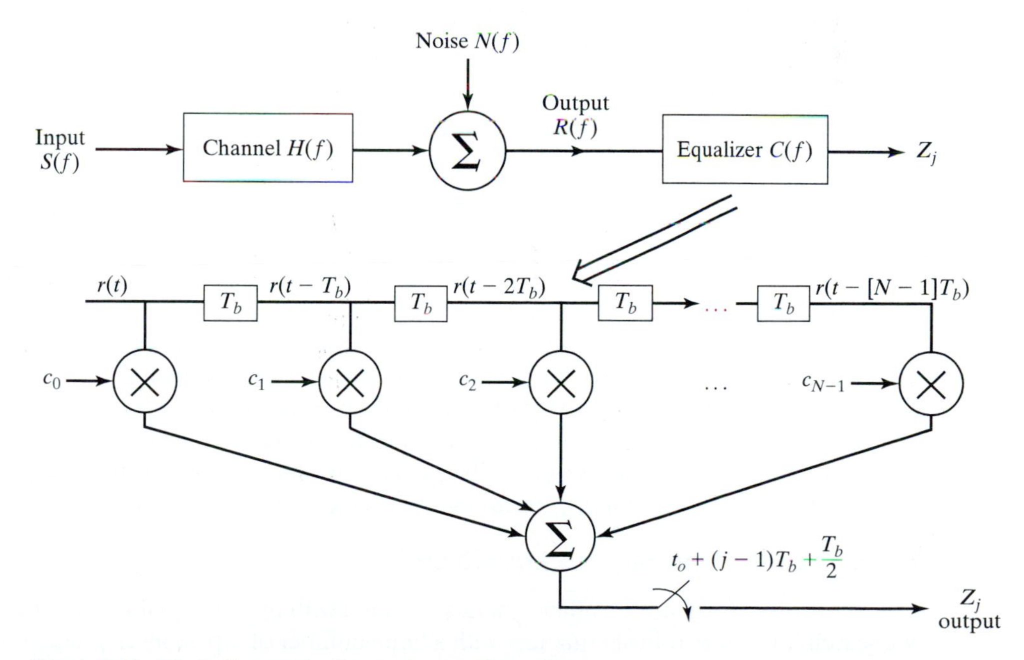

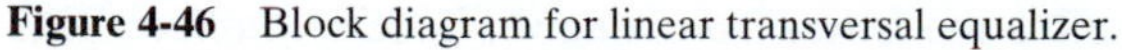
Figure 4-46 Block diagram for linear transversal equalizer.

implemented digitally, which means the output signal of the channel is sampled first and the tapped delay is simple shift registers. If the equalizer has N coefficients, $c_0, c_1, c_2, \ldots, c_{N-1}$, then the equalizer output for the jth bit, $z\{(j-1)T_b + T_b/2 + t_o\}$, can be expressed as

$$z\left\{(j-1)T_b + \frac{T_b}{2} + t_o\right\} = \sum_{k=0}^{N-1} c_k r\left\{[(j-1)-k]T_b + \frac{T_b}{2} + t_o\right\} \tag{4.131}$$

The equalizer coefficients, $c_0, c_1, c_2, \ldots, c_{N-1}$, are then chosen to force the impulse response of the combined channel and equalizer to have zero crossings at T_b-spaced intervals, in which case the ISI will disappear at the output of the equalizer. We call this a *zero-forcing (ZF) equalizer.*

To calculate the ZF equalizer coefficients, $c_0, c_1, c_2, \ldots, c_{N-1}$, we substitute for $r\{[(j-1)-i]T_b + T_b/2 + t_o\}$ in Equation (4.131) from Equation (4.127) and we obtain the following:

$$\begin{aligned} z\left\{(j-1)T_b + \frac{T_b}{2} + t_o\right\} &= \sum_{k=0}^{N-1} c_k d_j p(0) h(t_o) \\ &+ \sum_{k=0}^{N-1} c_k \sum_{\substack{i=1 \\ i\neq j}}^{n} d_i p\{(j-i-k)T_b\} h\{(j-i-k)T_b + t_o\} \\ &+ \sum_{k=0}^{N-1} c_k n\left\{(j-1-k)T_b + \frac{T_b}{2} + t_o\right\} \end{aligned} \tag{4.132}$$

The middle term in Equation (4.132) represents the ISI. We want a set of equalizer coefficients that make this middle term equal to zero, that is

$$\sum_{k=0}^{N-1} c_k \sum_{\substack{i=1 \\ i\neq j}}^{n} d_i p\{(j-i-k)T_b\} h\{(j-i-k)T_b + t_o\} = 0 \tag{4.133}$$

If we write Equation (4.132) for $j = 1, 2, \ldots, N-1$, we obtain N equations, which we can solve simultaneously to calculate the coefficients $c_0, c_1, c_2, \ldots, c_{N-1}$.

From Equation (4.131), we see that if the equalizer has an infinite number of tap delays, then it is simply an inverse for the channel-folded frequency response. In practice, the ZF equalizer will have a finite length and hence it can only approximate this inverse. In doing so, the ZF equalizer may enhance the noise excessively where the spectrum of the folded channel has high attenuation. This noise enhancement is an undesirable side effect, and it makes ZF equalizers less useful in practical applications. In fact, in the calculations of the coefficients of the ZF equalizer, we have neglected the effect of noise altogether. Let's try another approach that may have more practical applications.

4.4.3 Least-Mean-Square Equalizers

Given the problem of noise enhancement associated with the ZF equalizer, it is logical that we search for a more robust equalizer with a finite number of tap delay elements. One such equalizer that is commonly used is the *least-mean-square (LMS) equalizer.* This equalizer works by finding the coefficients $c_0, c_1, c_2, \ldots, c_{N-1}$ that minimize the sum of squares of all the ISI terms plus the noise power. Thus, the LMS equalizer considers the middle term plus the noise term in Equation (4.132). The sum of these two terms, e_j, is given by

$$e_j = \sum_{k=0}^{N-1} c_k \left\{ \sum_{\substack{i=1 \\ i \neq j}}^{n} d_i p\{(j-i-k)T_b\} h\{(j-i-k)T_b + t_o\} \right.$$
$$\left. + \, n\left\{(j-1-k)T_b + \frac{T_b}{2} + t_o\right\} \right\} \tag{4.134}$$

The LMS equalizer finds the set of coefficients $c_0, c_1, c_2, \ldots, c_{N-1}$ by minimizing each $e_j^2, j = 1, 2, 3, \ldots, N-1$. The specific optimization algorithm for this purpose can be found in a number of references (see Section 4.4.5).

4.4.4 Other Types of Equalizers

Several other equalizer structures have been developed to provide effective solutions to the ISI problem for specific types of channels. Examples of such equalization techniques include the following:

1. Adaptive equalization to deal with applications in which the channel characteristics vary with time. The equalizer has to track down such variations by continuously updating the equalizer coefficients.
2. Decision feedback equalizers to deal with channels that have severe amplitude distortion. A decision feedback equalizer has two parts: The forward part is similar to the linear transversal equalizer, while the feedback part is a transversal filter with a different set of coefficients. Decisions made in the forward loop are fed back through the second transversal filter. The equalized signal is the sum of the forward and the feedback parts of the equalizer. The forward and feedback coefficients may be adjusted simultaneously to minimize the mean-square error.
3. Fractionally spaced equalizers in which the delay line taps are spaced at an interval T' that is less than, or is a fraction of, the interval T_b. Fractionally spaced equalizers are employed in situations where sampling the signal at the rate $1/T_b$ causes spectral overlap or *aliasing* (see Chapter 8). When the phases of the overlapping components match, they add constructively, and when they are 180 degrees apart, they add destructively, which results in a cancellation or reduction in amplitude. In contrast, there is no spectral overlap when the signal is sampled at a fraction of T_b. That is why the fractionally spaced equalizer can adjust the channel spectrum at the two band-edge regions; that is, at frequency $1/(2T_b)$ and $-1/(2T_b)$.

4.4.5 Further Reading on Equalization

Several articles and publications are recommended for readers who want to cover the topic of signal equalization in more depth. Earlier publications by Lucky and Rudin [4.13], Salz [4.14], Forney [4.15], and Falconer and Ljung [4.16] provide strong foundation for equalization methods and techniques. Several books are also available that include detailed material on a number of equalization techniques (Lucky, Salz, and Weldon [4.17], Lee and Messerschmitt [4.18], Proakis [4.19], and Ziemer and Peterson [4.20]). Finally, a comprehensive survey of adaptive equalization techniques by Qureshi is presented [4.21].

4.5 Multi-Level PAM (M-ary PAM)

Until now, we have considered baseband pulse amplitude modulation techniques that transmit two different pulses: one pulse, $s_1(t)$, to signify a "1," and another pulse, $s_2(t)$, to signify a "0." In Chapter 3, we saw that different pulse shapes required different bandwidths, and one of our goals was to develop a pulse shape that required a minimum bandwidth for a particular transmission speed. Let's now examine another strategy for increasing the bandwidth efficiency of baseband communication systems, M-ary modulation.

Consider a modulation technique that transmits four different rectangular pulses, all of the same width but of different amplitude. These four pulses are shown in Figure 4-47. Each of the four pulses in Figure 4-47 can represent *two* bits of information from the source:

Let $s_1(t)$ signify "11" from the source.
Let $s_2(t)$ signify "10" from the source.
Let $s_3(t)$ signify "01" from the source.
Let $s_4(t)$ signify "00" from the source.

Each of the four different pulses is called a *symbol*. Generally speaking, if a transmission technique uses N different symbols, each symbol can represent $\log_2 N$ bits from the source. (The case above employs four different symbols, each representing $\log_2 4 = 2$ bits.) A modulation technique that uses two symbols is called a *binary* technique. A modulation technique that uses more than two symbols is called an *M-ary* technique.

What are the advantages and disadvantages of the four-symbol technique described above? One advantage is that transmission speed can be increased relative to a binary technique. From our work in Chapter 3 with binary PAM pulses, we remember that bandwidth is not affected by pulse amplitude, but rather by pulse shape and pulse width. Having four levels rather than two allows us to transmit two bits per pulse while retaining the same pulse width as the binary case, therefore allowing twice as much data to be transmitted in a given bandwidth. A disadvantage of M-ary techniques is that they are more complex and more susceptible to noise than comparable binary techniques.

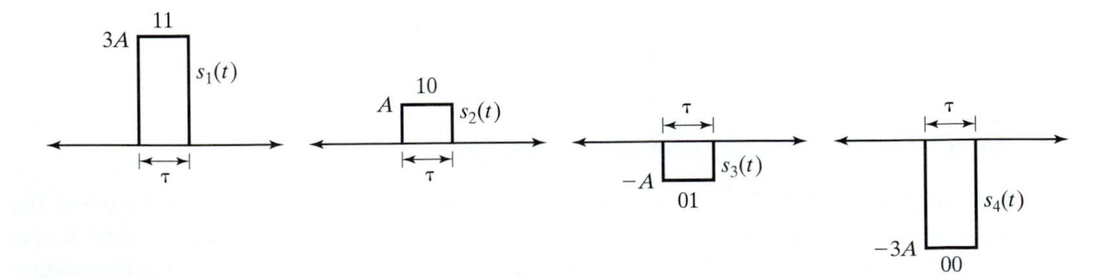

Figure 4-47 Four different pulse shapes, each capable of representing two bits of information from the source.

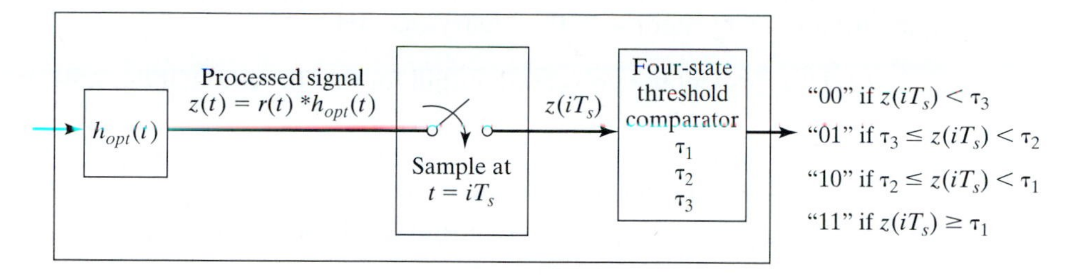

Figure 4-48 Receiver structure for a four-symbol system.

Consider a system using the four-symbol modulation technique just described. Let's assume that the bits output by the source are independent and equiprobable.[3] Figure 4-48 shows the optimal receiver for four-symbol PAM (this receiver design is merely an extension of the binary receiver developed in Section 4.2). T_s represents the symbol period—the amount of time required to transmit one symbol (note that $T_s = 2/r_b$). The symbol transmitted during the ith symbol period represents bits b_{2i-1} and b_{2i}.

Let's now evaluate the performance of the four-symbol system. If we define *probability of symbol error* (P_s) as the probability that the ith symbol is received in error, and if we assume that the source is outputting independent, equiprobable bits, then we can express the probability of symbol error as

$$\begin{aligned}
P_s &= P\{b_{2i-1}, b_{2i} = 11 \text{ and } z(iT_s) < \tau_1\} + P\{b_{2i-1}, b_{2i} = 10 \text{ and } z(iT_s) \ge \tau_1 \text{ or } < \tau_2\} \\
&\quad + P\{b_{2i-1}, b_{2i} = 01 \text{ and } z(iT_s) \ge \tau_2 \text{ or } < \tau_3\} + P\{b_{2i-1}, b_{2i} = 00 \text{ and } z(iT_s) \ge \tau_3\} \\
&= P\{b_{2i-1}, b_{2i} = 11\}P\{z(iT_s) < \tau_1 | b_{2i-1}, b_{2i} = 11\} \\
&\quad + P\{b_{2i-1}, b_{2i} = 10\}[P\{z(iT_s) \ge \tau_1 | b_{2i-1}, b_{2i} = 10\} + P\{z(iT_s) < \tau_2 | b_{2i-1}, b_{2i} = 10\}] \\
&\quad + P\{b_{2i-1}, b_{2i} = 01\}[P\{z(iT_s) \ge \tau_2 | b_{2i-1}, b_{2i} = 01\} + P\{z(iT_s) < \tau_3 | b_{2i-1}, b_{2i} = 01\}] \\
&\quad + P\{b_{2i-1}, b_{2i} = 00\}P\{z(iT_s) \ge \tau_3 | b_{2i-1}, b_{2i} = 00\} \\
&= \frac{1}{4}P\{z(iT_s) < \tau_1 | b_{2i-1}, b_{2i} = 11\} \\
&\quad + \frac{1}{4}[P\{z(iT_s) \ge \tau_1 | b_{2i-1}, b_{2i} = 10\} + P\{z(iT_s) < \tau_2 | b_{2i-1}, b_{2i} = 10\}] \\
&\quad + \frac{1}{4}[P\{z(iT_s) \ge \tau_2 | b_{2i-1}, b_{2i} = 01\} + P\{z(iT_s) < \tau_3 | b_{2i-1}, b_{2i} = 01\}] \\
&\quad + \frac{1}{4}P\{z(iT_s) \ge \tau_3 | b_{2i-1}, b_{2i} = 00\} \qquad (4.135)
\end{aligned}$$

[3]As mentioned in Section 4.1.1, these assumptions are consistent with a well-designed system.

Using notation consistent with the binary case, let

$a_1(iT_s)$ represent the processed output for the ith symbol if $b_{2i-1}, b_{2i} = 11$ and the receiver experiences no noise

$a_2(iT_s)$ represent the processed output for the ith symbol if $b_{2i-1}, b_{2i} = 10$ and the receiver experiences no noise

$a_3(iT_s)$ represent the processed output for the ith symbol if $b_{2i-1}, b_{2i} = 01$ and the receiver experiences no noise

$a_4(iT_s)$ represent the processed output for the ith symbol if $b_{2i-1}, b_{2i} = 00$ and the receiver experiences no noise

As with the binary case, we know that the received noise has a Gaussian probability density function with zero mean, and thus the processed signal plus noise will be a probabilistic signal with a Gaussian probability density function of variance, say, σ_o, and mean equal to $a_1(iT_s)$, $a_2(iT_s)$, $a_3(iT_s)$, or $a_4(iT_s)$, depending on whether the transmitted sequence was "11," "10," "01," or "00," respectively. Figure 4-49 shows the probability density functions and error regions for the four-symbol modulation technique. Total probability of error is the sum of the areas of the six error regions. As with the binary case, the thresholds are placed at the crossover points to minimize the sum of the error regions. Note that where there were only two error regions for a binary modulation technique, there are now six error regions for the four-symbol technique.

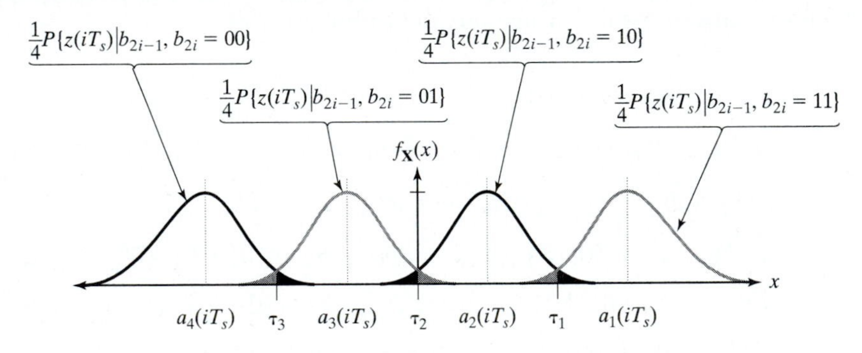

Figure 4-49 Probability distribution functions and error regions for a four-symbol modulation technique.

As we established for the binary modulation techniques, the optimum threshold values are the points where the probability density functions intersect. Thus,

$$\begin{aligned} \tau_{1,\,opt} &= \frac{a_1(iT_s) + a_2(iT_s)}{2} \\ \tau_{2,\,opt} &= \frac{a_2(iT_s) + a_3(iT_s)}{2} \\ \tau_{3,\,opt} &= \frac{a_3(iT_s) + a_4(iT_s)}{2} \end{aligned} \tag{4.136}$$

Let's now determine probability of symbol error by calculating the areas of each of the six error regions. Starting with the region to the right of τ_1

$$P\{z(iT_s) \geq \tau_1 | b_{2i-1}, b_{2i} = 10\} = \frac{1}{4}Q\left(\frac{\tau_1 - a_2(iT_s)}{\sigma_o}\right) \tag{4.137}$$

Substituting in the optimum threshold value for τ_1,

$$P\{z(iT_s) \geq \tau_1 | b_{2i-1}, b_{2i} = 10\} = \frac{1}{4}Q\left(\frac{\tau_1 - a_2(iT_s)}{\sigma_o}\right) = \frac{1}{4}Q\left(\frac{a_1(iT_s) - a_2(iT_s)}{2\sigma_o}\right) \tag{4.138}$$

Now observe that each of the six error regions has the same area. Thus, we can express probability of symbol error as

$$P_s = \frac{6}{4}Q\left(\frac{a_1(iT_s) - a_2(iT_s)}{2\sigma_o}\right) \tag{4.139a}$$

Using a correlation receiver and an approach similar to the one we used in Section 4.2.4.1 for the binary case, we can express Equation (4.139a) as

$$P_s = \frac{6}{4}Q\left(\frac{a_1(iT_s) - a_2(iT_s)}{2\sigma_o}\right) = \frac{6}{4}Q\left(\sqrt{\frac{E_{d,\,symbol}}{2N_o}}\right) \tag{4.139b}$$

where $E_{d,\,symbol}$ represents the component of the received signal's energy that is involved in making two adjacent symbols [$s_1(t)$ and $s_2(t)$; or $s_2(t)$ and $s_3(t)$; or $s_3(t)$ and $s_4(t)$] look different. For the four rectangular pulses in Figure 4-47, we can calculate $E_{d,\,symbol}$ as

$$E_{d,\,symbol} = \int_{(i-1)T_s}^{iT_s} [\gamma s_1(t) - \gamma s_2(t)]^2\, dt = 4\gamma^2 A^2 T_s \tag{4.140}$$

The pulses in Figure 4-47 exhibit a certain symmetry, and we can express $E_{d,\,symbol}$ in terms of the average received energy per symbol as

$$\begin{aligned} E_{ave,\,symbol} &= \frac{1}{4}\int_{(i-1)T_s}^{iT_s} (\gamma 3A)^2\, dt + \frac{1}{4}\int_{(i-1)T_s}^{iT_s} (\gamma A)^2\, dt \\ &\quad + \frac{1}{4}\int_{(i-1)T_s}^{iT_s} (-\gamma A)^2\, dt + \frac{1}{4}\int_{(i-1)T_s}^{iT_s} (-\gamma 3A)^2\, dt \\ &= 5\gamma^2 A^2 T_s \end{aligned} \tag{4.141}$$

Substituting into Equation (4.139b),

$$P_s = \frac{6}{4}Q\left(\sqrt{\frac{0.4E_{ave,\,symbol}}{N_o}}\right) \tag{4.142a}$$

How does the error rate for the four-symbol case compare with the binary case? For the binary case (assuming equiprobable bits and symmetric pulses) we know

$$P_b = Q\left(\sqrt{\frac{2E_b}{N_o}}\right) \tag{4.58R}$$

In order to compare the four-symbol and binary cases, we must do the following:

1. Determine the relationship between average energy per received symbol in the four-level case ($E_{ave,\, symbol}$) and average energy per received bit in the binary case (E_b).
2. Determine the relationship between a symbol error in the four-level case (related to P_s) and a bit error in the binary case (related to P_b).

For the first task, note that each symbol in the four-symbol case represents two bits so average energy per received bit in the four-symbol case is $0.5E_{ave,\, symbol}$. We can use the notation E_b to represent average energy per received bit for both the binary and four-symbol cases, and substitute $2E_b$ for $E_{ave,\, symbol}$ in Equation (4.142a), producing

$$P_s = \frac{6}{4}Q\left(\sqrt{\frac{0.4E_{ave,\, symbol}}{N_o}}\right) = \frac{6}{4}Q\left(\sqrt{\frac{0.8E_b}{N_o}}\right) \tag{4.142b}$$

Now we must address the second task—relating probability of bit error (binary case) and probability of symbol error (four-symbol case). Examining the probability density functions and error regions in Figure 4.49, we see that a vast majority of symbol errors are errors to adjacent symbols. Assuming for simplicity's sake that *all* symbol errors are errors to an adjacent symbol,

a. $P\{\text{“10” sent but “11” received}\} = \frac{1}{6}$

b. $P\{\text{“11” sent but “10” received}\} = \frac{1}{6}$

c. $P\{\text{“10” sent but “01” received}\} = \frac{1}{6}$

d. $P\{\text{“01” sent but “10” received}\} = \frac{1}{6}$

e. $P\{\text{“01” sent but “00” received}\} = \frac{1}{6}$

f. $P\{\text{“00” sent but “01” received}\} = \frac{1}{6}$

In cases a, b, e, and f, each symbol error corresponds to a one-bit error, but in cases c and d, each symbol error corresponds to a two-bit error. Thus, when the system transmits one symbol (two bits) and an error occurs, two-thirds of the time one bit will still be received correctly. Probability of bit error for the four-symbol case can be expressed as

$$\begin{aligned} P_{b,\,4\text{-}symbol} &= \frac{2}{3}(\text{one bit is in error; one bit is correct}) + \frac{1}{3}(\text{both bits are in error}) \\ &= \frac{2}{3}\left(\frac{1}{2}P_s\right) + \frac{1}{3}\left(P_s\right) = \frac{2}{3}P_s \end{aligned} \tag{4.143}$$

An examination of Figure 4-49 and Equation (4.143) suggests that a different bit assignment may produce a lower probability of bit error. Consider, for instance, the following assignment, in which adjacent symbols always differ in only one bit:

Let $s_1(t) = 11$
$s_2(t) = 10$
$s_3(t) = 00$
$s_4(t) = 01$

This type of assignment, shown in Figure 4-50, is called a *Gray code*. Assuming that all symbol errors are to adjacent symbols,

$$P_{b,\,\text{4-symbol, Gray code}} = \frac{1}{2}P_s \tag{4.144}$$

Comparing accuracy for a four-symbol, Gray-coded system versus a binary system,

$$P_{b,\,\text{4-symbol, Gray code}} = \frac{1}{2}P_{s,\,\text{4-symbol, Gray code}} = 0.75Q\left(\sqrt{\frac{0.8E_b}{N_o}}\right)$$

versus

$$P_{b,\,binary} = Q\left(\sqrt{\frac{2E_b}{N_o}}\right) \tag{4.145}$$

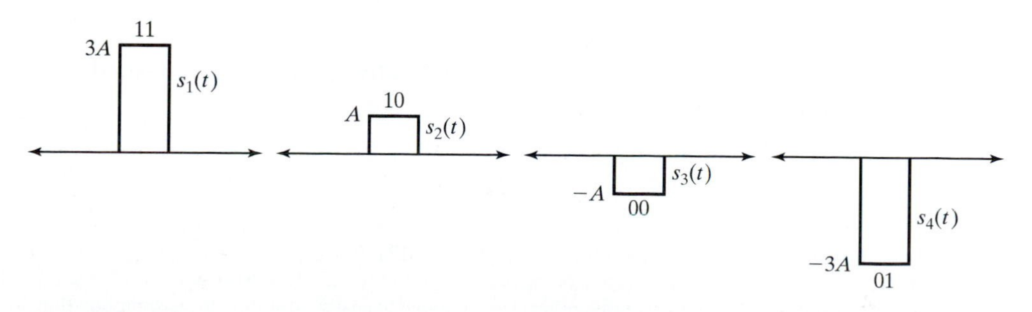

Figure 4-50 Gray code assignment for four different symbols.

We now have enough information to compare the error rates of our binary and four-symbol PAM systems. For practical values of P_b, the four-symbol case will require higher values of E_b/N_o than the binary case, and thus for a given average energy per received bit, the binary case will be more accurate.[4]

Note how average energy per received bit is related to average transmitted power. For both the binary and four-level cases,

$$P_{ave} = \frac{E_b r_b}{\gamma^2} \tag{4.146}$$

If both the binary and four-symbol systems are transmitting the same number of bits per second, the systems will be using the same average transmitted power to produce the same average received energy per bit. In this case, the advantage of the four-level system is that it requires only half the bandwidth of the binary system. If, however, the four-level system uses the same bandwidth as the binary system and is therefore transmitting at twice the speed of the binary system, a given average transmitted power level will produce only half as much energy per received bit for the four-symbol system.

As a final thought, let's very briefly consider a PAM system with eight distinct levels or symbols (this is called an *8-ary PAM system*). Each symbol is capable of transmitting three bits of information, thus tripling transmission speed relative to a binary system, or else reducing required bandwidth by a factor of three. However, with eight different symbols and a given average transmitted power, the energy difference between adjacent received symbols is even smaller than for a four-symbol system. Thus, for a given transmitted power level, P_s is higher for an eight-symbol system than for a four-symbol system. Additionally, the assumption that errors occur to adjacent symbols becomes less valid (again, since the energy difference between adjacent symbols is lower), and thus, even with a Gray code, one symbol error is more likely to cause multiple bits to be in error, increasing P_b.

We can generalize our observations, creating an *M-ary PAM system*—a PAM system with M distinct levels (in practical systems, M is usually a power of 2). For a given transmitted power level, the larger the value of M, the smaller the energy difference between adjacent received symbols and thus the larger the probability of symbol error. Also, the larger the value of M, the less valid the assumption that all errors occur to adjacent symbols. Increasing the number of different symbols in the system thus increases bandwidth efficiency at the cost of equipment complexity and accuracy. Upper and lower bounds on P_b relative to P_s are established in Chapter 5, at the end of section 5.8.3.1.

[4]You can prove this by plotting $P_{b,\,4\text{-symbol, Gray code}}$ versus $P_{b,\,binary}$ for various values of E_b/N_o. You will find that for values of $E_b/N_o > 0.235$ (approximately -6.3 dB), $P_{b,\,binary} < P_{b,\,4\text{-symbol, Gray code}}$. You will also find that $E_b/N_o = 0.235$ corresponds to a bit error rate of roughly 0.32. This error rate is too large for virtually all practical applications. This error rate is also large enough to easily invalidate the assumption that all errors in the four-symbol case occur to adjacent symbols, thereby making the 0.75 constant in the calculation of $P_{b,\,4\text{-symbol, Gray code}}$ overly optimistic.

PROBLEMS

4.1 Derive the equation for probability of bit error in a binary PAM system using the simple PAM receiver shown in Figure 4-1. Assume thermal noise, symmetric pulses, and equiprobable bits.

4.2 Discuss the differences between deterministic mathematics and stochastic (probabilistic) mathematics. Why is stochastic mathematics used to describe noise?

4.3 Show, either geometrically or mathematically, that

$$\int_a^{\infty} \frac{1}{\sqrt{2\pi}\sigma} e^{-\left(\frac{(x-\mu)^2}{2\sigma^2}\right)} dx = Q\left(\frac{a-\mu}{\sigma}\right)$$

where

$$Q(y) \equiv \int_y^{\infty} \frac{1}{\sqrt{2\pi}} e^{-\left(\frac{z^2}{2}\right)} dz$$

4.4 You are given the following two facts concerning the noise at the input of a receiver:

Fact 1: The noise can be modeled as zero-mean additive white Gaussian noise.

Fact 2: There is a 30.85% probability that at a given time the noise voltage will be greater than or equal to 0.4 volts.

Determine the average normalized noise power at the input to the receiver.

4.5 Example 4.6 transmits raised cosine pulses with amplitudes of +1 volt and −1 volt and achieves a probability of bit error of 0.38%, which is too high for most practical applications (see the comments at the end of Section 4.1.5 concerning probability of bit error values for typical systems). To what minimum value must the pulse amplitudes be increased to reduce the probability of bit error below 0.1%?

4.6 Repeat Problem 4.5 if the channel attenuates *power* such that the average normalized power of the received signal is one third the average normalized power of the transmitted signal.

4.7 Using a mathematical approach (rather than the graphical approach given in Section 4.2.3), prove that for independent, equiprobable bits

$$\tau_{opt} = \frac{a_1(iT_b) + a_2(iT_b)}{2}$$

4.8 A PAM system is transmitting rectangular pulses at 10,000 bits/sec. A correlation receiver is employed. Noise power spectral density at the receiver is $N_o/2 = 10^{-6}$ volts^2, and the channel attenuates the transmitted signal by 75% (so that only 25% of the transmitted signal arrives at the receiver). Assuming perfect synchronization at the receiver, what is the minimum pulse amplitude needed at the transmitter to guarantee a probability of bit error less than or equal to 10^{-3}?

4.9 Repeat Problem 4.8 for a probability of bit error less than or equal to 10^{-5}.

4.10 Repeat Problem 4.8 for a probability of bit error less than or equal to 10^{-7}.

4.11 Repeat Problem 4.8 if the channel attenuates the transmitted signal's *power* by 75%.

4.12 Prove that the matched filter receiver in Figure 4-12 and the correlation receiver in Figure 4-13 are equivalent.

4.13 Transmit data at 100,000 bit/sec using bipolar PAM with rectangular pulses. Suppose that the channel does not distort the transmitted signal, but that the channel attenuates the signal so that only 30% of the transmitted signal power reaches the receiver. A correlation receiver is used to demodulate the signal, and the noise at the receiver input has a two-sided power spectral density of 0.4×10^{-6} volts^2 per Hz. What is the minimum height needed for the pulses (i.e., the minimum value of A) in order to attain a probability of bit error of 10^{-5} or less?

4.14 Using numerical integration, plot $Q(z)$ versus the approximation in Equation (4.31) for $3 \leq z \leq 5$.

4.15 For asymmetric pulses, discuss why probability of bit error is related to the energy difference between the received signals for a "1" and a "0" rather than to the total energy in the symbols.

4.16 The correlation receiver in Figure 4-13 is used to demodulate and decode a binary PAM signal. Let $s_1(t)$ and $s_2(t)$ represent the two transmitted pulses, let $a_1(iT_b)$ represent the output of the integrator for the ith bit if $s_1(t)$ was transmitted in a noiseless environment, and let $a_2(iT_b)$ represent the output of the integrator for the ith bit if $s_2(t)$ was transmitted in a noiseless environment.

a. Assume that the source outputs bits that are independent but not necessarily equiprobable (let z represent the probability that a "1" was transmitted). Derive the expression for the optimal threshold setting in the presence of additive, white Gaussian noise.

b. Let the optimum threshold determined in Part a be represented as τ_{opt}. Let σ_o^2 represent the average normalized noise power at the output of the integrator. Derive the expression for probability of bit error in terms of the Q function.

4.17 Explain the role of each of the following components in the basic structure of the PLL:

a. The phase detector
b. The loop filter
c. The VCO

4.18 Derive the PLL phase transfer function given in Equation (4.97) by combining Equations (4.94)–(4.96).

4.19 Consider the PLL block diagram of Figure 4-28 and assume a simple loop filter for which $F(s) = 3.5$.

a. Determine the lock range.
b. State the transfer function.
c. Plot the transfer function in the time domain.
d. Is the PLL stable for $F(s) = -1$? Justify your answer.

4.20 Consider the PLL block diagram of Figure 4-28 with a loop filter having the transfer function

$$F(s) = K\left[\frac{s + 1.1}{s - 0.5}\right]$$

a. Determine the value of K that would yield a lock range with an upper bound of 2π.
b. Determine if the PLL circuit is stable.
c. Calculate the PLL bandwidth.

4.21 Consider the frequency sythesizer circuit shown in Figure 4-30. The input frequency is 2 GHz and the output frequency is 50 kHz with increments of 10 kHz.

a. Determine suitable values for n and m and calculate the corresponding upper limit for the PLL bandwidth.

b. What changes would you make to this PLL structure if you want to raise its bandwidth to 100 kHz while at the same time retaining the same values for the input and output frequencies and for the frequency increment?

4.22 For the timing recovery circuit shown in Figure 4-35,

a. Explain the role of the PLL.
b. Determine the range of values for the delay τ when the input bit stream has a transmission rate of 1.5 Mbit/sec.
c. Under what conditions will bit timing capture fail in this circuit?

4.23 **a.** State the Nyquist criterion for zero ISI in the time domain.
b. State the Nyquist criterion in the frequency domain.

4.24 For the power delay spread profile shown in Figure P4.24,

a. Find the rms delay spread.[5]

b. Estimate the maximum data symbol rate achievable without equalization.

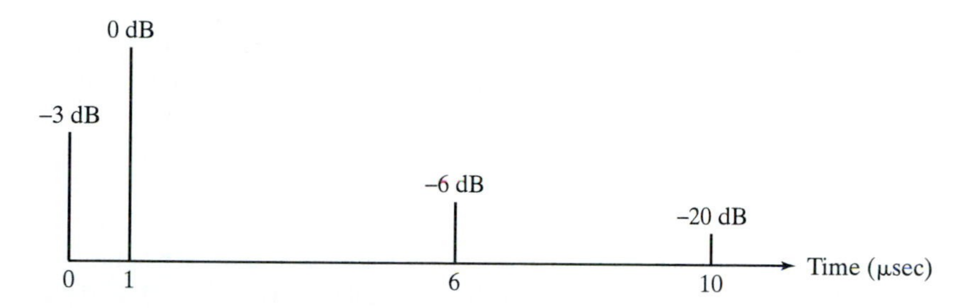

Figure P4.24 Power delay spread profile.

4.25 The output of a channel, sampled at time kT is

$$y_k = h_0 a_k + h_1 a_{k-1} + v_k$$

where the data symbols $a_k = \pm 1$ with equal probability and are uncorrelated, and the v_k are independent, zero-mean samples of noise with variance σ^2. A linear equalizer with three tap coefficients is used to process the channel's output. The equalizer output at time k is

$$z_k = \sum_{n=0}^{2} w_n^* y_{k-n}$$

The three tap coefficients are required to minimize the mean squared value of the error between the kth equalizer output sample and the $k - 1$st data symbol a_{k-1}.

a. Derive expressions for the optimum tap coefficients and minimum mean squared error as functions of the above parameters.

b. Determine the coefficients and minimum MSE for $h_0 = 1 + j$, $h_1 = 0.3 - j\,0.1$, and $\sigma^2 = 0.01$.

4.26 In one current digital cellular system (known as the North American IS-136 system), the modulation is QPSK (Quadrature Phase Shift Keying, which will be discussed in detail in Chapter 5), in which the bit rate is about 48 kb/sec (symbol rate is 24 ksymbols/sec), and an adaptive equalizer is usually used to combat delay spread of up to several tens of μs. In another current system (the Japanese digital cellular system), the same modulation is used, the expected delay spread is similar, the bit rate is about 42 kb/sec, and the system usually uses antenna (space) diversity instead of adaptive equalization. In antenna diversity, the same signal is received by two or more antennae spaced appropriately apart and then a signal-combining approach is used to determine the best signal version to be selected at each time instant. Referring to the effects of delay spread on transmission performance, and to principles of diversity and equalization, explain the rationale for each of these two approaches.

[5]RMS delay spread is defined as

$$RMS\ delay\ spread = \sqrt{\frac{\sum_{i=1}^{n} \tau_i^2 P_i}{\sum_{i=1}^{n} P_i} - \left(\frac{\sum_{i=1}^{n} \tau_i P_i}{\sum_{i=1}^{n} P_i}\right)^2}$$

where τ_i is the ith time delay and P_i is the corresponding avg. normalized power in volts^2.

4.27 Explain the role of the equalizer when the channel impulse response $h(t)$ does not satisfy Nyquist's first criterion. Explain how the equalizer known as the linear transversal filter fulfills this role and provide justification for the "tapped delay" structure of such an equalizer.

4.28 Discuss the significance of using Gray codes when transmitting with M-ary modulation techniques.

4.29 Explain the trade-offs involved in using M-ary PAM rather than binary PAM.

4.30 Consider a quaternary (4-ary) PAM system using rectangular pulses and transmitting 60,000 bits/sec.

a. Determine the channel bandwidth required for 90% in-band power.

b. What are the optimum amplitudes of the four different symbols if average normalized transmitted power must be ≤ 15 volts^2?

c. Determine a Gray code to assign two-bit sequences to each of the four different symbols.

d. Suppose the channel attenuates the signal so that only 50% of the transmitted signal's power arrives at the receiver. The two-sided average normalized power spectral density of the noise at the receiver is 5×10^{-6} volts^2/Hz. If a correlation receiver with three thresholds is employed, determine the probability of symbol error.

e. Translate your results in Part d into a probability of bit error, assuming that the Gray code from Part b is used and that all errors occur to adjacent symbols.

f. Suppose that the assumption that all errors occur to adjacent symbols is not valid. Translate your results in Part d to probability of bit error, assuming that errors to all symbols are equally likely.

g. Discuss how you can use the values you obtained in Parts e and f to evaluate a system where many but not all errors are to adjacent symbols.

4.31 Compare the performance of the system in Problem 4.30 with a binary PAM system transmitting rectangular pulses at the same bit rate and using the same average normalized power. Be sure to consider all performance-versus-cost parameters.

4.32 Consider an 8-ary PAM system using rectangular pulses and transmitting 60,000 bits/sec.

a. Determine the channel bandwidth required for 90% in-band power.

b. What are the optimum amplitudes of the eight different symbols if average normalized transmitted power must be ≤ 15 volts^2?

c. Determine a Gray code to assign three-bit sequences to each of the eight different symbols.

d. Extending the concepts used to determine the probability of symbol error for a quaternary (4-ary) PAM system, determine the probability of symbol error for an 8-ary PAM system.

e. Suppose the channel attenuates the signal so that only 50% of the transmitted signal's power arrives at the receiver. The two-sided average normalized power spectral density of the noise at the receiver is 5×10^{-6} volts^2/Hz. If a correlation receiver with seven thresholds is employed, determine the probability of symbol error.

f. Translate your results in Part e into a probability of bit error, assuming that the Gray code from Part c is used and that all errors are to adjacent symbols.

g. Suppose that the assumption that all errors occur to adjacent symbols is not valid. Translate your results in Part e to probability of bit error, assuming that errors to all symbols are equally likely.

h. Explain how you can use the values you obtained in Parts f and g to evaluate a system where many, but not all, errors occur to adjacent symbols.

4.33 The quaternary system in Problem 4.30, the binary system in Problem 4.31, and the 8-ary system in Problem 4.32 all transmit information at the same rate (i.e., the same number of bits/sec) and use the same average normalized power. Compare the performance of the three systems. Be sure to consider all performance-versus-cost parameters.

Chapter 5

Digital Bandpass Modulation and Demodulation Techniques (and Stochastic Mathematics, Part II)

IN Chapters 3 and 4 we developed pulse amplitude modulation (PAM) transmitters and receivers to efficiently send information across baseband channels. Can we also use PAM to efficiently send information across bandpass channels? Recall that in Chapter 3 we described baseband and bandpass channels as follows:

> *Baseband:* A *baseband channel* efficiently passes frequency components from dc to f_c Hz, where f_c is called the *cutoff frequency*. Frequency components above f_c are not efficiently passed by baseband channels. Wires and coaxial cables are examples of baseband channels.
>
> *Bandpass:* A *bandpass channel* efficiently passes frequency components within a certain band, say, between f_1 and f_2 Hz. Components outside the frequency band ($<f_1$ or $>f_2$) are not efficiently passed by bandpass channels. Airwaves and fiber optics are examples of bandpass channels.

As we established in Chapter 3, PAM signals were designed to produce an energy spectral density with all its significant components in the dc- to low-frequency range (which is necessary for efficient transmission across baseband channels). Unfortunately, these components will *not* be efficiently passed through a bandpass channel, so a PAM signal will experience significant distortion if transmitted through a bandpass channel. In order to efficiently transmit information through a bandpass channel, we need to develop a modulation technique that produces signals with an energy spectral density in the higher frequency band passed by the channel.

How then can we transmit information across a bandpass channel? We can *modulate a carrier signal*. What does "modulate a carrier signal" mean? Let's look at each of the key words.

- As we discussed in Chapter 1, in conventional English *modulate* means "to change."
- A *carrier signal* is a sinusoid with a frequency that can be efficiently passed over a particular channel. For example, if a bandpass channel efficiently passes frequencies in the band 50 kHz $\leq f \leq$ 250 kHz, a 150 kHz sine wave would be an appropriate carrier signal.

Therefore, "modulating a carrier signal" means that we are going to transmit information across a bandpass channel by changing a sinusoid that efficiently passes across the channel. There are many different ways to modulate (or change) a carrier signal. Let's begin by considering a technique known as binary *amplitude shift keying* (ASK).

5.1 Binary Amplitude Shift Keying

Binary ASK is best illustrated by example.

Example 5.1 Generating a binary ASK signal

Suppose a channel can efficiently pass frequencies in the range 50 kHz $\leq f \leq$ 250 kHz. Since PAM signals have very strong frequency components down to 0 Hz, we cannot send the information over the channel using PAM without serious attenuation and distortion. Let's consider instead the following scheme for carrier modulation:

- Let the transmitter send the carrier signal $A\sin(2\pi\times150{,}000t)$ volts if the information from the source is a "1."
- Let the transmitter send the signal 0 volts if the information from the source is a "0."

In other words, a 150 kHz sine wave is our carrier, and we turn it on when we want to send a "1" and off when we want to send a "0." Figure 5-1 shows the binary ASK signal corresponding to the data sequence "101101" sent at a transmission speed of 50,000 bit/sec.

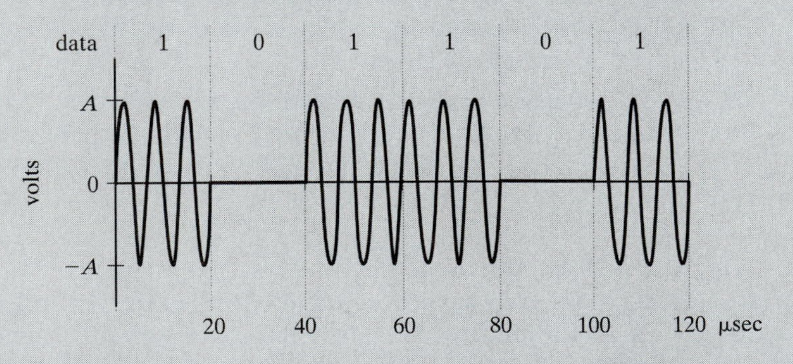

Figure 5-1 Binary ASK signal corresponding to "101101" transmitted at 50,000 bits/sec.

Examine Figure 5-1 carefully. Be sure you see why the bit period is 20 μsec and why the 150 kHz carrier signal completes exactly three cycles per bit period.

Intuitively, the binary ASK waveform in Example 5.1 seems as though it should pass efficiently across a bandpass channel with 50 kHz $\leq f \leq$ 250 kHz. All we're doing is turning on and off a 150 kHz sinusoid, and so we might believe that the average normalized power spectrum of the ASK signal is as shown in Figure 5-2. Unfortunately, we will soon see that our intuition has failed us.

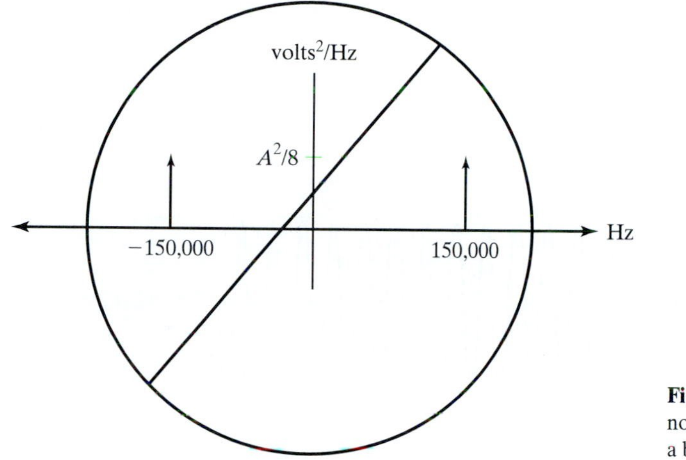

Figure 5-2 ***Not*** the average normalized power spectrum of a binary ASK signal.

To correctly determine the average normalized power spectrum of an ASK signal (and to refine our intuition), let's go back to our mathematical tools. We can express a typical binary ASK signal (such as the one in Example 5.1) as

$$s_{\text{ASK}}(t) = s_{\text{baseband}}(t)\sin(2\pi f_c t)$$
$$\text{where } s_{\text{baseband}}(t) = \begin{cases} A \text{ when the data from the source is a "1"} \\ 0 \text{ when the data from the source is a "0"} \end{cases} \quad (5.1)$$
and f_c represents the carrier frequency (150 kHz in Example 5.1)

Equation (5.1) is illustrated in Figure 5-3. Let's now determine the Fourier transform of $s_{\text{ASK}}(t)$.

$$\begin{aligned} S_{\text{ASK}}(f) &= \mathcal{F}\{s_{\text{ASK}}(t)\} \\ &= \mathcal{F}\{s_{\text{baseband}}(t)\sin(2\pi f_c t)\} \\ &= \int_{-\infty}^{\infty} s_{\text{baseband}}(t)\sin(2\pi f_c t)e^{-j2\pi f t}\,dt \quad (5.2) \\ &= \int_{-\infty}^{\infty} s_{\text{baseband}}(t)\left[\frac{e^{j2\pi f_c t} - e^{-j2\pi f_c t}}{2j}\right]e^{-j2\pi f t}dt \\ &= \frac{1}{2j}[S_{\text{baseband}}(f - f_c) - S_{\text{baseband}}(f + f_c)] \end{aligned}$$

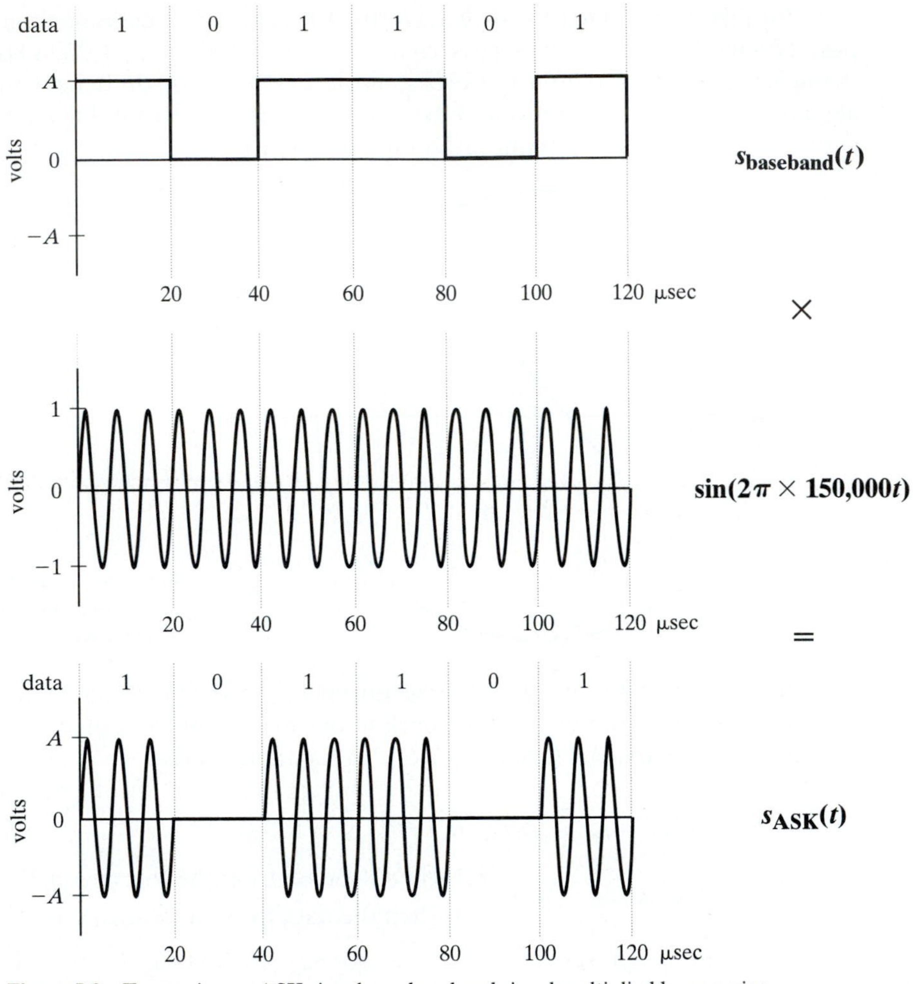

Figure 5-3 Expressing an ASK signal as a baseband signal multiplied by a carrier.

Now let's determine $S_{\text{baseband}}(f)$. As shown in Figure 5-4, we can express $s_{\text{baseband}}(t)$ as a rectangular PAM waveform plus a dc component.

$$s_{\text{baseband}}(t) = s_{\text{PAM}}(t) + \frac{A}{2} \tag{5.3}$$

Taking the Fourier transform,

$$S_{\text{baseband}}(f) = \mathcal{F}\left\{ s_{\text{PAM}}(t) + \frac{A}{2} \right\} = S_{\text{PAM}}(f) + \frac{A}{2}\delta(f) \tag{5.4}$$

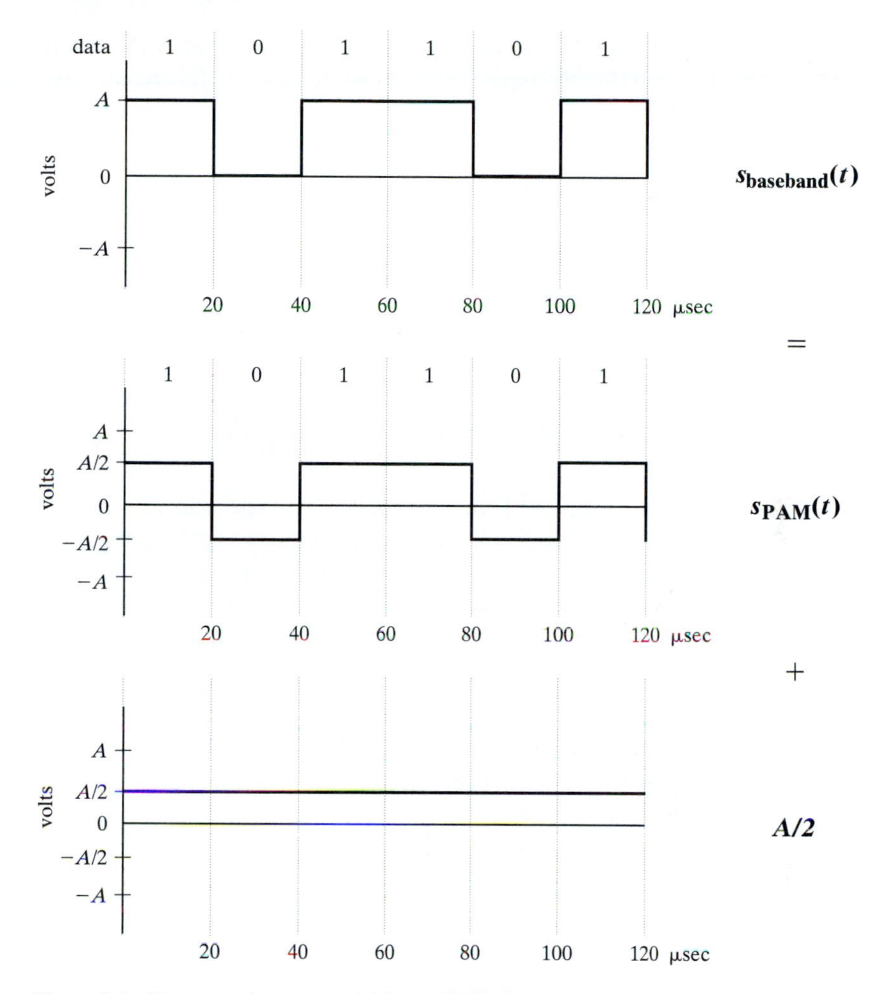

Figure 5-4 Decomposing $s_{\text{baseband}}(t)$ into a PAM signal plus a dc term.

Substituting Equation (5.4) into Equation (5.2) produces

$$S_{\text{ASK}}(f) = \frac{1}{2j}\left[S_{\text{PAM}}(f - f_c) + \frac{A}{2}\delta(f - f_c) - S_{\text{PAM}}(f + f_c) - \frac{A}{2}\delta(f + f_c) \right] \tag{5.5}$$

Note that the first and third terms of Equation (5.5) are continuous and that the second and fourth terms are discrete components of separate frequencies. The energy spectral density of a typical binary ASK signal is thus

$$\Psi_{\text{ASK}}(f) = |S_{\text{ASK}}(f)|^2$$

$$= \frac{1}{4}\left| S_{\text{PAM}}(f - f_c) - S_{\text{PAM}}(f + f_c) \right|^2 + \frac{A^2T}{16}\delta(f - f_c) + \frac{A^2T}{16}\delta(f + f_c) \tag{5.6a}$$

where T represents the total time required to transmit the ASK signal. Assuming that the magnitude spectra of $S_{\text{PAM}}(f - f_c)$ and $S_{\text{PAM}}(f + f_c)$ do not overlap (a reasonable assumption), then

$$\Psi_{\text{ASK}}(f) = \frac{1}{4}\left|S_{\text{PAM}}(f - f_c) - S_{\text{PAM}}(f + f_c)\right|^2 + \frac{A^2T}{16}\delta(f - f_c) + \frac{A^2T}{16}\delta(f + f_c)$$

$$= \frac{1}{4}\Psi_{\text{PAM}}(f - f_c) + \frac{1}{4}\Psi_{\text{PAM}}(f + f_c) + \frac{A^2T}{16}\delta(f - f_c) + \frac{A^2T}{16}\delta(f + f_c) \quad (5.6\text{b})$$

Similarly, we can express the average normalized power spectral density of a binary ASK signal as

$$G_{\text{ASK}}(f) = \frac{1}{4}G_{\text{PAM}}(f - f_c) + \frac{1}{4}G_{\text{PAM}}(f + f_c) + \frac{A^2}{16}\delta(f - f_c) + \frac{A^2}{16}\delta(f + f_c) \quad (5.7)$$

All we need now to complete Equation (5.7) is an expression for $G_{\text{PAM}}(f)$, the average normalized power spectral density of a typical binary PAM signal composed of rectangular pulses with pulse width $\tau = 1/r_b$. We calculated such a power spectral density in Equation (3.2). Given pulses of amplitude $\pm A/2$ and a transmission speed of r_b bits/sec,

$$G_{\text{PAM}}(f) = \frac{(A/2)^2}{r_b}\,\text{sinc}^2(\pi f/r_b) \quad (5.8)$$

Substituting Equation (5.8) into Equation (5.7), the average normalized power spectral density of a typical binary ASK signal is

$$G_{\text{ASK}}(f) = \frac{1}{4}\left\{\frac{A^2}{4r_b}\,\text{sinc}^2[\pi(f - f_c)/r_b] + \frac{A^2}{4r_b}\,\text{sinc}^2[\pi(f + f_c)/r_b] + \frac{A^2}{4}\delta(f - f_c) + \frac{A^2}{4}\delta(f + f_c)\right\}$$

$$= \frac{A^2}{16}\left\{\frac{1}{r_b}\,\text{sinc}^2[\pi(f - f_c)/r_b] + \frac{1}{r_b}\,\text{sinc}^2[\pi(f + f_c)/r_b] + \delta(f - f_c) + \delta(f + f_c)\right\} \quad (5.9)$$

which is plotted in Figure 5-5.

In observing Equation (5.9) and Figure 5-5, we note that the average normalized power spectral density for binary ASK is related to the average normalized power spectral density for binary PAM with rectangular pulses in the following ways:

1. Both densities contain sinc-squared-shaped curves with a main lobe width of $2r_b$. The PAM power spectral density has a single sinc-squared-shaped curve centered at 0 Hz, while the ASK power spectral density has two such curves—one centered at f_c (the carrier frequency) and one centered at $-f_c$. This result can be anticipated by observing the decomposition of the ASK signal shown in Figure 5-3 coupled with the modulation property of the Fourier transform given in Equation (2.56).
2. The ASK power spectral density also contains impulses at f_c and $-f_c$. This result can be anticipated by observing the decomposition of the baseband signal in Figure 5-4.

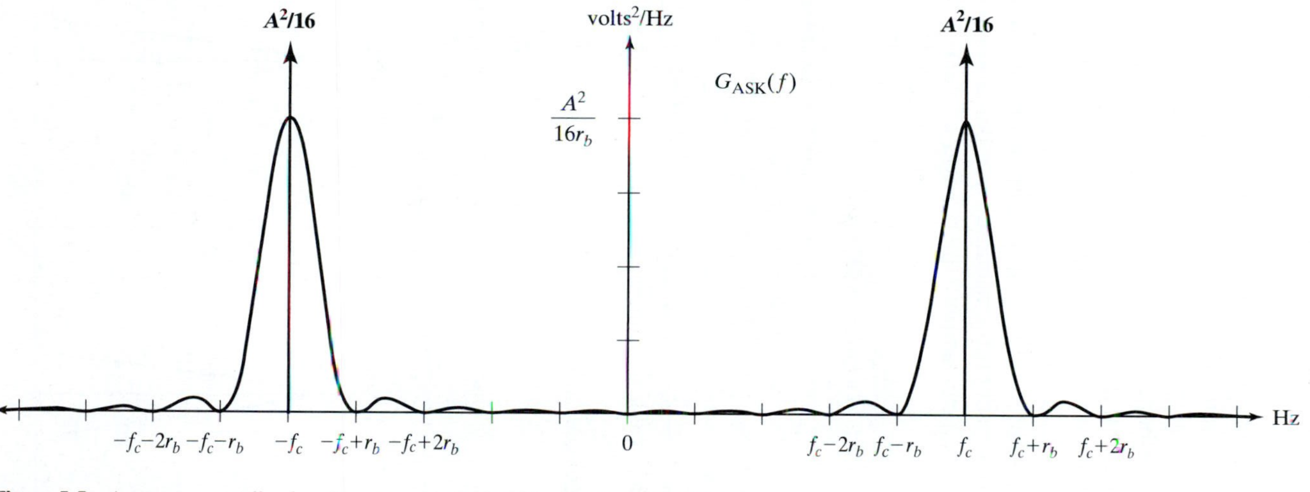

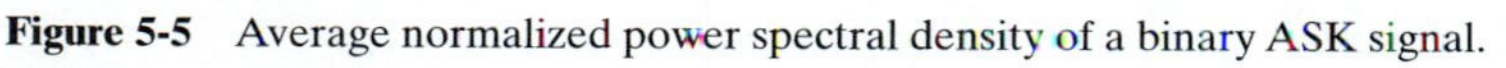

Figure 5-5 Average normalized power spectral density of a binary ASK signal.

3. The bandwidth of both the PAM and ASK signals depends on the percentage of in-band power required to maintain a certain degree of accuracy. As established in Chapter 3, a bandwidth of r_b for the PAM signal contains 90% of the signal's power, while a bandwidth of $2r_b$ contains 95% of the signal's power. For the ASK signal, however, bandwidth must be *doubled* relative to the PAM signal because symmetric components that occupied frequencies to the left of 0 Hz in the PAM spectrum are now shifted to the right of 0 Hz in the ASK spectrum. Thus, in order to obtain 95% in-band power for an ASK signal, the channel must be capable of reliably passing all frequencies from $f_c - 2r_b$ to $f_c + 2r_b$, which is a bandwidth of $4r_b$. We do not want to consider the power contained in the impulses—just as we proved that transmitting dc power has no impact on the accuracy of a PAM system (review Example 4.11), transmitting a discrete sinusoid at the carrier frequency has no impact on the accuracy of a bandpass system (although as we will see in Section 5.5, it can simplify the receiver circuit).

Example 5.2 Bandwidth of a binary ASK signal

In Example 5.1 the bandpass channel passes frequencies in the range 50 kHz $\leq f \leq$ 250 kHz, and we want to use binary ASK with a carrier frequency of 150 kHz to transmit data at a speed of 50,000 bits/sec. Suppose an application requires at least 95% in-band power.

a. Draw the average normalized power spectral density of a typical transmitted signal.

b. What bandwidth is required if at least 95% of the signal's power must be in-band?

c. Is the signal satisfactorily designed for the channel?

Solutions

a. The average normalized power spectral density is shown in Figure 5-6.

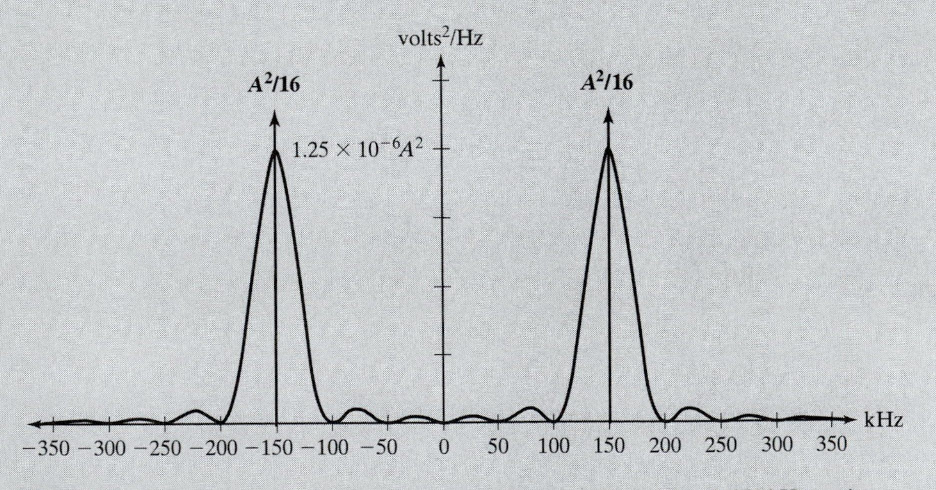

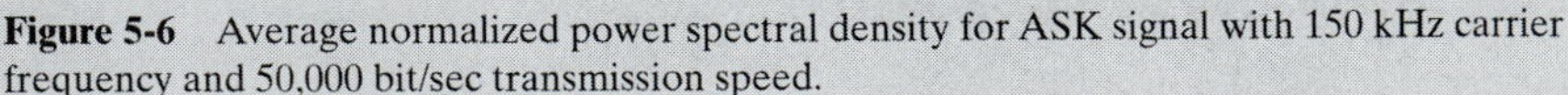

Figure 5-6 Average normalized power spectral density for ASK signal with 150 kHz carrier frequency and 50,000 bit/sec transmission speed.

b. Bandwidth $= 4r_b = 4(50{,}000 \text{ bits/sec}) = 200$ kHz.

c. From the calculations in Parts a and b of this example, we know that in order for the channel to be satisfactory, it must pass frequencies in the range from 50 kHz to 250 kHz. Since the channel can do this, it is satisfactory.

The concept we've just developed—decomposing a bandpass signal into a baseband signal multiplied by a carrier signal—can be used to build our intuition concerning the frequency domain representation of the bandpass signal. This decomposition was illustrated in the time domain in Figure 5-3. Figure 5-7 shows the decomposition in both the time domain (the left side of the figure) and in the frequency domain (the right side of the figure). Remembering that multiplication in the time domain corresponds to convolution in the frequency domain (see Equation (2.61)), we see that multiplying a baseband signal by a carrier in the time domain corresponds in the frequency domain to a scaling by 0.5 and a frequency shift of $\pm f_c$ in the baseband signal's magnitude spectrum. All the intuition that we developed for baseband signals can therefore be translated into bandpass signals, as long as we remember that the frequency shift effectively doubles bandwidth because symmetric components of the baseband magnitude spectrum that occupied frequencies to the left of 0 Hz are moved to the right of 0 Hz after the frequency shift.

5.2 Other Binary Bandpass Modulation Techniques

We now know that we can transmit information across a bandpass channel by modulating a carrier signal (i.e., changing a sinusoid that efficiently passes across the channel). Binary ASK uses different amplitudes of the carrier signal to indicate "1"s and "0"s ("1"s are transmitted by sending the carrier signal with an amplitude of A volts, and "0"s are transmitted by sending the carrier signal with an amplitude of 0 volts). Are there other methods of bandpass modulation besides binary ASK? To answer this question, let's examine a carrier signal to see if there are other parameters besides amplitude that might be modulated (changed) to indicate "1"s and "0"s. Consider a typical sinusoidal carrier signal:

$$s(t) = A\sin(2\pi f_c t + \theta) \tag{5.10}$$

This sinusoid has three independent parameters: amplitude, frequency, and phase. We've already developed a bandpass modulation technique based on changing amplitude. Let's now consider bandpass modulation techniques based on modulating the other two parameters: frequency and phase.

5.2.1 Binary Frequency Shift Keying

Binary frequency shift keying (FSK) uses changes in the frequency of the carrier signal to transmit digital information. Consider the general carrier signal $A\sin(2\pi f_c t + \theta)$.

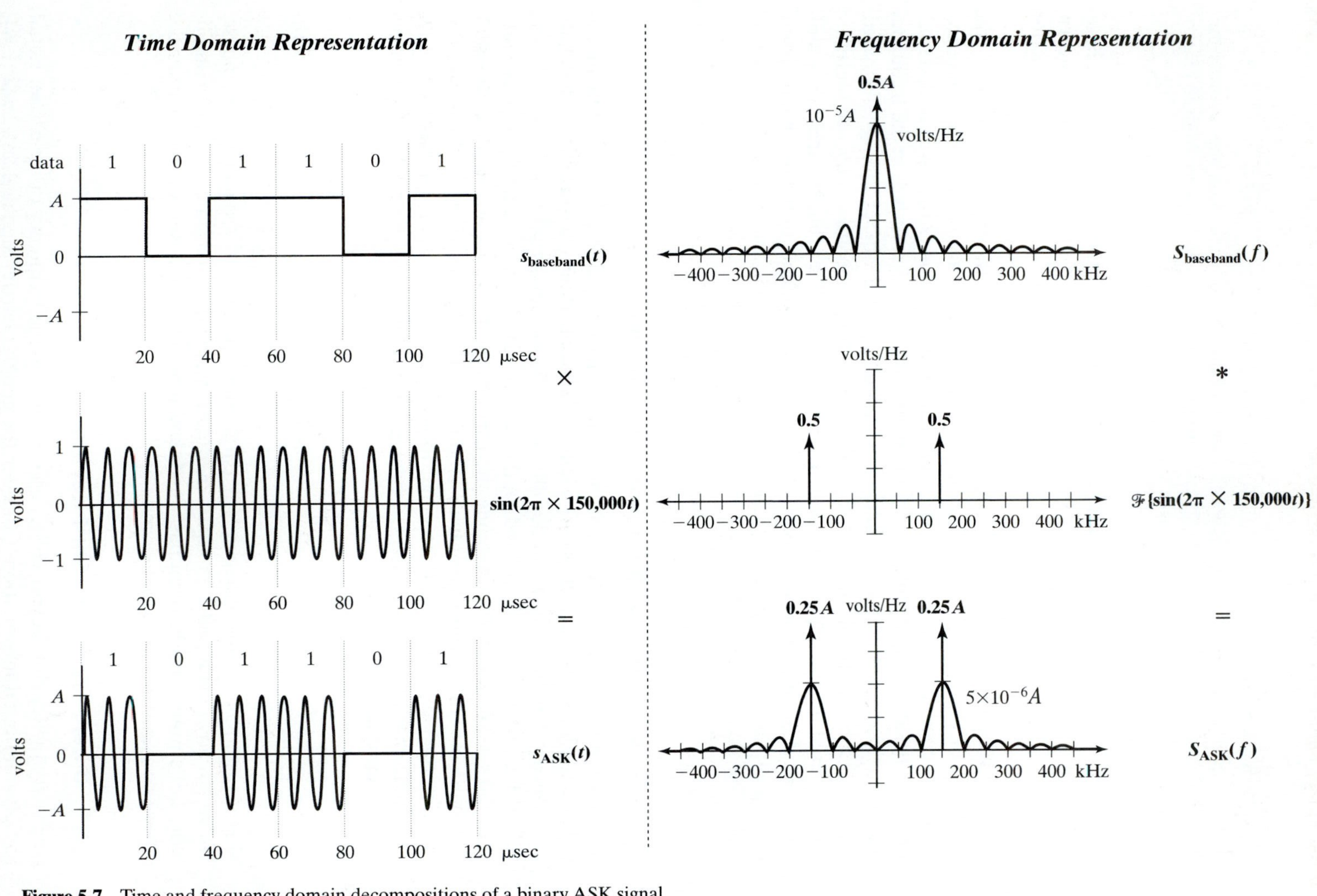

Figure 5-7 Time and frequency domain decompositions of a binary ASK signal.

Binary FSK can be described as follows:

- Let the transmitter send the signal $A\sin[2\pi(f_c + \Delta f)t + \theta]$ if the data from the source is a "1." Δf is a constant known as the *frequency offset.*
- Let the transmitter send the signal $A\sin[2\pi(f_c - \Delta f)t + \theta]$ if the information from the source is a "0."

With binary FSK, the carrier is shifted up in frequency by Δf Hz (a constant) to signify a "1" and is shifted down in frequency by Δf Hz to signify a "0." Figure 5-8 shows the binary FSK signal corresponding to the data sequence "101101" sent at a transmission speed of 50,000 bit/sec, with a carrier frequency of 150 kHz, an initial phase angle of zero, and a frequency offset (Δf) of 50 kHz. Examine Figure 5-8 carefully. Be sure you understand why the bit period is 20 μsec and why the transmitted signal completes four cycles per bit period (corresponding to 200 kHz) when the data is a "1" and two cycles per bit period (corresponding to 100 kHz) when the data is a "0."

5.2.2 Binary Phase Shift Keying

Binary phase shift keying (PSK) uses changes in the phase of the carrier signal to convey digital information. Consider the general carrier signal $A\sin(2\pi f_c t + \theta)$. Binary PSK can be described as follows:

- Let the transmitter send the signal $A\sin(2\pi f_c t + \theta)$ if the data from the source is a "1."
- Let the transmitter send the signal $A\sin(2\pi f_c t + \theta + \pi)$ if the information from the source is a "0."

With binary PSK, the carrier is transmitted without change to signify a "1" and is transmitted with an additional 180 degrees of phase to signify a "0."

Figure 5-9 shows the binary PSK signal corresponding to the data sequence "101101" sent at a transmission speed of 50,000 bit/sec, with a carrier frequency of 150 kHz and an initial phase angle of zero.

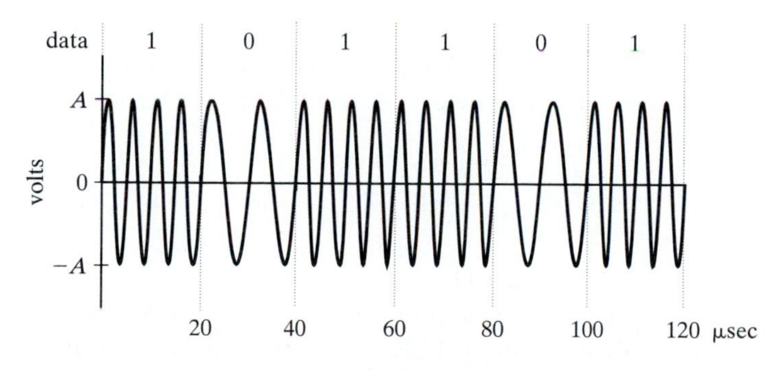

Figure 5-8 Binary FSK signal corresponding to "101101" transmitted at 50,000 bits/sec, $f_c = 150$ kHz, $\Delta f = 50$ kHz.

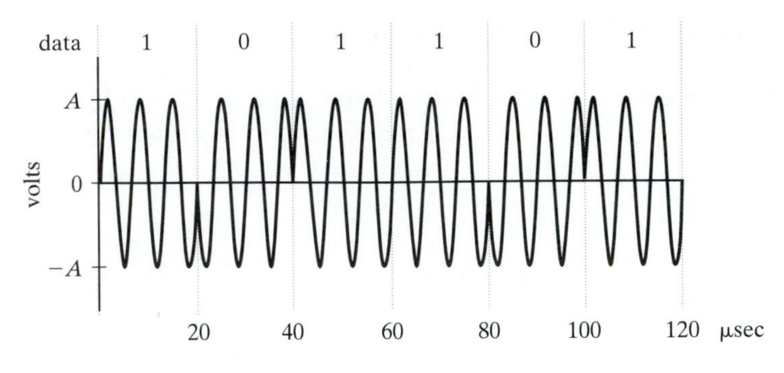

Figure 5-9 Binary PSK signal corresponding to "101101" transmitted at 50,000 bits/sec, $f_c = 150$ kHz.

5.2.3 Calculating Average Normalized Power Spectral Density for Binary FSK and Binary PSK

Let's now determine the average normalized power spectral density for binary PSK and binary FSK. We can use the same procedure we developed for binary ASK: decomposing the transmitted signal into a baseband signal multiplied by a carrier signal. We'll start with binary PSK.

The key to decomposing a binary PSK signal is to note that a 180 degree phase shift corresponds to multiplying the carrier by −1 (be sure you understand why). With this observation in mind, we can express the binary PSK signal as

$$s_{\text{PSK}}(t) = s_{baseband}(t)\sin(2\pi f_c t + \theta)$$

$$\text{where } s_{baseband}(t) = \begin{cases} A & \text{when the data from the source is a "1"} \\ -A & \text{when the data from the source is a "0"} \end{cases} \tag{5.11}$$

The decomposition of a typical PSK waveform (such as Figure 5-9) is shown in Figure 5-10. Observe that for binary PSK,

$$s_{baseband}(t) = s_{\text{PAM}}(t) \tag{5.12}$$

We can use a procedure similar to the one outlined for binary ASK [see Equations (5.4–5.8)] to establish the power spectral density for binary PSK:

$$G_{\text{PSK}}(f) = \frac{A^2}{4r_b}\left\{\text{sinc}^2[\pi(f - f_c)/r_b] + \text{sinc}^2[\pi(f + f_c)/r_b]\right\} \tag{5.13}$$

which is plotted in Figure 5-11.

From Equation (5.13) and by comparing Figure 5-11 with Figure 5-5, we can see that binary ASK and binary PSK have the same bandwidth and occupy the same frequency band (we'll discuss the significance of having the impulses only in the ASK power spectral density later in this chapter).

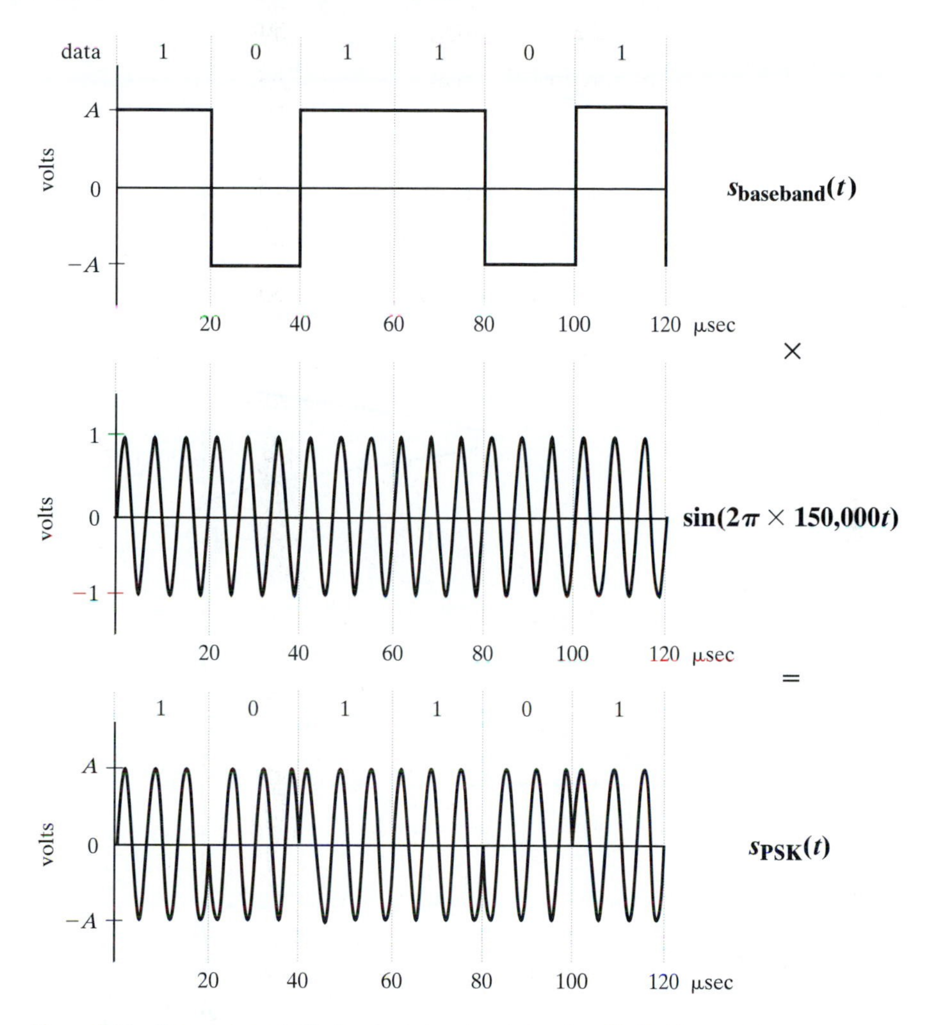

Figure 5-10 Expressing a PSK signal as a baseband signal multiplied by a carrier.

Developing the average normalized power spectral density for binary FSK is more difficult. We can decompose the binary FSK waveform as

$$s_{\text{FSK}}(t) = s_{\text{baseband 1}}(t)\sin[2\pi(f_c+\Delta f)t + \theta] + s_{\text{baseband 2}}(t)\sin[2\pi(f_c-\Delta f)t + \theta]$$

$$\text{where } s_{\text{baseband 1}}(t) = \begin{cases} A \text{ when the data from the source is a "1"} \\ 0 \text{ when the data from the source is a "0"} \end{cases} \tag{5.14}$$

$$\text{and} \quad s_{\text{baseband 2}}(t) = \begin{cases} 0 \text{ when the data from the source is a "1"} \\ A \text{ when the data from the source is a "0"} \end{cases}$$

We thus see that the binary FSK signal is essentially the sum of two ASK signals: one with a carrier frequency of $f_c + \Delta f$ Hz, the other with a carrier frequency of

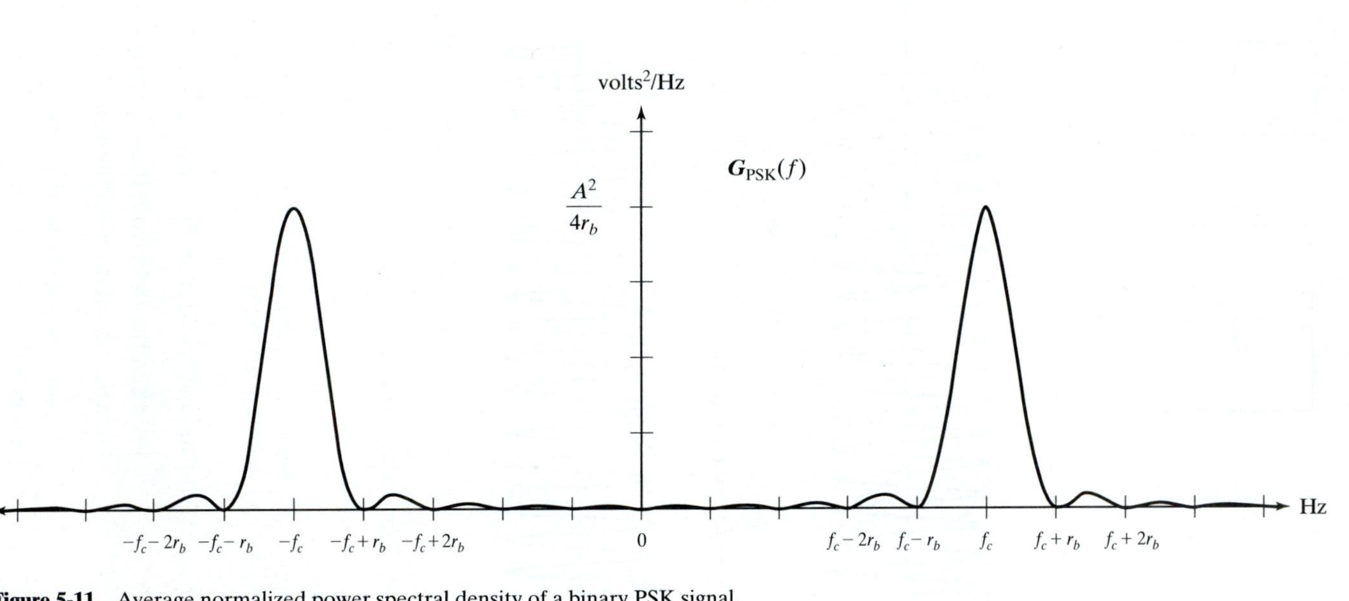

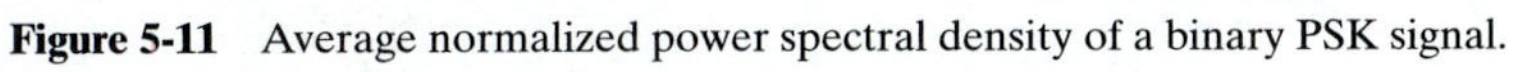

Figure 5-11 Average normalized power spectral density of a binary PSK signal.

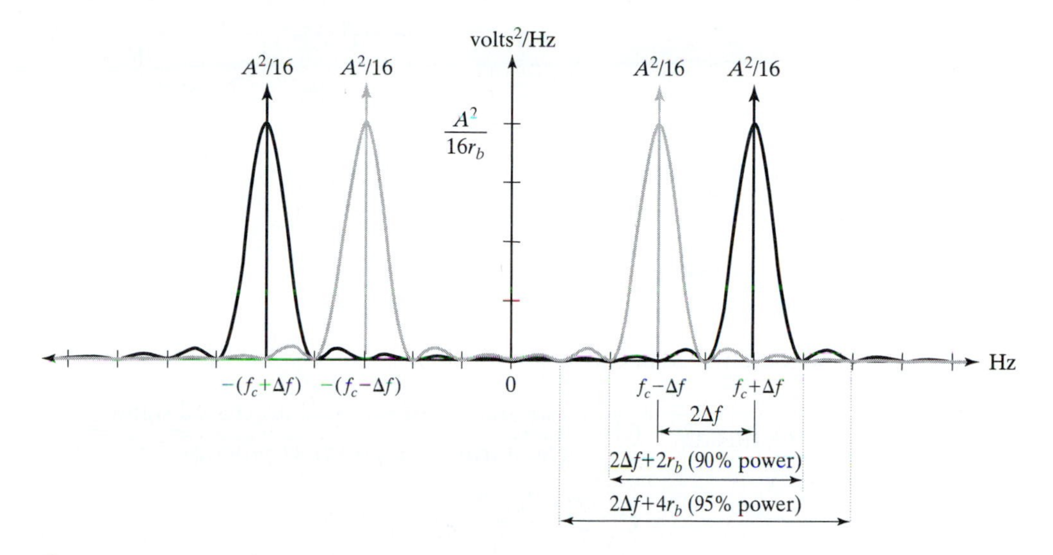

Figure 5-12 Conceptualizing the bandwidth of a binary FSK signal.

$f_c - \Delta f$ Hz. This concept is shown in Figure 5-12, where the average normalized power spectral density of the $f_c + \Delta f$ ASK signal is shown in black and the average normalized power spectral density of the $f_c - \Delta f$ ASK signal is shown in gray. Although the approach is not rigorous, we can see that a bandwidth of approximately $2\Delta f + 4r_b$ will be needed for 95% of the power to be in-band. (A rigorous approach requires that we add the spectra of the two ASK signals, using complex numbers, prior to squaring to determine energy and power. Such an approach will not significantly change the bandwidth calculations.)

5.3 Coherent Demodulation of Bandpass Signals

Let's develop optimum receivers for the three bandpass modulation techniques—PSK, ASK, and FSK. We can follow the same approach we used in Chapter 4, in Sections 4.2 and 4.3 for baseband receivers; that is, we can use Figure 4-10*R* to describe the general bandpass demodulation process, producing a general receiver as shown in Figure 4-11*R*.[1] Our challenge for PSK, ASK, and FSK, as it was for baseband systems, is to find the linear process $H(f)$ that produces the lowest probability of bit error.

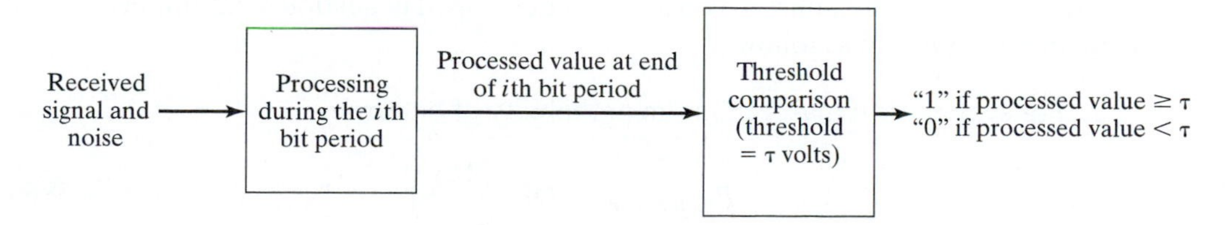

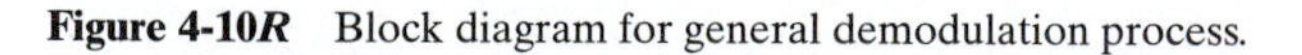
Figure 4-10*R* Block diagram for general demodulation process.

[1] In Figures 4-10*R* and 4-11*R*, note that the thresholds are generalized as τ volts rather than 0 volts, as was used in Figures 4-10 and 4-11.

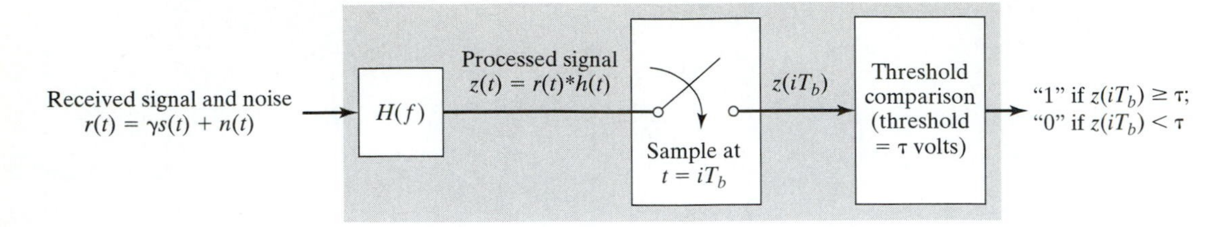

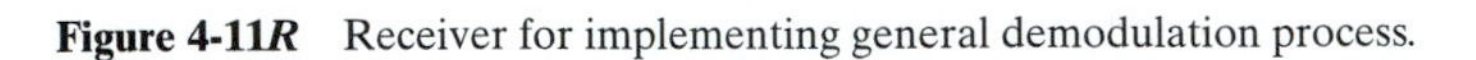
Figure 4-11*R* Receiver for implementing general demodulation process.

As we determined in Chapter 4,

$$P_{b,\text{ Fig. 4-11}R} = Q\left(\frac{\text{Noise margin of sampled processed signal}}{\sqrt{\text{avg. normalized power of processed noise}}}\right)$$

$$= Q\left(\frac{|\gamma s(iT_b) * h(iT_b)|}{\sigma_p}\right) \tag{4.47R}$$

where

σ_p^2 represents the average normalized power of the processed noise
$s(iT_b)$ represents the transmitted signal
γ represents the attenuation of the channel
$h(t)$ is the impulse response corresponding to the transfer function $H(f)$

$$h(t) = \mathcal{F}^{-1}[H(f)] \tag{4.45R}$$

As we also determined in Section 4.2.1, for symmetric signals the threshold should be set to 0 volts and the probability of bit error, as represented in Equation (4.47*R*), is minimized when

$$H(f) = k[S(f)e^{j2\pi fiT_b}]^* \tag{4.55R}$$

meaning that the optimal transfer function is

$$H_o(f) = kS^*(f)e^{-j2\pi fiT_b} \tag{4.56R}$$

where k is an arbitrary nonzero constant. As developed in Section 4.2.2, implications of Equation (4.56*R*) are as follows:

1. For symmetric signals, minimum probability of bit error is

$$P_{b,\,minimum} = Q\left(\sqrt{\frac{2E_b}{N_o}}\right) \tag{4.58R}$$

where E_b is the normalized energy of the signal at the receiver during the ith bit period and $N_o/2$ is the normalized power spectral density of the noise at the input to the receiver.

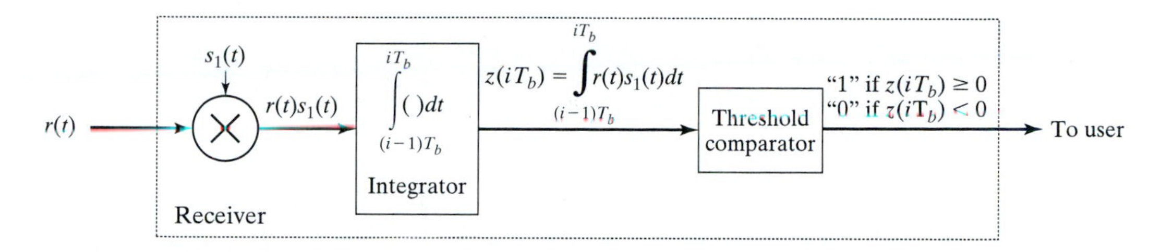

Figure 4-13*R* The correlation receiver for symmetric signals.

2. For symmetric signals, the optimum processing described in Equation (4.56*R*) can be practically implemented using a correlation receiver, as shown in Figure 4-13*R*. Note that $s_1(t)$ is the signal transmitted to represent a "1."

5.3.1 Developing a Coherent PSK Receiver

PSK signals are symmetric; assuming an initial phase angle of zero, the signal transmitted to represent a "1" is

$$s_1(t) = A\sin(2\pi f_c t + 0) \tag{5.15}$$

while the signal transmitted to represent a "0" is

$$s_2(t) = A\sin(2\pi f_c t + \pi) = -A\sin(2\pi f_c t) \tag{5.16}$$

Using Figure 4-13*R* as a starting point, the optimum receiver for PSK is therefore the correlation receiver shown in Figure 5-13. Note that there is no need for the constant A in the $s_1(t)$ term at the input to the mixer, since multiplying by A effectively amplifies both the received signal and the noise, thereby having no effect on accuracy.

Since the PSK signals are symmetric, the probability of bit error can be calculated as

$$P_{b,\,\text{PSK},\,correlation\ receiver} = Q\left(\sqrt{\frac{2E_{b,\,PSK}}{N_o}}\right) \tag{4.58R}$$

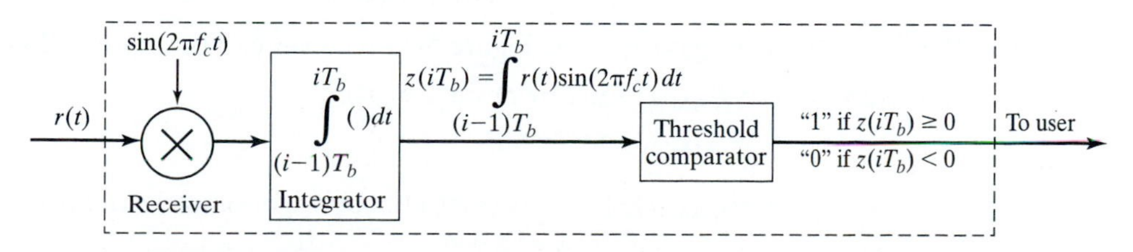

Figure 5-13 Optimum (coherent) receiver for binary PSK.

where

$$E_{b,\,\text{PSK}} = \int_{(i-1)T_b}^{iT_b} [\gamma A \sin(2\pi f_c t)]^2 dt \tag{5.17}$$

$$= \frac{(\gamma A)^2}{2} \int_{(i-1)T_b}^{iT_b} [1 - \cos(4\pi f_c t)]dt = \frac{(\gamma A)^2 T_b}{2} - \frac{(\gamma A)^2}{2} \int_{(i-1)T_b}^{iT_b} \cos(4\pi f_c t)\,dt$$

Assuming that the carrier produces a whole number of cycles within one bit period (a good practice, which, if the initial phase angle is zero, promotes smooth transitions in the signal between the end of one bit period and the beginning of the next bit period), the second term goes to zero and E_b reduces to

$$E_{b,\,\text{PSK}} = \frac{(\gamma A)^2 T_b}{2} \tag{5.18}$$

Substituting in,

$$P_{b,\,\text{PSK},\,\textit{correlation receiver}} = Q\left(\sqrt{\frac{2E_b}{N_o}}\right) = Q\left(\sqrt{\frac{(\gamma A)^2 T_b}{N_o}}\right) \tag{5.19}$$

The receiver in Figure 5-13 is called a *coherent receiver* because it must be capable of internally producing a reference signal, $\sin(2\pi f_c t)$, which is in exact phase and frequency synchronization with the carrier signal. (In practice, a phase shift must be added to the reference signal equal to the delay introduced by the channel. This concept was discussed in Chapter 4, concerning equalization.) As established in Problems 5.9 and 5.10, if the reference is not phase synchronized with the carrier signal, the received signal will be attenuated by the cosine of the phase difference, and if the reference is not frequency synchronized with the carrier signal, the received signal will oscillate with a frequency equal to the frequency difference. Both phenomena can significantly increase probability of bit error. The requirement for a synchronized reference signal increases the complexity of the receiver.

Let's reinforce our intuition concerning the PSK correlation receiver by tracing the received signal through each component in Figure 5-13. In Section 5.2.3 we saw that we could decompose a PSK signal into a baseband PAM signal multiplied by a carrier. In other words, combining Equations (5.11) and (5.12) and setting θ to zero, we can express a binary PSK signal as

$$s_{\text{PSK}}(t) = s_{\text{PAM}}(t)\sin(2\pi f_c t) \tag{5.20}$$

Thus, the input to the integrator in the Figure 5-13 receiver can be expressed as

$$\begin{aligned} r(t)\sin(2\pi f_c t) &= \gamma s_{\text{PAM}}(t)\sin(2\pi f_c t)\sin(2\pi f_c t) \\ &= \gamma s_{\text{PAM}}(t)\sin^2(2\pi f_c t) \\ &= \gamma s_{\text{PAM}}(t)\frac{1}{2}[1 - \cos(4\pi f_c t)] \quad \text{by trigonometric identity} \\ &= \frac{1}{2}\gamma s_{\text{PAM}}(t) - \frac{1}{2}\gamma s_{\text{PAM}}(t)\cos(4\pi f_c t) \end{aligned} \tag{5.21}$$

The input to the integrator can thus be considered as two signals: a baseband signal $0.5\gamma s_{\text{PAM}}(t)$ and a bandpass signal $0.5\gamma s_{\text{PAM}}(t)\cos(4\pi f_c t)$. Each of these signals contains all the information transmitted in the original PSK signal. Let's track these two signals through the rest of the receiver. As established in Equation (2.60), integration in the time domain corresponds to multiplication by $1/(j2\pi f)$ in the frequency domain. The integrator effectively acts as a lowpass filter, eliminating the bandpass signal. For the baseband signal, the integrator and threshold comparator act like the optimal PAM receiver, demodulating the original information from the source with minimum effect from the noise.

Example 5.3 Accuracy of a Coherently Demodulated PSK Signal—Analysis

A PSK system transmits 25,000 bits/sec using a signal with a peak amplitude of 0.1 volt. By the time the signal arrives at the receiver, its voltage is only 30% of the transmitted voltage. The noise at the input to the correlation receiver is additive white Gaussian noise with an average normalized power spectral density of 1.6×10^{-9} volts^2/Hz. Assuming that the receiver is perfectly synchronized, determine the accuracy of the system.

Solution

$$A = 0.1 \text{ volt}, \gamma = 0.3, T_b = 40\ \mu\text{sec, and } \frac{N_o}{2} = 1.6 \times 10^{-9} \text{ volts}^2/\text{Hz}$$

$$P_{b,\,\text{PSK},\,correlation\ receiver} = Q\left(\sqrt{\frac{(\gamma A)^2 T_b}{N_o}}\right) = Q\left(\sqrt{\frac{(0.03)^2(40x10^{-6})}{3.2x10^{-9}}}\right) = Q(3.35)$$

$$= 4x10^{-4}, \text{ or } 0.04\%$$

Example 5.4 Accuracy of a Coherently Demodulated PSK Signal—Design

For the system in Example 5.3, determine the minimum average normalized power for the transmitted signal that will provide an average accuracy of one error or less per 100,000 bits transmitted.

Solution

An average of one error per 100,000 bits corresponds to $P_b = .00001$. Thus,

$$P_{b,\,\text{PSK},\,correlation\ receiver} = Q\left(\sqrt{\frac{2E_b}{N_o}}\right) = Q\left(\sqrt{\frac{(\gamma A)^2 T_b}{N_o}}\right) = .00001$$

Using Table 4-1,

$$Q(a) = .00001 \text{ when } a = 4.27$$

So

$$\sqrt{\frac{2E_b}{N_o}} = \sqrt{\frac{E_b}{1.6x10^{-9}}} \geq 4.27$$

$$E_b \geq 2.917x10^{-8} \text{ volts}^2\text{-sec}$$

$$\text{Received average normalized power} = E_b r_b = 7.293x10^{-4} \text{ volts}^2$$

$$\text{Average normalized transmitted power} = \frac{7.293x10^{-4}}{(0.3)^2} = 8.103x10^{-3} \text{ volts}^2$$

5.3.2 Developing a Coherent ASK Receiver

ASK signals are not symmetric; assuming an initial phase angle of zero, the signal transmitted to represent a "1" is

$$s_1(t) = A\sin(2\pi f_c t+0) \tag{5.22}$$

while the signal transmitted to represent a "0" is

$$s_2(t) = 0 \tag{5.23}$$

Since the signals are nonsymmetric, we need to employ the general correlation receiver developed in Chapter 4, and reproduced as Figure 4.20*R*.

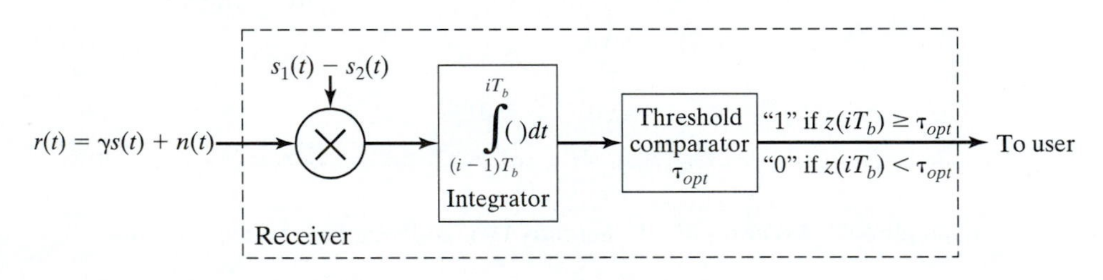

Figure 4-20*R* Optimum receiver for asymmetric signals.

As established in Section 4.2.4.1, the optimum threshold is

$$\tau_{opt} = \frac{a_1(iT_b) + a_2(iT_b)}{2} \tag{4.71R}$$

where

$$a_1(iT_b) = \int_{(i-1)T_b}^{iT_b} \gamma s_1(t)[s_1(t) - s_2(t)]\,dt,$$ the processed, received signal if a "1" was transmitted and there was no nois

and

$$a_2(iT_b) = \int_{(i-1)T_b}^{iT_b} \gamma s_2(t)[s_1(t) - s_2(t)]\,dt,$$ the processed, received signal if a "0" was transmitted and there was no noise

As also established in Chapter 4, for nonsymmetric signals

$$P_{b,\,minimum} = Q\left(\sqrt{\frac{E_d}{2N_o}}\right) \tag{4.78R}$$

where

$$E_d = \int_{(i-1)T_b}^{iT_b} [\gamma s_1(t) - \gamma s_2(t)]^2\, dt \tag{4.79R}$$

E_d is the total energy over the bit period of the *difference* between the two possible received signals.

The optimum receiver for ASK is therefore the correlation receiver shown in Figure 5-14. Note that although $s_1(t) - s_2(t) = A\sin(2\pi f_c t)$, there is no need for the constant A in the input to the mixer, since multiplying by A would effectively amplify both the received signal and the noise, and therefore have no effect on accuracy.

Let's now calculate the optimum threshold and probability of bit error for the coherent ASK receiver.

$$a_1(iT_b) = \int_{(i-1)T_b}^{iT_b} \gamma A\sin^2(2\pi f_c t)\, dt = \frac{\gamma A}{2}\int_{(i-1)T_b}^{iT_b} [1 - \cos(4\pi f_c t)]\, dt$$

$$= \frac{\gamma A T_b}{2} - \frac{\gamma A}{2}\int_{(i-1)T_b}^{iT_b} \cos(4\pi f_c t)\, dt \tag{5.24}$$

Let's assume that the carrier produces a whole number of cycles within one bit period (again, a good practice, which, if the initial phase angle is zero, promotes smooth transitions in the signal between the end of one bit period and the beginning of the next bit period). The second term goes to zero and $a_1(iT_b)$ reduces to

$$a_1(iT_b) = \frac{\gamma A T_b}{2} \tag{5.25}$$

Since $s_2(t) = 0$, $a_2(iT_b) = 0$. Thus,

$$\tau_{opt,\,\text{ASK}} = \frac{a_1(iT_b) + a_2(iT_b)}{2} = \frac{\gamma A T_b}{4} \tag{5.26}$$

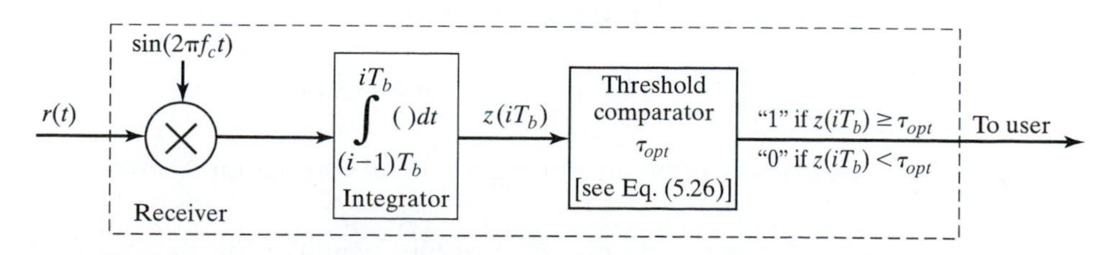

Figure 5-14 Optimum (coherent) receiver for ASK.

Applying Equation (4.79R) to ASK signals,

$$E_{d,\,\text{ASK}} = \int_{(i-1)T_b}^{iT_b} [\gamma s_1(t) - \gamma s_2(t)]^2\, dt = \int_{(i-1)T_b}^{iT_b} [\gamma A\sin(2\pi f_c t) - 0]^2\, dt \tag{5.27}$$

$$= \frac{\gamma^2 A^2}{2} \int_{(i-1)T_b}^{iT_b} [1 - \cos(4\pi f_c t)]\, dt = \frac{\gamma^2 A^2 T_b}{2} - \frac{\gamma^2 A^2}{2} \int_{(i-1)T_b}^{iT_b} \cos(4\pi f_c t)\, dt$$

Again, assuming that the carrier produces a whole number of cycles within one bit period, the second term goes to zero and

$$E_{d,\,\text{ASK}} = \frac{(\gamma A)^2\, T_b}{2} \tag{5.28}$$

Therefore, probability of bit error for ASK using the correlation receiver is

$$P_{b,\,\text{ASK},\,correlation\ receiver} = Q\left(\sqrt{\frac{E_d}{2N_o}}\right) = Q\left(\sqrt{\frac{(\gamma A)^2\, T_b}{4N_o}}\right) \tag{5.29}$$

In comparing Equations (5.19) and (5.29), we see that in order to produce the same probability of bit error, an ASK signal needs to be transmitted with twice the average power of a PSK signal (Problem 5.12). This observation makes intuitive sense because the ASK signal can be decomposed into a unipolar PAM signal multiplied by a carrier (Figure 5-3), and the PSK signal can be decomposed into a bipolar PAM signal times a carrier (Figure 5-10). As we established in Example 4.11, unipolar signals need to be transmitted with twice the average power of bipolar signals in order to achieve the same accuracy.

5.3.3 Developing a Coherent FSK Receiver

FSK signals are not symmetric; assuming an initial phase angle of zero, the signal transmitted to represent a "1" is

$$s_1(t) = A\sin[2\pi(f_c + \Delta f)t] \tag{5.30a}$$

while the signal transmitted to represent a "0" is

$$s_2(t) = A\sin[2\pi(f_c - \Delta f)t] \tag{5.30b}$$

Calculating probability of bit error using a general correlation receiver,

$$P_{b,\,minimum} = Q\left(\sqrt{\frac{E_d}{2N_o}}\right) \tag{4.78R}$$

where, for FSK

$$
\begin{aligned}
E_{d,\,\mathrm{FSK}} &= \int_{(i-1)T_b}^{iT_b} [\gamma s_1(t) - \gamma s_2(t)]^2\, dt \\
&= (\gamma A)^2 \int_{(i-1)T_b}^{iT_b} \{\sin[2\pi(f_c + \Delta f)t] - \sin[2\pi(f_c - \Delta f)t]\}^2\, dt \\
&= (\gamma A)^2 \int_{(i-1)T_b}^{iT_b} \{\sin^2[2\pi(f_c + \Delta f)t] \\
&\qquad - 2\sin[2\pi(f_c + \Delta f)t]\sin[2\pi(f_c - \Delta f)t] \\
&\qquad + \sin^2[2\pi(f_c - \Delta f)t]\}\, dt \\
&= \frac{(\gamma A)^2}{2} \int_{(i-1)T_b}^{iT_b} \{1 - \cos[4\pi(f_c + \Delta f)t]\}\, dt \\
&\quad - (\gamma A)^2 \int_{(i-1)T_b}^{iT_b} \sin[2\pi(f_c + \Delta f)t]\sin[2\pi(f_c - \Delta f)t]\, dt \\
&\quad + \frac{(\gamma A)^2}{2} \int_{(i-1)T_b}^{iT_b} \{1 - \cos[4\pi(f_c - \Delta f)t]\}\, dt \\
&= (\gamma A)^2\, T_b - \frac{(\gamma A)^2}{2} \int_{(i-1)T_b}^{iT_b} \cos[4\pi(f_c + \Delta f)t]\, dt \\
&\quad - (\gamma A)^2 \int_{(i-1)T_b}^{iT_b} \sin[2\pi(f_c + \Delta f)t]\sin[2\pi(f_c - \Delta f)t]\, dt \\
&\quad - \frac{(\gamma A)^2}{2} \int_{(i-1)T_b}^{iT_b} \cos[4\pi(f_c - \Delta f)t]\, dt
\end{aligned} \tag{5.31}
$$

To minimize P_b, we want to maximize $E_{d,\,\mathrm{FSK}}$. Generally speaking, we can maximize $E_{d,\,\mathrm{FSK}}$ if we select Δf such that the frequencies $f_c + \Delta f$ and $f_c - \Delta f$ produce a whole number of cycles within a bit period. This restriction forces the second, third, and fourth terms of Equation (5.31) to zero ($f_c + \Delta f$ and $f_c - \Delta f$ are now orthogonal). Therefore, for coherent FSK,

$$
P_{b,\,\mathrm{FSK},\,correlation\ receiver} = Q\left(\sqrt{\frac{E_{d,\,\mathrm{FSK}}}{2N_o}}\right) = Q\left(\sqrt{\frac{(\gamma A)^2 T_b}{2N_o}}\right) \tag{5.32}
$$

The correlation receiver for FSK is shown in Figure 5-15.

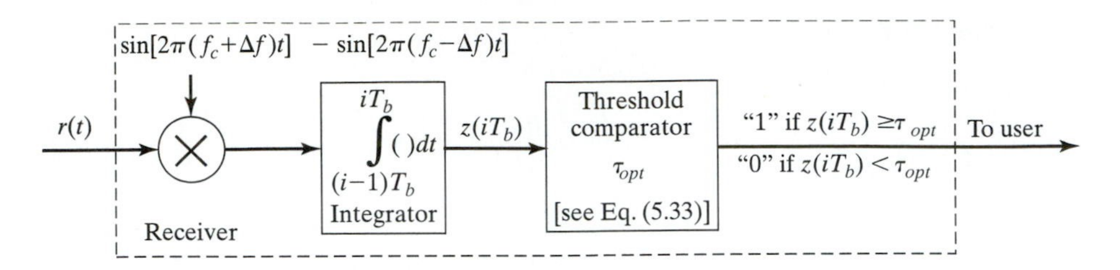

Figure 5-15 Optimum (coherent) receiver for FSK.

As you will establish in Problem 5.14, given the above restriction on Δf, the optimum threshold for the FSK correlation receiver is

$$\tau_{opt,\,\mathrm{FSK}} = 0 \tag{5.33}$$

We can reinforce our intuition concerning the FSK receiver and the restrictions on Δf by recalling that the correlation receiver can also be designed with two mixers and two integrators; that is, two correlators (see Figures 4-20 and 4-21). Figure 5-16 is therefore an equivalent design for the Figure 5-15 coherent FSK receiver. If $s_1(t)$ and $s_2(t)$ are orthogonal, then when $s_1(t)$ is transmitted, the output of the lower integrator has only a noise component while the output from the upper integrator has a signal component of $\gamma A T_b/2$ plus a noise component. Similarly, when $s_2(t)$ is transmitted, the output of the upper integrator has only a noise component while the output from the lower integrator has a signal component of $\gamma A T_b/2$ plus a noise component (note that the output of the lower correlator is inverted before summing and inputting to the threshold comparator). Orthogonality thus eliminates the effects of the received signal on the output of the correlator not corresponding to the transmitted data.

Practically speaking, the Figure 5-15 receiver is a better design (fewer components), but the dual-correlator design in Figure 5-16 helps reinforce basic concepts and will be needed when we examine M-ary bandpass modulation techniques later in Section 5.8.

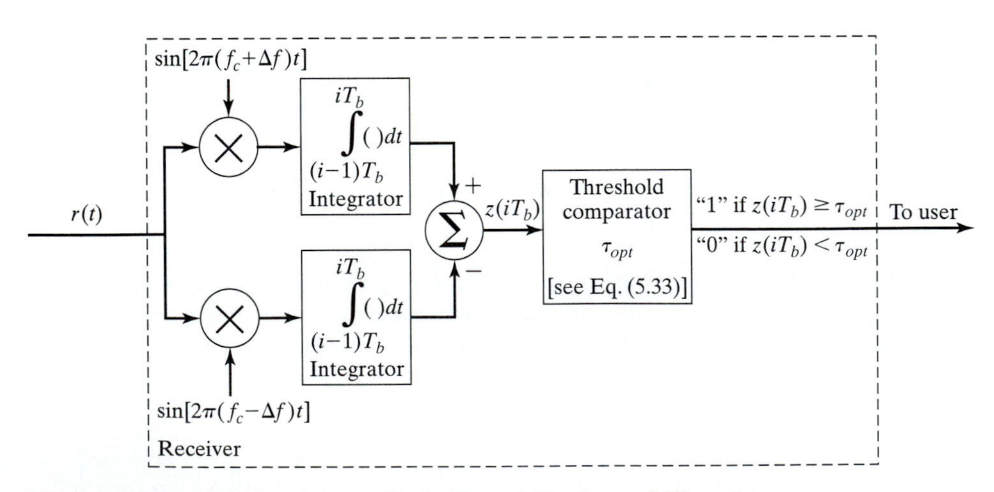

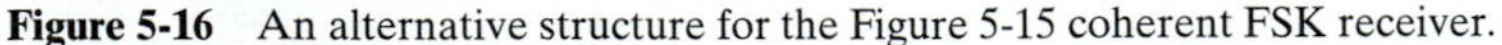
Figure 5-16 An alternative structure for the Figure 5-15 coherent FSK receiver.

5.3.4 Comparing Coherent PSK, FSK, and ASK

Table 5-1 shows many of the performance-versus-cost trade-offs for coherent PSK, FSK, and ASK. Note that equipment complexity is relatively high for all three systems, since the receivers must provide a reference signal that is in phase and frequency synchronization with the transmitted carrier signal.

Equations (5.19), (5.29), and (5.32) express probability of bit error for coherent PSK, ASK, and FSK, respectively, and are included in Table 5-1. Note, however, that for a fair comparison we must examine accuracy with all three systems transmitting with the same average normalized power or average normalized energy per bit. Thus, we need to express probability of bit error for the three systems in terms of E_b rather than amplitude of the received signal, γA. For PSK, using Equation (5.19),

$$P_{b,\text{ PSK},\, correlation\ receiver} = Q\left(\sqrt{\frac{(\gamma A)^2 T_b}{N_o}}\right) = Q\left(\sqrt{\frac{2E_b}{N_o}}\right) \tag{5.19R}$$

Expressing the probability of bit error in terms of E_b requires more work for FSK and ASK, since accuracy in nonsymmetric signaling is expressed in terms of energy difference between signals, E_d, instead of E_b. Normalized energy per bit for FSK is the same as for PSK. Thus, using Equation (5.18),

$$E_{b,\text{ FSK}} = E_{b,\text{ PSK}} = \frac{(\gamma A)^2 T_b}{2} \tag{5.34}$$

and therefore, substituting into Equation (5.32),

$$P_{b,\text{ FSK},\, correlation\ receiver} = Q\left(\sqrt{\frac{(\gamma A)^2 T_b}{2N_o}}\right) = Q\left(\sqrt{\frac{E_b}{N_o}}\right) \tag{5.35}$$

Table 5-1 A Comparison of Coherently Demodulated PSK, FSK, and ASK

Binary Modulation/ Demodulation Technique	Bandwidth (90% In-Band Power)	Relative Equipment Complexity	Accuracy (Bit Error Rate)
Coherent phase shift keying (Coherent PSK)	$2r_b$	High—tightly synchronized reference signal is needed at receiver	$P_b = Q(\sqrt{(\gamma A)^2 T_b/N_o}) = Q(\sqrt{2E_b/N_o})$ Most accurate of the techniques
Coherent frequency shift keying (Coherent FSK)	$2r_b + 2\Delta f$ Larger than ASK or PSK	High—tightly synchronized reference signal is needed at receiver	$P_b = Q(\sqrt{(\gamma A)^2 T_b/2N_o}) = Q(\sqrt{E_b/N_o})$ 3 dB less accurate than PSK
Coherent amplitude shift keying (Coherent ASK)	$2r_b$	High—tightly synchronized reference signal is needed at receiver	$P_b = Q(\sqrt{(\gamma A)^2 T_b/4N_o}) = Q(\sqrt{E_b/N_o})$ 3 dB less accurate than PSK

For ASK,

$$E_{b,\,\mathrm{ASK}} = \frac{1}{2}\int_{(i-1)T_b}^{iT_b} (\gamma A \sin 2\pi f_c t)^2\, dt + \frac{1}{2}\int_{(i-1)T_b}^{iT_b} 0^2\, dt = \frac{\gamma^2 A^2 T_b}{4} \tag{5.36}$$

Substituting into Equation (5.29),

$$P_{b,\,\mathrm{ASK},\,correlation\ receiver} = Q\left(\sqrt{\frac{(\gamma A)^2 T_b}{4N_o}}\right) = Q\left(\sqrt{\frac{E_b}{N_o}}\right) \tag{5.37}$$

Equations (5.19*R*), (5.35), and (5.37) are included in Table 5-1.

Accuracy of different systems can be compared in two ways. One way is to use a given average normalized power level (or average energy per bit) and calculate the difference in probability of bit error between the systems. This straightforward approach is suggested by Equations (5.19*R*), (5.35), and (5.37). Another approach, which is often more useful, is to start with a given probability of bit error and calculate the power required by each system to produce the given error rate. As discussed in Section 5.3.2 and supported by Equations (5.19*R*) and (5.37), coherent ASK requires twice as much average normalized power as coherent PSK to produce the same accuracy. Thus, we can state in Table 5-1 that coherent ASK is 3 dB less accurate than coherent PSK. Similarly, in comparing Equations (5.19*R*) and (5.35), we see that in order to produce the same probability of bit error, an FSK signal also needs to be transmitted with twice the average power of a PSK signal.

In examining Table 5-1, note that there are no practical advantages for coherent FSK or coherent ASK. Coherent FSK requires more bandwidth and the same equipment complexity as coherent PSK and yet is less accurate. Coherent ASK requires the same bandwidth and equipment complexity as coherent PSK and is also less accurate. Thus, if a designer is choosing a coherent receiver (i.e., if equipment complexity at the receiver is not an issue), coherent PSK is the only practical choice.

Suppose, however, that complexity of the receiver is important for a certain application (e.g., a mass consumer application where the dollar cost of the receiver may well determine whether or not the system can be successfully marketed). How can we design a receiver that does not require a coherent reference signal? How does the accuracy of such a receiver compare with a coherent PSK system? Before we develop noncoherent receivers for ASK, FSK, and PSK, we need to establish a few more mathematical tools (in particular, autocorrelation and the concept of the random process) in order to be able to assess the accuracy of noncoherent systems.

5.4 Stochastic Mathematics—Part II (Random Processes)

In Chapter 4 we discussed the need for stochastic (or probabilistic) mathematics to describe the behavior of noise in communication systems. We noted that we could not describe noise deterministically (i.e., we could not determine a function that would tell us the exact value of the noise at future times), but we could describe the *tendencies* of the noise. We used the variable **X** to denote the voltage of the noise at a particular time (we called **X** a *random variable*), and we established the following stochastic mathematical tools to describe the tendencies of **X**:

1. Probability distribution function

$$F_{\mathbf{X}}(a) \equiv P(\mathbf{X} \leq a) \tag{4.8R}$$

$F_{\mathbf{X}}(a)$ is the probability that the random variable **X** is less than or equal to some specific value a.

2. Probability density function

$$f_{\mathbf{X}}(x) \equiv \frac{dF_{\mathbf{X}}(x)}{dx} \tag{4.13R}$$

$f_{\mathbf{X}}(x)$ can be used to express the probability that the random variable **X** lies between two values, say, a and b.

$$P(a < \mathbf{X} \leq b) = \int_{a}^{b} f_{\mathbf{X}}(x)\, dx \tag{4.14R}$$

3. Mean (or expected value)

$$\mu_{\mathbf{X}} \equiv \int_{-\infty}^{\infty} x f_{\mathbf{X}}(x)\, dx \tag{4.20R}$$

$\mu_{\mathbf{X}}$ is the average value of the random variable **X**. The mean is often called the *expected value* of **X** and is symbolized as $E\{\mathbf{X}\}$. Using the notation for expected value,

$$E\{\mathbf{X}\} \equiv \mu_{\mathbf{X}} \equiv \int_{-\infty}^{\infty} x f_{\mathbf{X}}(x)\, dx \tag{4.20R}$$

4. Variance

$$\sigma_{\mathbf{X}}^2 = \int_{-\infty}^{\infty} (x - \mu_{\mathbf{X}})^2 f_{\mathbf{X}}(x)\, dx = E\{(\mathbf{X} - \mu_{\mathbf{X}})^2\} \tag{4.21R}$$

Variance is a measure of how much the values of the random variable **X** fluctuate (or *vary*) around its average. Thus, variance is a measure of the *unpredictability* of **X**.

5.4.1 Random Processes

In addition to measuring the effects of noise, stochastic (probabilistic) mathematics has another application in communication systems. In Chapter 1, we defined *information* as data that the user did not know prior to communicating with the source. Thus, even in the absence of noise, the signal arriving at the receiver is not deterministic from the receiver's point of view. In other words, the received signal has two different types of uncertainty or randomness:

1. Uncertainty due to noise
2. Uncertainty due to the information content of the signal

In order to handle signals with two degrees of uncertainty, we must develop another stochastic mathematical tool—the *random process*. To illustrate a random process, consider the following experiment:

> Being able to generate noise at a controlled average power level is a valuable capability for many laboratory procedures. Test equipment companies therefore manufacture a device known as a noise generator. Suppose we have ten noise generators, all placed on a lab bench, all turned on, and all adjusted to the same average power level. The output from each noise generator is a random variable (also known as a *sample function*). The entire group of ten outputs is called an *ensemble*. Now suppose we perform the following procedure:
>
> **a.** We pick a specific time t_k.
>
> **b.** At time t_k we randomly pick one of the noise generators and measure its voltage.

Do you understand how this experiment (or process) has two degrees of uncertainty?

Figure 5-17 illustrates the experiment. Each noise generator's output is a sample function, and the collection of all ten sample functions is an ensemble. To perform the experiment, time t_k is chosen, one noise generator (Noise Generator Three) is then chosen, and its output is measured as 0.5 volts. Again, be sure you understand how the measured value experiences two degrees of uncertainty. Figure 5-17 also introduces new notation for the sample functions. Each noise generator's output is a random variable, so we must use a subscript to identify the particular generator. Each generator's output can also change with time, so we must include time as a variable. $\mathbf{X}_3(t)$ is a random variable representing the output from the third signal generator, and $\mathbf{X}_{10}(t_o)$ represents the output from the tenth signal generator at a specific time t_o. We can designate the entire ensemble as $\mathbf{X}$, and we can use the notation $\mathbf{X}(t_k)$ to designate the random process described by the above experiment (i.e., $\mathbf{X}(t_k)$ means "pick a time t_k and then pick a noise generator and measure its output").

Example 5.5 Representing a received signal as a random process

Consider a bandpass communication system transmitting 100,000 bits/sec using FSK with a 250 kHz carrier and a frequency offset of 50 kHz. Peak voltage of the transmitted signal is 2 volts, the signal is attenuated so that the received signal has a peak

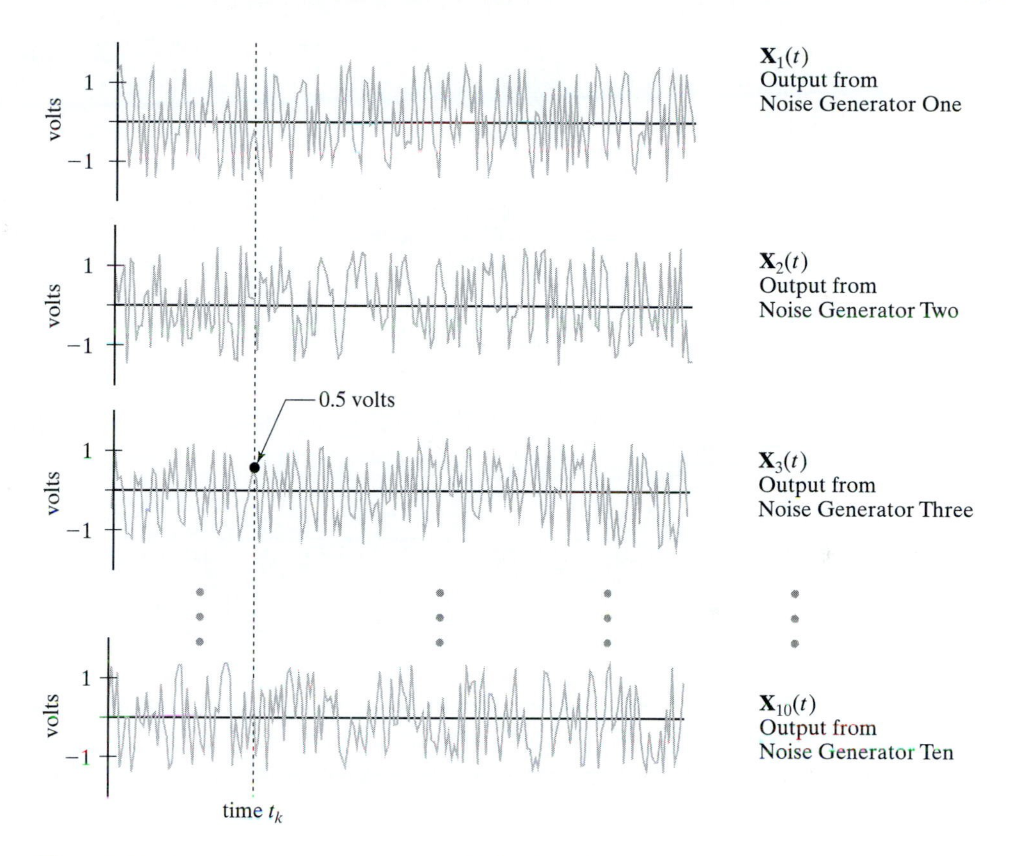

Figure 5-17 A random process with ten sample functions (outputs from ten noise generators). A particular time, t_k, is selected, one of the noise generators is then picked at random, and its output at time t_k is measured.

voltage of only 50% of the transmitted signal's peak. The noise is additive white Gaussian noise with an average normalized noise power of $\sigma^2 = 2.25$ volts^2 at the input to the receiver. Represent the received signal for the ith bit as a random process.

Solution

The transmitted signal for a particular bit is either a 300 kHz sinusoid (if the information from the source was a "1") or a 200 kHz sinusoid (if the information from the source was a "0"). As the signal is transmitted across the channel, it is corrupted by noise. The received signal can be represented as an ensemble containing two sample functions: one being a noise-corrupted, attenuated version of the 300 kHz sinusoid that would have been transmitted to represent a "1"; the second sample function being a noise-corrupted, attenuated version of the 200 kHz sinusoid that would have been transmitted to represent a "0." The two sample functions, designated $\mathbf{Y}_1(t)$ and $\mathbf{Y}_2(t)$, are shown in Figure 5-18. A voltage measurement taken by the receiver 1.5 μsec into the bit period would be designated as $\mathbf{Y}(.0000015)$.

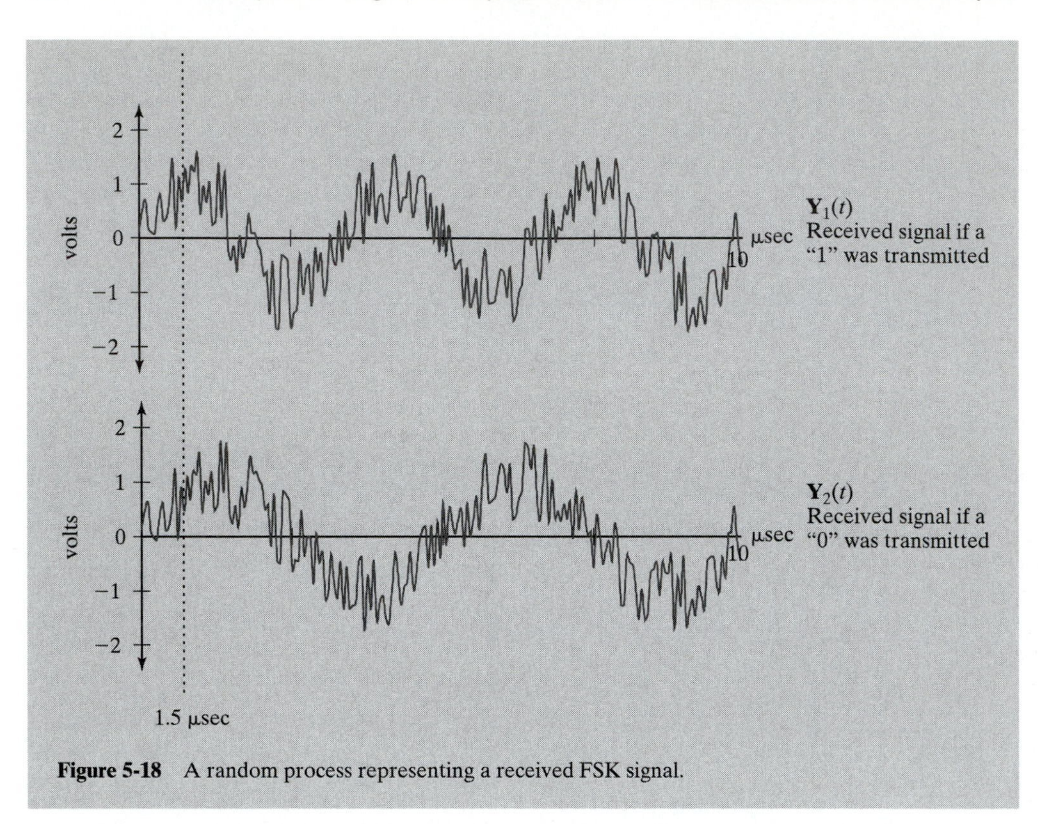

Figure 5-18 A random process representing a received FSK signal.

Now that we see how random processes are used to describe uncertainty in communication systems, let's develop the mathematical parameters we need to evaluate a random process.

5.4.1.1 Mean At a given time, the random process $\mathbf{X}(t_k)$ can be considered a random variable (be sure you understand why). It therefore makes sense to describe the mean of the random process at time t_k as

$$\mu_{\mathbf{X}}(t_k) \equiv E\{\mathbf{X}(t_k)\} \tag{5.38}$$

Unlike random variables, the mean of a random process may be time-dependent.

Example 5.6 Calculating the mean of a random process

Consider the random process established in Example 5.5. Assuming that a "1" and a "0" are equally probable from the source:

a. Determine the mean of the random process at $t = 1.5$ μsec.
b. Determine the mean of the random process at $t = 4.5$ μsec.

Solution

As established in Example 5.5, the received signal is either a 1 volt peak 300 kHz sinusoid corrupted by noise if the source output was a "1" or a 1 volt peak 200 kHz sinusoid corrupted by noise if the source output was a "0." We can therefore express the received signal representing the ith bit as

$$r_i(t) = \begin{cases} \sin[2\pi(300{,}000)t] + n(t) & \text{if } b_i = 1 \\ \sin[2\pi(200{,}000)t] + n(t) & \text{if } b_i = 0 \end{cases}$$

where $n(t)$ represents the additive white Gaussian noise at the receiver.

a. If $b_i = 1$, the received signal at time $t = 1.5$ μsec is

$$r_i(1.5\ \mu\text{sec}) = \sin(0.9\pi) + n(t) = 0.309 + n(t).$$

If $b_i = 0$, the received signal at time $t = 1.5$ μsec is

$$r_i(1.5\ \mu\text{sec}) = \sin(0.6\pi) + n(t) = 0.951 + n(t).$$

Since a "1" and a "0" are equally likely for the ith bit, the mean or expected (or *average*) value of $r_i(t)$ at time $t = 1.5$ μsec is

(0.5)(expected, or average, value of received signal given that a "1" was transmitted) +
(0.5)(expected, or average, value of received signal given that a "0" was transmitted) =
(0.5)(0.309 volts plus the expected value of the noise at $t = 1.5$ μsec) +
(0.5)(0.951 volts plus the expected value of the noise at $t = 1.5$ μsec) =
0.1545 volts + (0.5)(expected value of the noise at $t = 1.5$ μsec) +
0.4755 volts + (0.5)(expected value of the noise at $t = 1.5$ μsec) =
0.630 volts + (expected value of the noise at $t = 1.5$ μsec) =
0.630 volts + 0 volts = 0.630 volts, since the noise has zero mean.

Using mathematical notation, we can express the mean of the random process at $t = 1.5$ μsec as

$$\begin{aligned}
\mu_{\mathbf{X}}(1.5\ \mu\text{sec}) &= E\{\mathbf{X}(1.5\ \mu\text{sec})\} \\
&= P\{b_1 = 1\}E\{0.309 + n(t)\} + P\{b_1 = 0\}E\{0.951 + n(t)\} \\
&= (0.5)E\{0.309 + n(t)\} + (0.5)E\{0.951 + n(t)\} \\
&= (0.5)[(0.309) + E\{n(t)\}] + (0.5)[(0.951) + E\{n(t)\}] \\
&= 0.630 + E\{n(t)\} \\
&= 0.630 + 0 \\
&= 0.630 \text{ volts}
\end{aligned}$$

b.
$$\begin{aligned}
\mu_{\mathbf{X}}(4.5\ \mu\text{sec}) &= E\{\mathbf{X}(4.5\ \mu\text{sec})\} \\
&= P\{b_1 = 1\}E\{0.809 + n(t)\} + P\{b_1 = 0\}E\{-0.588 + n(t)\} \\
&= (0.5)E\{0.809 + n(t)\} + (0.5)E\{-0.588 + n(t)\} \\
&= (0.5)[(0.809) + E\{n(t)\}] + (0.5)[(-0.588) + E\{n(t)\}] \\
&= 0.111 + E\{n(t)\} \\
&= 0.111 + 0 \\
&= 0.111 \text{ volts}
\end{aligned}$$

The most difficult thing about working with random processes is the mathematical notation. Don't let the notation or equations confuse you—you should always be able to look at a particular step in the mathematical process and be able to explain it in English.

5.4.1.2 Autocorrelation

The *autocorrelation* of a random process $\mathbf{X}(t)$ is defined as

$$R_{\mathbf{XX}}(t_1, t_2) \equiv E\{\mathbf{X}(t_1)\mathbf{X}(t_2)\} \tag{5.39}$$

Autocorrelation is a measure of the degree to which two samples from the same random process (one taken at time t_1, the other taken at time t_2) are related. A few comments concerning autocorrelation:

1. The concept of multiplying two signals (or samples) and then determining their average is related to the term "correlation" as we discussed in Chapter 4 (regarding the correlation receiver).
2. The term "autocorrelation" is appropriate for Equation (5.39), since the equation determines how much a random process is correlated with itself (in conventional English "auto" often means "self").
3. Remember that the random process has two degrees of uncertainty (there are a number of sample functions from which to randomly choose, and the value of each sample function at any particular instant of time is random). Thus, when considering autocorrelation, remember that the sample taken at time t_2 may be from a different sample function than the sample taken at time t_1. Be sure you understand this point.
4. The higher the value of autocorrelation—that is, the larger the value of $R_{\mathbf{XX}}(t_1, t_2)$—the more the two time samples are related. In practical terms, a large value of $R_{\mathbf{XX}}(t_1, t_2)$ means that if we know the value of the random process at time t_1, we have a high probability of predicting its value at t_2.
5. Additive white Gaussian noise is *uncorrelated*. Knowing its value at any particular instant in time is of no use in predicting its value at any other instant in time.

Example 5.7 Calculating and interpreting the autocorrelation of a random process

Consider a bandpass communication system transmitting 100,000 bits/sec using PSK with a 400 kHz carrier. Peak voltage of the transmitted signal is 2 volts, the signal is attenuated so that the received signal has a peak voltage of only 50% of the transmitted signal's peak, and the noise is additive white Gaussian noise with an average normalized noise power of $\sigma^2 = 2.25$ volts^2. The data from the source is equiprobable.

a. Determine the correlation between two measurements at the receiver taken at times $t = 2.0$ μsec and $t = 6.0$ μsec.

b. Determine the correlation between two measurements at the receiver taken at times $t = 2.0$ μsec and $t = 11.0$ μsec.

c. Interpret the results of Parts a and b.

Solution

a.
$$R_{\mathbf{XX}}(2.0\ \mu\text{sec}, 4.0\ \mu\text{sec}) \equiv E\{\mathbf{X}(2.0\ \mu\text{sec})\mathbf{X}(4.0\ \mu\text{sec})\}$$

Note that $t = 2.0$ μsec and $t = 4.0$ μsec both occur within the first bit period. Therefore, the first bit is the only data bit that affects the autocorrelation.

$$
\begin{aligned}
R_{XX}(2.0\,\mu\text{sec}, 4.0\,\mu\text{sec}) &\equiv E\{\mathbf{X}(2.0\,\mu\text{sec})\mathbf{X}(4.0\,\mu\text{sec})\} \\
&= P\{b_1 = 1\}E\{[r(2.0\,\mu\text{sec})|b_1 = 1]r[(4.0\,\mu\text{sec})|b_1 = 1]\} + \\
&\quad P\{b_1 = 0\}E\{[r(2.0\,\mu\text{sec})|b_1 = 0]r[(4.0\,\mu\text{sec})|b_1 = 0]\} \\
&= (0.5)E\{[\sin(1.6\pi) + n(2.0\,\mu\text{sec})][\sin(3.2\pi) + n(4.0\,\mu\text{sec})]\} + \\
&\quad (0.5)E\{[\sin(2.6\pi) + n(2.0\,\mu\text{sec})][\sin(4.2\pi) + n(4.0\,\mu\text{sec})]\} \\
&= (0.5)E\left\{\begin{matrix}\sin(1.6\pi)\sin(3.2\pi) + \sin(1.6\pi)n(4.0\,\mu\text{sec}) + \\ n(2.0\,\mu\text{sec})\sin(3.2\pi) + n(2.0\,\mu\text{sec})n(4.0\,\mu\text{sec})\end{matrix}\right\} + \\
&\quad (0.5)E\left\{\begin{matrix}\sin(2.6\pi)\sin(4.2\pi) + \sin(2.6\pi)n(4.0\,\mu\text{sec}) + \\ n(2.0\,\mu\text{sec})\sin(4.2\pi) + n(2.0\,\mu\text{sec})n(4.0\,\mu\text{sec})\end{matrix}\right\} \\
&= (0.5)\left[\begin{matrix}E\{\sin(1.6\pi)\sin(3.2\pi)\} + E\{\sin(1.6\pi)n(4.0\,\mu\text{sec})\} + \\ E\{n(2.0\,\mu\text{sec})\sin(3.2\pi)\} + E\{n(2.0\,\mu\text{sec})n(4.0\,\mu\text{sec})\}\end{matrix}\right] + \\
&\quad (0.5)\left[\begin{matrix}E\{\sin(2.6\pi)\sin(4.2\pi)\} + E\{\sin(2.6\pi)n(4.0\,\mu\text{sec})\} + \\ E\{n(2.0\,\mu\text{sec})\sin(4.2\pi)\} + E\{n(2.0\,\mu\text{sec})n(4.0\,\mu\text{sec})\}\end{matrix}\right] \\
&= (0.5)\left[\begin{matrix}(-0.951)(-0.588) + (-0.951)E\{n(4.0\,\mu\text{sec})\} + \\ (-0.588)E\{n(2.0\,\mu\text{sec})\} + E\{n(2.0\,\mu\text{sec})n(4.0\,\mu\text{sec})\}\end{matrix}\right] + \\
&\quad (0.5)\left[\begin{matrix}(0.951)(0.588) + (0.951)E\{n(4.0\,\mu\text{sec})\} + \\ (0.588)E\{n(2.0\,\mu\text{sec})\} + E\{n(2.0\,\mu\text{sec})n(4.0\,\mu\text{sec})\}\end{matrix}\right]
\end{aligned}
$$

We know that the noise has zero mean, so

$$E\{n(2.0\,\mu\text{sec})\} = E\{n(4.0\,\mu\text{sec})\} = 0$$

Furthermore, we know that any two noise samples are uncorrelated, so

$$E\{n(2.0\,\mu\text{sec})n(4.0\,\mu\text{sec})\} = 0.$$

Thus, six of the eight terms in the previous calculation reduce to zero, resulting in

$$
\begin{aligned}
R_{XX}(2.0\,\mu\text{sec}, 4.0\,\mu\text{sec}) &= (0.5)[(-0.951)(-0.588) + 0 + 0 + 0] \\
&\quad + (0.5)[(0.951)(0.588) + 0 + 0 + 0] \\
&= 0.559
\end{aligned}
$$

b. $$R_{XX}(2.0\,\mu\text{sec}, 11.0\,\mu\text{sec}) \equiv E\{\mathbf{X}(2.0\,\mu\text{sec})\mathbf{X}(11.0\,\mu\text{sec})\}$$

Unlike Part a, note that $t = 2.0\,\mu\text{sec}$ occurs within the first bit period and that $t = 11.0\,\mu\text{sec}$ occurs within the second bit period. Therefore, both the first and second data bits affect the autocorrelation.

$$
\begin{aligned}
R_{XX}(2.0\,\mu\text{sec}, 11.0\,\mu\text{sec}) &\equiv E\{\mathbf{X}(2.0\,\mu\text{sec})\mathbf{X}(11.0\,\mu\text{sec})\} \\
&= P\{b_1 = 1, b_2 = 1\}E\{[r(2.0\,\mu\text{sec})|b_1 = 1][r(11.0\,\mu\text{sec})|b_2 = 1]\} + \\
&\quad P\{b_1 = 1, b_2 = 0\}E\{[r(2.0\,\mu\text{sec})|b_1 = 1][r(11.0\,\mu\text{sec})|b_2 = 0]\} + \\
&\quad P\{b_1 = 0, b_2 = 1\}E\{[r(2.0\,\mu\text{sec})|b_1 = 0][r(11.0\,\mu\text{sec})|b_2 = 1]\} + \\
&\quad P\{b_1 = 0, b_2 = 0\}E\{[r(2.0\,\mu\text{sec})|b_1 = 0][r(11.0\,\mu\text{sec})|b_2 = 0]\} \\
&= (0.25)E\{[\sin(1.6\pi) + n(2.0\,\mu\text{sec})][\sin(8.8\pi) + n(11.0\,\mu\text{sec})]\} + \\
&\quad (0.25)E\{[\sin(1.6\pi) + n(2.0\,\mu\text{sec})][\sin(9.8\pi) + n(11.0\,\mu\text{sec})]\} + \\
&\quad (0.25)E\{[\sin(2.6\pi) + n(2.0\,\mu\text{sec})][\sin(8.8\pi) + n(11.0\,\mu\text{sec})]\} + \\
&\quad (0.25)E\{[\sin(2.6\pi) + n(2.0\,\mu\text{sec})][\sin(9.8\pi) + n(11.0\,\mu\text{sec})]\} \\
&= (0.25)\sin(1.6\pi)\sin(8.8\pi) + (0.25)\sin(1.6\pi)\sin(9.8\pi) + \\
&\quad (0.25)\sin(2.6\pi)\sin(8.8\pi) + (0.25)\sin(2.6\pi)\sin(9.8\pi) \\
&= 0
\end{aligned}
$$

c. Interpreting our results in Part a of this example, the received signal's voltages at times $t = 2.0\ \mu\text{sec}$ and $t = 4.0\ \mu\text{sec}$ are correlated; thus, knowing the value at time $t = 2.0\ \mu\text{sec}$ allows us to somewhat anticipate the value at time $t = 4.0\ \mu\text{sec}$. From a practical standpoint, we can make measurements at times $t = 2.0\ \mu\text{sec}$ and $t = 4.0\ \mu\text{sec}$, and if the measurement at $t = 4.0\ \mu\text{sec}$ is close to its predicted value, we have greater confidence that neither of the two measurements were badly affected by noise and that we can accurately predict whether the received data was a "1" or a "0."

Interpreting our results in Part b of this example, the received signal's voltages at times $t = 2.0\ \mu\text{sec}$ and $t = 11.0\ \mu\text{sec}$ are not correlated; thus, knowing the value at time $t = 2.0\ \mu\text{sec}$ gives us no additional information concerning the signal's anticipated value at $t = 11.0\ \mu\text{sec}$.

Autocorrelation calculations can be extremely tedious, but the concepts are not too difficult. Nevertheless, you need to stay extremely alert concerning your understanding of probabilities and your mathematical bookkeeping.

5.4.1.3 Wide-sense stationarity As we've just established, the mean of a random process generally varies with the time of the sample, and the autocorrelation of a random process generally varies with the time of both samples. This dependence on time can greatly reduce the practical benefits of knowing the probabilistic statistics of a random process. Let's investigate a special class of random processes in which the mean is not time-variant and the autocorrelation depends only on the time difference between the two samples.

A random process $\mathbf{X}(t)$ is defined as *wide-sense stationary* (WSS) if its mean and autocorrelation are not affected by a shift in the time origin. In other words, a random process $\mathbf{X}(t)$ is WSS if and only if

a. for any times t_1, t_2, t_3, etc., $\mu_{\mathbf{X}}(t_k) = \mu_{\mathbf{X}}(t_2) = \mu_{\mathbf{X}}(t_3) = \ldots = \mu_{\mathbf{X}}$, a constant

and

b. $R_{\mathbf{XX}}(t_1, t_2) = R_{\mathbf{XX}}(t_1 + \gamma, t_2 + \gamma)$ for all values of γ. In other words, the autocorrelation is a function only of the difference between the two sample times, not the actual times. We can write the autocorrelation of a WSS random process using only one variable: $R_{\mathbf{XX}}(t_2 - t_1)$, or $R_{\mathbf{XX}}(\tau)$, where $\tau = t_2 - t_1$.

Many random processes describing practical communication systems are WSS.

If a random process is WSS, its autocorrelation function exhibits the following properties:

$$R_{\mathbf{XX}}(\tau) = R_{\mathbf{XX}}(-\tau) \tag{5.40}$$

$$R_{\mathbf{XX}}(\tau) \leq R_{\mathbf{XX}}(0) \text{ for all } \tau \tag{5.41}$$

$$R_{\mathbf{XX}}(0) = E\{\mathbf{X}^2(t)\}, \text{ the average normalized power of the signal} \tag{5.42}$$

$$\mathcal{F}\{R_{\mathbf{XX}}(\tau)\} = G_{\mathbf{X}}(f) \tag{5.43}$$

The properties given in Equations (5.40), (5.41), and (5.42) can all be proven in a straightforward manner (see Problem 5.19). Equation (5.43), known as the Wiener-Khintchine theorem, is extremely important and will be discussed in detail (and proven) in the next section of this chapter. Note that Equations (5.42) and (5.43) allow us to calculate power and power spectral density for probabilistic functions, providing us with our first common thread between deterministic and probabilistic signals.

5.4.2 The Wiener-Khintchine Theorem

The Wiener-Khintchine theorem states that if $\mathbf{X}(t)$ is a wide sense stationary (WSS) random process and if $R_{\mathbf{XX}}(\tau)$ is sufficiently small for large values of τ so that $\int_{-\infty}^{\infty} |\tau R_{\mathbf{XX}}(\tau)| d\tau < \infty$, then

$$\mathrm{G}_{\mathbf{X}}(f) = \mathcal{F}\{R_{\mathbf{XX}}(\tau)\} = \int_{-\infty}^{\infty} R_{\mathbf{XX}}(\tau) e^{-2\pi f \tau}\, d\tau \tag{5.43R}$$

and

$$R_{\mathbf{XX}}(\tau) = \mathcal{F}^{1}\{\mathrm{G}_{\mathbf{X}}(f)\} \tag{5.44}$$

The restriction that $R_{\mathbf{XX}}(\tau)$ be sufficiently small for large values of t holds true for realistic random processes: It states that there is less correlation between samples as the time interval between the samples becomes larger, and that as the time interval becomes infinitely long, the correlation between the two samples goes to zero.

The Wiener-Khintchine theorem is extremely important in the analysis of communication systems because it provides a bridge between deterministic and random signals. We now have a relationship between autocorrelation, a parameter we've used to characterize probabilistic signals, and power spectral density, a parameter that, until now, we've used only to characterize deterministic signals. The capability to calculate the power spectral density of both probabilistic and deterministic signals as they pass through a communication system is extremely important for characterizing the effects of noise. We will also be able to determine bandwidth of probabilistic signals and evaluate how the signals are affected by various types of filtering and processing. In the remainder of this section, we will prove the Wiener-Khintchine theorem and provide examples showing its importance in communication system analysis.

First, let's establish a formal definition of what we mean by the power spectral density of a random process. Consider a random process $\mathbf{X}(t)$. We use the notation $\mathbf{X}_i(t)$ to represent the ith sample function of $\mathbf{X}(t)$. Let $\mathbf{X}_i(t)_T$ represent the *truncated* version of $\mathbf{X}_i(t)$. In other words,

$$\mathbf{X}_i(t)_T = \begin{cases} \mathbf{X}_i(t) & \text{if } \frac{-T}{2} \le t \le \frac{T}{2} \\ 0 & \text{if } |t| > \frac{T}{2} \end{cases} \tag{5.45}$$

$\mathbf{X}_i(t)_T$ has finite energy, so we can define the Fourier transform of $\mathbf{X}_i(t)_T$ as

$$\mathcal{F}\{\mathbf{X}_i(t)_T\} = \mathbf{X}_i(f)_T = \int_{-\infty}^{\infty} \mathbf{X}_i(t)_T \mathrm{e}^{-j2\pi f t} dt = \int_{\frac{-T}{2}}^{\frac{T}{2}} \mathbf{X}_i(t) \mathrm{e}^{-j2\pi f t} dt \tag{5.46}$$

We can denote the average normalized energy of $\mathbf{X}_i(t)_T$ as

$$\text{Avg. Norm. Energy}_{\mathbf{X}_i(t)_T} = \int_{\frac{-T}{2}}^{\frac{T}{2}} E\{\mathbf{X}_i^2(t)\}dt = \int_{-\infty}^{\infty} E\{\mathbf{X}_i^2(t)_T\}dt \tag{5.47}$$

Remember that E represents expected value. Be sure you understand Equation (5.47).

Now consider the truncated random process $\mathbf{X}(t)_T$ to be made up of an ensemble of truncated sample functions. The average normalized energy for the whole random process (i.e., averaged across all truncated sample functions) is

$$\begin{aligned}
\text{Avg. Norm. Energy}_{\mathbf{X}(t)_T} &= \int_{\frac{-T}{2}}^{\frac{T}{2}} E\{\mathbf{X}^2(t)\}dt \\
&= \int_{-\infty}^{\infty} E\{\mathbf{X}^2(t)_T\}dt \\
&= \int_{-\infty}^{\infty} E\{|\mathbf{X}(f)_T|^2\}\, df \text{ by Parseval's theorem (see Chapter 2)}
\end{aligned} \tag{5.48}$$

Now let's consider power. Average normalized power during the time interval $-\frac{T}{2}$ to $\frac{T}{2}$ is

$$P_{\mathbf{X}(t)_T} = \frac{1}{T}[\text{Avg. Norm. Energy}_{\mathbf{X}(t)_T}] = \frac{1}{T}\int_{\frac{-T}{2}}^{\frac{T}{2}} E\{\mathbf{X}^2(t)\}dt \tag{5.49}$$

Let's now stretch the limits of truncation to $\pm\infty$ in order to get the true definition of average normalized power for the entire random process.

$$\begin{aligned}
P_{\mathbf{X}(t)} &= \lim_{T\to\infty} \frac{1}{T}\int_{-\frac{T}{2}}^{\frac{T}{2}} E\{\mathbf{X}^2(t)\}dt = \lim_{T\to\infty} \frac{1}{T}\int_{-\frac{T}{2}}^{\frac{T}{2}} E\{\mathbf{X}^2(t)_T\}dt \\
&= \int_{-\infty}^{\infty} \lim_{T\to\infty} \frac{E\{\mathbf{X}^2(t)_T\}}{T}\, dt = \int_{-\infty}^{\infty} \lim_{T\to\infty} \frac{E\{|\mathbf{X}(f)_T|^2\}}{T}\, df
\end{aligned} \tag{5.50}$$

Thus, in a manner consistent with deterministic signals, we can define the power spectral density of a random process as

$$\begin{aligned}
G_{\mathbf{X}}(f) &\equiv \lim_{T\to\infty} \frac{E\left\{|\mathbf{X}(f)_T|^2\right\}}{T} \\
\text{where } \mathbf{X}(f)_T &= \int_{-\frac{T}{2}}^{\frac{T}{2}} \mathbf{X}(t)e^{-j2\pi ft}\, dt
\end{aligned} \tag{5.51}$$

We can now begin our proof of the Wiener-Khintchine theorem:

$$G_{\mathbf{X}}(f) \equiv \lim_{T\to\infty} \frac{E\{|\mathbf{X}(f)_T|^2\}}{T} = \lim_{T\to\infty} \frac{E\left\{\left|\int_{-\frac{T}{2}}^{\frac{T}{2}} \mathbf{X}(t)e^{-j2\pi ft}\,dt\right|^2\right\}}{T}$$

$$= \lim_{T\to\infty} \frac{1}{T} E\left\{\left(\int_{-\frac{T}{2}}^{\frac{T}{2}} \mathbf{X}(t)e^{-j2\pi ft}\,dt\right)^*\left(\int_{-\frac{T}{2}}^{\frac{T}{2}} \mathbf{X}(h)e^{-j2\pi fh}\,dh\right)\right\}$$

$$= \lim_{T\to\infty} \frac{1}{T} \int_{-\frac{T}{2}}^{\frac{T}{2}} \int_{-\frac{T}{2}}^{\frac{T}{2}} E\{\mathbf{X}^*(t)\mathbf{X}(h)e^{j2\pi ft}e^{-j2\pi fh}\}\,dtdh \tag{5.52}$$

Assuming $\mathbf{X}(t)$ is real (a practical assumption),

$$G_{\mathbf{X}}(f) = \lim_{T\to\infty} \frac{1}{T} \int_{-\frac{T}{2}}^{\frac{T}{2}} \int_{-\frac{T}{2}}^{\frac{T}{2}} E\{\mathbf{X}(t)\mathbf{X}(h)e^{j2\pi ft}e^{-j2\pi fh}\}\,dtdh$$

$$= \lim_{T\to\infty} \frac{1}{T} \int_{-\frac{T}{2}}^{\frac{T}{2}} \int_{-\frac{T}{2}}^{\frac{T}{2}} E\{\mathbf{X}(t)\mathbf{X}(h)\}e^{j2\pi ft}e^{-j2\pi fh}\,dtdh \tag{5.53a}$$

Now let $\tau = h - t$. As h ranges from $-\frac{T}{2}$ to $\frac{T}{2}$, τ ranges from $-\frac{T}{2} - t$ to $\frac{T}{2} - t$ and $dh = d\tau$. Our integrals can thus be expressed as

$$G_{\mathbf{X}}(f) = \lim_{T\to\infty} \frac{1}{T} \int_{-\frac{T}{2}}^{\frac{T}{2}} \int_{-\frac{T}{2}-t}^{\frac{T}{2}-t} E\{\mathbf{X}(t)\mathbf{X}(t+\tau)\}e^{j2\pi ft}e^{-j2\pi f(t+\tau)}\,d\tau dt$$

$$= \lim_{T\to\infty} \frac{1}{T} \int_{-\frac{T}{2}}^{\frac{T}{2}} \int_{-\frac{T}{2}-t}^{\frac{T}{2}-t} R_{\mathbf{XX}}(\tau)e^{-j2\pi f\tau}\,d\tau dt \tag{5.53b}$$

Let's evaluate the double integral in Equation (5.53b) graphically. The area of integration is shown in Figure 5-19a. As also shown in Figure 5-19a, integrating first with respect to τ forms a vertical strip. We then integrate with respect to t, which corresponds to sliding the strip from left to right. Suppose, however, we reverse the order of integration, integrating first with respect to t and then integrating with respect to τ. Integrating with respect to t first forms a horizontal strip as shown in Figure 5-19b. Integrating with respect to τ then corresponds to sliding this strip from the bottom to the top.

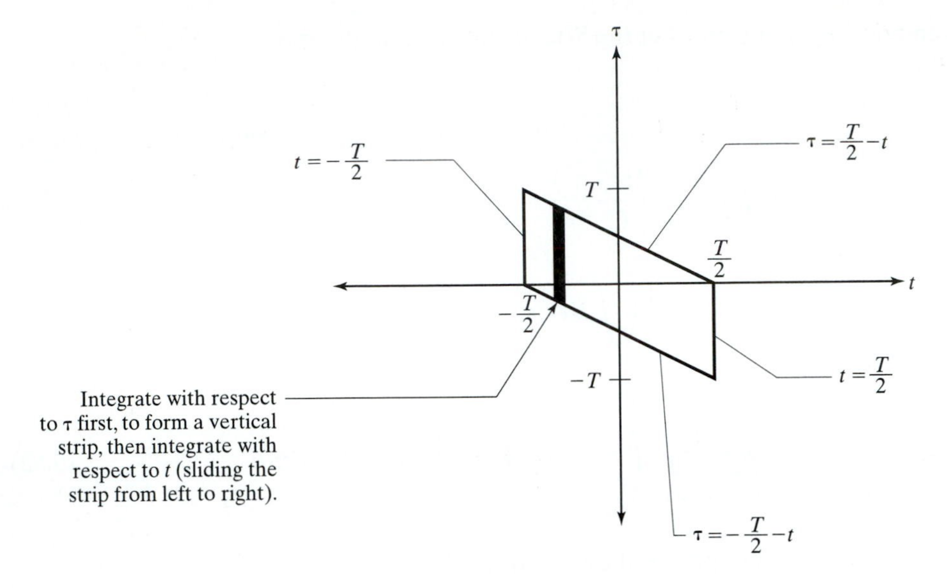

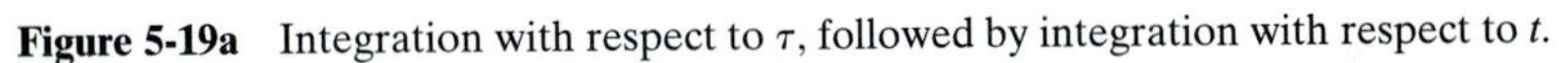

Figure 5-19a Integration with respect to τ, followed by integration with respect to t.

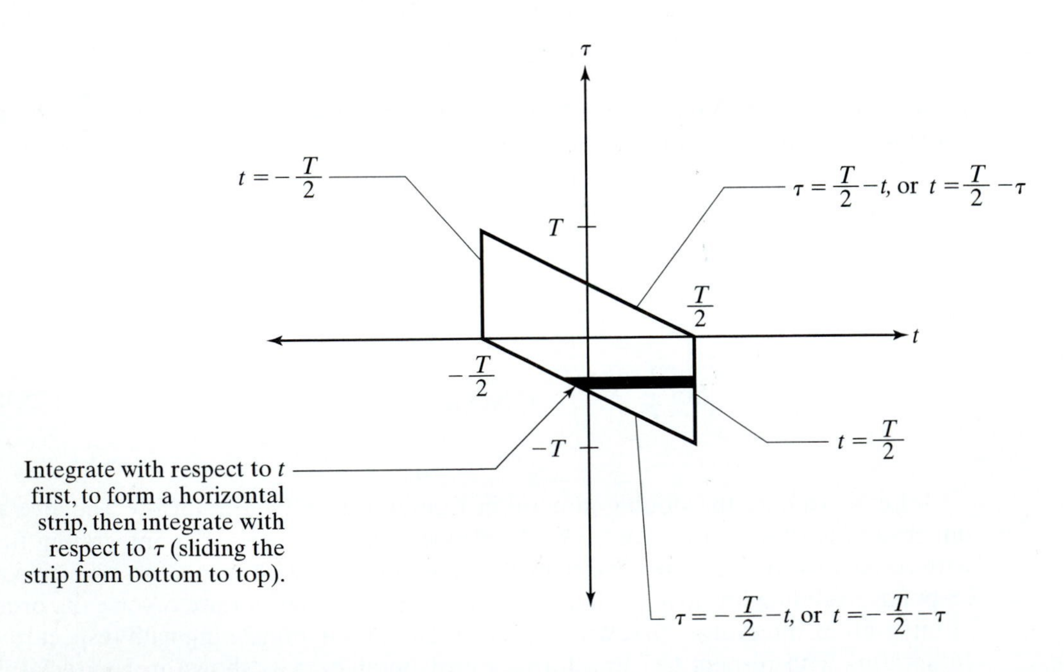

Figure 5-19b Integration with respect to t, followed by integration with respect to τ.

Reversing the order of integration, Equation (5.53b) thus becomes

$$G_{\mathbf{X}}(f) = \lim_{T\to\infty} \frac{1}{T} \int_{-\frac{T}{2}}^{\frac{T}{2}} \int_{-\frac{T}{2}-t}^{\frac{T}{2}-t} R_{\mathbf{XX}}(\tau) e^{-j2\pi f\tau}\, d\tau dt$$

$$= \lim_{T\to\infty} \frac{1}{T} \int_{-T}^{T} \int_{-\frac{T}{2}-\tau}^{\frac{T}{2}-\tau} R_{\mathbf{XX}}(\tau) e^{-j2\pi f\tau}\, dt d\tau$$

$$= \lim_{T\to\infty} \frac{1}{T} \left[\underbrace{\int_{-T}^{0} \int_{-\frac{T}{2}-\tau}^{\frac{T}{2}} R_{\mathbf{XX}}(\tau) e^{-j2\pi f\tau}\, dt d\tau}_{\text{sliding strip from bottom to } x \text{ axis}} + \underbrace{\int_{0}^{T} \int_{-\frac{T}{2}}^{\frac{T}{2}-\tau} R_{\mathbf{XX}}(\tau) e^{-j2\pi f\tau}\, dt d\tau}_{\text{sliding strip from } x \text{ axis to top}} \right]$$

$$= \lim_{T\to\infty} \frac{1}{T} \left[\int_{-T}^{0} R_{\mathbf{XX}}(\tau) e^{-j2\pi f\tau}\, d\tau \int_{-\frac{T}{2}-\tau}^{\frac{T}{2}} dt + \int_{0}^{T} R_{\mathbf{XX}}(\tau) e^{-j2\pi f\tau}\, d\tau \int_{-\frac{T}{2}}^{\frac{T}{2}-\tau} dt \right]$$

$$= \lim_{T\to\infty} \frac{1}{T} \left[\int_{-T}^{0} (T+\tau) R_{\mathbf{XX}}(\tau) e^{-j2\pi f\tau}\, d\tau + \int_{0}^{T} (T-\tau) R_{\mathbf{XX}}(\tau) e^{-j2\pi f\tau}\, d\tau \right] \tag{5.54}$$

Since $R_{\mathbf{XX}}(\tau) = R_{\mathbf{XX}}(-\tau)$ for WSS random processes, we can express Equation (5.54) as

$$G_{\mathbf{X}}(f) = \lim_{T\to\infty} \frac{1}{T} \left[\int_{-T}^{0} (T+\tau) R_{\mathbf{XX}}(\tau) e^{-j2\pi f\tau}\, d\tau + \int_{0}^{T} (T-\tau) R_{\mathbf{XX}}(\tau) e^{-j2\pi f\tau}\, d\tau \right]$$

$$= \lim_{T\to\infty} \frac{1}{T} \int_{-T}^{T} (T - |\tau|) R_{\mathbf{XX}}(\tau) e^{-j2\pi f\tau}\, d\tau$$

$$= \lim_{T\to\infty} \int_{-T}^{T} \frac{T-|\tau|}{T} R_{\mathbf{XX}}(\tau) e^{-j2\pi f\tau}\, d\tau$$

$$= \lim_{T\to\infty} \left[\int_{-T}^{T} R_{\mathbf{XX}}(\tau) e^{-j2\pi f\tau}\, d\tau - \int_{-T}^{T} \frac{|\tau|}{T} R_{\mathbf{XX}}(\tau) e^{-j2\pi f\tau}\, d\tau \right]$$

$$= \int_{-\infty}^{\infty} R_{\mathbf{XX}}(\tau) e^{-j2\pi f\tau}\, d\tau - \lim_{T\to\infty} \int_{-T}^{T} \frac{|\tau|}{T} R_{\mathbf{XX}}(\tau) e^{-j2\pi f\tau}\, d\tau \tag{5.55}$$

Initially, we specified that $R_{XX}(\tau)$ be sufficiently small for large values of t so that $\int_{-\infty}^{\infty} |\tau R_{XX}(\tau)| d\tau < \infty$, a restriction that holds true for realistic random processes. Thus,

$$\lim_{T\to\infty} \int_{-T}^{T} \frac{|\tau|}{T} R_{XX}(\tau) e^{-j2\pi f\tau}\, d\tau = 0 \tag{5.56}$$

and Equation (5.55) reduces to

$$\begin{aligned} G_X(f) &= \int_{-\infty}^{\infty} R_{XX}(\tau) e^{-j2\pi f\tau}\, d\tau - \lim_{T\to\infty} \int_{-T}^{T} \frac{|\tau|}{T} R_{XX}(\tau) e^{-j2\pi f\tau}\, d\tau \\ &= \int_{-\infty}^{\infty} R_{XX}(\tau) e^{-j2\pi f\tau}\, d\tau \equiv \mathcal{F}\{R_{XX}(\tau)\} \end{aligned} \tag{5.57}$$

Example 5.8 Formally determining the power spectral density of a binary PAM signal

In Chapter 3, we developed an informal approach to determining the power spectral density of a binary PAM signal. The signal was a series of rectangular pulses: a positive pulse of amplitude A representing each "1" output by the source and a negative pulse of amplitude $-A$ representing each "0." We now know that the transmitted signal can be represented as a random process. Let's use the Wiener-Khintchine theorem to formally determine the power spectral density of the transmitted signal.

Solution

In order to apply the Wiener-Khintchine theorem, we need to first establish that the signal is WSS. In Problem 5.20 you will determine that if the transmitted signal consists of a series of rectangular pulses of amplitude A and width γ, then $\mu_X = 0$ and autocorrelation is the triangular function plotted in Figure 5-20. Since the mean is a constant and the autocorrelation is only a function of the time between samples, the random process is WSS. Furthermore, in examining Figure 5-20 note that $\int_{-\infty}^{\infty} |\tau R_{XX}(\tau)| d\tau < \infty$. All the restrictions of the Wiener-Khintchine theorem are therefore met, and we can determine the power spectral density of the random process by taking the Fourier transform of the autocorrelation function.

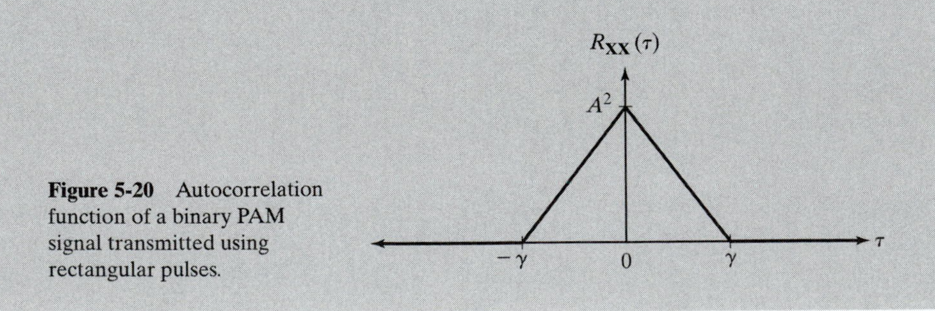

Figure 5-20 Autocorrelation function of a binary PAM signal transmitted using rectangular pulses.

$$
\begin{aligned}
G_X(f) = \mathcal{F}\{R_{XX}(\tau)\} &= \frac{1}{\gamma}\int_{-\gamma}^{0} A^2(\gamma+\tau)e^{-j2\pi f\tau}\,d\tau + \frac{1}{\gamma}\int_{0}^{\gamma} A^2(\gamma-\tau)e^{-j2\pi f\tau}\,d\tau \\
&= A^2\left\{\int_{-\gamma}^{0} e^{-j2\pi f\tau}\,d\tau + \frac{1}{\gamma}\int_{-\gamma}^{0} \tau e^{-j2\pi f\tau}\,d\tau + \int_{0}^{\gamma} e^{-j2\pi f\tau}\,d\tau - \frac{1}{\gamma}\int_{0}^{\gamma} \tau e^{-j2\pi f\tau}\,d\tau\right\} \\
&= A^2\left\{\int_{-\gamma}^{\gamma} e^{-j2\pi f\tau}\,d\tau + \frac{1}{\gamma}\int_{-\gamma}^{0} \tau e^{-j2\pi f\tau}\,d\tau - \frac{1}{\gamma}\int_{0}^{\gamma} \tau e^{-j2\pi f\tau}\,d\tau\right\} \\
&= \ldots \quad \text{(use integration by parts on the second and third terms)} \\
&= A^2\gamma\,\text{sinc}^2(\pi f\gamma)\ \text{volts}^2/\text{Hz}
\end{aligned}
$$

This confirms the results achieved using the less rigorous approach taken in Chapter 3, particularly Equation (3.2).

Example 5.9 Determining the autocorrelation of thermal noise

In Chapter 4, we stated that the probabilistic properties of thermal noise do not change with time. Thus, thermal noise is WSS.[2] We also established that the power spectral density of thermal noise is constant from dc to approximately 10^{12} Hz. The power spectral density of thermal noise is shown in Figure 4-4*R*.

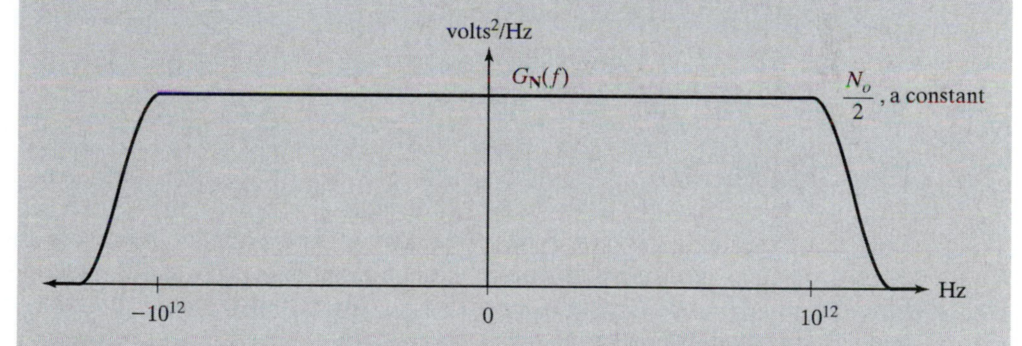

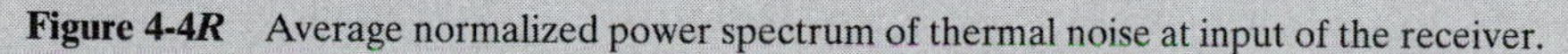

Figure 4-4*R* Average normalized power spectrum of thermal noise at input of the receiver.

If we approximate the power spectral density as

$$
G_N(f) = \begin{cases} N_o/2 & |f| \le 10^{12}\ \text{Hz} \\ 0 & |f| > 10^{12}\ \text{Hz} \end{cases}
$$

[2]In fact, we can categorize thermal noise as *strict-sense stationary* (a much more restrictive classification), meaning that *none* of its probabilistic properties change with respect to time.

then the autocorrelation function for the thermal noise is

$$R_{\mathbf{XX}}(\tau) = \mathcal{F}^{1}\{G_{\mathbf{N}}(f)\} = 10^{12}\, N_o\, \mathrm{sinc}[2\pi(10^{12})\tau]$$

The autocorrelation function is plotted in Figure 5-21. Note that it is an extremely tall, extremely thin sinc function.

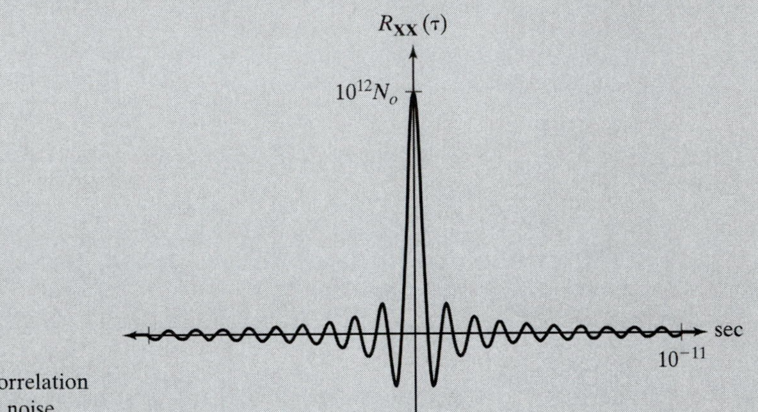

Figure 5-21 Autocorrelation function for thermal noise.

In examining Figure 5-21, note that autocorrelation for thermal noise is very close to zero for all but the shortest of time intervals between samples. Practically speaking, we can say that if the time between two samples is greater than 10^{-11} seconds, then the two samples are uncorrelated (i.e., knowing the value of the noise at any particular instant in time is of no use in predicting its value at any other instant in time more than 10^{-11} seconds away). Note that thermal noise is not, strictly speaking, white. It does not have frequency components above 10^{12} Hz. Practically speaking, as we've claimed earlier, we can model thermal noise as white and claim that samples are uncorrelated.

5.4.3 Ergodicity

Let's now consider a different type of average involving a random process. Instead of examining all of the sample functions within a random process, let's select one sample function and determine its statistics over time. These statistics are called *time averages* and are indicated by bracketing $< >$.

5.4.3.1 Time-averaged mean The *time-averaged mean* for the *i*th sample function, symbolized as $\langle \mu_{\mathbf{X}_i} \rangle$, is defined as

$$\langle \mu_{\mathbf{X}_i} \rangle = \lim_{T \to \infty} \frac{1}{T} \int_{-T/2}^{T/2} \mathbf{X}_i(t)\, dt \tag{5.58}$$

Note that $\langle \mu_{\mathbf{X}} \rangle$ may be different for different sample functions within the same random process.

5.4.3.2 Time-averaged autocorrelation

The *time-averaged autocorrelation* for the ith sample function, symbolized as $\langle R_{\mathbf{X}_i\mathbf{X}_i}(\tau)\rangle$, is defined as

$$\langle R_{\mathbf{X}_i\mathbf{X}_i}(\tau)\rangle \equiv \lim_{T\to\infty}\frac{1}{T}\int_{-T/2}^{T/2}\mathbf{X}_i(t)\mathbf{X}_i(t+\tau)\,dt \tag{5.59}$$

Again, generally speaking, the value of $\langle R_{\mathbf{XX}}(\tau)\rangle$ may be different for different sample functions within the same random process.

A random process is called *ergodic* if the values of its ensemble parameters are equal to the values of its time-averaged parameters. The WSS random process $\mathbf{X}(t)$ is *ergodic in its mean and autocorrelation* if $\mu_{\mathbf{X}} = \langle\mu_{\mathbf{X}_i}\rangle$ and $R_{\mathbf{XX}}(\tau) = \langle R_{\mathbf{X}_i\mathbf{X}_i}(\tau)\rangle$ for all sample functions within the ensemble (i.e., for all i).

What good are time-averaged statistics? In most practical cases, time averages are easier to measure than ensemble averages, and many random processes are ergodic in the mean and autocorrelation. If a signal can be represented by a random process that is ergodic in the mean and autocorrelation, then, as shown in Table 5-2, many significant properties of the signal are easy to measure and have solid physical and intuitive meaning.

Table 5-2 Physical Meaning of Stochastic Parameters for an Ergodic Random Process

Parameter		
Ensemble Average	Time Average	Physical Significance
$\mu_{\mathbf{X}}$	$\langle\mu_{\mathbf{X}_i}\rangle$	dc level of the signal
$R_{\mathbf{XX}}(0)$	$\langle R_{\mathbf{X}_i\mathbf{X}_i}(0)\rangle$	average normalized power of the signal
$\sqrt{R_{\mathbf{XX}}(0)}$	$\sqrt{\langle R_{\mathbf{X}_i\mathbf{X}_i}(0)\rangle}$	rms voltage or current of the signal
$\sigma^2_{\mathbf{X}}$	$\langle\sigma^2_{\mathbf{X}_i}\rangle$	average normalized ac power of the signal

5.5 Noncoherent Receivers for ASK and FSK

In Section 5.3, we developed coherent receivers for PSK, ASK, and FSK. We found that the coherent PSK system provided high accuracy and good bandwidth but required a high level of equipment complexity, since a coherent system requires a receiver that must independently produce a reference signal that is tightly synchronized in phase and frequency with the transmitted carrier signal. We also observed that coherent ASK and FSK systems required the same degree of receiver complexity and offered no improvements in bandwidth or accuracy relative to coherent PSK.

Suppose we have a certain application, such as an inexpensive mass-consumer product, for which a coherent system is too complex. How can we design a receiver that does not require a coherent reference signal? How does the accuracy of such a receiver compare with a coherent PSK system? Now that we've established the mathematical tools we'll need for analysis, let's develop noncoherent receivers—first for ASK and FSK, and then, in Section 5.6, for PSK.

5.5.1 The Envelope Detector

Consider a sine wave $x(t)$ input to the circuit shown in Figure 5-22a. During the first quarter cycle of the sine wave, the diode allows current to flow to the resistor and capacitor. As shown in Figure 5-22b, the capacitor voltage $y(t)$ rises with $x(t)$ (its value will actually be $x(t)$ minus a small voltage drop across the diode) and the capacitor starts charging. Early within the second quarter of the cycle, as $x(t)$ begins to drop, $y(t)$ becomes greater than $x(t)$, and the diode acts like an open circuit. This causes the capacitor to discharge energy through the resistor and, as shown in Figure 5-22b, the capacitor voltage $y(t)$ begins to drop. The diode continues to act as an open circuit during the third and fourth quarters of the sinusoid's cycle, causing $y(t)$ to continue to drop. As the sine wave $x(t)$ begins its second cycle, the input voltage catches up to the output voltage and the process repeats.

The significance of the circuit in Figure 5-22a is that if the values of R and C are chosen carefully, then the output $y(t)$ closely tracks the amplitude (or *envelope*) of the input $x(t)$. This is why the circuit is often called an *envelope detector*. As shown in Figure 5-23, if an ASK waveform is fed into the envelope detector, the detector's output approximates the envelope of the ASK signal. Furthermore, we can see from Figure 5-23 that the envelope of the ASK signal is itself a baseband unipolar PAM signal that represents the same data as the original ASK signal. In Figure 5-23, the frequency of the carrier is chosen as ten times the bit rate and the ASK signal represents the data "101101." For practical bandpass modulation systems, the frequency of the carrier signal is usually chosen to be much larger than the bit rate ($f_c \gg r_b$). If

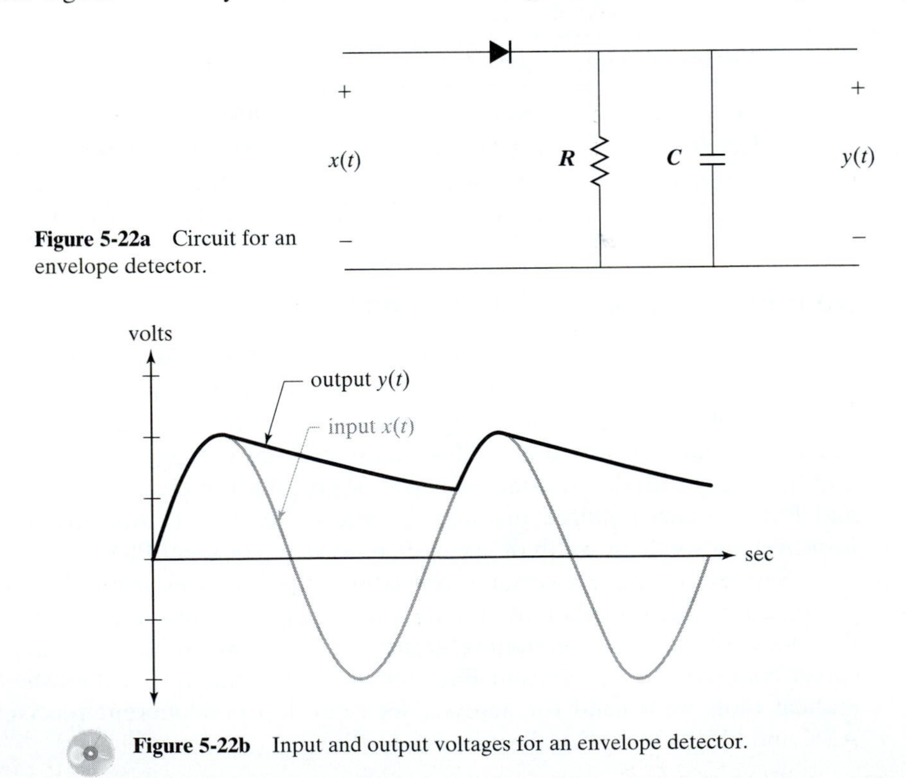

Figure 5-22a Circuit for an envelope detector.

Figure 5-22b Input and output voltages for an envelope detector.

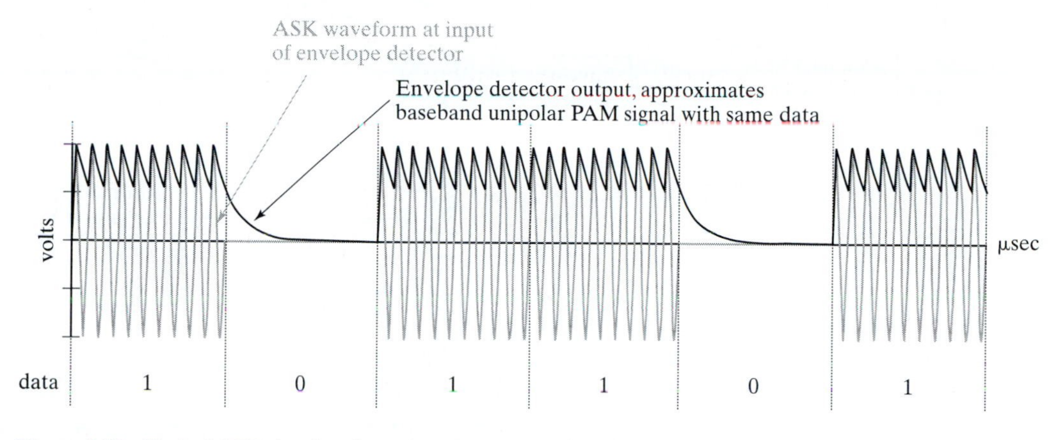

Figure 5-23 Typical ASK signal and envelope detector output.

$f_c \gg r_b$ and if values of R and C are chosen properly, then the envelope detector circuit can very closely track the envelope of the ASK signal.

5.5.2 Noncoherent Demodulation of ASK

We've just established that an envelope detector can convert an ASK signal into a unipolar PAM signal containing the same data. A noncoherent ASK receiver can therefore be constructed as shown in Figure 5-24. The receiver structure is similar to that of the simple PAM receiver (see Chapter 4), except that bandpass filtering is employed and that sampling occurs at the end of the bit period, allowing the envelope detector more time to accurately track the envelope of the received signal. When choosing the threshold τ, remember that the output from the envelope detector is unipolar.

5.5.3 Noncoherent Demodulation of FSK

In Section 5.2.3, we showed that we can decompose an FSK signal into two ASK signals: one ASK signal with a carrier frequency of $f_c + \Delta f$ Hz, turned on to represent a "1" and off to represent a "0," and the other ASK signal with a carrier frequency of $f_c - \Delta f$ Hz, turned on to represent a "0" and off to represent a "1." Let's call these two

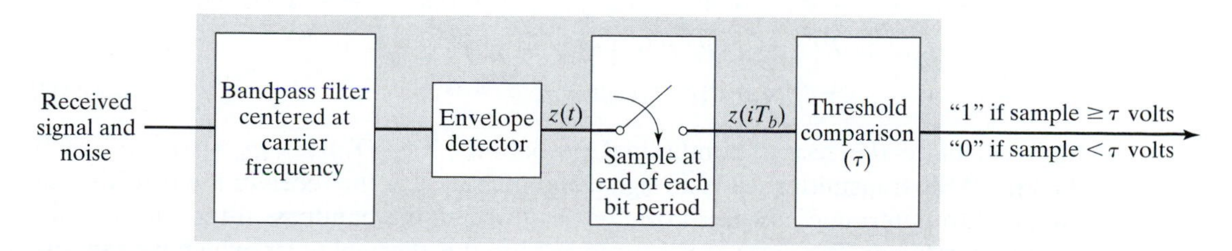

Figure 5-24 A noncoherent ASK receiver.

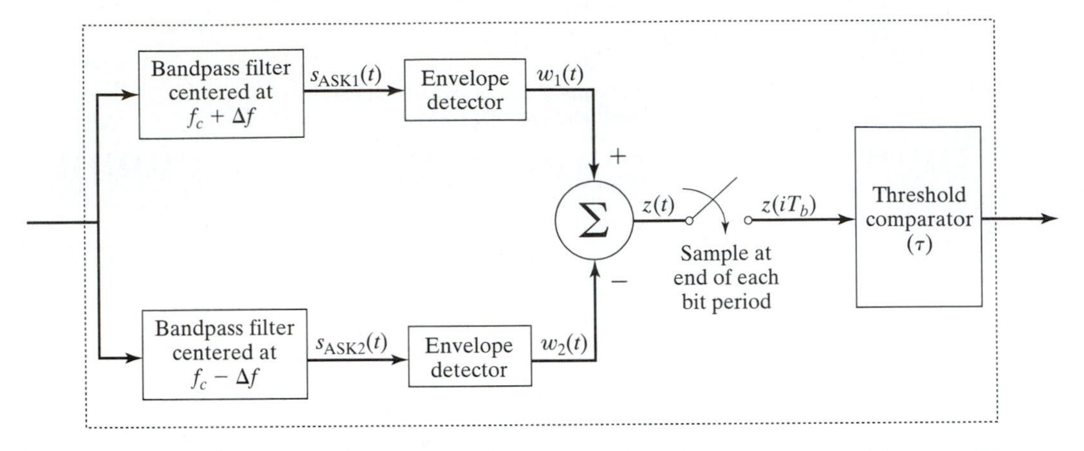

Figure 5-25 A noncoherent FSK receiver.

signals $s_{ASK1}(t)$ and $s_{ASK2}(t)$, respectively. As shown in Figure 5-25, we can use two bandpass filters to isolate $s_{ASK1}(t)$ and $s_{ASK2}(t)$. The output of the upper envelope detector, $w_1(t)$, is a unipolar PAM signal with a high voltage signifying a "1" and a low voltage signifying a "0." The output of the lower envelope detector, $w_2(t)$, is a unipolar PAM signal with a high voltage signifying a "0" and a low voltage signifying a "1." The output of the summer, $z(t) = w_1(t) - w_2(t)$, is therefore a bipolar PAM signal with a positive voltage signifying a "1" and a negative voltage signifying a "0."

5.5.4 Performance of Noncoherent ASK and FSK Receivers

5.5.4.1 Probability of bit error for a noncoherent ASK receiver Let's begin by determining probability of bit error for the noncoherent ASK receiver in Figure 5-24. We can use the same type of approach we used to determine P_b for our baseband and coherent bandpass systems, but we need to remember that, for the noncoherent receiver, sampling occurs at the end of the bit period and that the noncoherent ASK receiver uses an additional piece of circuitry—an envelope detector.

An error occurs if a "0" is transmitted but the sampled voltage at the end of the bit period is greater than or equal to the threshold, and an error also occurs if a "1" is transmitted but the sampled voltage at the end of the bit period is less than the threshold. Stating this mathematically,

$$\begin{aligned} P_{b,\,noncoherent\ \mathrm{ASK}} &= P\{b_i = 1 \text{ and } z(iT_b) < \tau\} + P\{b_i = 0 \text{ and } z(iT_b) \geq \tau\} \\ &= P\{b_i = 1\}P\{z(iT_b) < \tau | b_i = 1\} \\ &\quad + P\{b_i = 0\}P\{z(iT_b) \geq \tau | b_i = 0\} \end{aligned} \tag{5.60}$$

Let's calculate the second conditional probability, $P\{z(iT_b) \geq \tau | b_i = 0\}$. If $b_i = 0$, then the ASK transmitter sends no signal and the signal at the receiver is entirely noise. We need to determine how this noise passes through the bandpass filter and then the envelope detector. The noise at the receiver's input is thermal noise, which we can represent as a random process $\mathbf{X}(t)$. This thermal noise is additive, zero mean, stationary,

Gaussian noise with a flat power spectral density ($N_o/2$) from dc to approximately 10^{12} Hz, as shown in Figure 4-4*R*. The power spectral density of the noise after it has been bandpass filtered is shown in Figure 5-26, where f_c is the carrier frequency and B is the bandwidth of the filter (the same bandwidth as the ASK signal). Let the random process $\mathbf{Y}(t)$ represent the filtered noise. Since bandpass filtering is linear and time-invariant, the filtered noise is also additive, zero mean, and stationary, and can be modeled by a Gaussian probability density (as discussed in Section 4.1.3). The filtering merely changes the variance or average power of the noise.

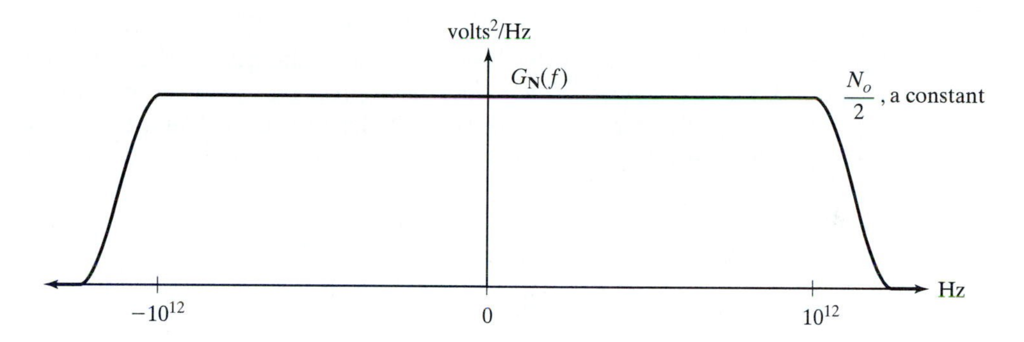

Figure 4-4*R* Average normalized power spectrum of thermal noise at input of the receiver.

We must now determine the characteristics of the filtered noise after it has passed through the envelope detector. The envelope detector is a nonlinear device (due to the diode), and so its processing of the noise signal is not linear. The noise at the output of the envelope detector is therefore not necessarily Gaussian. To understand the effects of the envelope detector on the noise, let's begin by determining the time domain characteristics of the noise at the output of the bandpass filter and at the output of the envelope detector. Because the bandpass filter is linear and time-invariant, we know that the noise at the output of the bandpass filter is Gaussian, zero mean, and stationary. Using the Wiener-Khintchine theorem, we can express the

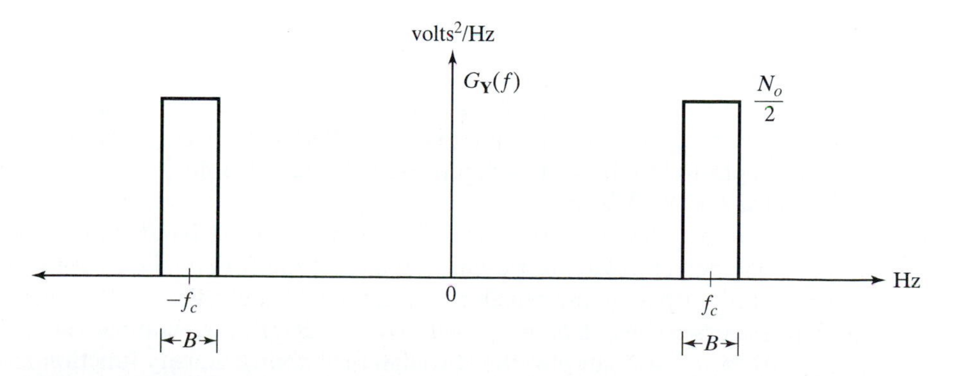

Figure 5-26 Average normalized power spectrum of thermal noise after bandpass filtering.

autocorrelation of the bandpass filtered noise as

$$R_{\mathbf{YY}}(\tau) = \mathcal{F}^{-1}\{G_{\mathbf{Y}}(f)\} = BN_o \operatorname{sinc}(\pi B\tau)\cos(2\pi f_c\tau) \tag{5.61}$$

Let's use $n_o(t)$ to represent the noise at the output of the bandpass filter (which we will call *narrowband noise*). For small values of τ, we can satisfy Equation (5.61) by expressing the narrowband noise as

$$n_o(t) = \mathbf{W}(t)\cos(2\pi f_c t) + \mathbf{Z}(t)\sin(2\pi f_c t) \tag{5.62}$$

where $\mathbf{W}(t)$ and $\mathbf{Z}(t)$ are uncorrelated, zero-mean, Gaussian random processes (Shanmugam, [5.2]). Equation (5.62) is often called the *quadrature* representation of narrowband noise. The term "quadrature" is appropriate because the sine and cosine components are 90 degrees (or a *quarter* of the unit circle) apart. We can express $n_o(t)$ in another form by adding the two quadrature components on the right side of Equation (5.62).

$$\begin{aligned} n_o(t) &= \mathbf{W}(t)\cos(2\pi f_c t) + \mathbf{Z}(t)\sin(2\pi f_c t) \\ &= \mathbf{R}(t)\cos(2\pi f_c t + \Theta(t)) \end{aligned} \tag{5.63}$$

where

$$\mathbf{R} = \sqrt{\mathbf{W}^2 + \mathbf{Z}^2}$$

and

$$\Theta = \arctan\left(\frac{\mathbf{Z}}{\mathbf{W}}\right)$$

Since $\mathbf{W}(t)$ and $\mathbf{Z}(t)$ are random variables, $\mathbf{R}(t)$ and $\Theta(t)$ are also random processes. Conversion from $\mathbf{W}$ and $\mathbf{Z}$ to $\mathbf{R}$ and Θ is the same as conversion from rectangular to polar coordinates. We can show that $\mathbf{R}(t)$ has the probability density function

$$f_{\mathbf{R}}(r) = \begin{cases} 0 & r < 0 \\ \dfrac{r}{\sigma_o^2} e^{-r^2/(2\sigma_o^2)} & r \geq 0 \end{cases} \tag{5.64}$$

where σ_o^2 is the average normalized power of the noise at the output of the bandpass filter (see Papoulis [5.1]). Equation (5.64) is called a *Rayleigh probability density function* and is plotted for $\sigma_o = 1$ in Figure 5-27. We can also show that Θ is uniformly distributed for $-\pi < \Theta \leq \pi$.

Equation (5.64) is perfect to describe how narrowband noise passes through an envelope detector. $\mathbf{R}(t)$ represents the magnitude (or *envelope*) of the narrowband noise, so the output of the envelope detector is merely $\mathbf{R}(t)$, a Rayleigh-distributed random process. Thus, if $b_i = 0$, then $z(iT_b) = \mathbf{R}(iT_b)$. Assuming that $P\{b_i = 1\} = P\{b_i = 0\} = 0.5$, we can plot the Rayleigh probability density function and the error region for the second term in Equation (5.60) as shown in Figure 5-28. The gray area

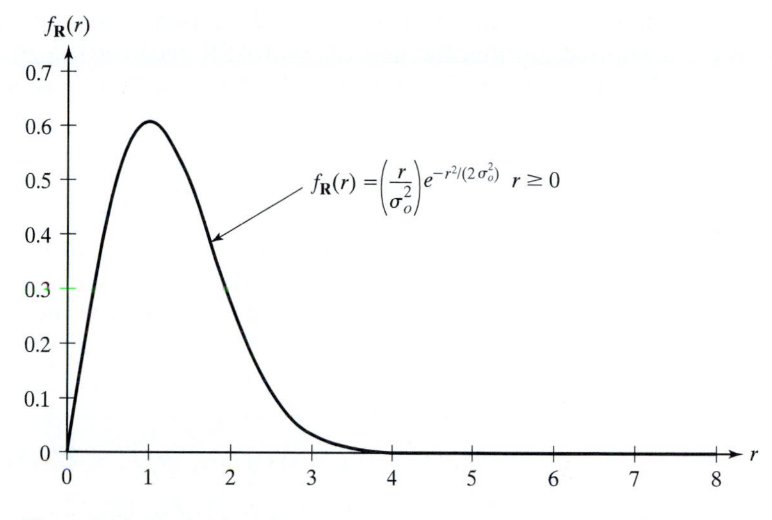

Figure 5-27 Rayleigh probability density function with $\sigma = 1$.

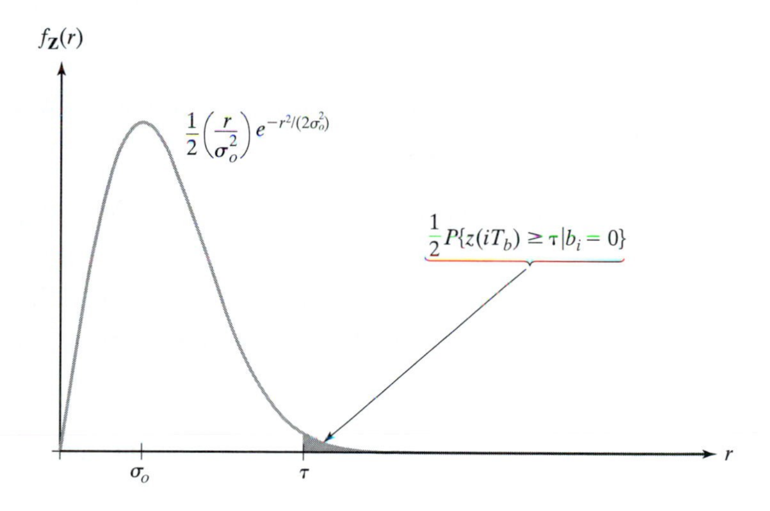

Figure 5-28 Error region for transmitting a "0" and having noncoherent ASK receiver demodulate it as a "1."

represents the probability that a "0" was transmitted but that the received signal was demodulated as a "1." Mathematically,

$$P\{z(iT_b) \geq \tau | b_i = 0\} = \int_{\tau}^{\infty} \frac{r}{\sigma_o^2} e^{-r^2/(2\sigma_o^2)} \, dr \tag{5.65}$$

Now we must examine the first term in Equation (5.60)—the probability that a "1" was transmitted but that the noncoherent ASK receiver incorrectly demodulated the received signal as a "0." If $b_i = 1$, then the ASK transmitter sends the carrier and, assuming zero initial phase, the signal at the receiver can be expressed as

$$r_{b_i=1}(t) = \gamma A \sin(2\pi f_c t) + n(t) \tag{5.66}$$

where γ represents the attenuation of the signal as it passes through the channel. Since the bandwidth of the bandpass filter is set to allow the signal to pass through undistorted, we can represent the received signal and noise at the output of the bandpass filter as

$$\begin{aligned} r_{b_i=1,\,filtered}(t) &= \gamma A \sin(2\pi f_c t) + n_o(t) \\ &= \gamma A \sin(2\pi f_c t) + \mathbf{W}(t)\cos(2\pi f_c t) + \mathbf{Z}(t)\sin(2\pi f_c t) \\ &= \mathbf{W}(t)\cos(2\pi f_c t) + (\gamma A + \mathbf{Z}(t))\sin(2\pi f_c t) \end{aligned} \tag{5.67}$$

We must now determine the characteristics of $r_{b_i=1,filtered}(t)$ after it passes through the envelope detector. Equation (5.67) is a quadrature representation of the received, filtered signal and noise. Adding the two quadrature components together produces

$$\begin{aligned} r_{b_i=1,\,filtered}(t) &= \mathbf{W}(t)\cos(2\pi f_c t) + (\gamma A + \mathbf{Z}(t))\sin(2\pi f_c t) \\ &= \mathbf{M}(t)\cos(2\pi f_c t + \theta) \end{aligned} \tag{5.68}$$

We can show (see Papoulis [5.1]) that $\mathbf{M}(t)$ has the Rician probability density function

$$f_{\mathbf{M}}(m) = \begin{cases} 0 & r < 0 \\ \dfrac{m}{\sigma_o^2} I_o\left(\dfrac{\gamma A m}{\sigma_o^2}\right) e^{-r[m^2+(\gamma A)^2]/(2\sigma_o^2)} & r \geq 0 \end{cases} \tag{5.69}$$

where $I_o(x)$ is the modified Bessel function of the first kind and zero order,

$$I_o(x) = \frac{1}{2\pi}\int_0^{2\pi} e^{x\cos\lambda}\, d\lambda \tag{5.70}$$

In a practical noncoherent ASK system, the received signal's amplitude, γA, must be significantly higher than the average magnitude of the noise; that is, $(\gamma A)^2 \gg \sigma_o^2$. When this is true, the Bessel function can be approximated by

$$I_o(x) = \frac{1}{2\pi}\int_0^{2\pi} e^{x\cos\lambda}\, d\lambda \approx \sqrt{\frac{1}{2\pi x}}e^{x} \tag{5.71}$$

Substituting this approximation into Equation (5.69) yields

$$
\begin{aligned}
f_{\mathbf{M}}(m) &\approx \begin{cases} 0 & r < 0 \\ \dfrac{m}{\sigma_o^2}\sqrt{\dfrac{\sigma_o^2}{2\pi\gamma Am}}\, e^{\gamma Am/\sigma_o^2} e^{-[m^2+(\gamma A)^2]/(2\sigma_o^2)} & r \geq 0 \end{cases} \\
&= \begin{cases} 0 & r < 0 \\ \sqrt{\dfrac{m}{2\pi\gamma A\sigma_o^2}}\, e^{-[m^2-2\gamma Am+(\gamma A)^2]/(2\sigma_o^2)} & r \geq 0 \end{cases} \\
&= \begin{cases} 0 & r < 0 \\ \sqrt{\dfrac{m}{2\pi\gamma A\sigma_o^2}}\, e^{-[m-\gamma A]^2/(2\sigma_o^2)} & r \geq 0 \end{cases}
\end{aligned}
\tag{5.72}
$$

The Rician distribution described in Equation (5.72) is plotted in Figure 5-29 for $\sigma_o = 1$ and $\gamma A = 4$. As we should expect, if the received signal amplitude is much greater than the noise amplitude, the mean of the distribution should be approximately the received signal amplitude.

Summarizing, if $b_i = 1$, then Equation (5.68) is a magnitude-phase representation of the received signal and noise at the input to the envelope detector, and the envelope detector output is the Ricean-distributed random process $\mathbf{M}(t)$. Thus, if $b_i = 1$, then $z(iT_b) = \mathbf{M}(iT_b)$. Assuming that $P\{b_i = 1\} = P\{b_i = 0\} = 0.5$, we can plot the Rician probability density function and the error region for the first term in Equation (5.60) as shown in Figure 5-30. The black area represents the probability that a "1" was transmitted but that the received signal was demodulated as a "0."

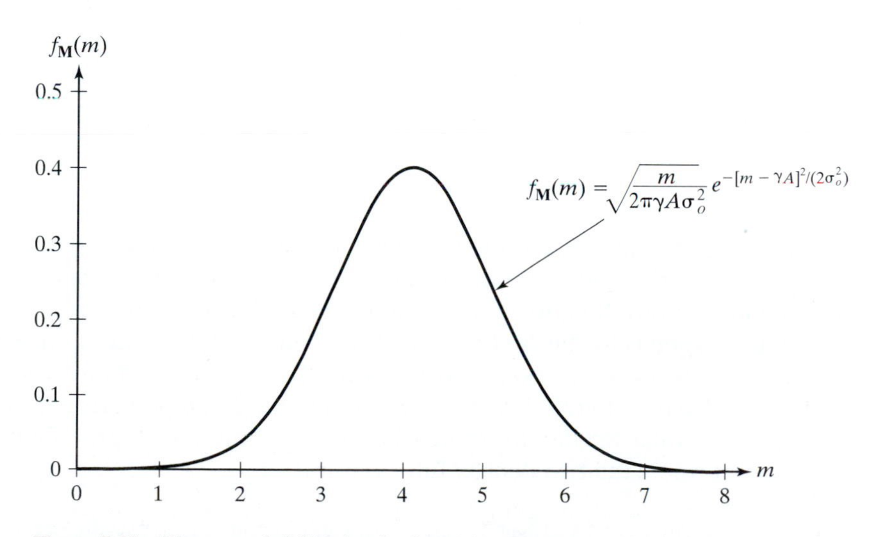

Figure 5-29 Ricean probability density function with $\sigma = 1$ and $\gamma A = 4$.

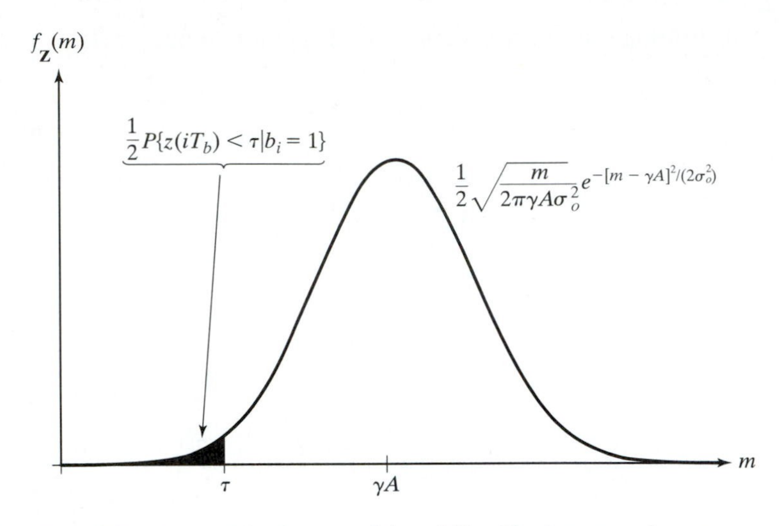

Figure 5-30 Error region for transmitting a "1" and having a noncoherent ASK receiver demodulate it as a "0" (assumes $\gamma A \gg \sigma_o$).

Mathematically,

$$P\{z(iT_b) < \tau | b_i = 1\} \approx \int_0^\tau \sqrt{\frac{m}{2\pi\gamma A\sigma_o^2}}e^{-[m-\gamma A]^2/(2\sigma_o^2)}\, dm \tag{5.73}$$

We now know all the probability distributions necessary in Equation (5.60) to determine the accuracy of the noncoherent ASK receiver. Assuming that the information from the source is represented using equiprobable, independent bits,

$$\begin{aligned} P_{b,\,noncoherent\ \text{ASK}} &= P\{b_i = 1\}P\{z(iT_b) < \tau | b_i = 1\} \\ &\quad + P\{b_i = 0\}P\{z(iT_b) \geq \tau | b_i = 0\} \\ &= \frac{1}{2}\int_0^\tau \sqrt{\frac{m}{2\pi\gamma A\sigma_o^2}}e^{-[m-\gamma A]^2/(2\sigma_o^2)}\, dm + \frac{1}{2}\int_\tau^\infty \frac{r}{\sigma_o^2}e^{-r^2/(2\sigma_o^2)}\, dr \end{aligned} \tag{5.74}$$

where the first term in Equation (5.74) is a Ricean distribution and the second term is a Rayleigh distribution. Total probability of error can be determined by plotting both probability density functions and combining the black error region of Figure 5-30, which corresponds to the first term of Equation (5.74), and the gray error region of Figure 5-28, which corresponds to the second term of Equation (5.74). Figure 5-31 shows the two probability density functions and the combined error regions. Note that the optimum value for the threshold (the value that minimizes probability of bit error) is the point where the two density functions intersect,

$$\sqrt{\frac{\tau_{opt}}{2\pi\gamma A\sigma_o^2}}e^{-[\tau_{opt}-\gamma A]^2/(2\sigma_o^2)} = \frac{\tau_{opt}}{\sigma_o^2}e^{-\tau_{opt}^2/(2\sigma_o^2)} \tag{5.75}$$

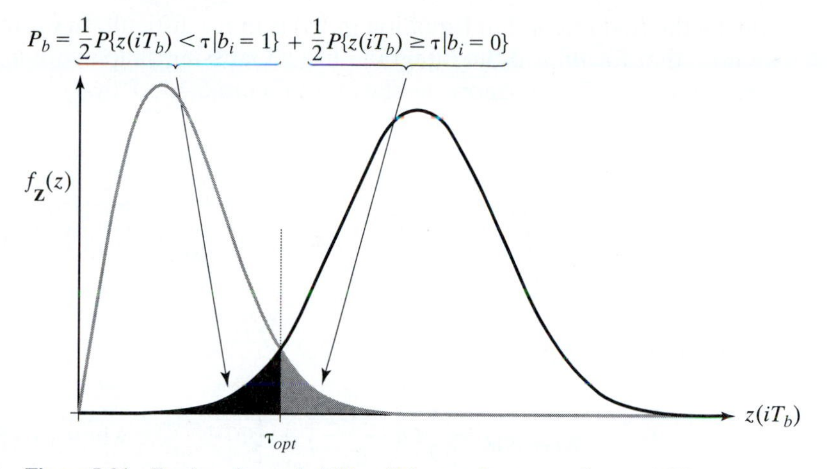

Figure 5-31 Regions for probability of bit error for a noncoherent ASK receiver.

As mentioned in Shanmugam [5.2], when $(\gamma A)^2 \gg \sigma_o^2$, we can approximate a solution to Equation (5.75) as

$$\tau_{opt} \approx \frac{\gamma A}{2}\sqrt{1 + \frac{8\sigma_o^2}{(\gamma A)^2}} \tag{5.76}$$

We've already noted that in a practical noncoherent ASK system the received signal's amplitude, γA, must be significantly higher than the average magnitude of the noise; that is, $(\gamma A)^2 \gg \sigma_o^2$. When this is true, we can further simplify Equation (5.76) to

$$\tau_{opt} \approx \frac{\gamma A}{2}\sqrt{1 + \frac{8\sigma_o^2}{(\gamma A)^2}} \approx \frac{\gamma A}{2} \text{ when } (\gamma A)^2 \gg \sigma_o^2 \tag{5.77}$$

Substituting into Equation (5.74),

$$P_{b,\,noncoherent\ \text{ASK}} = \frac{1}{2}\int_0^{\tau} \sqrt{\frac{m}{2\pi\gamma A\sigma_o^2}}\, e^{-[m-\gamma A]^2/(2\sigma_o^2)}\, dm + \frac{1}{2}\int_{\tau}^{\infty} \frac{r}{\sigma_o^2} e^{-r^2/(2\sigma_o^2)}\, dr$$

$$\approx \frac{1}{2}\int_0^{\gamma A/2} \sqrt{\frac{m}{2\pi\gamma A\sigma_o^2}}\, e^{-[m-\gamma A]^2/(2\sigma_o^2)}\, dm$$

$$+ \frac{1}{2}\int_{\gamma A/2}^{\infty} \frac{r}{\sigma_o^2} e^{-r^2/(2\sigma_o^2)}\, dr \text{ when } (\gamma A)^2 \gg \sigma_o^2 \tag{5.78}$$

We can solve the second integral in Equation (5.78) analytically:

$$\frac{1}{2}\int_{\gamma A/2}^{\infty} \frac{r}{\sigma_o^2} e^{-r^2/(2\sigma_o^2)}\, dr = \frac{1}{2} e^{-(\gamma A)^2/8\sigma_o^2} \tag{5.79}$$

Solving the first integral in Equation (5.78) is more difficult. In examining this integral, we note that for $m > 0$ the integrand has a Gaussian shape with a mean of γA and a variance of σ_o^2. Furthermore, as shown in Figure 5-31, if $(\gamma A)^2 \gg \sigma_o^2$, then the area under the curve in the region very close to zero is negligible. For these reasons (Shanmugam [5.2]), we can approximate the first integral in Equation (5.78) as

$$\frac{1}{2}\int_0^{\gamma A/2} \sqrt{\frac{m}{2\pi\gamma A\sigma_o^2}}\, e^{-[m-\gamma A]^2/(2\sigma_o^2)}\, dm \approx \frac{1}{2}Q\left(\frac{\gamma A/2}{\sigma_o}\right) \tag{5.80}$$

Substituting into Equation (5.78),

$$P_{b,\, noncoherent\ \mathrm{ASK}} \approx \frac{1}{2}Q\left(\frac{\gamma A/2}{\sigma_o}\right) + \frac{1}{2}e^{-(\gamma A)^2/8\sigma_o^2} \text{ when } (\gamma A)^2 \gg \sigma_o^2 \tag{5.81}$$

Example 5.10 Determining the accuracy of an ASK receiver

An ASK system transmits 25,000 bits/sec using a signal with a peak amplitude of 0.2 volts. By the time the signal arrives at the receiver, its voltage is only 30% of the transmitted voltage. The noise at the input to the receiver is additive white Gaussian noise with an average normalized power spectral density of 1.6×10^{-9} volts ^{2}/Hz.

a. Determine the accuracy of the system if a correlation receiver is used.

b. Determine the accuracy of the system if a noncoherent receiver is used.

Solution

a. For the correlation receiver,

$$P_{b,\,\mathrm{ASK},\,correlation\ receiver} = Q\left(\sqrt{\frac{E_b}{N_o}}\right) = Q\left(\sqrt{\frac{(\gamma A)^2 T_b}{4N_o}}\right)$$

$$= Q\left(\sqrt{\frac{(0.06)^2(4 \times 10^{-5})}{4(3.2x10^{-9})}}\right) = Q(3.35) = 0.0004$$

b. The noncoherent receiver is shown in Figure 5-24. Setting the bandwidth of the bandpass filter to $B = 2r_b$ will allow the ASK signal to satisfactorily pass but will eliminate out-of-band noise.[3] As shown in Figure 5-26 (including components to the left of zero), the average normalized noise power at the output of the bandpass filter is

$$\sigma_o^2 = 2B\frac{N_o}{2} = (2)(50{,}000)(1.6x10^{-9}) = 1.6 \times 10^{-4} \text{ volts}^2$$

[3] There is disagreement among texts concerning the width of the bandpass filters. Shanmugam [5.2] and Taub and Schilling [5.7] set $B = 2r_b$, while Lindsey [5.3] and Sklar [5.8] reduce bandwidth to $B = r_b$, the minimum value in a bandpass filter that will not cause ISI in the filtered, received signal (see discussion of Nyquist's first criterion in Chapter 4, Sections 4.4.1 and 4.4.2). In our analysis we will use the more conservative bandwidth.

Noting that $(\gamma A)^2 = 0.0036$ volts^2, we see that $(\gamma A)^2 \gg \sigma_o^2$ so we can use the approximation in Equation (5.81).

$$\begin{aligned} P_{b,\, noncoherent\ \mathrm{ASK}} &\approx \frac{1}{2}Q\left(\frac{\gamma A/2}{\sigma_o}\right) + \frac{1}{2}e^{-(\gamma A)^2/8\sigma_o^2} \text{ when } (\gamma A)^2 \gg \sigma_o^2 \\ &= \frac{1}{2}Q(2.37) + \frac{1}{2}e^{-2.8125} \\ &= 0.0045 + 0.0300 = 0.0345 \end{aligned}$$

As expected, the correlation receiver in Example 5.10 is more accurate than the noncoherent receiver, but at the cost of additional equipment complexity, since the correlation receiver must generate a coherent reference signal. Example 5.10 also illustrates another point. As established in Equation (5.81), calculation of $P_{b,\, noncoherent\ \mathrm{ASK}}$ involves two terms: the first term corresponding to $P\{b_i = 1\}P\{z(iT_b) < \tau | b_i = 1\}$, the probability that a "1" was transmitted but was demodulated as a "0"; and the second term corresponding to $P\{b_i = 0\}P\{z(iT_b) \geq \tau | b_i = 0\}$, the probability that a "0" was transmitted but was demodulated as a "1." As shown in Example 5.10, Part b, the values of the two terms are not equal. In fact, the probability of a "0" being incorrectly demodulated as a "1" is far greater than the probability of a "1" being incorrectly demodulated as a "0."

5.5.4.2 Probability of bit error for a noncoherent FSK receiver In examining Figure 5-25, we can see by symmetry that the optimum threshold for the noncoherent FSK receiver is zero. Therefore, assuming that the data bits are independent and equiprobable,

$$\begin{aligned} P_{b,\, noncoherent\ \mathrm{FSK}} &= P\{b = 1 \text{ and } z(iT_b) < 0\} + P\{b = 0 \text{ and } z(iT_b) \geq 0\} \\ &= P\{b_i = 1\}P\{z(iT_b) < 0 | b_i = 1\} \\ &\quad + P\{b_i = 0\}P\{z(iT_b) \geq 0 | b_i = 0\} \\ &= \frac{1}{2}P\{z(iT_b) < 0 | b_i = 1\} + \frac{1}{2}P\{z(iT_b) \geq 0 | b_i = 0\} \end{aligned} \tag{5.82}$$

In reexamining Figure 5-25, we see that $z(iT_b)$ is the difference in the output of the two envelope detectors; $z(iT_b) = w_1(iT_b) - w_2(iT_b)$. If a "1" is transmitted, the upper envelope detector will detect signal plus noise while the lower envelope detector will detect only noise; if a "0" is transmitted, the lower envelope detector will detect signal plus noise while the upper envelope detector will detect only noise. The subtractor and threshold comparator in Figure 5-25 determine which of the two envelope detectors has a larger output. Examining the first term in Equation (5.82)

$$P\{z(iT_b) < 0 | b_i = 1\} = P\{w_2(iT_b) > w_1(iT_b) | b_i = 1\} \tag{5.83}$$

From our analysis of the ASK receiver, we know that if a "1" is transmitted, then $w_1(iT_b)$ is a Rician-distributed random variable (signal plus noise at the output of the upper envelope detector), and $w_2(iT_b)$ is a Rayleigh-distributed random variable (noise at the output of the lower envelope detector). Therefore,

$$P\{w_2(iT_b) > w_1(iT_b)|b_i = 1\} = \int_0^\infty f_{\mathbf{M}}(m) \int_m^\infty f_{\mathbf{R}}(r)\, drdm \tag{5.84}$$

where $f_{\mathbf{R}}(r)$ is the Rayleigh probability density function given in Equation (5.64) and $f_{\mathbf{M}}(m)$ is the Rician probability density function given in Equation (5.69). Performing the integration in Equation (5.84) (see Shanmugam [5.2]).

$$P\{w_2(iT_b) > w_1(iT_b)|b_i = 1\} = \int_0^\infty f_{\mathbf{M}}(m) \int_m^\infty f_{\mathbf{R}}(r)\, drdm = \frac{1}{2}e^{-(\gamma A)^2/4\sigma_o^2} \tag{5.85}$$

By symmetry,

$$P\{w_1(iT_b) \geq w_2(iT_b)|b_i = 0\} = P\{w_2(iT_b) > w_1(iT_b)|b_i = 1\} \tag{5.86}$$

and therefore, substituting into Equation (5.82),

$$\begin{aligned} P_{b,\, noncoherent\ \mathrm{FSK}} &= \frac{1}{2}P\{z(iT_b) < 0|b_i = 1\} + \frac{1}{2}P\{z(iT_b) \geq 0|b_i = 0\} \\ &= \frac{1}{2}e^{-(\gamma A)^2/4\sigma_o^2} \end{aligned} \tag{5.87}$$

Example 5.11 Comparing the accuracy of ASK and FSK

In Example 5.10 we transmitted 25,000 bits/sec using an ASK system with a noncoherent receiver. Peak amplitude was 0.2 volts and by the time the signal arrived at the receiver, its voltage was only 30% of the transmitted voltage. The noise at the input to the receiver was additive white Gaussian noise with an average normalized power of 1.6×10^{-9} volts²/Hz. Employing a noncoherent ASK receiver, we determined $P_{b,\, noncoherent\ \mathrm{ASK}} = 0.0345$. Suppose that instead of ASK, we transmit FSK and use a noncoherent FSK receiver to demodulate. Using the same average transmitted signal power, determine differences in system performance between the noncoherent ASK and noncoherent FSK systems.

Solution

For noncoherent FSK,

$$P_{b,\, noncoherent\ \mathrm{FSK}} = \frac{1}{2}e^{-(\gamma A)^2/4\sigma_o^2}$$

The bandpass filters in the noncoherent FSK receiver each have the same bandwidth as the filter in the noncoherent ASK receiver, so

$$\sigma_o^2 = 2B\frac{N_o}{2} = (2)(50{,}000)(1.6x10^{-9}) = 1.6x10^{-4}\ \text{volts}^2$$

In Example 5.10, average normalized power of the transmitted signal was

$$P_{ave,\,\text{ASK}} = \frac{1}{2}\left(\frac{(0.2)^2}{2}\right) + \frac{1}{2}(0)^2 = 0.01 \text{ volts}^2$$

For FSK,

$$P_{ave,\,\text{FSK}} = \frac{1}{2}\left(\frac{A^2}{2}\right) + \frac{1}{2}\left(\frac{(-A)^2}{2}\right) = \frac{A^2}{2}$$

Therefore, for the FSK signal to have the same average transmitted power, $A = \sqrt{0.02} = 0.141$ volts. Probability of bit error for the noncoherent FSK system is

$$P_{b,\,noncoherent\ \text{FSK}} = \frac{1}{2}e^{-(\gamma A)^2/4\sigma^2} = \frac{1}{2}e^{-2.8125} = 0.0300$$

The noncoherent FSK system is slightly more accurate than the noncoherent ASK system, but requires more bandwidth. The noncoherent FSK receiver requires more parts than the noncoherent ASK receiver (two filters and envelope detectors versus a single filter and envelope detector) and also requires a summer, but the equipment is not more complex.

5.6 Differential (Noncoherent) PSK

Unlike ASK and FSK, an envelope detector cannot be used to noncoherently demodulate PSK, since the envelope of the PSK signal is the same whether a "1" or a "0" is transmitted. Consider instead the following strategy, known as binary *differential phase shift keying* (DPSK):

- To signify a "1," transmit the carrier signal with the same phase as used for the previous bit.
- To signify a "0," transmit the carrier signal with its phase shifted 180° relative to the previous bit.

Let b_i signify the ith bit and let b_{i-1} signify the i-1st bit. The modulation strategy for DPSK can be summarized as: "Shift the phase of the transmitted signal by 180 degrees if and only if $b_i = 0$." Figure 5-32a shows the binary PSK signal corresponding to the data sequence "101101" sent at a transmission speed of 50,000 bit/sec with a carrier frequency of 150 kHz. Figure 5-32b shows the same information transmitted using binary DPSK.

The term "differential" is appropriate because the information is transmitted using the phase *difference* between the present and previous bits. With PSK, information is based on the phase of the transmitted signal compared to the carrier; with DPSK, information is based on the phase of the transmitted signal relative to the previous transmitted bit. Note that DPSK requires a one-bit initialization period at the beginning of each transmission.

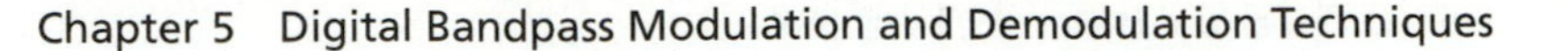

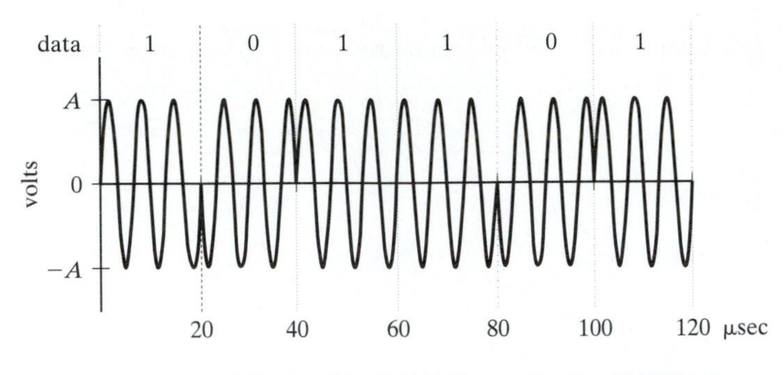

Figure 5-32a Binary PSK signal for "101101" transmitted at 50,000 bits/sec.

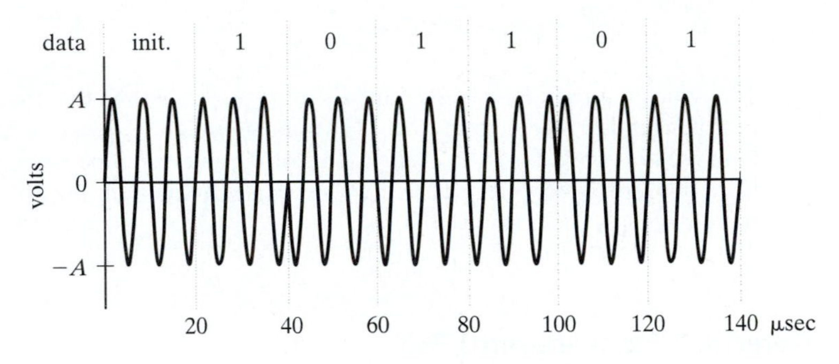

Figure 5-32b Binary DPSK signal corresponding to the same information in Figure 5-32a. Note the need for a one-bit initialization period.

5.6.1 Demodulation of Binary DPSK

The block diagram for a binary DPSK receiver is shown in Figure 5-33. The received signal corresponding to the present bit is multiplied by the received signal corresponding to the previous bit, and the product is integrated over the bit period. The result of the integration is then compared with a threshold to determine whether the transmitted bit was a "1" or a "0."

Let's analyze the operation of the DPSK receiver. Let $r_i(t)$ signify the received signal during the ith bit period, and let $r_{i-1}(t)$ signify the received signal during the

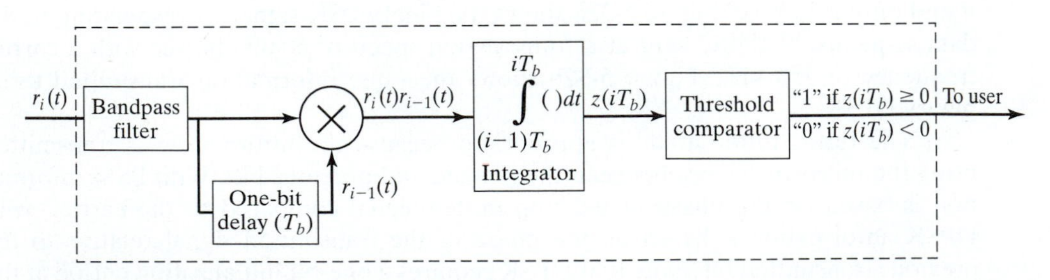

Figure 5-33 A receiver for binary DPSK.

i-1st bit period. Without loss of generality (and, for now, assuming noiseless operation), let

$$r_{i-1}(t) = \gamma A \sin[2\pi f_c(t - T_b) + \theta] \tag{5.88}$$

Assuming that the carrier frequency is an integral multiple of the bit rate (a good design practice lending to smooth transitions between bits),

$$r_{i-1}(t) = \gamma A \sin[2\pi f_c(t - T_b) + \theta] = \gamma A \sin(2\pi f_c t + \theta) \tag{5.89}$$

If $b_i = 0$, then

$$\begin{aligned} r_i(t)|_{b_i=0} &= \gamma A \sin(2\pi f_c t + \theta + \pi) \\ &= -\gamma A \sin(2\pi f_c t + \theta) \end{aligned} \tag{5.90}$$

The output of the integrator is therefore

$$\begin{aligned} z(iT_b)\Big|_{b_i=0} &= \int_{(i-1)T_b}^{iT_b} \gamma A \sin(2\pi f_c t + \theta + \pi)\gamma A \sin(2\pi f_c t + \theta)\, dt \\ &= \int_{(i-1)T_b}^{iT_b} -\gamma^2 A^2 \sin^2(2\pi f_c t + \theta)\, dt \\ &= -\frac{(\gamma A)^2 T_b}{2} \end{aligned} \tag{5.91}$$

and the output from the threshold comparator is a "0." If, on the other hand, $b_i = 1$, then

$$r_i(t)|_{b_i=1} = \gamma A \sin(2\pi f_c t + \theta) \tag{5.92}$$

and the output of the integrator is

$$\begin{aligned} z(iT_b)|_{b_i=1} &= \int_{(i-1)T_b}^{iT_b} \gamma A \sin(2\pi f_c t + \theta)\gamma A \sin(2\pi f_c t + \theta)\, dt \\ &= \int_{(i-1)T_b}^{iT_b} \gamma^2 A^2 \sin^2(2\pi f_c t + \theta)\, dt \\ &= \frac{(\gamma A)^2 T_b}{2} \end{aligned} \tag{5.93}$$

and the output from the threshold comparator is a "1."

5.6.2 Probability of Bit Error for a DPSK Receiver

This analysis starts the same way as for previous receivers:

$$\begin{aligned} P_{b,\,\mathrm{DPSK}} &= P\{b = 1 \text{ and } z(iT_b) < 0\} + P\{b = 0 \text{ and } z(iT_b) \geq 0\} \\ &= P\{b_i = 1\}P\{z(iT_b) < 0 | b_i = 1\} \\ &\quad + P\{b_i = 0\}P\{z(iT_b) \geq 0 | b_i = 0\} \end{aligned} \tag{5.94}$$

Adding the effects of noise to our analysis of the DPSK receiver, without loss of generality let

$$r_{i-1}(t) = \gamma A \sin[2\pi f_c(t - T_b) + \theta] + n(t - T_b) \tag{5.95}$$

Again, assuming that the carrier frequency is an integral multiple of the bit rate,

$$\begin{aligned} r_{i-1}(t) &= \gamma A \sin[2\pi f_c(t - T_b) + \theta] + n(t - T_b) \\ &= \gamma A \sin(2\pi f_c t + \theta) + n(t - T_b) \end{aligned} \tag{5.96}$$

If $b_i = 1$, then

$$r_i(t)|_{b_i=1} = \gamma A \sin(2\pi f_c t + \theta) + n(t) \tag{5.97}$$

and

$$\begin{aligned} z(iT_b)|_{b_i=1} &= \int_{(i-1)T_b}^{iT_b} [\gamma A \sin(2\pi f_c t + \theta) + n(t - T_b)][\gamma A \sin(2\pi f_c t + \theta) + n(t)]\, dt \\ &= \int_{(i-1)T_b}^{iT_b} \gamma^2 A^2 \sin^2(2\pi f_c t + \theta)\, dt + \int_{(i-1)T_b}^{iT_b} \gamma A \sin(2\pi f_c t + \theta) n(t)\, dt \\ &\quad + \int_{(i-1)T_b}^{iT_b} \gamma A \sin(2\pi f_c t + \theta) n(t - T_b)\, dt + \int_{(i-1)T_b}^{iT_b} n(t - T_b) n(t)\, dt \\ &= \frac{\gamma^2 A^2 T_b}{2} + \int_{(i-1)T_b}^{iT_b} \gamma A \sin(2\pi f_c t + \theta) n(t)\, dt \\ &\quad + \int_{(i-1)T_b}^{iT_b} \gamma A \sin(2\pi f_c t + \theta) n(t - T_b)\, dt + \int_{(i-1)T_b}^{iT_b} n(t - T_b) n(t)\, dt \\ &= \frac{\gamma^2 A^2 T_b}{2} + \mathbf{V}(t) \end{aligned} \tag{5.98}$$

where $\mathbf{V}(t)$ is a random process representing the last three terms. It has been shown that

$$P\{\mathbf{V}(\mathrm{t}) < -\frac{\gamma^2 A^2 T_b}{2} | b_i = 1\} = \frac{1}{2} e^{-(\gamma A)^2/\sigma_o^2} \tag{5.99}$$

(Shanmugam [5.2]), and thus

$$P\{z(iT_b) < 0 | b_i = 1\} = \frac{1}{2} e^{-(\gamma A)^2/\sigma_o^2} \tag{5.100}$$

By symmetry we can show that

$$P\{z(iT_b) \geq 0 | b_i = 0\} = \frac{1}{2} e^{-(\gamma A)^2/\sigma_o^2} \tag{5.101}$$

and so

$$\begin{aligned} P_{b,\,\mathrm{DPSK}} &= P\{b_i = 1\} P\{z(iT_b) < 0 | b_i = 1\} + P\{b_i = 0\} P\{z(iT_b) \geq 0 | b_i = 0\} \\ &= \frac{1}{2} e^{-(\gamma A)^2/\sigma_o^2} \end{aligned} \tag{5.102}$$

5.7 A Comparison of Binary Bandpass Systems

As with the coherent systems, it is often useful to express probability of bit error for the noncoherent systems in terms of E_b/N_o. In Example 5.10, we established that setting the bandwidth of the bandpass filters of the noncoherent receivers to $2r_b$ allows the transmitted signal to pass satisfactorily but eliminates out-of-band noise. (As discussed in the footnote to Example 5.10, we have selected a conservative value for bandwidth.) For noncoherent receivers, the average normalized noise power at the output of the bandpass filter is thus

$$\sigma_o^2 = 2B\frac{N_o}{2} = 2r_b N_o = \frac{2N_o}{T_b} \tag{5.103}$$

In Section 5.3.4 we established that

$$E_{b,\,\mathrm{FSK}} = E_{b,\,\mathrm{PSK}} = \frac{(\gamma A)^2 T_b}{2} \tag{5.34R}$$

and

$$E_{b,\,\mathrm{ASK}} = \frac{(\gamma A)^2 T_b}{4} \tag{5.36R}$$

Therefore,

$$\begin{aligned} P_{b,\,noncoherent\ \mathrm{ASK}} &\approx \frac{1}{2} Q\left(\frac{\gamma A/2}{\sigma_o}\right) + \frac{1}{2} e^{-(\gamma A)^2/8\sigma_o^2} \\ &= \frac{1}{2} Q\left(\sqrt{\frac{E_b}{2N_o}}\right) + \frac{1}{2} e^{-E_b/4N_o} \text{ when } (\gamma A)^2 \gg \sigma_o^2 \end{aligned} \tag{5.81R}$$

$$P_{b,\,noncoherent\ \mathrm{FSK}} = \frac{1}{2}e^{-(\gamma A)^2/4\sigma_o^2} = \frac{1}{2}e^{-E_b/4N_o} \tag{5.87R}$$

$$P_{b,\,\mathrm{DPSK}} = \frac{1}{2}e^{-(\gamma A)^2/\sigma_o^2} = \frac{1}{2}e^{-E_b/N_o} \tag{5.102R}$$

Table 5-3 compares the six binary bandpass systems we have analyzed: coherent and noncoherent ASK, FSK, and PSK. Of the six systems, two offer no apparent advantages in performance-versus-cost trade-offs: coherent FSK (more bandwidth, less accurate, and more complex than DPSK), and coherent ASK (as complex as coherent PSK and considerably less accurate). Practically speaking, our choices are therefore reduced to coherent PSK, DSPK, noncoherent FSK, and noncoherent ASK. These four systems are compared in Table 5-4. Remember that the noncoherent ASK system requires $(\gamma A)^2 \gg \sigma_o^2$ for reasonable accuracy. Figure 5-34 plots probability of bit error versus E_b/N_o for the four practical systems.

Table 5-3 A Comparison of Coherently and Noncoherently Demodulated PSK, FSK, and ASK

Binary Modulation/ Demodulation Technique	Bandwidth (90% In-Band Power)	Relative Equipment Complexity	Accuracy (Bit Error Rate)
Coherent Phase Shift Keying (Coherent PSK)	$2r_b$	High—tightly synchronized reference signal is needed at receiver	$P_b = Q(\sqrt{(\gamma A)^2 T_b/N_o}) = Q(\sqrt{2E_b/N_o})$ Most accurate of the techniques
Differential Phase Shift Keying (DPSK)	$2r_b$	Medium—synchronized reference signal not needed at receiver, but delay unit is required	$P_b = 0.5\exp(-(\gamma A)^2/\sigma_o^2)$ $= 0.5\exp(-E_b/N_o)$ Approx. 0.5 − 1 dB less accurate than coherent PSK for $10^{-3} < P_b < 10^{-7}$
Coherent Frequency Shift Keying (Coherent FSK)	$2r_b + 2\Delta f$ Larger than ASK or PSK	High—tightly synchronized reference signal is needed at receiver	$P_b = Q(\sqrt{(\gamma A)^2 T_b/2N_o}) = Q(\sqrt{E_b/N_o})$ 3 dB less accurate than PSK
Coherent Amplitude Shift Keying (Coherent ASK)	$2r_b$	High—tightly synchronized reference signal is needed at receiver	$P_b = Q(\sqrt{(\gamma A)^2 T_b/N_o}) = Q(\sqrt{E_b/N_o})$ 3 dB less accurate than PSK
Noncoherent FSK	$2r_b + 2\Delta f$ Larger than ASK or PSK	Moderately simple, same as noncoherent ASK except two filters and envelope detectors are required instead of one	$P_b = 0.5\exp(-(\gamma A)^2/4\sigma_o^2)$ $= 0.5\exp(-E_b/4N_o)$ Approx. 6.5 − 7 dB less accurate than coherent PSK for $10^{-3} < P_b < 10^{-7}$
Noncoherent ASK	$2r_b$	Moderately simple	$P_b = 0.5\exp(-(\gamma A)^2/8\sigma_o^2)$ $+0.5Q(\gamma A/2\sigma_o)$ $= 0.5\exp(-E_b/4N_o)$ $+0.5Q(\sqrt{E_b/2N_o})$ $(\gamma A)^2 \gg \sigma_o^2$ Approx. 6.5 − 7 dB less accurate than coherent PSK for $10^{-3} < P_b < 10^{-7}$

Table 5-4 A Comparison of the Four Practical Binary Bandpass Systems—Coherent PSK, DPSK, Noncoherent FSK, and Noncoherent ASK

Binary Modulation/ Demodulation Technique	Bandwidth (90% In-Band Power)	Relative Equipment Complexity	Accuracy (Bit Error Rate)
Coherent Phase Shift Keying (Coherent PSK)	$2r_b$	High—tightly synchronized reference signal is needed at receiver	$P_b = Q(\sqrt{(\gamma A)^2 T_b/N_o}) = Q(\sqrt{2E_b/N_o})$ Most accurate of the techniques
Differential Phase Shift Keying (DPSK)	$2r_b$	Medium—synchronized reference signal not needed at receiver, but delay unit is required	$P_b = 0.5\exp(-(\gamma A)^2/\sigma_o^2)$ $= 0.5\exp(-E_b/N_o)$ Approx. 0.5 − 1 dB less accurate than coherent PSK for $10^{-3} < P_b < 10^{-7}$
Noncoherent FSK	$2r_b + 2\Delta f$ Larger than PSK or ASK	Moderately simple, uses filters and envelope detectors	$P_b = 0.5\exp(-(\gamma A)^2/4\sigma_o{}^2)$ $= 0.5\exp(-E_b/4N_o)$ Approx. 6.5 − 7 dB less accurate than coherent PSK for $10^{-3} < P_b < 10^{-7}$
Noncoherent ASK	$2r_b$	Moderately simple	$P_b = 0.5\exp(-(\gamma A)^2/8\sigma_o^2)$ $+0.5Q(\gamma A/2\sigma_o)$ $= 0.5\exp(-E_b/4N_o)$ $+0.5Q(\sqrt{E_b/2N_o})$ $(\gamma A)^2 \gg \sigma_o^2$ Approx. 6.5 − 7 dB less accurate than coherent PSK for $10^{-3} < P_b < 10^{-7}$

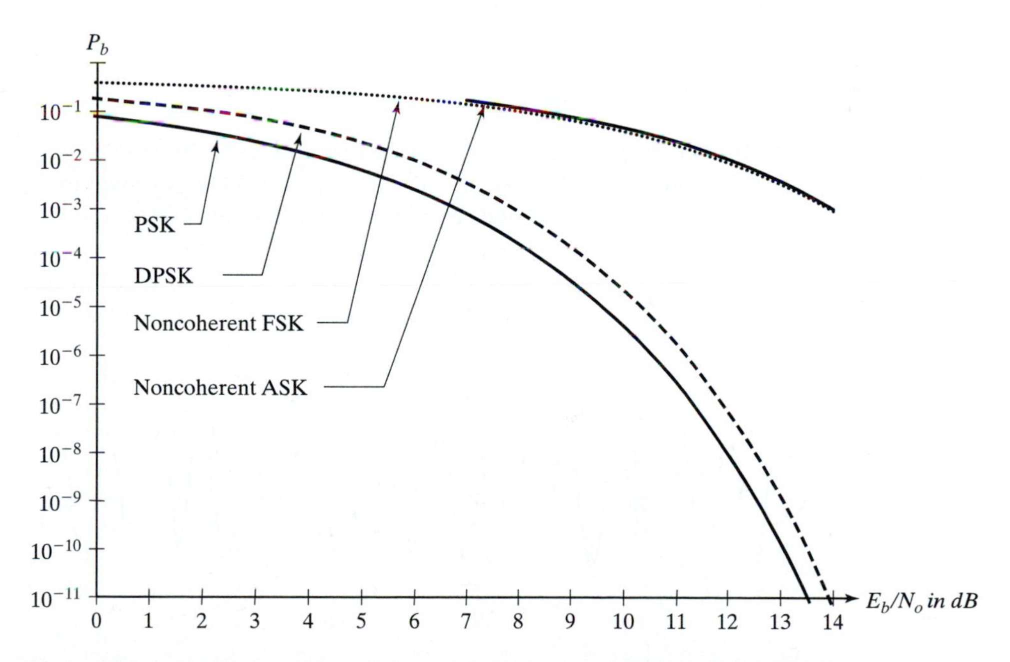

Figure 5-34 Probabilities of bit error for practical binary bandpass modulation/demodulation techniques.

5.8 M-ary Bandpass Techniques

In Chapter 4 we introduced the concept of M-ary PAM; that is, transmitting pulses with M possible different amplitudes and allowing each pulse to represent $\log_2 M$ bits. M-ary PAM significantly increases spectral efficiency for baseband systems at a cost of increased equipment complexity and reduced accuracy. Can we develop a similar concept for bandpass systems? Let's explore the possibilities and their trade-offs.

5.8.1 Quaternary Phase Shift Keying

Let's consider a modulation technique where we group the outputs from the source into bit pairs $b_{i-1}b_i$, and then

- Transmit $A\sin(2\pi f_c t + 45°)$ if $b_{i-1}, b_i =$ "11."
- Transmit $A\sin(2\pi f_c t + 135°)$ if $b_{i-1}, b_i =$ "10."
- Transmit $A\sin(2\pi f_c t + 225°)$ if $b_{i-1}, b_i =$ "00."
- Transmit $A\sin(2\pi f_c t + 315°)$ if $b_{i-1}, b_i =$ "01."

This modulation technique is called *quaternary phase shift keying* (QPSK).

Consistent with the terminology we developed for M-ary PAM, the four possible signals for QPSK are called *symbols*, and we can label them $s_1(t)$, $s_2(t)$, $s_3(t)$, and $s_4(t)$, with

$$\begin{aligned} s_1(t) &= A\sin(2\pi f_c t + 45°) \\ s_2(t) &= A\sin(2\pi f_c t + 135°) \\ s_3(t) &= A\sin(2\pi f_c t + 225°) \\ s_4(t) &= A\sin(2\pi f_c t + 315°) \end{aligned} \tag{5.104}$$

A typical QPSK waveform, representing the data sequence "1101100001" transmitted at a speed of 100,000 bits/sec with a carrier frequency of 150 kHz, is shown in Figure 5-35. Note that 100,000 bits/sec corresponds to 50,000 two-bit symbols/sec, and that the symbol period (T_s) is twice the bit period.

It is often convenient and helpful to graphically show the relationship between the different symbols in a modulation technique. One way to do this is to draw a

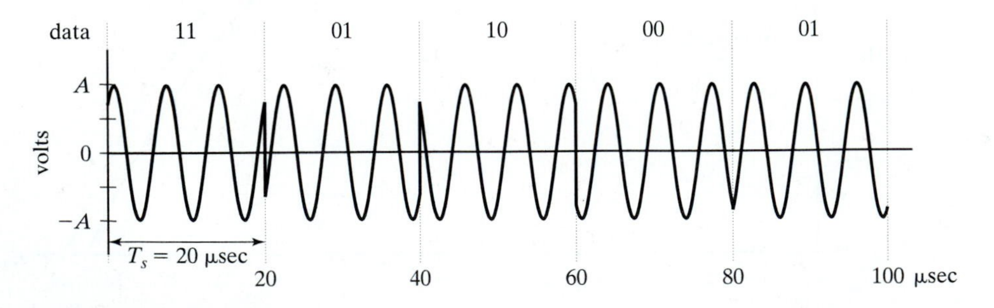

Figure 5-35 QPSK signal corresponding to "1101100001" transmitted at 100,000 bits/sec (50,000 symbols/sec).

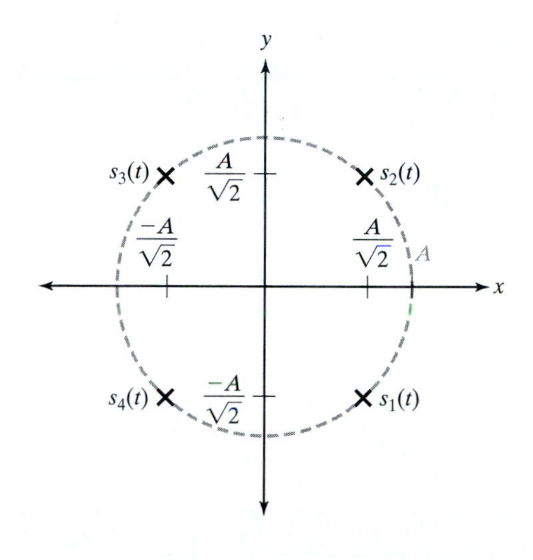

Figure 5-36 Constellation for QPSK.

constellation. Figure 5-36 shows a constellation for QPSK. Four points, corresponding to the tips of the phasors representing each of the four symbols $s_1(t)$, $s_2(t)$, $s_3(t)$, and $s_4(t)$ at time $t = 0$, are drawn in the x-y plane. (You may wish to review phasors in Section 2.2.2.) Note that all four points have the same magnitude (A) and that adjacent points are separated by 90° phase.

In examining the QPSK constellation, we see that each of the four symbols can be decomposed into two signals: one corresponding to a sine at the carrier frequency and the other corresponding to a cosine at the carrier frequency (remember that sine lags cosine by 90°).

$$
\begin{aligned}
s_1(t) &= A\sin(2\pi f_c t + 45^\circ) &&= \frac{A}{\sqrt{2}}\sin(2\pi f_c t) + \frac{A}{\sqrt{2}}\cos(2\pi f_c t) \\
s_2(t) &= A\sin(2\pi f_c t + 135^\circ) &&= -\frac{A}{\sqrt{2}}\sin(2\pi f_c t) + \frac{A}{\sqrt{2}}\cos(2\pi f_c t) \\
s_3(t) &= A\sin(2\pi f_c t + 225^\circ) &&= -\frac{A}{\sqrt{2}}\sin(2\pi f_c t) - \frac{A}{\sqrt{2}}\cos(2\pi f_c t) \\
s_4(t) &= A\sin(2\pi f_c t + 315^\circ) &&= \frac{A}{\sqrt{2}}\sin(2\pi f_c t) - \frac{A}{\sqrt{2}}\cos(2\pi f_c t)
\end{aligned} \tag{5.105}
$$

We can thus decompose a QPSK signal into the sum of two PSK signals with the same carrier frequency: one corresponding to a sine carrier (the *in-phase component*) and the other corresponding to a cosine carrier (the *quadrature component*). Figure 5-37 shows the decomposition of the QPSK waveform in Figure 5-35. Quaternary phase shift keying is often called quadrature phase shift keying because of the ability to decompose the QPSK signal into an in-phase component and a quadrature component.

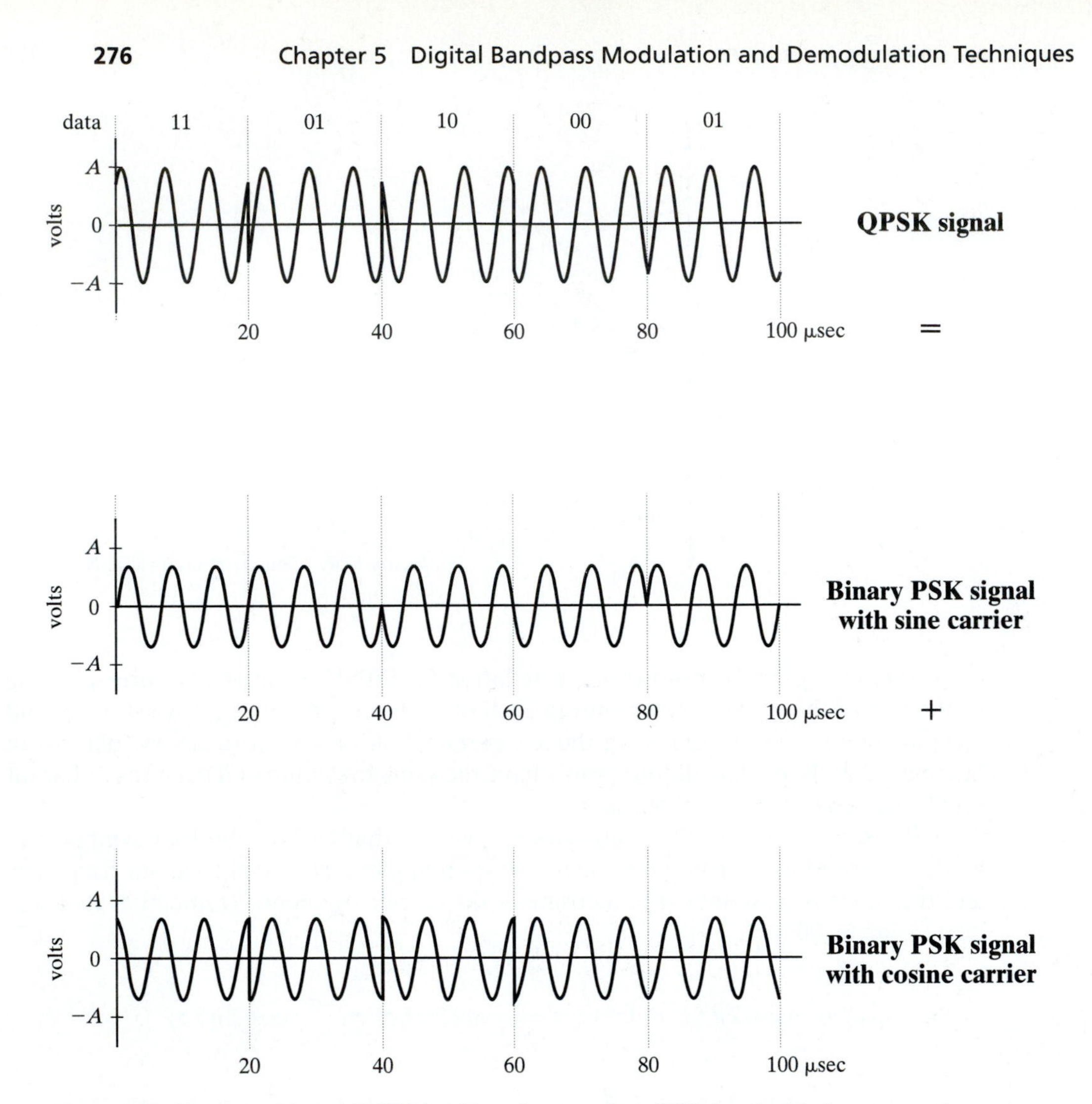

Figure 5-37 Decomposition of QPSK signal into the sum of two PSK signals with the same carrier frequency.

The decomposition in Figure 5-37 into in-phase and quadrature components shows many important characteristics of the QPSK waveform, including the following:

1. The in-phase and quadrature components are binary PSK waveforms with the same carrier frequency. The components experience phase shifts no more often than once every $T_s = 2/r_b$ seconds. The average normalized power spectral density of each component is therefore as shown in Figure 5-38, and the bandwidth is related to $r_b/2$. Since the QPSK signal is the sum of the two components, it has the same bandwidth as the components; in other words, its bandwidth is also related to $r_b/2$. This means that a QPSK signal, transmitted at the same bit rate as a binary PSK signal, requires only half the bandwidth. It also means that we can use

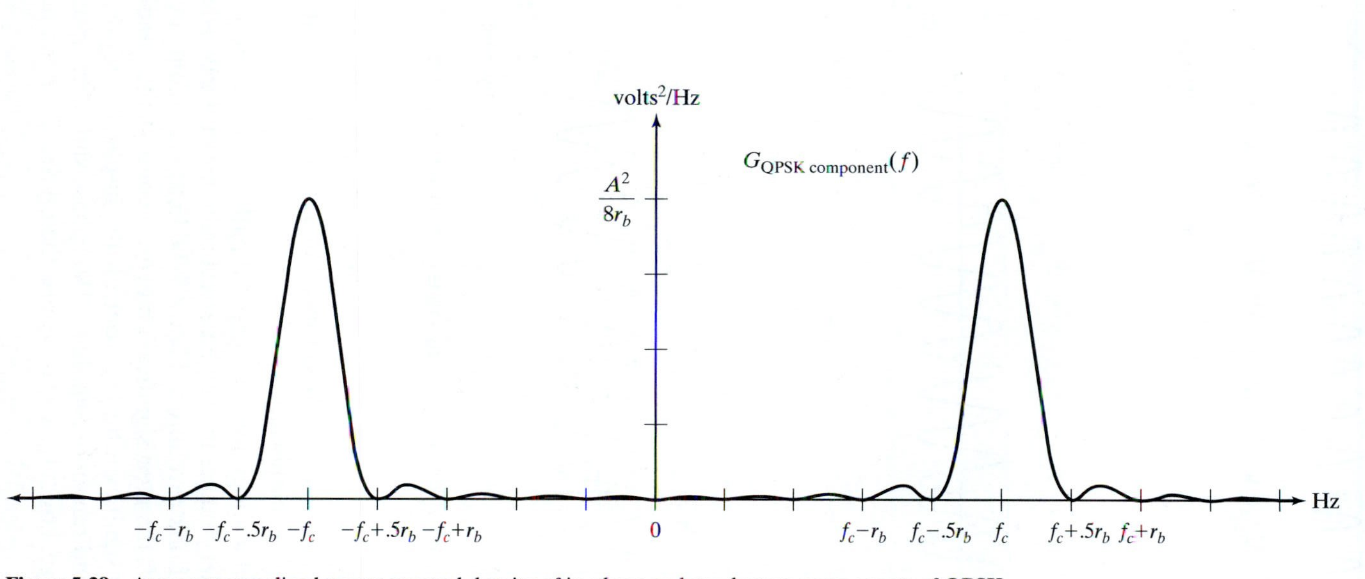

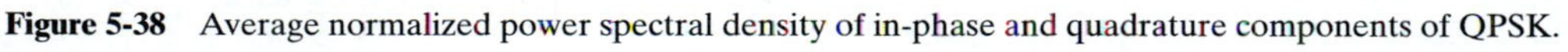

Figure 5-38 Average normalized power spectral density of in-phase and quadrature components of QPSK.

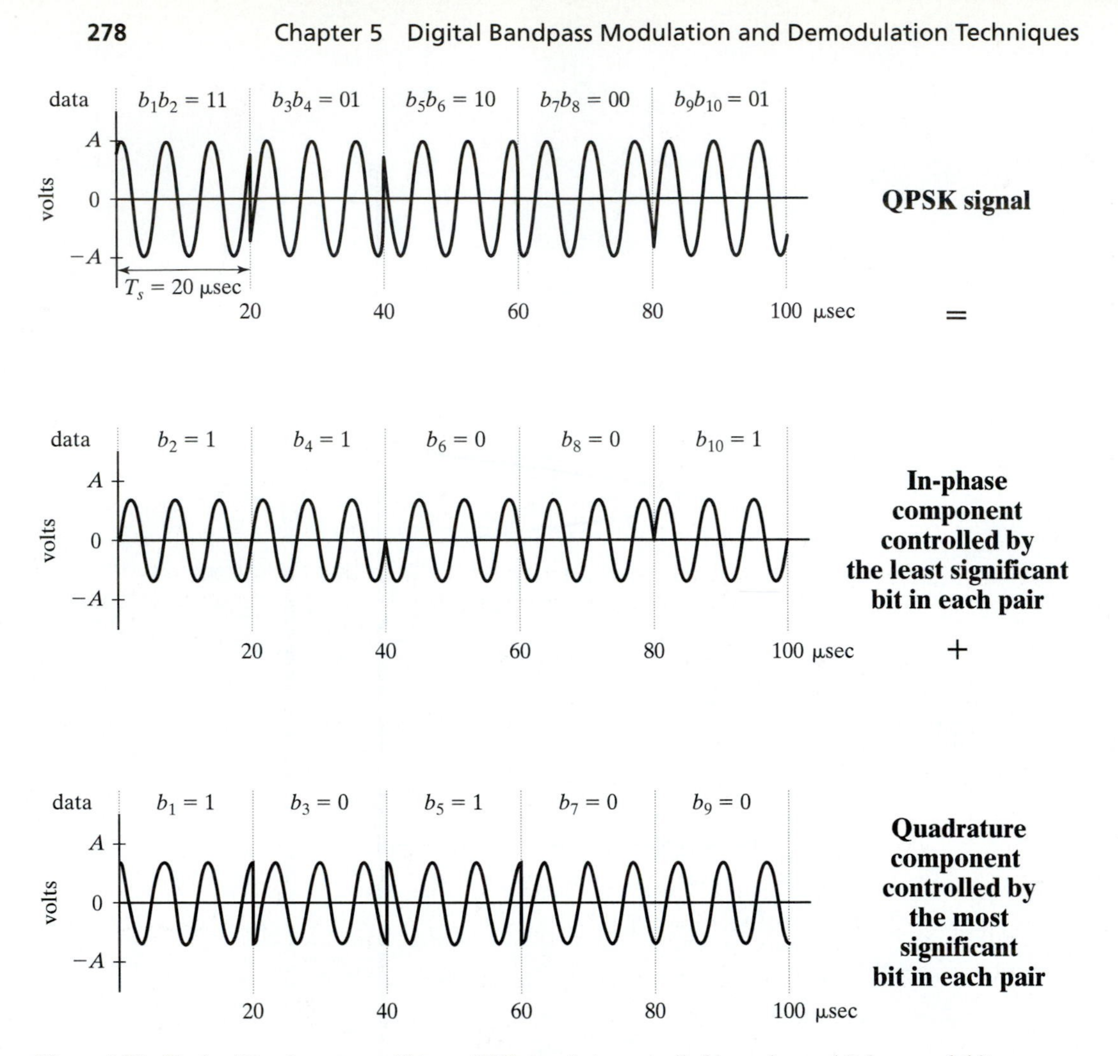

Figure 5-39 Each of the decomposed binary PSK signals is controlled by only one bit from each bit pair.

QPSK to double the transmission speed of a system relative to binary PSK without increasing the bandwidth.

2. If we are careful about how we assign bit patterns to each of the four QPSK symbols, we can let changes in each bit affect only one of the two decomposed binary PSK signals. Look carefully at Figure 5-39. Letting i assume even values, in each bit pair b_{i-1}, b_i the most significant bit (b_{i-1}) controls the quadrature component and the least significant bit (b_i) controls the in-phase component. Using this bit assignment and remembering that a 180° phase shift corresponds to multiplying the carrier signal by -1, we can express the QPSK signal mathematically as:

$$s_{\text{QPSK}}(t) = d_I \frac{A}{\sqrt{2}} \sin(2\pi f_c t) + d_Q \frac{A}{\sqrt{2}} \cos(2\pi f_c t) \qquad (5.106)$$

where

$$d_I = \begin{cases} 1 & \text{if } b_i = 1 \\ -1 & \text{if } b_i = 0 \end{cases}$$

and

$$d_Q = \begin{cases} 1 & \text{if } b_{i-1} = 1 \\ -1 & \text{if } b_{i-1} = 0 \end{cases}$$

The symbol d in Equation (5.106) is chosen to represent the word "data," with the I subscript used for the data corresponding to the in-phase component and the Q subscript used for the data corresponding to the quadrature phase component.

3. The decomposition of the QPSK signal into in-phase and quadrature components shows how to build a simple coherent QPSK receiver. If we choose the values of the carrier frequency f_c and the symbol period T_s so that a whole number of cycles of the carrier occur per symbol period (a good practice for reasons we discussed when examining binary PSK), then the in-phase and quadrature components are orthogonal and so

$$\int_0^{T_s} \sin(2\pi f_c t) \cos(2\pi f_c t)\, dt = 0 \tag{5.107}$$

This orthogonality allows us to isolate the in-phase and quadrature components at the receiver using mixers as shown in Figure 5-40. Let's consider the nth symbol

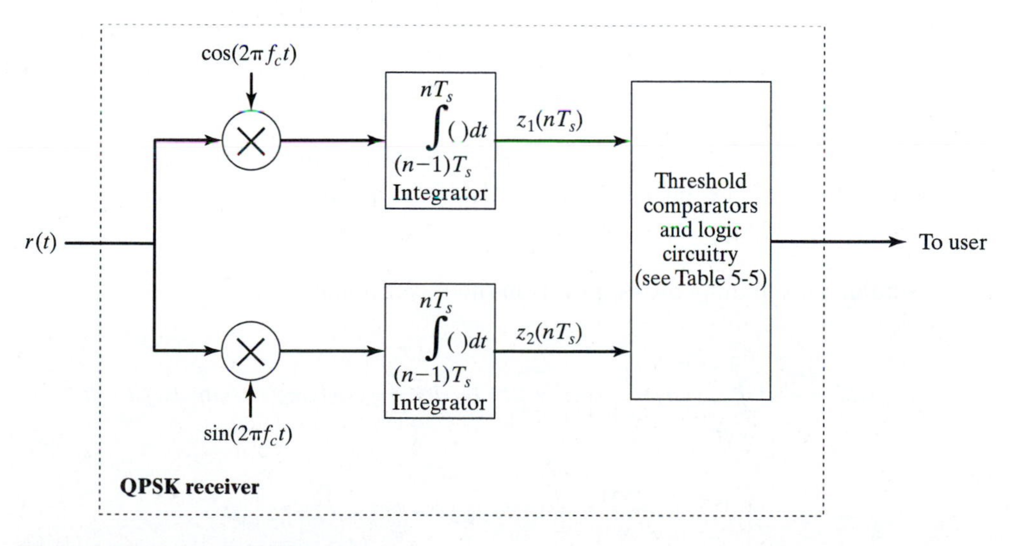

Figure 5-40 Coherent QPSK receiver.

period, which contains the bit pair b_{2n-1}, b_{2n}, and let's start our evaluation of the Figure 5-40 receiver by examining the output of the upper integrator at the end of the symbol period, $z_1(nT_s)$. Assuming no noise, the received QPSK signal is an attenuated version of the transmitted QPSK signal

$$\begin{aligned} r(t) &= \gamma s_{\text{QPSK}}(t) \\ &= \gamma\left[d_I \frac{A}{\sqrt{2}} \sin(2\pi f_c t) + d_Q \frac{A}{\sqrt{2}} \cos(2\pi f_c t)\right] \end{aligned} \tag{5.108}$$

where γ represents the attenuation in the channel. The output of the upper integrator is

$$\begin{aligned} z_1(nT_s) &= \int_{(n-1)T_s}^{nT_s} \gamma\left[d_I \frac{A}{\sqrt{2}} \sin(2\pi f_c t) + d_Q \frac{A}{\sqrt{2}} \cos(2\pi f_c t)\right]\cos(2\pi f_c t)\, dt \\ &= \frac{\gamma A}{\sqrt{2}} \int_{(n-1)T_s}^{nT_s} d_I \sin(2\pi f_c t) \cos(2\pi f_c t)\, dt \\ &\quad + \frac{\gamma A}{\sqrt{2}} \int_{(n-1)T_s}^{nT_s} d_Q \cos(2\pi f_c t) \cos(2\pi f_c t)\, dt \\ &= 0 + \frac{\gamma A}{\sqrt{2}} \int_{(n-1)T_s}^{nT_s} d_Q \cos^2(2\pi f_c t)\, dt \\ &= \frac{\gamma A T_s}{2\sqrt{2}} d_Q \end{aligned} \tag{5.109}$$

Thus, a positive output from the upper integrator indicates that the most significant bit in the bit pair, $b_{2n-1} = 1$, while a negative output indicates that $b_{2n-1} = 0$.

Similarly, examining the output from the lower integrator,

$$\begin{aligned} z_2(nT_s) &= \int_{(n-1)T_s}^{nT_s} \gamma\left[d_I \frac{A}{\sqrt{2}} \sin(2\pi f_c t) + d_Q \frac{A}{\sqrt{2}} \cos(2\pi f_c t)\right]\sin(2\pi f_c t)\, dt \\ &= \frac{\gamma A}{\sqrt{2}} \int_{(n-1)T_s}^{nT_s} d_I \sin(2\pi f_c t)\sin(2\pi f_c t)\, dt \end{aligned}$$

$$+\frac{\gamma A}{\sqrt{2}}\int_{(n-1)T_s}^{nT_s} d_Q\cos(2\pi f_c t)\sin(2\pi f_c t)\,dt$$

$$=\frac{\gamma A}{\sqrt{2}}\int_{(n-1)T_s}^{nT_s} d_I\sin^2(2\pi f_c t) + 0$$

$$=\frac{\gamma A T_s}{2\sqrt{2}}d_I \tag{5.110}$$

A positive output from the lower integrator indicates that the least significant bit $b_{2n} = 1$, while a negative output indicates that $b_{2n} = 0$. Combining the results of the upper integrator (which determines b_{2n-1}) and the lower integrator (which determines b_{2n}) produces the logic shown in Table 5-5.

Table 5-5 Relating Integrator Outputs to Received Data

$z_1(nT_s)$	$z_2(nT_s)$	Data to user $b_{2n-1}b_{2n}$
≥ 0	≥ 0	11
≥ 0	<0	10
<0	<0	00
<0	≥ 0	01

Let's now see how well our coherent QPSK receiver performs in the presence of noise. The signal and noise at the input to the receiver can be represented as

$$\begin{aligned} r_{\text{QPSK}}(t) &= \gamma s_{\text{QPSK}}(t) + n(t) \\ &= \gamma\left[d_I\frac{A}{\sqrt{2}}\sin(2\pi f_c t) + d_Q\frac{A}{\sqrt{2}}\cos(2\pi f_c t)\right] + n(t) \end{aligned} \tag{5.111}$$

and the output of the lower integrator is

$$\begin{aligned} z_2(nT_s) &= \int_{(n-1)T_s}^{nT_s}\left\{\gamma\left[d_I\frac{A}{\sqrt{2}}\sin(2\pi f_c t) + d_Q\frac{A}{\sqrt{2}}\cos(2\pi f_c t)\right] + n(t)\right\}\sin(2\pi f_c t)\,dt \\ &= \int_{(n-1)T_s}^{nT_s}\left[\gamma d_I\frac{A}{\sqrt{2}}\sin(2\pi f_c t) + n(t)\right]\sin(2\pi f_c t)\,dt \\ &\quad + \int_{(n-1)T_s}^{nT_s} d_Q\frac{A}{\sqrt{2}}\cos(2\pi f_c t)\sin(2\pi f_c t)\,dt \end{aligned} \tag{5.112a}$$

As established in Equation (5.107), the second term in Equation (5.112) is zero due to orthogonality, leaving

$$z_2(nT_s) = \int_{(n-1)T_s}^{nT_s} \left[\gamma d_I \frac{A}{\sqrt{2}} \sin(2\pi f_c t) + n(t)\right] \sin(2\pi f_c t)\, dt \tag{5.112b}$$

This is exactly the integrator output we would expect for a coherent binary PSK receiver. The probability that the noise causes the lower integrator to produce an error (i.e., output a signal less than 0 volts even though $b_{2n} = 1$ or output a signal greater than or equal to 0 volts even though $b_{2n} = 0$) is thus the same as the probability of bit error for a coherent binary PSK receiver. Therefore,

$$P\{b_{2n} \text{ is demodulated in error}\} = Q\left(\sqrt{\frac{2E_b}{N_o}}\right) \tag{5.113}$$

where

$$E_b = \frac{\left(\frac{\gamma A}{\sqrt{2}}\right)^2 T_s}{2} = \frac{(\gamma A)^2 T_s}{4} \tag{5.114}$$

E_b represents the energy per bit associated with the in-phase portion of the received signal.

Similarly, the output of the upper integrator in the presence of noise is

$$\begin{aligned} z_1(nT_s) &= \int_{(n-1)T_s}^{nT_s} \left\{\gamma\left[d_I \frac{A}{\sqrt{2}} \sin(2\pi f_c t) + d_Q \frac{A}{\sqrt{2}} \cos(2\pi f_c t)\right] + n(t)\right\} \cos(2\pi f_c t)\, dt \\ &= \int_{(n-1)T_s}^{nT_s} \gamma d_I \frac{A}{\sqrt{2}} \sin(2\pi f_c t)\cos(2\pi f_c t)\, dt \\ &\quad + \int_{(n-1)T_s}^{nT_s} \left[\gamma d_Q \frac{A}{\sqrt{2}} \cos(2\pi f_c t) + n(t)\right] \cos(2\pi f_c t)\, dt \\ &= \int_{(n-1)T_s}^{nT_s} \left[\gamma d_Q \frac{A}{\sqrt{2}} \cos(2\pi f_c t) + n(t)\right] \cos(2\pi f_c t)\, dt \end{aligned} \tag{5.115}$$

which again represents the output of a correlation receiver given a noisy binary PSK signal (albeit, with a cosine carrier instead of a sine carrier). As with the lower integrator, the probability that the noise causes the upper integrator to produce an error (i.e., output a signal less than 0 volts even though $b_{2n-1} = 1$ or output a signal greater than or equal to 0 volts even though $b_{2n-1} = 0$), is

$$P\{b_{2n-1} \text{ is demodulated in error}\} = Q\left(\sqrt{\frac{2E_b}{N_o}}\right) \tag{5.116}$$

where

$$E_b = \frac{\left(\frac{\gamma A}{\sqrt{2}}\right)^2 T_s}{2} = \frac{(\gamma A)^2 T_s}{4} \tag{5.114R}$$

representing the energy per bit associated with the quadrature portion of the received signal. Equations (5.112)–(5.116) show that the probability of bit error for QPSK is

$$P_{b,\, coherent\ \text{QPSK}} = Q\left(\sqrt{\frac{2E_b}{N_o}}\right) \tag{5.117}$$

where

$$E_b = \frac{(\gamma A)^2 T_s}{4} \tag{5.114R}$$

From our analysis and Equation (5.117), we see that a QPSK signal has the same bit error rate as a binary PSK signal, *provided that the two signals are transmitted using the same energy per bit*. In examining Equation (5.114R) and remembering that

$$E_b = \frac{\text{Average normalized signal power}}{\text{number of bits transmitted per second}} \tag{5.118}$$

note that if a binary PSK signal and a QPSK signal are transmitted with the same amplitude and the same number of bits per second, then the two signals have the same energy per bit and hence the same bit error rate. If, however, the two signals are transmitted at the same amplitude but the QPSK signal is transmitting twice as many bits per second as the binary PSK signal, then the QPSK signal has only half the energy per bit as the binary PSK signal and its bit error rate will be higher. In many practical cases, QPSK is used instead of binary PSK to provide an increase in transmission speed within a fixed bandwidth, so this second example (same average normalized power but faster transmission speed) is often encountered in practice. Once again, note that it is very important to keep track of all system parameters and to have a solid physical grasp of what our analysis means.

As a final exercise, we can calculate the probability of symbol error for QPSK.

$$\begin{aligned} P_{s,\, coherent\ \text{QPSK}} &= P\{b_{2n-1} \text{ is received in error or } b_{2n} \text{ is received in error}\} \\ &= 1 - P\{b_{2n-1} \text{ is accurately received and } b_{2n} \text{ is} \\ &\qquad \text{accurately received}\} \end{aligned} \tag{5.119}$$

The quadrature and in-phase components of the QPSK signal are independent. Since b_{2n-1} corresponds solely to the quadrature component and b_{2n} corresponds solely to the in-phase component,

$$\begin{aligned} &P\{b_{2n-1} \text{ is accurately received and } b_{2n} \text{ is accurately received}\} \\ &\quad = P\{b_{2n-1} \text{ is accurately received}\}P\{b_{2n} \text{ is accurately received}\} \\ &\quad = [1 - P_{b\,coherent\text{ QPSK}}][1 - P_{b\,coherent\text{ QPSK}}] \\ &\quad = 1 - 2P_{b\,coherent\text{ QPSK}} + P^2_{b\,coherent\text{ QPSK}} \end{aligned} \tag{5.120}$$

Substituting into Equation (5.119) yields

$$\begin{aligned} P_{s,\,coherent\text{ QPSK}} &= P\{b_{2n-1} \text{ is received in error or } b_{2n} \text{ is received in error}\} \\ &= 1 - P\{b_{2n-1} \text{ is accurately received and } b_{2n} \text{ is accurately received}\} \\ &= 1 - \{1 - 2P_{b\,coherent\text{ QPSK}} + P^2_{b\,coherent\text{ QPSK}}\} \\ &= 2P_{b\,coherent\text{ QPSK}} - P^2_{b\,coherent\text{ QPSK}} \end{aligned} \tag{5.119R}$$

For most practical applications, $P_{b\,coherent\text{ QPSK}}$ is quite small and we can approximate the probability of symbol error as

$$\begin{aligned} P_{s,\,coherent\text{ QPSK}} &= 2P_{b\,coherent\text{ QPSK}} - P^2_{b\,coherent\text{ QPSK}} \\ &\approx 2P_{b\,coherent\text{ QPSK}} \text{ for } P_{b\,coherent\text{ QPSK}} \ll 1 \end{aligned} \tag{5.121}$$

5.8.2 Differential Quaternary Phase Shift Keying

In our study of binary bandpass systems, we developed both coherent and noncoherent receivers. We saw that coherent receivers provide good performance but that they require a high level of equipment complexity because they must independently produce a reference signal that is tightly synchronized in phase and frequency with the transmitted carrier. Noncoherent receivers are simpler because they do not require a tightly synchronized reference signal, but noncoherent receivers are less accurate. System designers choose coherent or noncoherent receivers for a particular system by evaluating the performance-versus-cost trade-offs and determining which receiver is optimal for their particular application.

In Section 5.8.1, we developed and analyzed a coherent QPSK receiver and determined its complexity and accuracy. Let's now develop a noncoherent QPSK system. Our starting point will be the noncoherent binary PSK system (binary differential PSK) that we developed in Section 5.6. In binary differential PSK we conveyed information through the relative phasing of the transmitted signal: To signify a "1," we transmitted the carrier signal with the same phase used for the previous bit; to signify a "0," we transmitted the carrier signal with its phase shifted 180° relative to the previous bit. In other words, information was transmitted by the phase of the signal relative to its phase in the previous bit. We can extend this concept straightforwardly for differential QPSK (DQPSK):

- If b_{i-1}, b_i = "11", transmit the carrier signal with the same phase as used for the previous bit pair.
- If b_{i-1}, b_i = "10", shift the carrier's phase 90° relative to the phase used for the previous bit pair.

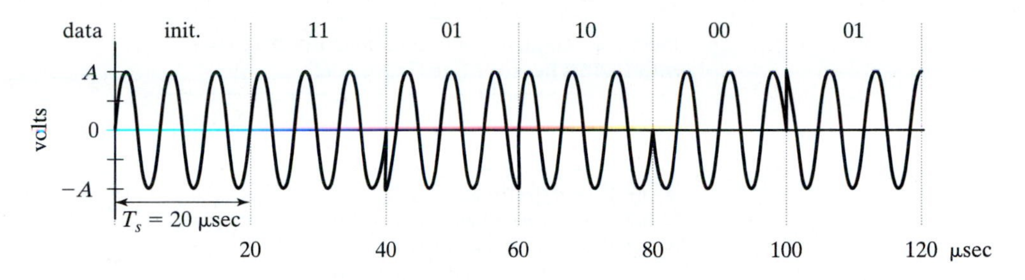

Figure 5-41 Differential QPSK signal corresponding to "1101100001" transmitted at 100,000 bits/sec (50,000 symbols/sec). Note initialization period.

- If b_{i-1}, b_i = "00", shift the carrier's phase 180° relative to the phase used for the previous bit pair.
- If b_{i-1}, b_i = "01", shift the carrier's phase 270° relative to the phase used for the previous bit pair.

A typical DQPSK waveform, representing the data sequence "1101100001" transmitted at a speed of 100,000 bits/sec with a carrier frequency of 150 kHz, is shown in Figure 5-41. As with binary DPSK, note that a one-symbol initialization period is needed.

From our analysis of QPSK, we know that our transmitted signal can be decomposed into an in-phase component and a quadrature component. Furthermore, we know that these two components are orthogonal. From our analysis of binary differential PSK, we know that we can demodulate a differentially phase-shifted signal by multiplying the received signal with a delayed version of the same received signal. Using all of these observations, we can develop a DQPSK receiver as shown in Figure 5-42.

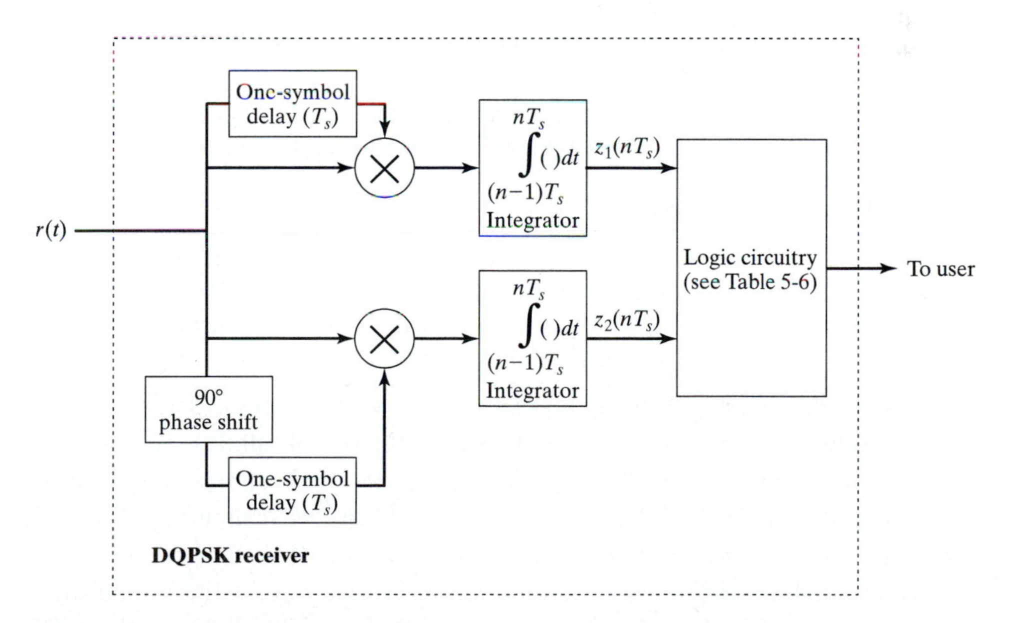

Figure 5-42 Differential QPSK receiver.

Table 5-6 Logic for Determining Outputs to User with DQPSK Receiver (see Figure 5-42)

State of Inputs to Logic Circuit	Outputs to User b_{i-1}, b_i
$\lvert z_1(nT_s)\rvert \geq \lvert z_2(nT_s)\rvert$ and $z_1(nT_s) \geq 0$	11
$\lvert z_1(nT_s)\rvert \geq \lvert z_2(nT_s)\rvert$ and $z_1(nT_s) < 0$	00
$\lvert z_2(nT_s)\rvert > \lvert z_1(nT_s)\rvert$ and $z_2(nT_s) \geq 0$	10
$\lvert z_2(nT_s)\rvert > \lvert z_1(nT_s)\rvert$ and $z_2(nT_s) < 0$	01

In a noiseless environment, the output of the upper integrator will be positive for a phase shift of 0°, negative for a phase shift of 180°, and zero for phase shifts of 90° and 270° (due to the orthogonality of sine and cosine). The output of the lower integrator will be positive for a phase shift of 90°, negative for a phase shift of 270°, and zero for phase shifts of 0° and 180°. Practical receivers must, of course, operate in environments with noise, so the logic circuit in the Figure 5-42 receiver must expect nonzero outputs from both integrators. Nevertheless, if the received signal is not overwhelmed by the noise, one of the two integrator outputs will be considerably larger in magnitude than the other. The logic circuit can demodulate the received signal by comparing the magnitudes of the two integrator outputs, determining which output has a larger magnitude, and then determining whether that output is positive or negative. Operation of the logic circuit is summarized in Table 5-6.

Determining the symbol error rate for DQPSK is rather complicated (see Lindsey and Simon [5.3] for a derivation of the approximation in Equation (5.122); see Simon, Hinedi, and Lindsey [5.4] for a derivation of an equality).

$$P_{s,\,\mathrm{DQPSK}} \approx 2Q\left(\sqrt{\frac{8E_b}{N_o}\sin^2\frac{\pi}{8}}\right) = 2Q\left(\sqrt{\frac{2E_b}{N_o}(0.586)}\right) \tag{5.122}$$

From Equation (5.122) we see that a DQPSK signal requires approximately 2.3 dB more power than a QPSK signal to achieve the same level of accuracy.

5.8.3 M-ary Phase Shift Keying

We've seen that quaternary PSK provides more spectral efficiency than binary PSK by modulating the carrier using four distinct phases instead of two. We can create even greater spectral efficiency by using even more distinct phases. Consider, for example, an eight-phase system where each symbol can represent three bits:

- Transmit $A\sin(2\pi f_c t + 0°)$ to represent "000"; we will call this symbol $s_1(t)$.
- Transmit $A\sin(2\pi f_c t + 45°)$ to represent "001"; we will call this symbol $s_2(t)$.
- Transmit $A\sin(2\pi f_c t + 90°)$ to represent "011"; we will call this symbol $s_3(t)$.
- Transmit $A\sin(2\pi f_c t + 135°)$ to represent "010"; we will call this symbol $s_4(t)$.
- Transmit $A\sin(2\pi f_c t + 180°)$ to represent "110"; we will call this symbol $s_5(t)$.
- Transmit $A\sin(2\pi f_c t + 225°)$ to represent "111"; we will call this symbol $s_6(t)$.
- Transmit $A\sin(2\pi f_c t + 270°)$ to represent "101"; we will call this symbol $s_7(t)$.
- Transmit $A\sin(2\pi f_c t + 315°)$ to represent "100"; we will call this symbol $s_8(t)$.

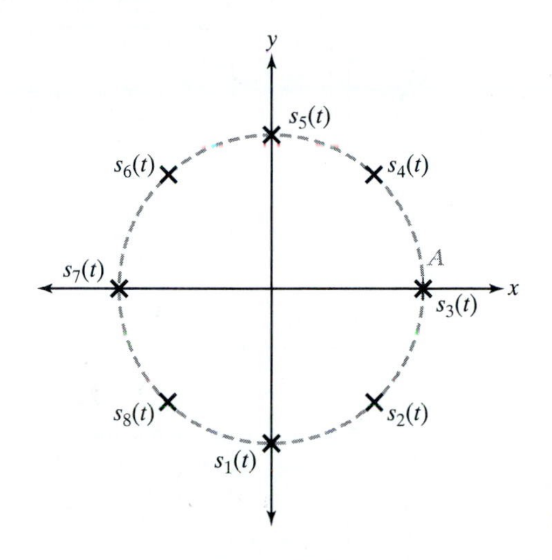

Figure 5-43 Constellation for 8-ary PSK.

Note that we've used a Gray code for the bit assignments to minimize the effects of errors to adjacent symbols (see Section 4.5 for a more detailed discussion of Gray codes and why they are used). The constellation for our 8-ary PSK modulation technique is shown in Figure 5-43.

As with binary PSK and QPSK, the bandwidth of 8-ary PSK is related to the symbol rate r_s. Since 8-ary PSK transmits three bits per symbol, it is three times as bandwidth-efficient as binary PSK and 1.5 times as bandwidth-efficient as QPSK. This means that within a given bandwidth, 8-ary PSK can transmit three times as many bits per second as binary PSK, or at a given bit rate, 8-ary PSK requires only one third as much bandwidth as binary PSK.

5.8.3.1 Coherent demodulation of M-ary PSK We can coherently demodulate 8-ary PSK using a bank of four correlators as shown in Figure 5-44. Note that n represents the nth symbol period, which includes bits b_{3n-2}, b_{3n-1}, and b_{3n}. As shown in Figures 5-44 and 5-45, the reference signals for the correlators are chosen to have phase angles exactly halfway between adjacent symbols in the constellation.

To see how the coherent 8-ary PSK demodulator works, let's consider a noiseless received signal $r(t)$, which we can represent as

$$r(t) = \gamma A \sin(2\pi f_c t + \theta), \text{ where } \theta = \begin{cases} 0^\circ & \text{if transmitted data is "000"} \\ 45^\circ & \text{if transmitted data is "001"} \\ 90^\circ & \text{if transmitted data is "011"} \\ 135^\circ & \text{if transmitted data is "010"} \\ 180^\circ & \text{if transmitted data is "110"} \\ 225^\circ & \text{if transmitted data is "100"} \\ 270^\circ & \text{if transmitted data is "101"} \\ 315^\circ & \text{if transmitted data is "111"} \end{cases} \tag{5.123}$$

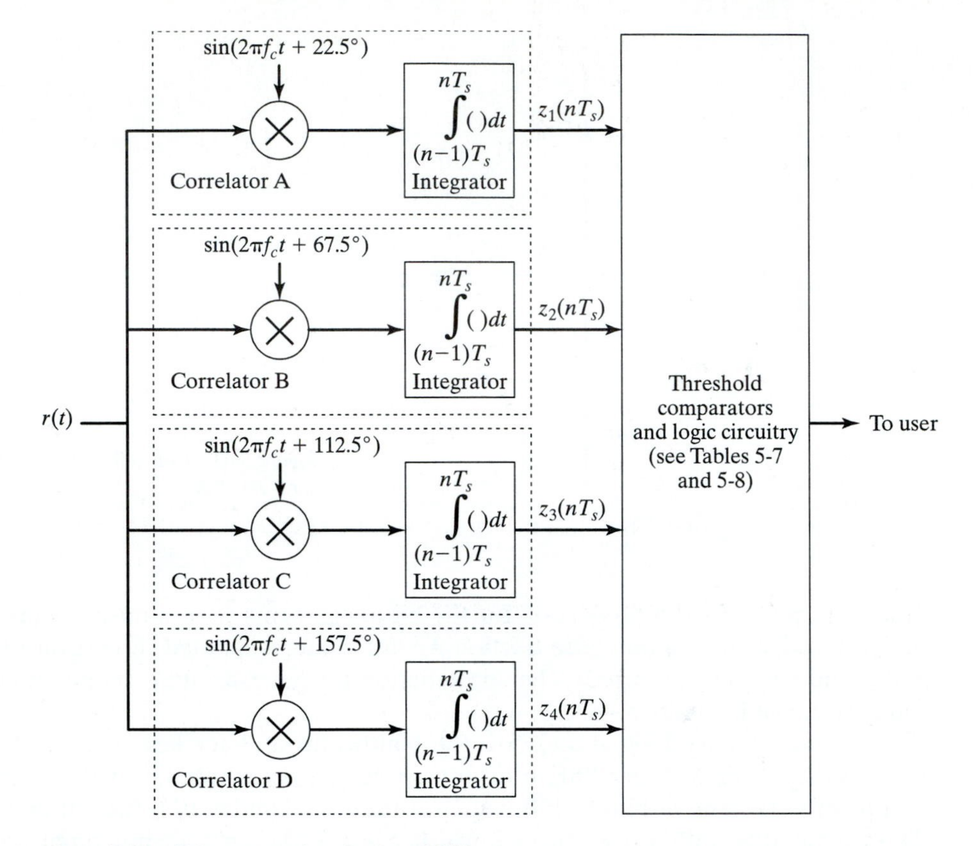

Figure 5-44 Coherent receiver for 8-ary PSK.

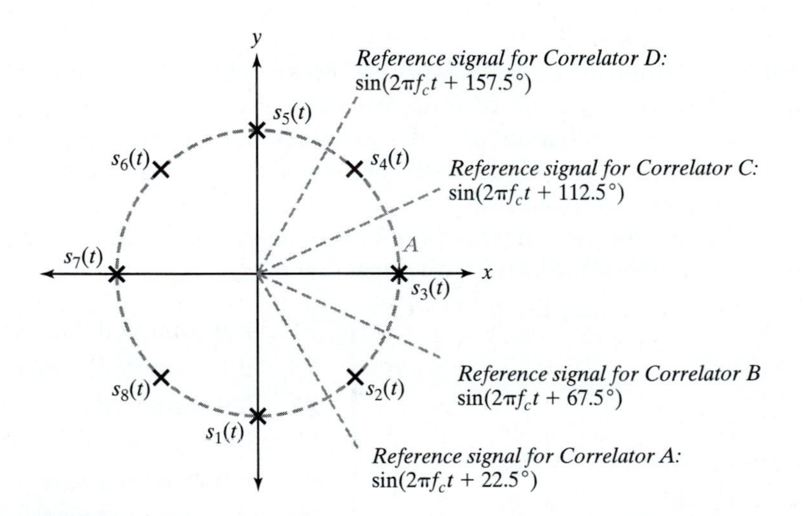

Figure 5-45 Reference signals for correlators in coherent 8-ary PSK receiver.

Let $s_{ref}(t)$ represent the reference signal from a particular correlator. We can represent $s_{ref}(t)$ as

$$s_{ref}(t) = \sin(2\pi f_c t + \phi) \text{ where } \phi = \begin{cases} 22.5^\circ & \text{for Correlator A} \\ 67.5^\circ & \text{for Correlator B} \\ 112.5^\circ & \text{for Correlator C} \\ 157.5^\circ & \text{for Correlator D} \end{cases} \quad (5.124)$$

Let $z(nT_s)$ represent the output from a particular correlator. Using trigonometric identities,

$$\begin{aligned} z(nT_s) &= \int_{(n-1)T_s}^{nT_s} \gamma A \sin(2\pi f_c t + \theta)\sin(2\pi f_c t + \phi)\, dt \\ &= \frac{\gamma A}{2} \int_{(n-1)T_s}^{nT_s} [\cos(\theta - \phi) - \cos(4\pi f_c t + \theta + \phi)]\, dt \\ &= \frac{\gamma A}{2} \int_{(n-1)T_s}^{nT_s} \cos(\theta - \phi)\, dt - \frac{\gamma A}{2} \int_{(n-1)T_s}^{nT_s} \cos(4\pi f_c t + \theta + \phi)\, dt \\ &= \frac{\gamma A T_s}{2} \cos(\theta - \phi) \end{aligned} \quad (5.125)$$

where θ is the phase angle of the received signal and ϕ is the phase angle of the correlator's reference signal (22.5° for Correlator A, 67.5° for Correlator B, 112.5° for Correlator C, and 157.5° for Correlator D). Note that $(\gamma A/2)\int_{(n-1)T_s}^{nT_s} \cos(4\pi f_c t + \theta + \phi)\, dt = 0$ because the integration is occurring over a whole number of cycles of the sinusoid.

Since $\gamma A T_s$ is positive, Equation (5.125) shows that the output of a particular correlator will be positive if the phase angle of the received signal (θ) is within $\pm 90^\circ$ of the phase angle of the reference signal for the correlator (ϕ), otherwise the correlator output will be negative. For example, the output of Correlator A will be positive if and only if $|\theta - 22.5^\circ| < 90^\circ$, corresponding to $-67.5^\circ < \theta < 112.5^\circ$. We have carefully chosen the phase angles for all of the correlator reference signals in the receiver so that each of the eight possible received symbols produces a unique combination of positive and negative outputs from the four correlators. These combinations are shown in Table 5-7.

Now let's consider how noise will affect the operation of the receiver. As long as the received signal is sufficiently stronger than the noise, the receiver can determine the transmitted symbol (and hence the transmitted bits) by examining the outputs of the four correlators. Table 5-8 shows the operation of the comparator and logic circuits in the 8-ary coherent PSK receiver (Figure 5-44).

Table 5-7 Each Received Symbol Creates a Different Combination of Correlator Outputs

Received Symbol	θ	Correlator Outputs			
		A	B	C	D
$s_1(t)$	0°	+	+	−	−
$s_2(t)$	45°	+	+	+	−
$s_3(t)$	90°	+	+	+	+
$s_4(t)$	135°	−	+	+	+
$s_5(t)$	180°	−	−	+	+
$s_6(t)$	225°	−	−	−	+
$s_7(t)$	270°	−	−	−	−
$s_8(t)$	315°	+	−	−	−

Table 5-8 Associating the Correlator Outputs with the Transmitted Information

Correlator Outputs				Bits to User		
A	B	C	D	b_{3n-2}	b_{3n-1}	b_{3n}
+	+	−	−	0	0	0
+	+	+	−	0	0	1
+	+	+	+	0	1	1
−	+	+	+	0	1	0
−	−	+	+	1	1	0
−	−	−	+	1	0	0
−	−	−	−	1	0	1
+	−	−	−	1	1	1

Now that we understand how a coherently demodulated 8-ary PSK system works, we can extend this concept to create an M-ary PSK system using M distinct phases (for most practical systems, M is a power of 2). We begin by creating a constellation where the M distinct phases are separated by angles of $360°/M$. For example, for a 16-ary system the distinct phases would be $0°, 22.5°, 45°, 67.5°$, etc. Since the system can produce M distinct symbols, each symbol can represent $\log_2 M$ bits. The system is M times as spectrally efficient as a binary PSK system, meaning that for a given bandwidth it can transmit M times as many bits per second, or for a given transmission speed it requires only $1/M$ times as much bandwidth as binary PSK. A coherent M-ary PSK receiver can be constructed using $M/2$ correlators, each employing a reference signal with a different phase. The phases of the reference signals for the $M/2$ correlators should be chosen halfway between adjacent symbols in the half-plane of the constellation corresponding to $0° < \theta < 180°$. Again using the 16-ary system as an example, a coherent receiver will employ eight correlators. The phases of the reference signals for the correlators will be $11.25°, 33.75°, 56.25°, 78.75°$, $101.25°$, $123.75°$, $146.25°$, and $168.75°$. Each possible received symbol will produce a different combination of positive and negative outputs among the correlators, so the received signal can be demodulated by examining the correlators' outputs.

Increasing the value of M increases the spectral efficiency of the system, but at the cost of additional equipment complexity and reduced accuracy. Probability of symbol error for coherently demodulated M-ary PSK is

$$P_{s,\,coherent\ \text{M-ary PSK}} \approx 2Q\left(\sqrt{\frac{A^2 T_s}{N_o}\sin^2\frac{\pi}{M}}\right) \quad M \geq 4$$

$$= 2Q\left(\sqrt{2\log_2 M\left[\frac{E_b}{N_o}\right]\sin^2\frac{\pi}{M}}\right) \quad M \geq 4 \qquad (5.126)$$

A derivation of Equation (5.126) is given in Carlson, Crilly, and Rutledge [5.5]. Figure 5-46 provides a plot of $P_{s,\,coherent\ \text{M-ary PSK}}$ versus E_b/N_o for various values of M.

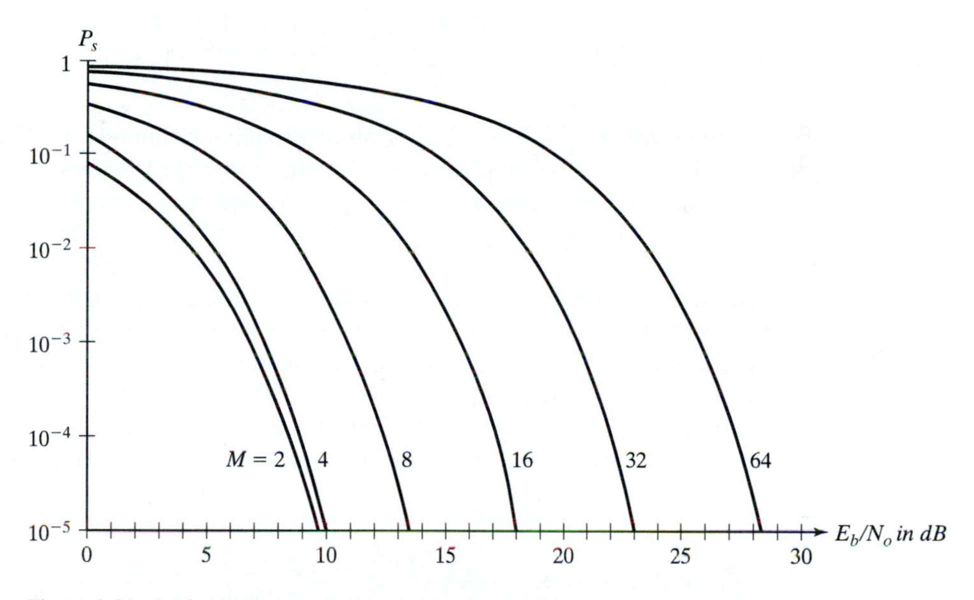

Figure 5-46 Probability of symbol error versus E_b/N_o for coherently demodulated M-ary PSK.

Intuitively, we know that accuracy is reduced as M increases because as we try to crowd more symbols into the PSK constellation, the phase difference between adjacent symbols becomes smaller since there is only a fixed 360° of possible phase. As two symbols become closer in phase, distinguishing between the two symbols becomes harder and the probability of error at the receiver increases.

To complete our analysis, we need to relate the probability of symbol error with the probability of bit error. If we use a Gray code to assign the bit patterns to each symbol (as we just did in the 8-ary PSK case) and if we assume that all errors will be to an adjacent symbol, then each symbol error produces only one erroneous bit and $\log_2 M - 1$ correct bits. Thus

$$P_{b,\,errors\ to\ adjacent\ symbol} = \frac{1}{\log_2 M}P_s \qquad (5.127)$$

Unfortunately, for larger values of M the phase difference between adjacent symbols becomes small and the assumption that all errors are to adjacent symbols becomes less valid. Let's consider the worst-case relationship between P_b and P_s, where an erroneous symbol has no relationship to the correct symbol. For this worst case, there are $M - 1$ incorrect symbols and in $M/2$ of these symbols a particular bit will differ from the same bit in the correct symbol. The worst-case probability of bit error is therefore

$$P_{b,\,worst\ case} = \frac{M/2}{M-1}P_s = \frac{M}{2(M-1)}P_s \tag{5.128}$$

We have now created an upper and lower bound on probability of bit error relative to P_s.

$$\frac{1}{\log_2 M}P_s \leq P_b \leq \frac{M}{2(M-1)}P_s \qquad M \geq 4 \tag{5.129}$$

5.8.3.2 *Differential M-ary PSK* Using an approach similar to DQPSK, we can also create a differential M-ary PSK system. As with coherently demodulated M-ary PSK, we create M symbols using M distinct phases separated by angles of $360°/M$. As with DQPSK, we transmit information in the nth symbol by changing the phase relative to the phase of the n-1st symbol. The differential M-ary PSK receiver has a structure similar to the DQPSK receiver shown in Figure 5-42, except that we now need a series of $M/2$ circuits that apply appropriate phase shifts and delays to the received signal. The differential M-ary PSK receiver uses less complex equipment than the coherent M-ary PSK receiver (since the differential receiver does not need a reference signal that is phase and frequency synchronized with the carrier), but it provides less accuracy. Probability of symbol error for a differential M-ary PSK system is

$$\begin{aligned} P_{s,\,differential\ \text{M-ary PSK}} &\approx 2Q\left(\sqrt{\frac{2A^2T_s}{N_o}\sin^2\frac{\pi}{2M}}\right) \qquad M \geq 4 \\ &= 2Q\left(\sqrt{4\log_2 M\left[\frac{E_b}{N_o}\right]\sin^2\frac{\pi}{2M}}\right) \qquad M \geq 4 \end{aligned} \tag{5.130}$$

A derivation of Equation (5.130) is given in Lindsey and Simon [5.3].

Figure 5-47 provides a plot of $P_{s,\,differential\ \text{M-ary PSK}}$ versus E_b/N_o for various values of M. As with coherently demodulated M-ary PSK,

$$\frac{1}{\log_2 M}P_s \leq P_b \leq \frac{M}{2(M-1)}P_s \qquad M \geq 4 \tag{5.129R}$$

5.8.4 M-ary Frequency Shift Keying

We can also create a series of symbols by modulating the frequency of the carrier. Consider, for instance, the following technique, which produces four distinct symbols and

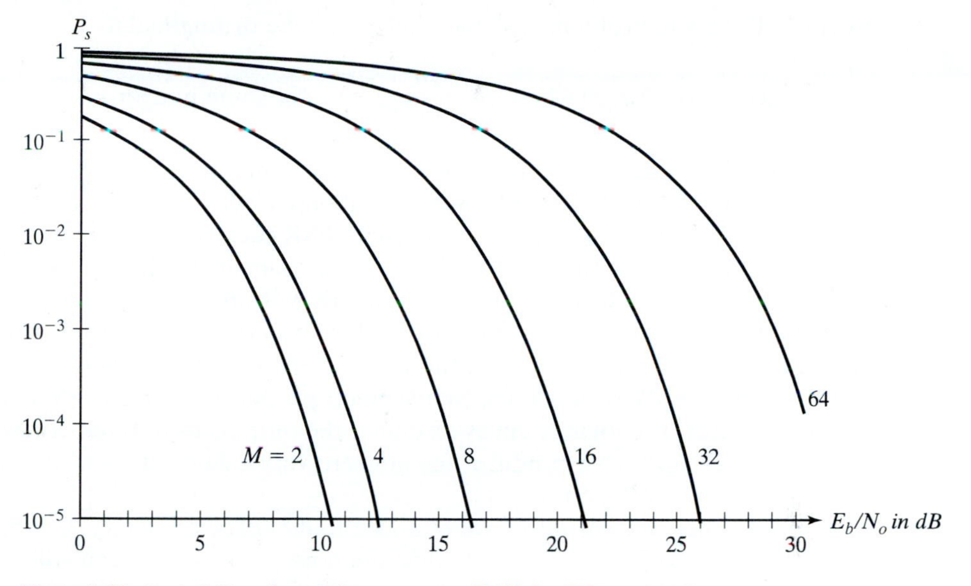

Figure 5-47 Probability of symbol error versus E_b/N_o for differential M-ary PSK.

hence can transmit two bits per symbol:

- Transmit $A\sin[2\pi(f_c+3\Delta f)t]$ to represent "11"; we will call this symbol $s_1(t)$.
- Transmit $A\sin[2\pi(f_c+\Delta f)t]$ to represent "10"; we will call this symbol $s_2(t)$.
- Transmit $A\sin[2\pi(f_c-\Delta f)t]$ to represent "00"; we will call this symbol $s_3(t)$.
- Transmit $A\sin[2\pi(f_c-3\Delta f)t]$ to represent "01"; we will call this symbol $s_4(t)$.

Figure 5-48 shows a typical quaternary FSK waveform representing the data sequence "1101100001" transmitted at a speed of 100,000 bits/sec (50,000 symbols/sec) with $f_c = 175$ kHz and $\Delta f = 25$ kHz.

If we choose the values of f_c and Δf such that all four symbols are orthogonal, then we can use a correlation receiver to coherently demodulate quaternary FSK.

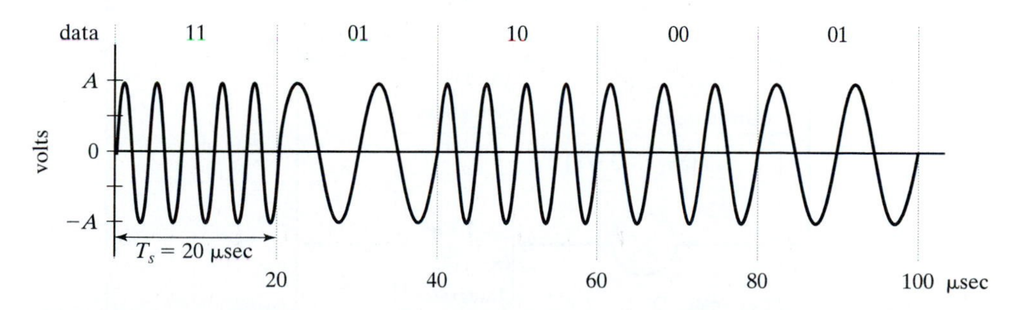

Figure 5-48 Quaternary FSK signal corresponding to "1101100001" transmitted at 100,000 bits/sec (50,000 symbols/sec).

As you will show in Problem 5.39, the symbols will be orthogonal if

$$s_i(t) = \sin(2\pi f_i t) \text{ where } f_i = \frac{(n+i)\pi}{2T_s} \quad \text{for some integer } n \tag{5.131}$$

Equation (5.131) states that the symbols are orthogonal if each one produces a whole number of half-cycles of sinusoid within a symbol period.

Figure 5-49 shows a coherent quaternary FSK receiver that exploits the orthogonality of the symbols. Suppose symbol $s_2(t)$ is transmitted. In a noiseless environment, the outputs of Correlators A, C, and D will all be zero and the output of Correlator B will be $0.5\gamma AT_s$. Generalizing this observation, for each of the four possible transmitted symbols, three of the four correlator outputs will be zero and the fourth will be $0.5\gamma AT_s$. Each of the four symbols produces a nonzero value from a different correlator. The logic circuitry examines the outputs of the four correlators and, based on which correlator produces the nonzero value, determines which symbol was transmitted (see Table 5-9).

In a practical system, noise will be added to the received signal. As long as the received signal is significantly stronger than the noise, the receiver can still demodulate the received signal by examining the four correlator outputs and choosing the symbol associated with the correlator having the largest output.

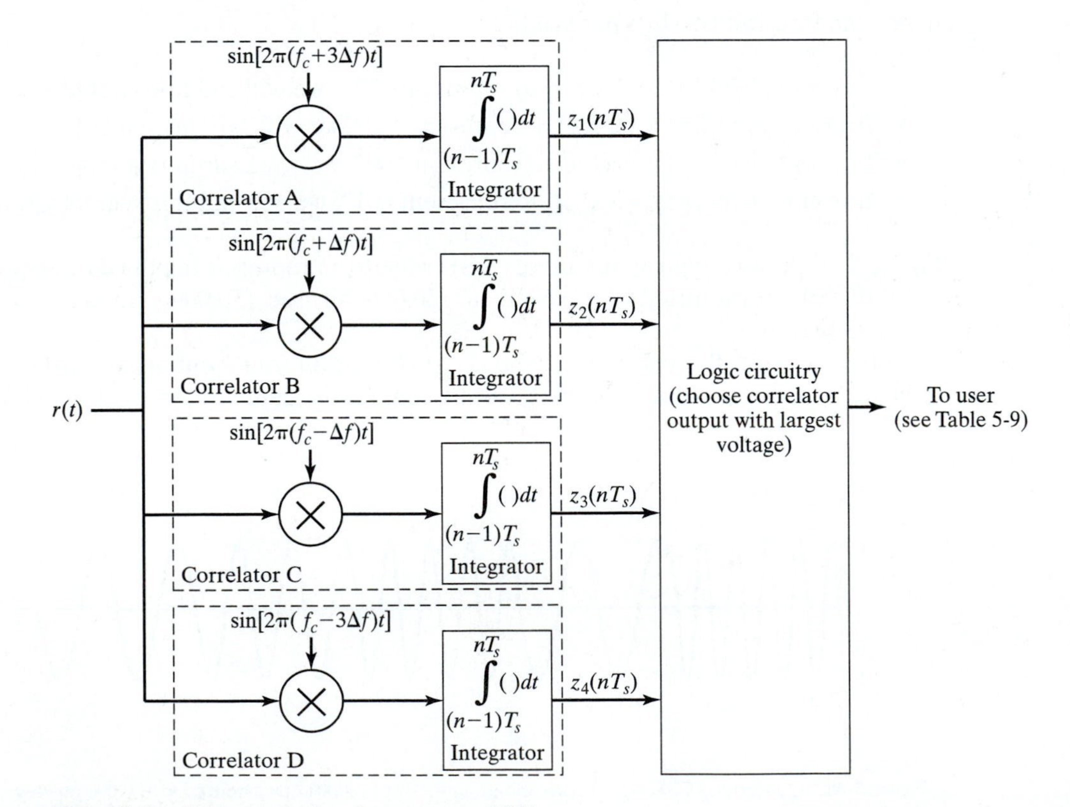

Figure 5-49 Coherent receiver for quaternary FSK.

Table 5-9 Logic for Coherent Quaternary FSK Receiver

Correlator with Largest Output	Outputs to User
A	11
B	10
C	00
D	01

As with the binary FSK signal, we can determine the bandwidth of the quaternary FSK signal by considering it the sum of ASK signals, one at each of the orthogonal frequencies. Each of the four ASK signals will be transmitting only one fourth of the time (as opposed to two ASK signals each transmitting half the time with binary FSK). Using a conceptual approach, Figure 5-50 shows the power spectral densities of each of the four ASK signals, one centered at $f_c-3\Delta f$, one centered at $f_c-\Delta f$, a third centered at $f_c+\Delta f$, and a fourth centered at $f_c+3\Delta f$. Although the approach is not rigorous, we can see that a bandwidth of approximately $6\Delta f+4r_s$ is needed for 95% of the power to be in-band. (As mentioned in Section 5.2.3, a rigorous approach requires that we add the spectra of the ASK signals, using complex numbers, prior to squaring to determine energy and power.)

We can extend the concept of quaternary FSK to develop an M-ary FSK system, creating M distinct symbols by using M mutually orthogonal frequencies. Such a system requires a bandwidth of $(M-1)\Delta f + 4r_s$ for 95% in-band power. A coherent M-ary FSK receiver can be constructed with M correlators, each using one of the orthogonal frequencies as its reference signal. Probability of symbol error for the coherently demodulated M-ary FSK system is the probability that noise causes the output of one of the $M-1$ correlators corresponding to the incorrect symbols to be greater than the output of the correlator corresponding to the correct symbol. As shown in Simon, Hinedi, and Lindsey [5.4], direct calculation of this probability is difficult but its value is upper bounded by

$$P_{s,\,coherent\ \text{M-ary FSK}} \leq (M-1)Q\left(\sqrt{\frac{A^2 T_s}{2N_o}}\right)$$

$$= (M-1)Q\left(\sqrt{\log_2 M\left[\frac{E_b}{N_o}\right]}\right) \qquad M \geq 4 \tag{5.132}$$

Figure 5-51 provides a plot of the upper bound of $P_{s,\,coherent\ \text{M-ary FSK}}$ versus E_b/N_o for various values of M.

The noise causes symbol errors to be equally likely among all $M-1$ correlators corresponding to incorrect symbols, so there is no advantage to Gray coding the bits, and probability of bit error for coherently demodulated M-ary FSK is related to symbol error as

$$P_{b,\,coherent\ \text{M-ary FSK}} = \frac{M}{2(M-1)} P_{s,\,coherent\ \text{M-ary FSK}} \tag{5.133}$$

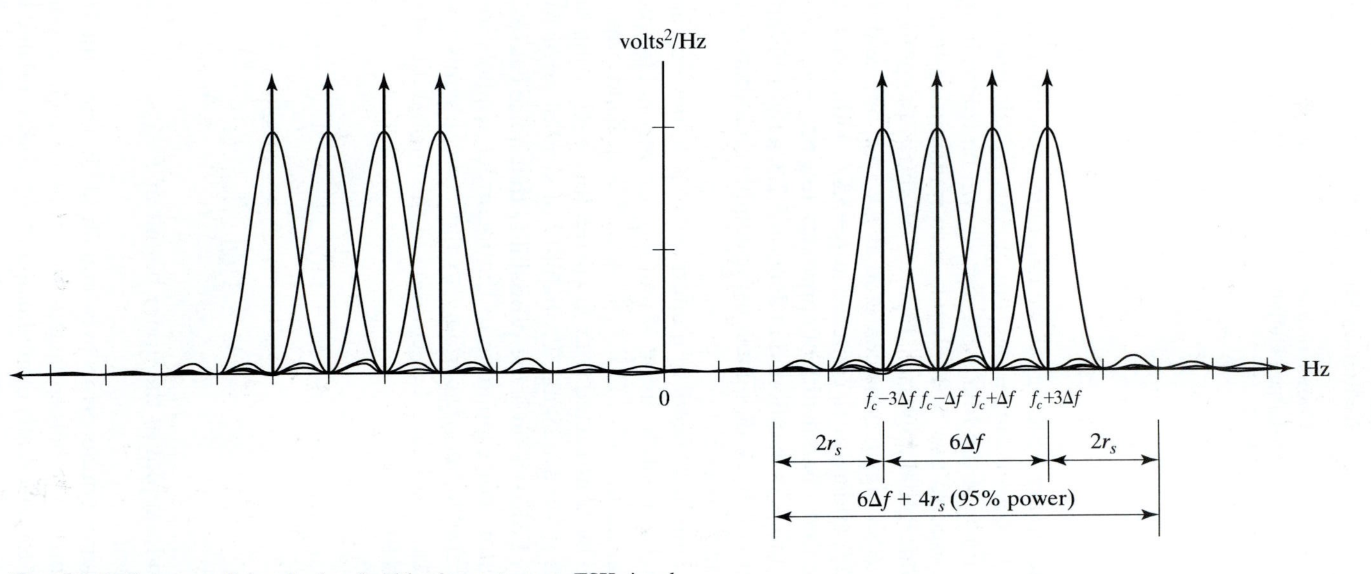

Figure 5-50 Conceptualizing the bandwidth of a quaternary FSK signal.

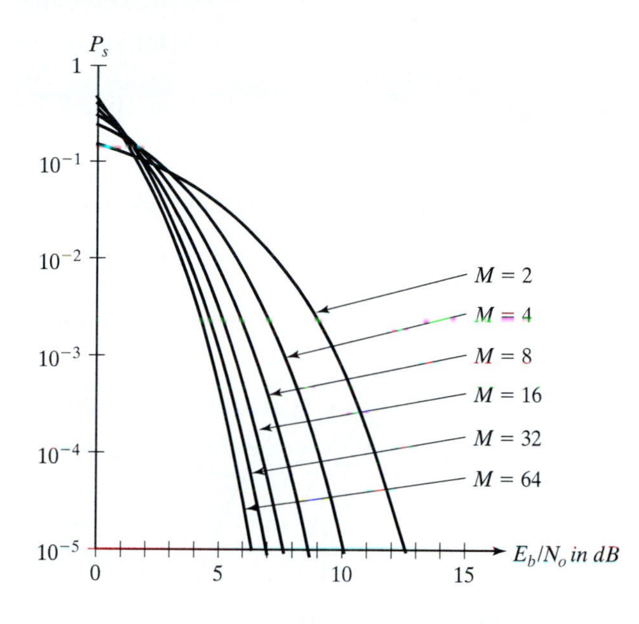

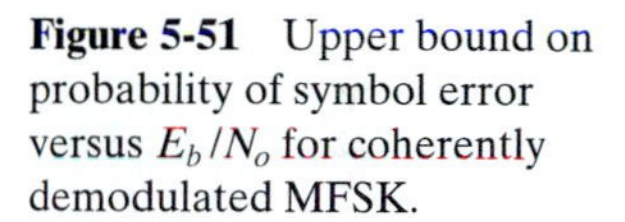
Figure 5-51 Upper bound on probability of symbol error versus E_b/N_o for coherently demodulated MFSK.

What are the benefits of M-ary FSK relative to binary FSK? In examining Figure 5-50, we see that increasing the value of M increases the bandwidth in an almost linear fashion, so M-ary FSK is not significantly more spectrally efficient than binary FSK. We can show, however, that an M-ary FSK system is more accurate than a series of binary FSK systems using separate carrier frequencies but which, taken together, transmit at the same effective bit rate as the M-ary system (see Problem 5.30).

We can also develop a noncoherent receiver for M-ary FSK. Such a receiver will use the same principles as used for the noncoherent binary FSK receiver (see Section 5.5.3 and Figure 5-25), employing a series of M bandpass filters, one centered at each of the possible transmitted frequencies, a series of M envelope detectors, and logic circuitry to determine which envelope detector has the highest output at the end of the symbol period.

In inspecting Figure 5-50 and Equation (5.131), we see that if we choose the frequencies for the M symbols the same way as we did for coherent M-ary FSK, then there will be considerable overlap between the lobes in the M-ary FSK spectrum. This overlap will increase the symbol error rate because, depending on the bandwidth of the receiver's filters, filtering will reduce the in-band power of the correct symbol and/or increase the in-band power of adjacent symbols. For practical noncoherent M-ary FSK systems, the frequency difference between adjacent symbols should be at least $1/T_s$. Using orthogonal symbols with a frequency difference between adjacent symbols of at least $1/T_s$, and using a filter bandwidth of $B = r_s$, the probability of symbol error for noncoherently demodulated M-ary FSK is

$$
\begin{aligned}
P_{s,\,noncoherent\ \text{M-ary FSK}} &= \frac{1}{M}e^{-\left(\frac{A^2T_s}{2N_o}\right)}\sum_{n=2}^{M}(-1)^n\frac{(M)!}{n!(M-n)!}e^{\left(\frac{A^2T_s}{n2N_o}\right)} \\
&= \frac{1}{M}e^{-\left(\frac{(\log_2 M)E_b}{N_o}\right)}\sum_{n=2}^{M}(-1)^n\frac{(M)!}{n!(M-n)!}e^{\left(\frac{(\log_2 M)E_b}{nN_o}\right)} \qquad (5.134)
\end{aligned}
$$

Figure 5-52 provides a plot of $P_{s,\,noncoherent\ \text{M-ary FSK}}$ versus E_b/N_o for various values of M. See Simon, Hinedi, Lindsey [5.4] for a derivation of $P_{s,\,noncoherent\ \text{M-ary FSK}}$. Note that the filter bandwidth used in Equation (5.134), Figure 5-52, and Simon, Hinedi, Lindsey [5.4] is not as conservative as the bandwidth we used in our analysis of noncoherent binary FSK (see the footnote to Example 5.10). For $M = 2$, Equation (5.134) will therefore produce E_b/N_o values that are 3 dB less than Equation (5.87), Table 5-3, and Figure 5-34.

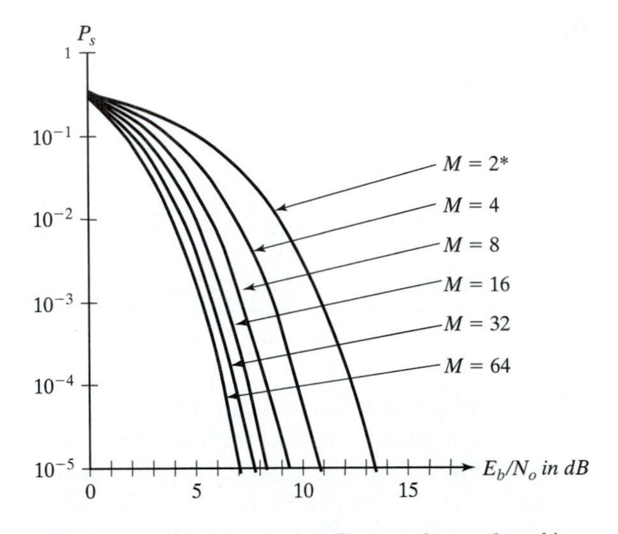

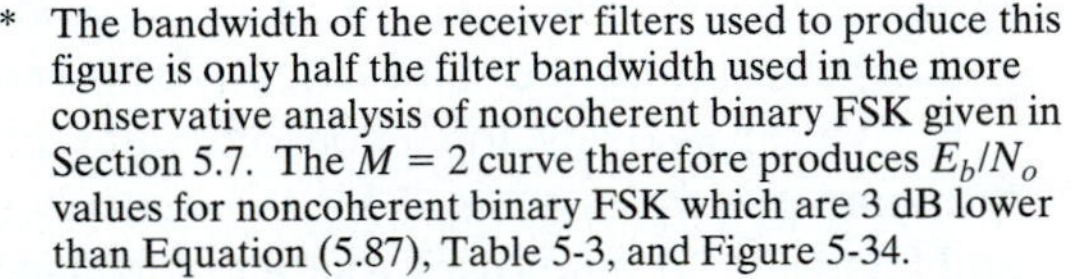

Figure 5-52 Probability of symbol error versus E_b/N_o for noncoherently demodulated M-ary FSK with orthogonal symbols.

* The bandwidth of the receiver filters used to produce this figure is only half the filter bandwidth used in the more conservative analysis of noncoherent binary FSK given in Section 5.7. The $M = 2$ curve therefore produces E_b/N_o values for noncoherent binary FSK which are 3 dB lower than Equation (5.87), Table 5-3, and Figure 5-34.

5.8.5 Multiparameter M-ary Bandpass Signaling

The two M-ary modulation techniques we've examined so far—M-ary FSK and M-ary PSK—provide an interesting set of performance-versus-cost trade-offs. M-ary FSK allows us to reduce bit error rate relative to a series of binary FSK signals transmitted with different carriers. Because of the need for orthogonality (for coherent detection) or because of the nature of the FSK spectrum and the practical limitations of filters (for noncoherent detection), M-ary FSK requires approximately the same bandwidth as a series of individual FSK binary signals and hence does not increase spectral efficiency. M-ary PSK allows us to increase spectral efficiency by increasing M, but for larger values of M accuracy is significantly reduced. Are there other M-ary modulation techniques that we can develop? If so, do these techniques provide a set of trade-offs that might be better than M-ary PSK or M-ary FSK for certain applications?

So far we have developed M-ary bandpass modulation techniques that vary only one parameter of the carrier signal (amplitude, frequency, or phase). Is it possible to simultaneously vary more than one parameter in order to produce M different symbols? Of course. Perhaps with a multiparameter approach we can create M distinct symbols that are spectrally efficient and yet are sufficiently different from each other

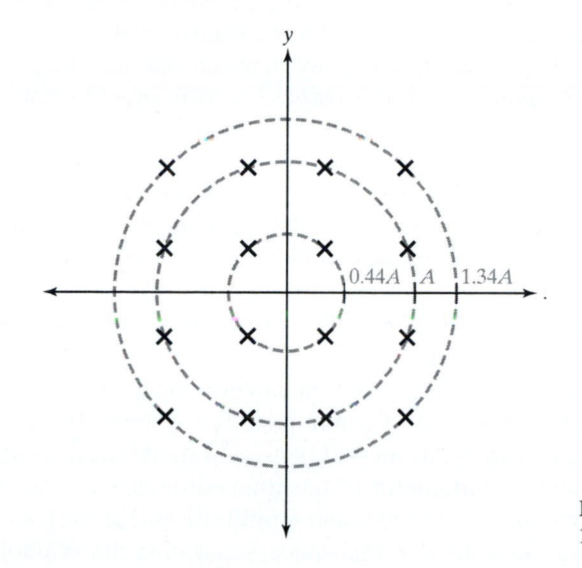

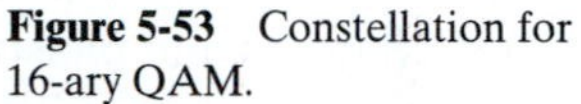
Figure 5-53 Constellation for 16-ary QAM.

to provide sufficient accuracy. Let's consider, for example, simultaneously varying both the amplitude and the phase of the carrier signal. Figure 5-53 shows a constellation with 16 distinct symbols. This modulation technique is called *16-ary quadrature amplitude modulation* (16-ary QAM).

The 16-ary constellation in Figure 5-53 uses three different amplitudes and 12 different phases. Before we develop a receiver and analytically calculate spectral efficiency and symbol error rate, let's see if we can intuitively understand why this constellation might produce a more accurate signal than, say, a constellation using one amplitude and 16 distinct phases. Figure 5-54 shows two constellations: one for 16-ary PSK and one for 16-ary QAM. Both constellations represent the same average transmitted signal power

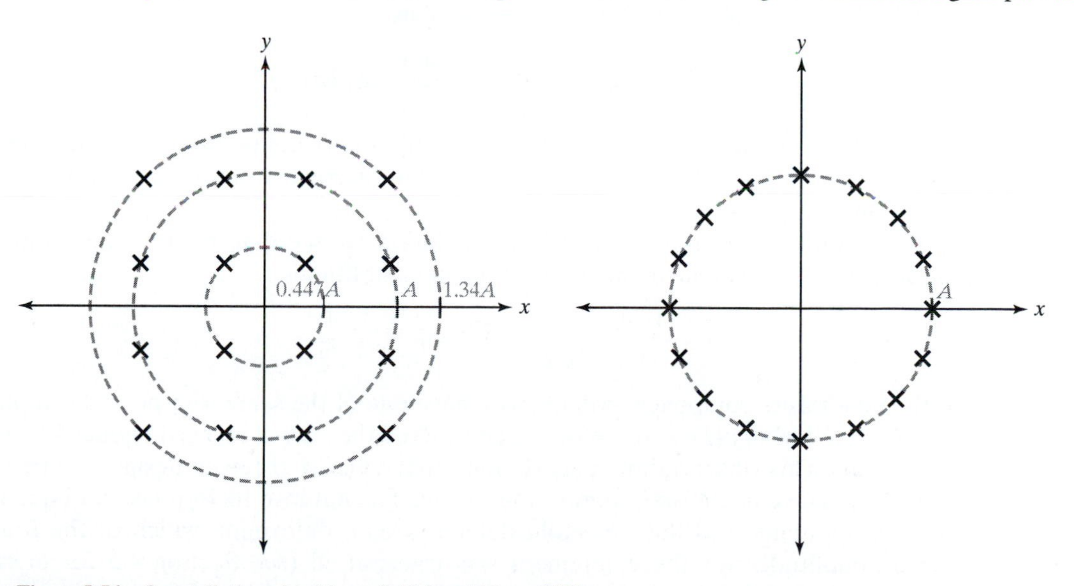

Figure 5-54 Constellations for 16-ary QAM and 16-ary PSK.

(you'll prove this in Problem 5.38). Let's consider the transmitted signal as a phasor. As it is sent across the channel, it is corrupted by noise, which changes the phasor's amplitude and its phase. The receiver takes the corrupted phasor and determines the transmitted information by finding the symbol in the constellation that is *geometrically closest* to the corrupted phasor. In looking closely at Figure 5-54, we can see that the symbols within the 16-ary QAM constellation are separated farther from each other than the symbols within the 16-ary PSK constellation. This greater separation means that it is less likely that the received phasor will be closer to an incorrect symbol than the correct symbol. The greater separation therefore means that for a given average transmitted power level, there is less probability of symbol error in the 16-ary QAM system than in the 16-ary PSK system.

Again we see that the use of a constellation and graphical techniques allows us to gain further insight into the operation of communication systems. In generalizing our geometric interpretation, we can create an n-dimensional signal space in which each dimension represents a different parameter of the transmitted signal (two-dimensional constellations and phasors allow us to represent amplitude and phase). We then create a series of symbols within the n-dimensional space, separating the symbols from each other as much as possible (this is equivalent to saying that we want to make the symbols "look as different from each other" as possible). The receiver then takes the received signal, places it at the appropriate point within the n-dimensional space, and then finds the symbol that is geometrically closest to it. (This is equivalent to saying that the receiver finds the symbol that "looks most like" the received signal.)

Let's now develop a QAM receiver. In inspecting the QAM constellation, we see that each of the 16 symbols can be decomposed into two signals: one corresponding to a sine at the carrier frequency (the in-phase component) and the other corresponding to a cosine at the carrier frequency (the quadrature component). For example, the symbol in the lower left corner of the 16-ary QAM constellation in Figure 5-54, $1.34A\sin(2\pi f_c t + 315°)$, can also be represented as

$$\frac{1.34A}{\sqrt{2}}\sin(2\pi f_c t) - \frac{1.34A}{\sqrt{2}}\cos(2\pi f_c t)$$

As with QPSK, we can decompose any QAM symbol into an in-phase component and a quadrature component. This decomposition is the reason for the word "quadrature" in the term "quadrature amplitude modulation."

Continuing to examine the QAM constellation, we see that the in-phase component of any symbol will have one of four possible amplitudes,

$$\frac{1.34A}{\sqrt{2}}, \frac{A}{\sqrt{2}}, -\frac{A}{\sqrt{2}}, \text{ or } -\frac{1.34A}{\sqrt{2}}$$

and the quadrature component will likewise have one of the same four possible amplitudes. A 16-QAM signal can therefore be considered the sum of two orthogonal 4-level ASK signals. This observation suggests the 16-ary QAM receiver design shown in Figure 5-55. The correlators separate the received signal into its in-phase and quadrature components, and the threshold detectors then determine which of the four possible amplitudes for the component was transmitted (see Section 4.5 for more details concerning the threshold detectors). After the amplitudes of the in-phase and

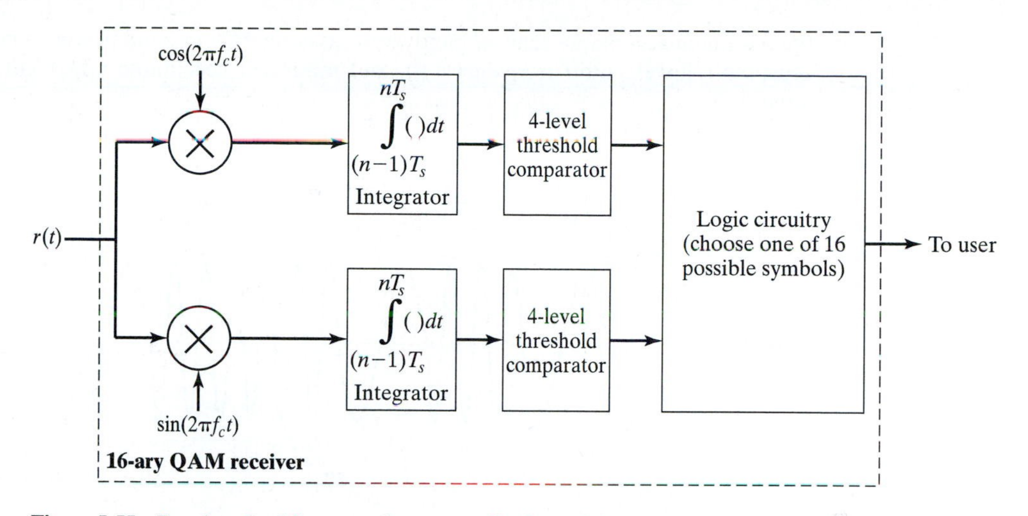

Figure 5-55 Receiver for 16-ary quadrature amplitude modulation.

quadrature components are known, the logic circuitry can select the appropriate symbol to be output to the user.

Assuming a rectangular shape for the constellation, a tight upper bound for the probability of symbol error for coherently demodulated M-ary QAM can be established (Proakis [5.6]).

$$P_{s,\,coherent\ \text{M-ary QAM}} \leq 4Q\left(\sqrt{\frac{3E_s}{(M-1)N_o}}\right) \tag{5.135}$$

where E_s represents the average normalized energy of one symbol of the received signal.

PROBLEMS

5.1 Explain the concept of creating a signal suitable for bandpass channels by modulating a carrier signal.

5.2 Consider a bandpass digital communication system using a carrier frequency of 600 kHz and a transmission speed of 100,000 bits/sec.

a. Draw the waveform corresponding to the data sequence "1101001" if ASK is chosen as the modulation technique.

b. Draw the waveform corresponding to the data sequence "1101001" if PSK is chosen as the modulation technique.

c. Draw the waveform corresponding to the data sequence "1101001" if FSK is chosen as the modulation technique and a frequency offset of 100 kHz is used.

5.3 Using the transmission speeds, carrier frequencies, and frequency offset from Problem 5.2, draw the average normalized power spectra for ASK, PSK, and FSK waveforms.

5.4 Each waveform in Figure P5-4 is transmitting data at a rate of 100,000 bits/sec. For each waveform, state

a. the modulation technique

b. the carrier frequency

c. the minimum acceptable channel bandwidth (assume 90% in-band power), maximum acceptable lower cutoff frequency (f_l), and minimum acceptable upper cutoff frequency (f_h)

d. the transmitted data

e. Δf, where applicable

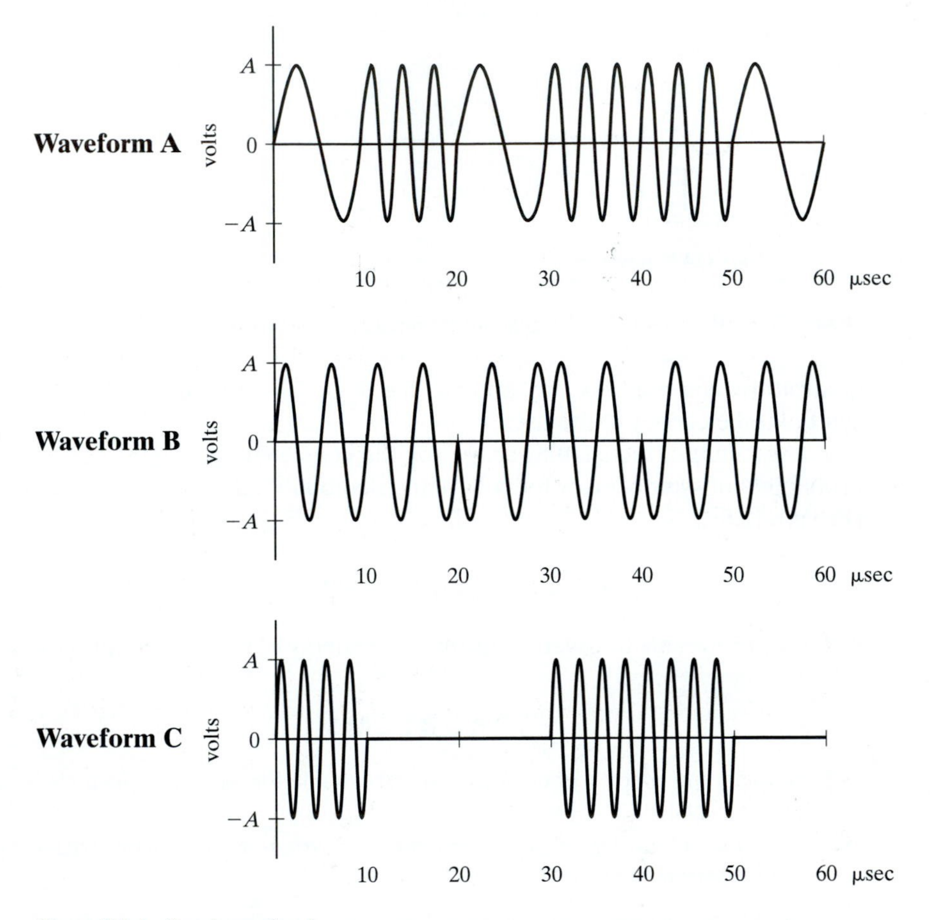

Figure P5-4 Bandpass signals.

5.5 A coherent PSK system using a correlation receiver is transmitting 100,000 bits/sec. Noise power spectral density at the receiver is $N_o/2 = 2.5 \times 10^{-7}$ volts^2 and the channel attenuates the transmitted signal by 75% (so that only 25% of the transmitted signal arrives at the receiver). The amplitude of the transmitted carrier signal is 3.0 volts and the carrier frequency is 800 kHz.

a. Determine the channel bandwidth required for the transmitted signal to have 95% in-band power. Also state the maximum acceptable lower cutoff frequency of the channel (f_l) and the minimum acceptable upper cutoff frequency of the channel (f_h).

b. Determine the average normalized power of the transmitted signal.

c. Determine the probability of bit error of the received signal.
d. Suppose we want to improve accuracy by raising the power of the transmitted signal. What is the minimum average normalized power of the transmitted signal that will produce a probability of bit error of 10^{-5} or less?

5.6 Repeat Problem 5.5, except the system is now transmitting FSK with $\Delta f = 100$ kHz.

5.7 Repeat Problem 5.5, except the system is now transmitting ASK. Comment on the "fairness" of comparing the results of Part c of Problems 5.5 and 5.6 to Part c of this problem. Comment on the "fairness" of comparing the results of Part d of Problems 5.5 and 5.6 to Part d of this problem.

5.8 Using the results of Problems 5.5–5.7, and using your knowledge of the structure of coherent PSK, ASK, and FSK receivers, explain the performance-versus-cost trade-offs associated with coherent PSK, ASK, and FSK.

5.9 Show that a binary PSK signal can be decomposed into the product of a baseband signal and a carrier. Discuss how this decomposition is useful in determining the power spectral density of binary PSK. Show how bandwidth and transmission speed are related for binary PSK.

5.10 Suppose a coherent PSK receiver is used in a communication system but the reference signal in the receiver is out of frequency synchronization with the transmitted carrier by a small amount, say Δf Hz. Derive an expression showing how the output of the receiver's mixer is affected by the frequency difference.

5.11 Suppose a coherent PSK receiver is used in a communication system. The reference signal in the receiver is in frequency synchronization with the transmitted carrier, but out of phase synchronization by $\Delta\theta$ degrees. Derive an expression showing how the output of the receiver's mixer is affected by the frequency difference.

5.12 Show that in order to obtain the same level of accuracy, a coherently demodulated ASK signal needs to be transmitted with twice the average normalized power of a coherently demodulated PSK signal.

5.13 Derive the equation for probability of bit error for a coherent PSK system.

5.14 Using the coherent FSK receiver shown in Figure 5-16, show that if we select Δf such that the frequencies $f_c+\Delta f$ and $f_c-\Delta f$ produce a whole number of cycles within a bit period, the optimum threshold for the receiver is 0 volts.

5.15 In your own words, define the term "random process." Explain the difference between a random process and a random variable. Explain how random processes are used in analyzing communication systems.

5.16 In your own words, define the terms "mean" and "autocorrelation" of a random process. Explain the significance of these two parameters.

5.17 State the Wiener-Khintchine theorem and discuss its significance.

5.18 Explain the difference between ensemble-averaged statistics and time-averaged statistics. When and why are time-averaged statistics useful?

5.19 Prove the following properties of a wide-sense stationary (WSS) random process.

a. $\mu_{\mathbf{X}}$ = the dc level of the signal described by the random process
b. $R_{\mathbf{XX}}(0)$ = average normalized power of **X**
c. $R_{\mathbf{XX}}(\tau) = R_{\mathbf{XX}}(-\tau)$
d. if $\mu_{\mathbf{X}} = 0$, then the variance is equal to the average normalized power of **X**

5.20 Consider a PAM signal using rectangular pulses of amplitude A and width γ. Show that $\mu_{\mathbf{X}} = 0$ and autocorrelation is the triangular function shown in Figure 5-20.

5.21 Explain the differences between coherent demodulation and noncoherent demodulation.

5.22 Consider a bandpass digital communication system using a carrier frequency of 600 kHz and a transmission speed of 100,000 bits/sec.

a. Draw the waveform corresponding to the data sequence "1101001" if phase shift keying is chosen as the modulation technique.

b. Draw the waveform corresponding to the data sequence "1101001" if differential phase shift keying is chosen as the modulation technique.

5.23 Repeat Problem 5.5, except the system is now transmitting DPSK and using a noncoherent receiver.

5.24 Repeat Problem 5.5, except the system is now transmitting FSK and using a noncoherent receiver.

5.25 Repeat Problem 5.5, except the system is now transmitting ASK and using a noncoherent receiver. Comment on the "fairness" of comparing the results of Part c of Problems 5.23 and 5.24 to Part c of this problem. Comment on the "fairness" of comparing the results of Part d of Problems 5.23 and 5.24 to Part d of this problem.

5.26 Explain the performance-versus-cost trade-offs in choosing among coherently demodulated PSK, DPSK, and noncoherently demodulated FSK systems.

5.27 Consider a system in which the average normalized power at the transmitter must be 6 volts^2 or less. The system must be designed to transmit over a bandpass channel that passes frequencies in the range 200 kHz $\leq f \leq$ 600 kHz. The attenuation in the channel is such that only 80% of the transmitted signal's power is present at the input to the receiver. The system operates in thermal noise that has a power spectral density as shown in Figure P5-27. The system requirements are a transmission speed of 50,000 bits/sec and a probability of bit error of 0.0025 or less. *Additionally, the system should be as inexpensive to build as possible.*

a. What binary modulation technique do you recommend? Justify your answer.

b. What is the probability of bit error for your recommended system?

c. Draw a block diagram for the receiver of your recommended system.

d. Can the transmission speed of the system be increased without exceeding the bandwidth of the channel (assume at least 95% in-band power), adding complexity to the transmitter or receiver, or causing the probability of bit error to exceed the maximum requirement? If so, what would be the maximum transmission speed? Justify your answer.

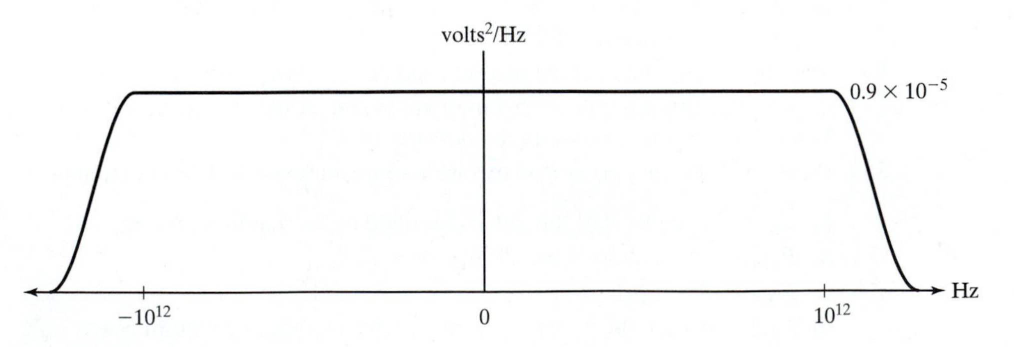

Figure P5-27 Power spectral density for noise in Problem 5.27.

5.28 Derive the equation for probability of bit error for a DPSK system.

5.29 Derive the equation for probability of bit error for a noncoherent FSK system.

5.30 Suppose you are given a choice of implementing binary PSK, DPSK, or noncoherent FSK in a communication system with additive white Gaussian noise.

a. Draw a block diagram of a receiver for each of the three systems.

b. Give the expression for probability of bit error for each of the three systems.

c. If a bit error rate of 10^{-4} is required for each system, determine how much average normalized received power is required for each of the three systems.

d. Explain advantages and disadvantages of using each of the three systems.

5.31 Consider the data sequence "1011001001," which needs to be transmitted at 25,000 bits/sec.

a. Draw the quaternary PSK (QPSK) waveform for the data, assuming that the transmitted signal has an amplitude of 2 volts.

b. Draw the differential quaternary PSK (DQPSK) waveform for the data, assuming that the transmitted signal has an amplitude of 2 volts.

c. The channel attenuates the signal so that only 20% of its amplitude arrives at the receiver. The two-sided power spectral density of the noise at the receiver is 8×10^{-7} volts^2. Determine the probability of symbol error for the waveforms in Parts a and b of this problem.

5.32 Explain the differences between QPSK and DQPSK, and discuss the trade-offs that need to be considered when choosing between the two for a particular design.

5.33 Design a 16-ary coherent PSK system.

a. What are the phases for the 16 symbols?

b. How many correlators are needed, and what are their reference angles?

c. Suppose we transmit at 800,000 bits/sec. What is the transmitted signal's bandwidth (assume a requirement for 95% in-band power)?

d. Assume that the channel attenuation is such that only 10% of the transmitted signal's power arrives at the receiver. The two-sided power spectral density for the noise is 10^{-6} volts^2. If the system requires a symbol error rate of 10^{-5}, what is the minimum required transmitted signal power?

e. How much would the transmitted signal power need to be increased if we wanted to use a noncoherent receiver and therefore needed to transmit 16-ary differential PSK? What advantages would such a system have over the coherent system?

5.34 Using coherent demodulation, and aiming for an accuracy of $P_s \leq 10^{-5}$, what is the difference in average normalized transmitted signal power needed for a 16-ary PSK system, as opposed to an 8-ary PSK system?

5.35 Explain the trade-offs in choosing the value of M for an M-ary PSK system.

5.36 Show that an 8-ary FSK system is more accurate than a series of three binary FSK systems transmitting at the same effective bit rate. How can you generalize this conclusion?

5.37 Explain the trade-offs involved in choosing the value of M for an M-ary FSK system.

5.38 Show that the two constellations in Figure 5-54 produce signals with the same average transmitted power.

5.39 Show that Equation (5.131) produces orthogonal symbols for M-ary FSK.

Chapter 6

Analog Modulation and Demodulation

Consider the analog signal $s(t)$ shown in Figure 6-1a. Suppose this signal represents music and we want to transmit it across a bandpass channel, such as the airwaves. The magnitude spectrum of $s(t)$, shown in Figure 6-1b, tells us that $s(t)$ has significant low-frequency components and therefore cannot be efficiently transmitted directly across a bandpass channel. As discussed in Chapter 1, we will therefore need to modulate the signal prior to transmission. One approach would be to convert the signal to a digital format (see Chapter 8), then transmit the digital information across the channel using a digital bandpass modulation technique, such as PSK or FSK (as we discussed in Chapter 5), and then convert the received digital signal back to analog format. This approach has many advantages (see Chapter 8), but it can involve significant equipment complexity. In this chapter we will investigate analog modulation techniques. These techniques modulate the signal directly, without analog-to-digital conversion.

6.1 Transmitting an Amplitude Modulated Signal

In Chapters 2 and 5 we established that multiplying a baseband signal, such as $s(t)$, by the sinusoid $A\cos(2\pi f_c t)$ merely shifts the spectrum of $s(t)$ by the frequency f_c. In other words,

$$\mathcal{F}\{s(t)\cos(2\pi f_c t)\} = \frac{1}{2}[S(f - f_c) + S(f + f_c)] \tag{2.56R}$$

We can use this concept to shift the spectrum of $s(t)$ into the frequency band appropriate to a particular bandpass channel. For example, consider the signal $s(t)$ shown in Figures 6-1a and b. Suppose we want to transmit the signal over a bandpass channel that can efficiently pass frequencies in the range 90 kHz $\leq f \leq$ 110 kHz. Multiplying $s(t)$ by $\cos(200{,}000\pi t)$

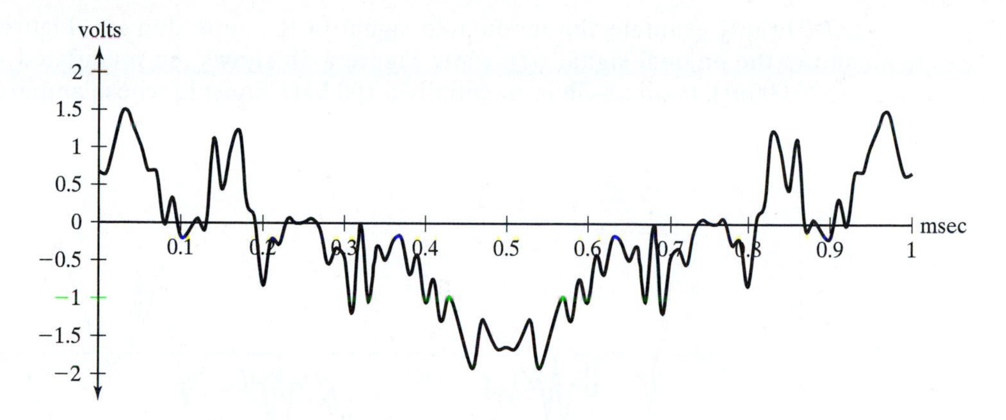

Figure 6-1a Typical baseband signal $s(t)$.

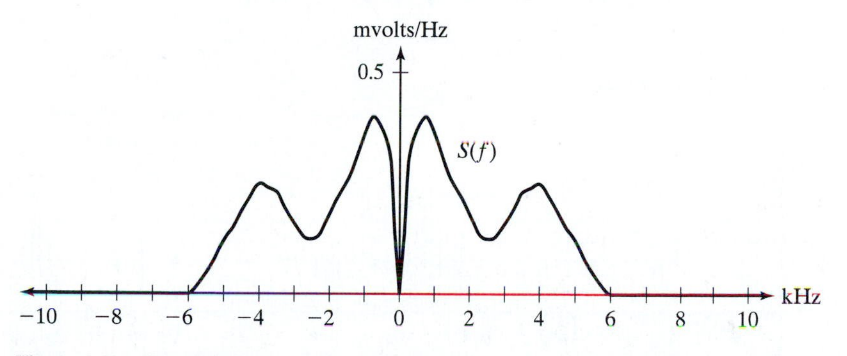

Figure 6-1b Two-sided magnitude spectrum of $s(t)$.

shifts its frequency by 100 kHz, producing the spectrum shown in Figure 6-2. This shifted spectrum is now within the frequency band that can be efficiently passed by the channel. Note that the bandwidth of the modulated signal is double the bandwidth of $s(t)$, since symmetric components of the baseband signal's magnitude spectrum that were originally to the left of 0 Hz are moved to the right of 0 Hz after the frequency shift. The portion of the spectrum in Figure 6-2 corresponding to the "positive" frequency components of $s(t)$ is called the *upper sideband*, and the portion corresponding to the "negative" frequency components of $s(t)$ is called the *lower sideband*. Later in this chapter we will discuss techniques for reducing the bandwidth of the modulated signal.

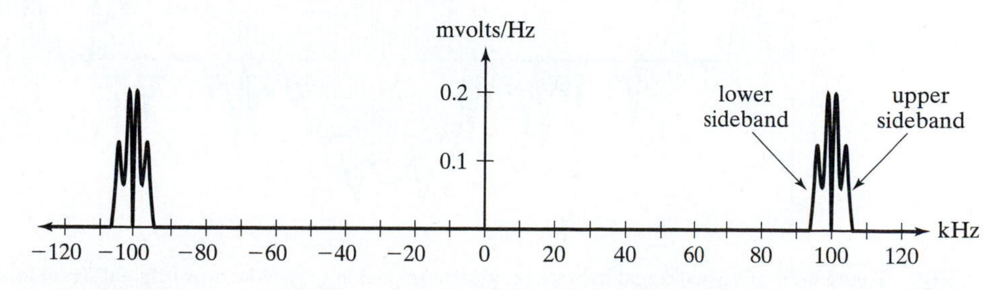

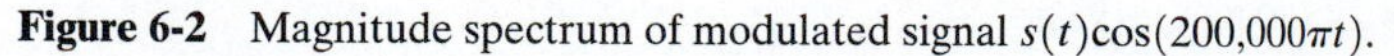
Figure 6-2 Magnitude spectrum of modulated signal $s(t)\cos(200{,}000\pi t)$.

Let's briefly examine the modulated signal in the time domain. Figure 6-3a again shows the original signal $s(t)$, while Figure 6-3b shows the modulated signal $s(t)\cos(200{,}000\pi t)$. Figure 6-3b is essentially a 100 kHz sinusoid whose amplitude is changed (or *modulated*) according to our original signal $s(t)$. Figure 6-3c shows the

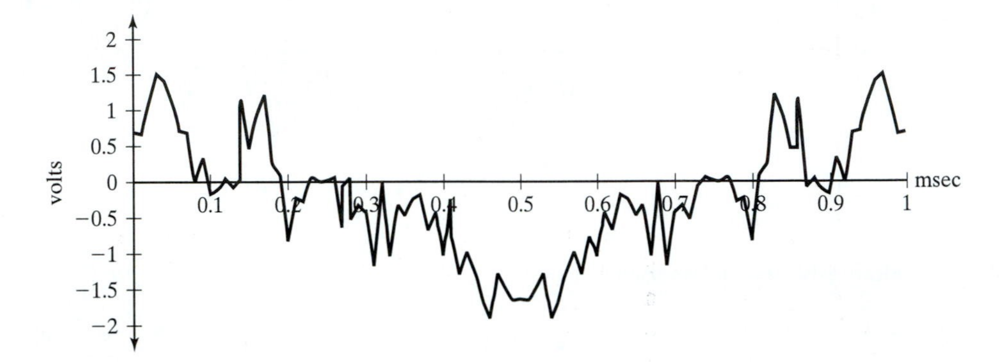

Figure 6-3a Original signal $s(t)$.

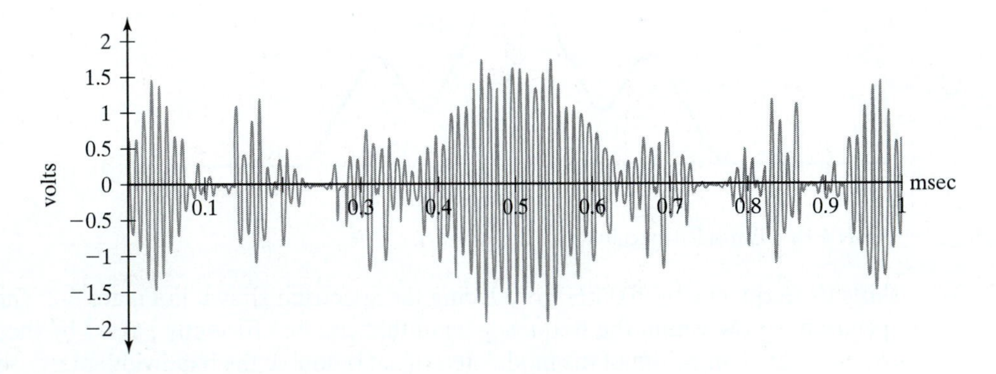

Figure 6-3b Amplitude-modulated signal $s(t)\cos(2\pi f_c t)$ $f_c = 100$ kHz.

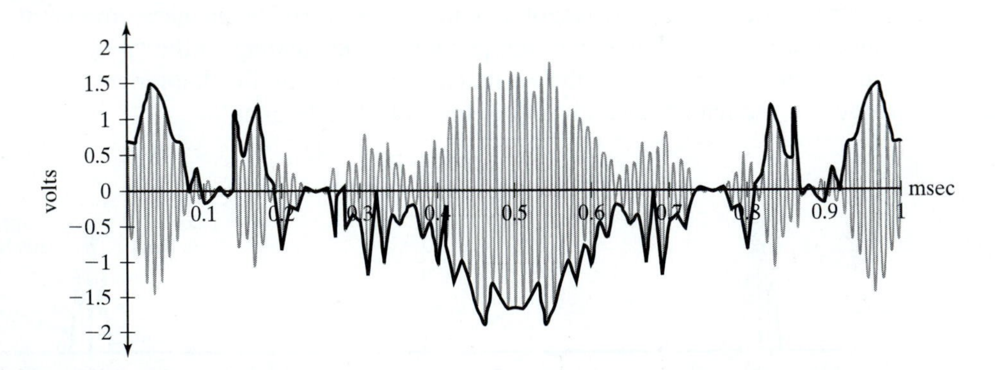

Figure 6-3c Original signal in black, modulated signal in gray. Note how original signal follows envelopes of the modulated signal.

original and modulated signals (note that when $s(t)$ is positive, it follows the upper envelope of the modulated signal; when $s(t)$ is negative, it follows the lower envelope of the modulated signal).

The modulation technique we have just developed is called *amplitude modulation* (AM) because we are transmitting information by modulating (or changing) the amplitude of a carrier signal. (Remember that a carrier signal was defined in Chapter 5 as a sinusoid that efficiently passes across a particular channel, such as a bandpass channel.) Generalizing, we can express an amplitude-modulated signal $s_{AM}(t)$ as

$$s_{AM}(t) = As(t)\cos(2\pi f_c t) \tag{6.1}$$

where $s(t)$ is the analog baseband signal we want to modulate, f_c is the frequency of the carrier signal chosen to allow $s_{AM}(t)$ to be efficiently transmitted across the bandpass channel, and A is a constant (called the *scaling factor*) that we can select to produce the appropriate power level for the transmitted signal.

There are many variations on AM—each with its advantages and disadvantages—which we will explore shortly. The type of AM just described is called *amplitude modulation-double sideband-suppressed carrier* (AM-DSB-SC). The term "double sideband" is appropriate because, as shown in Figure 6-2, the transmitted signal has both an upper sideband and a lower sideband. (Later in this chapter we will see why the term "suppressed carrier" is appropriate.)

Amplitude modulation is very similar to amplitude shift keying (ASK), a digital bandpass modulation technique we studied in Chapter 5. Both techniques transmit information by modulating the amplitude of a carrier signal. The difference between the techniques is that in AM, we allow the carrier amplitude to take on a continuum of possible values (reflecting the fact that the original signal $s(t)$ is analog); with ASK, the carrier amplitude is allowed to change only among a discrete number of values (A volts and 0 volts, for instance, for a binary system), reflecting the digital nature of the original signal. In general, we can say that AM is related to ASK in the same way that analog signals are related to digital signals. The similarities between AM and ASK will help us develop appropriate demodulation techniques for AM.

6.2 Coherent Demodulation of AM Signals

We now know that an AM signal $s_{AM}(t)$ can be produced by multiplying the analog baseband signal $s(t)$ by a carrier signal

$$s_{AM}(t) = As(t)\cos(2\pi f_c t) \tag{6.1R}$$

This multiplication shifts the spectrum of $s(t)$ into the frequency band efficiently passed by the channel. Once the signal is received, we need to demodulate it to reproduce the original signal $s(t)$ for the user. We can do this by shifting the spectrum of the received signal back into the original frequency band of $s(t)$.

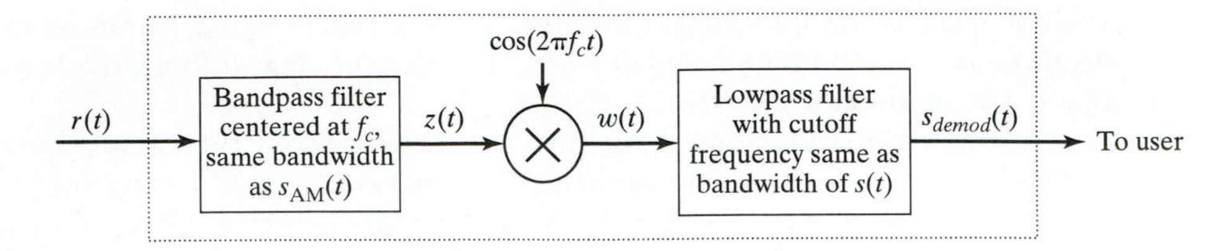

Figure 6-4 Coherent AM receiver.

Consider the circuit shown in Figure 6-4. As we established in Section 4.1, we can model the received signal as an attenuated transmitted signal plus noise

$$r(t) = \gamma s_{\mathrm{AM}}(t) + n(t) \tag{6.2}$$

where γ is a constant representing the attenuation of the channel and $n(t)$ represents additive white Gaussian noise.[1] The received signal first passes through a bandpass filter at the front end of the receiver. Since the bandwidth of the filter is the same as the bandwidth of the transmitted signal, the attenuated signal passes through the filter unaffected, but the out-of-band noise is eliminated. Thus, the signal at the output of the filter is

$$z(t) = \gamma s_{\mathrm{AM}}(t) + n_o(t) \tag{6.3}$$

where $n_o(t)$ represents the noise at the filter output. As we determined in Section 4.1.3, $n_o(t)$ has a Gaussian distribution, a flat power spectral density of $N_o/2$ volts^2/Hz within the passband of the filter, and zero power spectral density outside the passband of the filter. The power spectral density of $n_o(t)$ is shown in Figure 6-5. B represents the bandwidth of the original signal $s(t)$.

The filtered signal is then multiplied by the carrier reference, producing

$$\begin{aligned} w(t) &= [\gamma s_{\mathrm{AM}}(t) + n_o(t)]\cos(2\pi f_c t) \\ &= [\gamma A s(t)\cos(2\pi f_c t) + n_o(t)]\cos(2\pi f_c t) \\ &= \gamma A s(t)\cos^2(2\pi f_c t) + n_o(t)\cos(2\pi f_c t) \\ &= \frac{1}{2}\gamma A s(t) + \frac{1}{2}\gamma A s(t)\cos(4\pi f_c t) + n_o(t)\cos(2\pi f_c t) \end{aligned} \tag{6.4}$$

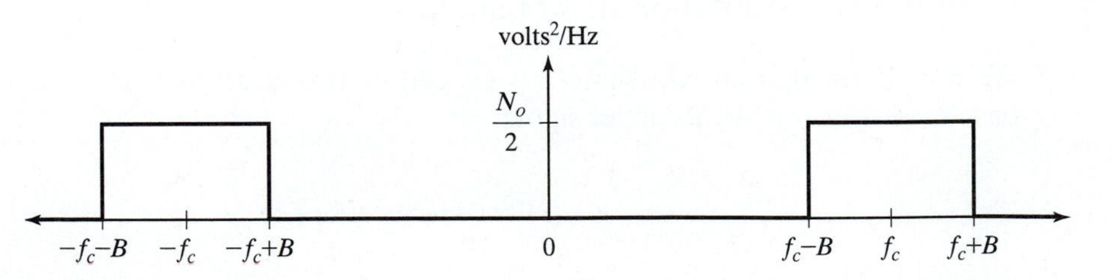

Figure 6-5 Average normalized power spectrum of thermal noise after bandpass filtering at the receiver.

[1]If you are studying analog systems before digital systems, you may want to pause here and read Sections 4.1.2 and 4.1.3, which discuss random variables and thermal noise.

The first term in Equation (6.4) is an attenuated version of original analog signal $s(t)$. This is exactly what we want: We can amplify the signal to remove the constant $A\gamma/2$. The second term is an attenuated version of the original analog signal frequency-shifted by $2f_c$. We can eliminate this term using a lowpass filter with the same bandwidth as $s(t)$. The third term represents the effects of noise; we will now examine this term in more detail.

As we established in Section 5.5.4, we can express the narrowband noise at the output of the bandpass filter using a quadrature representation

$$n_o(t) = \mathbf{W}(t)\cos(2\pi f_c t) + \mathbf{Z}(t)\sin(2\pi f_c t) \tag{5.62R}$$

where $\mathbf{W}(t)$ and $\mathbf{Z}(t)$ are uncorrelated, zero-mean, Gaussian random processes.[2] The power spectral densities of $\mathbf{W}(t)$ and $\mathbf{Z}(t)$ are shown in Figure 6-6.

Substituting Equation (5.62R) into the third term in Equation (6.4) and applying the appropriate trigonometric identities,

$$\begin{aligned} n_o(t)\cos(2\pi f_c t) &= [\mathbf{W}(t)\cos(2\pi f_c t) + \mathbf{Z}(t)\sin(2\pi f_c t)]\cos(2\pi f_c t) \\ &= \mathbf{W}(t)\cos^2(2\pi f_c t) + \mathbf{Z}(t)\sin(2\pi f_c t)\cos(2\pi f_c t) \\ &= \frac{1}{2}\mathbf{W}(t) + \frac{1}{2}\mathbf{W}(t)\cos(4\pi f_c t) + \frac{1}{2}\mathbf{Z}(t)\sin(0) + \frac{1}{2}\mathbf{Z}(t)\sin(4\pi f_c t) \\ &= \frac{1}{2}\mathbf{W}(t) + \frac{1}{2}\mathbf{W}(t)\cos(4\pi f_c t) + \frac{1}{2}\mathbf{Z}(t)\sin(4\pi f_c t) \end{aligned} \tag{6.5}$$

Substituting Equation (6.5) into Equation (6.4) produces

$$\begin{aligned} w(t) &= [\gamma s_{\text{AM}}(t) + n_o(t)]\cos(2\pi f_c t) \\ &= \frac{1}{2}\gamma A s(t) + \frac{1}{2}\gamma A s(t)\cos(4\pi f_c t) \\ &\quad + \frac{1}{2}\mathbf{W}(t) + \frac{1}{2}\mathbf{W}(t)\cos(4\pi f_c t) + \frac{1}{2}\mathbf{Z}(t)\sin(4\pi f_c t) \end{aligned} \tag{6.6}$$

As shown in Figure 6-4, this signal is then sent through a lowpass filter, eliminating the second, fourth, and fifth terms of Equation (6.6), producing

$$s_{demod}(t) = \frac{1}{2}\gamma A s(t) + \frac{1}{2}\mathbf{W}(t) \tag{6.7}$$

The analysis we've just completed is similar to the work we did in Chapters 4 and 5 when we developed and evaluated receivers for digital signals. The processed received signal $s_{demod}(t)$ has two terms: an attenuated version of the signal from the source,

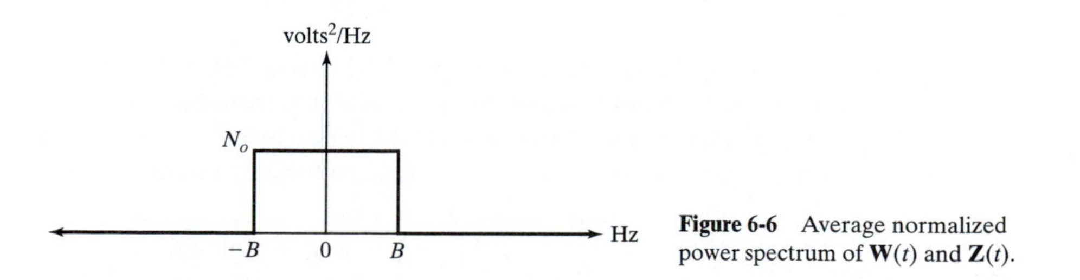

Figure 6-6 Average normalized power spectrum of $\mathbf{W}(t)$ and $\mathbf{Z}(t)$.

[2] If you are studying analog systems before digital systems, you may want to pause here and read Section 5.4, which discusses random processes.

$\frac{1}{2}\gamma As(t)$, which is what we want to present to the user; and a second term, $\frac{1}{2}\mathbf{W}(t)$, representing the processed noise. In our analysis of digital systems, the problem with the processed noise was that it was occasionally strong enough to overwhelm the signal, causing the receiver to output a "1" when a "0" had been transmitted or to output a "0" when a "1" had been transmitted. We measured the accuracy of a digital system by its probability of bit error, which was dependent on the modulation technique, the type of receiver employed, and the relative strength of the attenuated transmitted signal and the noise at the receiver. Measuring the effects of noise in an analog system requires a slightly different approach. Since the signal presented to the user is an analog waveform rather than a stream of "1"s and "0"s, we cannot use bit error rate to measure the accuracy of the received signal. Rather, we want to develop a measure of signal fidelity—a measure showing how badly the noise affects the perceived quality of the received signal. The parameter we will use to measure the quality of the received, processed analog signal is the *signal-to-noise ratio* (SNR), the ratio of the power of the demodulated signal in a noiseless environment to the power of the processed noise. SNR is not a new concept; in Chapters 4 and 5 we determined that this ratio was one of the key factors in determining bit error rate for digital systems.

Let's start by calculating the average normalized power of the demodulated signal in a noiseless environment. Let P_{trans} represent the average normalized power of the transmitted signal

$$s_{\text{AM}}(t) = As(t)\cos(2\pi f_c t) \tag{6.1R}$$

We can express P_{trans} as

$$\begin{aligned} P_{trans} &= \int_{-\infty}^{\infty} |S_{\text{AM}}(f)|^2\,df \\ &= \int_{-\infty}^{\infty} |\mathcal{F}\{As(t)\cos(2\pi f_c t)\}|^2\,df \\ &= A^2 \int_{-\infty}^{\infty} \left|\frac{1}{2}[S(f - f_c) + S(f + f_c)]\right|^2 df \\ &= \frac{A^2}{4}\int_{-\infty}^{\infty} |S(f - f_c) + S(f + f_c)|^2\,df \end{aligned} \tag{6.8}$$

where $S(f)$ represents the Fourier transform of the original signal from the source $s(t)$. In practical systems, the bandwidth of the original signal from the source is much lower than the carrier frequency, and so the spectra of the two terms $S(f - f_c)$ and $S(f + f_c)$ do not overlap. In this case, the cross-terms are zero for all frequencies and

$$\begin{aligned} P_{trans} &= \frac{A^2}{4}\int_{-\infty}^{\infty} |S(f - f_c) + S(f + f_c)|^2\,df \\ &= \frac{A^2}{2}P_s \end{aligned} \tag{6.9}$$

where P_s represents the average normalized power of the original signal from the source,

$$P_s = \int_{-\infty}^{\infty} |S(f)|^2\, df = \int_{-\infty}^{\infty} |S(f - f_c)|^2\, df = \int_{-\infty}^{\infty} |S(f + f_c)|^2\, df \tag{6.10}$$

If the AM-DSB-SC signal is transmitted through a noiseless channel, then the power of the demodulated signal would be the power associated with the first term of Equation (6.7), which we can calculate as

$$P_{demod,\ noiseless\ signal} = \frac{1}{4}\gamma^2 A^2 P_s = \frac{\gamma^2}{2} P_{trans} \tag{6.11}$$

Now let's calculate the average normalized power of the noise after it has passed through the receiver. This is the average normalized power associated with the second term of Equation (6.7). In examining Figure 6-6 and Equation (6.7), we can see that the average normalized power of the processed noise is $\frac{1}{4}N_o(2B)$. The SNR for the coherent AM receiver is thus

$$SNR_{coherent\ \mathrm{AM}} = \frac{\frac{\gamma^2}{2}P_{trans}}{\frac{1}{4}N_o(2B)} = \frac{\gamma^2 P_{trans}}{N_o B} \tag{6.12}$$

Coherent demodulation requires that we maintain a reference signal at the receiver that is in phase and frequency synchronization with the carrier signal at the transmitter. As we discussed in Chapter 5, this increases the complexity of the receiver. In Chapter 5 we also examined how accuracy was reduced if the receiver's reference was at a slightly different frequency than the carrier, or if the reference was out of phase with the carrier. Let's now examine the effects of loss of synchronization for a coherently demodulated analog AM system.

Suppose that the receiver's reference signal is at the same frequency as the carrier but is not phase synchronized. The reference signal can be represented as $\cos(2\pi f_c t + \phi)$, where ϕ represents the difference in phase between the carrier and the reference signal. This out-of-phase receiver is shown in Figure 6-7. The received signal is, of course, unchanged and is still represented as

$$r(t) = \gamma s_{\mathrm{AM}}(t) + n(t) \tag{6.2R}$$

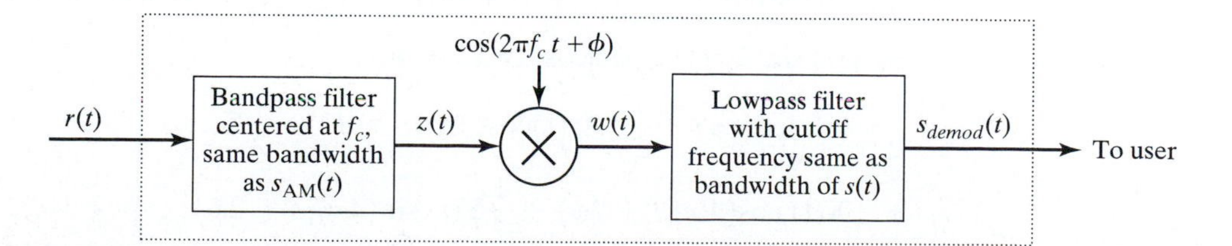

Figure 6-7 AM receiver with nonsynchronized phase.

Similarly, the output of the bandpass filter is unchanged and is still represented as

$$z(t) = \gamma s_{\text{AM}}(t) + n_o(t) \tag{6.3R}$$

The filtered signal is now multiplied by the out-of-phase carrier reference, producing

$$\begin{aligned} w(t) &= [\gamma s_{\text{AM}}(t) + n_o(t)]\cos(2\pi f_c t + \phi) \\ &= [\gamma A s(t)\cos(2\pi f_c t) + n_o(t)]\cos(2\pi f_c t + \phi) \\ &= \gamma A s(t)\cos(2\pi f_c t)\cos(2\pi f_c t + \phi) + n_o(t)\cos(2\pi f_c t + \phi) \\ &= \frac{1}{2}\gamma A s(t)\cos(\phi) + \frac{1}{2}\gamma A s(t)\cos(4\pi f_c t + \phi) + n_o(t)\cos(2\pi f_c t + \phi) \end{aligned} \tag{6.13}$$

The first term in Equation (6.13) is an attenuated version of original analog signal $s(t)$, similar to the first term in Equation (6.4), except that it is further attenuated by $\cos(\phi)$, the cosine of the phase difference between the carrier and the reference. The second term is an attenuated version of the original analog signal frequency-shifted by $2f_c$ and phase-shifted by ϕ. This term will be eliminated using the lowpass filter in the receiver. Let's now examine the third term, which represents the effects of noise.

As discussed in Chapter 5 and previously in this section, we can express the narrowband noise at the output of the bandpass filter using a quadrature representation

$$n_o(t) = \mathbf{W}(t)\cos(2\pi f_c t) + \mathbf{Z}(t)\sin(2\pi f_c t) \tag{5.62R}$$

where $\mathbf{W}(t)$ and $\mathbf{Z}(t)$ are uncorrelated, zero-mean, Gaussian random processes with power spectral densities as shown in Figure 6-6. We can shift the phase of the two components of $n_o(t)$, producing

$$n_o(t) = \mathbf{X}(t)\cos(2\pi f_c t + \phi) + \mathbf{Y}(t)\sin(2\pi f_c t + \phi) \tag{6.14}$$

The random processes $\mathbf{X}(t)$ and $\mathbf{Y}(t)$ are different from $\mathbf{W}(t)$ and $\mathbf{Z}(t)$, but as long as we retain quadrature (i.e., as long as the two components are 90° out of phase with each other), $\mathbf{X}(t)$ and $\mathbf{Y}(t)$ will have the same probability distributions and power spectral densities as $\mathbf{W}(t)$ and $\mathbf{Z}(t)$. Substituting Equation (6.14) into Equation (6.13),

$$\begin{aligned} w(t) &= [\gamma s_{\text{AM}}(t) + n_o(t)]\cos(2\pi f_c t + \phi) \\ &= \frac{1}{2}\gamma A s(t)\cos(\phi) + \frac{1}{2}\gamma A s(t)\cos(4\pi f_c t + \phi) + n_o(t)\cos(2\pi f_c t + \phi) \\ &= \frac{1}{2}\gamma A s(t)\cos(\phi) + \frac{1}{2}\gamma A s(t)\cos(4\pi f_c t + \phi) + [\mathbf{X}(t)\cos(2\pi f_c t + \phi) \\ &\quad + \mathbf{Y}(t)\sin(2\pi f_c t + \phi)]\cos(2\pi f_c t + \phi) \\ &= \frac{1}{2}\gamma A s(t)\cos(\phi) + \frac{1}{2}\gamma A s(t)\cos(4\pi f_c t + \phi) + \frac{1}{2}\mathbf{X}(t) \\ &\quad + \frac{1}{2}\mathbf{X}(t)\cos(4\pi f_c t + 2\phi) + \frac{1}{2}\mathbf{Y}(t)\sin(4\pi f_c t + 2\phi) \end{aligned} \tag{6.15}$$

As shown in Figure 6-7, this signal is then sent through a lowpass filter, eliminating the second, fourth, and fifth terms of Equation (6.15), producing

$$S_{demod,\ out\ of\ phase}(t) = \frac{1}{2}\gamma A s(t)\cos(\phi) + \frac{1}{2}\mathbf{X}(t) \tag{6.16}$$

Evaluating Equation (6.16),

$$\begin{aligned} SNR_{coherent\ \mathrm{AM},\ out\ of\ phase} &= \frac{\gamma^2\cos^2(\phi)P_{trans}}{N_o B} \\ &= [SNR_{coherent\ \mathrm{AM}}]\cos^2(\phi) \end{aligned} \tag{6.17}$$

We see that loss of phase synchronization reduces the SNR by the square of the cosine of the phase difference between the carrier and the reference.

Loss of frequency synchronization produces even worse effects. As you will show in Problem 6.4, if the reference signal is out of frequency synchronization by Δf,

$$\begin{aligned} S_{demod,\ not\ in\ freq.\ synch}(t) &= \frac{1}{2}\gamma A s(t)\cos(2\pi\Delta f t) + \frac{1}{2}\mathbf{X}(t)\cos(2\pi\Delta f t) \\ &\quad + \frac{1}{2}\mathbf{Y}(t)\sin(2\pi\Delta f t) \end{aligned} \tag{6.18}$$

The noise will retain the same power spectral density as it would with the synchronized receiver, but the amplitude of the signal will be distorted by the function $\cos(2\pi\Delta f t)$.

6.3 Noncoherent Demodulation of AM Signals

As we saw in Section 6.2, coherent demodulation of AM signals requires that the reference signal in the receiver be tightly frequency- and phase-synchronized with the carrier signal. This synchronization increases the complexity of the receiver. Is it possible to develop a noncoherent AM receiver, a receiver that does not need the synchronized reference signal?

In Figures 6-3a–c we observed that the original signal $s(t)$ is related to the envelopes of the AM signal: When $s(t)$ is positive, it follows the upper envelope of the modulated signal; when $s(t)$ is negative, it follows the lower envelope of the modulated signal. In Section 5.5.1 we developed a circuit for an envelope detector, a device capable of following the upper envelope of an ASK signal. Can the envelope detector be used to noncoherently demodulate an AM signal? The answer, as we will soon see, is yes, if we modify the AM signal.

The envelope detector circuit is shown in Figure 5-22a*R*. Briefly reviewing its theory of operation, consider a sine wave input to the circuit. During the first quarter cycle of the sine wave, the diode allows current to flow to the resistor and capacitor. As shown in Figure 5-22b*R*, the capacitor voltage $y(t)$ rises with $x(t)$ (its value will actually be $x(t)$ minus a small voltage drop across the diode) and the capacitor

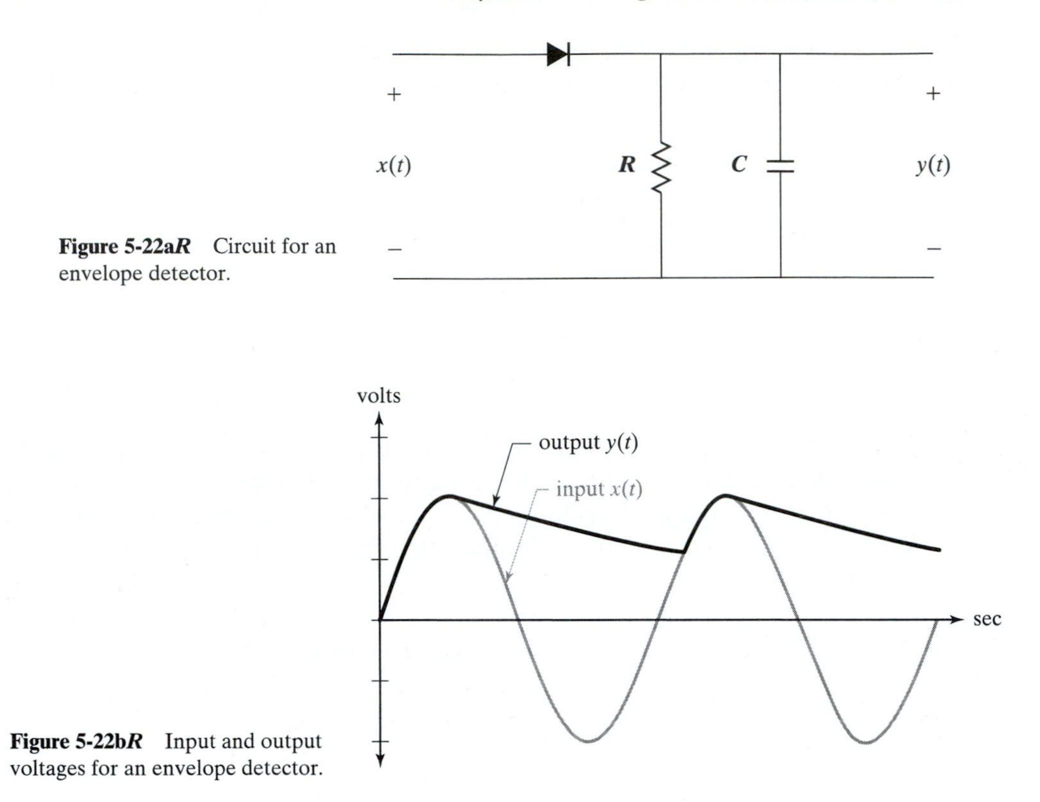

Figure 5-22a*R* Circuit for an envelope detector.

Figure 5-22b*R* Input and output voltages for an envelope detector.

starts charging. Early in the second quarter of the cycle, as $x(t)$ begins to drop, $y(t)$ becomes greater than $x(t)$, and the diode acts like an open circuit. This causes the capacitor to discharge energy through the resistor, so the capacitor voltage $y(t)$ begins to drop. The diode continues to act as an open circuit during the third and fourth quarters of the sinusoid's cycle, causing $y(t)$ to continue to drop. As the sine wave $x(t)$ begins its second cycle, the input voltage catches up to the output voltage and the process repeats.

If the carrier frequency is much larger than the highest significant frequency components in the original signal $s(t)$, then the envelope will not undergo large changes in amplitude between cycles of the carrier, and, with proper values of R and C, the envelope detector will be able to accurately track the upper envelope of the AM signal.

Unfortunately, as we saw in Figure 6-3c, the upper envelope follows the original signal $s(t)$ only when $s(t)$ is positive. When $s(t)$ is negative, the upper envelope follows $-s(t)$. Suppose, however, that prior to modulation we add a sufficiently large dc offset to the original signal so that the sum is always positive. Figure 6-8a shows the original signal $s(t)$. Note that its value is never less than -2 volts. Figure 6-8b shows the signal $s(t) + 2$ volts, which is always positive. Figure 6-8c shows the amplitude modulated signal $[s(t) + 2]\cos(200{,}000\pi t)$. As shown in Figure 6-8d, the upper envelope of this AM signal is always $s(t) + 2$ volts.

Suppose we now pass this AM signal through an envelope detector. The output of the envelope detector, shown as the thick black line in Figure 6-9a, approximates the upper envelope $s(t) + 2$ volts. The two-volt dc offset can be removed using an appropriate

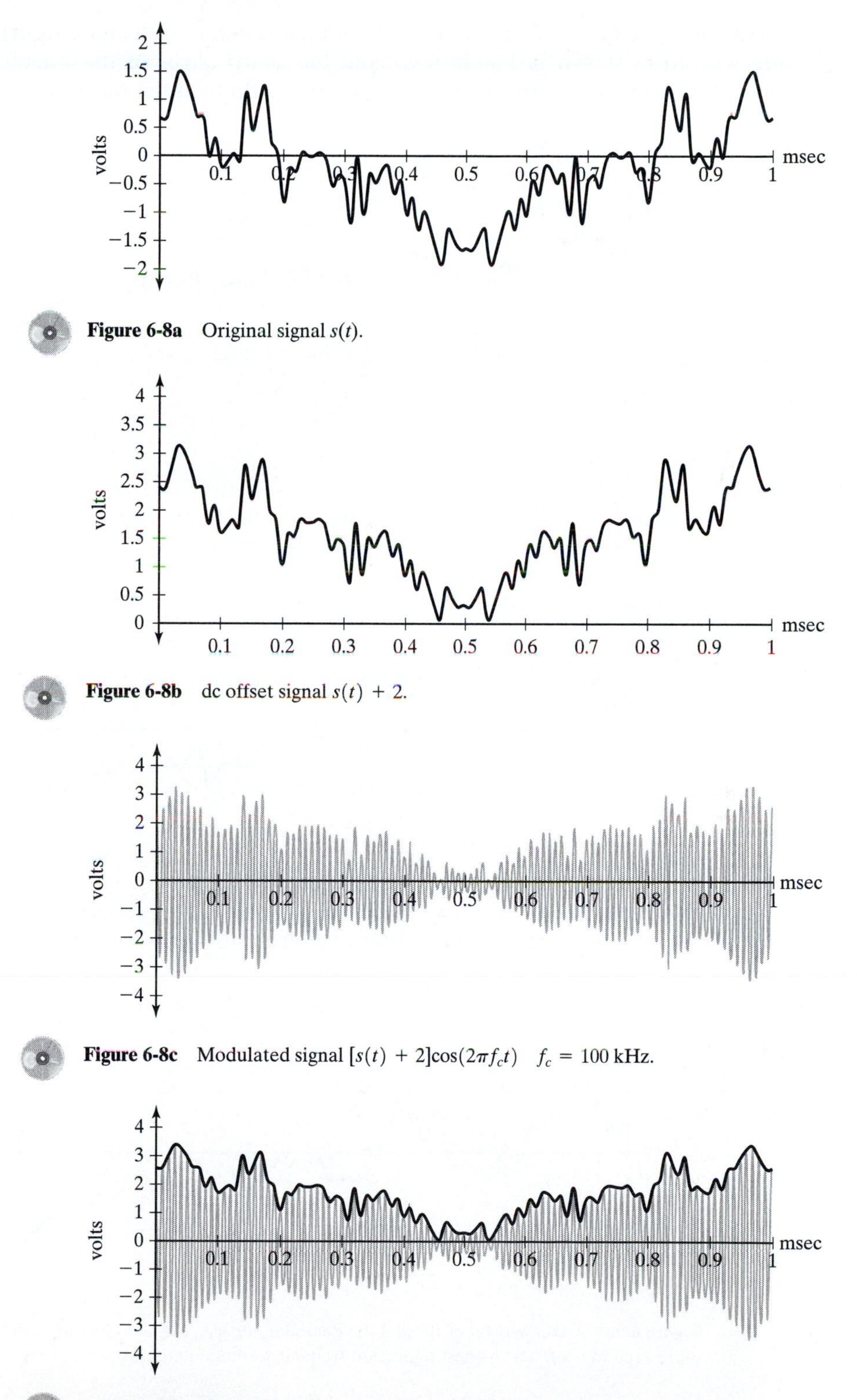

Figure 6-8a Original signal $s(t)$.

Figure 6-8b dc offset signal $s(t) + 2$.

Figure 6-8c Modulated signal $[s(t) + 2]\cos(2\pi f_c t)$ $f_c = 100$ kHz.

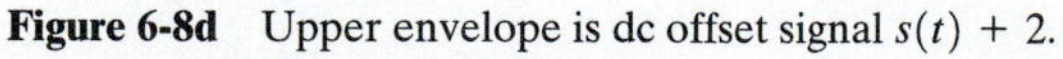

Figure 6-8d Upper envelope is dc offset signal $s(t) + 2$.

blocking capacitor (a capacitor in series with the envelope detector output), producing the waveform shown in Figure 6-9b, and the sharp edges of the waveform can be smoothed using an appropriate lowpass filter, resulting in the black waveform shown in Figure 6-9c. The original waveform $s(t)$ is shown in gray in Figure 6-9c. We see that

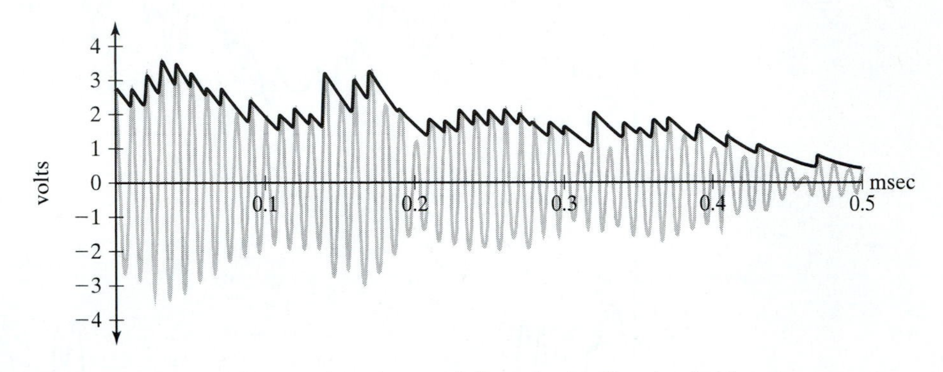

Figure 6-9a Output of the envelope detector follows the dc offset signal $s(t) + 2$.

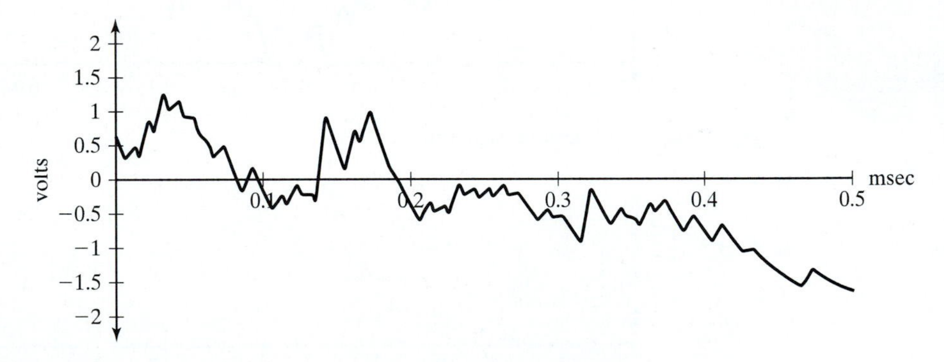

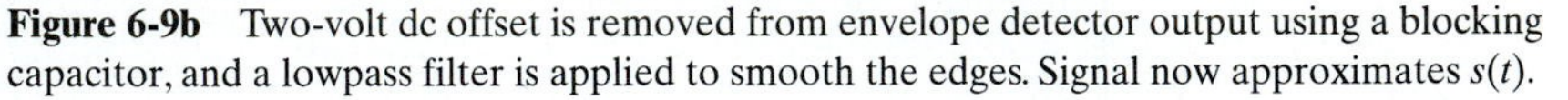

Figure 6-9b Two-volt dc offset is removed from envelope detector output using a blocking capacitor, and a lowpass filter is applied to smooth the edges. Signal now approximates $s(t)$.

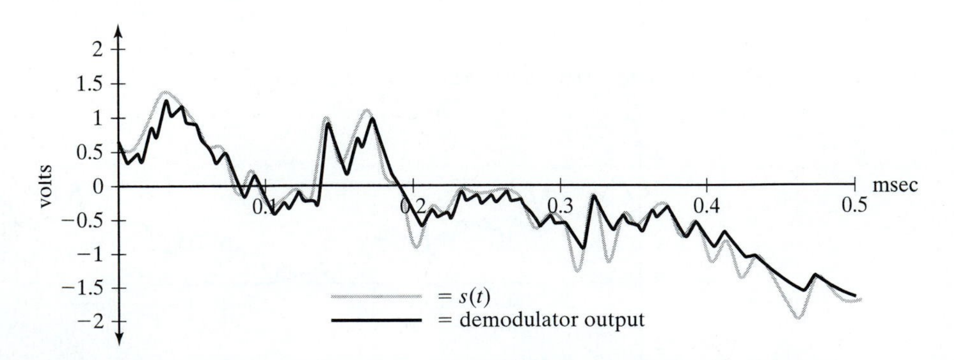

Figure 6-9c A comparison of $s(t)$ and the demodulated signal. Demodulation will be even more accurate if $f_c \gg$ highest significant frequency component of $s(t)$.

the envelope detector, coupled with a blocking capacitor and lowpass filter, provides effective demodulation for the dc-offset AM signal $[s(t) + 2]\cos(200{,}000\pi t)$. The demodulated signal will closely resemble $s(t)$, provided that the carrier frequency is much higher than the highest significant frequency components of $s(t)$. A circuit for a noncoherent AM demodulator is shown in Figure 6-10.

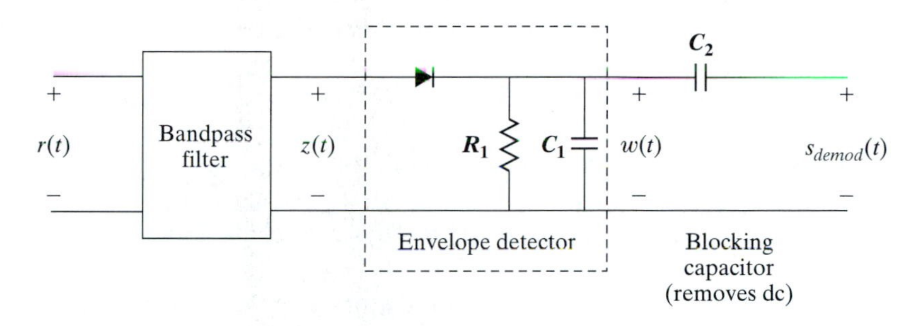

Figure 6-10 A noncoherent AM demodulator.

Let's investigate the trade-offs between coherent and noncoherent AM systems. By adding a dc offset prior to modulating the signal, we've simplified the receiver by not requiring synchronization. There is, however, a fairly significant cost to this dc offset. Figure 6-11 shows the spectrum of the transmitted signal $[s(t) + 2]\cos(200{,}000\pi t)$. Note that the spectrum is the same as the spectrum of the AM-DSB-SC signal in Figure 6-2 with an added discrete component at the carrier frequency. We can see the additional discrete component by decomposing the transmitted signal as

$$[s(t) + 2]\cos(200{,}000\pi t) = s(t)\cos(200{,}000\pi t) + 2\cos(200{,}000\pi t) \quad (6.19)$$

where the first term is the same as the AM-DSB-SC signal and the second term is a sinusoid at the carrier frequency with an amplitude equal to the dc offset. We call our new modulation technique *amplitude modulation-double sideband-carrier* (AM-DSB-C).

The discrete carrier component means that the AM-DSB-C signal requires more transmitted power than the AM-DSB-SC signal. As we shall soon see, this additional

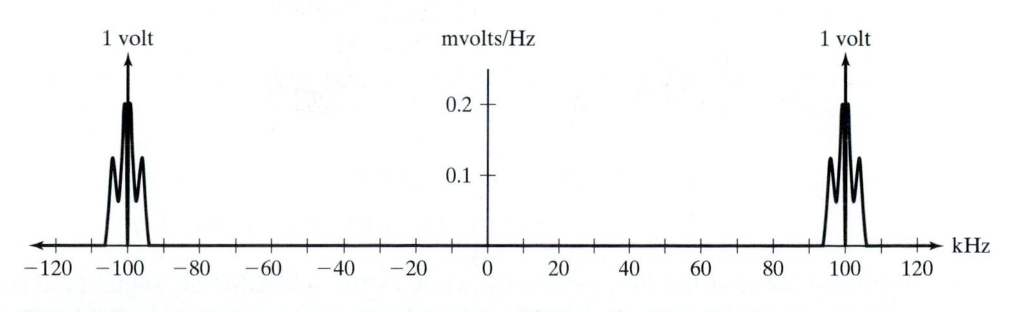

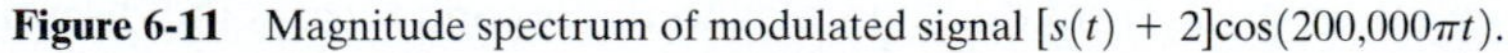

Figure 6-11 Magnitude spectrum of modulated signal $[s(t) + 2]\cos(200{,}000\pi t)$.

transmitted power, while necessary, does not improve the SNR of the demodulated signal. (We've seen this concept of "wasted" power before, when we examined unipolar versus bipolar PAM in Chapter 4 and ASK versus PSK in Chapter 5.) AM-DSB-C allows a simpler receiver but at the cost of requiring additional transmitted power. For many applications, including commercial AM radio systems, this is a very acceptable trade-off. The commercial AM station makes its money by selling advertising. The more listeners it has, the more the AM station can charge for a commercial. By making the receivers as inexpensive to purchase as possible, the AM station attracts more listeners.

Let's formalize our study of AM-DSB-C. Let $s(t)$ represent a general analog signal that we want to transmit. Let c be a constant representing a dc offset. The AM-DSB-C signal can be represented as

$$s_{\text{AM-DSB-C}}(t) = A[s(t) + c]\cos(2\pi f_c t) \tag{6.20}$$

where f_c is the frequency of the carrier signal, and A, again, is a scaling factor that we can choose to appropriately set the power level of the transmitted signal. The frequency domain representation of the transmitted signal is

$$\begin{aligned} S_{\text{AM-DSB-C}}(f) &= \mathcal{F}\{A[s(t) + c]\cos(2\pi f_c t)\} \\ &= A\mathcal{F}\{[s(t) + c]\cos(2\pi f_c t)\} \\ &= A\mathcal{F}\{s(t)\cos(2\pi f_c t)\} + A\mathcal{F}\{c\cos(2\pi f_c t)\} \\ &= \frac{A}{2}[S(f - f_c) + S(f + f_c)] + \frac{Ac}{2}[\delta(f - f_c) + \delta(f + f_c)] \end{aligned} \tag{6.21}$$

We obtain the power spectral density by squaring the magnitude of Equation (6.21), producing

$$G_{\text{AM-DSB-C}}(f) = \frac{A^2}{4}\left[|S(f - f_c)|^2 + |S(f + f_c)|^2\right] + \frac{A^2c^2}{4}[\delta(f - f_c) + \delta(f + f_c)] \tag{6.22}$$

The first term is the same as the power spectral density of the AM-DSB-SC signal, and the second term is the power produced by the discrete carrier.

We know that the addition of the discrete carrier allows us to use a simpler receiver (i.e., the envelope detector) but requires additional transmitted power compared with AM-DSB-SC. Let's now determine how much additional power is needed. Decomposing Equation (6.20),

$$\begin{aligned} s_{\text{AM-DSB-C}}(t) &= A[s(t) + c]\cos(2\pi f_c t) \\ &= As(t)\cos(2\pi f_c t) + Ac\cos(2\pi f_c t) \end{aligned} \tag{6.20R}$$

we again see that the first term is the same as the AM-DSB-SC signal (and therefore contains all the information from the source), and the second term is the discrete carrier signal. Let's call the first term the *information term* and the second term the *carrier term*. Since the information term is the same as the AM-DSB-SC signal, we can apply Equations (6.8) and (6.9) to determine the average normalized transmitted power of the information term,

$$P_{info\ term} = \frac{A^2}{2} P_s \tag{6.23}$$

Average normalized transmitted power of the carrier term is

$$P_{carrier\ term} = \frac{1}{T} \int_0^T [Ac \cos(2\pi f_c t)]^2 \, dt$$

$$= \frac{(Ac)^2}{2} \tag{6.24}$$

As we already discussed, $s(t) + c$ must always be nonnegative in order to avoid distortion. We can express this requirement as

$$c \geq |\min[s(t)]| \tag{6.25}$$

In a well-designed system, $s(t)$ will be conditioned prior to transmission so that it has no dc component and $|\min[s(t)]| \approx |\max[s(t)]|$. This conditioning is the analog equivalent of employing data compression prior to transmission (see Chapter 9). For such a signal,

$$c^2 \geq s^2(t) \quad \text{for all } t \tag{6.26}$$

and therefore

$$c^2 \geq P_s \tag{6.27}$$

Equations (6.23), (6.24), and (6.27) tell us that for AM-DSB-C

$$P_{carrier\ term} \geq P_{info\ term} \tag{6.28}$$

This means that in an AM-DSB-C system, the average power associated with transmitting the discrete carrier is at least 50% of the average power of the entire transmitted signal. The exact percentage of transmitted power required by the discrete carrier will depend on the signal $s(t)$ and the value of the dc offset c. Let's define the power efficiency η of an AM-DSB-C system as

$$\eta = \frac{P_{info\ term}}{P_{carrier\ term} + P_{info\ term}} = \frac{P_{info\ term}}{P_{trans\ \text{AM-DSB-C}}} \leq 0.5 \tag{6.29}$$

where $P_{trans\ \text{AM-DSB-C}}$ represents the total transmitted power of the AM-DSB-C signal. In one respect we can say that an AM-DSB-C system "wastes" at least half its transmitted power because the power associated with the carrier term does not carry any information. However, as we have seen, transmitting this extra power allows us to use much simpler equipment at the receiver.

The *modulation index, m*, is defined as

$$m = \frac{\max[s(t) + c] - \min[s(t) + c]}{\max[s(t) + c] + \min[s(t) + c]} \tag{6.30}$$

Modulation index is often used to describe the power efficiency of a particular AM-DSB-C system. As we have seen, the smaller the dc offset c, the less power is used transmitting the discrete carrier. However, we must choose c large enough to ensure that $s(t) + c$ is always nonnegative. As you will show in Problem 6.12, the larger the value of m, the greater the power efficiency; but if $m > 1$, then $\min[s(t) + c] < 0$, producing distortion. We therefore want to design our AM-DSB-C system so that m is relatively large but does not exceed 1.

In addition to requiring transmission of a discrete carrier component, noncoherent demodulation of AM-DSB-C differs from coherent demodulation of AM-DSB-SC because the envelope detector and the coherent receiver process the received noise differently. For the coherent AM receiver, we previously determined that

$$SNR_{coherent\ \mathrm{AM}} = \frac{\gamma^2 P_{trans}}{N_o B} \tag{6.12R}$$

Let's now calculate the SNR for the noncoherent receiver.

Referring back to Figure 6-10, we can express the received signal as

$$r(t) = \gamma s_{\mathrm{AM\text{-}DSB\text{-}C}}(t) + n(t) \tag{6.31}$$

where $n(t)$ is additive white Gaussian noise with a flat power spectral density of $N_o/2$. As we established in Section 6.2, the narrowband noise at the output of the bandpass filter can be represented in quadrature form as

$$n_o(t) = \mathbf{W}(t)\cos(2\pi f_c t) + \mathbf{Z}(t)\sin(2\pi f_c t) \tag{5.62R}$$

where $\mathbf{W}(t)$ and $\mathbf{Z}(t)$ are uncorrelated, zero-mean, Gaussian random processes with power spectral densities as shown in Figure 6-6. The signal at the input to the envelope detector is thus

$$\begin{aligned} z(t) &= \gamma s_{\mathrm{AM\text{-}DSB\text{-}C}}(t) + n_o(t) \\ &= \gamma s_{\mathrm{AM\text{-}DSB\text{-}C}}(t) + \mathbf{W}(t)\cos(2\pi f_c t) + \mathbf{Z}(t)\sin(2\pi f_c t) \\ &= \gamma A[s(t) + c]\cos(2\pi f_c t) + \mathbf{W}(t)\cos(2\pi f_c t) + \mathbf{Z}(t)\sin(2\pi f_c t) \\ &= \{\gamma A[s(t) + c] + \mathbf{W}(t)\}\cos(2\pi f_c t) + \mathbf{Z}(t)\sin(2\pi f_c t) \end{aligned} \tag{6.32}$$

Combining the sine and cosine terms to produce a single sinusoidal term with magnitude and phase,

$$\begin{aligned} z(t) &= \{\gamma A[s(t) + c] + \mathbf{W}(t)\}\cos(2\pi f_c t) + \mathbf{Z}(t)\sin(2\pi f_c t) \\ &= \mathbf{M}(t)\cos(2\pi f_c t + \Phi(t)) \end{aligned} \tag{6.33}$$

where

$$\mathbf{M}(t) = \sqrt{\{\gamma A[s(t) + c] + \mathbf{W}(t)\}^2 + \mathbf{Z}(t)^2} \tag{6.34}$$

and

$$\Phi(t) = \arctan\frac{\mathbf{Z}(t)}{\gamma A[s(t) + c] + \mathbf{W}(t)} \tag{6.35}$$

The envelope detector has been designed to output the envelope, or magnitude, of the input signal. Referring back to Figure 6-10,

$$w(t) = |\mathbf{M}(t)| \tag{6.36}$$

We can also use phasors to illustrate the effects of noise. Figure 6-12a shows the phasor representing the received signal in the absence of noise. For this noiseless case, the output of the envelope detector will be the magnitude of the phasor $\gamma A[s(t) + c]$. Figure 6-12b shows the effects of adding noise. As we've just discussed, the noise can be

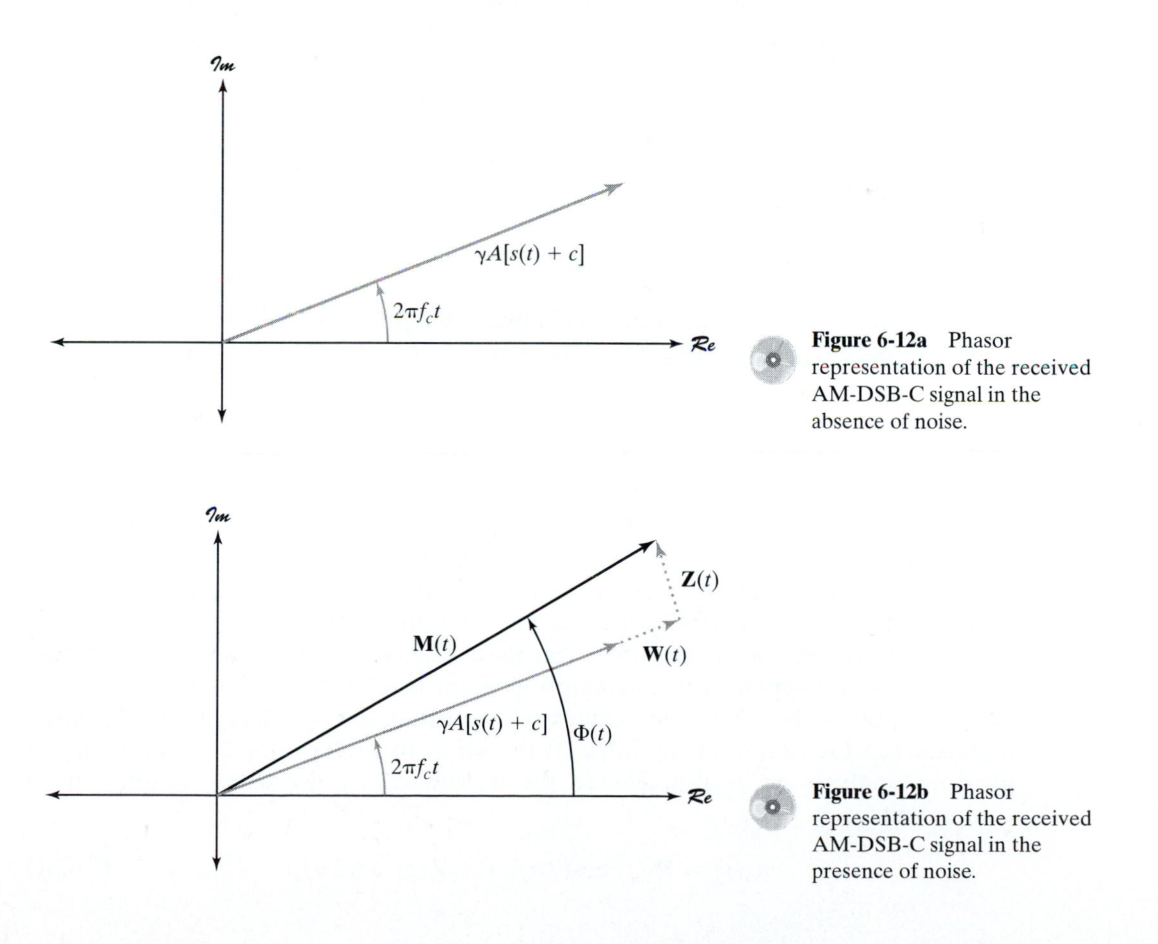

Figure 6-12a Phasor representation of the received AM-DSB-C signal in the absence of noise.

Figure 6-12b Phasor representation of the received AM-DSB-C signal in the presence of noise.

represented as two quadrature components, $\mathbf{W}(t)\cos(2\pi f_c t) + \mathbf{Z}(t)\sin(2\pi f_c t)$, shown in Figure 6-12b as phasors $\mathbf{W}(t)$ and $\mathbf{Z}(t)$. The phasor representing the received, noisy signal is the sum of the phasor representing the noiseless signal and the phasors representing the noise. This phasor is shown in Figure 6-12b as $\mathbf{M}(t)$. For a noisy signal, the output of the envelope detector will be the magnitude of the phasor $\mathbf{M}(t)$.

In practical AM-DSB-C systems, the signal is much stronger than the noise and we can approximate the envelope as

$$w(t) = |\mathbf{M}(t)| \approx \gamma A[s(t) + c] + \mathbf{W}(t) \tag{6.37}$$

(In other words, the $\mathbf{Z}(t)$ phasor, being much smaller than $A[s(t) + c]$ and acting at right angles to $\gamma A[s(t) + c] + \mathbf{W}(t)$, has little effect on the magnitude of the vector sum.) Referring again to Figure 6-10, the blocking capacitor C_2 removes the dc component, and thus

$$s_{demod}(t) \approx \gamma A s(t) + \mathbf{W}(t) \tag{6.38}$$

The first term in Equation (6.38) corresponds to the receiver output if there was no noise, and the second term corresponds to the processed noise. The power associated with the demodulated signal in a noiseless environment is

$$P_{demod,\, noiseless\ signal} = (\gamma A)^2 P_s \tag{6.39a}$$

Using Equations (6.9) and (6.23), and noting that the power in the information term of the AM-DSB-C signal is the same as the transmitted power of the AM-DSB-SC signal,

$$P_{demod,\, noiseless\ signal} = A^2\gamma^2 P_s = 2\gamma^2 P_{info\ term} \tag{6.39b}$$

The power associated with the processed noise is the power associated with the second term in Equation (6.38). Figure 6-6 shows that this power is $N_o(2B)$. Thus,

$$SNR_{\text{AM-DSB-C}} = \frac{2\gamma^2 P_{info\ term}}{2N_o B} = \frac{\gamma^2 P_{info\ term}}{N_o B} = \left(\frac{\gamma^2 P_{trans\ \text{AM-DSB-C}}}{N_o B}\right)\eta \tag{6.40}$$

Equation (6.40) makes intuitive sense: The bracketed term of Equation (6.40) looks like the SNR of the AM-DSB-SC system, and η represents the power efficiency, which, as we know from Equation (6.29), is the percentage of total transmitted power in an AM-DSB-C system corresponding to the information signal.

In our analysis we've assumed that the received signal is much stronger than the noise. What happens if this assumption is not true? Let's now consider how the system operates when the noise is much stronger than the signal. Figure 6-13a shows a phasor representation of the noise at the input to the envelope detector. In our previous analyses we've represented the narrowband noise using its quadrature components:

$$n_o(t) = \mathbf{W}(t)\cos(2\pi f_c t) + \mathbf{Z}(t)\sin(2\pi f_c t) \tag{5.62R}$$

As shown in Section 5.5.4, we can combine the quadrature terms, producing

$$\begin{aligned} n_o(t) &= \mathbf{W}(t)\cos(2\pi f_c t) + \mathbf{Z}(t)\sin(2\pi f_c t) \\ &= \mathbf{R}(t)\cos(2\pi f_c t + \Theta(t)) \end{aligned} \tag{5.63R}$$

where

$$\mathbf{R}(t) = \sqrt{\mathbf{W}(t)^2 + \mathbf{Z}(t)^2}$$

and

$$\Theta(t) = \arctan\left(\frac{\mathbf{Z}(t)}{\mathbf{W}(t)}\right)$$

We also saw in Section 5.5.4 that $\mathbf{R}(t)$ is Rayleigh distributed and that $\Theta(t)$ is uniformly distributed for $-\pi < \Theta \leq \pi$.

Referring back to Figure 6-10, the input to the envelope detector is

$$\begin{aligned} z(t) &= \gamma s_{\text{AM-DSB-C}}(t) + n_o(t) \\ &= \gamma A[s(t) + c]\cos(2\pi f_c t) + n_o(t) \end{aligned} \tag{6.41}$$

where the first term represents the signal and the second term represents the noise. Figure 6-13b shows the phasor representation of the envelope detector's input when the noise is much stronger than the signal. For this case, note that the envelope detector's output can be approximated as

$$w(t) = |\mathbf{M}(t)| \approx |\mathbf{R}(t)| + \gamma A[s(t) + c]\cos\Theta(t) \tag{6.42}$$

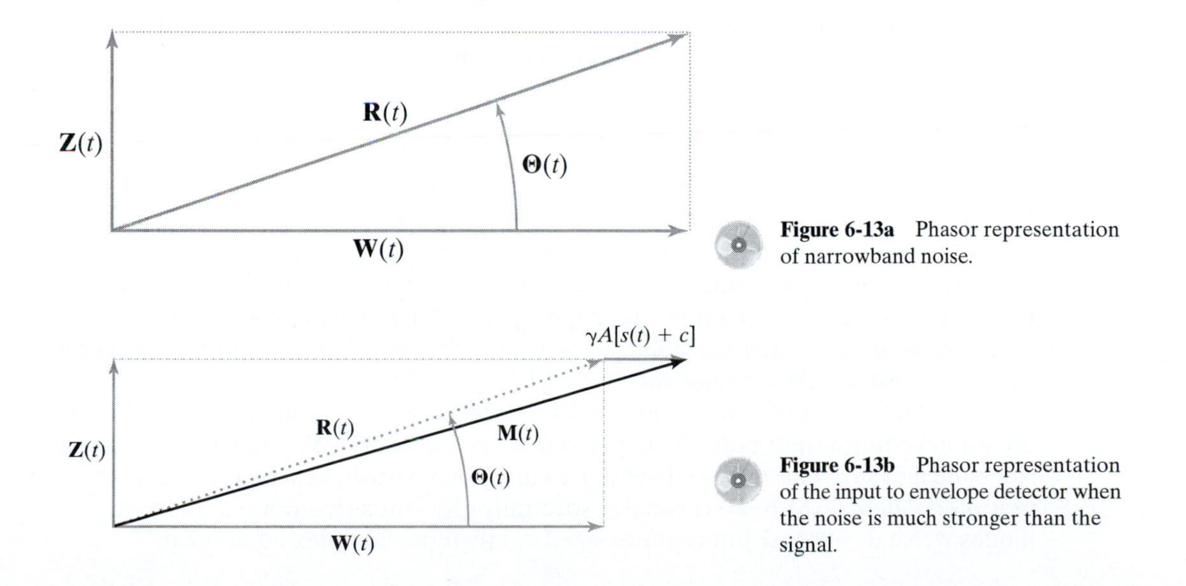

Figure 6-13a Phasor representation of narrowband noise.

Figure 6-13b Phasor representation of the input to envelope detector when the noise is much stronger than the signal.

Referring again to Figure 6-10, the blocking capacitor C_2 removes the dc component, and thus

$$s_{demod}(t) \approx |\mathbf{R}(t)| + \gamma As(t)\cos\Theta(t) \qquad (6.43)$$

In examining Equation (6.43), we note that the signal is badly affected by the noise in two different ways. First, the signal is overwhelmed by the noise, since the magnitude of the noise, represented by the first term in Equation (6.42), is much greater than the processed signal $\gamma As(t)$. Second, the amplitude of the processed signal is distorted because it is multiplied by the random factor $\cos\Theta(t)$, where $\Theta(t)$ is the phase of the noise. This second phenomenon is known as *capture*.

The combination of noise both overwhelming and capturing the processed signal produces a *threshold effect* that is not present in coherent demodulation. When the SNR at the receiver is above a certain level (called the *threshold*), noise causes a slight degrading of the demodulated signal, as described in Equation (6.40). However, once the SNR drops below the threshold, the demodulated signal degrades rapidly as the noise gets relatively stronger. This degradation occurs because the assumptions in Equation (6.40) no longer hold, and because capture begins to take effect. This threshold is normally defined as the SNR producing a 99% probability that the magnitude of the noise (described by the Rayleigh-distributed random variable $\mathbf{R}(t)$) is less than the magnitude of the received signal. This definition corresponds to an SNR of approximately 10 dB [Shanmugam (6.1)]. For many applications, including commercial music and voice transmission, SNR must be considerably higher than the threshold to produce a demodulated signal of acceptable quality.

We can now see the trade-offs involved in coherent-versus-noncoherent AM systems more clearly. The noncoherent systems require considerably more transmitted signal power, both because of the discrete carrier component of the AM-DSB-C signal and because of the threshold effect at the receiver. The noncoherent systems, however, have much simpler receivers. Let's now explore single sideband and vestigial sideband AM systems. These systems require more complex equipment than AM-DSB-SC or AM-DSB-C, but significantly less power and bandwidth.

6.4 Single Sideband and Vestigial Sideband AM Systems

The amplitude modulation techniques we've studied so far—AM-DSB-SC and AM-DSB-C—require twice the bandwidth of the original analog information signal. As we saw in Figures 6-2 and 6-11, these modulation techniques produce magnitude spectra with two symmetric sidebands (an upper sideband and a lower sideband). These sidebands are redundant: If we know the shape of one sideband, then we can determine the shape of the other. Either sideband, all by itself, contains all of the information necessary to reconstruct the original analog signal.

Theoretically, our observation that the sidebands are symmetric means that we do not have to transmit both the upper and lower sidebands: We can transmit just one sideband and save half the bandwidth. This approach, known as *single sideband amplitude modulation* (AM-SSB) is twice as spectrally efficient as the double sideband techniques we've developed, but requires significantly more complex equipment.

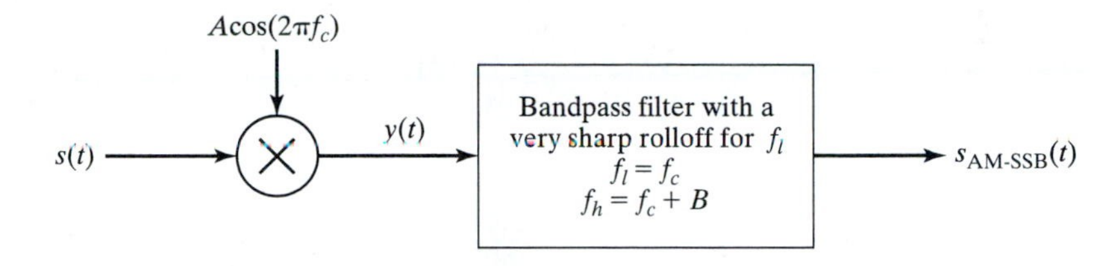

Figure 6-14 AM-SSB transmitter (upper sideband).

Figure 6-14 shows an AM-SSB transmitter. The analog information signal $s(t)$ is mixed with the carrier to produce an AM-DSB-SC signal $y(t)$, which is then passed through a bandpass filter ($f_l = f_c, f_h = f_c + B$) with a very sharp low-frequency rolloff to eliminate the lower sideband. The rolloff for f_h does not need to be sharp, since the upper sideband is transmitted. (Note that the decision to transmit the upper sideband instead of the lower sideband was chosen arbitrarily. We could just as easily eliminate the upper sideband and transmit the lower sideband by changing the cutoff frequencies of the bandpass filter and requiring the upper frequency rolloff to be sharp.) Figures 6-15a, b, and c show the spectra of a typical analog information signal $s(t)$, the output of the mixer $y(t)$, and the AM-SSB signal, respectively. For Figure 6-15c, $f_c = 50$ kHz, $f_l = 50$ kHz, and $f_h = 56$ kHz.

Figure 6-16 shows a coherent AM-SSB receiver. To understand how this receiver works, let's consider the transmitted AM-SSB signal whose magnitude spectrum is shown in Figure 6-15c. This signal arrives at the receiver and is bandpass filtered to eliminate out-of-band noise. The filtered signal $z(t)$ is then multiplied by a reference signal that is phase- and frequency-synchronized with the carrier, producing the signal $w(t)$, whose spectrum is shown in Figure 6-15d. Note that the spectrum of $w(t)$ includes an attenuated spectrum of the original signal $s(t)$ and higher frequency components in the frequency band around $2f_c$. Using a lowpass filter to eliminate the higher frequency components in $z(t)$, the output of the receiver, $s_{demod}(t)$, has the spectrum shown in Figure 6-15e. In comparing Figure 6-15a (the original analog signal) and Figure 6-15e, we see that the demodulated signal is an attenuated version of the original analog signal. SNR for an AM-SSB system is the same as for an AM-DSB-SC system using coherent demodulation, since the AM-SSB system eliminates both the signal power and the noise power from one of the sidebands.

As we've seen, AM-SSB systems require only half the bandwidth of AM-DSB-SC and AM-DSB-C systems, but require more complex equipment. The filter required for the AM-SSB modulator must have a very sharp rolloff, since one of the sidebands must be completely eliminated, while the other must pass through the filter.[3] The nature of the AM-SSB signal also makes it difficult for the receiver to generate a reference signal that is phase- and frequency-synchronized to the carrier.

Vestigial sideband amplitude modulation (AM-VSB) offers a compromise between the bandwidth efficiency of AM-SSB and the simpler equipment used in

[3] Other techniques exist for generating AM-SSB signals, but these techniques also have drawbacks. Many are based on observing that an AM-SSB signal can be deconstructed into in-phase and quadrature components. See Shanmugam [6.1] and Taub and Schilling [6.2].

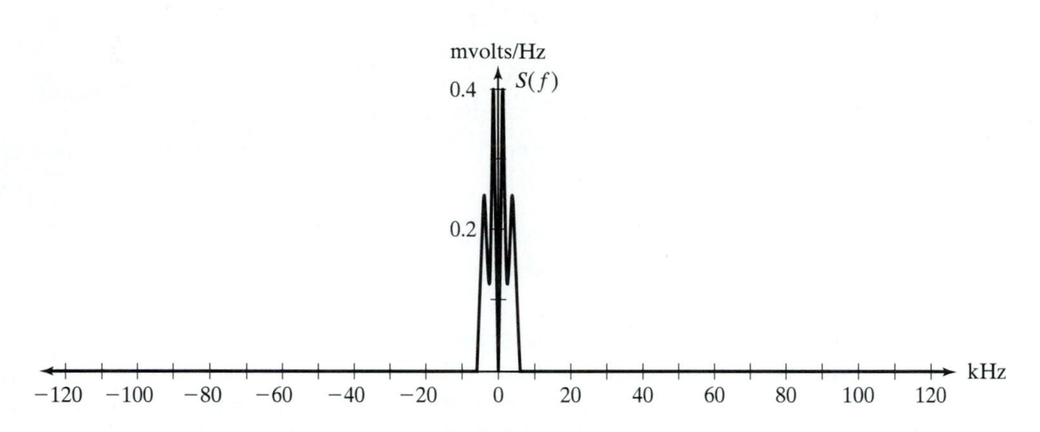

Figure 6-15a Magnitude spectrum of analog signal from source.

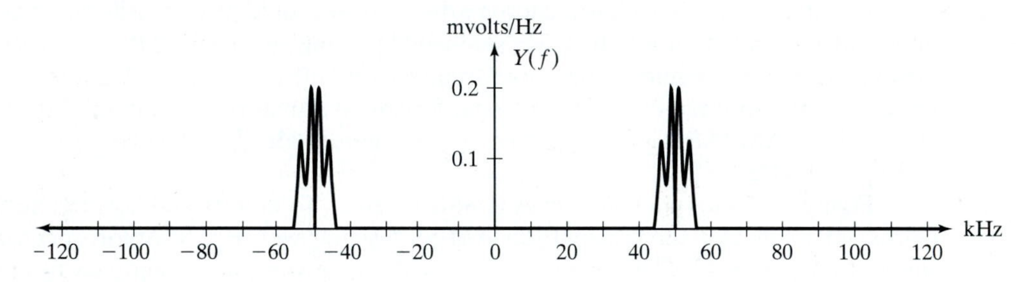

Figure 6-15b Spectrum after mixing with a 50 kHz carrier (an AM-DSB-SC signal).

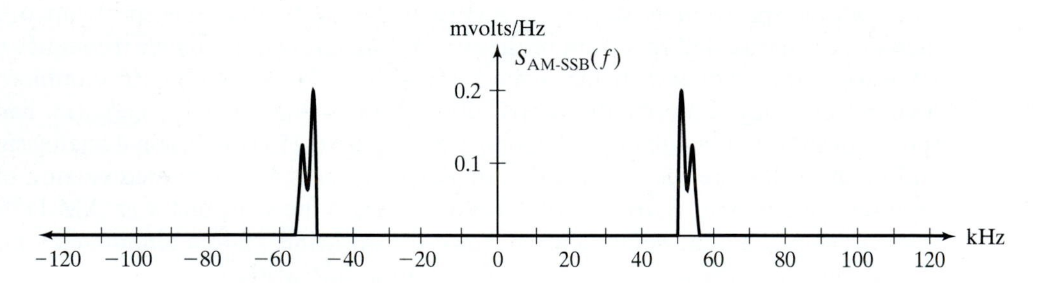

Figure 6-15c Spectrum of transmitted AM-SSB signal.

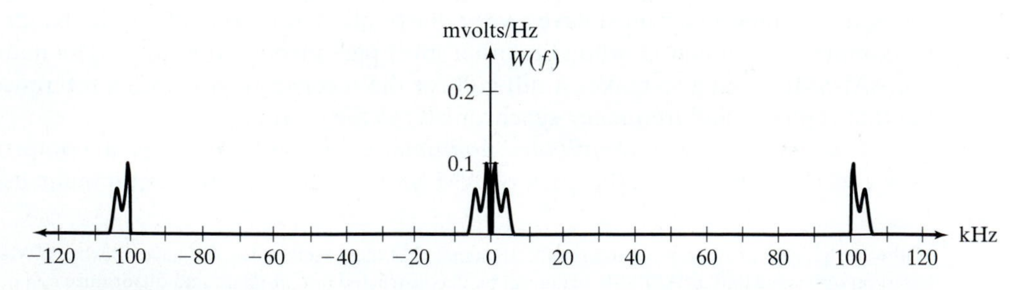

Figure 6-15d Spectrum of received signal after bandpass filtering and mixing.

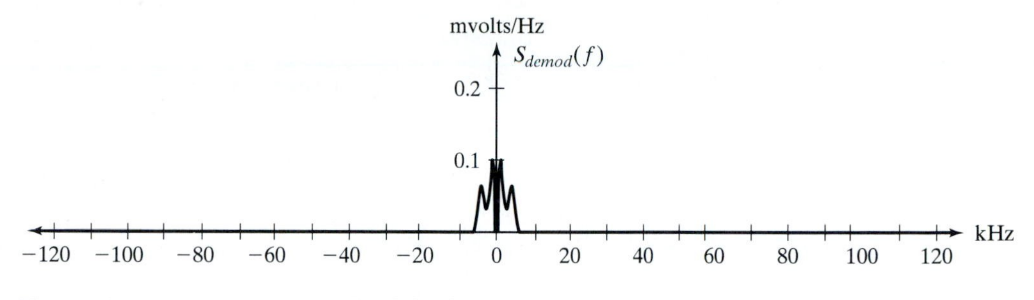

Figure 6-15e Spectrum of demodulated signal.

double sideband AM systems (AM-DSB-SC and AM-DSB-C). Consider the AM-SSB transmitter shown in Figure 6-14, which transmits only the upper sideband. Suppose we relax the requirements on the filter so that a steep rolloff at the lower cutoff frequency ($f_l = f_c$) is no longer needed. All that we will require is a transfer function with a linear phase response and a symmetric rolloff around f_c over the frequency band $f_c - B < f < f_c + B$, where B is the bandwidth of the original analog signal $s(t)$. Such a filter is much simpler to implement than a filter with a very sharp rolloff.

A typical transfer function for a VSB filter is shown in Figure 6-17. The value of Δ is chosen to be a small percentage of the bandwidth B of the original analog signal

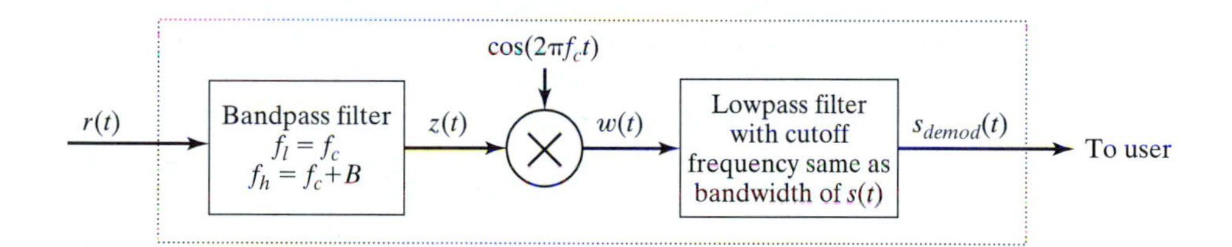

Figure 6-16 Coherent AM-SSB receiver (upper sideband).

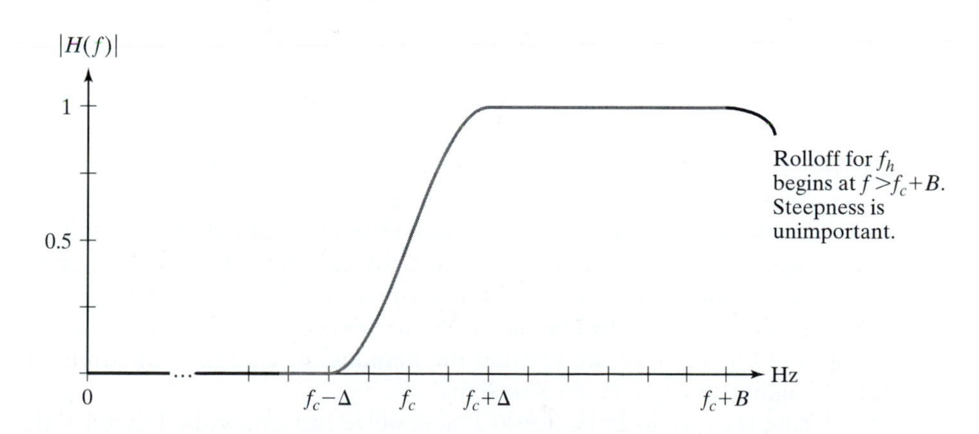

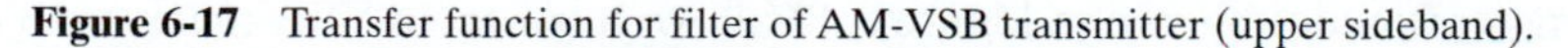
Figure 6-17 Transfer function for filter of AM-VSB transmitter (upper sideband).

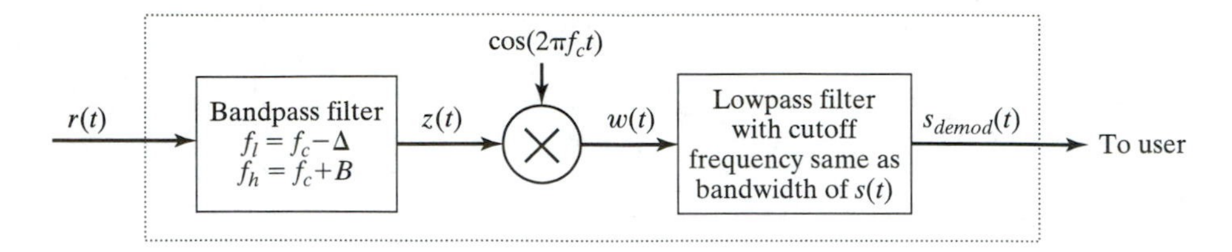

Figure 6-18 Coherent AM-VSB receiver (upper sideband).

(trade-offs in selecting the value of Δ will be discussed shortly). Using this filter, the AM-VSB transmitter transmits a small portion (or *vestige*) of the lower sideband along with the upper sideband, which is why the modulation technique is named "vestigial sideband." Note that the upper sideband is distorted for $f_c < f < f_c + \Delta$ and that the transmitted signal has a bandwidth of $B + \Delta$, larger than an AM-SSB signal but smaller than a double sideband signal.

Suppose we use the coherent receiver shown in Figure 6-18 to demodulate the VSB signal. Note that this receiver is similar to the coherent AM-SSB receiver shown in Figure 6-16, except that the bandpass filter at the input is adjusted for the wider bandwidth of the AM-VSB signal.

We may initially believe that the receiver will produce a distorted replica of the original information signal $s(t)$ because the filter in Figure 6-17 introduces a vestige of the lower sideband and distorts the upper sideband before transmission. If we look closer, however, we will see that the symmetry of the VSB filter allows the receiver to compensate for the distortion. To see this compensation, consider an information signal $s(t)$ with a magnitude spectrum as shown in Figure 6-19a. The AM-VSB transmitter multiplies the signal $s(t)$ with the carrier, producing the AM-DSB-SC signal whose magnitude spectrum is shown in Figure 6-19b (for this example, $f_c = 50$ kHz). This signal then passes through the VSB filter shown in Figure 6-17, producing a transmitted signal with the magnitude spectrum shown in Figure 6-19c (the reason for the shading in Figure 6-19c will be apparent shortly). The receiver eliminates the out-of-band noise and then multiplies the received signal with a reference signal (phase- and frequency-synchronized with the carrier), producing a signal $w(t)$ with the spectrum shown in Figure 6-19d. In inspecting Figure 6-19d, we see that the multiplication has caused two portions of the frequency spectrum to overlap in the frequency band corresponding to the original signal (dc-20 kHz in this example). Adding the two overlapping components (and remembering that the VSB filter has a linear phase response), we can see that the vestige of the lower sideband compensates exactly for the distortion in the upper sideband: When the two portions of the spectra in the dc-20 kHz frequency band are added together, the sum produces the spectrum of the original analog information signal (with attenuation). Using a lowpass filter to eliminate the higher frequency components in $w(t)$, the output of the receiver, $s_{demod}(t)$, has the spectrum shown in Figure 6-19e. In comparing Figure 6-19a (the original analog information signal) and Figure 6-19e, we see that the demodulated signal is an attenuated version of the original analog information signal.

Using the spectra in Figures 6-19a–e, we've just established that a VSB filter with the transfer function shown in Figure 6-17 allows a coherent VSB receiver to reconstruct

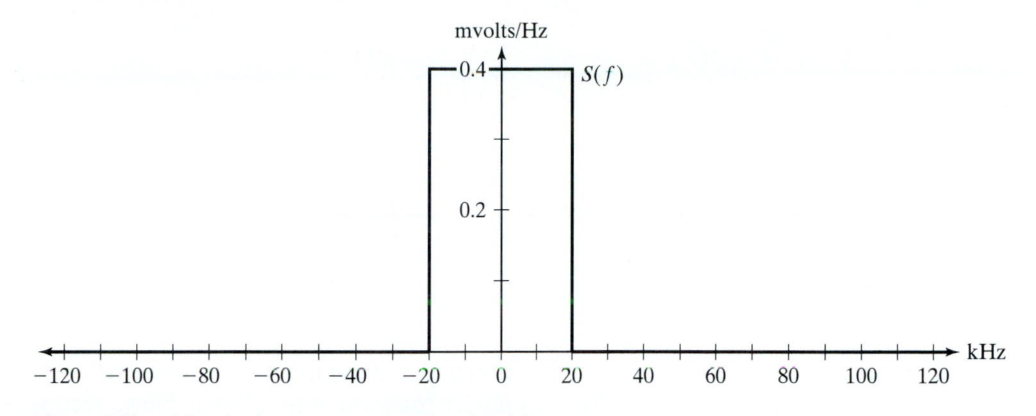

Figure 6-19a Magnitude spectrum of analog signal from source.

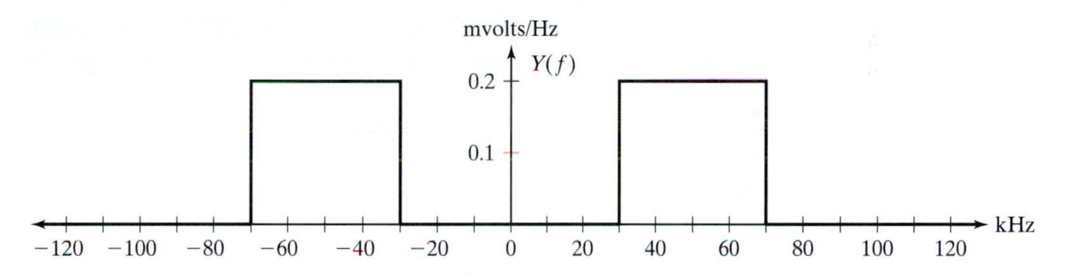

Figure 6-19b Spectrum after mixing with a 50 kHz carrier (an AM-DSB-SC signal).

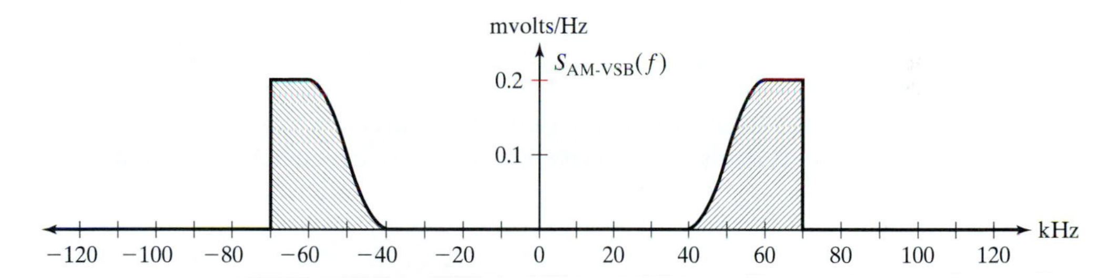

Figure 6-19c Spectrum of transmitted AM-VSB signal.

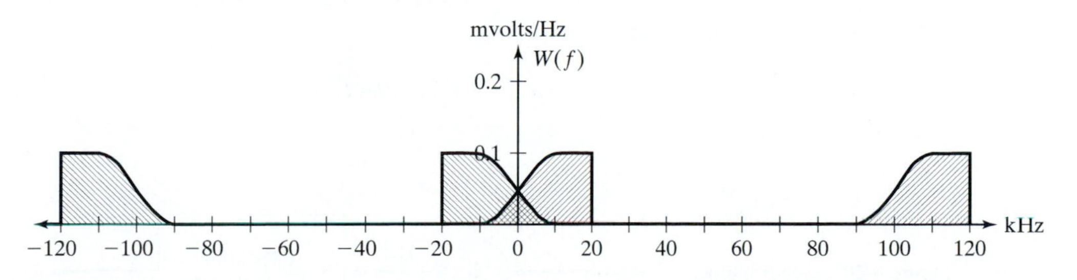

Figure 6-19d Spectrum of received signal after bandpass filtering and mixing. Note overlap of low frequency components.

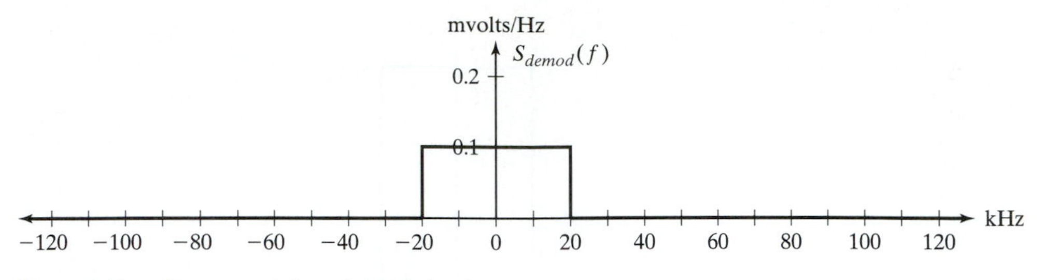

Figure 6-19e Spectrum of demodulated signal.

the original waveform without distortion. Let's mathematically prove that the more general conditions for a VSB filter (a transfer function with a linear phase response and a symmetric rolloff around f_c over the frequency band $f_c - B < f < f_c + B$) are sufficient for distortionless demodulation of the received signal.

Figure 6-20 shows the general block diagram of an AM-VSB transmitter. The analog information signal from the source, $s(t)$, is mixed with the carrier, producing

$$y(t) = s(t)\cos(2\pi f_c t) \tag{6.44}$$

Expressing $y(t)$ in the frequency domain,

$$Y(f) = \mathcal{F}\{s(t)\cos(2\pi f_c t)\} = \frac{1}{2}[S(f - f_c) + S(f + f_c)] \tag{6.45}$$

The signal is now passed through the VSB filter, producing the transmitted signal

$$S_{\text{AM-VSB}}(f) = \frac{1}{2}[S(f - f_c) + S(f + f_c)]H(f) \tag{6.46}$$

where $H(f)$ is the transfer function of the VSB filter.

Assuming no noise, the received signal is an attenuated version of the transmitted signal

$$r(t) = \gamma s_{\text{AM-VSB}}(t) \tag{6.47}$$

In examining Figure 6-18, the schematic for a coherent VSB receiver, we see that the signal is filtered and then mixed with a reference signal that is phase- and frequency-synchronized with the carrier, producing

$$w(t) = r(t)\cos(2\pi f_c t) \tag{6.48}$$

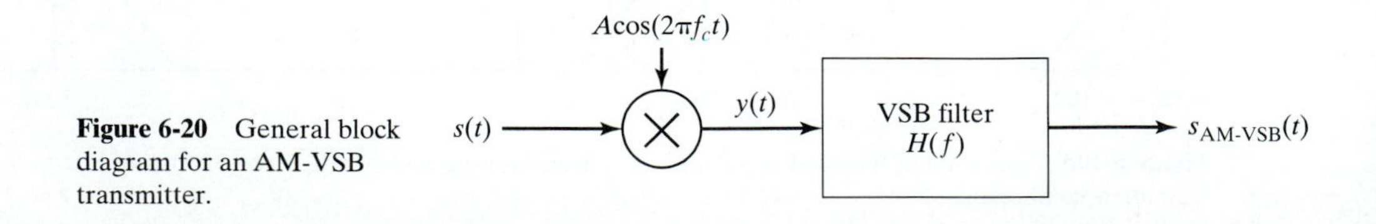

Figure 6-20 General block diagram for an AM-VSB transmitter.

Expressing $w(t)$ in the frequency domain,

$$\begin{aligned}
W(f) &= \mathcal{F}\{r(t)\cos(2\pi f_c t)\} \\
&= \frac{1}{2}[R(f - f_c) + R(f + f_c)] \\
&= \frac{\gamma}{2}[S_{\text{AM-VSB}}(f - f_c) + S_{\text{AM-VSB}}(f + f_c)] \\
&= \frac{\gamma}{4}\{[S(f - 2f_c) + S(f)]H(f - f_c) \\
&\quad + [S(f) + S(f + 2f_c)]H(f + f_c)\} \\
&= \frac{\gamma}{4}\{S(f - 2f_c)H(f - f_c) + S(f)[H(f - f_c) + H(f + f_c)] \\
&\quad + S(f + 2f_c)H(f + f_c)\}
\end{aligned} \tag{6.49}$$

This signal is then lowpass filtered, removing the first and third terms and leaving

$$S_{demod}(f) = \frac{\gamma}{4}S(f)[H(f - f_c) + H(f + f_c)] \tag{6.50}$$

SNR for an AM-VSB system with coherent demodulation is the same as SNR for an AM-DSB-SC system or an AM-SSB system, since the filtering has the same effect on both signal and noise. Distortionless demodulation is accomplished if

$$S_{demod}(f) = kS(f) \tag{6.51}$$

where k is any positive constant (i.e., the demodulated signal is a scaled version of the transmitted signal). Equation (6.51) is satisfied if

$$H(f - f_c) + H(f + f_c) = k \quad 0 \leq |f| \leq B \tag{6.52}$$

where k is any positive constant and B is the bandwidth of $S(f)$. Equation (6.52) corresponds to the condition that $H(f)$ be symmetric around f_c over the frequency band $f_c - B < f < f_c + B$).

As we've seen, AM-VSB is simpler to implement than AM-SSB because the filter in the AM-VSB transmitter is easier to build, but AM-VSB requires more bandwidth. It is also possible to insert a carrier in the transmitted AM-VSB signal (in a manner similar to AM-DSB-C) and then use a simple envelope detector instead of a coherent receiver for demodulation. This further simplifies the equipment, but it introduces distortion. The distortion can be reduced by increasing Δ (thereby increasing bandwidth) or by increasing the strength of the carrier relative to the rest of the transmitted signal (thereby increasing transmitted power). As always, we have trade-offs (for more details, see Shanmugam [6.1]). Television stations in North America transmit their signals using AM-VSB with a carrier.

6.5 Frequency Modulation and Phase Modulation

Are there other methods of analog bandpass modulation besides AM? If so, what are their trade-offs? Might there be better methods than AM for certain applications? Consider a typical sinusoidal carrier signal.[4]

$$x(t) = A\cos(2\pi f_c t + \phi) \tag{5.10R}$$

This sinusoid has three independent parameters: amplitude, frequency, and phase angle. We've already developed analog bandpass modulation techniques based on changing amplitude. Let's now consider the other two parameters. We can transmit a baseband analog signal $s(t)$ across a bandpass channel by modulating phase angle

$$s_{\mathrm{PM}}(t) = A\cos[2\pi f_c t + \alpha s(t)] \tag{6.53}$$

where α is a constant. This modulation technique is called *phase modulation* (PM). We can also transmit a baseband analog signal $s(t)$ across a bandpass channel by modulating frequency

$$s_{\mathrm{FM}}(t) = A\cos\{2\pi[f_c + ks(t)]t + \phi\} \tag{6.54}$$

where k is a constant. This modulation technique is called *frequency modulation* (FM).

In order to examine PM and FM further, we need to carefully define a few terms. We already know some of these terms, but we need to be sure that we understand exactly what they mean.

The *phase angle* of a sinusoidal signal is the angle of the argument of the sinusoid at time $t = 0$. We use the symbol ϕ to denote phase angle. As you can see, we used the proper notation in Equation (5.10R).

The *instantaneous phase* of a sinusoidal signal is the angle of the argument of the sinusoid at time t. We will use the symbol $\Psi(t)$ to denote instantaneous phase. As an example, the instantaneous phase of the unmodulated carrier signal in Equation (5.10R) is

$$\Psi_{carrier}(t) = 2\pi f_c t + \phi \tag{6.55}$$

The instantaneous phase of the PM signal in Equation (6.53) is

$$\Psi_{\mathrm{PM}}(t) = 2\pi f_c t + \alpha s(t) \tag{6.56}$$

The instantaneous phase of the FM signal in Equation (6.54) is

$$\Psi_{\mathrm{FM}}(t) = 2\pi[(f_c + ks(t)]t + \phi \tag{6.57}$$

[4] In Chapter 5 we expressed the carrier signal as $A\sin(2\pi f_c t + \theta)$. The form shown here is equivalent, since both θ and ϕ represent arbitrary phase angles.

Often, the instantaneous phase is just called the *angle* of the signal.

The *instantaneous frequency* of a sinusoidal signal is the time rate of change of the signal's instantaneous phase. Denoting instantaneous frequency as $f(t)$,

$$f(t) = \frac{d\Psi(t)}{dt} \tag{6.58}$$

The instantaneous frequency of the unmodulated carrier signal in Equation (5.10R) is

$$\begin{aligned} f_{carrier}(t) &= \frac{d\Psi_{carrier}(t)}{dt} = \frac{d}{dt}\{2\pi f_c t + \phi\} \\ &= 2\pi f_c \text{ radians per second, or } f_c \text{ Hz} \end{aligned} \tag{6.59}$$

Using Equation (6.58), we can also express instantaneous phase as

$$\Psi(t) = \int_{-\infty}^{t} f(\lambda)d\lambda = \int_{0}^{t} f(\lambda)d\lambda + \phi \tag{6.60}$$

where λ is just a variable of integration. Remember that ϕ represents phase angle, which can also be defined as $\Psi(0)$.

We need to impose a few practical limits on instantaneous frequency and instantaneous phase. To avoid ambiguity and distortion in FM signals, we need

$$ks(t) \leq f_c \text{ for all } t \tag{6.61}$$

This restriction ensures that the instantaneous frequency is always positive. To avoid ambiguity and distortion in PM signals, we need

$$-\pi < as(t) \leq \pi \text{ radians for all } t \tag{6.62}$$

Phase can be resolved only within a range of 2π radians (we can't tell the difference between a phase angle of 2.5π radians and a phase angle of 0.5π radians at the receiver), so the restriction in Equation (6.62) ensures that the phase detected by the receiver is always directly related to the unmodulated signal. We satisfy the above restrictions on instantaneous phase and instantaneous frequency by carefully choosing values for constants α in Equation (6.53) and k in Equation (6.54).

Using our new definitions, we can see that FM and PM are related. They both change the instantaneous phase (or angle) of the carrier signal as a function of the analog information signal $s(t)$. For this reason, the two techniques are often grouped together and called *angle modulation*. The distinction between FM and PM is that in PM, instantaneous phase is a linear function of $s(t)$; while in FM, instantaneous frequency minus carrier frequency is a linear function of $s(t)$.

As a final comment, let's examine the signal

$$x(t) = A\cos\left\{2\pi f_c t + \int_{-\infty}^{t} ks(\lambda)d\lambda + \phi\right\} \tag{6.63}$$

Is this signal FM, PM, or neither? The instantaneous phase of the signal is

$$\Psi_x(t) = 2\pi f_c t + \int_{-\infty}^{t} ks(\lambda)d\lambda + \phi$$

which is not a linear function of $s(t)$, so the signal is not PM. The instantaneous frequency of the signal is

$$f_x(t) = \frac{d\Psi_x(t)}{dt} = 2\pi f_c + ks(t)$$

so $f(t) - f_c = ks(t)$, which is a linear function of $s(t)$. The signal $x(t)$ in Equation (6.63) is therefore an FM signal. In many applications and in much of the literature, you will see FM signals represented in the form of Equation (6.63), often with ϕ set to zero for simplification.

Example 6.1 PM and FM signals

The analog information signal $s(t) = 5 \sin 1000\pi t$ is to be transmitted across a bandpass channel.

a. If $s(t)$ is to be transmitted using PM, write an equation for the transmitted signal. Are there any practical limitations on the constant α in the PM signal?

b. If $s(t)$ is to be transmitted using FM, write an equation for the transmitted signal. Assume that the initial phase angle is zero. Are there any practical limitations on the constant k in the FM signal?

Solution

a. If $s(t)$ is phase modulated, we can express the transmitted signal as

$$s_{PM}(t) = A\cos(2\pi f_c t + 5\alpha \sin 1000\pi t)$$

where α is a constant. From Equation (6.62), we require $|\alpha| < 0.2\pi$ in order to unambiguously demodulate the signal.

b. If $s(t)$ is frequency modulated, we can express the transmitted signal as

$$s_{FM}(t) = A\cos\{2\pi[f_c + 5k \sin 1000\pi t]t\}$$

where k is a constant. From Equation (6.61), we require $|k| \leq 0.2f_c$ in order to unambiguously demodulate the signal.

Let's now introduce some terminology that will be useful in evaluating FM and PM signals. *Maximum phase deviation* (sometimes called just *phase deviation*) is the maximum instantaneous phase difference between the transmitted signal and an unmodulated carrier signal. Maximum phase deviation for the PM signal in Equation (6.53) is $\max|\alpha s(t)|$ radians. As shown in Equation (6.62), maximum phase deviation for a practical PM signal must be less than or equal to π radians. *Maximum frequency deviation* (sometimes called just *frequency deviation*) is the maximum instantaneous frequency difference between the transmitted signal and an unmodulated carrier. Maximum frequency deviation for the FM signal in Equation (6.54) is $\max|2\pi ks(t)|$

radians/sec, or $\max|ks(t)|$ Hz. Maximum frequency deviation is symbolized as Δf, As shown in Equation (6.61), maximum frequency deviation for a practical FM signal must be less than or equal to f_c, the frequency of the carrier.

Example 6.2 Maximum phase deviation and maximum frequency deviation—Part 1

The analog information signal $s(t) = 4\cos 2000\pi t$ is to be transmitted across a bandpass channel using PM.

a. What is the maximum phase deviation of the transmitted signal?
b. Are there any practical limitations on the value of α in the PM signal?
c. What is the maximum frequency deviation of the transmitted signal?

Solution

a. The transmitted PM signal is

$$s_{\text{PM}}(t) = A\cos(2\pi f_c t + 4\alpha \cos 2000\pi t)$$

where α is a constant. Maximum phase deviation is $\max|4\alpha \cos 2000\pi t| = 4\alpha$ radians.

b. Since maximum phase deviation must be less than π radians, $|\alpha| < 0.25\pi$ radians.
c. We can express instantaneous frequency as

$$f(t) = \frac{d}{dt}(2\pi f_c t + 4\alpha \cos 2000\pi t) = 2\pi f_c - 8000\pi\alpha \sin 2000\pi t$$

Maximum frequency deviation is therefore

$$\begin{aligned}\Delta f &= \max|(2\pi f_c - 8000\pi\alpha \sin 2000\pi t) - 2\pi f_c| = \max|-8000\pi\alpha \sin 2000\pi t| \\ &= 8000\pi\alpha \text{ radians/sec} = 4000\alpha \text{ Hz}\end{aligned}$$

Example 6.3 Maximum phase deviation and maximum frequency deviation—Part 2

In generalizing Example 6.2, consider an information signal $s(t) = W\sin 2\pi f_m t$, where W and f_m are constants. If this signal is phase modulated, the transmitted signal is

$$s_{\text{PM}}(t) = A\sin(2\pi f_c t + \alpha W \sin 2\pi f_m t)$$

Note that in this example we are using a sine carrier instead of a cosine carrier. This is not a problem, since the general form of the carrier has an arbitrary phase angle. The maximum phase deviation is αW radians, and the maximum frequency deviation is $\Delta f = \alpha W f_m$ Hz. If we want, we can express the PM signal as

$$\begin{aligned}s_{\text{PM}}(t) &= A\sin(2\pi f_c t + \alpha W \sin 2\pi f_m t) \\ &= A\sin\left[2\pi f_c t + \frac{\Delta f}{f_m}\sin 2\pi f_m t\right]\end{aligned}$$

Since maximum phase deviation is restricted to π radians, in a practical system $|\alpha| < \pi/W$.

In order to assess the costs and performance of PM and FM, we need to determine what PM and FM signals look like in the frequency domain. This will help us determine bandwidth and how well the signals perform in the presence of noise. Consider a signal representing angle modulation of a single sinusoid of frequency f_m

$$v(t) = A\sin(2\pi f_c t + \beta \sin 2\pi f_m t) \tag{6.64}$$

Both PM and FM signals can take the form of $v(t)$ in Equation (6.64); $v(t)$ may be a PM signal in the form of Equation (6.53), or it may be an FM signal in the form of Equation (6.63) after the integration is performed.

Determining the frequency domain representation of $v(t)$ is not simple. $v(t)$ is periodic with finite energy per period, so we need to determine its Fourier series. How can we determine the Fourier series of a *sinusoid of a sinusoid*? Let's try to manipulate $v(t)$ into a form we can more easily use. Applying Euler's identity, note that

$$\begin{aligned} e^{j2\pi f_c t} e^{j\beta \sin 2\pi f_m t} &= e^{j(2\pi f_c t + \beta \sin 2\pi f_m t)} \\ &= \cos(2\pi f_c t + \beta \sin 2\pi f_m t) + j \sin(2\pi f_c t + \beta \sin 2\pi f_m t) \end{aligned} \tag{6.65}$$

We can therefore express $v(t)$ as

$$\begin{aligned} v(t) &= A\sin(2\pi f_c t + \beta \sin 2\pi f_m t) \\ &= \text{Im}\{A e^{j2\pi f_c t} e^{j\beta \sin 2\pi f_m t}\} \end{aligned} \tag{6.66}$$

where $\text{Im}\{x\}$ means "the imaginary part of x." Equation (6.66) is somewhat messy, but let's keep going. Note that $e^{j\beta \sin 2\pi f_m t}$ is periodic with finite energy per period, so we can determine its Fourier series. Using the two-sided form and noting that the period of the signal is $1/f_m$,

$$e^{j\beta \sin 2\pi f_m t} = \sum_{n=-\infty}^{\infty} c_n e^{j2\pi n f_m t} \tag{6.67}$$

where

$$\begin{aligned} c_n &= f_m \int_{t_o}^{t_o+1/f_m} e^{j\beta \sin 2\pi f_m t} e^{-j2\pi n f_m t}\, dt \\ &= f_m \int_{t_o}^{t_o+1/f_m} e^{j(\beta \sin 2\pi f_m t - 2\pi n f_m t)}\, dt \end{aligned} \tag{6.68}$$

Performing a change of variable, let $\theta = 2\pi f_m t$. We can now simplify the notation in Equation (6.68):

$$c_n = f_m \int_{t_o}^{t_o+1/f_m} e^{j(\beta \sin 2\pi f_m t - 2\pi n f_m t)}\, dt$$

$$= \frac{1}{2\pi} \int_{-\pi}^{\pi} e^{j(\beta \sin \theta - n\theta)}\, d\theta \tag{6.69}$$

The final form of Equation (6.69) can be recognized as $J_n(\beta)$, the Bessel function of the first kind and order n. Bessel functions cannot be solved in closed form, but they are tabulated in many handbooks (see Table 6-1) and are also available as standard functions in mathematical computation packages, such as MATLAB. Substituting into Equation (6.69),

$$c_n = \frac{1}{2\pi} \int_{-\pi}^{\pi} e^{j(\beta \sin \theta - n\theta)}\, d\theta = J_n(\beta) \tag{6.70}$$

and so, substituting into Equation (6.67),

$$e^{j\beta \sin 2\pi f_m t} = \sum_{n=-\infty}^{\infty} c_n e^{j2\pi n f_m t} = \sum_{n=-\infty}^{\infty} J_n(\beta) e^{j2\pi n f_m t} \tag{6.71}$$

Table 6-1 Values of Bessel Function of the First Kind $J_n(\beta)$ for Various Values of n and β

n	$\beta = 1$	$\beta = 2$	$\beta = 3$	$\beta = 4$	$\beta = 5$	$\beta = 6$	$\beta = 7$	$\beta = 8$	$\beta = 9$
0	0.7652	0.2239	−0.2601	−0.3971	−0.1776	0.1506	0.3001	0.1717	−0.0903
1	0.4401	0.5767	0.3391	−0.0660	−0.3276	−0.2767	−0.0047	0.2346	0.2453
2	0.1149	0.3528	0.4861	0.3641	0.0466	−0.2429	−0.3014	−0.1130	0.1448
3	0.0196	0.1289	0.3091	0.4302	0.3648	0.1148	−0.1676	−0.2911	−0.1809
4	0.0025	0.0340	0.1320	0.2811	0.3912	0.3576	0.1578	−0.1054	−0.2655
5	0.0002	0.0070	0.0430	0.1321	0.2611	0.3621	0.3479	0.1858	−0.0550
6	*	0.0012	0.0114	0.0491	0.1310	0.2458	0.3392	0.3376	0.2043
7	*	0.0002	0.0025	0.0152	0.0534	0.1296	0.2336	0.3206	0.3275
8	*	*	0.0005	0.0040	0.0184	0.0565	0.1280	0.2235	0.3051
9	*	*	0.0001	0.0009	0.0055	0.0212	0.0589	0.1263	0.2149
10	*	*	*	0.0002	0.0015	0.0070	0.0235	0.0608	0.1247
11	*	*	*	*	0.0004	0.0020	0.0083	0.0256	0.0622
12	*	*	*	*	0.0001	0.0005	0.0027	0.0096	0.0274
13	*	*	*	*	*	0.0001	0.0008	0.0033	0.0108
14	*	*	*	*	*	*	0.0002	0.0010	0.0039
15	*	*	*	*	*	*	0.0001	0.0003	0.0013
16	*	*	*	*	*	*	*	0.0001	0.0004
17	*	*	*	*	*	*	*	*	0.0001
18	*	*	*	*	*	*	*	*	*
19	*	*	*	*	*	*	*	*	*

*The values designated by * are all $<10^{-4}$. The values in each column below the bar are all less than 0.1. Practically speaking, such values produce negligible spectral components.

Substituting Equation (6.71) back into Equation (6.66),

$$\begin{aligned}
v(t) &= A\sin(2\pi f_c t + \beta \sin 2\pi f_m t) \\
&= \mathrm{Im}\{Ae^{j2\pi f_c t}e^{j\beta \sin 2\pi f_m t}\} \\
&= \mathrm{Im}\left\{Ae^{j2\pi f_c t}\sum_{n=-\infty}^{\infty} J_n(\beta)e^{j2\pi n f_m t}\right\} \\
&= \mathrm{Im}\left\{A\sum_{n=-\infty}^{\infty} J_n(\beta)e^{j2\pi n f_m t}e^{j2\pi f_c t}\right\} \\
&= \mathrm{Im}\left\{A\sum_{n=-\infty}^{\infty} J_n(\beta)e^{j2\pi (n f_m + f_c) t}\right\} \\
&= \mathrm{Im}\left\{A\sum_{n=-\infty}^{\infty} J_n(\beta)\{\cos[2\pi(nf_m + f_c)t] + j\sin[2\pi(nf_m + f_c)t]\}\right\} \\
&= A\sum_{n=-\infty}^{\infty} J_n(\beta)\sin[2\pi(nf_m + f_c)t] \qquad (6.72a)
\end{aligned}$$

We finally have what we wanted: $v(t)$ expressed as a summation of sinusoids. Restating Equation (6.72),

$$v(t) = A\sum_{n=-\infty}^{\infty} J_n(\beta)\sin[2\pi(nf_m + f_c)t] \qquad (6.72b)$$

where $J_n(\beta)$ is the Bessel function of the first kind and order n.

Bessel functions have many useful properties:

1. For even values of n, $J_{-n}(\beta) = J_n(\beta)$.
2. For odd values of n, $J_{-n}(\beta) = -J_n(\beta)$.
3. $J_{n-1}(\beta) + J_{n+1}(\beta) = (2n/\beta)J_n(\beta)$. This property helps us construct tables such as Table 6-1.

Figure 6-21 shows plots of $J_n(\beta)$ versus β for several integer values of n.

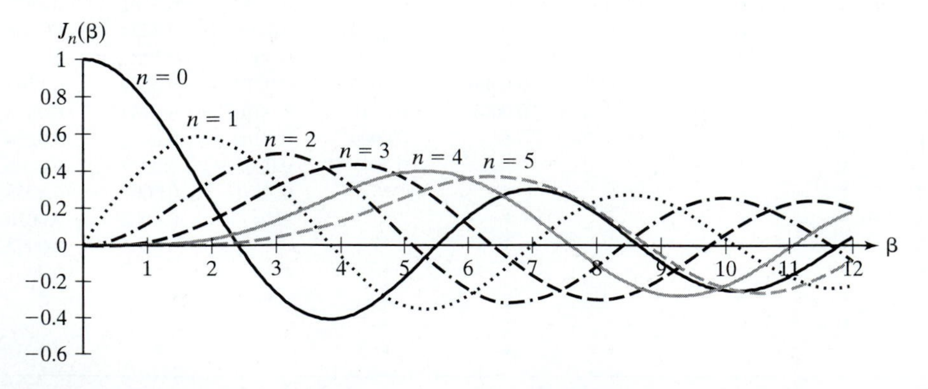

Figure 6-21 Bessel function of the first kind for various values of n (order).

Example 6.4 Magnitude spectrum of an angle-modulated signal

Determine the magnitude spectrum of the angle-modulated signal

$$x(t) = 4\sin(200{,}000\pi t + 2\sin 20{,}000\pi t)$$

Solution

We can express $x(t)$ as

$$x(t) = A\sin(2\pi f_c t + \beta \sin 2\pi f_m t)$$

where $A = 4$, $\beta = 2$, $f_c = 100{,}000$, and $f_m = 10{,}000$. Applying Equation (6.72b),

$$\begin{aligned} x(t) &= A \sum_{n=-\infty}^{\infty} J_n(\beta)\sin[2\pi(nf_m + f_c)t] \\ &= 4 \sum_{n=-\infty}^{\infty} J_n(2)\sin[2\pi(100{,}000 + 10{,}000n)t] \end{aligned}$$

Using Table 6-1 we can determine the values of $J_n(2)$. Expanding the summation for $x(t)$ and noting from Table 6-1 that $J_n(2)$ is negligibly small for $|n| > 3$,

$$\begin{aligned} x(t) &= 4 \sum_{n=-\infty}^{\infty} J_n(2)\sin[2\pi(100{,}000 + 10{,}000n)t] \\ &= 4\{\ldots + J_{-3}(2)\sin[2\pi(70{,}000)t] + J_{-2}(2)\sin[2\pi(80{,}000)t] \\ &\quad + J_{-1}(2)\sin[2\pi(90{,}000)t] + J_0(2)\sin[2\pi(100{,}000)t] \\ &\quad + J_1(2)\sin[2\pi(110{,}000)t] + J_2(2)\sin[2\pi(120{,}000)t] \\ &\quad + J_3(2)\sin[2\pi(130{,}000)t] + \ldots\} \\ &= 4\{\ldots - 0.1289\sin[2\pi(70{,}000)t] + 0.3528\sin[2\pi(80{,}000)t] \\ &\quad - 0.5767\sin[2\pi(90{,}000)t] + 0.2239\sin[2\pi(100{,}000)t] \\ &\quad + 0.5767\sin[2\pi(110{,}000)t] + 0.3528\sin[2\pi(120{,}000)t] \\ &\quad + 0.1289\sin[2\pi(130{,}000)t] + \ldots\} \\ &= \ldots - 0.5156\sin[2\pi(70{,}000)t] + 1.4112\sin[2\pi(80{,}000)t] \\ &\quad - 2.3068\sin[2\pi(90{,}000)t] + 0.8956\sin[2\pi(100{,}000)t] \\ &\quad + 2.3068\sin[2\pi(110{,}000)t] + 1.4112\sin[2\pi(120{,}000)t] \\ &\quad + 0.5156\sin[2\pi(130{,}000)t] + \ldots \end{aligned}$$

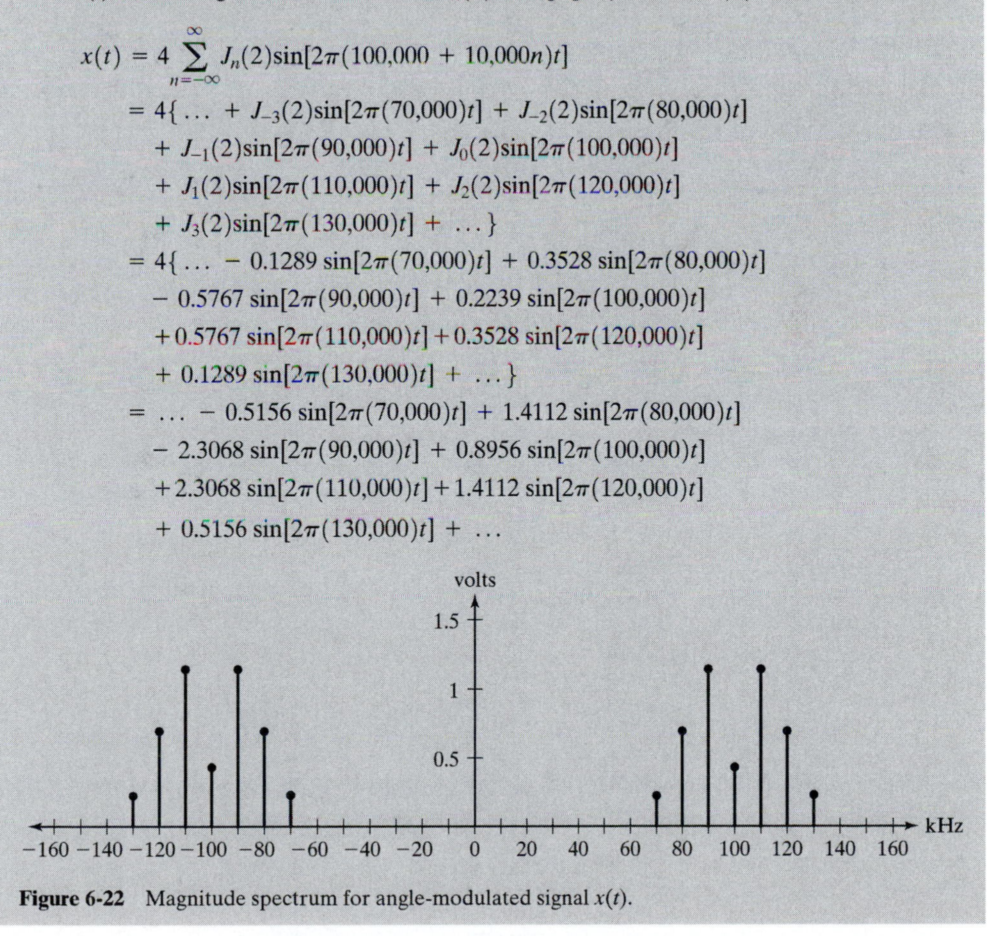

Figure 6-22 Magnitude spectrum for angle-modulated signal $x(t)$.

The magnitude spectrum for $x(t)$ is plotted in Figure 6-22. Theoretically, the angle-modulated signal $x(t)$ has infinite bandwidth because $J_n(2)$ is nonzero for all values of n. Practically speaking, however, $J_n(2)$ is negligibly small for $|n| > 3$, so $x(t)$ has a practical bandwidth of 60 kHz.

We can generalize our observations concerning the bandwidth of an angle-modulated sinusoid. In looking closely at Table 6-1, we can see that, practically speaking, values of $J_n(1)$ are negligibly small for $|n| > 2$, values of $J_n(2)$ are negligibly small for $|n| > 3$, values of $J_n(3)$ are negligibly small for $|n| > 4$, etc. For any value of n, values of $J_n(\beta)$ are negligibly small for $|n| > \beta + 1$. This means that any angle-modulated sinusoid of the form

$$v(t) = A\sin(2\pi f_c t + \beta \sin 2\pi f_m t) \qquad (6.64R)$$

will have a practical bandwidth of

$$B = 2(\beta + 1)f_m \text{ Hz} \qquad (6.73)$$

We can also relate bandwidth to maximum frequency deviation Δf. In inspecting Equation (6.64R), we can see that the maximum frequency deviation of $v(t)$ is βf_m. We can therefore rewrite Equation (6.73) as

$$B = 2(\beta + 1)f_m = 2\beta f_m + 2f_m = 2\Delta f + 2f_m = 2(\Delta f + f_m) \qquad (6.74)$$

Equations (6.73) and (6.74), relating the bandwidth of an angle-modulated sinusoid to its frequency and its maximum frequency deviation, are known as *Carson's rule.*[5]

To determine how useful Carson's rule will be to us, we need to answer two questions:

1. How good of an approximation is Carson's rule?
2. How do we calculate bandwidth for a phase-modulated or frequency-modulated general information signal? Remember that all we've done so far is determine the frequency spectrum for an information signal consisting of a single sinusoid.

Addressing the first question, let's calculate the percentage of an angle-modulated sinusoid's power that lies within the bandwidth determined by Carson's rule. In Example 6.4, we saw that

$$x(t) = 4\sin(200{,}000\pi t + 2\sin 20{,}000\pi t)$$

This signal has a constant envelope of amplitude 4 volts, so the average normalized power of the transmitted signal is $4^2/2 = 8$ volts^2. From Example 6.4 and Figure 6-22,

[5]Named in honor of J. R. Carson, co-author of Carson and Fry [6.3].

the average normalized power that lies in the frequency band from 70–130 kHz is

$$\begin{aligned} P_{in\text{-}band}(t) &= 2(0.2578^2 + 0.7056^2 + 1.1534^2 + 0.4478^2 + 1.1534^2 \\ &\quad + 0.7056^2 + 0.2578^2) \\ &= 7.98 \text{ volts}^2 \end{aligned}$$

For Example 6.4, the bandwidth given by Carson's rule contains 99.75% of the modulated sinusoid's power. As discussed in Chapter 2, such a high percentage means that the modulated signal can be reconstructed from the in-band components with very little distortion. For the more general case where the modulated sinusoid is given in Equation (6.64), the power within the bandwidth determined by Carson's rule is

$$P_{in\text{-}band,\ sinusoid}(t) = \frac{A^2}{2} \sum_{n=-(\beta+1)}^{\beta+1} J_n^2(\beta) \tag{6.75}$$

Be sure you understand how Equation (6.75) was derived. In the general case of a modulated sinusoid, the in-band power computed by Equation (6.75) will be at least 98% of the modulated sinusoid's power, so Carson's rule is valid.

Let's now address the second question concerning Carson's rule: How do we calculate bandwidth for a general phase-modulated or frequency-modulated information signal? Let $s(t)$ represent a general analog information signal from the source. If $s(t)$ is periodic, we know that $s(t)$ can be represented in the frequency domain as a summation of sinusoids. We therefore know that the angle-modulated signal corresponding to $s(t)$ can be represented as

$$s(t)_{angle\ modulated,\ periodic} = A\sin(2\pi f_c t + \beta \sum X_i \cos(2\pi i f_o t + \phi_i) \tag{6.76}$$

where the values X_i are constants. As we established in Example 6.1, the X_i values and the sinusoids in the summation of Equation (6.76) will correspond to the Fourier series coefficients and sinusoids of $s(t)$ if the angle-modulated signal is PM. If the angle-modulated signal is FM, the X_i values and sinusoids in the summation will correspond to the integral of the Fourier series components of $s(t)$. Either way, the sinusoids in the summation will have the same frequencies as the Fourier series components of $s(t)$. Bandwidth for a general FM signal can be conservatively expressed as

$$B = 2(\Delta f + f_{max}) \tag{6.77}$$

where f_{max} is the maximum frequency component in the original analog signal $s(t)$ and Δf represents maximum frequency deviation. Equation (6.77) is Carson's rule in a more generalized form.

6.6 Generating and Demodulating FM and PM Signals

6.6.1 FM and PM Modulators

Voltage-controlled oscillators (VCOs) can be used to create simple FM and PM modulators. As briefly discussed in Section 4.3.1, the output of a VCO is a signal whose instantaneous frequency is dependent on the voltage of an input signal applied to the

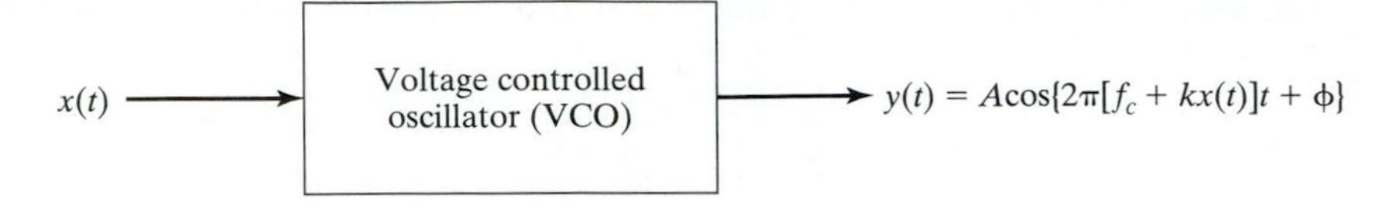

Figure 6-23 The input–output characteristics of a voltage-controlled oscillator. Note that a VCO can be used directly as an FM modulator.

VCO. With a zero volt input, the VCO outputs a sinusoid of a certain constant frequency (say, f_c). Increasing the input voltage linearly increases the frequency of the sinusoidal output of the VCO, and decreasing the input voltage linearly decreases the frequency of the sinusoidal output of the VCO. As shown in Figure 6-23, for an input signal $x(t)$, the output of the VCO is a sinusoid $A\cos\{2\pi[f_c + kx(t)]t + \phi\}$ where A, f_c, and k are constants and ϕ is a fixed, arbitrary phase angle. VCOs are available as inexpensive integrated circuits with a wide variety of user-adjustable ranges for f_c and k.

In inspecting Equation (6.54), we see that if an information signal $s(t)$ is input to a VCO, the VCO output will be an FM signal directly related to $s(t)$. With appropriate amplification circuitry, a VCO can be used as an FM transmitter.

We can also use a VCO to generate PM signals. Consider the circuit shown in Figure 6-24 and remember that a differentiator can be easily constructed using an op amp, resistors, and capacitors. Let $s(t)$ represent the information signal. The output of the differentiator is

$$w(t) = \frac{ds(t)}{dt} \tag{6.78}$$

and the output of the VCO is

$$\begin{aligned} y(t) &= A\cos\{2\pi[f_c + kw(t)]t + \phi\} \\ &= A\cos\left\{2\pi\left[f_c + k\frac{ds(t)}{dt}\right]t + \phi\right\} \end{aligned} \tag{6.79}$$

In inspecting Equation (6.79), we see that the instantaneous frequency of the VCO output is

$$f(t) = 2\pi\left[f_c + k\frac{ds(t)}{dt}\right] \text{ radians/sec} \tag{6.80}$$

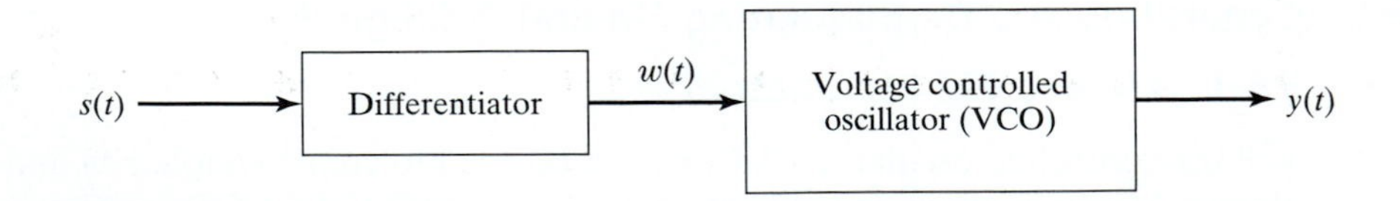

Figure 6-24 Phase modulation using a differentiator and a VCO.

and the instantaneous phase is

$$\Psi(t) = \int_{-\infty}^{t} f(\lambda)\,d\lambda = 2\pi f_c t + ks(t) \text{ radians} \tag{6.81}$$

which is a phase-modulated signal related to $s(t)$.

6.6.2 FM and PM Demodulators

A discriminator is a device that, when given a sinusoidal input, outputs a voltage that is directly proportional to how much the frequency of the sinusoidal input differs from a fixed frequency f_c. The operation of a discriminator is shown in Figure 6-25. A discriminator is basically a frequency-to-voltage converter. Since the discriminator performs the inverse function of a VCO, it is well suited for demodulation of FM and PM signals. Discriminators are available as inexpensive integrated circuits with a wide variety of user-adjustable ranges for f_c and k.

Discriminators are not difficult to build. Consider the circuit in Figure 6-26, a differentiator followed by an envelope detector and a blocking capacitor. Let the input signal to the differentiator be a typical FM signal

$$r(t) = \gamma A\cos\{2\pi[f_c + ks(t)]t + \phi\} \tag{6.82}$$

where $s(t)$ is an analog information signal. If the circuit in Figure 6-26 operates like a discriminator, then it will output a voltage proportional to $2\pi ks(t)$, the difference between the instantaneous frequency, $2\pi[f_c + ks(t)]$, and the carrier frequency, $2\pi f_c$.

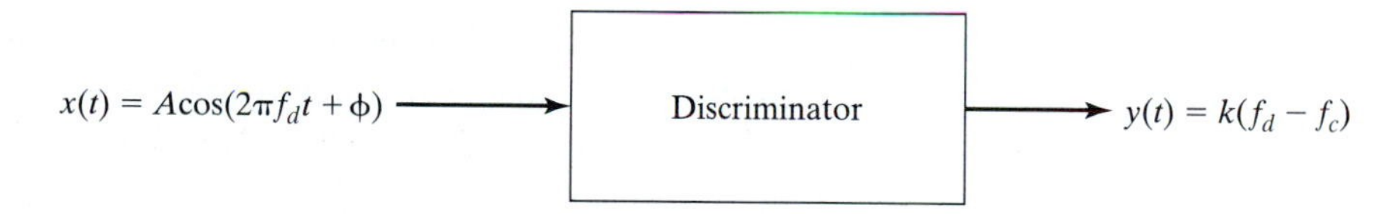

Figure 6-25 The input–output characteristics of a discriminator.

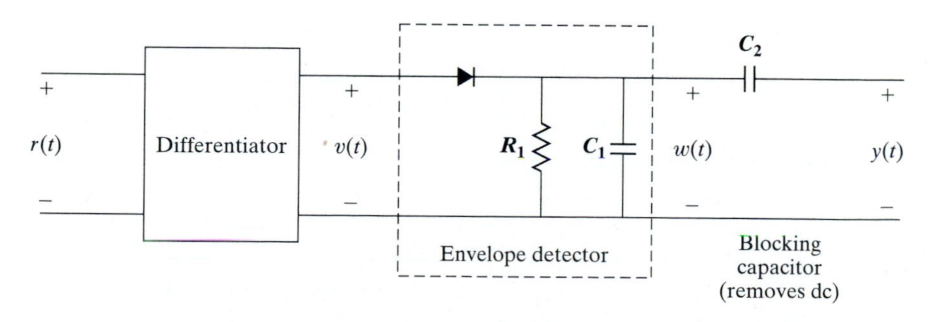

Figure 6-26 A discriminator circuit.

In examining Figure 6-26, the output of the differentiator is

$$v(t) = \frac{dr(t)}{dt} = -\gamma A2\pi[f_c + ks(t)]\sin\{2\pi[f_c + ks(t)]t + \phi\} \tag{6.83}$$

As long as $f_c > ks(t)$, a condition equal to satisfying the practical restriction of Equation (6.61), the output of the envelope detector is

$$w(t) = -\gamma A2\pi[f_c + ks(t)] = -\gamma A2\pi f_c - \gamma A2\pi ks(t) \tag{6.84}$$

The first term of Equation (6.82) is a constant and is removed using a blocking capacitor, yielding

$$y(t) = -\gamma A2\pi ks(t) \tag{6.85}$$

The output of the circuit is directly proportional to the information signal $s(t)$. The minus sign is unimportant because either $y(t)$ can be inverted or k can be set as a negative value in the modulator.

We can build an FM receiver from a discriminator as shown in Figure 6-27. The transmitted FM signal has a constant envelope, but noise will cause the envelope of the received signal to fluctuate. The received signal $r_{\mathrm{FM}}(t)$ is first passed through a limiter, a device that uses clipping to keep the maximum received signal from exceeding a fixed value. The output of the limiter is then input to a bandpass filter whose characteristics have been designed to pass the frequency components within the bandwidth of the transmitted signal, but to eliminate out-of-band noise and some of the harmonics caused by the clipping. The filtered signal is then input to the discriminator, which outputs a voltage proportional to the difference between the carrier frequency and the instantaneous frequency of the received signal. This signal will be a scaled version of the original information signal $s(t)$. The discriminator output is passed through a lowpass filter (called a *post-detection filter*) to eliminate any demodulated noise components lying outside the frequency band of $s(t)$.

PM signals are demodulated in a similar way, using a phase detector instead of a discriminator. Phase detection can be accomplished using phase-locked loops (PLLs), discussed in Section 4.3. Note that PLLs can also be used for FM demodulation, since the voltage being input to the VCO in the PLL will be proportional to the change in frequency of the received signal. Figure 6-28 provides a block diagram of a PM receiver.

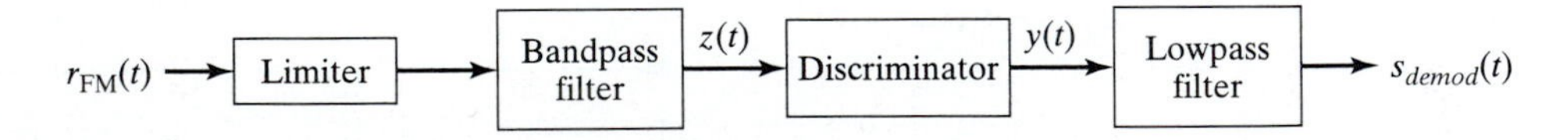

Figure 6-27 Block diagram of an FM receiver.

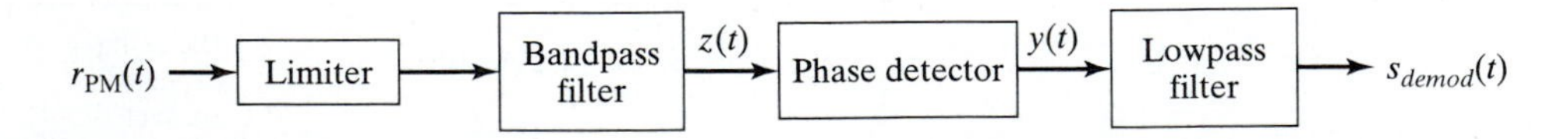

Figure 6-28 Block diagram of a PM receiver.

6.6.3 Noise in FM and PM Systems

Let's evaluate how FM and PM receivers operate in the presence of noise. We will start by examining a general angle-modulated signal

$$s_{angle\text{-}modulated}(t) = A\cos[\Psi(t)] \tag{6.86}$$

where $\Psi(t)$ represents the instantaneous phase. This signal may be FM or PM, depending on how the information signal is related to the instantaneous phase. Transmitting this angle-modulated signal across a channel, it will be attenuated and corrupted by noise. The received signal can be represented as

$$\begin{aligned} r_{angle\text{-}modulated}(t) &= \gamma s_{angle\text{-}modulated}(t) + n(t) \\ &= \gamma A\cos[\Psi(t)] + n(t) \end{aligned} \tag{6.87}$$

where $n(t)$ is additive white Gaussian noise with a flat power spectral density of $N_o/2$. As shown in Figures 6-27 and 6-28, regardless of whether we are using an FM or PM demodulator, the first two steps of demodulation are the same: limiting and bandpass filtering. The bandpass filter is designed to effectively pass the angle-modulated signal, but to eliminate out-of-band noise. The output of the bandpass filter can be expressed as

$$z(t) = \gamma A\cos[\Psi(t)] + n_o(t) \tag{6.88}$$

where $n_o(t)$ represents narrowband noise. As we discussed in our analysis of noise in AM systems, narrowband noise can be represented as

$$\begin{aligned} n_o(t) &= \mathbf{W}(t)\cos(2\pi f_c t) + \mathbf{Z}(t)\sin(2\pi f_c t) \\ &= \mathbf{R}(t)\cos(2\pi f_c t + \Theta(t)) \end{aligned} \tag{5.63R}$$

where $\mathbf{W}(t)$ and $\mathbf{Z}(t)$ are uncorrelated, zero-mean, Gaussian random variables with the power spectral densities shown in Figure 6-6, and where

$$\mathbf{R}(t) = \sqrt{\mathbf{W}(t)^2 + \mathbf{Z}(t)^2}$$

and

$$\Theta(t) = \arctan\left(\frac{\mathbf{Z}(t)}{\mathbf{W}(t)}\right)$$

We saw in Sections 5.5.4 and 6.3 that $\mathbf{R}(t)$ is Rayleigh distributed and that $\Theta(t)$ is uniformly distributed for $-\pi < \Theta(t) \leq \pi$. Rewriting Equation (6.88), the input to the discriminator can be expressed as

$$\begin{aligned} z(t) &= \gamma A\cos[\Psi(t)] + n_o(t) \\ &= \gamma A\cos[\Psi(t)] + \mathbf{R}(t)\cos(2\pi f_c t + \Theta(t)) \end{aligned} \tag{6.89}$$

We now need to determine the instantaneous phase of the signal $z(t)$. This is most easily understood if we use phasors to visualize Equation (6.89). Figure 6-29a is the

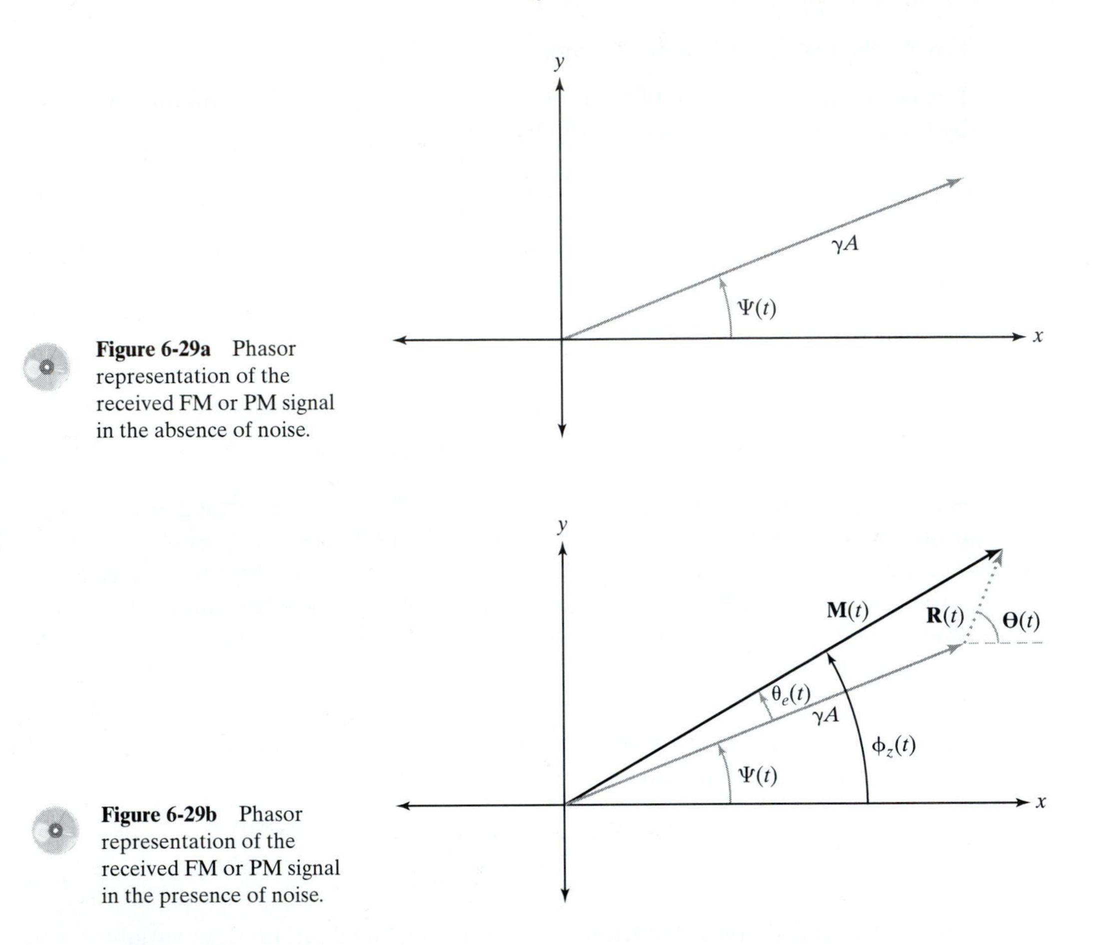

Figure 6-29a Phasor representation of the received FM or PM signal in the absence of noise.

Figure 6-29b Phasor representation of the received FM or PM signal in the presence of noise.

phasor representation of the bandpass filter output signal $z(t)$ in a noiseless environment. The instantaneous phase of the signal is $\Psi(t)$. Since $\Psi(t)$ is the instantaneous phase of the transmitted signal, $z(t)$ contains all of the information of the original source signal. The discriminator (if the signal is FM) or phase detector (if the signal is PM) can then reconstruct the information signal $s(t)$ with no distortion or corruption. In a noisy environment, however, the phasor representing the bandpass filter output $z(t)$ is the sum of the phasor of the received noiseless signal and the phasor of the noise. This is shown in Figure 6-29b, where $\mathbf{R}(t)$ is the phasor representing the noise and $\mathbf{M}(t)$ is the phasor representing the input signal to the discriminator (if the signal is FM) or the phase detector (if the signal is PM). As shown in Figure 6-29b, the instantaneous phase of this noisy signal is the angle of $\mathbf{M}(t)$, labeled $\phi_z(t)$. Note that $\phi_z(t)$ differs from the instantaneous phase of the transmitted signal $\Psi(t)$. This difference, labeled $\phi_e(t)$ in Figure 6-29b, produces distortion and corruption in the demodulated signal. In carefully examining Figure 6-29b, we can see that

$$\phi_e(t) = \arctan\frac{\mathbf{R}(t)\sin[\Theta(t) - \Psi(t)]}{\gamma A + \mathbf{R}(t)\cos[\Theta(t) - \Psi(t)]} \tag{6.90}$$

In practical systems, the average normalized power of the received signal will be significantly greater than the average normalized power of the noise. Under these conditions $\gamma A \gg \mathbf{R}(t)$, and we can approximate the corruption due to noise as

$$\begin{aligned}\phi_e(t) &= \arctan\frac{\mathbf{R}(t)\sin[\Theta(t) - \Psi(t)]}{\gamma A + \mathbf{R}(t)\cos[\Theta(t) - \Psi(t)]} \\ &\approx \arctan\frac{\mathbf{R}(t)\sin[\Theta(t) - \Psi(t)]}{\gamma A} \\ &\approx \frac{\mathbf{R}(t)\sin[\Theta(t) - \Psi(t)]}{\gamma A}\end{aligned} \tag{6.91}$$

Using the approximation in Equation (6.91), the instantaneous phase of $z(t)$ is

$$\phi_z(t) = \Psi(t) + \phi_e(t) \approx \Psi(t) + \frac{\mathbf{R}(t)}{\gamma A}\sin[\Theta(t) - \Psi(t)] \tag{6.92}$$

The first term of Equation (6.92) is the component due to the transmitted information signal and the second term is the component due to the noise.

6.6.3.1 Signal-to-noise ratio for PM

Consider a typical PM signal

$$s_{\text{PM}}(t) = A\cos[2\pi f_c t + \alpha s(t)] \tag{6.53R}$$

This signal is transmitted across a channel where it is attenuated and corrupted by noise. Using Figure 6-28 (the block diagram of a PM receiver) and Equation (6.89), the received signal after it has passed through the limiter and the bandpass filter is

$$\begin{aligned}z(t) &= \gamma A\cos[\Psi(t)] + n_o(t) \\ &= \gamma A\cos\{2\pi f_c t + \alpha s(t)\} + n_o(t) \\ &= \gamma A\cos\{2\pi f_c t + \alpha s(t)\} + \mathbf{R}(t)\cos(2\pi f_c t + \Theta(t))\end{aligned} \tag{6.93}$$

This signal has an instantaneous phase of

$$\phi_z(t) = \Psi(t) + \phi_e(t) \approx 2\pi f_c t + \alpha s(t) + \frac{\mathbf{R}(t)}{\gamma A}\sin[\Theta(t) - \Psi(t)] \tag{6.94}$$

The phase detector will output a voltage proportional to the instantaneous phase minus the carrier component $2\pi f_c t$

$$y(t) = \lambda\left\{\alpha s(t) + \frac{\mathbf{R}(t)}{\gamma A}\sin[\Theta(t) - \Psi(t)]\right\} = \lambda\alpha s(t) + \frac{\lambda\mathbf{R}(t)}{\gamma A}\sin[\Theta(t) - \Psi(t)]) \tag{6.95}$$

where λ is a constant. We will call the first term in Equation (6.95) the *signal term* because it is directly proportional to the analog information signal and is, therefore, the desired term for accurate demodulation of the PM signal. We will call the second term in Equation (6.95) the *noise term* because it is unwanted and causes distortion and corruption in the demodulated signal. Note that the value of the noise term is actually a function of both the Gaussian noise at the input of the receiver and the instantaneous phase of the transmitted signal.

We can now determine the SNR of a demodulated PM signal. Since the signal term passes unaffected through the post-detection filter, the average normalized power of the signal term at the post-detection filter output is

$$P_{signal,\ \mathrm{PM}} = (\lambda\alpha)^2 P_s \tag{6.96}$$

where P_s is the average normalized power of the analog information signal. Calculating the average normalized power of the noise term appears to be more difficult, since, as we just mentioned, the noise term is a function of both the Gaussian noise at the input of the receiver and the instantaneous phase of the transmitted signal. Fortunately, the instantaneous phase of the transmitted signal has little effect on average normalized noise power.[6] We can therefore set $\Psi(t)$ to a convenient value (say, zero) when calculating the average normalized power of the noise term.

Setting $\Psi(t)$ to zero produces a noise term of $(\lambda\mathbf{R}(t)/(\gamma A))\sin\Theta(t)$. From Section 5.5.4, Equation (5.63*R*) and earlier in this section, we can recognize the term $\mathbf{R}(t)\sin\Theta(t)$ as the quadrature magnitude term of narrowband noise. This quadrature magnitude term, represented by the random variable $\mathbf{Z}(t)$ in Equation (5.63*R*), has a flat power spectral density $G_{\mathbf{Z}}(f) = N_o$ volts²/Hz over the bandwidth from dc to B, where B is the bandwidth of the bandpass filter in the receiver. This power spectral density was shown in Figure 6-6. Noting the effects of the lowpass post-detection filter, the average normalized noise power at the post-detection filter output is

$$P_{noise,\,PM} = \left(\frac{\lambda}{\gamma A}\right)^2 \int_{-f_{\max}}^{f_{\max}} N_o\, df = 2f_{\max}N_o\left(\frac{\lambda}{\gamma A}\right)^2 \tag{6.97}$$

Note that the limits on the integral in Equation (6.97) are $\pm f_{\max}$, the maximum frequency within the original analog signal $s(t)$ and the bandwidth of the post-detection filter. SNR for the demodulated PM signal is thus

$$SNR_{\mathrm{PM}} = \frac{(\lambda\alpha)^2 P_s}{2f_{\max}N_o\left(\dfrac{\lambda}{\gamma A}\right)^2} = \frac{(\alpha\gamma A)^2 P_s}{2N_o f_{\max}} \tag{6.98}$$

For a given analog information signal $s(t)$, Equation (6.98) shows us that there are two ways of increasing the SNR (and therefore the quality) of the demodulated PM signal. As with every other modulation technique we've studied, we can improve

[6]See Downing [6.4]. The reason $\Psi(t)$ has little effect on average normalized noise power is that it produces frequency components in the second term of Equation (6.95) that are above the frequency band of the analog information signal, and can thus be filtered out by the post-detection lowpass filter.

SNR by increasing A, the amplitude of the transmitted signal. The cost of doing this is an increase in the transmitted power. As we've previously discussed, this trade-off is suitable for some applications, but for others (such as cellular telephone systems), governmental regulations and practical considerations (such as battery size) may restrict the maximum transmitted power. A second way to improve SNR for a PM system is by increasing α. Note, however, that increasing α increases maximum frequency deviation Δf and therefore increases the bandwidth of the transmitted signal. PM therefore provides a way to achieve increases in received signal quality at the cost of increases in bandwidth—an interesting trade-off. As mentioned previously, we also need to remember that in practical systems the maximum value of α is restricted by

$$-\pi < \alpha s(t) \leq \pi \text{ radians for all } t \tag{6.62R}$$

6.6.3.2 Signal-to-noise ratio for FM We know from Equation (6.63) that we can express an FM signal as

$$s_{\mathrm{FM}}(t) = A\cos\left\{2\pi f_c t + \int_{-\infty}^{t} ks(\lambda)d\lambda + \phi\right\} \tag{6.63R}$$

This FM signal has an instantaneous phase of

$$\Psi(t) = 2\pi f_c t + \int_{-\infty}^{t} ks(\lambda)d\lambda + \phi \tag{6.99}$$

The FM signal is transmitted across a channel where it is attenuated and corrupted by noise. Using Figure 6-27 (the block diagram of an FM receiver) and Equation (6.92), the instantaneous phase of the received signal after it has passed through the limiter and bandpass filter is

$$\phi_z(t) = \Psi(t) + \phi_e(t) \approx 2\pi f_c t + \int_{-\infty}^{t} ks(\lambda)d\lambda + \phi + \frac{\mathbf{R}(t)}{\gamma A}\sin[\Theta(t) - \Psi(t)] \tag{6.100}$$

The instantaneous frequency of this signal can be expressed as

$$\begin{aligned} f_z(t) &= \frac{d\phi_z(t)}{dt} \\ &\approx \frac{d}{dt}\left\{2\pi f_c t + \int_{-\infty}^{t} ks(\lambda)d\lambda + \phi + \frac{\mathbf{R}(t)}{\gamma A}\sin[\Theta(t) - \Psi(t)]\right\} \\ &= 2\pi f_c + ks(t) + \frac{d}{dt}\left\{\frac{\mathbf{R}(t)}{\gamma A}\sin[\Theta(t) - \Psi(t)]\right\} \end{aligned} \tag{6.101}$$

The third term of Equation (6.101), which looks somewhat unwieldy, will be analyzed in more detail shortly.

The output of a discriminator is proportional to the instantaneous frequency of its input minus a reference frequency (in this case, the reference is the carrier frequency f_c). In Figure 6-27, the discriminator output can therefore be expressed as

$$\begin{aligned} y(t) &= c(f_z(t) - 2\pi f_c) \\ &\approx kcs(t) + c\frac{d}{dt}\left\{\frac{\mathbf{R}(t)}{\gamma A}\sin[\Theta(t) - \Psi(t)]\right\} \end{aligned} \tag{6.102}$$

where c is a constant. The first term in Equation (6.102) is the signal term because it is directly proportional to the analog information signal and is, therefore, the desired term for accurate demodulation of the FM signal. The second term in Equation (6.102) is the noise term because it is unwanted and causes distortion and corruption in the demodulated signal. As with PM, note that the value of the noise term is a function of both the Gaussian noise at the input of the receiver and the instantaneous phase of the transmitted signal.

We can now determine the SNR of a demodulated FM signal. Since the signal term passes unaffected through the post-detection filter, the average normalized power of the signal term at the post-detection filter output is

$$P_{signal,\,\mathrm{FM}} = (kc)^2 P_s \tag{6.103}$$

where P_s is the average normalized power of the analog information signal. As with PM, calculating the average normalized power of the noise term appears to be difficult, since the noise term is a function of both the Gaussian noise at the input of the receiver and the instantaneous phase of the transmitted signal. In practical systems, the average normalized power of the received signal will be significantly greater than the average normalized power of the noise, and under these conditions, the instantaneous phase of the transmitted FM signal has little effect on average normalized noise power (again, see Downing [6.4]). As we did with our analysis of SNR for PM, we can therefore set $\Psi(t)$ to a convenient value (say, zero). Again, as we saw in our PM analysis, the resulting noise term contains $\mathbf{R}(t)\sin\Theta(t)$, which is the quadrature magnitude term of narrowband noise. We can therefore represent the noise term as

$$\begin{aligned} y_{noise}(t) &= c\frac{d}{dt}\left\{\frac{\mathbf{R}(t)}{\gamma A}\sin[\Theta(t) - \Psi(t)]\right\} \\ &\approx c\frac{d}{dt}\left\{\frac{\mathbf{Z}(t)}{\gamma A}\right\} \\ &= \frac{c}{\gamma A}\frac{d\mathbf{Z}(t)}{dt} \end{aligned} \tag{6.104}$$

where $\mathbf{Z}(t)$ is a random variable with flat power spectral density $G_{\mathbf{Z}}(f) = N_o$ volts^2/Hz over the frequency band from dc to B, where B is the bandwidth of the bandpass filter in the receiver. This power spectral density was shown in Figure 6-6.

How do we take the derivative of a random variable? Instead of answering this question directly, let's examine how differentiation affects power spectral density. Consider a signal $n(t)$. As was shown in Equation (2.59), differentiation in the time domain corresponds to multiplication by $j2\pi f$ in the frequency domain, so

$$\mathcal{F}\left\{\frac{dx(t)}{dt}\right\} = j2\pi f X(f) \tag{6.105}$$

The power spectral density of a signal is the square of its magnitude spectrum, so differentiating a signal is equivalent to multiplying its power spectral density by $(2\pi f)^2$. The power spectral density of the noise term in Equation (6.104) is therefore

$$G_{noise,\, discriminator\ output}(f) = \left|\left\{\frac{c}{\gamma A}\right\}^2 (j2\pi f)^2\right| G_{\mathbf{Z}}(f) \tag{6.106}$$

The output of the discriminator is passed through a post-detection filter, so the average normalized power of the noise term at the post-detection filter output is

$$\begin{aligned} P_{noise,\,\mathrm{FM}} &= \left(\frac{c}{\gamma A}\right)^2 \int_{-f_{\max}}^{f_{\max}} (2\pi f)^2 N_o\, df \\ &= \left(\frac{2\pi c}{\gamma A}\right)^2 \frac{2N_o f_{\max}^3}{3} \end{aligned} \tag{6.107}$$

SNR for the demodulated FM signal is thus

$$SNR_{\mathrm{FM}} = \frac{(kc)^2 P_s}{\left(\dfrac{2\pi c}{\gamma A}\right)^2 \dfrac{2N_o f_{\max}^3}{3}} = \frac{3}{2}\left(\frac{k\gamma A}{2\pi}\right)^2 \frac{P_s}{N_o f_{\max}^3} \tag{6.108}$$

As with PM, Equation (6.108) shows us that for a given analog information signal $s(t)$, there are two ways of increasing the SNR (and therefore the quality) of the demodulated FM signal. As with every other modulation technique we've studied, we can improve SNR by increasing A, the amplitude of the transmitted signal. The cost of doing this is an increase in the transmitted power, which, as we've previously discussed, is suitable for some applications but not for others. A second way to improve SNR for an FM system is by increasing k. Note, however, that increasing k increases maximum frequency deviation Δf and therefore increases the bandwidth of the transmitted signal. Just like PM, FM provides a way to achieve increases in received signal quality at the cost of increases in bandwidth. We need to remember that in practical systems the maximum value of k is restricted by

$$ks(t) \leq f_c \text{ for all } t \tag{6.61R}$$

For practical carrier frequencies, this restriction is much more relaxed than the restriction placed on α in PM systems. Generally speaking, FM systems therefore provide much more flexibility than PM systems in trade-offs concerning signal quality versus bandwidth.

6.6.3.3 *Pre-emphasis/de-emphasis in FM signals* Using Equation (6.106), the power spectral density of the noise at the output of the post-detection filter is

$$\begin{aligned} G_{noise\ post\ detection}(f) &= \left| \left\{ \frac{c}{\gamma A} \right\}^2 (j2\pi f)^2 \right| G_{\mathbf{Z}}(f) \\ &= \begin{cases} \left\{ \frac{c}{\gamma A} \right\}^2 N_o (2\pi f)^2 & |f| \leq f_{\max} \\ 0 & \text{elsewhere} \end{cases} \end{aligned} \tag{6.109}$$

This power spectral density is plotted in Figure 6-30. In inspecting Equation (6.109) and Figure 6-30, we see that the power of the noise increases as a function of f^2. Thus, higher frequency components of the analog information signal $s(t)$ will experience more corruption during demodulation than lower frequency components.

Let's consider a way to combat this noise. Suppose that prior to modulation we pass the analog information signal $s(t)$ through a circuit with a particular transfer function, say $H_T(f)$, which increases the magnitudes of the signal's high frequency

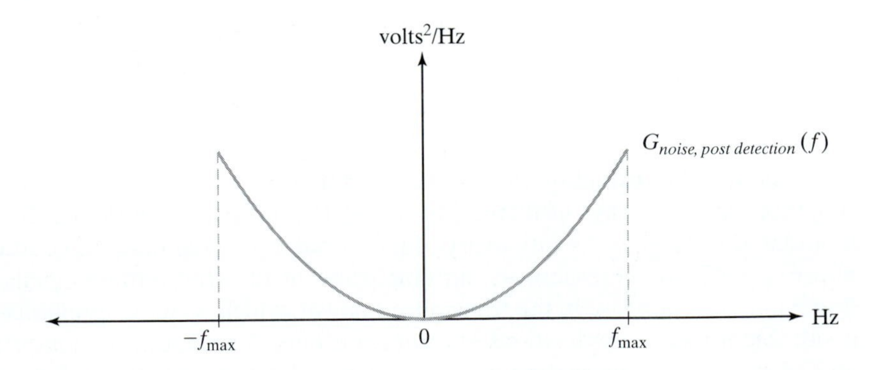

Figure 6-30 Power spectral density of FM noise at the output of the post-detection filter.

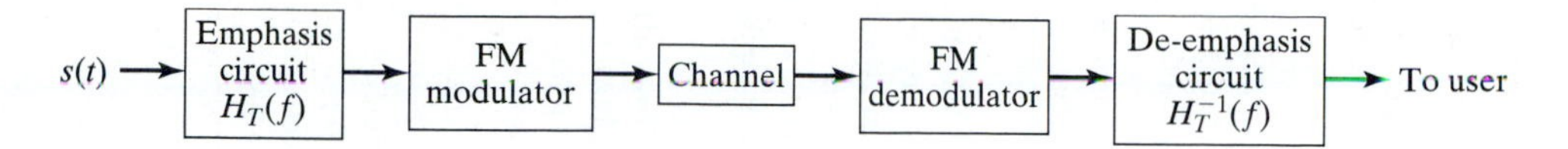

Figure 6-31 Pre-emphasis and de-emphasis in an FM system.

components. This process is called *pre-emphasis* because it emphasizes the higher frequency components of the information signal. The demodulated signal will now be less susceptible to noise, but it will be distorted because of the pre-emphasis. This distortion is removed by passing the demodulated signal through a circuit with a transfer function of $\frac{1}{H_T(f)}$. The processing of the demodulated signal is called *de-emphasis*. Figure 6-31 shows a block diagram of an FM system with pre-emphasis and de-emphasis. The concept of emphasis/de-emphasis is very similar to companding, which we will study in Chapter 8.

As an example of pre-emphasis/de-emphasis, consider a commercial FM broadcasting system. The system needs only one transmitter, but every listener must have a receiver. We therefore want the receivers to be inexpensive. To reduce expense, FM receivers use a very simple de-emphasis circuit: a series RC circuit. The transfer function of the de-emphasis circuit is

$$\frac{1}{H_T(f)} = \frac{1}{1 + j2\pi fRC} \tag{6.110}$$

This type of de-emphasis circuit is consistent with our intuition: It attenuates the high-frequency components more than the low-frequency components. Solving Equation (6.110) for the transfer function of the emphasis circuit yields

$$H_T(f) = 1 + j2\pi fRC \tag{6.111}$$

This circuit can also be simply realized using resistors and capacitors.

6.7 A Comparison of Analog Modulation Techniques

Table 6-2 provides a comparison of the performance and costs of AM-DSB-SC, AM-DSB-C, AM-SSB, AM-VSB, PM, and FM. Among the AM techniques, AM-SSB is the most bandwidth efficient, but at the cost of very high equipment complexity. AM-VSB represents a compromise between the bandwidth efficiency of AM-SSB and the reduced equipment complexity of other AM techniques. FM and PM provide a way to increase SNR at the cost of increased bandwidth. The trade-off is limited by practical restrictions on α (for PM) and k (for FM). Generally speaking, the restrictions on FM are not as severe as the restrictions on PM.

Table-2 A Comparison of Analog Modulation Techniques

Analog Modulation/ Demodulation Technique	Bandwidth	Relative Equipment Complexity	Signal-to-Noise Ratio	Comments
AM-DSB-SC	$2B$	High, requires coherent demodulation	$\frac{\gamma^2 P_{trans}}{N_o B}$	
AM-DSB-C	$2B$	Low, uses envelope detector	$\left(\frac{\gamma^2 P_{trans}}{N_o B}\right)\eta$	Simpler equipment than AM-DSB-SC, but considerably lower SNR because discrete carrier term must be transmitted.
AM-SSB	B	Very high, requires sharp filtering and coherent demodulation	$\frac{\gamma^2 P_{trans}}{N_o B}$	Same SNR as AM-DSB-SC, uses more complex equipment but less bandwidth.
AM-VSB	$B + \Delta$	High if coherent demodulation is used, lower if discrete carrier is also transmitted (allowing demodulation using an envelope detector)	$\frac{\gamma^2 P_{trans}}{N_o B}$ if discrete carrier is not transmitted. Multiply SNR by η if discrete carrier is transmitted.	$\Delta$is a parameter of the AM-VSB transmitter filter, $0 < \Delta < B$. AM-VSB represents a compromise between the bandwidth efficiency of AM-SSB (at the cost of equipment complexity) and the simplicity of other AM modulation techniques.
PM	$2(\Delta f + f_{max})$	Moderate	$\frac{(\alpha\gamma A)^2 P_s}{2N_o f_{max}}$	PM provides technique for tradeoffs between bandwidth and SNR. Note, however, that the extent of the tradeoff is limited by the practical restriction $-\pi < \alpha s(t) \leq \pi$ for all t.
FM	$2(\Delta f + f_{max})$	Moderate	$\frac{3}{2}\left(\frac{k\gamma A}{2\pi}\right)^2 \frac{P_s}{N_o f_{max}^3}$	FM provides technique for tradeoffs between bandwidth and SNR. Note, however, that extent of the tradeoff is limited by the practical restriction $ks(t) \leq f_c$ for all t. Generally speaking, this restriction is not as severe as the restriction on PM.

$s(t)$ = original analog signal from the source
B = bandwidth of the original analog signal from the source
P_{trans} = average normalized power of transmitted signal
η = power efficiency of AM-DSB-C. See Equation (6.29).
The transmitted PM signal is represented as $s_{PM}(t) = A\cos[2\pi f_c t + \alpha s(t)]$. See Equation (6.53).
f_{max} = maximum frequency component in original analog signal from source. For a baseband source signal, $f_{max} = B$.
Δf = maximum frequency deviation
The transmitted FM signal is represented as $s_{FM}(t) = A\cos\{2\pi[f_c + ks(t)]t + \phi\}$. See Equation (6.54).

PROBLEMS

6.1 Consider an information signal shown in Figure P6-1. Draw the transmitted signal if AM-DSB-SC is employed with a carrier frequency of 100 kHz and a scaling factor of 3.

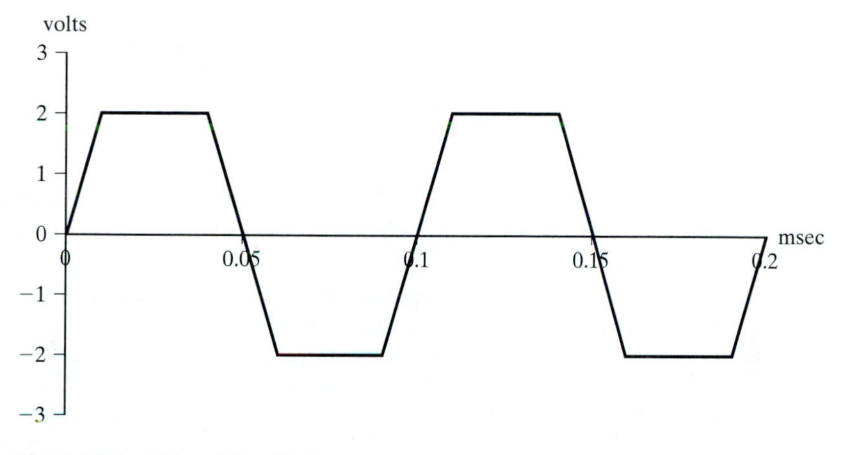

Figure P6-1 Information signal.

6.2 Consider an information signal with the magnitude spectrum shown in Figure P6-2. Suppose we have a channel capable of passing frequencies in the range $300\,\text{kHz} \leq f \leq 320\,\text{kHz}$ and we want to transmit the signal across the channel using AM-DSB-SC with a scaling factor of 1.

a. Determine the carrier frequency.

b. Draw the magnitude spectrum of the transmitted signal.

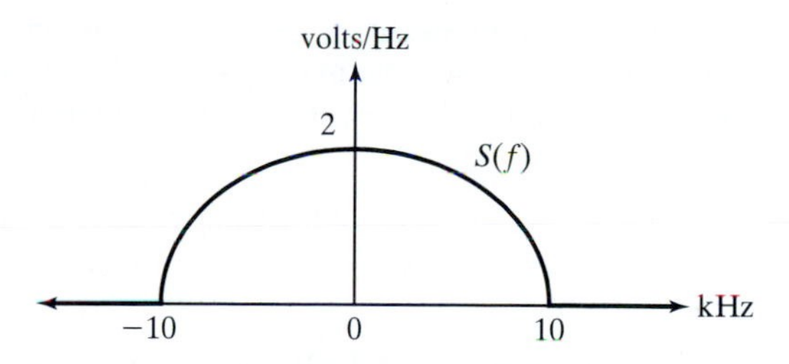

Figure P6-2 Magnitude spectrum of information signal.

6.3 **a.** Draw a block diagram for a communication system using AM-DSB-SC. Draw as much detail as you can in the transmitter and receiver, leaving enough room in your diagram for Part c of this problem.

b. Does AM-DSB-SC use coherent or noncoherent demodulation? Explain your answer.

c. Given the magnitude spectrum of the information signal in Figure P6-3, and given a carrier frequency f_c, draw the magnitude spectrum of the signal at each point in your block diagram. (Note that if you have more than one component within a block, you need to draw the magnitude spectrum of the signal at the output of each component.)

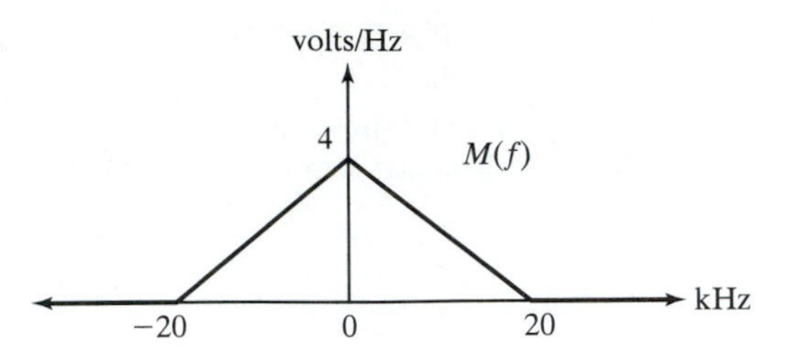

Figure P6-3 Magnitude spectrum of information signal.

6.4 For an AM-DSB-SC system, derive Equation (6.18), which shows the effect on the receiver's output signal if the transmitter and receiver carriers are not perfectly frequency synchronized.

6.5 Consider an AM-DSB-SC system transmitting a signal across a channel that attenuates the transmitted signal by 75% (so that only 25% of the transmitted signal arrives at the receiver). Noise power spectral density at the receiver is $N_o/2 = 2.5 \times 10^{-7}$ volts^2 and the channel attenuates the transmitted signal by 75% (so that only 25% of the transmitted signal arrives at the receiver). The bandwidth of the analog signal from the source is 10 kHz and the carrier frequency is 800 kHz. Determine the average normalized transmitted signal power necessary to attain an SNR of 12 dB for the demodulated signal.

6.6 Derive the equation for SNR in AM-DSB-SC systems corrupted by additive white Gaussian noise.

6.7 Consider the information signal shown in Figure P6-1. Draw the transmitted signal if AM-DSB-C is employed with a carrier frequency of 100 kHz, a scaling factor of 2, and a dc offset (c) of 3 volts.

6.8 Determine the modulation index of the signal in Problem 6.7.

6.9 Consider an information signal with the magnitude spectrum shown in Figure P6-2. Suppose we have a channel capable of passing frequencies in the range 300 kHz $\leq f \leq$ 320 kHz and we want to transmit the signal across the channel using AM-DSB-C with a scaling factor of 1. Further suppose that the maximum amplitude of the information signal is +2 volts, the minimum amplitude is −2 volts, and a modulation index of 0.667 is used.

- **a.** Determine the carrier frequency.
- **b.** Draw the magnitude spectrum of the transmitted signal.
- **c.** Compare the SNR of the demodulated signal with the SNR for the AM-DSB-SC system in Problem 6.2.

6.10 Draw a block diagram of an AM-DSB-C receiver and explain in detail how it operates.

6.11 **a.** Draw the circuit for an envelope detector and explain its operation in as much detail as possible.

- **b.** Which amplitude modulation techniques use an envelope detector in their receiver circuit?
- **c.** Explain in detail what happens if the values chosen for the resistor and capacitor in the envelope detector are too large.
- **d.** Explain in detail what happens if the values chosen for the resistor and capacitor in the envelope detector are too small.

6.12 For AM-DSB-C signals, show that the larger the modulation index, the greater the power efficiency; but if $m > 1$, then $\min\,[s(t) + c] < 0$, producing distortion.

6.13 Consider an AM-DSB-C system transmitting a signal across a channel that attenuates the transmitted signal by 75% (i.e., only 25% of the transmitted signal arrives at the receiver). Noise power spectral density at the receiver is $N_o/2 = 2.5 \times 10^{-7}$ volts^2 and the channel attenuates the transmitted signal by 75% (so that only 25% of the transmitted signal arrives at the receiver). The bandwidth of the analog signal from the source is 10 kHz and the carrier frequency is 800 kHz. The modulation index has been set so that the power of the transmitted discrete carrier component is twice the power of the transmitted information component.

a. Determine the power efficiency of the system.
b. Determine the average normalized transmitted signal power necessary to attain an SNR of 12 dB for the demodulated signal.
c. Compare your answer in Part b with the answer to Problem 6.5.

6.14 Derive the equation for SNR in AM-DSB-C systems corrupted by additive white Gaussian noise. Be sure to explain the threshold effect.

6.15 Explain the performance-versus-cost trade-offs between AM-DSB-C and AM-DSB-SC.

6.16 Consider an information signal with the magnitude spectrum shown in Figure P6-2. Suppose we have a channel capable of passing frequencies in the range 300 kHz $\leq f \leq$ 310 kHz and we want to transmit the signal across the channel using AM-SSB with a scaling factor of 1.

a. You may choose to transmit either the upper or lower sideband. State which sideband you are choosing and determine the appropriate carrier frequency.
b. Draw the magnitude spectrum of the transmitted signal.

6.17 **a.** Draw a block diagram for a communication system using AM-SSB. Draw as much detail as you can in the transmitter and receiver, leaving plenty of room in your diagram for Part c of this problem. You may choose to transmit either the upper or lower sideband.
b. Does AM-SSB use coherent or noncoherent demodulation? Explain your answer.
c. Given the magnitude spectrum of the information signal in Figure P6-3, and given a carrier frequency f_c, draw the magnitude spectrum of the signal at each point in your block diagram. (Note that if you have more than one component within a block, you will need to draw the magnitude spectrum of the signal at the output of each component.)

6.18 A certain symmetry is required for the transfer function of the filter in an AM-VSB transmitter. Describe the symmetry and explain why it is required.

6.19 Consider an information signal with the magnitude spectrum shown in Figure P6-2. Suppose we have a channel capable of passing frequencies in the range 300 kHz $\leq f \leq$ 315 kHz and we want to transmit the signal across the channel using AM-VSB with a scaling factor of 1.

a. You may choose to transmit either the upper or lower sideband. State which sideband you are choosing and draw an appropriate transfer function for the filter of the AM-VSB transmitter.
b. Determine the carrier frequency.
c. Draw the magnitude spectrum of the transmitted signal.

6.20 **a.** Draw a block diagram for a communication system using AM-VSB. Draw as much detail as you can in the transmitter and receiver, leaving plenty of room in your diagram for Part c of this problem. You will want to design the transfer function for the filter of the AM-VSB transmitter to accommodate an analog information signal with a bandwidth of 20 kHz.

b. Does AM-VSB use coherent or noncoherent demodulation? Explain your answer.

c. Given the magnitude spectrum of the information signal in Figure P6.3, and given a carrier frequency f_c, draw the magnitude spectrum of the signal at each point in your block diagram. (Note that if you have more than one component within a block, you will need to draw the magnitude spectrum of the signal at the output of each component.)

6.21 Discuss the performance-versus-cost trade-offs for AM-DSB-SC, AM-DSB-C, AM-SSB, and AM-VSB.

6.22 A source produces an analog output signal $s(t)$. The signal is angle-modulated, and the modulated signal $y(t) = A\cos[2\pi f_c t + \zeta s(t)]$ is transmitted across the channel (ζ is a constant).

a. Is the transmitted signal PM, FM, or neither?

b. State any practical restrictions on the value of the constant ζ and explain the reasons for the restrictions.

6.23 A source produces an analog output signal $s(t)$. The signal is angle-modulated, and the modulated signal $z(t) = A\cos\{2\pi[(f_c + ms(t))]t + \phi\}$ is transmitted across the channel (m is a constant).

a. Is the transmitted signal PM, FM, or neither?

b. State any practical restrictions on the value of the constant m and explain the reasons for the restrictions.

6.24 Explain why both Equations (6.54) and (6.63) produce FM signals.

6.25 Consider the signal $v(t) = 5\sin[2\pi f_c t + 3\cos 2\pi f_m t + 2\cos 4\pi f_m t]$ where $f_c = 250$ kHz and $f_m = 20$ kHz.

a. Determine mathematical expressions for the instantaneous phase and frequency of $v(t)$.

b. Determine mathematical expressions for the phase and frequency deviation of $v(t)$.

c. What is the maximum phase deviation for $v(t)$?

d. What is the maximum frequency deviation for $v(t)$?

e. What is the practical bandwidth of $v(t)$?

6.26 Consider the signal $v(t) = 8\cos\{2\pi[10^5 + 4\sin(5 \times 10^3 t)]t\}$ volts. Suppose $v(t)$ is passed through an ideal bandpass filter centered at 100 kHz.

a. Draw the amplitude spectrum of $v(t)$.

b. What percentage of the power of $v(t)$ will pass through the filter if the filter's bandwidth is:

i. 2 kHz
ii. 12 kHz
iii. 22 kHz
iv. 32 kHz
v. 42 kHz
vi. 52 kHz
vii. 62 kHz
viii. 72 kHz

6.27 The amplitude spectrum of an angle-modulated signal is shown in Figure P6-27.

a. What is the carrier frequency of the transmitted signal?
b. The information signal was a single sinusoid. What is its frequency?
c. Write the time domain expression for the transmitted signal.

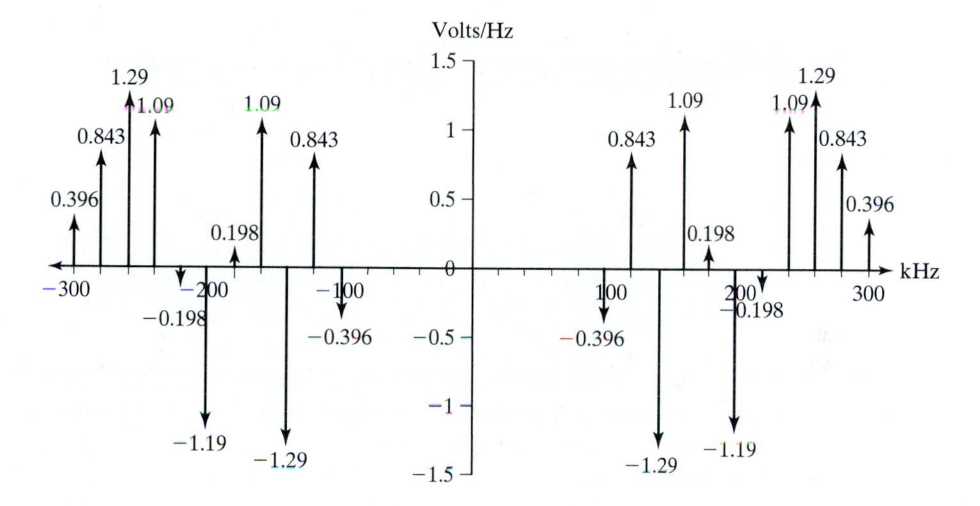

Figure P6-27 Amplitude spectrum of angle-modulated signal.

6.28 Consider the FM signal $v(t) = 8 \cos [2\pi \times 10^5 t + 3\int_{-\infty}^{t} m(t)dt]$ volts.

a. What is the instantaneous frequency of $v(t)$?
b. What is the frequency deviation of $v(t)$?
c. Suppose $\max|m(t)| = 4000\pi$. What is the maximum frequency deviation of $v(t)$?
d. If $m(t)$ has a bandwidth of 5 kHz, what is the bandwidth of $v(t)$?

6.29 Derive the equation for SNR in FM systems corrupted by additive white Gaussian noise.

6.30 Derive the equation for SNR in PM systems corrupted by additive white Gaussian noise.

6.31 Explain the performance-versus-cost trade-offs involved in selecting an analog modulation technique. Be sure to consider AM-DSB-SC, AM-DSB-C, AM-SSB, AM-VSB, PM, and FM.

Chapter 7

Multiplexing Techniques

So far, our examples have been systems with a single source and a single user transmitting across a single communication channel. In reality, you're more likely to find *networks* of multiple sources and users sharing multiple channels. In such cases, we pool resources for efficiency.

Imagine if we all had to run our own private telephone lines every place we wanted to call. Not only would we be crushed under the hardware, but the waste of resources would be ruinously expensive. Building your own telephone company to call out for a pizza once a week is ridiculous. Before long, you'd figure out how to share the expense by letting other users in, and the more the better. Next, you'd have to control the traffic; no bottlenecks and no dead time. Clearly, sharing existing lines is better, and using existing lines "to the max" is ideal. *Multiplexing* is just wringing the most use out of the fewest resources, with the least hassle.

To better understand multiplexing, consider a channel with a bandwidth sufficient to support a transmission rate of 64 kbits/sec. The channel could be used to transmit a single telephone conversation, but good-quality digital voice communication requires only a transmission rate of about 8 kbits/sec. So, when a source speaks to a user over this channel, it will be using only one eighth of the channel's capacity. We need to use eight eighths of the capacity for economy's sake, but we need to keep the messages straight or we'll lose paying customers.

Methods that allow a group of independent sources and users to share a common communication channel are called *multiplexing techniques*. Messages that do not interfere with one another are said to be *orthogonal to one another in signal space*. There are three principal ways to separate the messages and achieve orthogonality: Separate them in time; separate them in frequency; and separate them in characteristics. More specifically:

1. The messages can be separated in time, meaning that different sources transmit at different times. This technique is called *time division multiplexing*.
2. The messages can be separated in frequency, meaning that different sources use different portions of the channel's frequency band. This technique is called *frequency division multiplexing*.

3. The messages can be transmitted at the same time and in the same frequency band, but they can be made orthogonal by using special coding. This technique is called *code division multiplexing*.

Before we explore the basics of each of the multiplexing methods, let's understand the physical shape of the medium, or *media topology*. Figures 7-1a–f show a few of the most common multiplexing topologies, the ones used by cell phones, cable TVs,

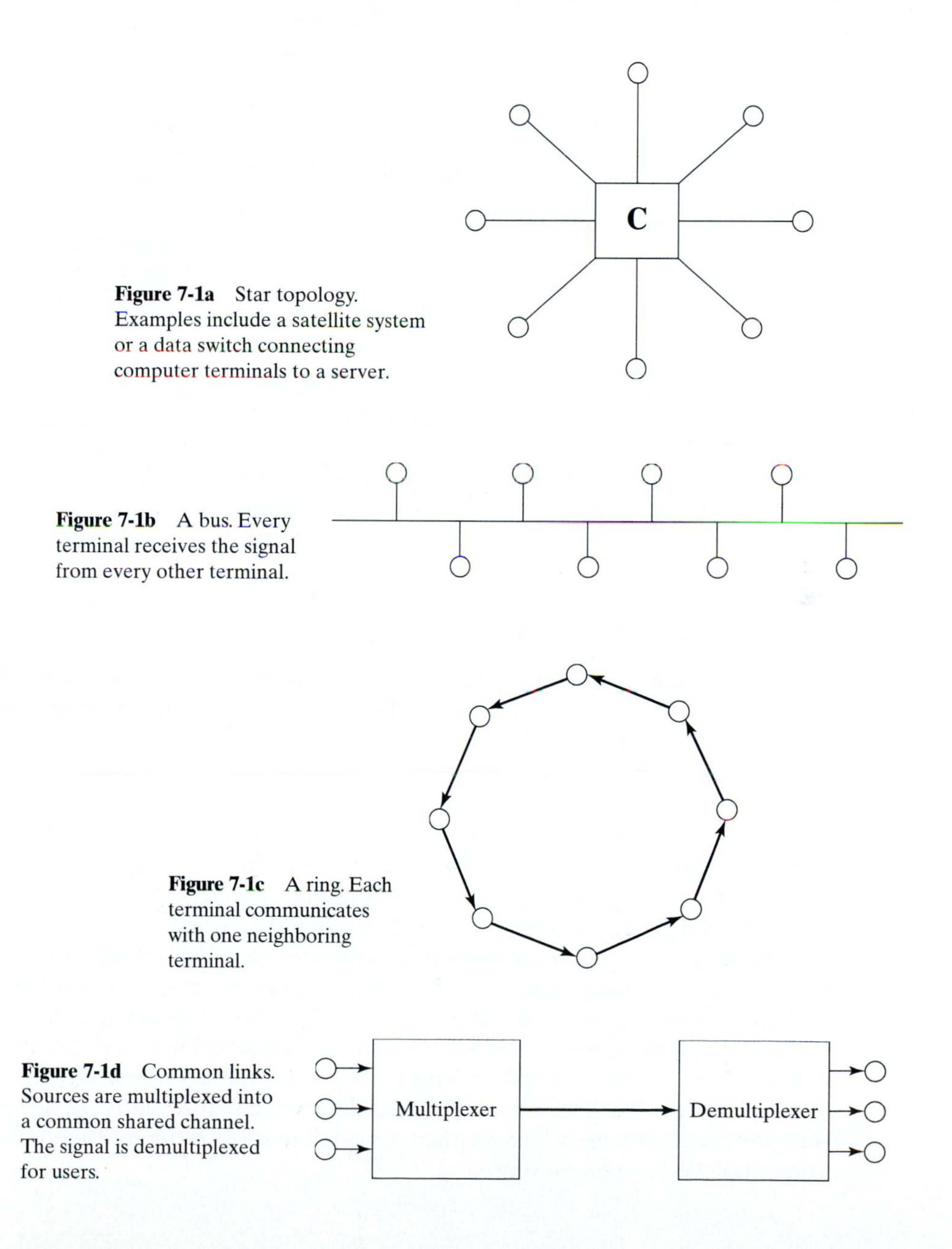

Figure 7-1a Star topology. Examples include a satellite system or a data switch connecting computer terminals to a server.

Figure 7-1b A bus. Every terminal receives the signal from every other terminal.

Figure 7-1c A ring. Each terminal communicates with one neighboring terminal.

Figure 7-1d Common links. Sources are multiplexed into a common shared channel. The signal is demultiplexed for users.

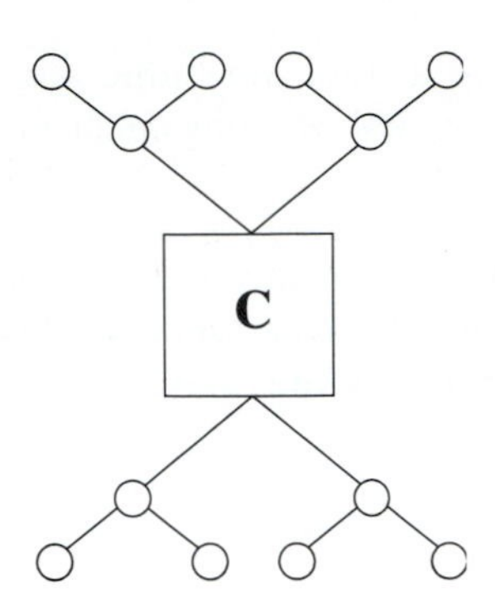

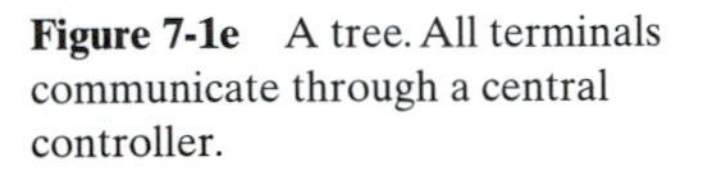
Figure 7-1e A tree. All terminals communicate through a central controller.

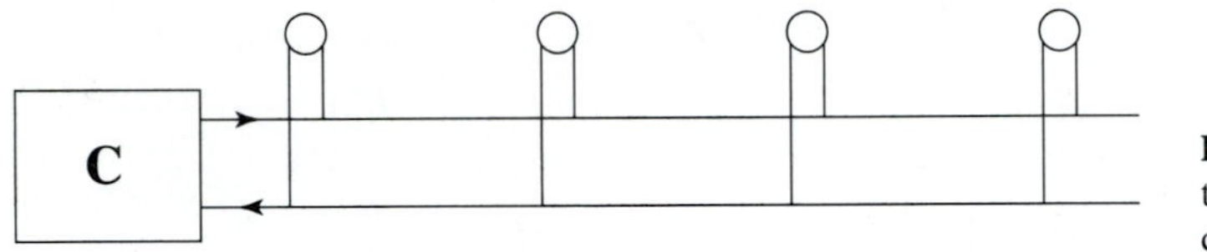

Figure 7-1f A common two-way bus with centralized control.

and computers linked to a local network. In these figures, each line represents a communication link. The link is one-way if the line has a single arrowhead and two-way if it does not. The link may be wireless (such as a radio channel) or a wireline (such as cable or fiber). A box with a letter C represents a controller (a device that controls access to the shared channel). A circle signifies a transmitter or receiver. Now that we've reviewed some familiar topologies, let's examine each of the three multiplexing methods.

7.1 Time Division Multiplexing

It's often practical to combine a set of low-bit-rate streams, each with a fixed and predefined bit rate, into a single high-speed bit stream that can be transmitted over a single channel. This technique is called *time division multiplexing* (TDM) and has many applications, including wireline telephone systems and some cellular telephone systems. The main reason to use TDM is to take advantage of existing transmission lines. It would be very expensive if each low-bit-rate stream were assigned a costly physical channel (say, an entire fiber optic line) that extended over a long distance.

Consider, for instance, a channel capable of transmitting 192 kbit/sec from Chicago to New York. Suppose that three sources, all located in Chicago, each have 64 kbit/sec of data that they want to transmit to individual users in New York. As shown in Figure 7-2, the high-bit-rate channel can be divided into a series of *time slots*, and the time slots can be alternately used by the three sources. The three sources are thus capable of transmitting all of their data across the single, shared channel. Clearly, at the other end of the channel (in this case, in New York), the process must be reversed (i.e., the system must divide the 192 kbit/sec multiplexed data stream back into the original three 64 kbit/sec data streams, which are then provided to three different users). This reverse process is called *demultiplexing*.

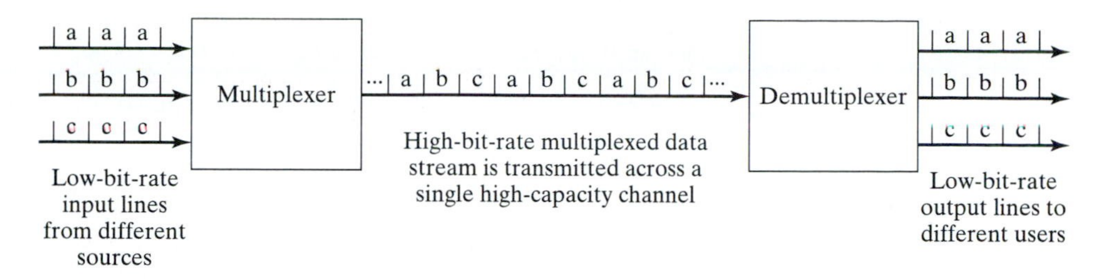

Figure 7-2 Time division multiplexing.

Choosing the proper size for the time slots involves a trade-off between efficiency and delay. If the time slots are too small (say, one bit long) then the multiplexer must be fast enough and powerful enough to be constantly switching between sources (and the demultiplexer must be fast enough and powerful enough to be constantly switching between users). If the time slots are larger than one bit, data from each source must be stored (buffered) while other sources are using the channel. This storage will produce delay. If the time slots are too large, then a significant delay will be introduced between each source and its user. Some applications, such as teleconferencing and videoconferencing, cannot tolerate long delays.

As shown in Example 7.2, the sources that are multiplexed may have different bit rates. When this occurs, each source is assigned a number of time slots in proportion to its transmission rate.

Example 7.1 The T1 system for wireline telephone networks

The T1 system is used for wireline long-distance service in North America and is an excellent example of TDM. Speech from a telephone conversation is sampled once every 125 μsec and each sample is converted into eight bits of digital data (see Chapter 8 for more details). Using this technique, a transmission speed of 64,000 bits/sec is required to transmit the speech. A T1 line is essentially a channel capable of transmitting at a speed of 1.544 Mbit/sec. This is a much higher transmission speed than a single telephone conversation needs, so TDM is used to allow a single T1 line to carry 24 different speech signals between, say, two different telephone substations (called *central offices*) within a city. As shown in Figure 7-3, the 1.544 Mbit/sec bit stream is divided into 193-bit frames. The duration of each frame is

$$\frac{193 \text{ bits per frame}}{1.544 \text{ Mbit/sec}} = 125 \ \mu\text{sec}$$

corresponding to the period between samples of the speech. Each frame is divided into 24 slots, which are each eight bits wide (corresponding to the number of bits needed to digitize a speech sample). One additional bit at the end of the frame is used for signaling. The eight bits of data corresponding to a sample of the speech are placed into one of the 24 slots in the frame.

For longer distances (say, between two large cities) higher-capacity channels are used and multiple T1 lines are time division multiplexed onto the new channels. A T3

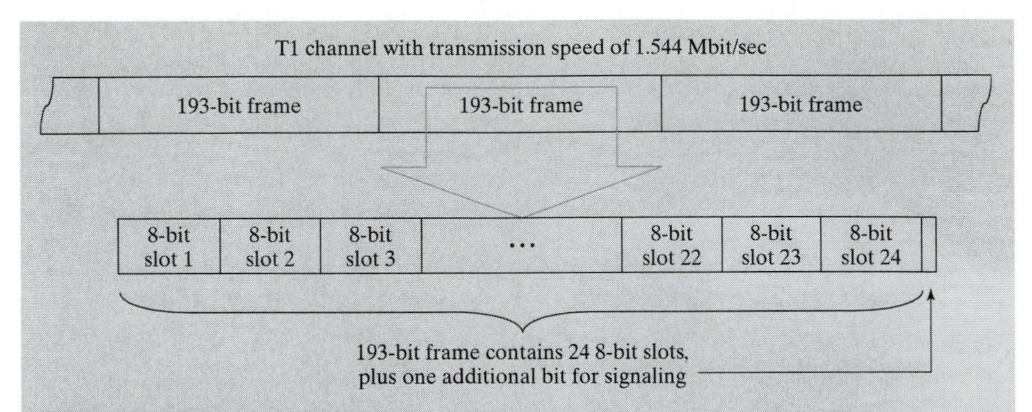

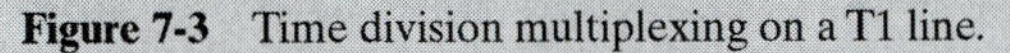

Figure 7-3 Time division multiplexing on a T1 line.

channel, for example, has a transmission speed of 44.736 Mbit/sec and uses TDM to carry 28 T1 lines (a total of 672 different speech signals) plus signaling. For more information on this hierarchical multiplexing system, see Bellamy [7.1].

Example 7.2 TDM with sources having different data rates

Consider the case of three streams with bit rates of 8 kbit/sec, 16 kbit/sec, and 24 kbit/sec, respectively. We want to combine these streams into a single high-speed stream using TDM. The high-speed stream in this case must have a transmission rate of 48 kbit/sec, which is the sum of the bit rates of the three sources. To determine the number of time slots to be assigned to each source in the multiplexing process, we must reduce the ratio of the rates, 8:16:24, to the lowest possible form, which in this case is 1:2:3. The sum of the reduced ratio is 6, which will then represent the minimum length of the repetitive cycle of slot assignments in the multiplexing process. The solution is now readily obtained: In each cycle of six time slots we assign one slot to Source A (8 kbit/sec), two slots to Source B (16 kbit/sec), and three slots to Source C (24 kbit/sec). Figure 7-4 illustrates this assignment, using "a" to indicate data from Source A, "b" to indicate data from Source B, and "c" to indicate data from Source C.

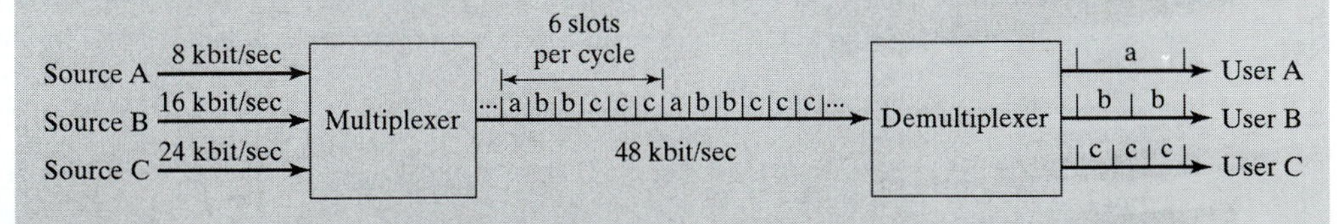

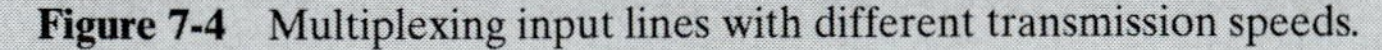

Figure 7-4 Multiplexing input lines with different transmission speeds.

Example 7.3 A more complex TDM system

Consider a system with four low-bit-rate sources of 10 kbit/sec, 15 kbit/sec, 20 kbit/sec, and 30 kbit/sec. Determine the slot assignments when the data streams are combined using TDM.

Solution

The rate ratio 10:15:20:30 reduces to 2:3:4:6. The length of the cycle is therefore $2 + 3 + 4 + 6 = 15$ slots. Within each cycle of 15 slots, we assign two slots to the 10 kbit/sec source, three slots to the 15 kbit/sec source, four slots to the 20 kbit/sec source, and six slots to the 30 kbit/sec source.

So far we have considered a form of TDM that is based on fixed slot assignments to each of the low-bit-rate data streams. In other words, each stream has predefined slot positions in the combined stream, and the receiver must be aware which slots belong to which input stream. Both transmission ends, the transmitter and the receiver, must be perfectly synchronized to the slot period. For this reason, the technique is usually called *synchronous TDM*.

There is another important version of TDM, usually referred to as *statistical TDM*. Statistical TDM is useful for applications in which the low-bit-rate streams have speeds that vary in time. For example, a low-bit-rate stream to a single terminal in a computer network may fluctuate between 2 kbit/sec and 50 kbit/sec during an active connection session (we've all seen variable speeds during Internet connections, for instance). If we assign the stream enough slots for its peak rate (that is, for 50 kbit/sec), then we will be wasting slots when the rate drops well below the peak value. This waste can be especially significant if the system has many variable-speed low-bit-rate streams.

Statistical TDM works by calculating the average transmission rates of the streams to be combined, and then uses a high-speed multiplexing link with a transmission rate that is equal to (or slightly greater than) the statistical average of the combined streams. Since the transmission rates from each source are variable, we no longer assign a fixed number of time slots to each data stream. Rather, we dynamically assign the appropriate number of slots to accommodate the current transmission rates from each stream. Because the combined rate of all the streams will also fluctuate in time between two extreme values, we need to buffer the output of the low-bit-rate streams when the combined rate exceeds the transmission rate of the high-speed link.

With statistical TDM, we are no longer relying on synchronized time slots with fixed assignments for each input stream, as we did with synchronous TDM. So how does the demultiplexer in statistical TDM know which of the received bits belongs to which data stream? Prior to transmission, we divide each stream of bits coming from a source into fixed-size *blocks*. We then add a small group of bits called a *header* to each block, with the header containing the addresses of the source and intended user for that block. The block and the header are then transmitted together across the channel. Combined, the block and header are called a *packet*.

Actually, the header may contain other information besides the source and user addresses, such as extra bits for error control (see Chapter 10) or additional bits for link control (used, for example, to indicate the position of a particular block in a sequence of blocks coming from the same user, or to indicate priority level for a particular message). Extra bits can also be added to the beginning and end of a block for synchronization; a particular pattern of bits, called a *start flag*, can be used in the header to mark the start of a block, and another particular pattern of bits, called an *end flag*, can

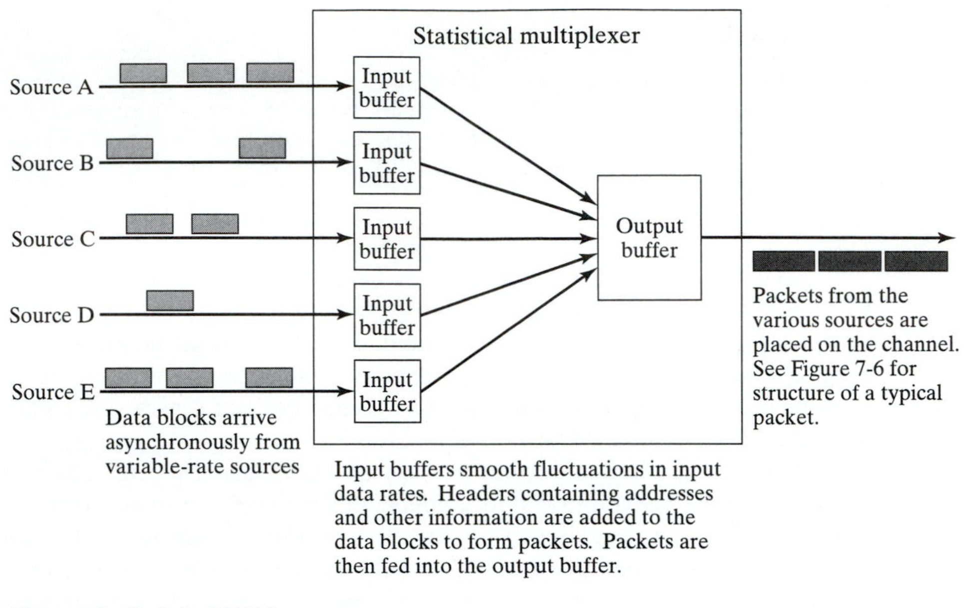

Figure 7-5 Statistical TDM.

Start flag	Address field	Control field	Information bits	Error control	End flag

Figure 7-6 Structure of a typical statistical TDM packet.

be used to conclude the block. Each block transmitted across the channel thus contains a group of information bits that the user wants, plus additional bits needed by the system to ensure proper transmission. These additional bits, while necessary to system operation, reduce the effective transmission rate on the channel. Figures 7-5 and 7-6 present the statistical TDM technique and the structure of a typical packet.

7.2 Frequency Division Multiplexing

In many communication systems, a single, large frequency band is assigned to the system and is shared among a group of users. Examples of this type of system include:

1. A microwave transmission line connecting two sites over a long distance. Each site has a number of sources generating independent data streams that are transmitted simultaneously over the microwave link.
2. AM or FM radio broadcast bands, which are divided among many channels or stations. The stations are selected with the radio dial by tuning a variable-frequency filter. (We examined AM and FM in Chapter 6.)

3. A satellite system providing communication between a large number of ground stations that are separated geographically but that need to communicate at the same time. The total bandwidth assigned to the satellite system must be divided among the ground stations.
4. A cellular radio system that operates in full-duplex mode over a given frequency band. The earlier cellular telephone systems, for example AMPS, used analog communication methods. The bandwidth for these systems was divided into a large number of channels. Each pair of channels was assigned to two communicating end-users for full-duplex communications.

Frequency division multiplexing (FDM) means that the total bandwidth available to the system is divided into a series of nonoverlapping frequency sub-bands that are then assigned to each communicating source and user pair. Figures 7-7a and 7-7b show how this division is accomplished for a case of three sources at one end of a system that are

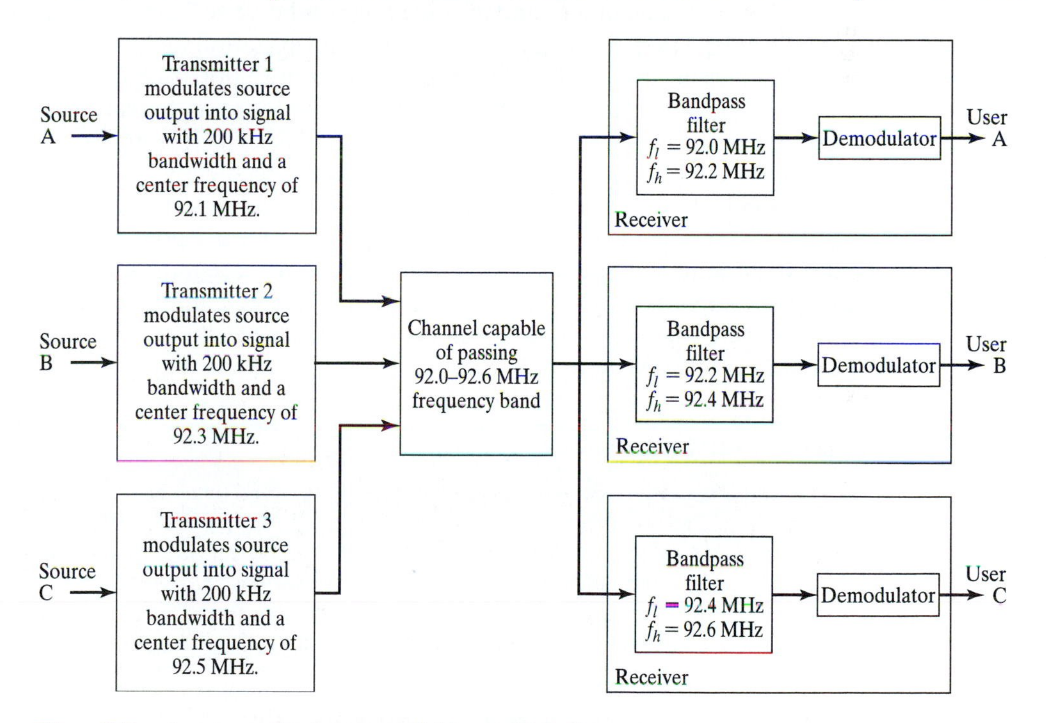

Figure 7-7a A system using frequency division multiplexing.

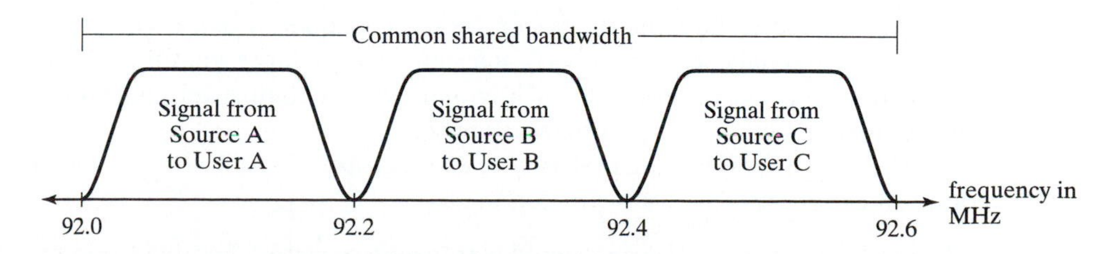

Figure 7-7b Spectral occupancy of signals in an FDM system.

communicating with three separate users at the other end. Note that each transmitter modulates its source's information into a signal that lies in a different frequency sub-band (Transmitter 1 generates a signal in the frequency sub-band between 92.0 MHz and 92.2 MHz, Transmitter 2 generates a signal in the sub-band between 92.2 MHz and 92.4 MHz, and Transmitter 3 generates a signal in the sub-band between 92.4 MHz and 92.6 MHz). The signals are then transmitted across a common channel.

At the receiving end of the system, bandpass filters are used to pass the desired signal (the signal lying in the appropriate frequency sub-band) to the appropriate user and to block all the unwanted signals. To ensure that the transmitted signals do not stray outside their assigned sub-bands, it is also common to place appropriate passband filters at the output stage of each transmitter. It is also appropriate to design an FDM system so that the bandwidth allocated to each sub-band is slightly larger than the bandwidth needed by each source. This extra bandwidth, called a *guardband*, allows systems to use less expensive filters (i.e., filters with fewer poles and therefore less steep rolloffs).

FDM has both advantages and disadvantages relative to TDM. The main advantage is that unlike TDM, FDM is not sensitive to propagation delays. Channel equalization techniques needed for FDM systems are therefore not as complex as those for TDM systems. Disadvantages of FDM include the need for bandpass filters, which are relatively expensive and complicated to construct and design (remember that these filters are usually used in the transmitters as well as the receivers). TDM, on the other hand, uses relatively simple and less costly digital logic circuits. Another disadvantage of FDM is that in many practical communication systems, the power amplifier in the transmitter has nonlinear characteristics (linear amplifiers are more complex to build), and nonlinear amplification leads to the creation of out-of-band spectral components that may interfere with other FDM channels. Thus, it is necessary to use more complex linear amplifiers in FDM systems.

Example 7.4 FDM for commercial FM radio

The frequency band from 88 MHz to 108 MHz is reserved over the public airwaves for commercial FM broadcasting. The 88–108 MHz frequency band is divided into 200 kHz sub-bands. As we saw in Chapter 6, the 200 kHz bandwidth of each sub-band is sufficient for high-quality FM broadcast of music. The stations are identified by the center frequency within their channel (e.g., 91.5 MHz, 103.7 MHz). This system can provide radio listeners with their choice of up to 100 different radio stations.

7.3 Code Division Multiplexing and Spread Spectrum

Code division multiplexing (CDM) allows signals from a series of independent sources to be transmitted at the same time over the same frequency band. This is accomplished by using orthogonal codes to spread each signal over a large, common frequency band. At the receiver, the appropriate orthogonal code is then used again to recover the particular signal intended for a particular user.

The key principle of CDM is *spread spectrum*. Spread spectrum is a means of communication with the following features:

1. Each information-bearing signal is transmitted with a bandwidth in excess of the minimum bandwidth necessary to send the information.

2. The bandwidth is increased by using a *spreading code* that is independent of the information.
3. The receiver has advance knowledge of the spreading code and uses this knowledge to recover the information from the received, spread-out signal.

Spread spectrum seems incredibly counterintuitive. We've spent most of this book studying ways to transmit information using a minimum amount of bandwidth. Why should we now study ways to intentionally increase the amount of bandwidth required to transmit a signal? By the end of this chapter you will see that spread spectrum is a good technique for providing secure, reliable, private communication in an environment with multiple transmitters and receivers. In fact, spread spectrum and CDM are currently being used in an ever-increasing number of commercial cellular telephone systems.

7.3.1 Direct Sequence Spread Spectrum

Let's begin by considering a technique for spreading the spectrum of an information signal. Suppose we have a series of information bits—say, "110110"—that we want to transmit at a particular speed—say, $r_b = 1000$ bits/sec. We know that the bandwidth required to transmit this signal will be proportional to the transmission speed (e.g., if PSK is used and we need 90% in-band power, the signal will require a bandwidth of $2r_b = 2$ kHz). Now consider the circuit shown in Figure 7-8, which exclusive-ORs[1] the information bits from the source with a second sequence of bits known as a spreading code. The spreading code is being clocked at a rate three times as fast as the source is outputting information. If the source information and the spreading code are synchronized, the sequence at the output of the exclusive-OR gate also has a rate of $3r_b$. Let's use r_{ss} to symbolize the rate of the spreading code, which we call the *chip rate*.

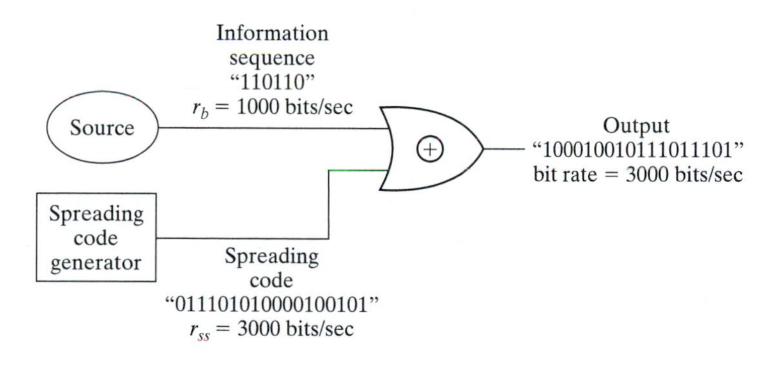

Figure 7-8 Spreading circuit.

[1]The exclusive-OR function is equivalent to modulo-2 addition, which is discussed in Chapter 10. Briefly, the exclusive-OR operator is symbolized by $\oplus$ and defined as $0 \oplus 0 = 0, 0 \oplus 1 = 1, 1 \oplus 0 = 1$, and $1 \oplus 1 = 0$.

The circuit in Figure 7-8 (which we can call a *spreader*) converts the information sequence, which is being output by the source at a rate of r_b, into a longer, faster sequence of bits being output at a rate of r_{ss}. This longer, faster sequence of bits will require more bandwidth to transmit (again, if PSK is used and we need 90% in-band power, the new sequence will require a bandwidth of $2r_{ss} = 6$ kHz). The technique of exclusive-ORing an information sequence with a faster spreading code is known as *direct sequence spread spectrum*.

As shown in Figure 7-8, suppose that the spreading code is the sequence "011101010000100101" and that it is being output at a chip rate of 3 kbit/sec. Table 7-1 shows the source information sequence, the spreading sequence, and the output of the spreader. The large numbers in the "source information" column indicate that the source is outputting only one bit per millisecond.

Table 7-1 Data Sequence for Spreader

Time (msec)	Source Information	Spreading Code	Spreader Output
0	1	0	1
0.33		1	0
0.67		1	0
1	1	1	0
1.33		0	1
1.67		1	0
2	0	0	0
2.33		1	1
2.67		0	0
3	1	0	1
3.33		0	1
3.67		0	1
4	1	1	0
4.33		0	1
4.67		0	1
5	0	1	1
5.33		0	0
5.67		1	1

Now suppose that the output of the spreader is transmitted across the channel at 3 kbit/sec. The user needs to extract the original 6-bit information sequence ("110110") from the received 18-bit sequence ("100010010111011101"). This is accomplished by exclusive-ORing the received sequence again by the spreading sequence. This process is called *despreading*, and it works because for any two binary variables X and Y,

$$X \oplus Y \oplus Y = X \tag{7.1}$$

Thus, exclusive-ORing the information bits with the spreading sequence twice (once prior to transmission and once after reception) reproduces the original information bits. The process of spreading and despreading is shown in Figure 7-9 and Table 7-2.

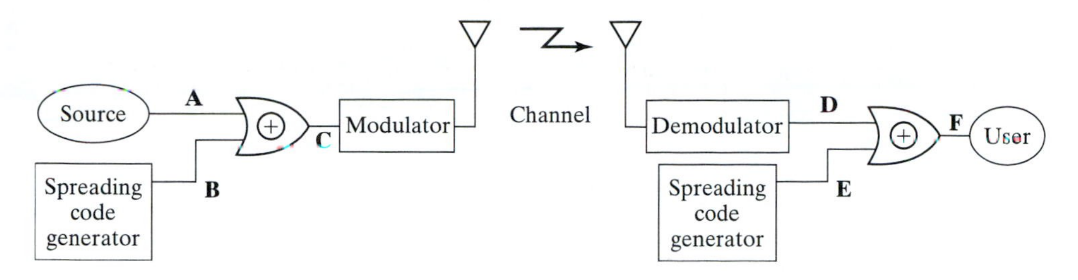

Figure 7-9 Information flow through the spreading and despreading circuits (letters correspond to entries in Table 7-2).

Table 7-2 Spreading and Despreading Operations

Time (msec)	Source Information **A**	Spreading Code **B**	Transmitted Sequence **C**	Received Sequence **D**	Spreading Code **E** (Same as **B**)	Despread Information **F**
0	1	0	1	1	0	1
0.33		1	0	0	1	
0.67		1	0	0	1	
1	1	1	0	0	1	1
1.33		0	1	1	0	
1.67		1	0	0	1	
2	0	0	0	0	0	0
2.33		1	1	1	1	
2.67		0	0	0	0	
3	1	0	1	1	0	1
3.33		0	1	1	0	
3.67		0	1	1	0	
4	1	1	0	0	1	1
4.33		0	1	1	0	
4.67		0	1	1	0	
5	0	1	1	1	1	0
5.33		0	0	0	0	
5.67		1	1	1	1	

We can illustrate the general spreading and despreading processes in the frequency domain as shown in Figure 7-10. Spreading increases the bandwidth by a factor of r_{ss}/r_b, and despreading reduces the bandwidth back to that of the original signal. This factor, r_{ss}/r_b, is known as the *processing gain* and is symbolized G_p

$$G_p = \frac{r_{ss}}{r_b} \tag{7.2}$$

We've now shown that spreading and despreading are mathematically valid operations, but we have yet to show why they are useful. In fact, as mentioned earlier, these

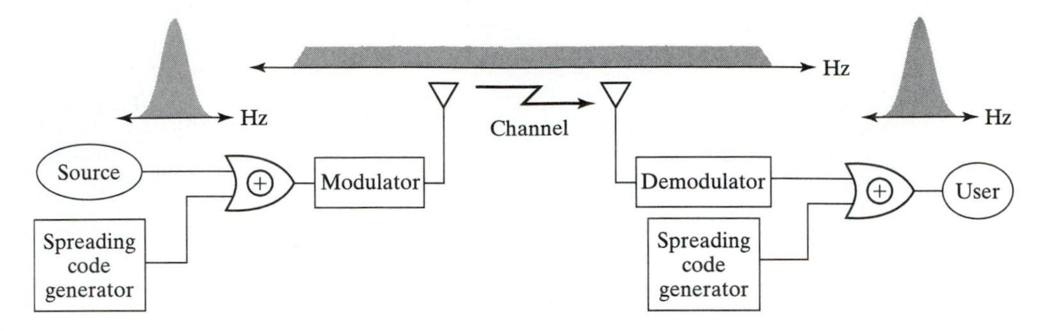

Figure 7-10 Spectrum of a signal as it passes through a spread spectrum system.

operations seem counterintuitive because they increase the bandwidth of the transmitted signal—something communication systems engineers instinctively avoid. In order to see practical applications for spreading and despreading, we must investigate the properties of the spreading code.

Although any sequence of "1"s and "0"s can be used as a spreading code, practical spreading codes must look like a sequence of random, independent, equiprobable bits. These codes are called *pseudo-random* (PN)—rather than truly random—because the transmitter and receiver must generate the same sequence (otherwise despreading will not work).[2]

The bits in a PN code are essentially uncorrelated. Thus, when a PN spreading code is applied to additive white Gaussian noise, the power spectral density of the noise remains flat and unchanged.

Figure 7-11 shows the spreading, transmitting, and despreading processes for a general signal corrupted in the channel by additive white Gaussian noise. A PN spreading code is used. Note that the spreading process flattens (as well as spreads) the spectrum of

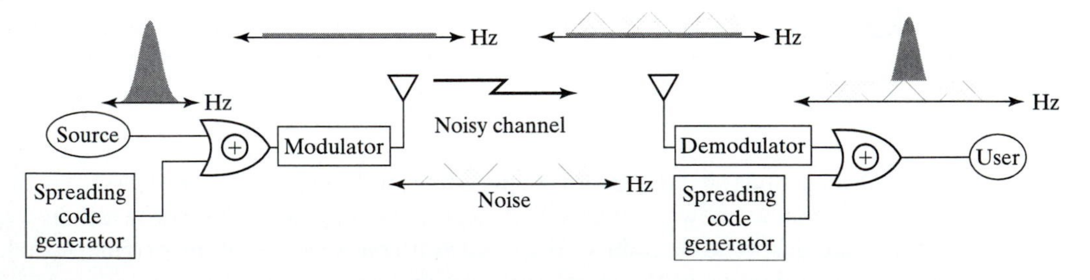

Figure 7-11 The effects of noise on a spread spectrum signal.

[2]For the transmitter and receiver to generate the same code, they must generate the code from a series of deterministic rules. The code is, therefore, technically not random. PN is a short form for "pseudo-random noise."

the information signal, and that the spread transmitted signal is essentially buried in the noise by the time it arrives at the receiver. The despreading process narrows and heightens the spectrum of the received signal, yet leaves the spectrum of the received noise essentially unchanged. The receiver can now extract the despread signal from the noise using a bandpass filter.

Despreading reduces the bandwidth of the signal by a factor of r_{ss}/r_b, a value we defined in Equation (7.2) as the processing gain G_p. The in-band signal-to-noise ratio (SNR) of the despread signal is therefore a factor of G_p greater than the in-band SNR of the received, spread signal. This is why G_p is called the processing gain.

One advantage of spread spectrum is that the transmitted signal is buried in the noise and is virtually undetectable by other receivers that do not know the spreading code. This provides security. Also, since a spread spectrum signal contains low energy spread over a large bandwidth, the signal can be transmitted on a channel containing other communications without causing significant interference to the other communications (i.e., the spread spectrum signal will simply look like a very small amount of additional noise). In some applications, this property can allow a spread spectrum system to share bandwidth with existing communication systems without significantly affecting the operation of the other system (Pickholtz [7.2]).

Let's now investigate the pseudo-random spreading code and its effects in more detail. As mentioned earlier, we want each bit in the spreading code to be random, independent, and equiprobable. These characteristics are achieved if the code exhibits the following three properties (Lee [7.3]):

1. *Balance*. The number of "1"s in the code should differ from the number of "0"s by no more than one. Balance is an important indicator that the bits are equiprobable.
2. *Run property*. Among the groups of "1"s and "0"s in the code sequence, half the groups of each type should be one bit long, one fourth of the groups of each type should be two bits long, one eighth of the groups of each type should be three bits long, etc. The run property is an important indicator that the bits are random and independent.
3. *Correlation*. Consider an n-bit binary sequence $b_1 b_2 b_3 \ldots b_n$. We can write this binary sequence in matrix form as

$$\mathbf{[b]} = [b_1 \quad b_2 \quad b_3 \quad b_4 \quad \ldots \quad b_n]$$

Now consider a second n-bit sequence, say, $c_1 c_2 c_3 \ldots c_n$. Again, we can write this binary sequence in matrix form as

$$\mathbf{[c]} = [c_1 \quad c_2 \quad c_3 \quad c_4 \quad \ldots \quad c_n]$$

We can use matrix notation to signify a bit-by-bit exclusive-ORing of the two sequences:

$$\mathbf{[b]} \oplus \mathbf{[c]} = [b_1 \oplus c_1 \quad b_2 \oplus c_2 \quad b_3 \oplus c_3 \quad b_4 \oplus c_4 \quad \ldots \quad b_n \oplus c_n] \tag{7.3}$$

Let's consider one other operation—a cyclic shift. Let the notation $\mathbf{[b]}^{(n)}$ signify the sequence of bits $\mathbf{[b]}$ cyclically shifted n places to the right. In other words,

$$[\boldsymbol{b}] = [b_1 \quad b_2 \quad b_3 \quad b_4 \quad \ldots \quad b_n]$$

$$[\boldsymbol{b}]^{(1)} = [b_n \quad b_1 \quad b_2 \quad b_3 \quad \ldots \quad b_{n-1}]$$

$$[\boldsymbol{b}]^{(j)} = [b_{(n-j)+1} \quad b_{(n-j)+2} \quad b_{(n-j)+3} \quad \ldots \quad b_n \quad b_1 \quad \ldots \quad b_{n-j}] \tag{7.4}$$

The third desirable property in a PN code is that the n-bit sequence produced by $[\boldsymbol{b}] \oplus [\boldsymbol{b}]^{(j)}$ should exhibit balance for all nonzero values of j less than n (the number of "1"s in $[\boldsymbol{b}] \oplus [\boldsymbol{b}]^{(j)}$ should differ from the number of "0"s by no more than one for any value $1 \le j \le n - 1$). This property also helps ensure independence and randomness of the bits in the PN code.

Why does the correlation property help ensure independence and randomness? Because exclusive-ORing two bits produces a "0" if both bits are the same (if both bits are "0"s or if both bits are "1"s) and produces a "1" if the two bits are different (if one of the bits is a "0" and the other is a "1"). If a sequence of bits is truly random and independent, then cyclically shifting the sequence by an arbitrary number of places and performing a bit-by-bit comparison of the original and shifted sequences should produce the same number of agreements (the values of the two bits are the same) as disagreements (the values of the two bits are different). Of course, if the sequence contains an odd number of bits, the number of agreements and disagreements will have to differ by at least one.

Example 7.5 Evaluating a pseudo-random code

Evaluate the 15-bit spreading code "100110101111000" for balance, run property, and correlation.

Solution

Balance: The code has eight "1"s and seven "0"s. Balance is achieved.

Run property: The code has two one-bit runs of "1"s and two one-bit runs of "0"s; one two-bit run of "1"s and one two-bit run of "0"s; no three-bit runs of "1"s and one three-bit run of "0"s; one four-bit run of "1"s and no four bit runs of "0"s. For a 15-bit sequence, this code comes as close to satisfying the run property as is possible.

Correlation: We will investigate the sequences produced by $[\boldsymbol{b}] \oplus [\boldsymbol{b}]^{(1)}$ and $[\boldsymbol{b}] \oplus [\boldsymbol{b}]^{(2)}$ here. The evaluations are rather tedious, and evaluating some of the remaining 13 sequences is left as an exercise in Problem 7.20.

$$[\boldsymbol{b}] \oplus [\boldsymbol{b}]^{(1)} = [1\,0\,0\,1\,1\,0\,1\,0\,1\,1\,1\,1\,0\,0\,0] \oplus [0\,1\,0\,0\,1\,1\,0\,1\,0\,1\,1\,1\,1\,0\,0]$$

$$= [1\,1\,0\,1\,0\,1\,1\,1\,1\,0\,0\,0\,1\,0\,0],\ \text{a balanced sequence (eight "1"s and seven "0"s)}$$

The mathematics is simpler to perform if we stack the two sequences in rows and then perform an exclusive-OR on each column, in much the same way we add two large numbers. In other words, we can calculate $[\boldsymbol{b}] \oplus [\boldsymbol{b}]^{(1)}$ as

$$\begin{array}{rl} & [1\,0\,0\,1\,1\,0\,1\,0\,1\,1\,1\,1\,0\,0\,0] \\ \oplus & [0\,1\,0\,0\,1\,1\,0\,1\,0\,1\,1\,1\,1\,0\,0] \\ \hline & [1\,1\,0\,1\,0\,1\,1\,1\,1\,0\,0\,0\,1\,0\,0] \end{array}$$

Similarly, $[b] \oplus [b]^{(2)} =$

$$\begin{array}{rl} & [1\,0\,0\,1\,1\,0\,1\,0\,1\,1\,1\,1\,0\,0\,0] \\ \oplus & [0\,0\,1\,0\,0\,1\,1\,0\,1\,0\,1\,1\,1\,1\,0] \\ \hline & [1\,0\,1\,1\,1\,1\,0\,0\,0\,1\,0\,0\,1\,1\,0], \text{ another balanced sequence} \end{array}$$

We can similarly show that the sequences produced by $[b] \oplus [b]^{(j)}$ are balanced for $j = 3, 4, \ldots 14$.

We have shown that the 15-bit spreading code "100110101111000" exhibits balance, run property, and correlation. This code is therefore pseudo-random.

As discussed earlier, PN spreading codes are desirable because they allow a signal to be spread and despread but do not affect the power spectral density of the noise. PN codes can be generated rather easily using a series of shift registers and logic gates in feedback as shown in Figure 7-12.

Let's investigate the Figure 7-12 circuit in more detail. Initializing the shift registers with the values $X_1X_2X_3X_4 = 1001$ and clocking the circuit at 3000 cycles per second, Table 7-3 shows the changing values of the shift registers and the generator output. Note that the pseudo-random output repeats itself in a cycle every 15 bits.

Fifteen bits is the longest cycle that can be produced using four shift registers and logic gates. In reviewing Figure 7-12 and Table 7-3, we see that no longer cycle can be produced because the present output of the generator and the next states of the shift registers depend solely on the present values of the shift registers, and there are only 15 different nonzero values of $X_1X_2X_3X_4$. (The value $X_1X_2X_3X_4 = 0000$ must be avoided because this state will cause no changes to ever occur in the shift register values or in the output.) In general, a circuit with logic gates and N shift registers can produce at maximum a sequence of $2^N - 1$ bits before it starts repeating itself (note that there are ways to connect the feedback gates to produce a shorter sequence, but shorter sequences are less random because they repeat more often, and hence are less desirable). A circuit configured to produce the maximum sequence of nonrepeating bits for a given number of shift registers is called a *maximal length* PN code generator. The algorithm illustrated in Figure 7-12 can also, if desired, be easily implemented in software.

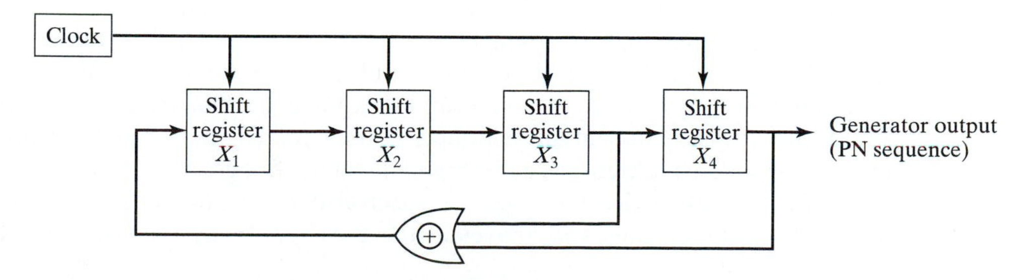

Figure 7-12 A pseudo-random sequence generator.

Table 7-3 Output Sequence and Internal Shift Register Values for the PN Generator Given in Figure 7-12

Time (msec)	X_1	X_2	X_3	X_4	Output	
0	1	0	0	1	1	
0.33	1	1	0	0	0	
0.67	0	1	1	0	0	
1	1	0	1	1	1	
1.33	0	1	0	1	1	
1.67	1	0	1	0	0	
2	1	1	0	1	1	
2.33	1	1	1	0	0	
2.67	1	1	1	1	1	
3	0	1	1	1	1	
3.33	0	0	1	1	1	
3.67	0	0	0	1	1	
4	1	0	0	0	0	
4.33	0	1	0	0	0	
4.67	0	0	1	0	0	
5	1	0	0	1	1	Sequence begins repeating (see t = 0, 0.33, etc.)
5.33	1	1	0	0	0	
5.67	0	1	1	0	0	
6	1	0	1	1	1	
•	•	•	•	•	•	
•	•	•	•	•	•	
•	•	•	•	•	•	

Spreading can be used as a multiplexing technique by developing a series of orthogonal spreading codes. These are PN codes with one additional feature: If any two different orthogonal spreading codes are exclusive-ORed bit by bit, the resulting series of bits will itself be a PN code. Thus, if a signal is spread using one code and then despread using another orthogonal code, the result will just look like a PN code and will have a power spectral density similar to wideband white Gaussian noise.[3]

Consider the system shown in Figure 7-13, with each of the three source/user pairs employing mutually orthogonal spreading codes to transmit information over a common channel. Source A uses one spreading code (let's call it Spreading Code A) and transmits the spread spectrum signal shown in medium gray in Figure 7-13. This signal is wideband and, from the viewpoint of the channel and any observer who doesn't know the code, the signal looks exactly like additive white Gaussian noise. Source B uses a second, orthogonal spreading code (Spreading Code B) and transmits another spread spectrum signal, shown in light gray in Figure 7-13. Again, from the viewpoint of the channel and any observer who doesn't know the code, this transmitted signal looks like additive white Gaussian noise. Source C uses a third, orthogonal spreading code (Spreading Code C) to transmit a message, shown in dark gray, across the channel. The sum of the signals (which has a power spectral density similar to wideband noise) carries across the channel and arrives at the receivers being employed by the three

[3]Developing a series of mutually orthogonal codes is not a simple task. See Simon [7.4].

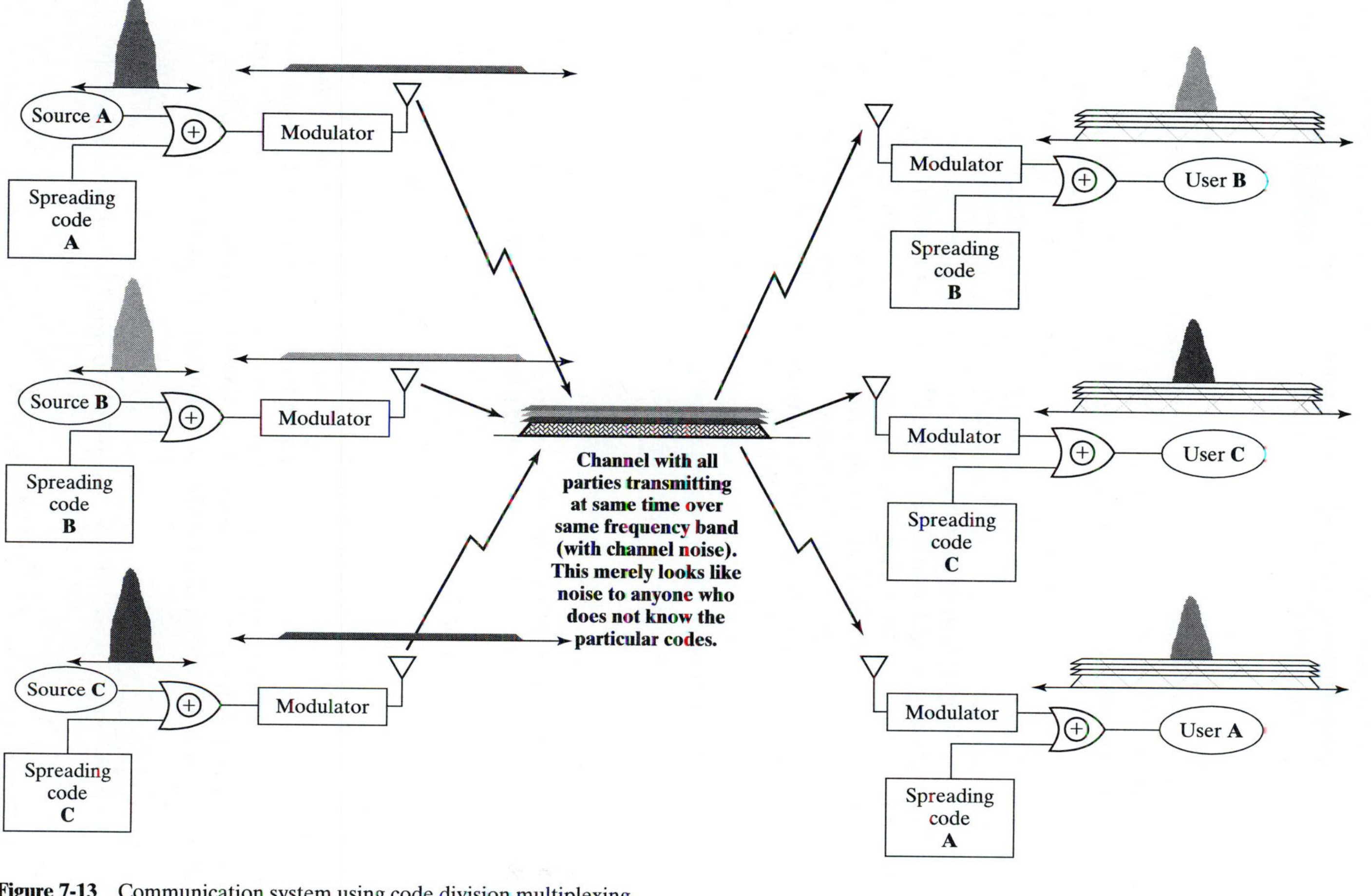

Figure 7-13 Communication system using code division multiplexing.

users. Some noise from the channel itself may also be added (this is the brick portion of the power spectral density in the channel in Figure 7-13).

Let's now consider the receivers. The receiver associated with User B (upper right corner in Figure 7-13) applies Spreading Code B to the total received signal. This despreads the portion of the signal transmitted from Source B (the light gray part of the spectrum in the upper right corner of Figure 7-13) but leaves all other portions of the received signal with a wideband noise-like power spectral density (the cross-hatched part of the spectrum). Using a narrow bandpass filter, User B may now extract the portion of the signal associated with Source B, with the channel noise and the interference from Users A and C significantly reduced. Similarly, User C may use Spreading Code C to extract its intended message from Source C (see the spectrum in the center right part of Figure 7-13), and User A may use Spreading Code A to extract its intended message from Source A (see the spectrum in the lower right corner of Figure 7-13).

An inexact (although illuminating) analogy to a spread spectrum system is a cocktail party where each conversing group is speaking a different language. At such a party the background noise may be much louder than the voice of the person speaking in your group, but if you listen carefully you can still understand your speaker.

Let's now consider a general system with M sources and M users employing code division multiplexing. Let the channel noise be additive white Gaussian noise with a two-sided average normalized power spectral density of $N_o/2$, and let each source transmit its spread message with an average normalized power of P. To simplify matters, let's assume there is no attenuation in the channel. Let BW_{ss} represent the bandwidth of the spread spectrum signal. The average normalized power of the total received signal is

$$P_{total\ received\ power} = MP + \left[\frac{N_o}{2}\right] BW_{ss} \tag{7.5}$$

where the first term (MP) is the power from all the transmitters in the system and the second term is the noise from the channel. The SNR at each receiver before despreading is

$$SNR_{before\ despreading} = \frac{P}{(M-1)P + \left[\frac{N_o}{2}\right] BW_{ss}} \tag{7.6}$$

where the first term in the denominator represents the interference from all of the other transmitters in the system and the second term represents the noise.

Despreading concentrates the desired signal's power into a bandwidth that is $1/G_p$ times the bandwidth of the spread signal, but despreading does not affect the power spectral density of the interfering signals from the other transmitters or the channel noise. Thus, after despreading and filtering, the SNR becomes

$$\begin{aligned} SNR_{after\ despreading} &= \frac{P}{\frac{1}{G_p}\left\{(M-1)P + \left[\frac{N_o}{2}\right] BW_{ss}\right\}} = \frac{G_p P}{(M-1)P + \left[\frac{N_o}{2}\right] BW_{ss}} \\ &= G_p SNR_{before\ despreading} \end{aligned} \tag{7.7}$$

We can now see why G_p is called "processing gain." In a practical system, the combined power of all the interferers is much greater than the channel noise, and hence the second term in the denominator of Equation (7.7) can be neglected, producing

$$SNR_{after\ despreading} \approx \frac{G_p P}{(M - 1)P} = \frac{G_p}{(M - 1)} \tag{7.8}$$

If we need a given SNR for a certain level of system performance, we can use Equation (7.8) to determine how much spreading is needed (a function of G_p) to accommodate a certain number of users in the system (M).

Code division multiplexing has many advantages. Individual parties have a high degree of security; as we've discussed, an individual's transmitted signal is virtually buried within the noise and the signals from other users, therefore the signal is virtually undetectable by anyone who does not know the particular spreading code. CDM also works well as a multiple access technique; there is no need for all users in the system to be synchronized, nor is there any need for a complex control protocol to assign resources (once a source and user agree on one of a series of orthogonal spreading codes, transmission can start any time). Statistical multiplexing can be accomplished with a minimum of overhead, and additional transmitters can be accommodated with only a gradual degradation in SNR.

A word of warning: You must be careful with the concept of gradual degradation, since it relates to SNR and not directly to either bit-error rate or to the quality of the received message. We've seen in previous chapters that bit-error rate and SNR are not linearly related. Thus, a small degradation in SNR may produce a significant increase in bit-error rate. Furthermore, in certain types of voice and image communication systems, once a threshold bit-error rate is reached, even a small increase in the error rate may significantly impact the quality of the received voice signal or image.

7.3.2 Frequency-Hopping Spread Spectrum

In Section 7.3.1, we developed a spread spectrum signal by exclusive-ORing a sequence of information bits with a faster spreading code. We called this technique direct sequence spread spectrum. Let's now consider another technique for spreading spectrum—*frequency hopping*, as illustrated in Figure 7-14. In Figure 7-14, the bandwidth of the signal is spread by pseudo-randomly changing the transmitter's carrier frequency many times during the transmission of a message. The carrier signal is thus "hopping" around over a wide frequency band. The spread received signal is decoded using the same pseudo-random sequence of carrier frequencies in the demodulator.

Much of the analysis and design we've done for direct sequence spread spectrum also applies to frequency hopping. The circuitry used to generate the pseudo-random code in Figure 7-14 is the same as for direct sequence spread spectrum (see Figure 7-12). The concepts of orthogonal codes and processing gain are also directly applicable. In a frequency-hopped spread spectrum system, processing gain is defined as

$$G_p = \frac{\text{bandwidth of spread spectrum signal}}{\text{bandwidth of despread signal}} \tag{7.9}$$

This definition is consistent with Equation (7.2).

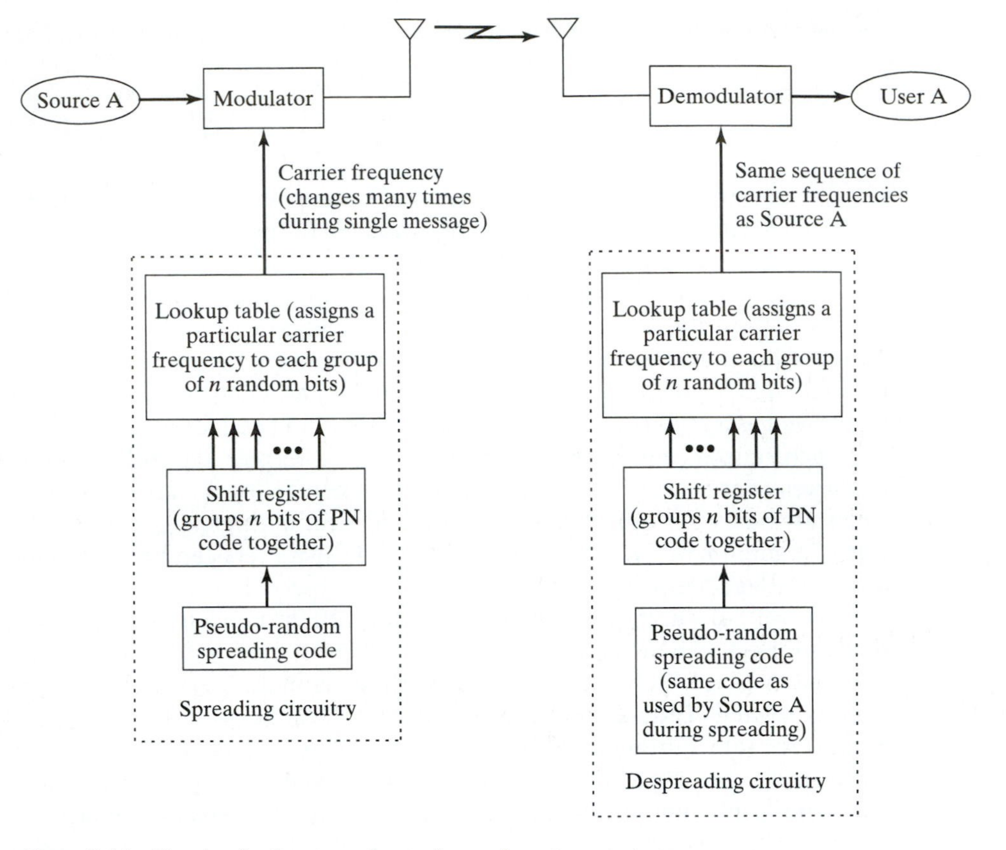

Figure 7-14 Circuitry for frequency-hopped spread spectrum system.

Frequency-hopping spread spectrum has many advantages over direct-sequence spread spectrum. First, current technology permits us to obtain a much larger bandwidth using frequency hopping than using direct-sequence spread spectrum. A frequency-hopped system can thus produce a significantly higher processing gain. Second, if hopping is fast enough, a frequency-hopped spread spectrum system can provide more immunity to certain types of channel distortion (Pickholtz [7.2]). Frequency-hopped spread spectrum also has disadvantages, chief among them being that phase modulation techniques such as PSK and DPSK are difficult to use because of problems maintaining phase coherence across frequency hops. The current technique of choice for frequency-hopped spread spectrum systems is noncoherent M-ary FSK.

As we discussed in Chapter 5, information is conveyed in FSK by shifting the frequency of the transmitted signal upward or downward by fixed amounts with respect to the carrier. For example, in a four-level FSK system, frequencies of $f_c - 3\Delta f$, $f_c - \Delta f$, $f_c + \Delta f$, and $f_c + 3\Delta f$ could be used to transmit the binary pairs 00, 01, 11, and 10, respectively. As also explained in Chapter 5, for noncoherent detection we need to carefully choose the four frequencies so that they are all integral multiples of the symbol rate. To implement frequency-hopped spread spectrum using noncoherent M-ary FSK, we need to select a series of different possible carrier frequencies, all of which should be integral multiples of Δf. As with direct-sequence spread spectrum, the key to frequency-hopped spread spectrum is to use orthogonal signaling. If the carrier frequencies, Δf, and the hop rate are chosen so that all the frequencies have an integral number of cycles either per bit period or per hop (whichever is shorter), then all the signals will be orthogonal and easily demodulated. Consider the following example.

Example 7.6 A spread spectrum system using frequency hopping

A frequency-hopped spread spectrum transmitter is using four-level FSK to transmit at a rate of 100,000 symbols/sec (i.e., 200,000 bits/sec). Let $\Delta f = 10$ MHz, and let $f_c - 3\Delta f, f_c - \Delta f, f_c + \Delta f$, and $f_c + 3\Delta f$ be used to transmit the binary pairs 00, 01, 11, and 10, respectively. Eight different possible carrier frequencies have been selected: 270 MHz, 350 MHz, 430 MHz, 510 MHz, 590 MHz, 670 MHz, 750 MHz, and 830 MHz. Let's refer to these eight frequencies as $f_1, f_2, f_3, f_4, f_5, f_6, f_7$, and f_8. The data to be transmitted is "1001001110," and the pseudo-random code has generated the random hopping sequence $f_2, f_6, f_7, f_1, f_3, f_8, f_4, f_5, f_7, f_3, f_1, f_4, \ldots$.

a. Plot the transmitted frequency versus time if hopping occurs once every 20 microseconds (i.e., once every two symbols).

b. Plot the transmitted frequency versus time if hopping occurs once every five microseconds (i.e., twice per symbol).

Solution

a. The data sequence "1001001110" corresponds to the frequencies $f_c + 3\Delta f$, $f_c - \Delta f, f_c - 3\Delta f, f_c + \Delta f, f_c + 3\Delta f$. The first symbol is transmitted during the time period $0 \le t < 10$ μsec, the second symbol is transmitted during $10 \le t < 20$ μsec, the third symbol during $20 \le t < 30$ μsec, the fourth symbol during $30 \le t < 40$ μsec, and the fifth symbol during $40 \le t < 50$ μsec. The hopping rate is once every two symbols, so $f_c = f_2 = 350$ MHz when $0 \le t < 20$ μsec, then $f_c = f_6 = 670$ MHz when $20 \le t < 40$ μsec, then $f_c = f_7 = 750$ MHz when $40 \le t < 60$ μsec. Table 7-4 shows the transmitted frequencies and Figure 7-15 provides a plot of transmitted frequency versus time. The value of Δf is larger than is necessary to ensure an integral number of cycles per bit period, but this will not be the case for Part b.

Table 7-4 Frequency Hopping in Example 7.6a

Time	Information	Modulation	f_c	Transmitted Frequency
$0 \le t < 10$ μsec	10	$f_c + 3\Delta f$	350 MHz	380 MHz
$10 \le t < 20$ μsec	01	$f_c - \Delta f$	350 MHz	340 MHz
$20 \le t < 30$ μsec	00	$f_c - 3\Delta f$	670 MHz	640 MHz
$30 \le t < 40$ μsec	11	$f_c + \Delta f$	670 MHz	680 MHz
$40 \le t < 50$ μsec	10	$f_c + 3\Delta f$	750 MHz	780 MHz

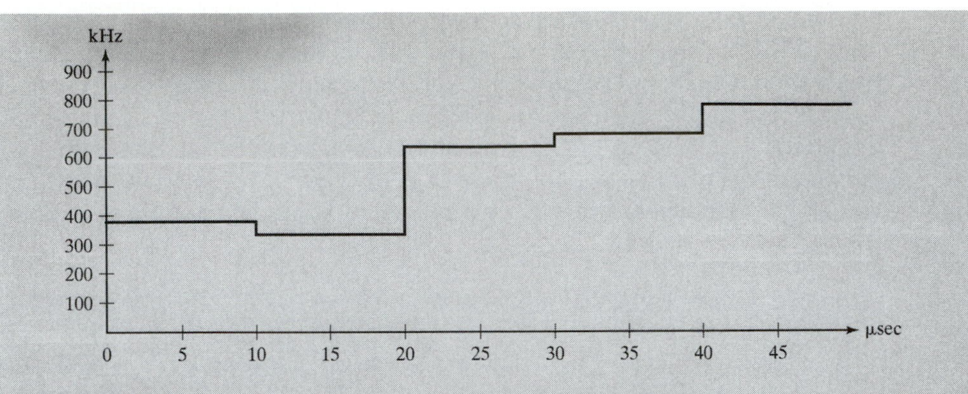

Figure 7-15 Transmitted frequency versus time for Example 7.6a.

b. As in Part a, the data sequence "1001001110" corresponds to the frequencies $f_c + 3\Delta f, f_c - \Delta f, f_c - 3\Delta f, f_c + \Delta f, f_c + 3\Delta f$, and the symbol period is ten microseconds. Unlike Part a, hopping now occurs twice per symbol (i.e., once every five microseconds). Table 7-5 shows the transmitted frequencies and Figure 7-16 provides a plot of transmitted frequency versus time.

Table 7-5 Frequency Hopping in Example 7.6b

Time (μsec)	Information	Modulation	f_c (MHz)	Transmitted Frequency (MHz)
$0 \le t < 5$	10	$f_c + 3\Delta f$	350	380
$5 \le t < 10$	10	$f_c + 3\Delta f$	670	700
$10 \le t < 15$	01	$f_c - \Delta f$	750	740
$15 \le t < 20$	01	$f_c - \Delta f$	270	260
$20 \le t < 25$	00	$f_c - 3\Delta f$	430	400
$25 \le t < 30$	00	$f_c - 3\Delta f$	830	800
$30 \le t < 35$	11	$f_c + \Delta f$	510	520
$35 \le t < 40$	11	$f_c + \Delta f$	590	600
$40 \le t < 45$	10	$f_c + 3\Delta f$	750	780
$45 \le t < 50$	10	$f_c + 3\Delta f$	430	460

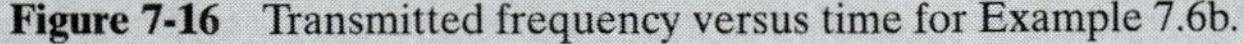

Figure 7-16 Transmitted frequency versus time for Example 7.6b.

As a final observation, we can classify frequency-hopping spread spectrum systems into two groups. If the hop rate is less than one hop per symbol (as in Example 7.6a), the system is *slow hopping*. If the hop rate is greater than one hop per symbol (as in Example 7.6b), the system is *fast hopping*. Fast hopping is more complex, but can provide more immunity to certain types of channel distortion (in particular, frequency-dependent fading (Pickholtz [7.2])). To ensure orthogonality, the hopping period in a slow-hopping system should be an integral multiple of the symbol period. In a fast-hopping system, to ensure orthogonality, the symbol period should be an integral multiple of the hop period.

PROBLEMS

7.1 Why are signals multiplexed?

7.2 Explain how a TDM system operates.

7.3 Information bits generated on a continuous basis by four different users are time division multiplexed and then transmitted over a channel. The first user's data rate is 100 kbit/sec, the second user's is 300 kbit/sec, the third user's is 500 kbit/sec and the fourth user's is 600 kbit/sec. The bit stream from each user is first arranged into 100-bit-long slots. The slots from the four streams are then multiplexed into 3000-bit-long frames. Thus, each frame contains 30 slots.

a. What is the minimum data rate required for the channel?
b. How many slots in each multiplexed frame will be assigned to each of the four bit streams?
c. Draw a sketch similar to Figure 7-3, showing the structure of the system's TDM frames.

7.4 Information bits generated on a continuous basis by three different users are time division multiplexed and then transmitted over a channel. The first user's data rate is 200 kbit/sec, the second user's is 400 kbit/sec, and the third user's is 600 kbit/sec. The bit stream from each user is first arranged into 50-bit-long slots, and the slots from the three streams are then multiplexed into frames.

a. What is the minimum data rate required for the channel?
b. What is the minimize size of the frame?
c. Using your answer from Part b, how many slots in each multiplexed frame will be assigned to each of the three bit streams?
d. Draw a sketch similar to Figure 7-3, showing the structure of the system's TDM frames.

7.5 Explain why slot synchronization is important in TDM systems.

7.6 Explain the trade-offs involved in choosing slot size and frame size for TDM systems.

7.7 Explain the differences between synchronous TDM and statistical TDM. Discuss the advantages and disadvantages of each, and give examples when each should be used.

7.8 Consider a statistical TDM system serving 100 users, each with an average transmission rate of 5000 bits/sec and a peak rate of 40,000 bits/sec.

a. If no control signals were needed (i.e., if the packets needed no headers), what would be the minimum data rate required by the channel?
b. What types of information are contained within a typical header?
c. Explain why buffering is required in a statistical TDM system.
d. Suppose that the information in the packet headers requires another 1000 bits/sec for each user. What would then be the minimum data rate required by the channel?

e. How much more efficient is statistical TDM than synchronous TDM for the system in this problem? Assume the statistical TDM system needs the packet headers discussed in Part d of this problem and the synchronous TDM system must accommodate each user's peak data rate.

7.9 Explain how an FDM system operates.

7.10 Discuss the differences between TDM and FDM and discuss the trade-offs associated with choosing between the two techniques.

7.11 An FDM system transmits signals from six users. The bandwidth of each signal is as follows:

Signal 1: 100 kHz
Signal 2: 150 kHz
Signal 3: 200 kHz
Signal 4: 300 kHz
Signal 5: 350 kHz
Signal 6: 400 kHz

The channel is assigned the frequency band 910 MHz–911.8 MHz. Draw a magnitude spectrum showing how each of the signals is multiplexed within the channel. Be sure to identify the center frequency of the sub-band for each signal. If possible, include guardbands between each of the signals and half-sized guardbands at 910 and 911.8 MHz.

7.12 Consider an AM-DSB-SC system used to frequency division multiplex four 100 kHz baseband signals into a channel with a frequency band centered at 2 MHz. Remembering that AM-DSB-SC is a *double sideband* system,

a. Determine the frequency band needed to transmit the signals if no guardbands are used.

b. Suppose we use a 20 kHz guardband between each of the signals (i.e., the highest frequency in the magnitude spectrum of the multiplexed signal from the first source is 20 kHz below the lowest frequency in the magnitude spectrum of the multiplexed signal from the second source). Determine the frequency band needed to transmit the signals and draw a magnitude spectrum showing how each user's signal is multiplexed within the channel. Be sure to identify the center frequency for each user's signal.

7.13 Repeat Problem 7.12 for an AM-SSB system.

7.14 Explain why frequency guardbands are important in FDM systems.

7.15 Research commercial AM broadcasting and explain how it employs FDM. Give technical details similar to those in Example 7.4, which examines commercial FM systems.

7.16 Explain how a CDM system operates.

7.17 Explain the differences between TDM, FDM, and CDM, and explain the trade-offs associated with choosing among the three techniques.

7.18 Consider a direct sequence spread spectrum system. The source outputs the information sequence "01101" at a rate of 100 kbit/sec. The spreader has a chip rate of four and uses the spreading code "10101000010010101110".

a. Determine the transmitted bit sequence (i.e., the output of the spreader) and the transmission speed.

b. Show that if the transmitted sequence is despread at the receiver using the same spreading code, the output is "01101"—the information sequence from the source.

7.19 Design a maximal-length PN code generator capable of outputting a seven-bit-long pseudo-random code. You can do this by drawing a circuit of the code generator, then stating the initial values you want to place in the shift registers (you can choose any values except all zeroes), and then determining the sequence of "1"s and "0"s in the PN sequence.

7.20 Evaluate $[b] \oplus [b]^{(3)}$, $[b] \oplus [b]^{(5)}$, and $[b] \oplus [b]^{(8)}$ in Example 7.5 to show that the correlation property is satisfied for these three cyclic shifts.

7.21 A binary PSK system using CDM requires a data rate for each user of 5 kbit/sec and has a spread spectrum bandwidth (90% in-band power) of 80 kHz.

a. Determine the processing gain of the system.
b. If an SNR of 2 dB is required for adequate performance, how many users can the system support?

7.22 Derive Equations (7.7) and (7.8) for a CDM system.

7.23 Explain the differences between a direct-sequence spread spectrum system and a frequency-hopping spread spectrum system.

7.24 Consider a slow-hopping binary FSK system with orthogonal signals transmitting at a rate of 10,000 bits/sec and hopping once every three bits.

a. Suppose one of the carrier frequencies used in the system is 225 kHz. Using minimal bandwidth and yet preserving orthogonality, determine Δf and the two frequencies used to transmit a "1" and a "0."
b. If the system is large enough that thermal noise is small relative to the interference from other users, how much bandwidth (90% in-band power) is needed to provide a system of 20 users with an SNR of 6 dB?
c. Let the hop sequence for one of the users produce the following sequence of carrier frequencies: 125 kHz, 215 kHz, 175 kHz. Plot transmitted frequency versus time if the user's data sequence is "101011101."

7.25 Suppose the binary FSK system in Problem 7.24 is converted to a fast-hopping system with a rate of three hops per bit.

a. How does the change in hop rate affect the answers to Problem 7.24a and 7.24b?
b. Let the hop sequence for one of the users produce the following sequence of carrier frequencies: 405 kHz, 225 kHz, 585 kHz, 465 kHz, 705 kHz, 345 kHz. Plot transmitted frequency versus time if the user's data sequence is "101."

Chapter 8

Analog-to-Digital and Digital-to-Analog Conversion

IN Chapters 3–5, we've assumed that the source produced information in a digital format (i.e., a series of "1"s and "0"s). While this is true for many applications, such as computer networks and automated teller machines, it is also true that many sources, such as voice and music, produce information in an analog format. How do we want to transmit this analog-formatted information? We have two principal choices: Either we can employ an analog modulation technique and transmit the information directly (see Chapter 6) or we can convert the information into a digital format prior to transmission, use a digital modulation technique to transmit the information across the channel, and then convert the received information back into analog format for the user. These two approaches are illustrated in Figure 8-1.

Although the analog transmission approach appears less complex and more straightforward, an ever-increasing number of new-generation communication and instrumentation systems, such as sound recording systems (CDs), video recording and transmitting systems (DVDs and HDTVs), and cellular telephones are designed using the digital transmission approach, even though these systems must start with analog signals and even though the end users require the signal to be presented in analog format. Compact discs are an excellent example: The music is originally in analog format and must be converted to digital format prior to being encoded onto the disc, and the CD player must then convert the digital information back to analog music before playing it to the listening audience. Why do engineers designing these new systems choose the additional complexity of converting the signal to digital prior to transmission and then

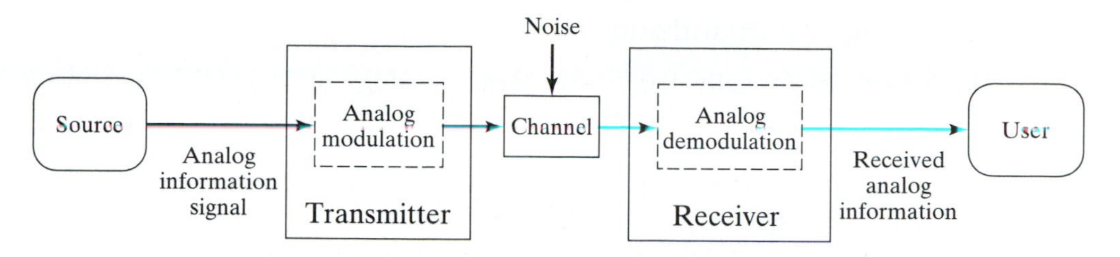

Figure 8-1a Analog transmission of analog-formatted information.

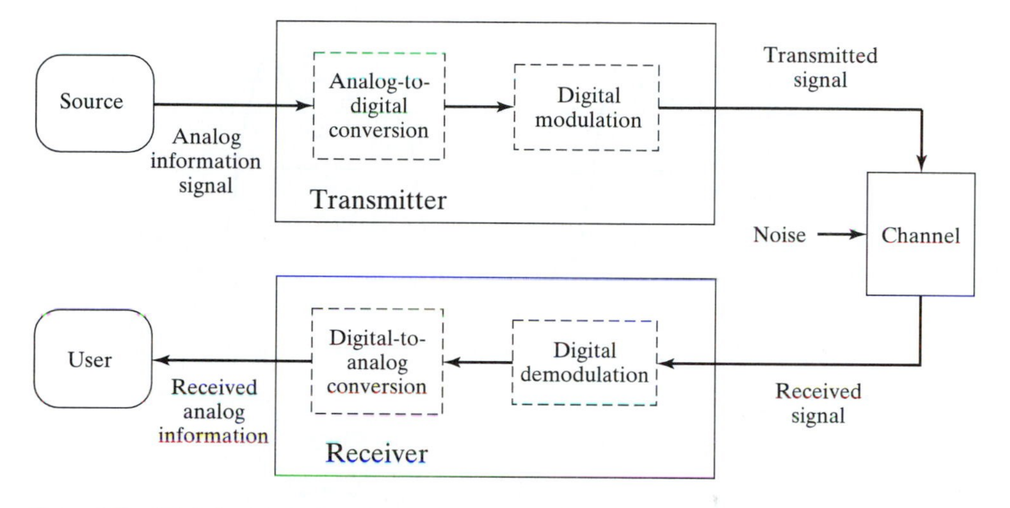

Figure 8-1b Digital transmission of analog-formatted information.

converting the received signal back to analog before the signal is played back or displayed? There are many reasons:

1. Digital transmission provides tighter control of accuracy and digital storage is more durable and more compact. (Imagine the appeal to manufacturers of digital entertainment media.)
2. Signal processing is simpler if the signal is in digital format. With a sound card, a PC, and the appropriate software, anyone can be a recording engineer. Straightforward digital algorithms can perform noise reduction, signal compression, graphic equalization, error detection and correction, and encryption. Using analog circuitry to perform similar processing on analog-formatted signals is much more difficult and much more expensive.
3. In a digital format, more flexible and powerful signal processing techniques can be used. Such techniques include adaptive and predictive processing. It is also simpler to incorporate new services, such as multimedia transmission, if the signal is in digital format.

In this chapter we will develop techniques for converting analog-formatted signals into digital format and vice versa.

8.1 Sampling and Quantizing

Let's begin by formally defining the terms "analog" and "digital." As shown in Figure 8-2a, an analog signal is continuous in time and continuous in amplitude. Such a signal cannot be directly converted into digital format because it has an infinite number of points (since it is continuous in time) and because each point requires infinite resolution (since the signal is continuous in amplitude).

Suppose we sample the signal at discrete intervals of time, say, every 125 μsec. This sampled signal, shown in Figure 8-2b, is now discrete in time and continuous in

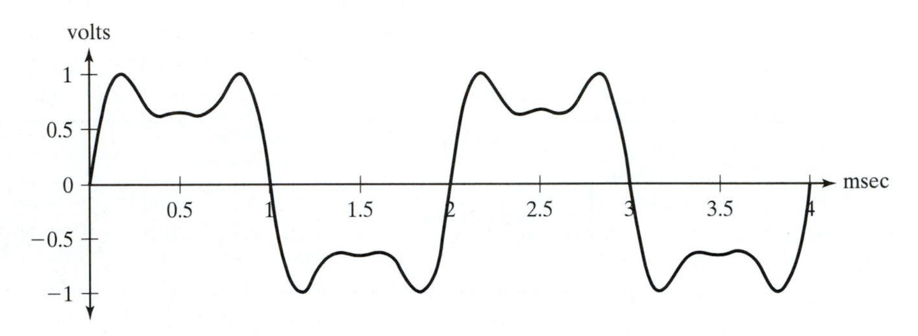

Figure 8-2a Analog signal—continuous in both time and amplitude.

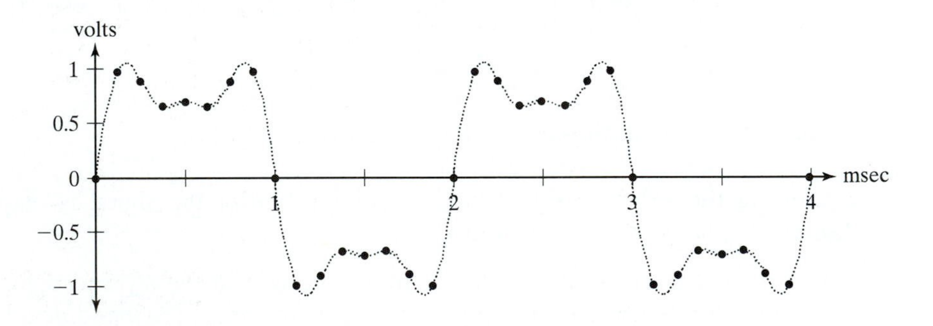

Figure 8-2b Sampled signal—discrete in time but continuous in amplitude.

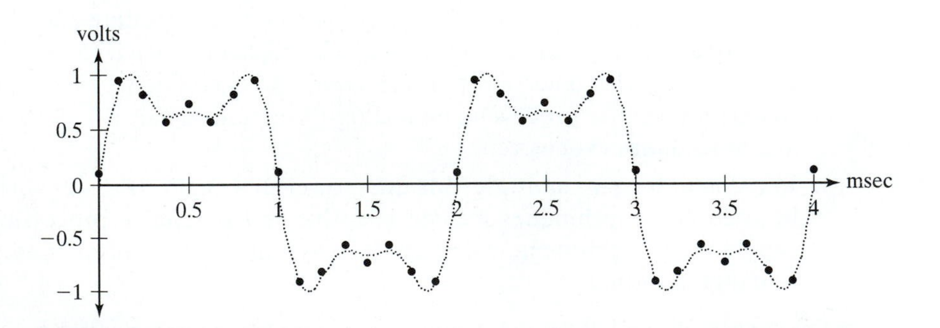

Figure 8-2c Sampled and quantized signal—discrete in time and discrete in amplitude. Note the roundoff in the sample points.

amplitude. Such a signal (often called a *discrete signal* or *sampled signal*) still cannot be directly converted into digital format because even though the signal has a finite number of points, the voltage represented by each point still requires infinite resolution (i.e., an infinite number of "1"s and "0"s are required to exactly represent the base-2 value of the voltage of each sample).

Suppose we now take the sampled signal and round the value of each sample to the nearest in a finite series of preset values. For example, as shown in Figure 8-2c, we can round each sample to the nearest of 16 preset values shown in Table 8-1. This roundoff process, called *quantization*, allows each sample to be represented in digital format using a finite number of "1"s and "0"s. In our example, we need four bits to uniquely identify each of the 16 different rounded-off values (see Table 8-1). The sampled and quantized signal in Figure 8-2c can be easily represented in digital format, using four bits for each sample to identify the rounded-off value of the sample. If the quantization levels are labeled as shown in Table 8-1, the sampled and quantized signal can be represented by the digital sequence 1000111111101100... where the first four bits (1000) represent the quantized value of the first sample, the second four bits (1111) represent the quantized value of the second sample, etc. The number of bits/sec required to represent the signal in digital format is known as the *encoding rate*. In our example, the encoding rate is (8000 samples/sec)(4 bits/sample) = 32,000 bits/sec.

Figures 8-2a–c show how an analog signal can be represented digitally using sampling and quantization. Intuitively, it appears that both sampling and quantizing will cause distortion, and therefore information present in the analog signal will be lost. In the remainder of this section we will investigate the distortion introduced by sampling and quantizing.

Table 8-1 Quantization Values and Binary Assignments for the Sampled Signal

Voltage Range of Signal $x(t)$	Rounded-off (Quantized) Value	Binary Assignment for Quantized Value
$-1.000 \le x(t) < -0.875$ v	−0.9375 v	0000
$-0.875 \le x(t) < -0.750$ v	−0.8125 v	0001
$-0.750 \le x(t) < -0.625$ v	−0.6875 v	0010
$-0.625 \le x(t) < -0.500$ v	−0.5625 v	0011
$-0.500 \le x(t) < -0.375$ v	−0.4375 v	0100
$-0.375 \le x(t) < -0.250$ v	−0.3125 v	0101
$-0.250 \le x(t) < -0.125$ v	−0.1875 v	0110
$-0.125 \le x(t) < 0.000$ v	−0.0625 v	0111
$0.000 \le x(t) < 0.125$ v	0.0625 v	1000
$0.125 \le x(t) < 0.250$ v	0.1875 v	1001
$0.250 \le x(t) < 0.375$ v	0.3125 v	1010
$0.375 \le x(t) < 0.500$ v	0.4375 v	1011
$0.500 \le x(t) < 0.625$ v	0.5625 v	1100
$0.625 \le x(t) < 0.750$ v	0.6875 v	1101
$0.750 \le x(t) < 0.875$ v	0.8125 v	1110
$0.875 \le x(t) \le 1.000$ v	0.9375 v	1111

8.1.1 Sampling Baseband Analog Signals

As shown in the left side of Figure 8-3, we can mathematically model the sampling process by multiplying the analog waveform by a periodic series of unit impulses. Let f_s represent the sampling rate (number of samples per second) and let $T_s = 1/f_s$ represent the time interval between samples. Representing the original signal as $x(t)$ and the sampled signal as $x_s(t)$,

$$x_s(t) = x(t) \sum_{k=-\infty}^{\infty} \delta(t - kT_s) \tag{8.1}$$

In Figure 8-3, the sampling rate is $f_s = 8000$ samples per second and the sampling interval is $T_s = 125$ μsec.

To understand how sampling may introduce distortion, we need to examine the original and sampled signals in the frequency domain. As we established in Equation (2.61), multiplication in the time domain corresponds to convolution in the frequency domain. Thus, the frequency domain representation of the sampled signal is

$$\begin{aligned} X_s(f) &= X(f) * \mathcal{F}\left\{ \sum_{k=-\infty}^{\infty} \delta(t - kT_s) \right\} = X(f) * \left\{ f_s \sum_{k=-\infty}^{\infty} \delta(f - kf_s) \right\} \\ &= f_s \sum_{k=-\infty}^{\infty} X(f - kf_s) \end{aligned} \tag{8.2}$$

The right side of Figure 8-3 shows the frequency domain representation of the original and sampled signals. Note that the magnitude spectrum of $X_s(f)$ consists of the spectrum of the original signal $X(f)$, plus copies of the original signal's spectrum (called *aliases*) centered at integral multiples of the sampling frequency, all scaled by the constant f_s. Be sure you understand how Equation (8.2) produces the spectrum of $X_s(f)$ shown in Figure 8-3.

We can develop more intuition concerning aliasing by examining its effects in the time domain. Consider a 2-volt, 500 Hz sinusoid sampled at 5000 samples/sec, as shown in Figure 8-4a. We can visualize the original sinusoid by "connecting the samples" as shown in Figure 8-4b. Note, however, that there are other ways to connect the samples. Figure 8-4c shows a 4500 Hz sinusoid and Figure 8-4d shows a 5500 Hz sinusoid, both of which can also be produced by connecting the samples. These higher-frequency sinusoids are aliases of the original 500 Hz signal, and their existence can easily be predicted if we examine the magnitude spectrum of the sampled signal, shown in Figure 8-4e. Note that there are also aliases at 9500 Hz, 10,500 Hz, etc. As an exercise, you may want to connect the samples in Figure 8-4a using these other aliases.

Let's reexamine Figure 8-3 in more detail. The magnitude spectrum of $X_s(f)$, reproduced in Figure 8-5, shows how we can reconstruct the original analog signal from the sampled signal. If the sampling frequency f_s is sufficiently high so that the aliases (shown in gray) do not overlap the spectrum of the original analog signal

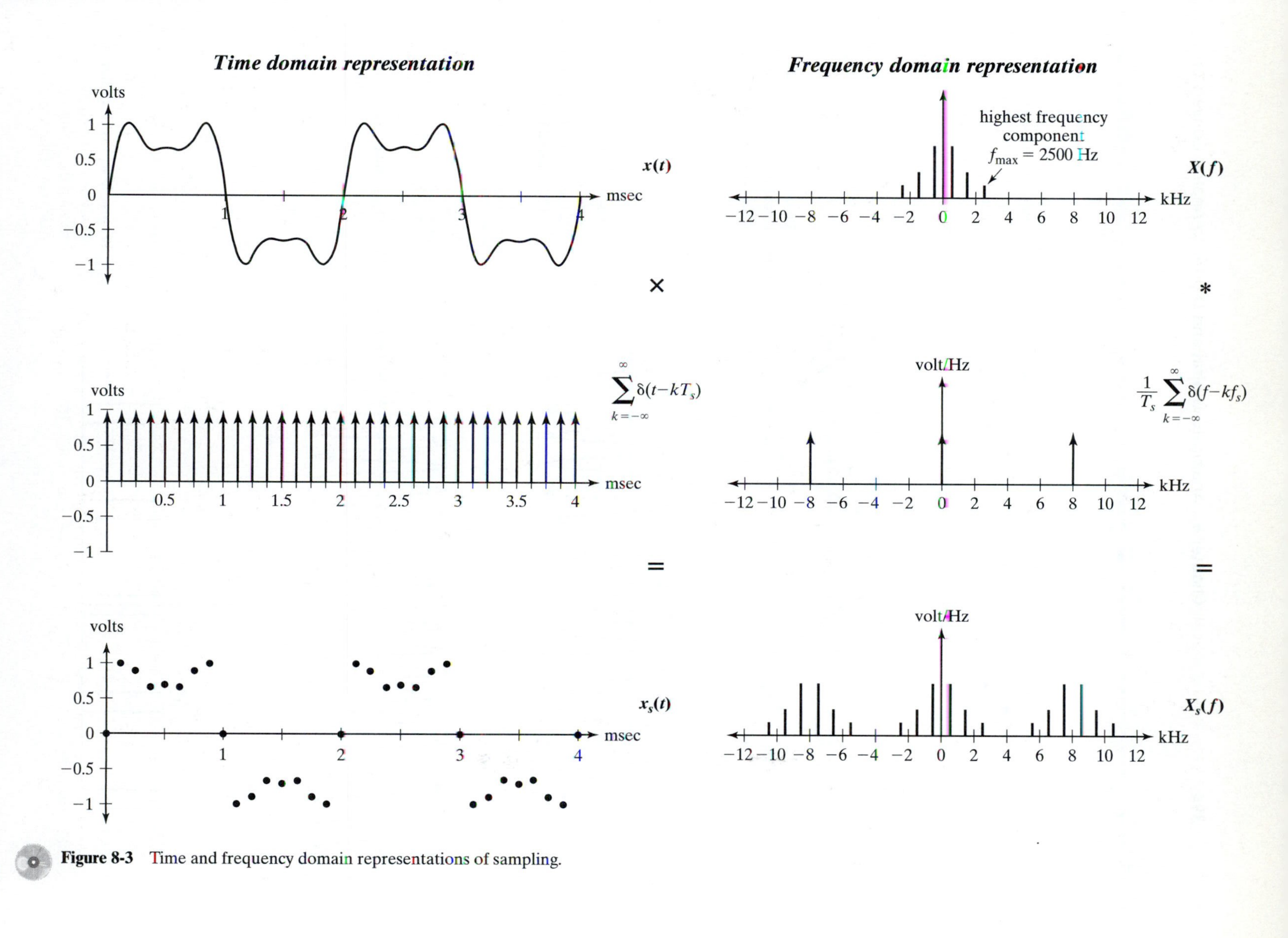

Figure 8-3 Time and frequency domain representations of sampling.

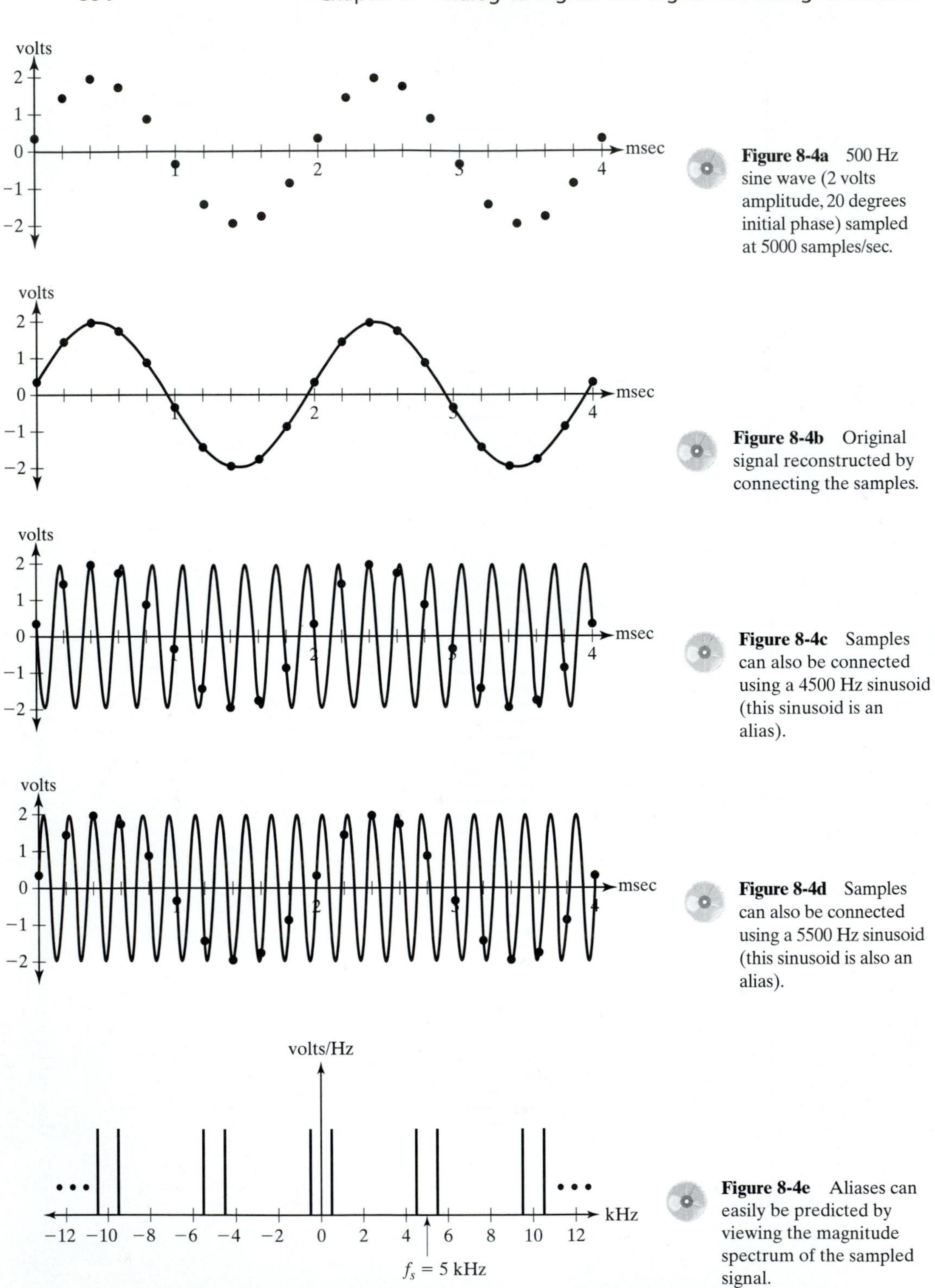

Figure 8-4a 500 Hz sine wave (2 volts amplitude, 20 degrees initial phase) sampled at 5000 samples/sec.

Figure 8-4b Original signal reconstructed by connecting the samples.

Figure 8-4c Samples can also be connected using a 4500 Hz sinusoid (this sinusoid is an alias).

Figure 8-4d Samples can also be connected using a 5500 Hz sinusoid (this sinusoid is also an alias).

Figure 8-4e Aliases can easily be predicted by viewing the magnitude spectrum of the sampled signal.

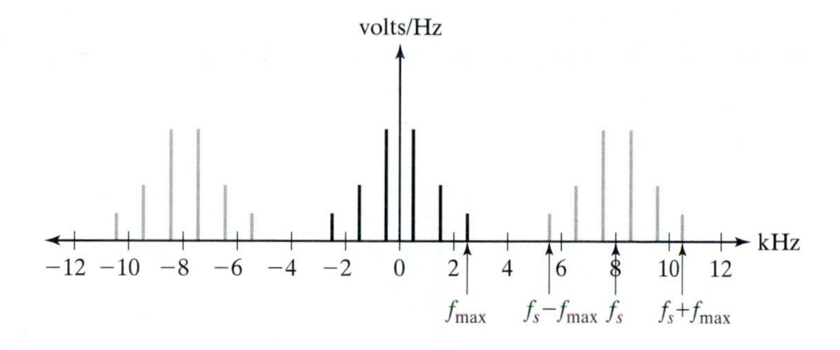

Figure 8-5 Magnitude spectrum of sampled signal (enlarged from Figure 8-3).

(shown in black), then we can pass the sampled signal through a lowpass filter with a cutoff frequency set to eliminate all the aliases but pass the spectrum of the original analog signal. After adjusting for the scaling factor f_s, the signal at the output of the lowpass filter will have the same spectrum as the original signal and will therefore be the same as the original analog signal. Reconstruction is shown in Figure 8-6.[1]

As we've just established, we can reconstruct an analog baseband signal from its samples with no distortion as long as the sampling rate is high enough to prevent the aliases from overlapping the spectrum of the original analog signal. In reviewing Figure 8-5, we see that the highest frequency in the original signal is f_{max} and the lowest frequency component in any of the aliases is $f_s - f_{max}$. Thus, distortionless reconstruction is possible if

$$f_s - f_{max} > f_{max} \tag{8.3a}$$

which we can rewrite as

$$f_s > 2f_{max} \tag{8.3b}$$

The value $2f_{max}$ is called the *Nyquist rate*.

Let's reflect on our discoveries. We have shown that for a baseband analog signal, we can sample the signal and subsequently reconstruct it *without any distortion* as long as our sampling rate is greater than $2f_{max}$. This seems counter-intuitive if we examine the signal in the time domain (after all, when sampling, we appear to be discarding the portions of the signal between the samples, so how can we reconstruct the original signal without distortion?). Yet, if we examine sampling in the frequency domain, we can easily see that distortionless reconstruction is possible as long as our sampling is faster than the Nyquist rate.

[1]Practical systems perform sampling by latching the analog signal at the appropriate times rather than by multiplying the analog signal by a periodic series of impulse functions. This practical approach does not produce a scaling factor in the spectrum of the sampled signal.

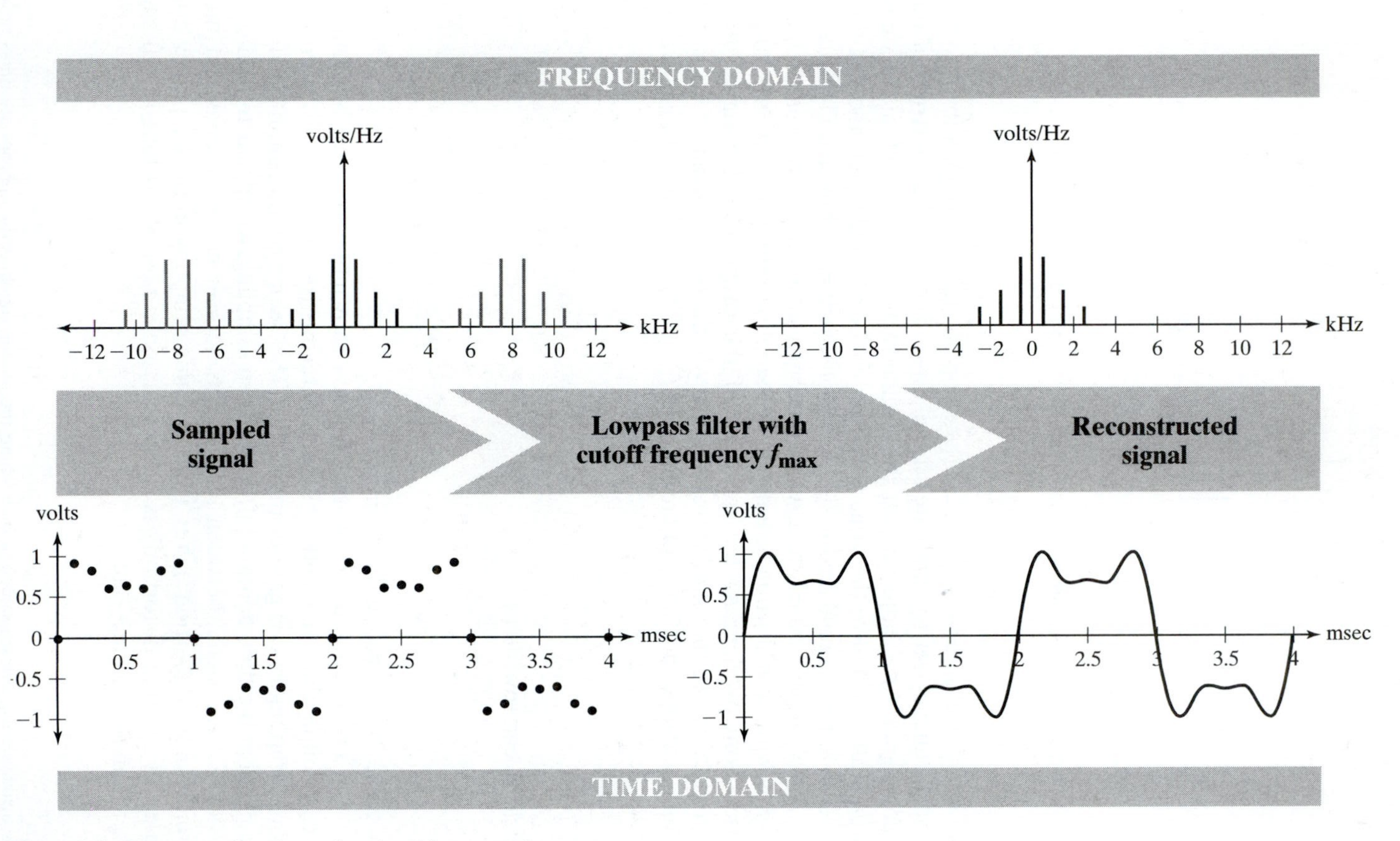

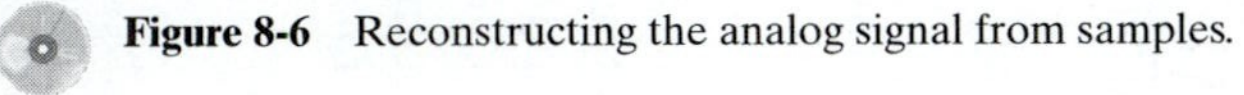

Figure 8-6 Reconstructing the analog signal from samples.

8.1.2 Practical Considerations in Baseband Sampling

8.1.2.1 Sampling at higher-than-Nyquist rates We've established that the Nyquist rate $2f_{max}$ represents the theoretical lower limit for distortionless sampling of baseband signals. In practice, however, a higher sampling rate is required because practical filters do not have infinitely steep rolloffs. As shown in Figure 8-7, in order to reconstruct the original signal without distortion, all frequencies up to f_{max} must lie within the flat passband region of the filter and all frequencies within the aliases must lie within the stopband region of the filter.[2] The region between the passband and stopband (i.e., the region where rolloff occurs) cannot contain any frequency components of the original signal because such components will be distorted by the filter. Similarly, the region cannot contain any frequency components of the aliases because such components will not

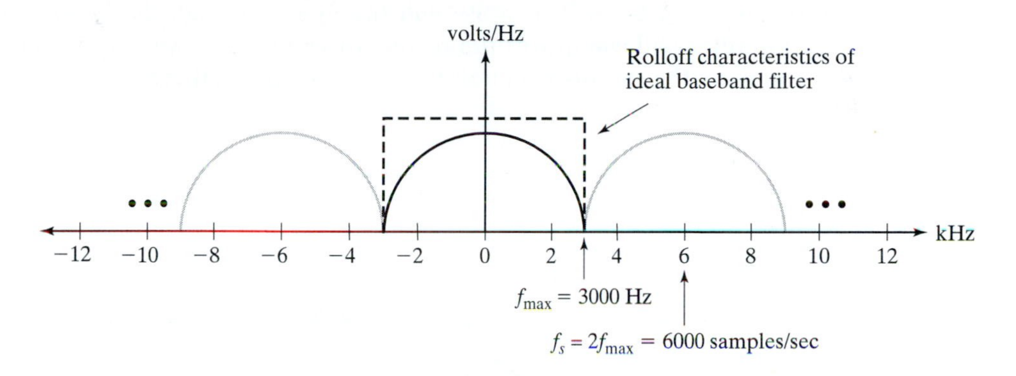

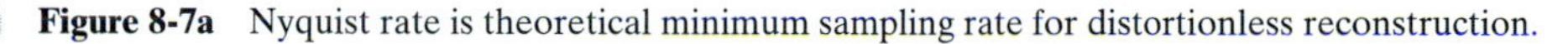

Figure 8-7a Nyquist rate is theoretical minimum sampling rate for distortionless reconstruction.

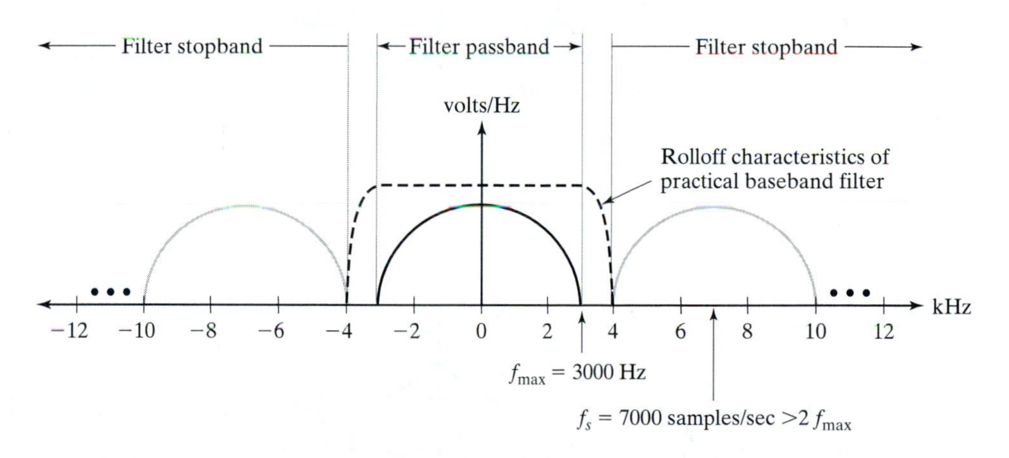

Figure 8-7b For practical systems, sampling rate must be a higher than Nyquist rate (produces guardbands to compensate for nonideal filters).

[2] In order to avoid distortion, the filter must also have a linear phase response within the passband region (see Section 4.4).

be completely stopped by the filter. The sampling rate must therefore be sufficiently high to produce *guardbands* between the spectrum of the original signal and the aliases. The size of these guardbands depends on the characteristics of the filter used for reconstruction. For example, in compact discs the original music signal may contain frequency components up to 20 kHz and the sampling rate is 44.1 ksamples/sec, producing a guardband of 4.1 kHz.

8.1.2.2 Anti-alias filtering Prior to sampling, the analog baseband signal is often passed through a lowpass filter with a cutoff frequency slightly above f_{max}. This filter, called an *anti-aliasing filter*, is necessary for the following two situations:

1. If the original analog signal is corrupted by out-of-band noise, sampling can cause aliases of the noise to fall within the frequency band of the original signal (see Figures 8-8a and 8-8b). This noise will then pass through the lowpass reconstruction filter and will distort the reconstructed signal . An anti-aliasing filter will eliminate the out-of-band noise prior to sampling, thereby preventing this problem.

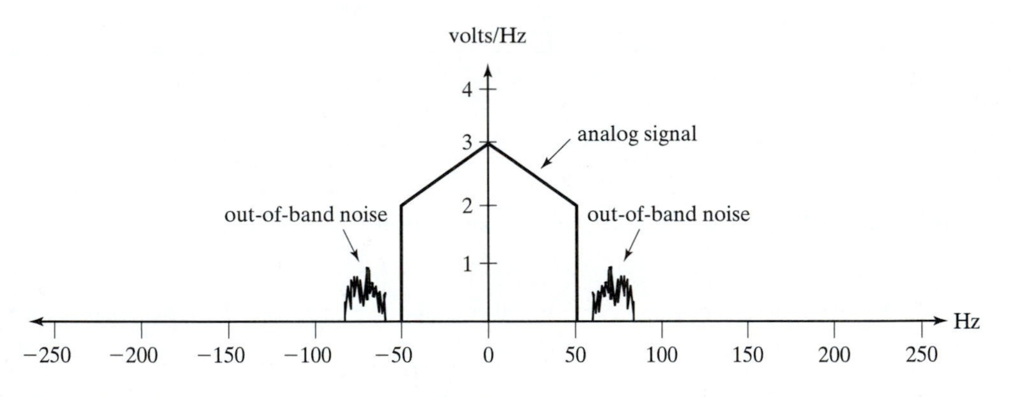

Figure 8-8a Baseband analog signal with out-of-band noise.

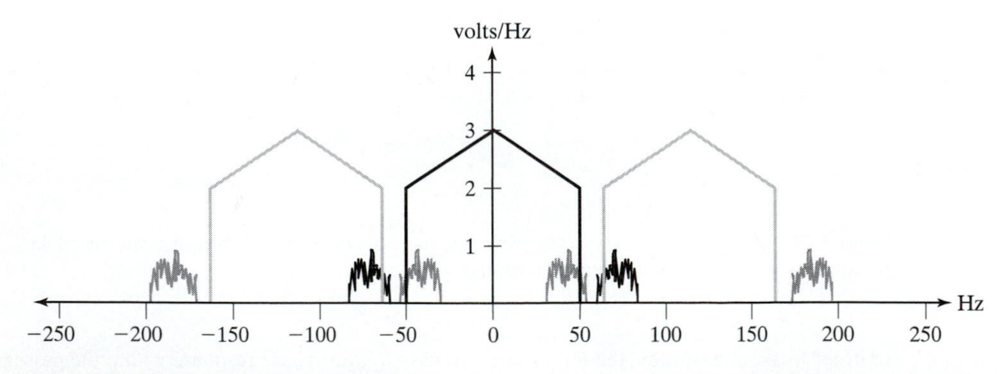

Figure 8-8b Sampled signal. Note that aliases of the noise now overlap the original spectrum. The out-of-band noise needs to be filtered prior to sampling.

2. If the original analog signal is not strictly band-limited to f_{max} Hz but still has all significant frequency components at or below f_{max} Hz, then anti-alias filtering will cause slight distortion (by eliminating insignificant frequency components) but will prevent aliases of the insignificant components from falling within the frequency band of the original signal.

8.1.3 Sampling Bandpass Analog Signals

Let's now consider sampling a bandpass analog signal, that is, a signal whose spectrum contains all its significant components in a frequency band between a nonzero lower limit—say, f_1 Hz—and an upper limit—say, f_2 Hz. The spectrum for a typical bandpass signal is shown in Figure 8-9a, where $f_1 = 8$ kHz and $f_2 = 10$ kHz. We know that sampling this signal at a rate faster than the Nyquist rate (20 ksamples/sec) will cause all the aliases to be at higher frequencies than the original signal (as shown in Figure 8-9b), and that the original signal can then be reconstructed without distortion by using a lowpass filter with a cutoff frequency of 10 kHz. Suppose, however, we sample at a slower rate, say, 7 ksamples/sec. The spectrum of the bandpass signal sampled at this slower rate is shown in Figure 8-9c. Note that aliases appear at frequencies lower

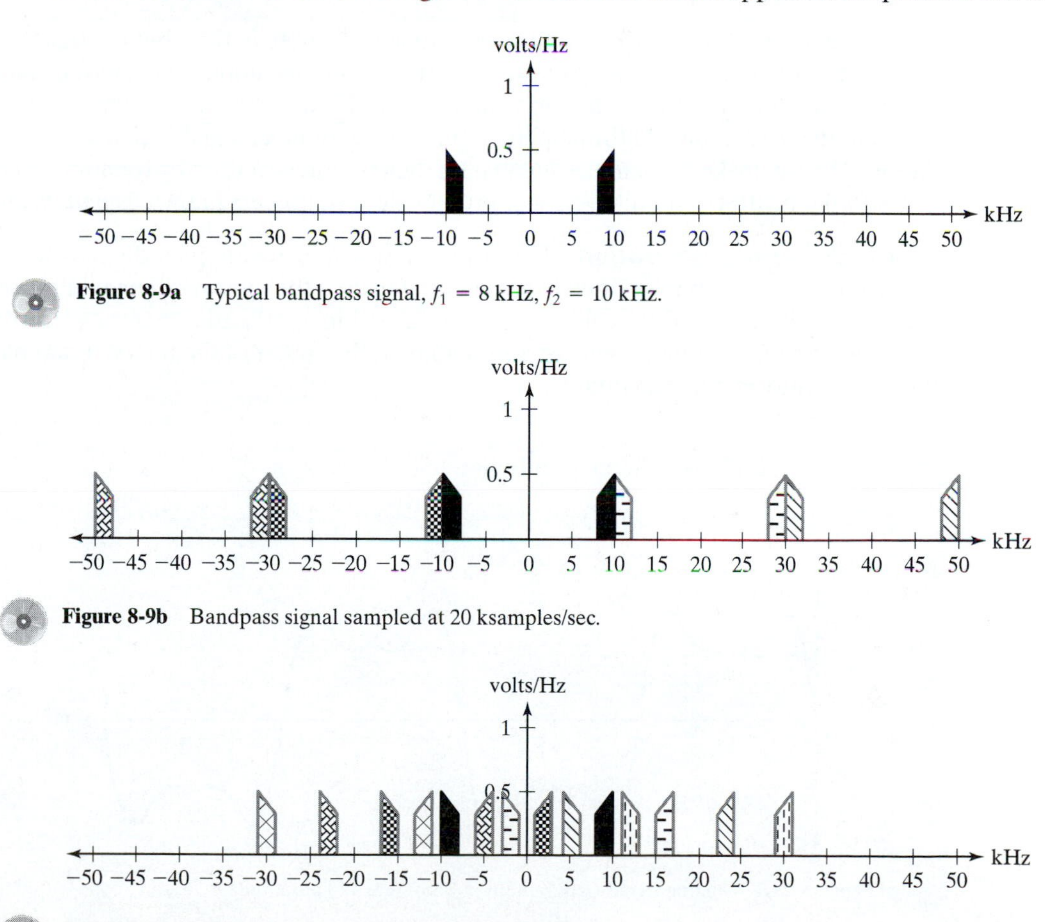

Figure 8-9a Typical bandpass signal, $f_1 = 8$ kHz, $f_2 = 10$ kHz.

Figure 8-9b Bandpass signal sampled at 20 ksamples/sec.

Figure 8-9c Bandpass signal sampled at 7 ksamples/sec.

than the spectrum of the original signal, but the aliases do not overlap the spectrum of the original signal. The original analog bandpass signal can therefore be reconstructed without distortion by passing the sampled signal through a bandpass filter with the same passband as the original signal (i.e., from 8 kHz to 10 kHz).

The technique of sampling bandpass signals at lower-than-Nyquist rates is called *bandpass sampling*. The success of the technique and the degree to which the sampling rate can be reduced below the Nyquist rate depends on the values of f_1 and f_2. For example, the technique will never work if the bandwidth of the signal is greater than f_1, since—in this case—there is insufficient room to fit an alias in the low-frequency region below f_1. In the best of cases, the bandpass sampling rate has a lower limit of twice the bandwidth of the signal. Mathematical formulae and plots have been generated to show possible under-sampling for various values of f_1 and f_2 (see, e.g., Lyons [8.1]), but we recommend that for each case you draw the spectrum of the sampled signal for various sampling rates and solve the problem geometrically. (Caution: Be sure you draw enough aliases to ensure there is no overlap.) As with the sampling of baseband signals, anti-alias filtering and guardbands must be considered in practical systems when sampling bandpass signals.

8.1.4 The Quantizing Process

Sampling converts a signal from continuous in time to discrete in time, but the signal is still continuous in amplitude. A sampled signal with continuous amplitude values cannot be represented digitally because an infinite number of "1"s and "0"s are required to exactly represent the base-2 value of the amplitude (i.e., the continuous signal requires infinite resolution). We can make the samples suitable for digital representation by rounding off their values to the nearest in a finite series of preset values (a process known as *quantization*).

8.1.4.1 Uniform quantization Uniform quantization techniques divide the range of possible voltages for the original analog signal (called the *dynamic range*) into a series of equal-width quantization regions. Any sample within a particular quantization region is rounded off to the voltage value corresponding to the center of the region. Example 8.1 illustrates uniform quantization.

Example 8.1 Applying uniform quantization

Suppose we want to convert the analog signal shown in Figure 8-2a into a digital format.

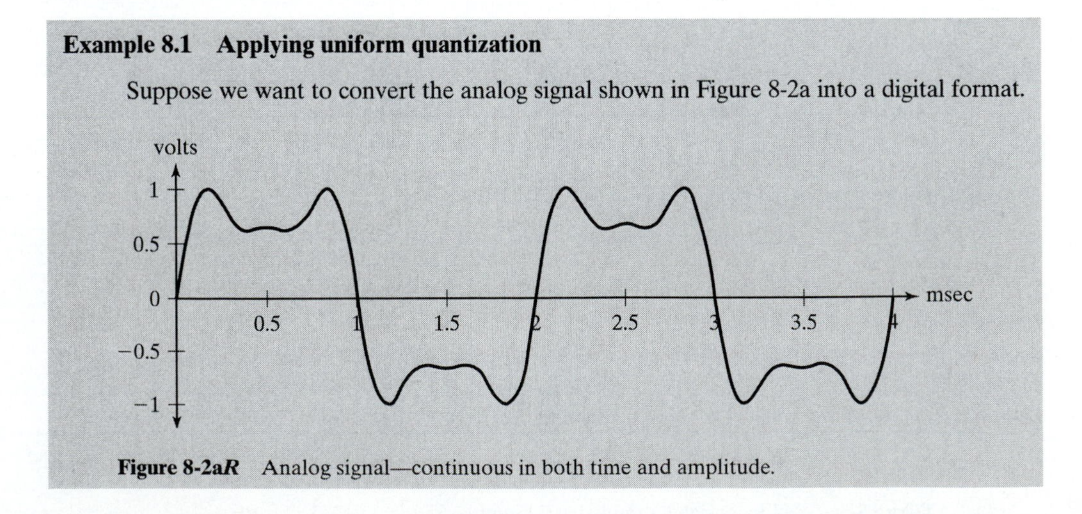

Figure 8-2a*R* Analog signal—continuous in both time and amplitude.

In Figure 8-3, we showed the spectrum of the signal and established that the signal has a maximum frequency component of 2500 Hz. Let's sample the signal at a rate of 8000 samples/sec, guaranteeing no distortion from sampling, and producing good guardbands. We saw the sampled signal in Figure 8-2b.

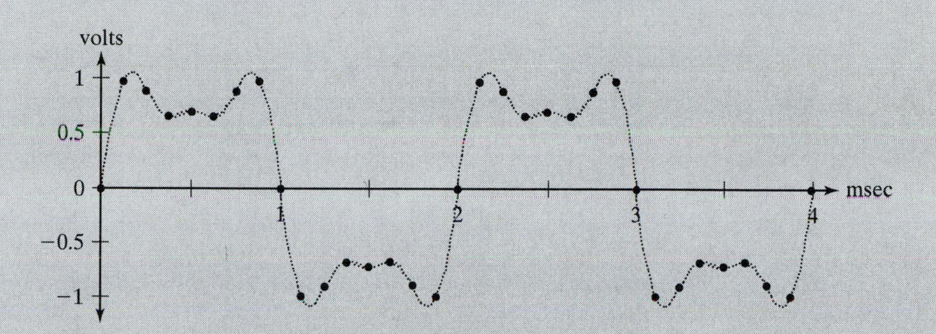

Figure 8-2b*R* Sampled signal—discrete in time but continuous in amplitude.

Our next step is to quantize each of the samples. Suppose the transmission speed of our system is limited to 32,000 bits/sec. We can quantize the samples as follows:

a. Since we are sampling at 8000 samples/sec and are limited to an encoding rate of 32,000 bits per second, we are limited to using only four bits or less to represent the quantized value of each sample.

b. If we use four bits to represent the quantized value of each sample, we can uniquely identify $2^4 = 16$ different quantization values (we can round off the sample to one of 16 different pre-assigned values and represent each value using a unique four-bit pattern).

c. The signal's amplitude always lies between −1 volt and +1 volt; the signal therefore has a dynamic range of 2 volts.

d. We can divide the range of possible voltages for the signal into 16 different quantization regions. If the regions are of equal size, the quantization process is known as *uniform quantization*. For uniform quantization, the width of each quantization region is

$$\text{uniform quantization region width} = \frac{\text{dynamic range of signal}}{\text{number of different quantization values}} \tag{8.4}$$

For our signal, if we use uniform quantization, the width of each quantization region is 0.125 volts.

e. Starting with the lowest voltage within the dynamic range of the signal (−1 volt) and ending with the highest voltage within the dynamic range of the signal (+1 volt), we divide the y-axis into quantization regions 0.125 volts wide. Within each region, let the voltage value at the *center* of the region be one of the quantization voltages and assign it a unique four-bit pattern. This process is shown in Figure 8-10.

f. Any sampled voltage that lies within a quantization region is rounded off to the quantization value in the center of the region and is represented by the four-bit pattern assigned to that quantization value. For example, the second sample ($t = 125$ μsec) has a value of 0.926459 volts, which is quantized (rounded off) to 0.9375 volts and represented by the binary sequence 1111.

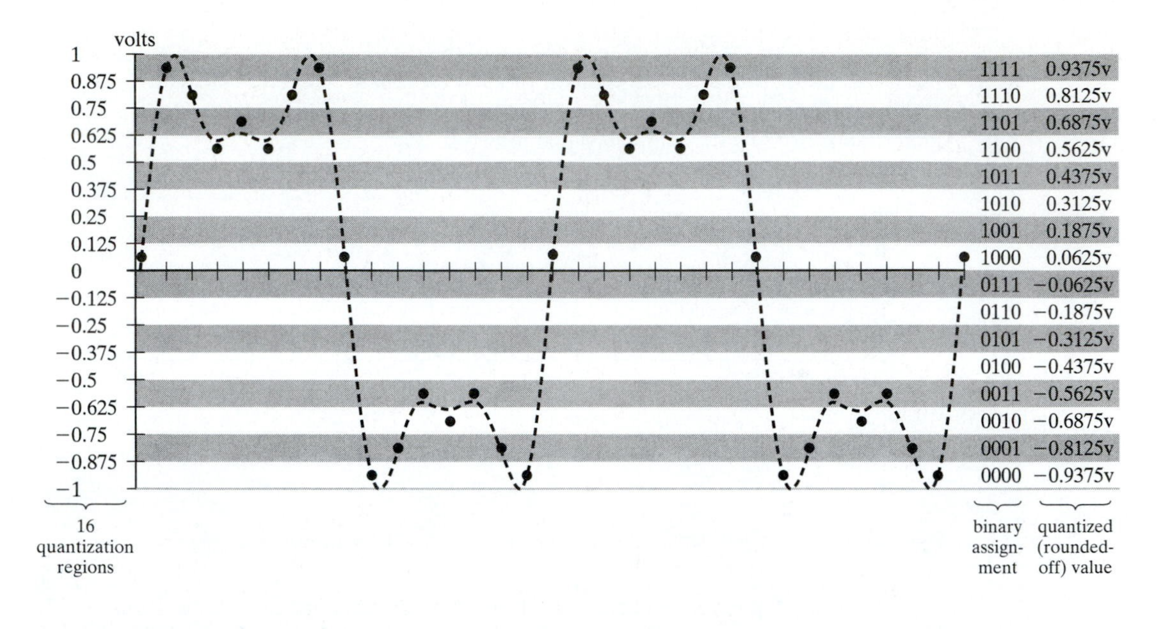

Figure 8-10 Dividing the dynamic range into 16 uniform regions, determining quantization value for each region, and assigning a unique four-bit sequence to each region.

g. The sampled and quantized signal is represented in digital form by sequentially transmitting the binary assignments representing the quantized values of each sample. The binary sequence for the sampled and quantized signal in this example is 1000111111101100110111001110....

8.1.4.2 Error associated with uniform quantization The error associated with quantizing is probabilistic. Since quantization is merely roundoff, if the analog signal has equal probability of being any value within a particular quantization region, then the error is equally likely to be any value ranging from zero to plus or minus half the quantization region. We can therefore represent error as a uniformly distributed random variable. For Example 8.1, the probability distribution for the quantization error **X** is shown in Figure 8-11.

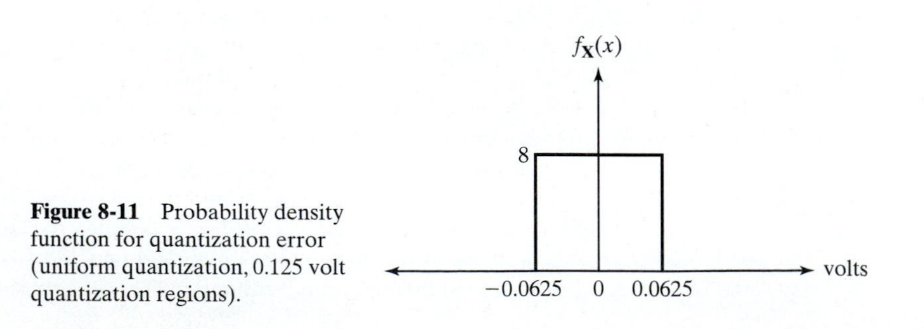

Figure 8-11 Probability density function for quantization error (uniform quantization, 0.125 volt quantization regions).

Quantization error is often called *quantization noise* because it affects the received signal in much the same way as noise. Let's examine some characteristics of quantization noise when uniform quantization is used.

Assuming that the analog signal is equally likely to be any value within the dynamic range,

1. The maximum error in using uniform quantization is

$$\begin{aligned}\mathbf{X}_{\max} &= \pm 0.5(\text{uniform quantization region width})\\ &= \pm 0.5\left(\frac{\text{dynamic range of signal}}{\text{number of different quantization values}}\right)\\ &= \pm 0.5\left(\frac{\text{dynamic range of signal}}{2^n}\right)\\ &= \pm\frac{\text{dynamic range of signal}}{2^{n+1}}\end{aligned} \tag{8.5}$$

   where n is the number of bits used to represent each quantization level. We often say that n is the number of *quantization bits*. We can reduce quantization error by increasing the number of quantization bits. The trade-off is that increasing the number of quantization bits requires the system to transmit at a faster speed.

2. Quantization noise can be represented as a uniformly distributed random variable whose probability density is

$$f_{\mathbf{X}}(x) = \begin{cases} \dfrac{1}{2\mathbf{X}_{\max}} & -\mathbf{X}_{\max} \le \mathbf{X} \le \mathbf{X}_{\max} \\ 0 & \text{elsewhere} \end{cases} \tag{8.6}$$

   You can verify this and build your intuition by reexamining Figure 8.11.

3. The mean of the quantization noise is zero.

4. The variance of the quantization noise is

$$\sigma_{\mathbf{X}}^2 = \int_{-\mathbf{X}_{\max}}^{\mathbf{X}_{\max}} (x-\mu_{\mathbf{X}})^2 f_{\mathbf{X}}(x)\,dx = \int_{-\mathbf{X}_{\max}}^{\mathbf{X}_{\max}} x^2\left(\frac{1}{2\mathbf{X}_{\max}}\right)dx = \frac{\mathbf{X}_{\max}^2}{3} \tag{8.7}$$

   Since we have shown that variance is equal to average normalized power for zero-mean random variables (see Section 4.1.2.5), the variance of the quantization noise can be used to express quantization error in terms of a signal-to-noise ratio.

Be sure you understand how Equations (8.5) and (8.7) are impacted by changes in dynamic range and by changes in the number of quantization bits used.

Some systems set the quantization voltage to the lowest voltage within the quantization region rather than to the value in the center of the region. Such a strategy involves truncation rather than rounding off. As you'll show in Problem 8.9, truncation produces a significantly larger average normalized quantization noise power.

Music CDs employ uniform quantization. As mentioned earlier, two analog channels (for stereo) are each sampled at 44,100 samples/sec, providing distortionless sampling and guardbands for frequencies up to 20 kHz. Each sample is then quantized using 16 bits. The encoding rate for music on CDs is thus

$$(44{,}100 \text{ samples/sec})(16 \text{ quantizing bits/sample})(2 \text{ channels for stereo}) = 1.411 \text{ Mbits/sec}$$

Compact discs can store up to 74 minutes of music and audio.

8.1.4.3 Nonuniform quantization As its name implies, a *nonuniform quantization* technique divides the dynamic range of a signal into a series of nonuniform quantization regions. For example, lower magnitudes may have smaller quantization regions than higher magnitudes. As with uniform quantization, any sample within a particular quantization region is rounded off to the voltage value corresponding to the center of the region.

Nonuniform quantization is employed in many applications where the user may require different resolutions for different regions of the signal's range. For example, in many applications the impact of an error depends on the error's size relative to the magnitude of the signal (i.e., error as a percentage of the signal's magnitude). One application for nonuniform quantization is in digitizing speech, since the human ear perceives volume change logarithmically rather than linearly.

Example 8.2 Uniform versus nonuniform quantization

Suppose we want to convert a typical analog speech signal, shown in Figure 8-12, into a digital format. As with Example 8.1, suppose we sample at 8,000 samples/sec and use four bits per sample for quantization.

Let's evaluate two possible quantization strategies: uniform quantization and a nonuniform quantization approach. Since we are using four bits per sample for quantization, we can divide the dynamic range of the signal into $2^4 = 16$ different quantization regions. For uniform quantization, each region has a width of 0.125 volts. For our nonuniform quantization approach, let's use small regions for the smaller signal magnitudes and let the regions become progressively larger as the signal's magnitudes increase. The uniform and nonuniform quantization regions are shown in Table 8-2.

Uniform quantization of the speech signal is shown in Figure 8-13a. Note the large roundoff errors associated with the smaller voltage values. Since these errors are a large percentage of the signal's magnitude, the errors will significantly impact the perceived quality of the signal—the speech will sound artificial. Nonuniform quantization of

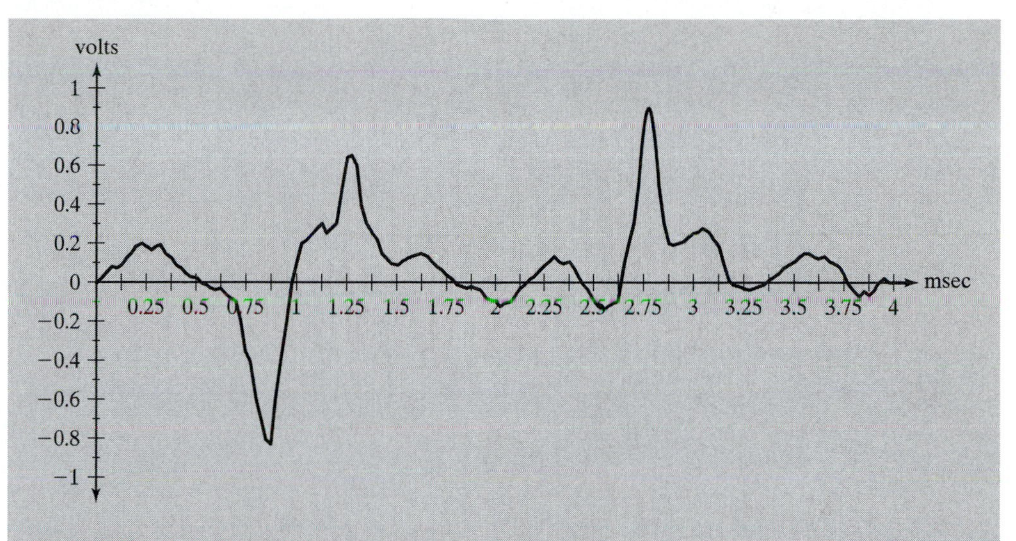

Figure 8-12 Typical speech signal.

Table 8-2 Uniform and Nonuniform Quantization Regions

Binary Assignment	Uniform Quantization	Nonuniform Quantization
0000	$-1.000 \le x(t) < -0.875$v	$-1.000 \le x(t) < -0.805$v (0.195v)
0001	$-0.875 \le x(t) < -0.750$v	$-0.805 \le x(t) < -0.630$v (0.175v)
0010	$-0.750 \le x(t) < -0.625$v	$-0.630 \le x(t) < -0.475$v (0.155v)
0011	$-0.625 \le x(t) < -0.500$v	$-0.475 \le x(t) < -0.340$v (0.135v)
0100	$-0.500 \le x(t) < -0.375$v	$-0.340 \le x(t) < -0.225$v (0.115v)
0101	$-0.375 \le x(t) < -0.250$v	$-0.225 \le x(t) < -0.130$v (0.095v)
0110	$-0.250 \le x(t) < -0.125$v	$-0.130 \le x(t) < -0.055$v (0.075v)
0111	$-0.125 \le x(t) < 0.000$v	$-0.055 \le x(t) < 0.000$v (0.055v)
1000	$0.000 \le x(t) < 0.125$v	$0.000 \le x(t) < 0.055$v (0.055v)
1001	$0.125 \le x(t) < 0.250$v	$0.055 \le x(t) < 0.130$v (0.075v)
1010	$0.250 \le x(t) < 0.375$v	$0.130 \le x(t) < 0.225$v (0.095v)
1011	$0.375 \le x(t) < 0.500$v	$0.225 \le x(t) < 0.340$v (0.115v)
1100	$0.500 \le x(t) < 0.625$v	$0.340 \le x(t) < 0.475$v (0.135v)
1101	$0.625 \le x(t) < 0.750$v	$0.475 \le x(t) < 0.630$v (0.155v)
1110	$0.750 \le x(t) < 0.875$v	$0.630 \le x(t) < 0.805$v (0.175v)
1111	$0.875 \le x(t) \le 1.000$v	$0.805 \le x(t) \le 1.000$v (0.195v)

the signal (using the quantization regions in the right column of Table 8-2) is shown in Figure 8-13b. The roundoff errors associated with the smaller voltage values have been considerably reduced, improving the quality of the speech. Although the roundoff errors associated with the larger voltage values have increased, these errors are still relatively small when viewed as a percentage of the signal's amplitude. Since the ear perceives volume changes logarithmically, the larger errors associated with the larger voltage values will not significantly impact perceived speech quality.

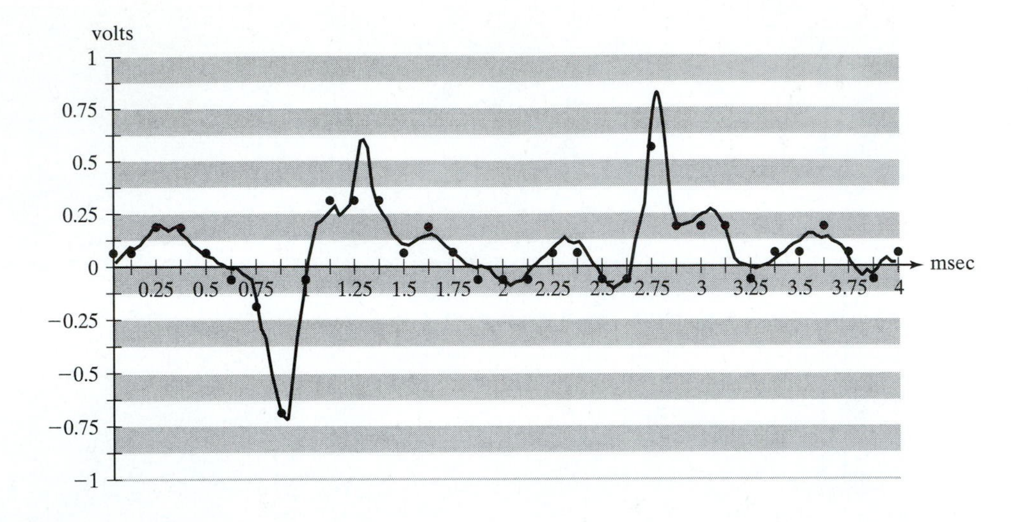

Figure 8-13a Uniform quantization of the speech signal. Note the relatively large amount of error for the smaller samples.

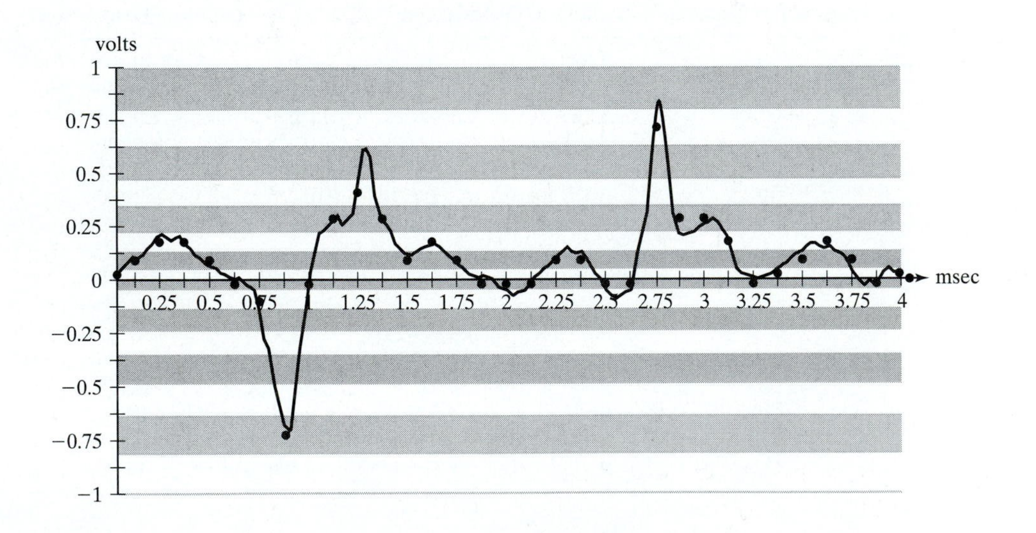

Figure 8-13b Nonuniform quantization of the speech signal. Error in the smaller samples is reduced, and error in the larger samples is still a small percentage of the sample's amplitude.

Although uniform quantization is simpler to implement, Example 8.2 shows that for certain applications, nonuniform quantization can significantly improve perceived quality. If we want to use nonuniform quantization, we can do so directly or we can

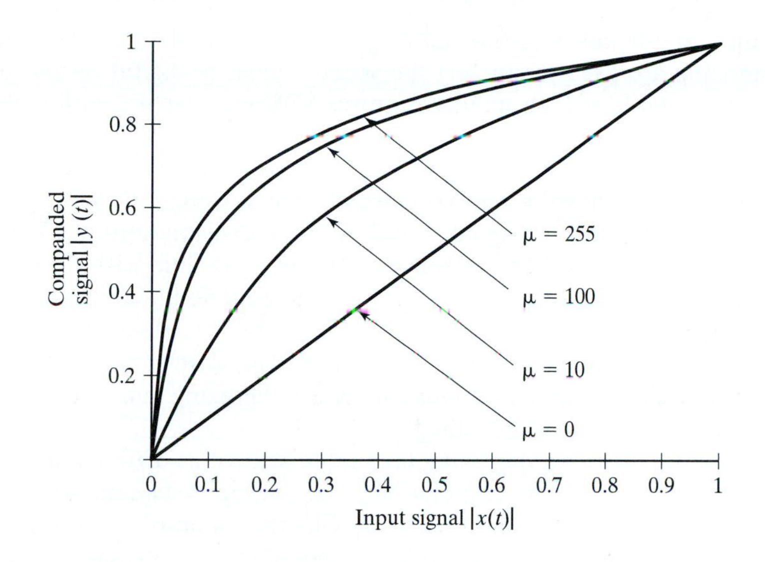

Figure 8-14 μ-law companding.

implement the following, mathematically equivalent, procedure:

1. At the transmitter, perform nonlinear compression on the analog signal and then digitize the compressed signal using uniform quantization. This step effectively "warps" the signal prior to quantizing.
2. At the receiver, convert the warped signal back into analog form and then decompress (expand) the signal using the inverse of the nonlinear operation employed in Step 1. This effectively "unwarps" the signal.

This procedure is referred to as *companding*, a term derived from "COMpressing and exPANDING." As an example of companding, consider the nonlinear function

$$|y(t)| = \frac{\ln(1 + \mu|x(t)|)}{\ln(1 + \mu)} \tag{8.8}$$

where $x(t)$ represents the input signal, $y(t)$ represents the output signal, and μ is a constant ($0 \le \mu \le 255$). Figure 8-14 plots Equation (8.8) for various values of μ. The larger the value of μ, the greater the warping, which emphasizes low amplitudes and de-emphasizes high amplitudes. Equation (8.8) is referred to as μ-law companding, and the value chosen for μ depends on the application. For example, wireline telephone systems in North America use μ-law companding with $\mu = 255$.[3]

8.2 Differential Pulse Coded Modulation

For many applications, limitations in transmission speed and channel capacity make sampling and quantizing a less-than-optimal method for analog-to-digital conversion.

[3]Another popular companding technique, called A-law companding, is described in Sklar [8.2].

As an example, consider transmitting speech over a telephone system. If sampling and quantizing are used to convert the speech signal to digital format prior to transmission, acceptable speech quality requires 8000 samples/sec with eight quantizing bits per sample. This technique, known as *pulse coded modulation* (PCM), requires an encoding rate of 64,000 bits/sec per user for each telephone conversation. What if it was possible to produce the same quality of speech using only half as many bits/sec per user? The telephone company could use time division multiplexing (see Section 7.1) to transmit twice as many telephone calls as is possible with PCM (and collect twice the revenue) without having to add more telephone lines. Can speech (or other analog signals) be digitized using a lower encoding rate than required by sampling and quantizing, without compromising quality? The answer is "yes," if we are willing to consider more complicated techniques for digitizing (another trade-off—increased complexity for reduced encoding rate).

Let's consider a digitizing technique known as *differential pulse coded modulation* (DPCM), which exploits the correlation between samples to reduce the encoding rate. As we remember from Chapters 4 and 5, correlation means that the present sample is not completely independent of previous samples. DPCM, illustrated in Figure 8-15, works as follows:

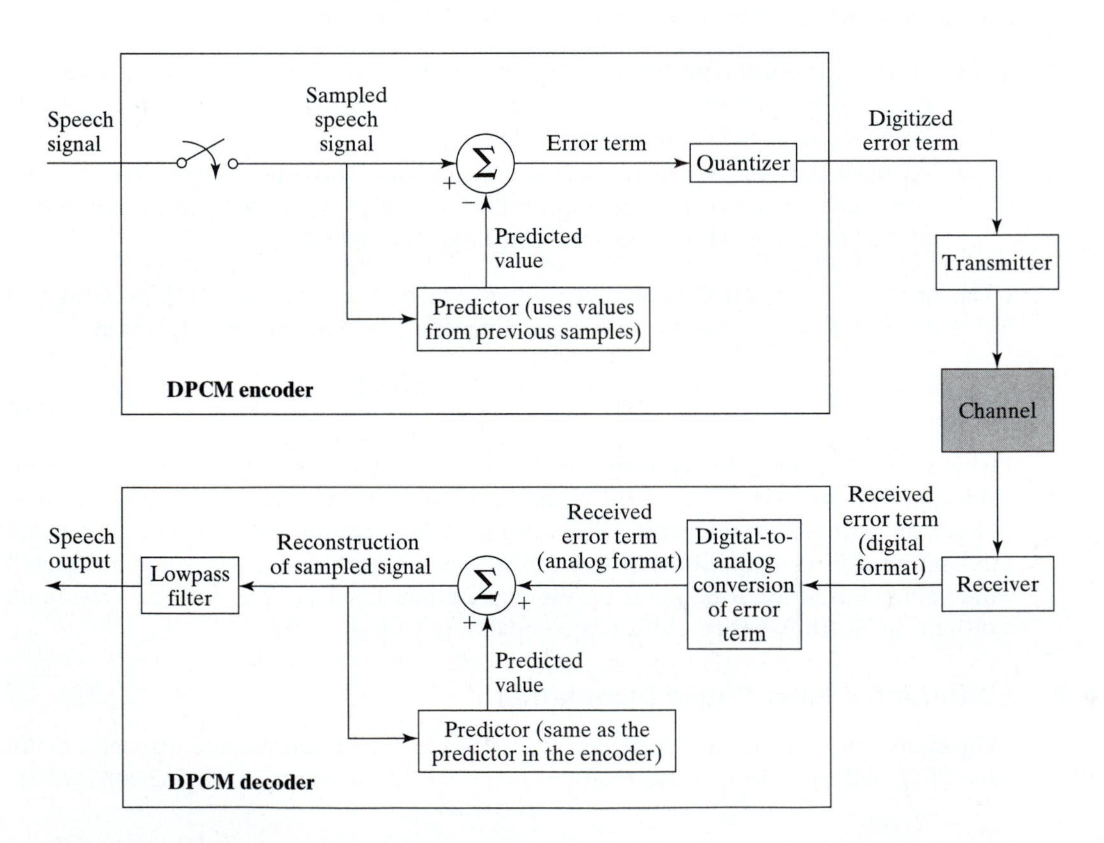

Figure 8-15 A differential pulse coded modulation encoder and decoder.

1. Based on the amplitude of the previous samples, the encoder (the analog-to-digital converter) predicts the value of the present sample.
2. The encoder compares the actual value of the present sample with the value that was predicted in Step 1. The difference between the actual and predicted values, called the *error term*, is then quantized.
3. Only the quantized error term is transmitted.
4. At the receiver, the DPCM decoder (the digital-to-analog converter) uses the same method as the encoder to predict the value of the present sample. After the error term is received and demodulated, it is added to the prediction, producing again the actual amplitude of the present sample.

The present sample is predicted based on a linear combination of the previous N samples. The predictor in the encoder in Figure 8-15 may convert the sampled signal to digital format so that the predictive algorithm can be implemented digitally. Such a prediction algorithm can be implemented using a tapped-delay line similar to the one shown in Figure 4-46.

As illustrated in Example 8.3, DPCM requires a significantly lower encoding rate than PCM because, if the prediction algorithm is chosen properly, the error term will have a significantly lower dynamic range than the original speech signal and will thus need fewer quantizing bits to achieve the same quantization error.

Example 8.3 Comparing DPCM with PCM

Suppose we want to convert the analog signal shown in Figure 8-16a into digital format. Sampling at 8000 samples/sec and using uniform quantization at 8 bits/sample, we can achieve a maximum quantization error of

$$\mathbf{X}_{\max} = \pm\frac{4 \text{ volts}}{2^9} = \pm 0.0078125 \text{ volts}$$

Now suppose we use DPCM to digitize the signal and suppose we predict the value of the present sample by the algorithm

$$\hat{s}(n) = 0.75s(n-1) + 0.20s(n-2) + 0.05s(n-3)$$

where $\hat{s}(n)$ represents the predicted value of the nth sample
$s(n-1)$ represents the actual value of the n-1st sample
$s(n-2)$ represents the actual value of the n-2nd sample, etc.

Predicted values and error terms for the first 10 samples are shown in Table 8-3. Note that initially there are no previous samples, so their values in the prediction algorithm are zero.

Figure 8-16b shows the actual signal, predicted signal, and error term using DPCM. Note that the dynamic range of the error term is only 0.5 volts, considerably less than the four-volt dynamic range of the original signal. As we've established, the original signal needs eight quantization bits to achieve a maximum quantization error of ±0.0078125 volts. This same maximum quantization error can be achieved for the error term using only five quantization bits. Using sampling and uniform quantization, the signal in Figure 8-16a required 64,000 bits/sec. Using DPCM, the same signal can be transmitted at the same high quality using only 40,000 bits/sec.

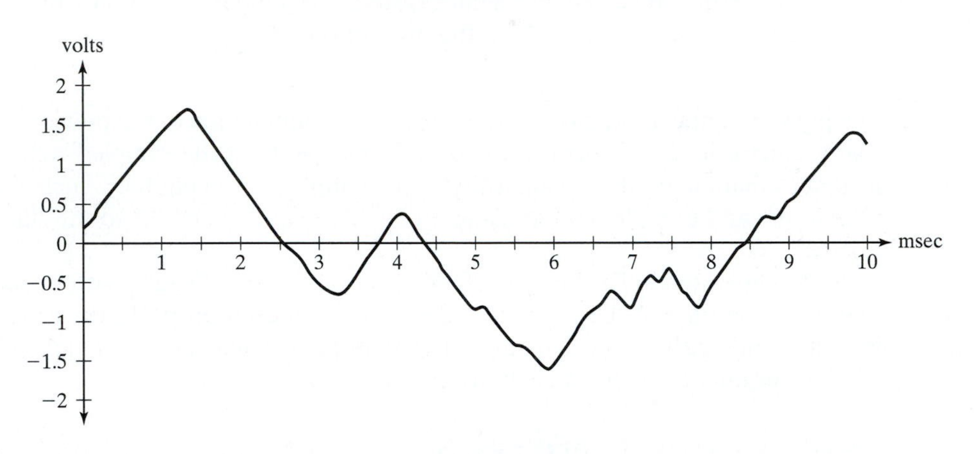

Figure 8-16a Typical signal to be encoded using DPCM.

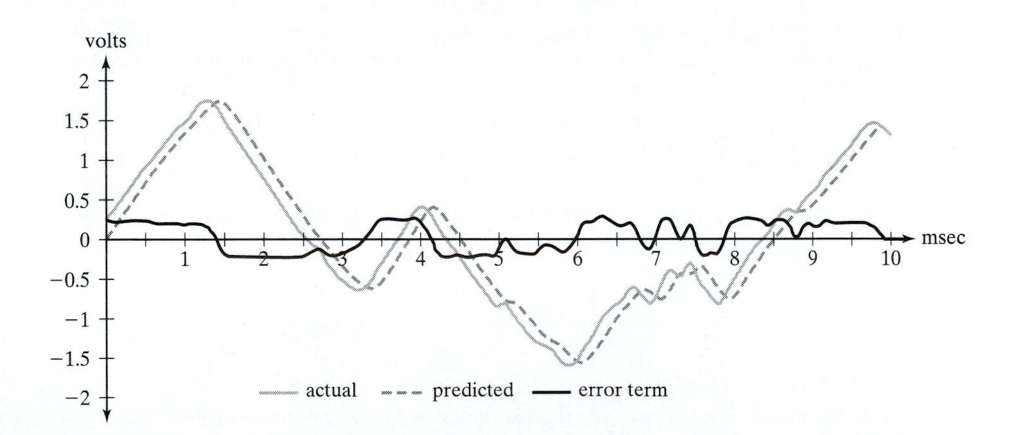

Figure 8-16b Actual signal, predicted signal, and error term using DPCM.

As a final improvement, the algorithm used to predict the present sample based on a linear combination of previous samples can be *adaptive*; that is, the algorithm can change the values of the linear coefficients based on long-term changes in the correlation of the samples. Adaptive prediction is more complex but it produces greater accuracy if the

Table 8-3 Actual Values, Predicted Values, and Error Terms for Figure 8-16a Using DPCM

Time (msec)	Predicted Values		Actual Values	Error Terms
0	0.75(0) + 0.20(0) + 0.05(0) =	0v	0.23v	0.23v
0.125	0.75(0.23) + 0.20(0) + 0.05(0) =	0.1725v	0.38v	0.2075v
0.25	0.75(0.38) + 0.20(0.23) + 0.05(0) =	0.33v	0.56v	0.229v
0.375	0.75(0.56) + 0.20(0.38) + 0.05(0.23) =	0.5075v	0.73v	0.2225v
0.5	0.75(0.73) + 0.20(0.56) + 0.05(0.38) =	0.6785v	0.90v	0.2215v
0.625	0.75(0.90) + 0.20(0.73) + 0.05(0.56) =	0.849v	1.05v	0.201v
0.7		1.004v	1.2v	0.196v
0.875		1.155v	1.35v	0.195v
1		1.305v	1.48v	0.175v
1.125		1.440v	1.62v	0.18v
etc.		etc.	etc.	etc.

characteristics of the signal change over the long term or if you are not completely confident in the initial values of your prediction coefficients. Many long-distance telephone services use a 32 kbit/sec adaptive differential pulse coded modulation (ADPCM), allowing a telephone system to double its capacity relative to 64 kbit/sec PCM.

8.3 Delta Modulation and Continuously Variable Slope Delta Modulation

8.3.1 Delta Modulation

As we've just seen, we can reduce the encoding rate for an analog signal by exploiting the correlation between adjacent samples. With DPCM, we used past samples to predict the present sample and we then encoded just the error term. Encoding only the error term required fewer quantizing bits than encoding the entire sample because the error term had a smaller dynamic range. Can we make further reductions in the encoding rate for an analog signal? Let's see.

If we increase the sampling rate, we increase the correlation between adjacent samples. Increased correlation should allow us to predict the present sample more accurately, which will reduce the dynamic range of the error term, meaning that fewer quantizing bits will be needed to encode the error term. Thus, given a certain acceptable quantization error, increasing the sampling rate should reduce the number of bits needed to reproduce each sample. Unfortunately, increasing the sampling rate also increases the number of samples/sec that must be digitized, since

$$\text{encoding rate} = (\text{sampling rate})(\text{number of quantizing bits per sample}) \tag{8.9}$$

Let's examine an extreme case to see if the benefit (reducing the number of quantizing bits per sample) outweighs the deficit (increasing the sampling rate).

Delta modulation (DM) is an extreme case of DPCM. Consider the prediction algorithm

$$\hat{s}(n) = \begin{cases} \hat{s}(n-1) + \Delta & \text{if } \hat{s}(n-1) \le s(n-1) \\ \hat{s}(n-1) - \Delta & \text{if } \hat{s}(n-1) > s(n-1) \end{cases} \tag{8.10}$$

where Δ is a constant. Translating Equation (8.10) into English: "If our prediction for the n-1st sample was too low, then our prediction for the nth sample is the same prediction plus a constant (Δ). If our prediction for the n-1st sample was too high, then our prediction for the nth sample is the same prediction minus a constant (Δ)." The prediction algorithm in Equation (8.10) mimics a guessing game. Suppose we ask you to guess a number between one and 10 and, after you guess, we tell you that your guess was too low. For your next guess you will increase your last guess by some increment.

Now suppose we quantize the error term using only a single bit. Let a "1" signify a positive error term (the prediction was too low) and a "0" signify a negative error term (the prediction was too high). This single bit is sufficient to predict the value of the n + 1st sample.

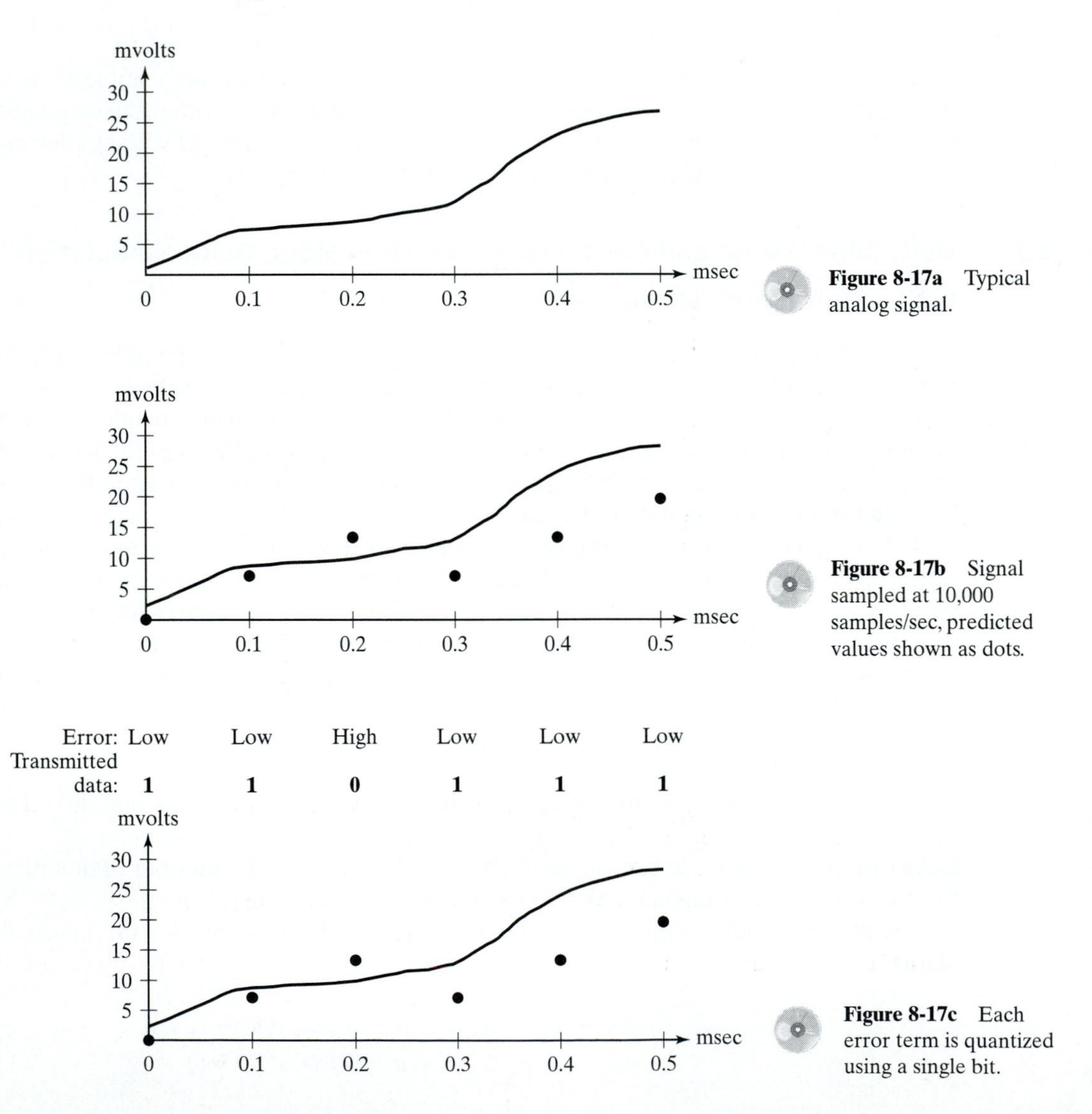

Figure 8-17a Typical analog signal.

Figure 8-17b Signal sampled at 10,000 samples/sec, predicted values shown as dots.

Figure 8-17c Each error term is quantized using a single bit.

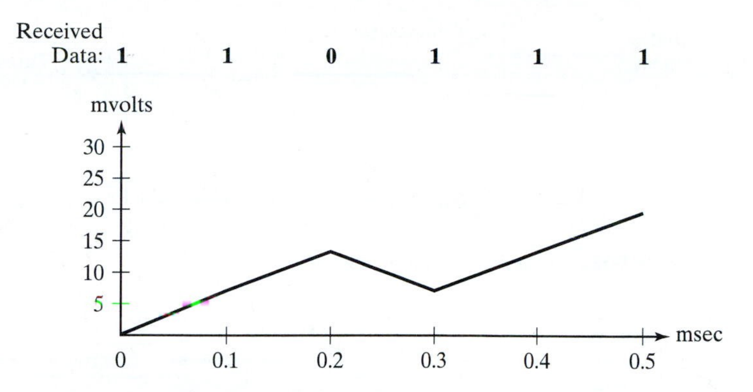

Figure 8-17d Signal is reconstructed by ramping voltage up for a "1" and down for a "0" (ramp rate = Δ/f_s).

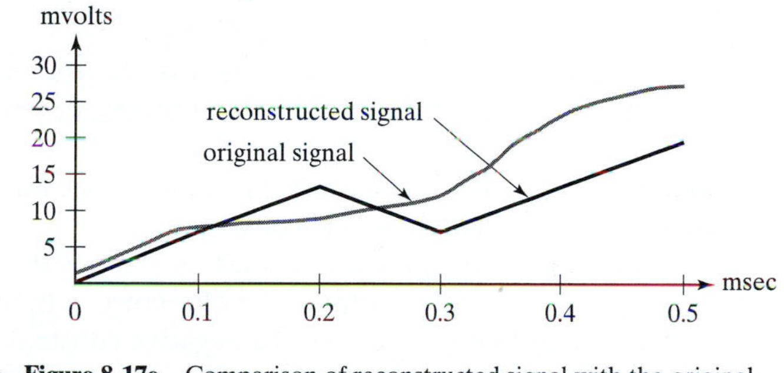

Figure 8-17e Comparison of reconstructed signal with the original signal.

Figure 8-17a shows a typical analog signal. Suppose we sample the signal at 10,000 samples/sec and use DM with $\Delta = 6.25$ mvolts. As shown in Figure 8-17b, our initial prediction (at $t = 0$) is 0 mvolts. Since the actual sample is 2 mvolts, the prediction is too low. The next prediction (for $t = 0.1$ msec) is therefore 6.25 mvolts. The actual sample at 0.1 msec is 7 mvolts, so the prediction is, again, too low. The prediction for $t = 0.2$ msec is therefore 12.5 mvolts. The actual sample at 0.2 msec is 9 mvolts, so the prediction is too high. The prediction for $t = 0.3$ msec is thus reduced by 6.25 mvolts. The process continues as illustrated in Figure 8-17b. Each time a prediction is too low, the error term is quantized as a single "1" and each time the prediction is too high, the error term is quantized as a single "0." As shown in Figure 8-17c, the encoded waveform is thus represented by the data pattern "110111."

Converting the encoded signal back into analog format is a straightforward procedure. Starting at 0 volts, each time a "1" is received, the signal is ramped upward at a rate of Δ/f_s (for our example, 62.5 volts per second). Each time a "0" is received, the signal is ramped downward at the same rate of Δ/f_s. The waveform produced by this conversion is shown in Figure 8-17d. Figure 8-17e compares the converted waveform with the original analog signal.

One of the major advantages of DM is its simplicity. Figure 8-18a shows the block diagram for a DM encoder. The analog signal is compared to its predicted value

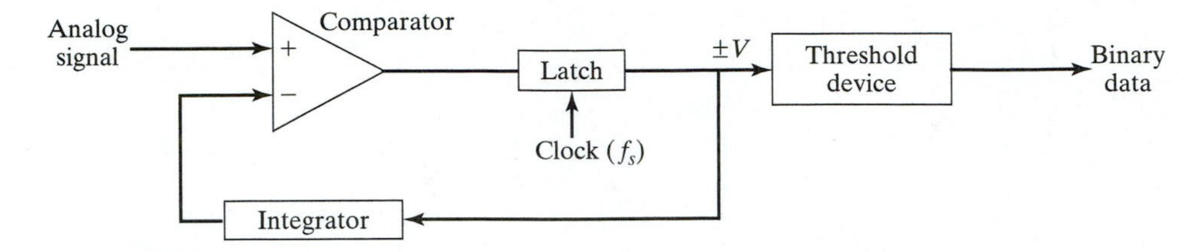

Figure 8-18a A delta modulation encoder.

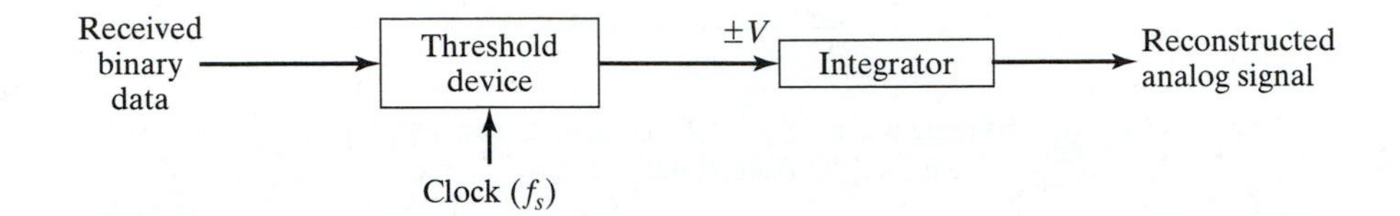

Figure 8-18b A delta modulation decoder.

(generated by the integrator, as we will discuss shortly), and the output of the comparator is either $+V$ volts if the signal is greater than its predicted value or $-V$ volts if the signal is less than its predicted value. The comparator output is then latched for the duration of the sample period, and the threshold device outputs either a "1" if its input is $+V$ volts or a "0" if its input is $-V$ volts.

The output of the integrator is a constant k times the integral of the integrator's input. At $t = 0$, the integrator output is 0 volts. For $t > 0$, the input to the integrator is either a positive constant ($+V$ volts) or a negative constant ($-V$ volts), so the integrator output either ramps upward at a slope kV if the input is positive, or downward at a slope $-kV$ if the input is negative. If the constant k is chosen so that the slope $kV = \Delta/f_s$, then the output of the integrator will be the reconstructed signal (similar to Figure 8-17d). At each instant of sampling, this output is equal to the predicted value of the analog signal (compare Figure 8-17d with 8-17c).

The DM decoder is shown in Figure 8-18b. The digital data is input to a threshold device that outputs $+V$ volts if the input was a "1" and $-V$ volts if the input was a "0." The output of the threshold device is then fed into an integrator similar to the one in the DM encoder. As with the encoder's integrator, the output of the decoder's integrator will be an upward ramp of slope kV if the integrator's input is $+V$ and a downward ramp if the integrator's input is $-V$. Choosing the constant k so that the slope $kV = \Delta/f_s$, the output of the integrator is the reconstructed signal.

To gauge the accuracy of DM, we need to examine the process for a longer period of time than we did in Figure 8-17. Figure 8-19 shows the same analog signal as in Figure 8-17 over a longer time period (2 msec rather than 0.5 msec) and also shows the reconstituted DM signal, sampled at 10,000 samples/sec with $\Delta = 6.25$ mvolts. The DM signal bears a fair resemblance to the original analog signal, and the resemblance can be improved by using a lowpass filter at the decoder's output to smooth out the sharp edges of the reconstituted signal. Since the sampling rate and encoding rate for a DM signal are the same, the signal in Figure 8-19 is being encoded using only 10,000 bits/sec—considerably lower than the encoding rate required by PCM or DPCM. Note, however, that the quality is not as high as would be obtained using PCM or DPCM. Increasing the

Data: 1 1 0 1 1 1 1 0 0 0 0 1 1 0 1 1 1 1 0 0 0

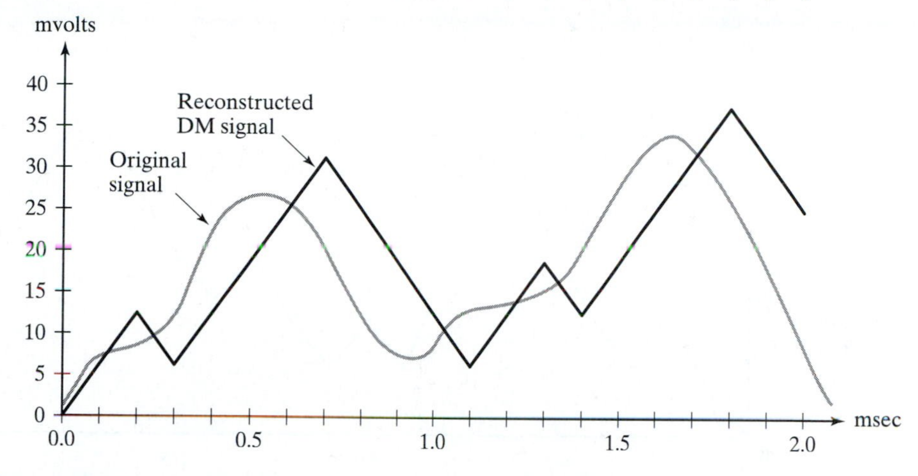

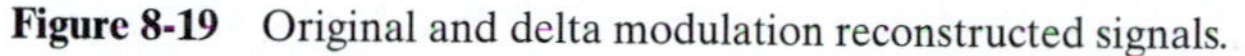

Figure 8-19 Original and delta modulation reconstructed signals.

sampling rate of the DM signal (up to about 20,000 samples/sec for speech signals) will produce marginal increases in signal quality at the cost of an increase in encoding rate, but the signal's quality will still not compare favorably with PCM or DPCM. DM thus offers a low-bit-rate, simple method for analog-to-digital and digital-to-analog conversion, but the resulting signals are of lower quality than with PCM or DPCM.

8.3.2 Continuously Variable Slope Delta Modulation

Can we improve the quality of DM without raising the encoding rate? Let's reexamine the original and DM reconstructed waveforms shown in Figure 8-19. Note the inaccuracies in the DM reconstructed signal occurring between 0.3 and 0.6 msec, and between 0.7 and 1.0 msec. In these regions the slope of the DM reconstructed signal is too shallow to "catch up" with the original signal. Increasing Δ (and hence increasing the slope of the DM reconstructed signal) will improve the accuracy in these two time intervals but it will cause overshoot problems in other intervals where the slope of the original signal is significantly smaller (for instance, $1.0 < t < 1.4$ msec). Instead of using a fixed value for Δ, suppose we develop a strategy in which we let Δ change as a function of the accuracy of previous predictions. This strategy is known as *continuously variable slope delta* (CVSD) modulation.

Consider, for example, the original analog signal in Figure 8-19 and let's use the following rules for changing the value of Δ:

a. Initially, let $\Delta = 5.0$ mvolts.

b. When the error signal is either positive for two consecutive samples or negative for two consecutive samples, increase Δ to 7.5 mvolts.

c. When the error signal is either positive for three consecutive samples or negative for three consecutive samples, increase Δ to 10.0 mvolts.

d. Any time the error signal changes polarity between two adjacent samples, reduce Δ to 5.0 mvolts.

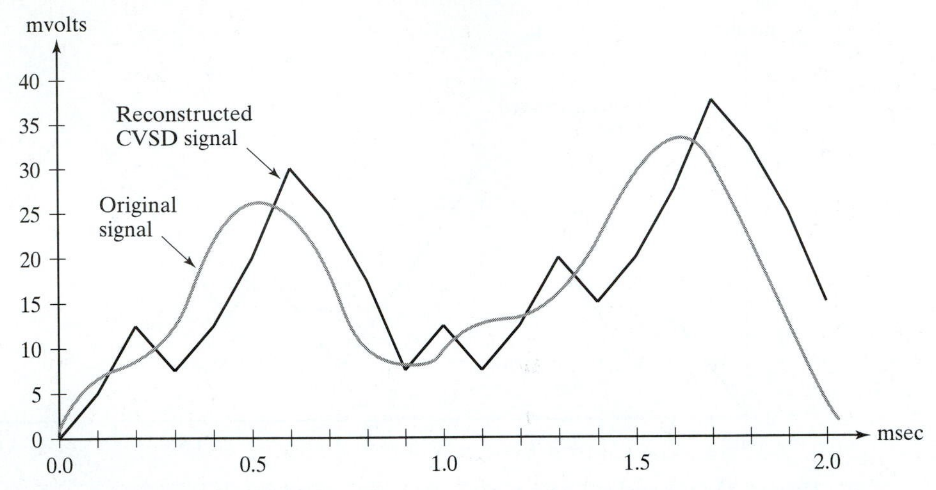

Figure 8-20 Original and continuously variable slope delta reconstructed signals.

Figure 8-20 shows the original signal and the reconstructed waveform when CVSD is applied using the Δ values given above. At $t = 0$, $\Delta = 5$ mvolts, but at $t = 0.1$ msec, when the second estimated value in a row is too low, Δ is increased to 7.5 mvolts. Since the estimated value for the third sample ($t = 0.2$ msec) is too high, Δ is reduced to 5 mvolts, where it remains until the fifth sample ($t = 0.4$ msec), when the estimated value is too low for the second time in a row and so Δ is increased to 7.5 mvolts. At $t = 0.5$ msec, the estimated value is too low for the third consecutive sample and so Δ is increased to 10.0 mvolts, etc.

Compare Figure 8-20 with Figure 8-19 and observe the increase in accuracy that CVSD provides over DM (again, a lowpass filter can be used to smooth the sharp edges of the reconstituted signal). CVSD encoders and decoders are more complex than their DM counterparts, but only slightly. Figures 8-21a and b show a CVSD encoder and decoder. The only additional components required are a small amount of memory and logic to keep track of the previous few estimates, and the capability to change the gain in the integrator. For many applications, CVSD provides an excellent compromise between

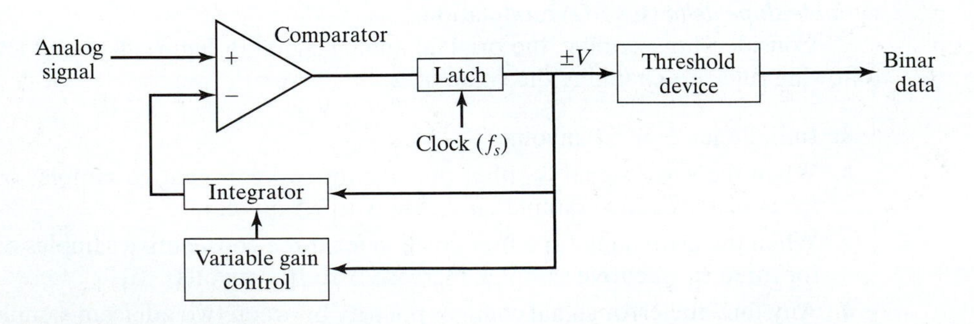

Figure 8-21a A CVSD encoder.

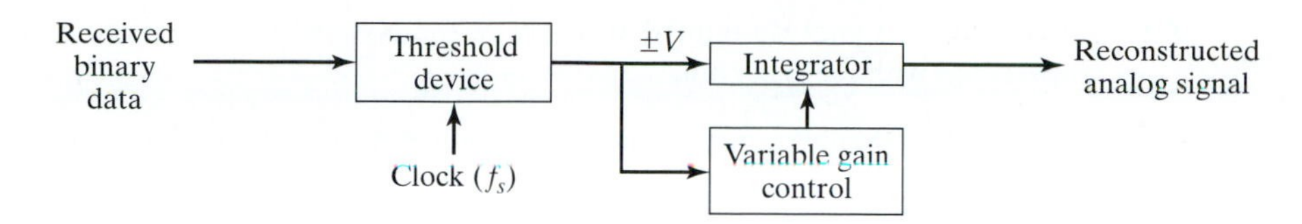

Figure 8-21b A CVSD decoder.

speech quality, complexity, and encoding rate. For example, many of the free Internet-based long-distance software packages use CVSD to encode speech. Also, many manufacturers provide single-chip CVSD encoders/decoders for less than $10.

8.4 Further Reading on Analog-to-Digital and Digital-to-Analog Conversion

Encoding rates can be further reduced, at the cost of increased complexity, by using more advanced techniques such as linear predictive coding (LPC) and vector quantization (VQ). For more information concerning these advanced encoding techniques, see Kondoz [8.3] and Deller, Hansen, and Proakis [8.4]. For more information concerning the different types of algorithms and integrated circuits used to perform sampling and quantizing, see Mitra [8.5].

PROBLEMS

8.1 The waveform in Figure P8-1 is $f(t) = 2\sin(2\pi 1000t) - 3\sin(2\pi 2000t)$.

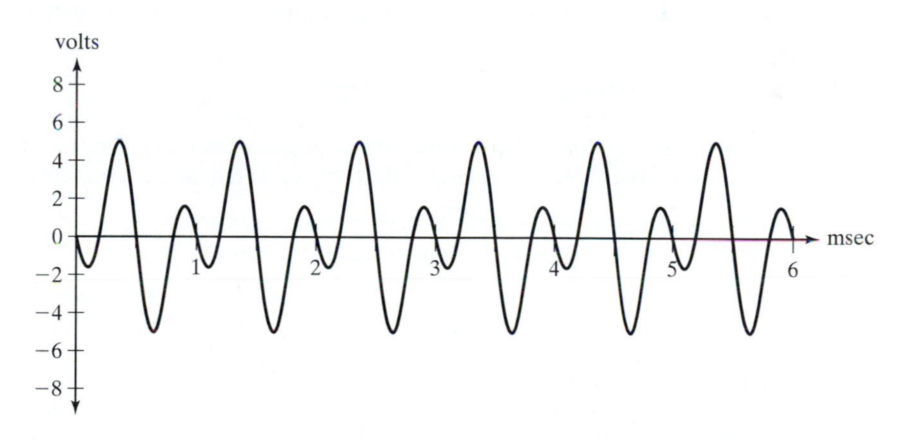

Figure P8-1 Analog signal requiring sampling.

a. Find the minimum sampling frequency theoretically required to reconstruct $f(t)$ without distortion.

b. Discuss any practical considerations that require the signal to be sampled at a rate above the theoretical minimum.

c. Draw the magnitude spectrum of the sampled signal if the sampling rate is 1.5 times the theoretical minimum.

8.2 Prove that if a signal is sampled at or above the Nyquist rate, then the signal can be reconstructed without distortion.

8.3 Explain why sampling is usually performed at rates significantly higher than the Nyquist rate.

8.4 Describe an anti-aliasing filter. Why is such a filter used prior to sampling?

8.5 Explain in detail how and why certain bandpass signals can be sampled at a rate below $2f_{max}$ and still be reconstructed without distortion.

8.6 Consider the bandpass signal whose magnitude spectrum is shown in Figure P8-6.

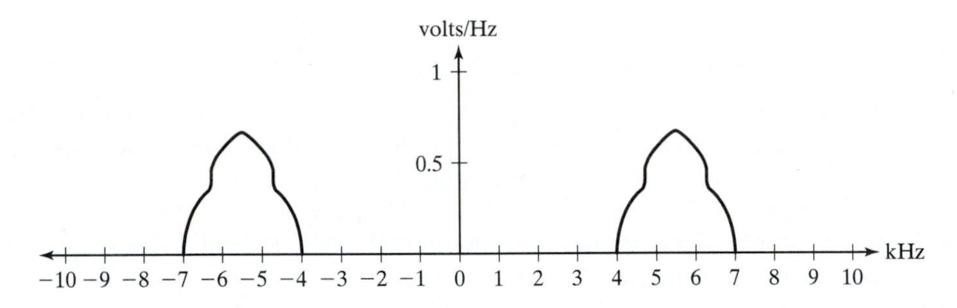

Figure P8-6 Magnitude spectrum of analog bandpass signal.

a. Can this signal be sampled at a rate lower than twice its maximum frequency component and then reconstructed without distortion? Justify your answer.

b. Draw the magnitude spectrum of the sampled signal if the sampling rate is 7000 samples/sec.

8.7 Suppose the signal in Problem 8.1 is sampled at 1.5 times the Nyquist rate and suppose each sample is uniformly quantized using four bits. Determine the maximum quantization error, the average normalized power of the quantization noise, and the number of bits per second required to store or transmit the sampled and quantized waveform.

8.8 In Problem 8.7 we digitized the waveform using a sampling rate of 1.5 times the theoretical minimum and four-bits-per-sample quantizing. Now,

a. Explain how we can increase the accuracy of the digitized waveform.

b. Discuss the costs involved in increasing the accuracy.

8.9 In Section 8.1.4.2, we claimed that setting quantization voltage to the value in the center of the quantization region produced lower quantization noise power than setting quantization voltage to the lowest voltage within the quantization region. Prove this claim.

8.10 Consider the signal shown in Figure P8-10. Suppose the signal is sampled 10,000 times per second and that each sample is uniformly quantized using six bits.

a. What is the dynamic range of the signal?
b. What is the width of the quantization region?
c. What is the maximum quantization error?
d. What is the average normalized power of the quantization noise?

e. How many bits per sample are needed to reduce the maximum quantization error to 12.5 millivolts?

f. What are the costs associated with the increase in accuracy described in Part e?

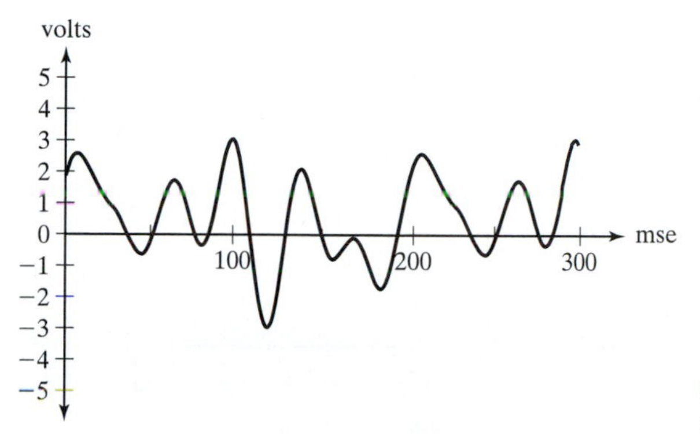

Figure P8-10 Analog signal requiring sampling and quantizing.

8.11 Describe DPCM and discuss how it differs from conventional sampling and quantizing. Discuss its advantages and disadvantages relative to conventional sampling and quantizing.

8.12 A DPCM system uses 8000 samples/sec and predicts the value of the present sample by the algorithm

$$\hat{s}(n) = 0.5s(n-1) + 0.3s(n-2) + 0.2s(n-3)$$

Determine the error terms for the first 10 samples of the waveform in Figure 8.16a. (Note that the values of the waveform samples are given in Table 8-3.) Compare the prediction algorithm in this problem with the algorithm in Example 8.3. Which is better for encoding the waveform in Figure 8-16a? Justify your answer in terms of quantization noise.

8.13 Consider the waveform shown in Figure 8-2a (a magnified version of the waveform is shown in Figure 8-10). The waveform is sampled at 8000 samples/sec and quantized at four bits/sample, with the bit assignments as shown in Figure 8-10.

a. Determine the data sequence for the first five samples if uniform quantization is used.

b. Determine the data sequence for the first five samples if μ-law companding is used with $\mu = 10$. You will have to estimate the original values of the samples by inspecting the waveform.

c. Determine the data sequence for the first five samples if μ-law companding is used with $\mu = 100$. You will have to estimate the original values of the samples by inspecting the waveform.

d. Compared with uniform quantization, what are the advantages and disadvantages of nonuniform quantization?

8.14 The waveform shown in Figure P8-14 is to be encoded using DM at a rate of 20,000 bits/sec using a slope of ±15 volts/sec.

a. Determine the digital output (string of "1"s and "0"s) of the DM encoder.

b. Draw the waveform that will appear at the output of the DM decoder prior to lowpass filtering.

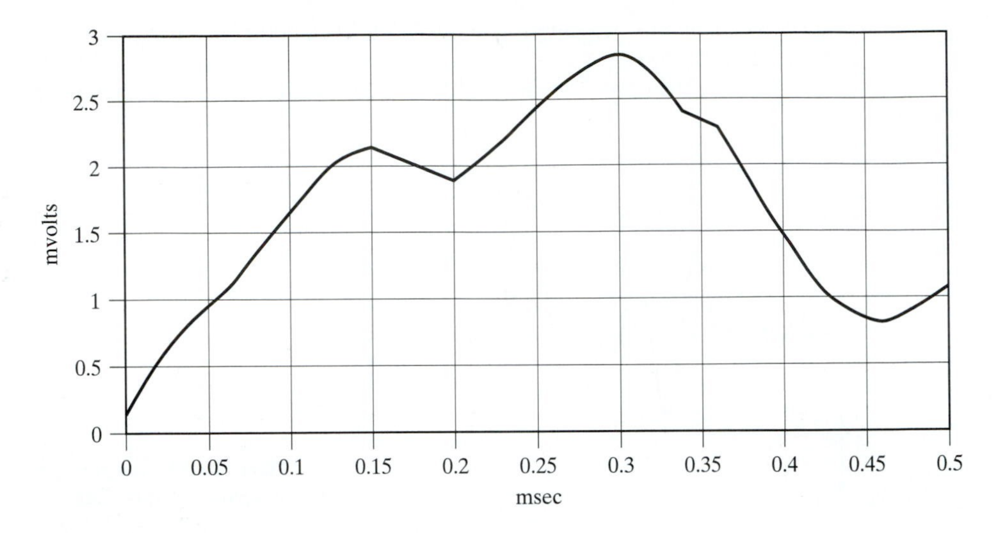

Figure P8-14 Analog signal requiring DM.

8.15 The waveform shown in Figure P8-14 is to be encoded using CVSD modulation at a rate of 20,000 bits/sec using slopes of ±10 volts/sec., ±15 volts/sec., and ±20 volts/sec.

a. Determine the digital output (string of "1"s and "0"s) of the CVSD encoder.

b. Draw the waveform that will appear at the output of the CVSD decoder prior to low-pass filtering.

8.16 Compare the results of DM encoding in Problem 8.14 with the results of CVSD encoding in Problem 8.15.

8.17 Discuss the performance-versus-cost trade-offs between standard sampling and quantizing (uniform and nonuniform), DPCM, and CVSD.

Chapter 9

Fundamentals of Information Theory, Data Compression, and Image Compression

IN Chapters 3–5 we developed techniques to transmit digital information across baseband and bandpass channels. In our performance-versus-cost trade-offs we used sophisticated waveforms (raised cosine pulse shapes, M-ary baseband and M-ary bandpass signaling, etc.) to increase transmission speed without increasing the required bandwidth. These trade-offs were our attempts to transmit more information across a fixed-bandwidth channel (in other words, to develop a more spectrally efficient communication system). In these trade-offs, our emphasis was on transmitting *faster*. We can also increase spectral efficiency and transmit more information across a fixed-bandwidth channel by transmitting *smarter*—by using data compression to ensure that the information from the source is represented using a shorter sequence of "1"s and "0"s. As an example, consider the transmission of a typical 3×4 inch color image across the Internet. Without data compression, representing this image might require 52 million bits (6.5 Mbytes)—much too large for efficient storage or fast transmission. Using a compression technique such as JPEG (discussed later in this chapter) we will need only about 1/20 as many bits to represent the same image. In this chapter we will develop efficient compression techniques for data, still images, and moving images.

9.1 Information Content, Entropy, and Information Rate of Independent Sources

Let's start by establishing a quantitative definition of information. In Chapter 1 we defined information qualitatively, saying that information is data that the user did not know prior to communicating with the source. Now let's determine a way to measure *how much* information is contained in any particular message from the source. To be most useful, our quantitative measure of information should have the following properties:

1. If a particular message is known by the user prior to being transmitted, the message contains zero information.
2. If potential messages from a source are all equally likely, then the information contained in each particular message should be equal to the number of "1"s and "0"s required to uniquely identify the message. For example, if the source flips a fair coin (a coin equally likely to be heads or tails), the message "the coin landed heads up" could be represented by the data sequence "1," while the message "the coin landed tails up" could be represented by the data sequence "0." The source needs one binary digit (one "1" or one "0") to represent the information. As another example, assume a person has eight shirts of different colors in a closet. If the person walks into the closet in the dark and blindly chooses one of the eight shirts, then the eight three-bit data sequences—"000," "001," "010," "011," "100," "101," "110," and "111"—can be used to represent which one of the eight shirts was chosen. The source in this case needs three binary digits to represent the information.
3. If two potential messages are not equally likely, the least likely message should contain the most information. As an example, suppose customers are nine times more likely to order a hamburger than a fish sandwich in a certain fast-food restaurant. The message "customer X ordered a hamburger" should contain less information than the message "customer X ordered a fish sandwich." A customer ordering a fish sandwich is more unusual, and hence less predictable, than a customer ordering a hamburger.

There is only one way to measure information that satisfies all three of these properties. This measure, developed by Claude Shannon [Shannon 9.1], states that if a certain event has a probability of occurrence of P, then a message stating that that event has occurred has an information content of

$$I = \log_2\left(\frac{1}{P}\right) = -\log_2 P \tag{9.1}$$

The unit for I is *bits*. This can cause some confusion, since "bits" is also the unit for the number of binary digits in a data sequence. In this chapter, we will use the term "bits" to specify information content and the term "binary digits" to specify the number of "0"s and "1"s in a data sequence.

Example 9.1 Measuring information content

a. Consider a source flipping a coin. How much information is contained in the message "the coin landed heads up"?
b. Consider a fast-food restaurant in which a customer is nine times as likely to order a hamburger as a fish sandwich. How much information is contained in the

message "the customer wants a hamburger"? How much information is contained in the message "the customer wants a fish sandwich"?

c. How much information is contained in the message "you are reading this textbook"?

Solution

a. The probability that the coin landed heads up is 0.5. Information content of the message is thus

$$I_{heads} = \log_2\left(\frac{1}{0.5}\right) = \log_2(2) = 1 \text{ bit}$$

b. The probability that the customer ordered a hamburger is 0.9. Information content of the message "the customer ordered a hamburger" is thus

$$I_{hamburger} = \log_2\left(\frac{1}{0.9}\right) = \log_2(1.11) = 0.152 \text{ bit}$$

Many calculators do not provide keys that calculate bases other than e and 10. You may therefore need to remember the identity

$$\log_2 x = \frac{\ln(x)}{\ln(2)} \tag{9.2}$$

Going back to our example, the information content of the message "the customer ordered a fish sandwich" is

$$I_{fish} = \log_2\left(\frac{1}{P_{fish}}\right) = \log_2(10) = 3.32 \text{ bits}$$

c. The probability that you are reading this textbook right now is 100%. Therefore

$$I_{textbook} = \log_2\left(\frac{1}{1}\right) = \log_2(1) = 0 \text{ bits}$$

This result corresponds with our qualitative definition of information, since the message "you are reading this textbook" did not tell you anything you did not already know.

Let's define *entropy* (symbolized H) as the average information content of a message from a particular source. Let's consider a source with n possible messages where the probability of the first message is P_1, the probability of the second message is P_2, etc. The entropy of the source (i.e., the average information content of a message from the source) can be calculated as

$$H = \sum_{j=1}^{n} P_j I_j = \sum_{j=1}^{n}\left[P_j \log_2\left(\frac{1}{P_j}\right)\right] = -\sum_{j=1}^{n} P_j \log_2(P_j) \tag{9.3}$$

Example 9.2 Measuring entropy

Suppose that in the fast-food restaurant mentioned previously each customer orders either one hamburger or one fish sandwich. The average information content in a customer's order is

$$H_{restaurant\ order} = 0.9I_{hamburger} + 0.1I_{fish} = 0.469 \text{ bits}$$

Let's compare this with the entropy of a coin flip

$$H_{coin\ flip} = 0.5I_{heads} + 0.5I_{tails} = 1 \text{ bit}$$

Entropy is a measure of the predictability of a source. Each of the two sources just described (the customer in the restaurant and the coin being flipped) has two possible outcomes, but we are far more likely to correctly guess a particular customer's order than to correctly guess whether a coin lands heads up or tails up. The more predictable a source, the lower its entropy.

Using the concept of entropy, we can calculate the average rate at which information flows from a source. Average information rate, symbolized as R, is the entropy of the source multiplied by the average number of messages output by the source per unit of time.

Example 9.3 Measuring information rate

Suppose our fast-food restaurant serves an average of eight customers per minute. The information rate of the food orders is

$$\begin{aligned} R = (0.469 \text{ bits per message})(8 \text{ messages per minute}) &= 3.75 \text{ bits per minute} \\ &= 0.0625 \text{ bits/sec} \end{aligned}$$

The average information rate tells us how much a source can benefit from data compression. Shannon has shown that if a source has an information rate of R bits per second, then there exists a data compression technique that can compress the bit stream coming from the source into a sequence averaging $R + \varepsilon$ binary digits per second, where ε is an arbitrarily small positive number. The information rate R therefore represents a lower limit on how many binary digits are needed, on the average, to represent the information flowing from the source. This principle is known as Shannon's first theorem.[1]

9.2 Variable-Length, Self-Punctuating Coding for Data Compression

Let's begin by considering an example to illustrate the principle of Shannon's first theorem.

[1]Although Shannon's first theorem guarantees that such a compression technique exists, it does not tell us how to develop the technique. Shannon's first theorem is useful in establishing a bound for the performance of a lossless compression code.

Example 9.4 A variable-length, self-punctuating code

Consider a system that transmits the weather from Los Angeles every day at 10:00 a.m. The system transmits one of four possible messages: "It's sunny in L.A. today," "It's raining in L.A. today," "It's foggy in L.A. today," or "It's too smoggy to determine what the weather is in L.A. today." A statistical study over a long period of time has shown that 50% of the time it's sunny at 10:00 a.m. in L.A., 25% of the time it's raining, 12.5% of the time it's foggy, and 12.5% of the time it's too smoggy to determine the weather. Let's assume that the weather each day is independent of any other day, and let's consider two possible codes for representing our L.A. weather information. These codes are shown in Table 9-1. Each entry in the Code 1 and Code 2 columns is a sequence of "1"s and "0"s representing a particular message. These sequences are called *codewords*. Code 1 is a straight binary code and requires two binary digits per message. Code 2 is a variable-length code that is completed when the user receives the first "1" or when the user receives three consecutive "0"s.

Table 9-1 Two Codes for L.A. Weather Information

Message	Probability of Occurrence	Code 1	Code 2
Sunny	0.5	00	1
Rainy	0.25	01	01
Foggy	0.125	10	001
Smoggy	0.125	11	000

The entropy of the L.A. weather information is

$$\begin{aligned} H_{L.A.\ weather} &= -\sum_{j=1}^{n} P_j \log_2(P_j) \\ &= -[(0.5)(-1) + (0.25)(-2) + (0.125)(-3) + (0.125)(-3)] \\ &= 1.75 \text{ bits per message} \end{aligned}$$

If the system transmits one message per day, then Shannon's first theorem tells us that there exists a code that can compress the L.A. weather information into an average arbitrarily close to 1.75 binary digits per message. Code 1 always uses two binary digits per message, but Code 2 uses an average of

(1 binary digit for "sunny")(sunny 50% of the time) +

(2 binary digits for "raining")(raining 25% of the time) +

(3 binary digits for "foggy")(foggy 12.5% of the time) +

(3 binary digits for "smoggy")(smoggy 12.5% of the time)

= 1.75 binary digits per message

Let $\overline{\ell}$ symbolize the length of the average compressed message, called the *average codeword length*. In the L.A. weather example, $\overline{\ell}_{Code\,1} = 2$ binary digits per message and $\overline{\ell}_{Code\,2} = 1.75$ binary digits per message. Shannon's first theorem establishes a lower limit on $\overline{\ell}$. For a code to be able to represent all of the information from a source

$\overline{\ell} \geq H$. This makes intuitive sense, as we need at least one binary digit to represent each bit of information. Since $\overline{\ell}_{Code\,2} = H$, we know that Code 2 is the most efficient way possible to encode all of the L.A. weather information.

Let's look at a third possible code for L.A. weather information, as shown in Table 9-2. This third code has an average length of only one binary digit per message, but it cannot convey all of the information regarding L.A. weather; Code 3 can tell us only whether or not it is sunny. This type of code is called a *lossy* code (as opposed to a *lossless* code), since it loses some of the source's information. Lossy codes are not necessarily bad—there are many applications in which the user either does not need or cannot use all of the information available from the source. As an example, Code 3 is acceptable if the user is a surfer who wants to go to the beach only when it's sunny. We will examine lossy codes later in this chapter, especially when we investigate image compression codes where, in many applications, the source image may contain more information (higher resolution and more contrast) than the human eye can discern.

Table 9-2 A Third Code for L.A. Weather Information

Message	Probability of Occurrence	Code 1	Code 2	Code 3
Sunny	0.5	00	1	1
Rainy	0.25	01	01	0
Foggy	0.125	10	001	0
Smoggy	0.125	11	000	0

For now, let's focus on lossless codes. In examining Code 2 in Example 9.4, we can see many desirable properties. First, the code is variable length with the shorter codewords assigned to the more probable messages. This property produces a shorter average codeword length. Second, the code is *self-punctuating*. We can tell when each codeword ends (we encounter either our first "1" or three consecutive "0"s) without requiring additional binary digits for demarcation. For example, if we store every day's L.A. weather information on a floppy disk, then the sequence "10110011000011" represents the series of messages "sunny, rainy, sunny, foggy, sunny, smoggy, rainy, sunny" and requires only 14 bits, as opposed to the 16 bits that would be required to store the messages in a straight binary assignment like Code 1.

9.2.1 Prefix Coding and the Tree Diagram

We see from Example 9.4 that a variable-length, self-punctuating code provides data compression. Can we produce such a code for any source with any set of messages? To find out, let's see what type of structure we need to make a code self-punctuating. Figure 9-1 represents L.A. Weather Code 2 in Example 9.4 using a *tree diagram*, a tool that can help us see certain features of a code geometrically.

As shown in Figure 9-2, tree diagrams are easy to construct. Starting at the left end, we create a node (Figure 9-2a). Next, we produce two branches radiating from the node, one pointing slightly upward, the other slightly downward. We then assign a binary "1" to one of the branches and a binary "0" to the other (Figure 9-2b). Consider each branch to have a node at its right end. We now go to one of these two new nodes

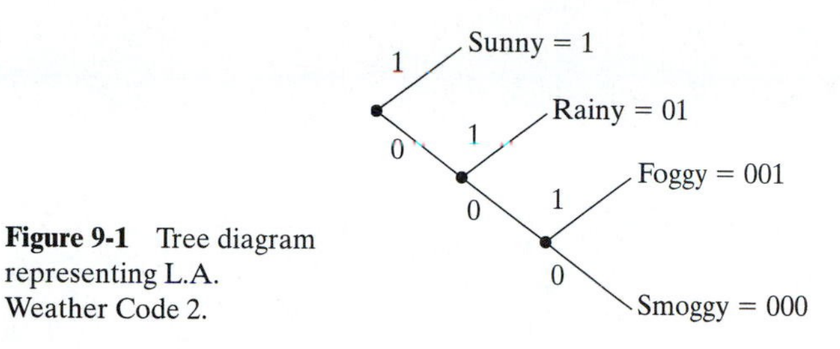

Figure 9-1 Tree diagram representing L.A. Weather Code 2.

and produce two more branches as just described (Figure 9-2c). The tree now has three nodes that have not produced branches. We call such nodes *end nodes*. We continue to grow the tree as desired by selecting an end node, producing two new branches (we now have four end nodes), selecting another end node, producing two more new branches, etc.

We can associate a sequence of binary digits with each node in the tree. Imagine tracing a path along the tree starting with the left node and ending at a particular node elsewhere in the tree. The path consists of a series of branches, each labeled with a "1" or a "0." We can uniquely identify any particular node by the sequence of "1"s and "0"s representing the path to that node. For example, the path in the Figure 9-1 tree to its lowest right node is represented by the sequence "000." A particular code can thus be represented by a tree that contains nodes associated with each of the possible codewords.

As we stated earlier, data compression can be achieved by creating a variable-length, self-punctuating code and assigning the shorter codewords to the more probable messages. In examining Figure 9-1, we see that the tree can indeed produce variable-length codewords. Furthermore, a careful look at the tree shows that if we assign all our codewords using only sequences associated with end nodes, then the code will be self-punctuating. Why is this true? If we use only the end nodes, then no path through the tree to a node representing one of the codewords will touch a node representing another codeword. This means that no sequence of "1"s and "0"s representing a particular codeword will have within it the sequence representing another codeword. A code that uses only sequences associated with the end nodes is also called a *prefix code*, since no valid codeword will be the prefix of another valid codeword.

Figure 9-2a Start the tree with a single node.

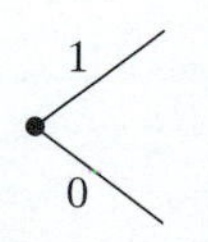

Figure 9-2b The tree grows by producing two branches from the node.

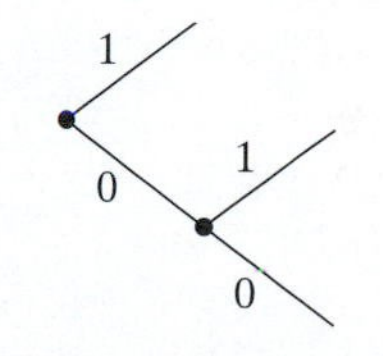

Figure 9-2c The tree continues to grow by producing two more new branches from an end node.

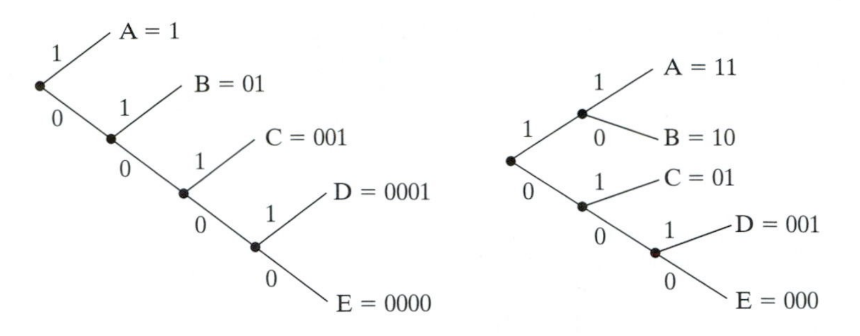

Figure 9-3a Tree 1.

Figure 9-3b Tree 2.

Table 9-3 Two Different Five-Message, Variable-Length, Self-Punctuating Codes

Message	Code from Tree 1	Code from Tree 2
A	1	11
B	01	10
C	001	01
D	0001	001
E	0000	000

We've now shown that we can construct a tree and, using only the end nodes, create a variable-length, self-punctuating code. Note, however, that there are often many different ways to construct the tree. Figure 9-3 shows two different trees, each with five end nodes, which are capable of producing a variable-length, self-punctuating code for five different messages. The two different codes are shown in Table 9-3. Which code is better? The answer, as illustrated in Examples 9.5 and 9.6, depends on the relative probabilities of the five different messages.

Example 9.5 Evaluating different variable-length codes—Part 1

Consider a source with five different, independent messages whose relative probabilities are given in Table 9-4. Which of the two codes in Table 9-3 is the more efficient way to represent the messages?

Table 9-4 Probabilities for Each of the Five Messages in Example 9.5

Message	Probability
A	0.46
B	0.23
C	0.15
D	0.10
E	0.06

Solution

Code 1 produces an average codeword length of

$$\overline{\ell}_{Code\,1} = (0.46)(1) + (0.23)(2) + (0.15)(3) + (0.10)(4) + (0.06)(4)$$
$$= 2.01 \text{ binary digits per message}$$

Code 2 produces an average codeword length of

$$\overline{\ell}_{Code\,2} = (0.46)(2) + (0.23)(2) + (0.15)(2) + (0.10)(3) + (0.06)(3)$$
$$= 2.16 \text{ binary digits per message}$$

Code 1 is therefore more efficient. Let's now calculate the entropy of the source.

$$H_{Example\ 9.5} = -\sum_{j=1}^{n} P_j \log_2(P_j) = -[(0.46)(-1.12) + (0.23)(-2.12)$$
$$+ (0.15)(-2.74) + (0.10)(-3.32) + (0.06)(-4.06)]$$
$$= 1.99 \text{ bits per message}$$

Not only is Code 1 more efficient than Code 2, but its average length is very close to the entropy of the source, making it an excellent compression code for this example. Using Code 1, each 2.01 binary digits transmitted will contain 1.99 bits of information. The extra 0.02 binary digits are *redundant*.

Example 9.6 Evaluating different variable-length codes—Part 2

Now consider a source with five different, independent messages whose relative probabilities are given in Table 9-5. Note that these probabilities are very different from those given in Example 9.5. Which of the two codes in Table 9-3 is the more efficient way to represent these messages?

Table 9-5 Probabilities for Each of the Five Messages in Example 9.6

Message	Probability
V	0.30
W	0.25
X	0.20
Y	0.15
Z	0.10

Solution

Code 1 produces an average codeword length of

$$\overline{\ell}_{Code\,1} = (0.30)(1) + (0.25)(2) + (0.20)(3) + (0.15)(4) + (0.10)(4)$$
$$= 2.40 \text{ binary digits per message}$$

Code 2 produces an average codeword length of

$$\overline{\ell}_{Code\,2} = (0.30)(2) + (0.25)(2) + (0.20)(2) + (0.15)(3) + (0.10)(3)$$
$$= 2.25 \text{ binary digits per message}$$

So, for the set of five messages with probabilities expressed in Table 9-5, Code 2 is more efficient. Again, let's calculate the entropy of the source in this example.

$$H_{Example\ 9.6} = -\sum_{j=1}^{n} P_j \log_2(P_j) = -[(0.30)(-1.74) + (0.25)(-2.00)$$
$$+ (0.20)(-2.32) + (0.15)(-2.74) + (0.10)(-3.32)]$$
$$= 2.23 \text{ bits per message}$$

Not only is Code 2 more efficient than Code 1 for Example 9.6, but its average length is very close to the entropy of the source, making it an excellent compression code for this example.

As shown in Examples 9.5 and 9.6, the compression capability of a particular code depends on the relative probabilities of the messages from the source. In Example 9.5, where the first message has a high probability and where the probabilities of the five messages are very different, Code 1 is very efficient and Code 2 is significantly less efficient. However, in Example 9.6, where the probabilities of the five messages are much closer, Code 2 is very efficient and Code 1 is significantly less efficient.

To see what makes a particular code good for a particular set of message probabilities, we need to examine the tree diagrams. The key to creating an efficient code is to balance the tree (as much as possible) at each of the nodes. To illustrate this concept, let's revisit Example 9.5. Figure 9-4a shows the Code 1 tree with each of the intermediate nodes labeled. At node α, the total probabilities of all codewords associated with the upper branch is $P_A = 0.46$, while the total probabilities of all codewords associated with the lower branch is $(P_B + P_C + P_D + P_E) = 0.54$. This balance is important because the first binary digit in Code 1 essentially conveys the information "the message is A" if the digit is a "1" and the information "the message is B, C, D, or E" if the digit is a "0." By balancing the tree, these two pieces of information have roughly equal probabilities and the binary digit conveying the information has maximum entropy. Similarly, at node β the total probabilities of all codewords corresponding to the upper branch is 0.23, while the total probabilities of all codewords associated with the lower

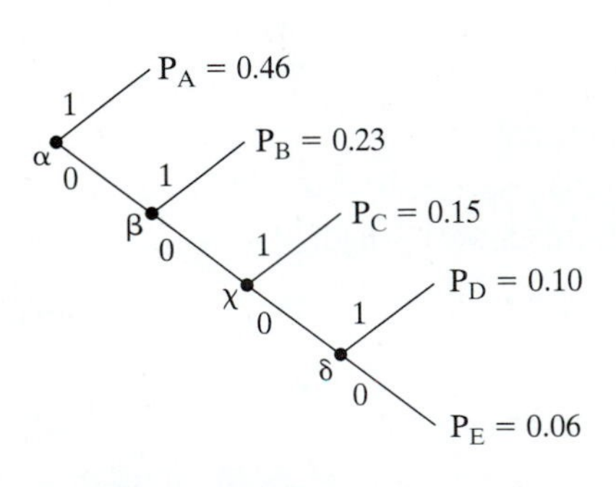

Figure 9-4a Branches and probabilities for Example 9.5 with Tree 1.

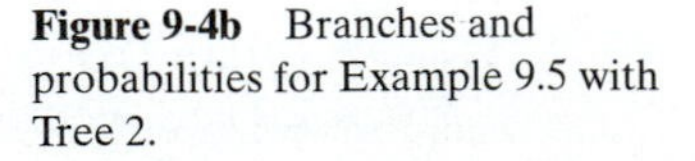

Figure 9-4b Branches and probabilities for Example 9.5 with Tree 2.

branch is 0.31. By achieving good balance at node β, Code 1 assures that if a second binary digit is needed, the digit will have maximum entropy.

Figure 9-4b shows that the Code 2 tree is badly out of balance for the message probabilities in Example 9.5. At node ω, the total probabilities of all codewords associated with the upper branch is $P_A + P_B = 0.69$, while the total probabilities of all codewords associated with the lower branch is $P_C + P_D + P_E = 0.31$. The tree is similarly out of balance at many of its other nodes.

Note, however, that the Code 2 tree achieves good balance for the messages in Example 9.6, while the Code 1 tree achieves poor balance. As you can calculate from Figures 9-5a and 9-5b, the average information content of the binary digit describing the path taken from node ω is 0.993 bits, while the average information content of the binary digit describing the path taken from node α is 0.881 bits.

We now know that we achieve the best coding efficiency when we balance the tree as much as possible at each node. This is an important analysis tool, but can it help us design the best code for a source outputting a set of independent messages with given probabilities? If the set of possible messages from the source is fairly large, there will be a prohibitively large number of different trees, and hence different variable-length, self-punctuating codes from which to choose. Given a set of independent messages and given their probabilities, can we develop a general method for building the most balanced tree and hence designing the most efficient code? As determined by David A. Huffman in 1952, the answer is "yes," and the key is to build the tree *backwards*, starting at the end nodes and working back toward the trunk. The Huffman coding algorithm, described in the next section, produces the most efficient code for compressing independent messages.

9.2.2 Huffman Coding

The Huffman coding algorithm (Huffman [9.2]) produces the most efficient prefix code for a given set of independent messages (Blahut [9.3]). The algorithm, which is simple and straightforward (albeit time-consuming), can be described as follows:

1. Arrange the messages into a column in order of their probability of occurrence, with the message that has the lowest probability at the bottom of the column. Also list the probability of occurrence of each message.

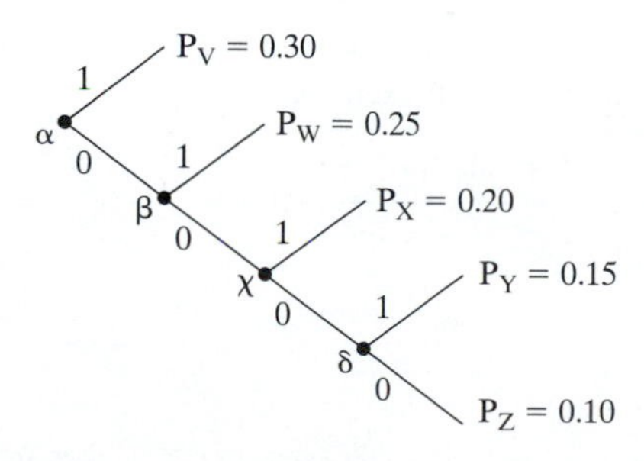

Figure 9-5a Branches and probabilities for Example 9.6 with Tree 1.

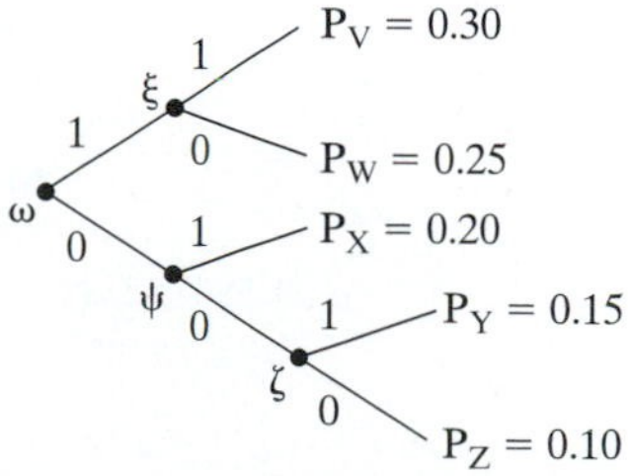

Figure 9-5b Branches and probabilities for Example 9.6 with Tree 2.

2. Combine the two messages that have the lowest probability of occurrence. Consider the combination to be a new message with a probability of occurrence equal to the sum of the two individual probabilities.
3. Reproduce the column of probabilities, deleting the two probabilities for the messages that were combined in Step 2 and adding the new probability that resulted from the combination. Remember to reorder the column, if necessary, so that the probabilities are in descending order. Note that the new column will have one less entry than the original column.
4. Draw a bracket around the two probabilities in the original column that were combined, then draw a line from the bracket to the new probability in the new column.
5. Using the new column constructed in Step 3, repeat Step 2 and then repeat Steps 3 and 4.
6. Repeat Step 5 until a column with only two probabilities is produced.
7. Assign a "1" to one of the probabilities in the final column. Assign a "0" to the other probability. Now trace back through the columns, using the brackets and lines described in Step 4 to form a tree structure. Each place where there is a branch, assign a "1" to one part of the branch and assign a "0" to the other part.
8. Starting at the left column, trace each message's path through all of the columns. Record each "0" and "1" encountered along the path. The sequence of "0"s and "1"s represents the codeword corresponding to that original message, with the most significant bit corresponding to the "1" or "0" in the right column.

   The tree structure of the Huffman algorithm guarantees a prefix code, and the recording of the characters within the column, described in Step 3, ensures optimum coding efficiency.

Example 9.7 illustrates Huffman coding.

Example 9.7 Developing a Huffman code

Consider a group of eight possible messages (each message may be a single character of text, a sentence, or the description of a possible outcome). Let's represent the messages using the characters A through H. The messages are independent and have the probabilities of occurrence listed in Table 9-6. Develop a Huffman code to efficiently represent the set of messages.

Table 9-6 Probabilities for Each Independent Message in Example 9.7

Message	Probability of Occurrence
A	.4
B	.2
C	.15
D	.13
E	.05
F	.04
G	.02
H	.01

Solution

Step 1 has already been performed. Steps 2, 3, and 4 are performed as follows:

A	.4	.4
B	.2	.2
C	.15	.15
D	.13	.13
E	.05	.05
F	.04	.04
G	.02	.03
H	.01	

Step 5 is now performed. Note the reordering that is necessary in the new column.

A	.4	.4	.4
B	.2	.2	.2
C	.15	.15	.15
D	.13	.13	.13
E	.05	.05	.07
F	.04	.04	.05
G	.02	.03	
H	.01		

Since the new column has more than two probabilities, Step 5 is performed again.

A	.4	.4	.4	.4
B	.2	.2	.2	.2
C	.15	.15	.15	.15
D	.13	.13	.13	.13
E	.05	.05	.07	.12
F	.04	.04	.05	
G	.02	.03		
H	.01			

The new column still has more than two probabilities, so Step 5 is repeated.

A	.4	.4	.4	.4	.4
B	.2	.2	.2	.2	.25
C	.15	.15	.15	.15	.2
D	.13	.13	.13	.13	.15
E	.05	.05	.07	.12	
F	.04	.04	.05		
G	.02	.03			
H	.01				

Still more than two probabilities . . .

A	.4	.4	.4	.4	.4	.4
B	.2	.2	.2	.2	.25	.35
C	.15	.15	.15	.15	.2	.25
D	.13	.13	.13	.13	.15	
E	.05	.05	.07	.12		
F	.04	.04	.05			
G	.02	.03				
H	.01					

One last time . . .

A	.4	.4	.4	.4	.4	.4	.6
B	.2	.2	.2	.2	.25	.35	.4
C	.15	.15	.15	.15	.2	.25	
D	.13	.13	.13	.13	.15		
E	.05	.05	.07	.12			
F	.04	.04	.05				
G	.02	.03					
H	.01						

Now perform Step 7.

A	.4	.4	.4	.4	.4	.4	.6 **1**
B	.2	.2	.2	.2	.25	.35 **1**	.4 **0**
C	.15	.15	.15	.15	.2 **1**	.25 **0**	
D	.13	.13	.13	.13 **1**	.15 **0**		
E	.05	.05	.07 **1**	.12 **0**			
F	.04	.04 **1**	.05 **0**				
G	.02 **1**	.03 **0**					
H	.01 **0**						

After completing Step 7, we can construct the tree for the Huffman code by reading the columns from right to left. The tree is shown in Figure 9-6. Be sure you understand how the tree is related to the series of columns, the brackets, and the lines produced at the end of Step 7.

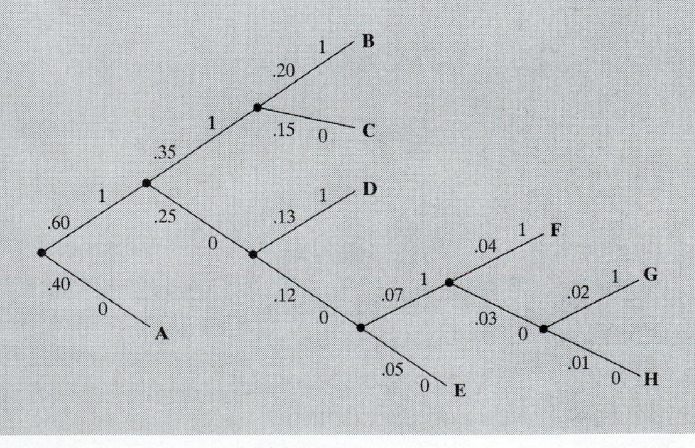

Figure 9-6 Tree for the Huffman code.

We don't need to actually draw the tree: Step 8 allows us to determine the code sequence directly from the columns by tracing the branches back to the original messages. In performing Step 8, we see that the path for message A produces the sequence 0. The codeword for message A is thus 0.

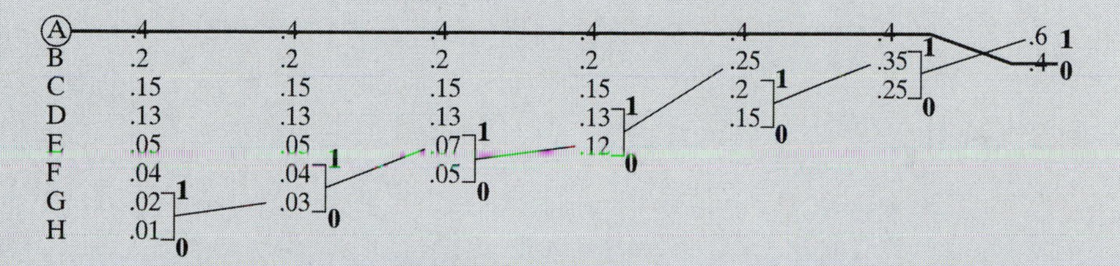

The path for B produces the sequence 111. The codeword for B is thus 111.

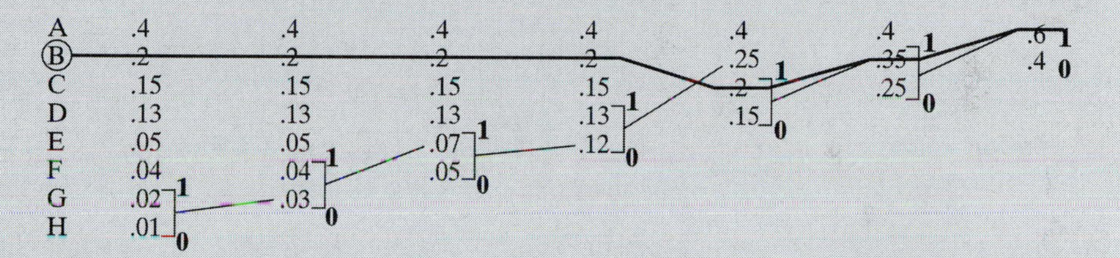

The path for C produces the sequence 011. The codeword for C is thus 110.

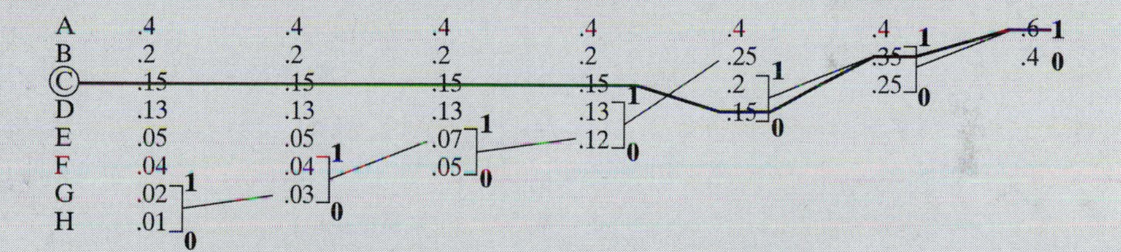

The path for D produces the sequence 101. The codeword for D is thus 101.

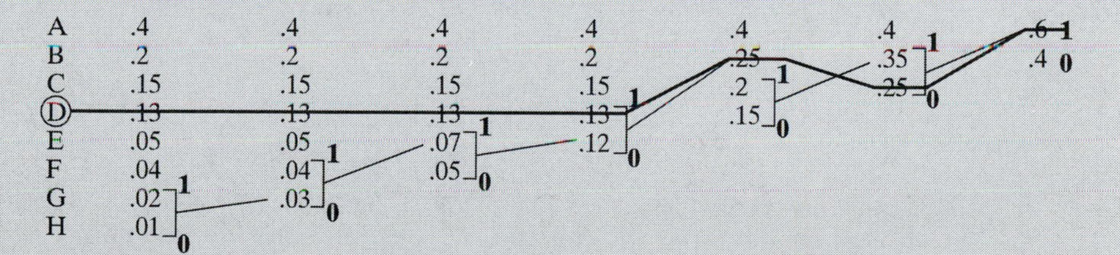

The path for E produces the sequence 0001. The codeword for E is thus 1000.

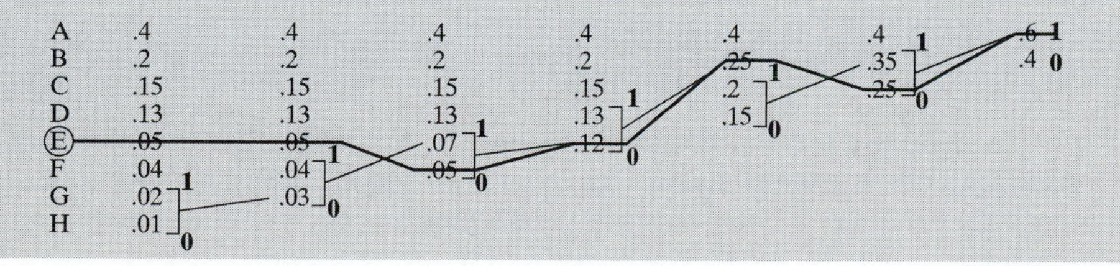

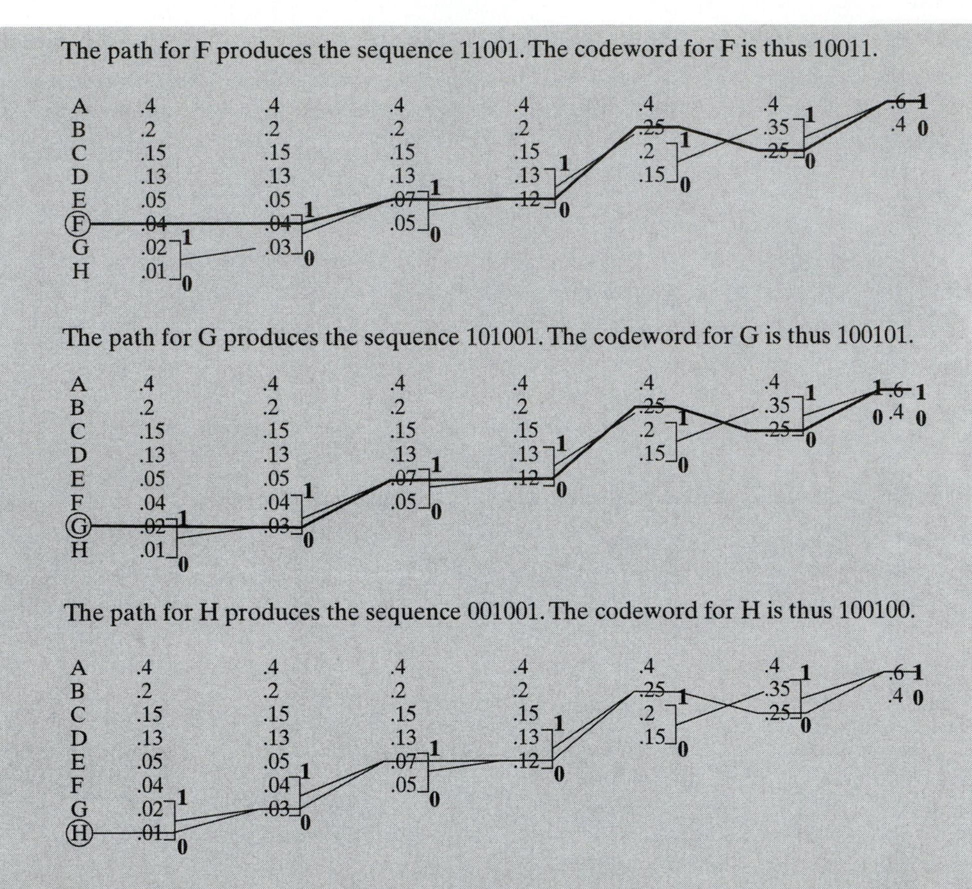

The path for F produces the sequence 11001. The codeword for F is thus 10011.
The path for G produces the sequence 101001. The codeword for G is thus 100101.
The path for H produces the sequence 001001. The codeword for H is thus 100100.

The Huffman code for the message set is thus

Message	Codeword
A	0
B	111
C	110
D	101
E	1000
F	10011
G	100101
H	100100

The average codeword length in this example is 2.42 binary digits, compared with three binary digits that would be needed for straight binary coding. With a fixed-capacity channel, the Huffman code thus allows transmission of 124% as much information as the straight binary code. The entropy of the message set is 2.37 bits per message.

In looking closely at Example 9.7, we can see that the Huffman coding algorithm produces a prefix code by using a tree structure with only end nodes as codewords. The algorithm produces a balanced tree (and hence an efficient code) by building the tree backwards—in each iteration the algorithm balances two branches as closely as possible and then connects them into a single node.

Huffman coding has limitations. Although the Huffman code produces the most efficient compression technique for transmitting single independent messages, there is often still some redundancy (i.e., average codeword length is greater than the entropy of the source). This redundancy is produced because the tree often cannot be perfectly balanced. Given a large number of possible messages and a diverse set of probabilities, as in Example 9.7, the redundancy is often small. This may not be true, however, given a small set of messages. Consider Example 9.8.

Example 9.8 Finding potential inefficiencies in a Huffman code

A source has three possible messages. The messages are independent and have the following probabilities of occurrence. Develop a Huffman code to represent the set of messages and evaluate the efficiency of the code.

Message	Probability of Occurrence
A	.75
B	.1875
C	.0625

Solution

Developing the code is straightforward; the results are as follows:

Message	Codeword
A	1
B	01
C	00

Evaluating the efficiency of the code,

$$\bar{\ell} = (0.75)(1) + (0.1875)(2) + (0.0625)(2) = 1.25 \text{ binary digits per message}$$

$$H = -\sum_{j=1}^{n} P_j \log_2(P_j) = -[(0.75)(-.415) + (0.1875)(-2.42) + (0.0625)(-4)]$$

$$= 1.015 \text{ bits per message}$$

Although the Huffman code produces the most efficient representation possible for each independent message, the code in this example is not very efficient, requiring an average of 1.25 binary digits for each 1.015 bits of information transmitted. We could have anticipated this lack of efficiency by examining the code's tree and seeing that, even though the tree is balanced as well as possible, it is still badly out of balance.

If the source is outputting multiple messages, one way to improve the efficiency of the code is to group the messages prior to encoding. In Example 9.8, for instance, if the source is outputting many messages, we could wait for the source to output two messages and then transmit a code representing each message pair. Since the messages are independent, the probability for each message pair is the product of the probabilities of each independent message. Pairing the messages produces a larger number of characters for the Huffman code and a more distributed set of probabilities. Consider Example 9.9.

Example 9.9 Improving code efficiency by grouping multiple independent messages

Suppose the source in Example 9.8 is outputting multiple, independent messages. Develop a Huffman code to represent each *pair* of possible outputs, and evaluate the efficiency of the code.

Solution

The table below shows the probabilities of occurrence for each message pair. Since the messages are independent, the probability for each message pair is the product of the probabilities of each independent message.

Message Pair	Probability of Occurrence
AA	.5625
AB	.1406
AC	.0469
BA	.1406
BB	.0352
BC	.0117
CA	.0469
CB	.0117
CC	.0039

A Huffman code for the nine message pairs is shown below (see Problem 9.9). Average codeword length (again, see Problem 9.9) for each message pair is 2.074 binary digits, or 1.037 binary digits per message. This value is much closer to the 1.015 bit/message entropy of the source.

Message Pair	Codeword
AA	1
AB	011
AC	0011
BA	010
BB	0001
BC	00001
CA	0010
CB	000001
CC	000000

We can see from Examples 9.8 and 9.9 that redundancy can be reduced by grouping messages together and developing a Huffman code for each group. We can make the amount of redundancy arbitrarily small if we create large enough groups (i.e., groups of three independent messages, four independent messages, etc.), thereby satisfying Shannon's first theorem. Unfortunately, grouping has disadvantages: The number of codewords grows exponentially, code development becomes more complex, and delay is introduced.

9.3 Sources with Dependent Messages

Let's now consider sources outputting a stream of characters that are not independent. As an example, suppose a source outputs English words and transmits three characters, the first being a "D" and the third being a "T." Do we know the second character? Yes: "O" is the only character that will make a valid English word. Since the second character was completely predictable, it contains no information.

Sources that produce interdependent outputs are said to have *memory*. As we've just shown, the interdependence makes the characters more predictable and therefore reduces the information content of the messages.[2]

Huffman coding is based on independent messages and hence is suboptimal when used to encode sources with memory. If each message is Huffman-encoded independently, then the redundancy inherent in the interdependency of the messages is not exploited at all. If messages are grouped together and then Huffman encoded, redundancy is reduced by exploiting the interdependency of messages within the group, but interdependency between groups is not used. Other problems with grouping include questions such as, "Where should group boundaries be formed?" and "How many groups can exist before the number is too large for the code to be practical?" Consider, for instance, encoding English. Should groups be all the same size, say, two characters long (a *bigram*), three characters long (a *trigram*), etc? Should groups be variable length; for example, should each word in the English language have its own code? Let's try another approach, called *dictionary encoding*.

9.3.1 Static Dictionary Encoding

English text consists of strings of characters that are often heavily interdependent and predictable (this is the basis for games such as *Wheel of Fortune*). These strings are not necessarily complete words (e.g., the string "ed" or "ing"). If we examine a large quantity of English text, we can see the same strings are often repeated. Suppose we use the following technique to compress the text:

1. Examine a large quantity of English text and determine the 2^n most commonly used strings of c characters or less (typically $10 \leq n \leq 14$, producing between 1024 and 16,384 strings, and typically $c \leq 8$ characters).

[2] Determining the entropy of a source with memory can be difficult. In general, the source needs to be modeled as a Markov process with a sufficient number of states, and one must determine the appropriate transition probabilities. See Shanmugam [9.4] for more details. Sometimes the entropy of a source with memory can be determined experimentally. For an interesting and entertaining example, see Shannon [9.5] in which Shannon experimentally measures the entropy of printed English by playing a form of the game "hangman."

2. Assign a unique sequence of binary digits (i.e., a unique codeword) to each of the strings.
3. Form complete messages by concatenating the strings (i.e., by piecing the strings together).

Example 9.10 Using a static dictionary

Suppose Table 9-7 shows the 256 most commonly used English text strings that are one-to-four characters long for a particular application (to save space, many of the entries have been left out).[3] The space character is symbolized using the character "-." Using the table, encode the sentence "The cat ate the fat rat."

Table 9-7 Most Commonly Used English Text Strings in a Particular Application.

Text String	Code
E	00000000 (0)
-	00000001 (1)
⋮	⋮
T	00000110 (6)
A	00000111 (7)
.	00001000 (8)
⋮	⋮
H	00001011 (11)
R	00001100 (12)
C	00001101 (13)
⋮	⋮
AN	00011000 (24)
TO	00011001 (25)
⋮	⋮
F	00101010 (42)
⋮	⋮
THE-	00110010 (50)
⋮	⋮
AT	00111110 (62)
⋮	⋮
ATI	11111110 (254)
AT-	11111111 (255)

Solution

Using the smallest number of text strings, "the cat ate the fat rat" compresses to the sequence 50, 13, 255, 62, 0, 1, 50, 42, 255, 12, 62, 8. Using a straight binary assignment with eight binary digits to represent each string, the sentence, which contains 24 English characters, can be transmitted using 96 binary digits.

[3]The strings in this example are for illustration only and are not representative of typical English text. To find the most commonly used strings for typical English text, see Storer [9.6].

Table 9-7 is called a *dictionary*, and the technique just described is known as *dictionary encoding*. Dictionary encoding is not restricted to text but can be used with any type of information comprised of a series of strings.

Dictionary encoding is a straightforward, efficient way to compress text or other information comprised of interdependent characters. The dictionary must contain all single-character strings, even if those strings are not among the most commonly used, because otherwise it would be possible for the source to output a sequence of characters that could not be encoded. This property of ensuring that all possible outputs from the source can be encoded is called *dictionary guaranteed progress*. Since the strings in the dictionary are not equally likely to appear in the source's output, we can use Huffman coding rather than a straight binary code to assign a sequence of binary digits to represent each string. This will further reduce the number of binary digits needed to represent the average text message.

Other important properties of a compression technique are how quickly it can encode messages and how much memory the technique requires for compression and decompression. If we store our dictionary in tree form, we can significantly improve encoding speed by reducing the time it takes to examine the entire dictionary to find the longest matching string. Figure 9-7 shows a tree structure used to represent the dictionary in Example 9.10. We can also reduce memory requirements for each branch in the tree by storing only the additional characters involved in that branch (e.g., in Figure 9-7, we store only "HE-" with the terminating branch for the string "THE-"; information associated with "T" is obtained from earlier branches).

Dictionaries for different applications will contain different sets of text strings. Furthermore, as applications and the usage of particular text strings change, the dictionary originally chosen by the designer may no longer be optimal. Dictionaries can be updated either off-line by temporarily halting transmission of information and then sending a new dictionary, or on-line by having both the source and user review previous transmissions and, according to a given set of rules, concurrently modify their copy of the dictionary. On-line techniques, known as *dynamic dictionary encoding*, are more efficient but are also more complex. Section 9.3.2 explores one of the most popular dynamic dictionary encoding techniques.

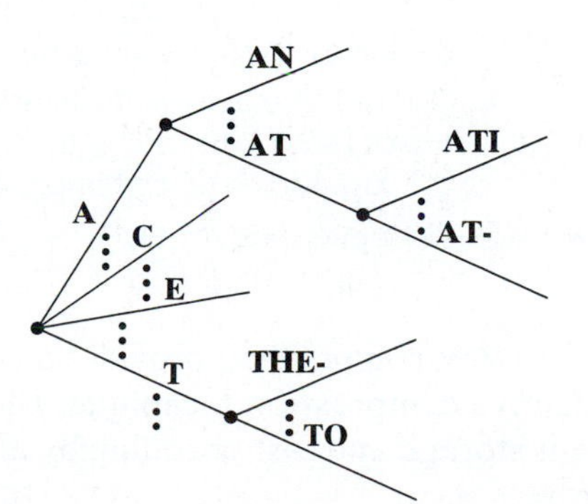

Figure 9-7 Using a tree structure to store strings in a dictionary.

9.3.2 LZW Compression—an Example of Dynamic Dictionary Encoding

Lempel-Ziv-Welch (LZW) encoding uses a dynamic dictionary to achieve flexibility and significant data compression (Ziv and Lempel [9.7], Welch [9.8]). LZW compression is used in many applications, including the ZIP programs that run on most PCs. The algorithm works as follows:

1. Begin by creating a large buffer; this will be our dynamic dictionary. Initialize the dictionary by storing all single characters in the first few buffer locations.
2. Starting at the beginning of the message, find the longest match in the buffer (this will, of course, be a single character).
3. Transmit the buffer location containing the match, then transmit the first character after the match.
4. Take the match and concatenate the first character after the match. Store this new string in the next location in the buffer (this new string is now part of the dynamic dictionary).
5. Starting at the next character in the message, find the longest match within the dynamic dictionary, then repeat Steps 3 and 4.
6. Repeat Step 5 until the message is completely encoded.

Example 9.11 Using LZW dynamic dictionary encoding

Consider the character set consisting of all capital letters, the space character (symbolized –), and the punctuation characters comma and period. Compress the message "ALL ALAN WANTS IS A PIE" using LZW compression.

1. Begin by initializing the buffer with all possible single characters, as shown in Figure 9-8a.
2. Since the longest match is the single character A, and the match is followed by the character L, transmit "1,12" and add the string AL to the bottom of the dictionary, as shown in Figure 9-8b.
3. The first two characters in the message have now been encoded. Starting with the third character in the message, the longest match is the single character L followed by the space character, so transmit "12,27" and add the string L- to the bottom of the dictionary, as shown in Figure 9-8c.
4. The fifth character in the message is now the first unencoded character. The longest match in the dictionary is the string AL followed by an A, so transmit "30,1" and add the string ALA to the bottom of the dictionary, as shown in Figure 9-8d.
5. Continue compressing.

LZW compression provides dictionary guaranteed progress and is an extremely effective compression technique. The structure of LZW compression provides efficient storage and fast encoding by allowing the dictionary to be represented in tree form.

1	**A**
2	**B**
3	**C**
4	**D**
5	**E**
6	**F**
7	**G**
8	**H**
9	**I**
10	**J**
11	**K**
12	**L**
13	**M**
14	**N**
15	**O**
16	**P**
17	**Q**
18	**R**
19	**S**
20	**T**
21	**U**
22	**V**
23	**W**
24	**X**
25	**Y**
26	**Z**
27	–
28	,
29	.

Figure 9-8a

1	**A**
2	**B**
3	**C**
4	**D**
5	**E**
6	**F**
7	**G**
8	**H**
9	**I**
10	**J**
11	**K**
12	**L**
13	**M**
14	**N**
15	**O**
16	**P**
17	**Q**
18	**R**
19	**S**
20	**T**
21	**U**
22	**V**
23	**W**
24	**X**
25	**Y**
26	**Z**
27	–
28	,
29	.
30	**AL**

Figure 9-8b

1	**A**
2	**B**
3	**C**
4	**D**
5	**E**
6	**F**
7	**G**
8	**H**
9	**I**
10	**J**
11	**K**
12	**L**
13	**M**
14	**N**
15	**O**
16	**P**
17	**Q**
18	**R**
19	**S**
20	**T**
21	**U**
22	**V**
23	**W**
24	**X**
25	**Y**
26	**Z**
27	–
28	,
29	.
30	**AL**
31	**L–**

Figure 9-8c

1	**A**
2	**B**
3	**C**
4	**D**
5	**E**
6	**F**
7	**G**
8	**H**
9	**I**
10	**J**
11	**K**
12	**L**
13	**M**
14	**N**
15	**O**
16	**P**
17	**Q**
18	**R**
19	**S**
20	**T**
21	**U**
22	**V**
23	**W**
24	**X**
25	**Y**
26	**Z**
27	–
28	,
29	.
30	**AL**
31	**L–**
32	**ALA**

Figure 9-8d

One problem with LZW compression is that, even using a tree representation, after a while the dictionary fills all available memory. At this point the designer has two options. The simplest option, which is to just stop adding new strings, has the disadvantage of reducing the flexibility, adaptability, and efficiency of the code if the characteristics of the source change over time. A second option is to monitor the performance of the code and, when the performance deteriorates below a certain level, to clear out the entire dictionary (except for all single-character strings) and start over.

9.4 Still-Image Compression

The key to digitally representing still images involves understanding how we see and then exploiting the physiology of the eye. The eye has limitations concerning resolution and color, so there is no need to transmit information that is beyond the eye's ability to discern. We can therefore use lossy compression techniques to there is no need to transmit information that is beyond the eye's ability to discern reduce the number of binary digits necessary to represent images, and, if we are careful, the information lost will not significantly impact the quality of the image.

Still images are represented using a grid of dots called *pixels* (short for *picture elements*). You can easily see this principle when you look closely at newspaper photos. If we represent an image using a sufficient number of pixels per inch (say, 600 horizontally and 600 vertically), then because of the limited resolution of the human eye, the individual pixels appear to blend and the image looks continuous. (Anyone who has shopped for an ink-jet printer or a scanner has likely been bombarded by the "dots per inch" resolution specs from the various manufacturers.)

Suppose we use a resolution of 600×600 (i.e., 600 pixels/inch horizontally and 600 pixels/inch vertically) to represent an image. The image requires 360,000 pixels per square inch. Now suppose we use a series of binary digits to represent the color and intensity of each pixel, and suppose we use 256 colors and 16 intensity levels (these values represent mid-level performance of a computer monitor). We will require 12 binary digits per pixel, or 51,840,000 binary digits to represent a single 3×4 inch color image. Clearly, we need to develop more efficient ways to represent our image.

Let's begin our study in image compression by examining facsimile, since facsimile transmits simple monochromatic, low-resolution images. After we understand facsimile compression, we will examine techniques to compress high-resolution, full-color still images.

9.4.1 Facsimile

Facsimile is the simplest case of a still image. Fax images are monochromatic (in this case only black and white), have no intensity levels, and are usually low resolution. Each pixel in a fax image can be represented using one binary digit (e.g., a "1" could represent that the pixel is black and a "0" could represent that the pixel is white). Consider the image in Figure 9-9, where the grid is used to subdivide the image into pixels. The image is 63 pixels horizontally and 31 pixels vertically (represented as 63×31). If the image was encoded using one binary digit per pixel, then 1953 binary digits would be required to represent the image. Using one binary digit per pixel can produce rather large files: An 8.5×11 inch facsimile with a standard horizontal and vertical resolution of 200 pixels per inch contains 3,740,000 pixels. Let's see if we can develop a more efficient way to represent the facsimile image.

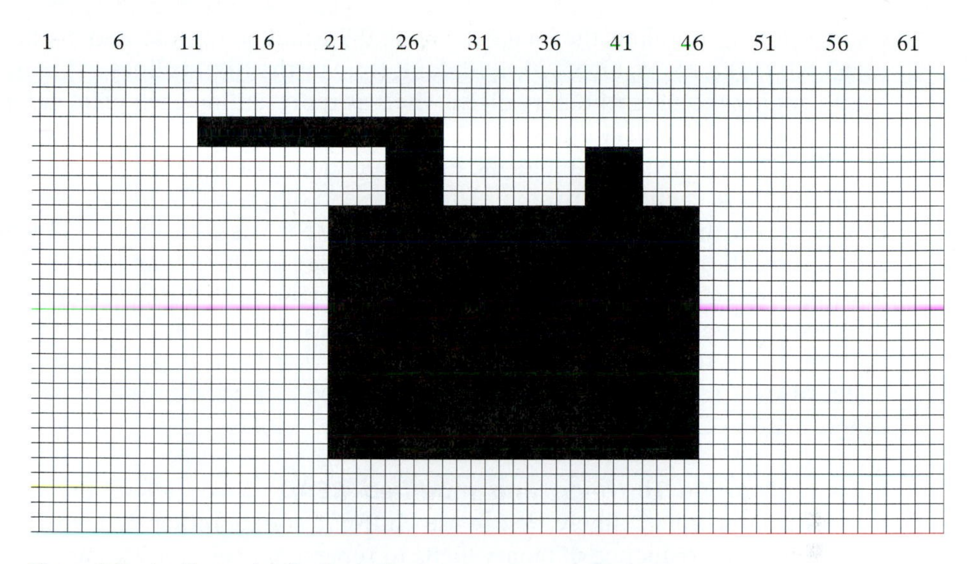

Figure 9-9 An illustration for facsimile.

9.4.1.1 Run-length coding Pixels along a row in a facsimile are not independent: Often the black pixels cluster together and the white pixels cluster together. For example, in the fourth row of Figure 9-9 we see that the row starts with a run of 11 consecutive white pixels followed by a run of 17 consecutive black pixels, followed by a run of 35 consecutive white pixels. Instead of transmitting the black/white status of each pixel in the row, suppose that we transmit the lengths of each run of like-colored pixels. If we assume that a row always starts with a white pixel, we can transmit all the image information in the fourth row using the sequence "11, 17, 35." Since the row is 63 pixels wide, and since we must also be able to transmit a run of zero (in case the row starts with a black pixel), we can use six binary digits to encode each run length. Such an approach is known as *run-length coding*. Using run-length coding, the Row 4 information can be represented in binary form using the 18-binary-digit sequence "001011010001100011," a pattern considerably shorter than the 63 binary digits that would be required to individually represent each pixel.

Let's formally establish the procedure for run-length coding.

1. Start with the first row in the facsimile image.
2. Starting with the first pixel in the row, transmit the number of white pixels before the first black pixel is encountered (this number may be 0 if the first pixel in the row is black).
3. Now transmit the number of black pixels before the next white pixel is encountered, then the number of white pixels before the next black pixel is encountered, etc.
4. Continue until the end of the row is reached, then go to Step 2 and begin again with the next row.

Using run-length coding, the image in Figure 9-9 could be represented by the sequence 63, 63, 63, 11, 17, 35, 11, 17, 35, 24, 4, 10, 4, 21, etc. For the first six lines of Figure 9-9, run-length coding requires 84 binary digits, whereas representing each pixel independently would require 378 binary digits.

Run-length coding exploits the interdependence of adjacent pixels in a row. Theoretically, there is no need to stop a run at the end of a row. For instance, in Figure 9-9, Row 4 ends with a run of 35 white pixels and Row 5 starts with a run of 11 white pixels. This information could be represented by a single run of 46 white pixels. In practice, however, runs are terminated at the end of each row so that if an error occurs in the received message, then the error will affect only a single row and will not propagate to the remaining rows in the image. *Error propagation* is an important, undesired side effect of data and image compression and will be discussed in the remainder of this chapter and in Chapter 10.

9.4.1.2 Modified Huffman and modified READ Since all of the possible run lengths are not equally likely, we can use Huffman coding rather than a straight binary code to assign a sequence of binary digits to represent each run. This approach significantly reduces average codeword length, but introduces another problem. A line of fax can contain up to 1728 pixels. Furthermore, the probabilities concerning white run lengths and black run lengths are different. We therefore appear to need 3458 different codewords (1729 for all possible white runs and an equal number for all possible black runs), producing a prohibitively large and cumbersome code. This problem has been addressed by considering a run length, N, to be the sum of two variables m and n

$$N = 64m + n \tag{9.4}$$

where $0 \leq m \leq 27$ and $0 \leq n \leq 63$. By encoding m and n separately, the number of possible codewords is kept manageable.

An international standards body known as the International Telecommunications Union (ITU) has established a technique for transmitting facsimile based on these concepts. Called the modified Huffman (MH) code, this technique is formalized in ITU Standard T.4 [9.9]. The standard includes separate codes for black and white runs and provides additional features, such as an end-of-line (EOL) character to reduce error propagation. For more information on the MH code, see [9.9] and Sayood [9.10].

Just as the pixels along a row in a facsimile image are not independent, pixels along a column are not independent. By exploiting this vertical interdependency as well as the horizontal interdependency, we can develop even more efficient data compression techniques for facsimile. Typical fax images tend to have sharp edges where black becomes white or vice versa. For example, note the black-to-white transition on the far right of Figure 9-9 in Rows 10 through 26. In all of these rows, the transition occurs on the 47th pixel. An examination of Figure 9-9 shows many other similar sharp transitions. For typical facsimile images, approximately 50% of all black-to-white or white-to-black transitions in a row echo a similar transition in exactly the same position in the previous row. These transitions produce a vertical edge. An additional 25% of the transitions in a row echo a similar transition in the previous row not immediately above, but offset one pixel to the left or right. These transitions produce an edge with a slight diagonal.

The ITU has developed a standard known as modified Relative Element Address Designate (modified READ) based on the horizontal and vertical dependencies in typical facsimile images (Sayood [9.10], ITU Standard T.6 [9.11]). This standard is used in most current facsimile transmissions and provides significant compression. A typical letter consisting of 4352 × 3072 pixels can be compressed into 14,290 bytes using modified READ (Sayood [9.10]).

9.4.2 Monochromatic Gray Scale Images

Facsimile has many applications, but it generally provides poor quality for images such as photographs. Based on our experience with newspaper photographs, we know that the quality of black-and-white images can be improved if we allow the pixels to assume various shades of gray as well as just black or just white. The use of shades or intensities is often called *gray scale*. We can represent the various shades of gray by assigning an intensity level to each pixel. For example, we can represent the intensity of a particular pixel by using an integer ranging from 0 to 63, with 0 representing white, increasing numbers representing increasingly darker shades of gray, and 63 representing black. The drawback of this approach is that run-length coding and modified READ are no longer efficient compression techniques because, with 64 different intensity levels, the lengths of the runs are much shorter. Also, since the runs no longer alternate between just black and just white, we must represent the intensity of each run. We need to use a more sophisticated compression technique, which we will develop in the next section when we discuss an even more complicated problem—compression of color images.

9.4.3 Color Images (the DCT and JPEG)

Now that we understand compression of facsimile images, let's examine techniques to compress high-resolution color images. Compression techniques for these images are considerably more complex than facsimile, since we must now develop methods to represent color and intensity.

Representation of color is straightforward. The cones in the human eye contain three pigments, each sensitive to a different color. These pigments are rhodopsin (sensitive to blue at a wavelength of 447 nm), chlorolabe (sensitive to green at a wavelength of 540 nm), and erythrolabe (sensitive to red at a wavelength of 557 nm). All visible colors are combinations of various intensities of these three "primary colors."[4] We can therefore represent any color image as the sum of three monochromatic images (red, green, and blue). This is the principle behind television screens and computer monitors.[5] As with facsimile, the pixels in the red, green, and blue monochromatic

[4]Almost all nonengineering books correctly claim that the primary colors are red, blue, and *yellow*, but almost every engineering and image-processing book will refer to green, not yellow, as the third primary color.
[5]Each pixel on the television or computer monitor screen consists of three separate dots of phosphor in close proximity, one that illuminates red, one that illuminates green, and one that illuminates blue when hit by an electron beam.

components of color images exhibit strong horizontal and vertical interdependencies. However, unlike facsimile, the red, green, and blue component images have many different intensity levels, so, as we discussed above, run-length coding for each component image is not efficient. We will employ a more sophisticated technique, based on the *discrete cosine transform* (DCT).

Let's consider a color image that is n pixels wide and n pixels long (for the remainder of this chapter we will say that such an image is $n \times n$). As discussed above, we can deconstruct the image into three monochromatic $n \times n$ images—one red, one green, and one blue. We can now represent each of the three monochromatic images using an $n \times n$ matrix, where the value of a particular element within the matrix refers to the intensity level of the corresponding pixel. Suppose, for example, that there are 256 different possible intensity levels and that in a particular image, the red component of the pixel in the third row and second column is very bright—say an intensity level of 248. Let **[R]** represent the matrix corresponding to the red component of the image and let $r_{i,j}$ correspond to the element in the ith row and jth column of **[R]**. In our example, $r_{3,2} = 248$. To keep matrix size from becoming unwieldy, most images are broken up into a series of smaller blocks, usually 8×8 or 16×16. Figure 9-10a shows an 8×8 matrix corresponding to the red monochromatic component of an 8×8 block of pixels within a particular standardized test image called CHEETAH.

Each element in the Figure 9-10a matrix corresponds to one particular pixel and contains three pieces of information: the x-coordinate of the pixel, the y-coordinate of the pixel, and the red intensity level of the pixel. The x and y coordinates for the pixel are determined by the matrix element's location within the matrix, and the intensity level is determined by the value of the matrix element.

140	144	147	140	140	155	179	175
144	152	140	147	140	148	167	179
152	155	136	167	163	162	152	172
168	145	156	160	152	155	136	160
162	148	156	148	140	136	147	162
147	167	140	155	155	140	136	162
136	156	123	167	162	144	140	147
148	155	136	155	152	147	147	136

Figure 9-10a Typical intensity matrix (from 8×8 block of CHEETAH).

186	−18	15	−9	23	−9	−14	19
21	−34	26	−9	−11	11	14	7
−10	−24	−2	6	−18	3	−20	−1
−8	−5	14	−15	−8	−3	−3	8
−3	10	8	1	−11	18	18	15
4	−2	−18	8	8	−4	1	−7
9	1	−3	4	−1	−7	−1	−2
0	−8	−2	2	1	4	−6	0

Figure 9-10b Shifted, DCT-transformed matrix (from 8×8 block of CHEETAH).

Like the Fourier transform, the DCT allows us to represent information in another domain. As we established in Chapter 2, the Fourier transform allows us to express a signal as the summation of its frequency components. The DCT of a monochromatic image allows us to express the image not in terms of the intensities of each of its pixels, but rather in terms of how rapidly the intensity changes from pixel to pixel within the image. This rapidity of change is called *frequency*, and this term can be problematic. This term should not be confused with frequency as we've previously defined it, or in terms of the wavelengths of particular colors. For the remainder of this chapter, we need to think of frequency as a measure of how rapidly the intensity of a monochromatic image changes.

Let **[C]** represent a particular monochromatic $n \times n$ intensity matrix with elements $c(i, j)$ and 2^m intensity levels. Before applying the two-dimensional DCT, we subtract 2^{m-1} from each element in **[C]** so that all intensity values are between -2^{m-1} and $2^{m-1} - 1$ rather than between 0 and $2^m - 1$. Let's call this new matrix $[\mathbf{C}_{shift}]$ with elements $c_{shift}(i, j)$. Now let **[D]** be the same matrix after the two-dimensional DCT has been applied. The elements in **[D]**, denoted $d(i, j)$ are

$$d(i, j) = \sqrt{\frac{2}{n}}\alpha_i\alpha_j\sum_{k=0}^{n-1}\sum_{l=0}^{n-1}c_{shift}(k, l)\cos\left(\frac{\pi(2k + 1)i}{2n}\right)\cos\left(\frac{\pi(2l + 1)j}{2n}\right) \tag{9.5}$$

$$\text{where } \alpha_z = \begin{cases} \frac{1}{\sqrt{2}} & \text{if } z = 0 \\ 1 & \text{if } z \neq 0 \end{cases}$$

Figure 9-10b shows the shifted, DCT-transformed matrix of Figure 9-10a, the red-component 8×8 block of the CHEETAH image.[6] There is no longer a one-to-one correspondence between elements in the matrix and pixels within the image's block. Rather, the position of a matrix element in a row represents the horizontal frequency (rapidity of change of intensity in horizontally adjacent rows) within the entire 8×8 block. The farther an element is toward the right side of the matrix, the greater the horizontal frequency it represents. The position of a matrix element in a column represents the vertical frequency (rapidity of change of intensity in vertically adjacent rows) within the entire 8×8 block. The further an element is toward the bottom of the matrix, the greater the vertical frequency it represents. The value of a matrix element represents the strength of the particular frequency component within the 8×8 block. The upper-left corner element represents the component corresponding to no intensity change (often called the dc component). The further an element is from the upper-left corner, the higher the frequency components it represents.

Initially, it appears that we've gained nothing by transforming the matrix with the DCT; after all, we still have an 8×8 matrix containing 256 values that we must

[6]You will verify the values in the shifted, DCT-transformed matrix in Problem 9.19. Note that many commercial software packages will compute the two-dimensional DCT, but you need to be careful in specifying the function. For example, in MATLAB the DCT function computes the one-dimensional DCT, which will be incorrect for a two-dimensional application. Instead, you will want to use the DCT2 function in the MATLAB signal processing toolbox.

represent using binary digits. The advantages of using the transformed matrix come from the following properties of still color images:

1. Most color images are composed primarily of low-frequency components. This is equivalent to saying that most color images exhibit high interdependence between adjacent pixels. We can see this tendency in Figure 9-10b by noticing that, in general, the closer a transformed matrix element is to the upper-left corner of the matrix, the higher its value.
2. From extensive subjective testing, we know that for the human eye, low-frequency information is perceptually more important to image quality than high-frequency information. This is equivalent to saying that the quality of an image depends more on accurately representing small changes in color (corresponding to shading or a change in the angle of reflection of light) than in accurately representing large changes in color (corresponding to a transition from one object to another).

The DCT allows us to represent a block from an image by its frequency components. From the two properties just described, we now see that all of the elements within the DCT-transformed matrix are not equally important. We can achieve great compression without significantly sacrificing image quality by using fewer binary digits to represent the higher-frequency components than to represent the lower-frequency components. This process is essentially nonuniform quantization, as discussed in detail in Chapter 8. The further a matrix element is from the upper-left corner of the DCT matrix, the less important its accurate representation is to the quality of the image, and the fewer binary digits we can use to represent it. Figure 9-11 shows a path within the DCT matrix transitioning through elements in their order of importance (with the most important element first).

The DCT and nonuniform quantization are key concepts used in the JPEG compression algorithm. The JPEG algorithm also uses Huffman and run-length coding to efficiently represent values within the DCT-transformed matrix (runs of zero are common in the higher-frequency components of the matrix after quantization). Named for the Joint Photographic Expert Group (the committee that established the JPEG standard), this standard is one of the most popular for image compression, especially for transmission of images over the Internet. Quantizing the values within the DCT-transformed matrix causes the code to be lossy, but perceived quality is preserved by using the DCT to isolate the high-frequency components and then by concentrating the losses on these components. The amount of compression achieved using JPEG depends on both the original image and the particular level of quality desired in the compressed image. For high-quality compression and a typical image, JPEG achieves a compression ratio of approximately 20:1, meaning that the JPEG image can be represented using 1/20 as many binary digits as would be required for the uncompressed image. (For more details on JPEG compression, see Gonzalez and Woods [9.12].)

9.5 Moving-Image Compression

As with still-image compression, the key to digitally representing moving images involves understanding how we see and then exploiting the physiology of the eye. For

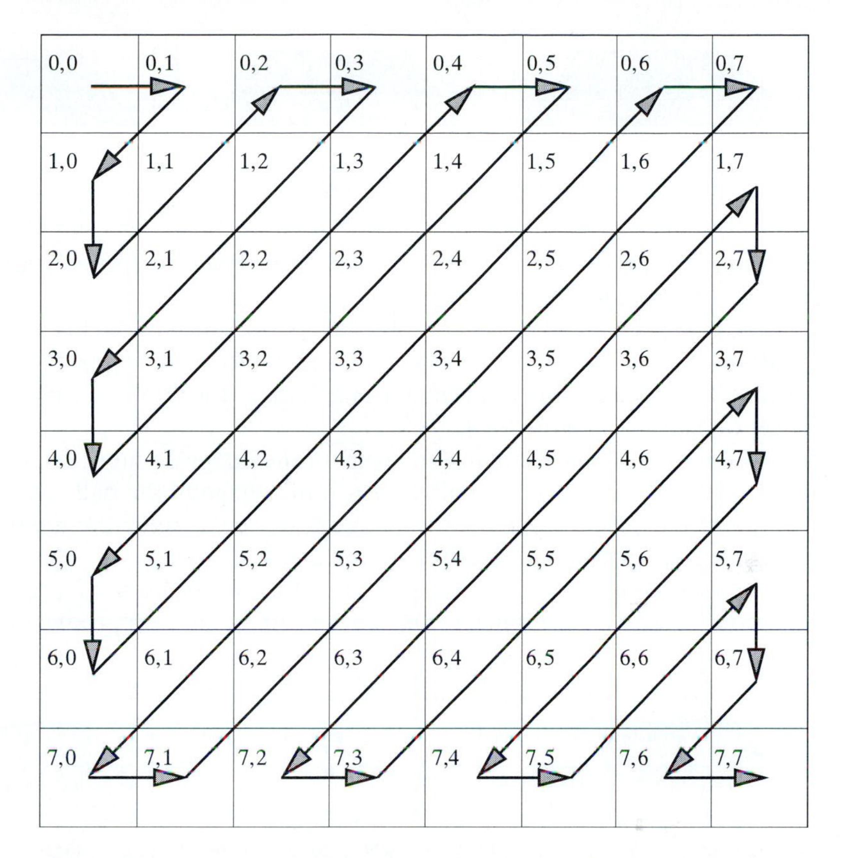

Figure 9-11 Path through DCT-transformed matrix showing order of importance of the elements.

still images, we exploited the eye's limitations concerning resolution and color. For moving images, we must also exploit the eye's limitations concerning persistence.

The eye retains an image for a fraction of a second after viewing it. Therefore, motion can be represented by sequencing through a series of still images and, if the sequence rate is fast enough, the images will appear to be moving fluidly. Movies use a sequence rate of 24 still images per second, and television in the United States uses a sequence rate of roughly 30 still images per second. Each still image is called a *frame*.

As we established earlier, still images have strong vertical and horizontal interdependencies between pixels. These *spatial dependencies* are, of course, also present in the frames of a moving image. Moving images additionally have strong interdependencies between pixels in adjacent frames. These *temporal dependencies*, combined with the spatial dependencies, can be exploited to effectively compress moving images.

Let's explore temporal dependencies further. Consider the sequence of three frames shown in Figure 9-12, representing a ball rolling on a table that contains other objects in the background. (The example in this figure is simplistic, but it will illustrate certain fundamental concepts; we will develop more sophisticated approaches shortly.)

In examining Figure 9-12, we can see that much of the picture does not change from frame to frame, and that the changes that do occur are often predictable. To illustrate the predictability, suppose that, as shown in Figure 9-13, we are given only the

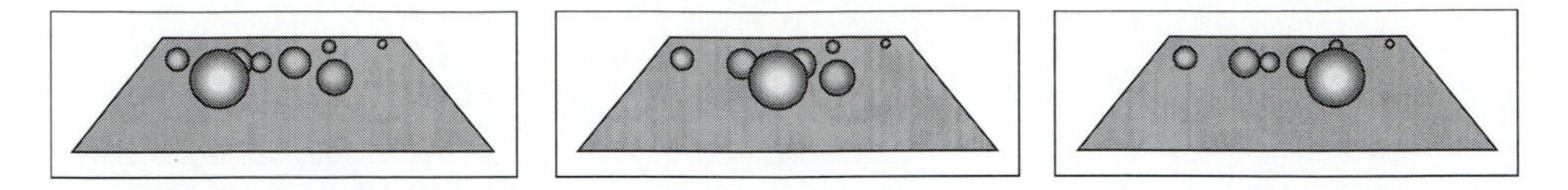

Figure 9-12 A sequence of frames in a moving image.

first and third frames and are then asked to reconstruct the second frame. How can we develop a method for reconstruction?

First, in reviewing Figure 9-13, we can see that much of the background in Frames 1 and 3 is the same. Let's divide Frames 1 and 3 into 16×16 pixel blocks and assume that if a block is the same in Frames 1 and 3, then it is also the same in Frame 2. This concept is shown in Figure 9-14.

Next, we can observe that some parts of the image do move from frame to frame, but do not change in shape or color. Thus, if the image of the ball in Frame 1 is merely shifted, say, 640 pixels to the right in Frame 3, we can assume that the image of the ball was shifted 320 pixels to the right in Frame 2.

Finally, we note that the movement of the ball in Frame 2 produces some new background relative to Frame 1. This background can be reproduced from Frame 3.

Figure 9-13 Given the first and third frames, how can we reconstruct the second frame?

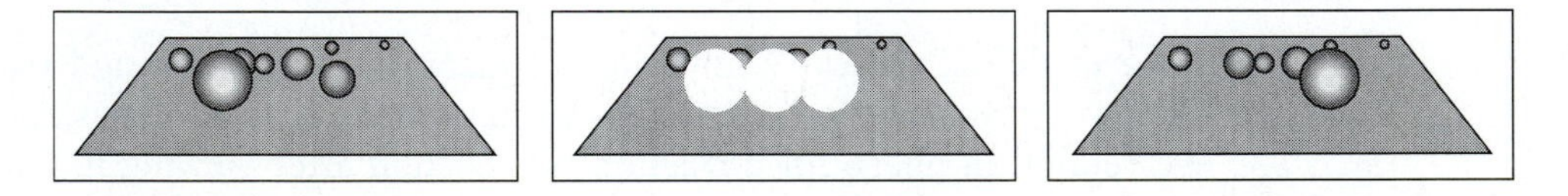

Figure 9-14 Many parts of the image do not change from frame to frame.

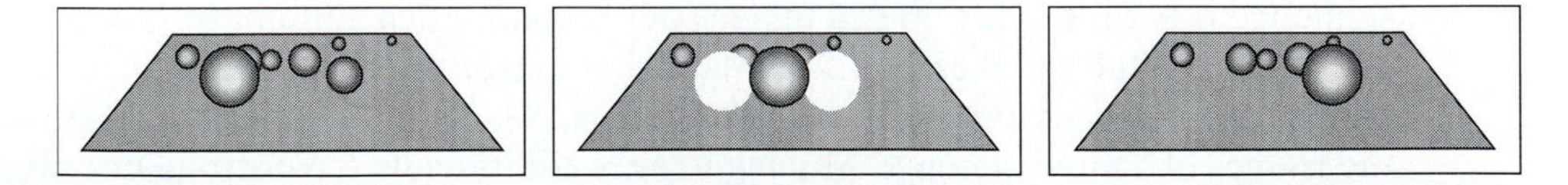

Figure 9-15 Some parts of the image do move from frame to frame, but do not change composition.

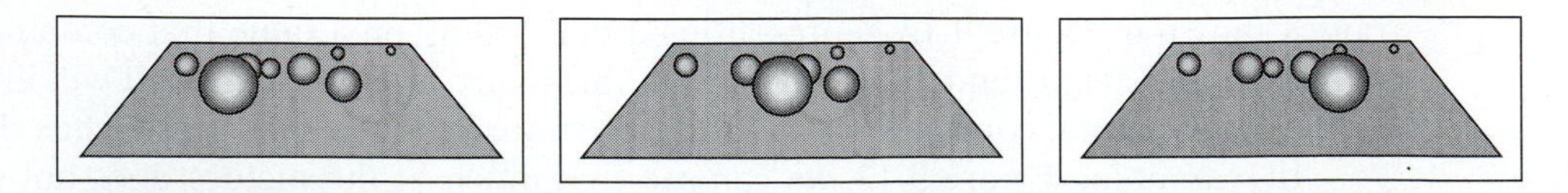

Figure 9-16 The second frame has been completely reconstructed.

Similarly, Frame 1 can be used to produce the background to the immediate right of the ball in Frame 2. Figure 9-16 shows the reconstructed Frame 2.

The observations we've just made can help us develop a good approach for compressing moving images. Suppose we want to send a sequence of three frames. Consider the following encoding process:

1. Represent the entire first and third frames using JPEG.
2. Represent the second frame as follows:
   a. Divide the first, second, and third frames into 16 × 16 pixel macroblocks. If a macroblock in Frame 2 is exactly the same as in Frame 1 or Frame 3, represent this fact using a short sequence of binary digits.
   b. If the macroblock is not the same, see if perhaps that portion of the image has moved from Frame 1 to Frame 2 but has not changed shape. Examining Frames 1 and 3, predict the location in Frame 2 where you expect the macroblock to have moved. Now search the area around the predicted location in Frame 2 and see if the macroblock is indeed present somewhere in that area. If so, use a short sequence of binary digits to indicate this fact and to give the new location of the macroblock relative to its predicted location. If the exact macroblock cannot be found in the area, but a "close fit" is found, use a sequence of binary digits to represent this fact, to give the location of the "close fit," and to state the differences in the "close fit." If neither the exact Frame 1 macroblock nor a "close fit" can be found in Frame 2, compress the Frame 2 macroblock using the DCT and JPEG.

This encoding process is complex and time-consuming, especially the searches for moving macroblocks (called *motion compensation*), but the process is usually extremely effective in compressing the second frame. (Note that this process is very much like the DPCM algorithm we studied in Chapter 8.) We can extrapolate the process to achieve even greater compression by using JPEG encoding for, say, the first and fifth frames and then using the above process for encoding the second, third, and fourth frames. Note that the JPEG-encoded frames do not exploit temporal dependencies and thus are less compressed. Nevertheless, they are needed to provide stability, to efficiently respond to scene changes, and to combat error propagation. The number of frames between JPEG-encoded frames is a parameter that the code designer can use to trade off various properties of the compression technique.

One of the most popular compression techniques for moving images is the MPEG series. Named for the Moving Pictures Expert Group (the committee that established the standard), the MPEG-1 standard allows VHS-quality representation of slow-to-medium-motion images at approximately 1.2 Mbit/sec. Consider a sequence of frames, as shown in Figure 9-17. Depending on a frame's relative position in the sequence, it will be compressed in one of three ways: interframe (I), predictive (P), or bidirectional (B). I frames are treated as still images and are essentially compressed using the DCT and JPEG. As mentioned earlier, although these frames do not exploit temporal dependency, they are needed to provide stability, to efficiently respond to scene changes, and to combat error propagation. P frames are encoded with respect to previous I frames and previous P frames, using a variation of the motion-prediction algorithm described earlier. Note that future frames are not used in the prediction.

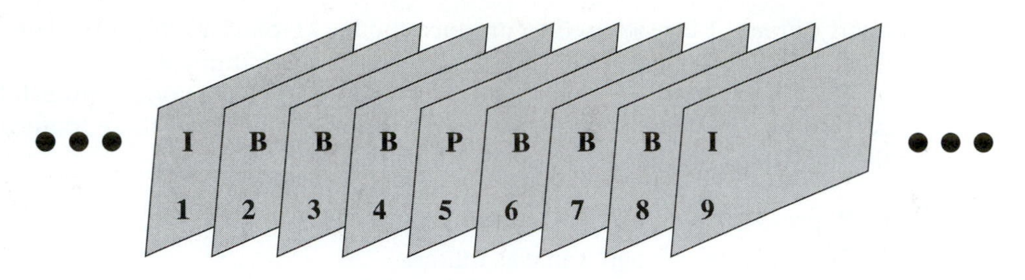

Figure 9-17 Typical sequence of frames for MPEG compression.

B frames are encoded using the same process as P frames, except that the motion-prediction algorithm uses both past and future P and I frames. Figure 9-17 shows a typical grouping of frames and a typical I, P, B sequence. Image quality and computational complexity can be traded off versus compression efficiency by altering the grouping of frames and the I, P, B sequence. MPEG-1 was established for digital storage media such as compact discs. MPEG-2 adds features such as support for interlacing (a requirement for television) and achieves a compression ratio greater than 75:1 for high-definition television applications.

PROBLEMS

9.1 Qualitatively and quantitatively define the term "information."

9.2 Discuss the process used to develop the quantitative definition of information

$$I = \log_2\left(\frac{1}{P}\right) = -\log_2 P$$

9.3 A bag contains one penny, four nickels, three dimes, and two quarters. The source takes a coin out of the bag and then transmits a message to the user concerning the coin. How much information is contained in each of the following messages?

a. "The coin was a nickel."
b. "The coin was a penny."
c. "The coin was a quarter."
d. "The coin was worth at least ten cents."
e. "The coin was not a penny."

9.4 Calculate the entropy of a message from the source in Problem 9.3.

9.5 Why does the Huffman-coding algorithm produce efficient codes for sources with independent messages?

9.6 Consider a sequence of messages using L.A. Weather Code 2, as shown in Table 9-1 and Figure 9-1. As discussed in the text, we can separate individual messages from the sequence because we know that each codeword is completed whenever the user receives the first "1" or when the user receives three consecutive "0"s. Explain in a similar way how messages can be separated using each of the two five-message codes described in Figure 9-3 and Table 9-3.

9.7 A candy maker has a jar that contains 100 jelly beans. Both the candy maker and his colleague (who is in another building) know that 35 of the jelly beans are licorice, 25 are grape, 18 are lime, 12 are pineapple, seven are lemon, and three are cherry. It's a very slow business day, and both the candy maker and his colleague are bored. They devise the following game: The candy maker closes his eyes, selects a jelly bean, opens his eyes, and then sends a message to his colleague stating which flavor jelly bean he has selected. The candy maker then returns the jelly bean to the jar, shakes the jar, and repeats the process.

a. Determine the information content of each of the following messages:

i. "I chose a licorice jelly bean."
ii. "I chose a lemon jelly bean."

b. What is the average information content of a message from the candy maker?
c. What system capacity is required if the candy maker wants to select a jelly bean and transmit a message once every second? Define system capacity as the number of bits/sec that the system can transmit *error free* (this is actually a crude definition of system capacity, but sufficient to illustrate the concept; a more precise definition will be given in Chapter 10).
d. Devise an efficient code for transmitting the jelly bean flavor information.
e. What is the average length of a message, using the code you created in Part d?
f. Describe the performance improvement of the system using the code in Part e as opposed to using a straight three-bit binary assignment for each jelly bean flavor.

9.8 Consider the following set of messages:

Message	Probability of Occurrence
A	0.35
B	0.30
C	0.20
D	0.08
E	0.034
F	0.016
G	0.012
H	0.008

Assume that the probability of occurrence of each message is independent and that each message is transmitted independently.

a. Calculate the entropy of the message set.
b. How many binary digits ("1"s and "0"s) are required to transmit each message if a straight binary assignment is used?
c. Suppose the communication system has the capability to transmit 6000 bits/sec error free across the channel. How many messages per second can be transmitted if a straight binary assignment is used?
d. Develop a Huffman code to represent the eight messages.
e. What is the average codeword length using the Huffman code?
f. On the average, how many messages per second could be transmitted across the 6000 bits/sec channel if the Huffman code is used?
g. On the average, how many messages per second could be transmitted across the 6000 bits/sec channel if an optimum code could be developed?

9.9 Develop a Huffman code for the message pairs in Example 9.9. Calculate the average codeword length and compare to the entropy of the source.

9.10 Using the Internet or other resources, find a product, design, or application that employs Huffman coding. Describe the product, design, or application and how Huffman coding is used. List any references you used to find this information.

9.11 Describe the differences between static and dynamic dictionary encoding. What are the advantages and disadvantages of each?

9.12 Consider the character set consisting of all capital letters, the space character, and the punctuation characters comma and period. Compress the message "DAD ADDED JIM TO THE TEAM" using LZW compression.

9.13 Using the Internet or other resources, determine how much compression LZW provides for standard English text when compared with conventional ASCII coding. List any references you used or procedures you followed to obtain this information.

9.14 Consider the image shown in Figure 9-9.

a. If each pixel is encoded independently, using one bit to indicate whether it is black or white, how many bits are required to represent the entire image?

b. As stated in the text, with run-length coding we can transmit all the information in the fourth row using the sequence "11, 17, 35." What sequence would be used to represent Row 7? What sequence would be used to represent Row 20? What sequence would be used to represent Row 30?

c. State the pattern of "1"s and "0"s that would be used in run-length coding to represent the first five rows of the image.

d. How many bits would be required to transmit the entire image using run-length coding? How does this compare with the number of bits required to transmit the unencoded image (see your answer to Part a)?

9.15 Consider the image shown in Figure P9-15.

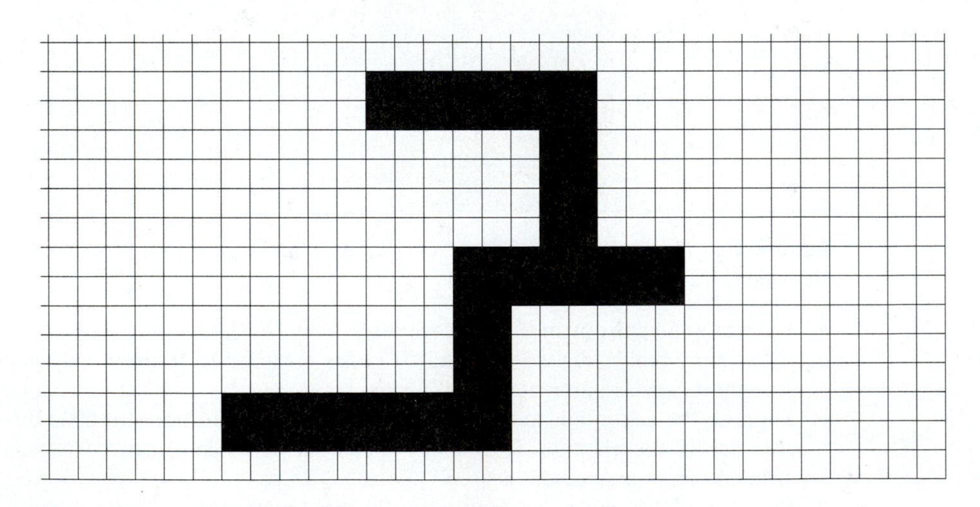

Figure P9-15 Image for facsimile transmission.

a. If each pixel is encoded independently, using one bit to indicate whether it is black or white, how many bits are required to represent the entire image?

b. If run-length coding is applied to the image, how many bits are required to represent each run length?

c. Apply run-length coding to the image and determine the sequence which would represent the image.

d. Compare the number of bits required to represent the unencoded image with the number of bits required to transmit the image if run-length coding is employed.

9.16 What are the advantages and disadvantages of using run-length coding, rather than encoding each pixel independently?

9.17 Describe the differences between run-length coding, modified Huffman coding, and modified READ coding. Explain the advantages and disadvantages of each technique.

9.18 Why are run-length coding and modified READ coding not efficient compression techniques for monochromatic gray scale images or for color images?

9.19 Perform the DCT on the 8×8 intensity matrix in Figure 9-10a and verify that your results match the values in Figure 9-10b.

9.20 Explain in general terms the principle behind the DCT and explain why the DCT provides a very efficient way to represent monochromatic and color images.

9.21 Define the term "*frequency*" when used in image compression. How is this definition different from the conventional definition of *frequency* in the rest of our studies?

9.22 Briefly describe the JPEG compression algorithm and explain its advantages and disadvantages relative to uncompressed representation of still color images.

9.23 Explain how the MPEG compression algorithm exploits temporal dependency to provide efficient compression of moving color images.

9.24 Why is motion compensation useful in MPEG compression?

9.25 Explain the trade-offs involved in determining how many B frames to place between P frames and I frames in an MPEG sequence.

Chapter 10

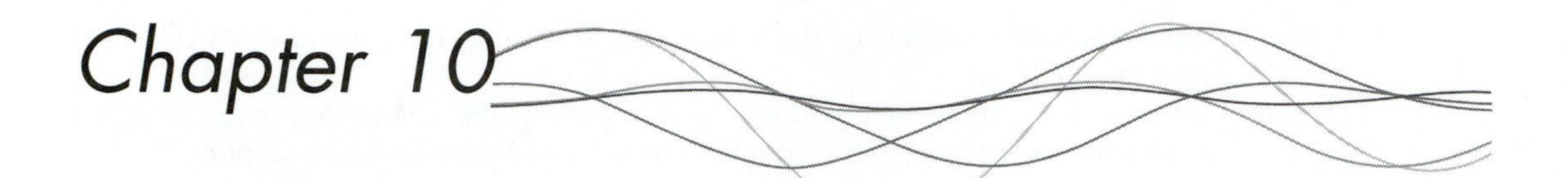

Basics of Error Control Coding

WE know that all channels have physical limitations that distort and attenuate transmitted signals, and we know that channels also corrupt the transmitted signal by adding noise. In our studies in Chapters 2–5, we saw how distortion, attenuation, and noise can cause errors in the received signal. We learned to combat distortion by designing signals with specified bandwidths and by using equalization. We developed various baseband and bandpass modulation and demodulation techniques to minimize the effects of noise, but we know that noise can still cause errors.

In Chapter 9 we established techniques for measuring the average amount of information flowing from a source. We also developed data compression techniques for removing redundancy prior to transmission. Suppose we now want to add some redundancy back into the compressed bit stream prior to transmission so that errors at the receiver can be detected and possibly even corrected (this process is called *error control coding*). We can consider the transmitter to have three parts, as shown in Figure 10-1: a source encoder, which removes redundancy from the source (i.e., compresses the source output); a channel encoder, which adds back a controlled amount of redundancy for error control; and a modulator, which shapes and formats the bit stream into a signal that can be efficiently transmitted across the channel.[1]

In inspecting Figure 10-1, it may initially seem that the source encoder and the channel encoder are working toward opposite goals: The source encoder extracts redundancy from the source and the channel encoder then adds redundancy back in order to provide a degree of error control for the received signal. Note, however, that

[1]In many of the newer, more advanced communication systems, the functions of channel encoding and modulation are often designed interdependently (e.g., intersymbol interference may be purposefully incorporated in the modulated signal to provide redundancy via interdependency between adjacent transmitted symbols). This topic is beyond the scope of this book, but to find out more, see Wilson [10.1].

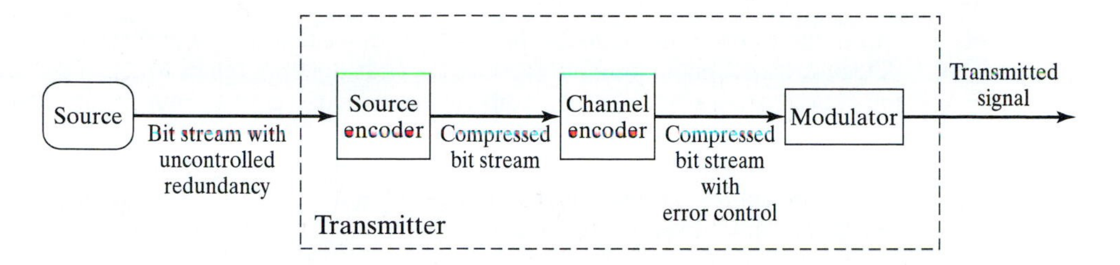

Figure 10-1 Block diagram showing the three functions of a transmitter.

the redundancy at the source output is uncontrolled—the designer cannot control how much redundancy is produced by the source, nor where the redundancy is located in the source's output. The key to error control is *controlled* redundancy, both in its amount and its location. The source encoder and channel encoder are really working together: The source encoder removes uncontrolled redundancy and then the channel encoder inserts the required amount of controlled redundancy prior to modulation. Note that error control can also help combat error propagation, which, as we saw in Chapter 9, is one of the major problems introduced by data compression.

There are two different approaches to error control coding:

1. We can add a small amount of controlled redundancy to a message to allow the receiver to detect errors. If an error is detected, the user can either ignore the message or request that the source resend the message (note that in order for the user to request retransmission, the communication link must be two-way). This strategy is known as *automatic retransmission request* (ARQ).
2. We can add a larger amount of controlled redundancy to a message to allow the receiver to both detect and correct errors. This approach initially requires the transmission of more redundant bits, but it allows for correction of erroneous messages without requiring retransmission. This strategy is known as *forward error correction* (FEC).

Example 10.1 illustrates an ARQ code (even parity) while Example 10.2 illustrates an FEC code (majority voting). These examples are not especially powerful or efficient codes (we will develop much stronger codes shortly), but they serve well to illustrate the differences between ARQ and FEC strategies.

Example 10.1 Even parity

Consider an n-bit message generated by a source. We add up the number of "1"s in the message, and if the sum is an odd number, we add one more "1" bit to the end of the message so that the sum becomes even. If the sum is an even number, we add one more "0" bit to the end of the message so that the sum remains even. This extra bit is called a *parity bit.*

The benefit of using an even parity code is that it allows the receiver to detect any one-bit error that may occur during reception, since such an error will cause the sum of "1"s in the received message to be an odd number. In fact, any odd number of bit errors within the received message can be detected. Note that although these errors can be detected, they cannot be corrected, since the code does not tell us which of the received bits was erroneous. Note also that this code cannot detect an even number of bit errors in the received message.

Even parity is a fairly weak ARQ code. It is simple to implement and requires only one redundant bit per message (or, if you want to divide a long message into a series of shorter *blocks*, one redundant bit per block).

Example 10.2 Majority voting

Consider an n-bit message generated by a source. We transmit this message three times. At the receiver we compare all three copies of the message bit by bit. If there is any disagreement in the bit-by-bit comparison, we assume that the majority is correct (e.g., if the first copy of bit four is 1 but the second and third copies of bit four are both 0, we assume that the fourth bit from the source was 0).

With majority voting we can detect and correct any single-bit error that occurs in the received bit stream. We can also detect and correct multiple errors if they correspond to different bits in the copies. Majority voting provides guaranteed single-bit error-detection and error-correction capability, but it requires a large amount of overhead (two redundant bits must be transmitted for each bit in the source's message; an overhead of 200%).

As a final point, note that if majority voting is used and an error occurs in the same bit of two copies, then the receiver will "correct" the bit to the *wrong* value and the user will have a false (and potentially dangerous) sense of confidence that the received message was correct. This is another issue we will address shortly.

10.1 Channel Capacity

Do the physical limitations of the channel (distortion, noise, and bandwidth) limit the amount of information that a system can accurately transmit in a given period of time? Claude Shannon addressed this question (Shannon [10.2]), determined that the answer is "yes," and developed an analytic expression for the limit. This limit is called the *channel capacity* and is symbolized by C. Shannon determined that for a channel corrupted by bandlimited additive white Gaussian noise, channel capacity is

$$C = BW \log_2\left(1 + \frac{S}{N}\right) \text{ bits/sec} \tag{10.1}$$

where BW represents the channel bandwidth and $\frac{S}{N}$ is the signal-to-noise ratio (SNR) at the receiver.

This concept, known as the Shannon-Hartley theorem, has many implications. Equation (10.1) makes intuitive sense: We know that greater bandwidth allows faster transmission speeds and therefore should allow faster flow of information. We also know that higher SNRs mean lower error rates and hence more accurate transmission. A deeper implication of the Shannon-Hartley theorem is that, coupled with

Shannon's first theorem (see Chapter 9), it tells us that for any source outputting information at a rate R and wanting to transmit this information across a channel with capacity C, there exists an error control code that can ensure accurate transmission of all of the source's information provided $C \geq R$. The Shannon-Hartley theorem does not tell us how to develop such a code—that is our job. The codes we need to develop are much stronger than the parity checking and majority voting codes illustrated in Examples 10.1 and 10.2. To develop these codes, we must first examine some new mathematical tools.

10.2 Field Theory and Modulo-2 Operators

Parity coding (Example 10.1) and majority voting (Example 10.2) are good codes for illustrating the concepts of error detection and error correction, but the codes themselves are rather weak or require excessive overhead. Parity, for instance, can detect only an odd number of bit errors in the group of protected bits. Any even number of bit errors will go undetected. Majority voting has the capability to detect and correct any single error in the protected group of bits, but it requires an overhead of 200%; in order to protect the information, we must send three copies of the information sequence. We will shortly examine much more powerful and efficient FEC and ARQ error control codes, but in order to understand these codes we must first develop some mathematical tools based on algebra theory.

10.2.1 Galois Field of Order 2

Let's start by examining a special mathematical system that contains only two elements, 0 and 1 (i.e., a *binary system*). Let's define two operations that we can perform on these elements: one that we will call *modulo-2 addition* and will symbolize as $\oplus$, and one that we will call *modulo-2 multiplication* and will symbolize as $\bullet$. Let's define modulo-2 addition as follows:

$$\begin{aligned} 0 \oplus 0 &= 0 \\ 0 \oplus 1 &= 1 \\ 1 \oplus 0 &= 1 \\ 1 \oplus 1 &= 0 \end{aligned} \tag{10.2}$$

Note that modulo-2 addition is equivalent to the exclusive-OR operation, which we learned about in our digital logic class (and discussed briefly in Chapter 7).

Let's define modulo-2 multiplication as follows:

$$\begin{aligned} 0 \bullet 0 &= 0 \\ 0 \bullet 1 &= 0 \\ 1 \bullet 0 &= 0 \\ 1 \bullet 1 &= 1 \end{aligned} \tag{10.3}$$

Note that modulo-2 multiplication is equivalent to the AND operation, which we also learned about in our digital logic class.[2]

Let the variables A and B represent elements in our new mathematical system (i.e., the value of A is 1 or 0 and the value of B is 1 or 0). We call A and B *binary variables*. The $\oplus$ operator has many significant properties, including the following:[3]

1. The system is *closed* under the $\oplus$ operator. For any two binary variables A and B, $A \oplus B$ is also an element in our system.
2. The $\oplus$ operator is *associative*. For any three binary variables A, B, and C,

$$(A \oplus B) \oplus C = A \oplus (B \oplus C) \tag{10.4}$$

3. An *identity element* exists for the $\oplus$ operator; it is the element 0. For any binary variable A,

$$A \oplus 0 = A \tag{10.5}$$

4. Each element has an *inverse*. We define the inverse of a binary variable A, symbolized $\overline{A}$, as the element that, when operated upon by A, produces the identity element. In other words,

$$\begin{gathered} A \oplus \overline{A} = 0 \\ 0 \oplus 0 = 0, \text{ so the inverse of 0 is 0 (i.e., } \overline{0} = 0) \\ 1 \oplus 1 = 0, \text{ so the inverse of 1 is 1 (i.e., } \overline{1} = 1) \end{gathered} \tag{10.6}$$

We see that for the $\oplus$ operator, the inverse of any binary variable A is itself.

$$\overline{A} = A \tag{10.7}$$

5. The $\oplus$ operator is *commutative*. For any two binary variables A and B,

$$A \oplus B = B \oplus A \tag{10.8}$$

The $\bullet$ operator has many significant properties, including the following:

1. The system is closed under the $\bullet$ operator. For any two binary variables A and B, $A \bullet B$ is also an element in our system.

[2]A more general way to define modulo-2 operations is to consider the modulus itself as an operator. Let a and m be integers. We define $a_{modulo\text{-}m} = r$, where r is the *remainder* resulting from dividing a by m. For example, $7_{modulo\text{-}3} = 1$, since 7 divided by 3 produces a quotient of 2 and a remainder of 1. Using this definition, the modulo-2 sum of 1 and 1, for instance, is $1 \oplus 1 = (1 + 1)_{modulo\text{-}2} = 2_{modulo\text{-}2} = 0$. We can, of course, create mathematical systems using a modulus other than 2, but such systems are only marginally useful (with the significant exception of moduli that are powers of 2, as we shall see in Section 10.7.2).

[3]A mathematical system consisting of elements and one operator that contains the first four of the following properties is called a *group*. If the system also contains the fifth property, it is called a *commutative group*.

2. The • operator is associative. For any three binary variables A, B, and C,

$$(A \bullet B) \bullet C = A \bullet (B \bullet C) \tag{10.9}$$

3. An identity element exists for the • operator; it is the element 1. For any binary variable A,

$$A \bullet 1 = A \tag{10.10}$$

4. The 1 element has an inverse, but the 0 element does not.
5. The • operator is commutative. For any two binary variables A and B,

$$A \bullet B = B \bullet A \tag{10.11}$$

We can also observe that the • operator is *distributive* over the ⊕ operator

$$A \bullet (B \oplus C) = (A \bullet B) \oplus (A \bullet C) \tag{10.12}$$

A mathematical system containing the above properties (a set of elements and two operators, both of which are closed, associative, commutative, and have an identity element, one of which has an inverse and the other of which has an inverse for all elements except 0; plus a distributive property) is called a *field*. The field we've just established, consisting of the elements $\{0, 1\}$ and the operators ⊕ and •, is called a *Galois field of order 2*, symbolized as $GF(2)$.

$GF(2)$ is a highly structured mathematical system, and we can develop powerful error detection and correction codes by exploiting this structure. Violations in the structure (resulting from incorrectly received bits) will alert us that errors have occurred and can even show us which bits were received in error, allowing us to correct them. We will begin developing our codes shortly, but let's first find a way to efficiently represent and manipulate a series of bits in $GF(2)$.

10.2.2 Matrix Representation and Manipulation

Matrix representation and manipulation in $GF(2)$ will be useful to us when we start describing some of the more complex error control codes. Using notation consistent with Chapter 7, we can represent a sequence of bits—say, $u_1, u_2, u_3, \ldots u_n$—as a $1 \times n$ matrix or *vector* $[\boldsymbol{u}]$

$$[\boldsymbol{u}] = [u_1 \; u_2 \; u_3 \; \ldots \; u_n]$$

We can add or multiply two matrices in $GF(2)$. Modulo-2 matrix addition is the same as conventional matrix addition, except that we use the ⊕ operator instead of adding. Consider adding a 2×3 matrix $[\boldsymbol{a}]$ with another 2×3 matrix $[\boldsymbol{b}]$

$$\begin{bmatrix} a_{11} & a_{12} & a_{13} \\ a_{21} & a_{22} & a_{23} \end{bmatrix} \oplus \begin{bmatrix} b_{11} & b_{12} & b_{13} \\ b_{21} & b_{22} & b_{23} \end{bmatrix} = \begin{bmatrix} a_{11} \oplus b_{11} & a_{12} \oplus b_{12} & a_{13} \oplus b_{13} \\ a_{21} \oplus b_{21} & a_{22} \oplus b_{22} & a_{23} \oplus b_{23} \end{bmatrix} \tag{10.13}$$

Modulo-2 matrix multiplication is the same as conventional matrix multiplication, except that we use the • and ⊕ operators instead of conventional multiplication and addition. Consider a 1×2 matrix **[*c*]** postmultiplied by a 2×3 matrix **[*b*]**. Using modulo-2 matrix multiplication, the product is defined as the 1×3 matrix

$$[c_{11} \quad c_{12}] \bullet \begin{bmatrix} b_{11} & b_{12} & b_{13} \\ b_{21} & b_{22} & b_{23} \end{bmatrix}$$

$$= [(c_{11} \bullet b_{11}) \oplus (c_{12} \bullet b_{21}) \quad (c_{11} \bullet b_{12}) \oplus (c_{12} \bullet b_{22}) \quad (c_{11} \bullet b_{13}) \oplus (c_{12} \bullet b_{23})] \tag{10.14}$$

Example 10.3 Modulo-2 matrix operations

Perform the following modulo-2 matrix operation

$$[1 \quad 1 \quad 0 \quad 0] \bullet \begin{bmatrix} 1 & 1 & 0 & 1 & 0 & 0 & 0 \\ 0 & 1 & 1 & 0 & 1 & 0 & 0 \\ 1 & 1 & 1 & 0 & 0 & 1 & 0 \\ 1 & 0 & 1 & 0 & 0 & 0 & 1 \end{bmatrix}$$

Solution

We know the product will be a 1×7 matrix **[*d*]**, where

$$\begin{aligned}
d_{11} &= (1 \bullet 1) \oplus (1 \bullet 0) \oplus (0 \bullet 1) \oplus (0 \bullet 1) = 1 \oplus 0 \oplus 0 \oplus 0 = 1 \\
d_{12} &= (1 \bullet 1) \oplus (1 \bullet 1) \oplus (0 \bullet 1) \oplus (0 \bullet 0) = 1 \oplus 1 \oplus 0 \oplus 0 = 0 \\
d_{13} &= (1 \bullet 0) \oplus (1 \bullet 1) \oplus (0 \bullet 1) \oplus (0 \bullet 1) = 0 \oplus 1 \oplus 0 \oplus 0 = 1 \\
d_{14} &= (1 \bullet 1) \oplus (1 \bullet 0) \oplus (0 \bullet 0) \oplus (0 \bullet 0) = 1 \oplus 0 \oplus 0 \oplus 0 = 1 \\
d_{15} &= (1 \bullet 0) \oplus (1 \bullet 1) \oplus (0 \bullet 0) \oplus (0 \bullet 0) = 0 \oplus 1 \oplus 0 \oplus 0 = 1 \\
d_{16} &= (1 \bullet 0) \oplus (1 \bullet 0) \oplus (0 \bullet 1) \oplus (0 \bullet 0) = 0 \oplus 0 \oplus 0 \oplus 0 = 0 \\
d_{17} &= (1 \bullet 0) \oplus (1 \bullet 0) \oplus (0 \bullet 0) \oplus (0 \bullet 1) = 0 \oplus 0 \oplus 0 \oplus 0 = 0
\end{aligned}$$

Thus

$$[1 \quad 1 \quad 0 \quad 0] \bullet \begin{bmatrix} 1 & 1 & 0 & 1 & 0 & 0 & 0 \\ 0 & 1 & 1 & 0 & 1 & 0 & 0 \\ 1 & 1 & 1 & 0 & 0 & 1 & 0 \\ 1 & 0 & 1 & 0 & 0 & 0 & 1 \end{bmatrix} = [1 \quad 0 \quad 1 \quad 1 \quad 1 \quad 0 \quad 0]$$

We are now ready to begin developing and analyzing effective error control codes.

10.3 Hamming Codes

Consider a channel encoder that takes a four-bit information sequence $(u_1\, u_2\, u_3\, u_4)$ and produces a seven-bit coded sequence $(v_1\, v_2\, v_3\, v_4\, v_5\, v_6\, v_7)$ as shown in Figure 10-2. Let the encoded bits be related to the unencoded bits as follows:

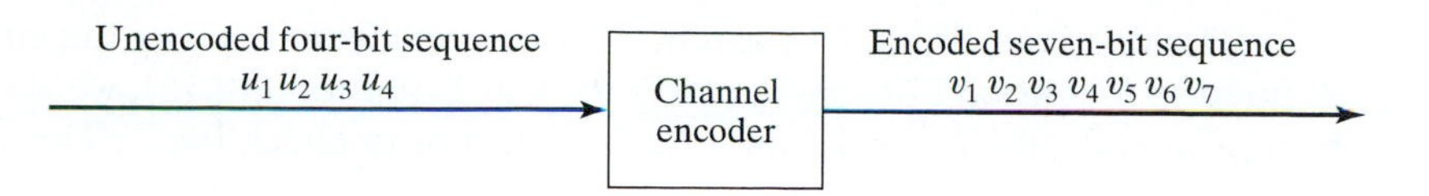

Figure 10-2 A channel encoder producing a seven-bit code to represent four information bits.

$$\begin{aligned} v_4 &= u_1 \\ v_5 &= u_2 \\ v_6 &= u_3 \\ v_7 &= u_4 \\ v_1 &= u_1 \oplus u_3 \oplus u_4 = v_4 \oplus v_6 \oplus v_7 \\ v_2 &= u_1 \oplus u_2 \oplus u_3 = v_4 \oplus v_5 \oplus v_6 \\ v_3 &= u_2 \oplus u_3 \oplus u_4 = v_5 \oplus v_6 \oplus v_7 \end{aligned} \tag{10.15}$$

Now let the seven-bit sequence be transmitted to the receiver. Let $\hat{v}_i$ represent the *i*th received bit. If no errors occurred, received bits $\hat{v}_4$, $\hat{v}_5$, $\hat{v}_6$, and $\hat{v}_7$ will be the same as the original four information bits u_1, u_2, u_3, and u_4. Suppose, however, that we suspect an error may have occurred. Single-bit errors can be detected and corrected in the received bit sequence by the following procedure:

1. Using the received bits $\hat{v}_4$, $\hat{v}_5$, $\hat{v}_6$, and $\hat{v}_7$ (called the *data bits*), determine the expected values for $\hat{v}_1$, $\hat{v}_2$, and $\hat{v}_3$ (called the *check bits*, or *parity bits*). Let's call these expected values $\hat{v}_{1ex}$, $\hat{v}_{2ex}$, and $\hat{v}_{3ex}$.

$$\begin{aligned} \hat{v}_{1ex} &= \hat{v}_4 \oplus \hat{v}_6 \oplus \hat{v}_7 \\ \hat{v}_{2ex} &= \hat{v}_4 \oplus \hat{v}_5 \oplus \hat{v}_6 \\ \hat{v}_{3ex} &= \hat{v}_5 \oplus \hat{v}_6 \oplus \hat{v}_7 \end{aligned} \tag{10.16}$$

2. Compare the calculated values for the check bits, $\hat{v}_{1ex}$, $\hat{v}_{2ex}$, and $\hat{v}_{3ex}$, with the actual received values $\hat{v}_1$, $\hat{v}_2$, and $\hat{v}_3$. As shown in Table 10-1, the relationship between the expected and received check bits will determine whether or not a single-bit error has occurred. If a single-bit error has occurred, Table 10-1 will also show which received bit is in error. Since the value of a bit is either 1 or 0, if we know that a particular bit has been received in error, we can correct its value by inverting it.

Table 10-1 Detecting (and Correcting) an Error by Comparing the Values of Expected and Received Check Bits

Bit Received in Error	$\hat{v}_{1ex}$ vs. $\hat{v}_1$	$\hat{v}_{2ex}$ vs. $\hat{v}_2$	$\hat{v}_{3ex}$ vs. $\hat{v}_3$
none	$=$	$=$	$=$
$\hat{v}_1$	$\neq$	$=$	$=$
$\hat{v}_2$	$=$	$\neq$	$=$
$\hat{v}_3$	$=$	$=$	$\neq$
$\hat{v}_4$	$\neq$	$\neq$	$=$
$\hat{v}_5$	$=$	$\neq$	$\neq$
$\hat{v}_6$	$\neq$	$\neq$	$\neq$
$\hat{v}_7$	$\neq$	$=$	$\neq$

The results in Table 10-1 can be seen directly from the way the encoded bits are defined in Equation (10.15). Looking back at Equation (10.15), we see that the data bit v_4 is a variable in check bits v_1 and v_2 but not in check bit v_3. Therefore, an error in the received value of v_4 will cause $\hat{v}_{1ex}$ and $\hat{v}_1$ to disagree (since the error will affect the calculation of $\hat{v}_{1ex}$ but will not affect the received bit $\hat{v}_1$). It will also cause $\hat{v}_{2ex}$ and $\hat{v}_2$ to disagree, but it will not cause $\hat{v}_{3ex}$ and $\hat{v}_3$ to disagree. Since there is no other data bit that is a variable in check bits v_1 and v_2 but not in check bit v_3, any time $\hat{v}_{1ex} \neq \hat{v}_1$ and $\hat{v}_{2ex} \neq \hat{v}_2$, but $\hat{v}_{3ex} = \hat{v}_3$, we can deduce that data bit v_4 has been received in error, as shown in the fifth row of Table 10-1. (This, of course, assumes only one error in the received sequence of seven bits.) We can show similar relationships for single-bit errors in the received values of data bits v_5 (see Row 6 in Table 10-1), v_6 (see Row 7), and v_7 (see Row 8). We can also see that each of the data bits (v_4, v_5, v_6, and v_7) is a variable in at least two of the check bit equations, so a disagreement between $\hat{v}_{1ex}$ and $\hat{v}_1$ but agreement between $\hat{v}_{2ex}$ and $\hat{v}_2$ and agreement between $\hat{v}_{3ex}$ and $\hat{v}_3$ means that $\hat{v}_1$ was received in error; disagreement only between $\hat{v}_{2ex}$ and $\hat{v}_2$ means that $\hat{v}_2$ was received in error; and disagreement only between $\hat{v}_{3ex}$ and $\hat{v}_3$ means that $\hat{v}_3$ was received in error. Finally, no disagreements means that all bits were correctly received.

The code we've just developed is known as a *Hamming code*, and it provides the capability to detect and correct one error in every four bits of data. The overhead required is three check bits per four data bits, which is considerably fewer than the overhead required by majority voting in Example 10.2.

Example 10.4 Using a Hamming Code

A communication system employs the Hamming code described in Equation (10.15). Suppose the seven-bit pattern 1011101 is received.

a. Did an error occur in reception?

b. What were the original four information bits from the source?

Solution

Calculating the expected values for the check bits,

$$\hat{v}_{1ex} = \hat{v}_4 \oplus \hat{v}_6 \oplus \hat{v}_7 = 1 \oplus 0 \oplus 1 = 0$$
$$\hat{v}_{2ex} = \hat{v}_4 \oplus \hat{v}_5 \oplus \hat{v}_6 = 1 \oplus 1 \oplus 0 = 0$$
$$\hat{v}_{3ex} = \hat{v}_5 \oplus \hat{v}_6 \oplus \hat{v}_7 = 1 \oplus 0 \oplus 1 = 0$$

Since $\hat{v}_1 = 1$, $\hat{v}_2 = 0$, and $\hat{v}_3 = 1$, we have disagreement between $\hat{v}_{1ex}$ and $\hat{v}_1$; agreement between $\hat{v}_{2ex}$ and $\hat{v}_2$; and disagreement between $\hat{v}_{3ex}$ and $\hat{v}_3$. Using Table 10-1, we see that an error occurred in $\hat{v}_7$. Correcting this error, the seven-bit encoded pattern should have been 1011100 and so the four information bits from the source are 1100.

To better understand the operation of a Hamming code, let's apply a matrix notation to the unencoded and encoded bits and see if we can describe the encoding and decoding processes using matrix manipulation. Consider the four unencoded information bits u_1, u_2, u_3, and u_4. We can represent these four bits using a 1×4 matrix, or *vector*

$$[\mathbf{u}] = [u_1 \quad u_2 \quad u_3 \quad u_4]$$

We can represent the seven encoded bits using a 1×7 vector

$$[\mathbf{v}] = [v_1 \quad v_2 \quad v_3 \quad v_4 \quad v_5 \quad v_6 \quad v_7]$$

We can now use modulo-2 matrix multiplication to represent the encoding process for our Hamming code

$$[u_1 \quad u_2 \quad u_3 \quad u_4] \bullet \begin{bmatrix} 1 & 1 & 0 & 1 & 0 & 0 & 0 \\ 0 & 1 & 1 & 0 & 1 & 0 & 0 \\ 1 & 1 & 1 & 0 & 0 & 1 & 0 \\ 1 & 0 & 1 & 0 & 0 & 0 & 1 \end{bmatrix} = [v_1 \quad v_2 \quad v_3 \quad v_4 \quad v_5 \quad v_6 \quad v_7] \tag{10.17}$$

where the 4×7 matrix is known as a *generator matrix* and is represented by $[\mathbf{G}]$. Examine Equation (10.17) carefully and make sure you see that it is consistent with Equation (10.15). (Be sure you remember how we earlier defined modulo-2 matrix multiplication—consistent with standard matrix multiplication, except using the $\bullet$ and $\oplus$ operators instead of conventional multiplication and conventional addition.) Equation (10.17) shows that we can represent the encoding process by postmultiplying our information vector $[\mathbf{u}]$ by a generator matrix $[\mathbf{G}]$ to obtain the encoded bit sequence $[\mathbf{v}]$. Expressed in matrix form,

$$[\mathbf{v}] = [\mathbf{u}][\mathbf{G}] \tag{10.18}$$

Example 10.5 Encoding by matrix multiplication

A source outputs the four information bits 1100. These four bits are to be encoded using the Hamming code just described. Using the generator matrix in Equation (10.17), determine the encoded seven-bit pattern to be transmitted.

Solution

$$[1 \quad 1 \quad 0 \quad 0] \bullet \begin{bmatrix} 1 & 1 & 0 & 1 & 0 & 0 & 0 \\ 0 & 1 & 1 & 0 & 1 & 0 & 0 \\ 1 & 1 & 1 & 0 & 0 & 1 & 0 \\ 1 & 0 & 1 & 0 & 0 & 0 & 1 \end{bmatrix} = [1 \quad 0 \quad 1 \quad 1 \quad 1 \quad 0 \quad 0]$$

The seven-bit sequence 1011100 is transmitted to represent the four data bits 1100. (Note that we performed this exact modulo-2 matrix multiplication in Example 10.3.)

Let's take a closer look at the generator matrix $[\mathbf{G}]$ for our Hamming code. From Equation (10.17),

$$[\mathbf{G}] = \begin{bmatrix} 1 & 1 & 0 & 1 & 0 & 0 & 0 \\ 0 & 1 & 1 & 0 & 1 & 0 & 0 \\ 1 & 1 & 1 & 0 & 0 & 1 & 0 \\ 1 & 0 & 1 & 0 & 0 & 0 & 1 \end{bmatrix} \tag{10.19}$$

We can see that the generator matrix has the following properties:

1. The matrix is k rows by n columns: where $k = 4$, the number of information bits from the source; and $n = 7$, the total number of encoded bits. (Be sure you see why this has to be true.)
2. The last four columns of **[*G*]** form an identity matrix. The identity matrix maps the four information bits into the last four bits of the encoded seven-bit pattern. A code in which the information bits are always repeated in a portion of the encoded sequence is called a *systematic code*. Systematic codes are desirable because they make it easy to extract the original data from the encoded sequence.
3. The first three columns of **[*G*]** are created by listing *as rows* different three-bit combinations of two or more "1"s. This ensures that each data bit is part of the modulo-2 sum of a unique combination of two or more check bits. We saw this property when we examined Table 10-1 and determined why the agreements and disagreements among the received and calculated check bits indicated which bit was received in error.
4. The generator matrix is not unique. For example, we can develop an equally valid Hamming code by switching the first and second columns in **[*G*]**.

Now that we've examined the properties of the generator matrix, let's deconstruct **[*G*]** into two submatrices as shown in Figure 10-3. The 4×4 *identity submatrix* **[*I*]** maps the four information bits into the last four bits of the encoded seven-bit pattern, ensuring a systematic code. The 4×3 *parity submatrix* **[*P*]** produces the three check bits (also known as *parity bits*) at the beginning of the encoded seven-bit pattern.

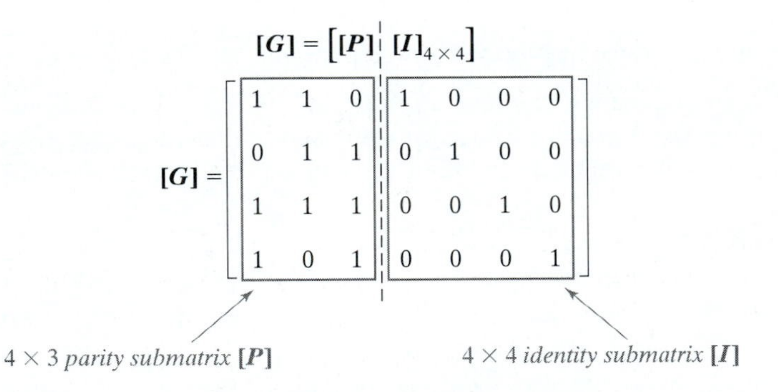

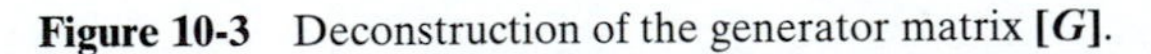

Figure 10-3 Deconstruction of the generator matrix **[*G*]**.

Before we examine Hamming and other error control codes in more detail, let's introduce some terminology. A *block code* divides the bit stream from the source encoder into k-bit blocks. Each k-bit block is then encoded into an n-bit block prior to transmission, where the values of the n bits depend on only the values of the k bits in the corresponding source block. The Hamming code we've just developed is a block code which takes the data stream from the source encoder, divides it into four-bit blocks, and then encodes each four-bit block into a seven-bit block prior to transmission.

Block codes are described in terms of n and k, stated as (n,k). The code we've developed is called a (7,4) Hamming code. The *code rate*, represented by r, is the ratio of k to n.

$$r = \frac{k}{n} \tag{10.20}$$

The Hamming code we've developed is termed a "rate four-sevenths" code.

Representing a Hamming code in matrix form and understanding the properties and structure of **[*G*]** allow us to see ways to develop similar codes. Suppose we want to develop a Hamming code to provide single-error detection and correction capabilities for a larger block of data. Using four parity bits, we can produce 11 different combinations of those four bits with two or more "1"s; we can therefore use four parity bits to provide single-error detection and correction capabilities for a block of 11 information bits. The first four columns of the 11×15 generator matrix for our new code are formed by listing as rows all possible four-bit combinations of two or more "1"s. The last 11 columns form an 11×11 identity matrix. Equation (10.21) shows a generator matrix for our (15,11) Hamming code.

$$[\boldsymbol{G}] = \begin{bmatrix} 1 & 1 & 0 & 0 & 1 & 0 & 0 & 0 & 0 & 0 & 0 & 0 & 0 & 0 & 0 \\ 1 & 0 & 1 & 0 & 0 & 1 & 0 & 0 & 0 & 0 & 0 & 0 & 0 & 0 & 0 \\ 1 & 0 & 0 & 1 & 0 & 0 & 1 & 0 & 0 & 0 & 0 & 0 & 0 & 0 & 0 \\ 0 & 1 & 1 & 0 & 0 & 0 & 0 & 1 & 0 & 0 & 0 & 0 & 0 & 0 & 0 \\ 0 & 1 & 0 & 1 & 0 & 0 & 0 & 0 & 1 & 0 & 0 & 0 & 0 & 0 & 0 \\ 0 & 0 & 1 & 1 & 0 & 0 & 0 & 0 & 0 & 1 & 0 & 0 & 0 & 0 & 0 \\ 1 & 1 & 1 & 0 & 0 & 0 & 0 & 0 & 0 & 0 & 1 & 0 & 0 & 0 & 0 \\ 1 & 1 & 0 & 1 & 0 & 0 & 0 & 0 & 0 & 0 & 0 & 1 & 0 & 0 & 0 \\ 1 & 0 & 1 & 1 & 0 & 0 & 0 & 0 & 0 & 0 & 0 & 0 & 1 & 0 & 0 \\ 0 & 1 & 1 & 1 & 0 & 0 & 0 & 0 & 0 & 0 & 0 & 0 & 0 & 1 & 0 \\ 1 & 1 & 1 & 1 & 0 & 0 & 0 & 0 & 0 & 0 & 0 & 0 & 0 & 0 & 1 \end{bmatrix} \tag{10.21}$$

This new (15,11) code isn't as powerful as our (7,4) code for protecting information; we can protect the information from only a single error in 11 information bits, whereas if we broke the 11 information bits into three four-bit blocks and used the (7,4) code, we could achieve more protection. Note, however, that our new code uses considerably less overhead: Only four parity bits are required for each 11 information bits, compared with nine parity bits needed for three smaller, four-bit blocks of information.

Let's now go back to our original (7,4) Hamming code, as previously shown in Figure 10-3, and see if we can use matrix operations to represent decoding, error detecting, and error correcting. As discussed previously, we can deconstruct the generator matrix **[*G*]** into a 4×3 parity matrix and a 4×4 identity matrix

$$[\boldsymbol{G}] = [[\boldsymbol{P}] \;\vdots\; [\boldsymbol{I}]_{4\times4}] \tag{10.22}$$

Let's consider forming a 3×7 matrix **[*H*]**

$$[\boldsymbol{H}] = [[\boldsymbol{I}]_{3\times3} \;\vdots\; [\boldsymbol{P}]^T] \tag{10.23}$$

where $[\boldsymbol{I}]_{3\times3}$ represents the 3 × 3 identity matrix and $[\boldsymbol{P}]^T$ represents the transpose of the parity submatrix. (Remember that to transpose a matrix, we turn the rows into columns and the columns into rows: The first row of $[\boldsymbol{P}]$ becomes the first column of $[\boldsymbol{P}]^T$, etc.) The matrix $[\boldsymbol{H}]$ is thus

$$[\boldsymbol{H}] = \left[\begin{array}{ccc|cccc} 1 & 0 & 0 & 1 & 0 & 1 & 1 \\ 0 & 1 & 0 & 1 & 1 & 1 & 0 \\ 0 & 0 & 1 & 0 & 1 & 1 & 1 \end{array}\right] \qquad (10.24)$$

3 × 3 *identity submatrix* $[\boldsymbol{I}]$ 3 × 4 *transpose of parity matrix* $[\boldsymbol{P}]^T$

Now let $[\boldsymbol{r}]$ represent the received codeword

$$[\boldsymbol{r}] = [\hat{v}_1 \quad \hat{v}_2 \quad \hat{v}_3 \quad \hat{v}_4 \quad \hat{v}_5 \quad \hat{v}_6 \quad \hat{v}_7]$$

Suppose we postmultiply the received codeword by the transpose of $[\boldsymbol{H}]$

$$[\boldsymbol{r}][\boldsymbol{H}]^T =$$

$$[\hat{v}_1 \quad \hat{v}_2 \quad \hat{v}_3 \quad \hat{v}_4 \quad \hat{v}_5 \quad \hat{v}_6 \quad \hat{v}_7] \bullet \begin{bmatrix} 1 & 0 & 0 \\ 0 & 1 & 0 \\ 0 & 0 & 1 \\ 1 & 1 & 0 \\ 0 & 1 & 1 \\ 1 & 1 & 1 \\ 1 & 0 & 1 \end{bmatrix} = \qquad (10.25)$$

$$[\hat{v}_1 \oplus \hat{v}_4 \oplus \hat{v}_6 \oplus \hat{v}_7 \quad \hat{v}_2 \oplus \hat{v}_4 \oplus \hat{v}_5 \oplus \hat{v}_6 \quad \hat{v}_3 \oplus \hat{v}_5 \oplus \hat{v}_6 \oplus \hat{v}_7] =$$

$$[\hat{v}_1 \oplus \hat{v}_{1ex} \quad \hat{v}_2 \oplus \hat{v}_{2ex} \quad \hat{v}_3 \oplus \hat{v}_{3ex}]$$

The resulting 1 × 3 vector, called the *syndrome*, tells us which expected check bits agree with their received counterparts and which check bits do not. ($\hat{v}_1 \oplus \hat{v}_{1ex} = 0$ if $\hat{v}_1 = \hat{v}_{1ex}$; $\hat{v}_1 \oplus \hat{v}_{1ex} = 1$ if $\hat{v}_1 \neq \hat{v}_{1ex}$. Similar results are obtained for $\hat{v}_2 \oplus \hat{v}_{2ex}$ and $\hat{v}_3 \oplus \hat{v}_{3ex}$.) As we've already seen, this agree/disagree information is exactly what we need to detect and correct a single-bit error in the received seven-bit sequence. Let's use $[\boldsymbol{s}]$ to represent the three-bit syndrome. We can now create a table similar to Table 10-1 to provide error correction based on the value of the syndrome. This information is provided in Table 10-2. Be sure you see how the information in Table 10-2 is derived from Table 10-1.

There is no magic in how we created $[\boldsymbol{H}]^T$ or in the fact that it produces the syndrome. As shown in Figure 10-4, $[\boldsymbol{H}]^T$ deconstructs into a 3 × 3 identity submatrix followed by the parity submatrix. The parity submatrix produces the calculated check bits $\hat{v}_{1ex}$, $\hat{v}_{2ex}$, and $\hat{v}_{3ex}$ when the modulo-2 multiplication involves the first, second, and third columns of $[\boldsymbol{H}]^T$, respectively, and the identity submatrix causes an exclusive-ORing of the first, second, or third received bit with the calculated check bit. This exclusive-ORing produces 0 if the received and calculated check bits

Table 10-2 Detecting (and Correcting) an Error by Evaluating the Syndrome [*s*]

[*s*]	Bit Received in Error
[0 0 0]	none
[0 0 1]	$\hat{v}_3$
[0 1 0]	$\hat{v}_2$
[0 1 1]	$\hat{v}_5$
[1 0 0]	$\hat{v}_1$
[1 0 1]	$\hat{v}_7$
[1 1 0]	$\hat{v}_4$
[1 1 1]	$\hat{v}_6$

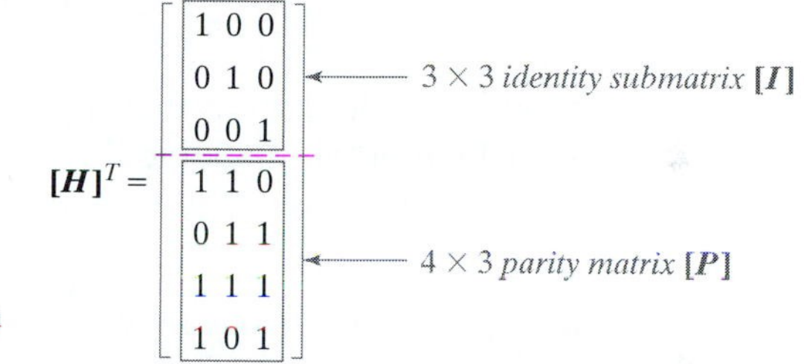

Figure 10-4 Deconstruction of decoding matrix $[H]^T$.

agree or 1 if they disagree, since

$$A \oplus B = \begin{cases} 0 \text{ if } A = B \\ 1 \text{ if } A \neq B \end{cases} \tag{10.26}$$

Matrix multiplication using $[H]^T$ followed by syndrome decoding using Table 10-2 is therefore equivalent to the decoding process we originally described and used in Example 10.4. $[H]^T$ is therefore called the *decoding matrix*.

Example 10.6 Using matrix operations to detect and correct an error—Part 1

As in Example 10.4, a communication system employs the (7,4) Hamming code described in Equation (10.15) and the seven-bit pattern 1011101 is received. Use matrix operations to determine if an error occurred and then determine the original four information bits from the source.

Solution

We calculate the syndrome by postmultiplying the received seven-bit pattern by the decoding matrix $[H]^T$. Note that $1 \oplus 1 \oplus 1 = 1$.

$$[1 \quad 0 \quad 1 \quad 1 \quad 1 \quad 0 \quad 1] \bullet \begin{bmatrix} 1 & 0 & 0 \\ 0 & 1 & 0 \\ 0 & 0 & 1 \\ 1 & 1 & 0 \\ 0 & 1 & 1 \\ 1 & 1 & 1 \\ 1 & 0 & 1 \end{bmatrix} = [1 \quad 0 \quad 1]$$

The syndrome is therefore $[s] = [1 \quad 0 \quad 1]$, which, from Table 10-2, corresponds to an error in the seventh received bit. The corrected received bit sequence is therefore 1011100 and, since the code is systematic, the information bits are 1100.

As a final observation, note that modulo-2 matrix addition can be used to represent the final step of error correction, inverting the erroneous bit once it has been identified. Suppose, as in Example 10.6, the seventh bit is received in error. Modulo-2 adding the received bit sequence with [0 0 0 0 0 0 1] will leave the first six received bits unchanged but will invert the seventh received bit, since

$$A \oplus 0 = A \tag{10.27}$$

and

$$A \oplus 1 = \begin{cases} 1 \text{ if } A = 0 \\ 0 \text{ if } A = 1 \end{cases} \tag{10.28}$$

Generalizing, once we determine that an error has occurred in the ith bit, we can correct the error by modulo-2 adding the received vector $[r]$ with an error vector $[e]$ that consists of all zeroes except for a 1 in its ith element.

Example 10.7 Using matrix operations to detect and correct an error—Part 2

Use modulo-2 matrix addition to correct the error in the received bit sequence in Example 10.6.

Solution

In Example 10.6 we determined that $[s] = [1 \quad 0 \quad 1]$, corresponding to an error in the seventh received bit. Correcting the error,

$$\begin{aligned} [r] \oplus [e] &= [1 \quad 0 \quad 1 \quad 1 \quad 1 \quad 0 \quad 1] \oplus [0 \quad 0 \quad 0 \quad 0 \quad 0 \quad 0 \quad 1] \\ &= [1 \quad 0 \quad 1 \quad 1 \quad 1 \quad 0 \quad 0] \end{aligned}$$

The transmitted bit sequence is 1011100 and the original information bits are 1100.

10.4 A Geometric Interpretation of Error Control Coding

In the previous section we used matrix notation and modulo-2 matrix multiplication to describe the encoding and decoding operations using a Hamming code. To more fully understand the structure of the Hamming code (and to develop even more powerful codes), we need to introduce a new concept: a geometric interpretation of error control coding.

Let's begin by considering an encoder that uses majority voting, as described previously in Example 10.2. Let's use one-bit blocks of information. We can represent the information sequence (one bit) as

$$[u] = [u_1]$$

and the encoded sequence (three bits) as

$$[v] = [v_1 \ v_2 \ v_3]$$

where

$$\begin{aligned} v_1 &= u_1 \\ v_2 &= u_1 \\ v_3 &= u_1 \end{aligned} \tag{10.29}$$

The encoder is illustrated in Figure 10-5.

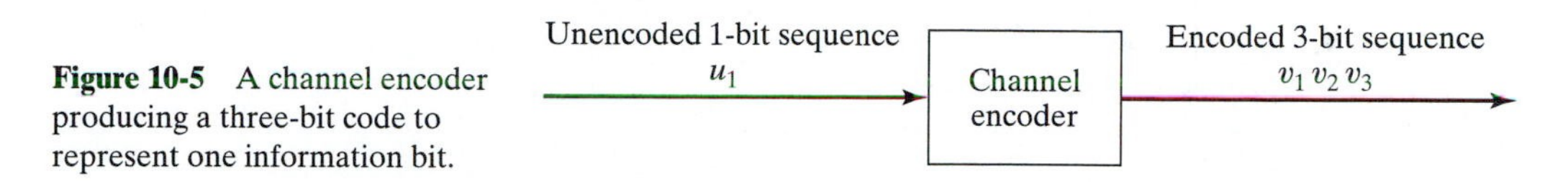

Figure 10-5 A channel encoder producing a three-bit code to represent one information bit.

Let's consider the information bit u_1 to be a point in a one-dimensional vector space, as shown in Figure 10-6a. There are only two points in this one-dimensional space that represent possible values for the information bit: 0 and 1. These two points are represented using dots in Figure 10-6a. Now let's represent the encoded three-bit sequence v_1, v_2, v_3 as a point in three-dimensional vector space, as shown in Figure 10-6b. Note that there are only two points in this three-dimensional space that represent possible values for the encoded three-bit sequence: (0,0,0) and (1,1,1). In examining Figures 10-6a and 10-6b, we can say that the encoding process maps a point from one-dimensional vector space (u_1) into a point in three-dimensional vector space (v_1, v_2, v_3).

Using this same geometric concept, we can also represent the received three-bit sequence as a point in three-dimensional vector space ($\hat{v}_1$, $\hat{v}_2$, $\hat{v}_3$). As shown in Figure 10-7, there are eight points in the three-dimensional vector space that represent possible received bit sequences. Two of these eight points, shown in black, represent valid codewords (i.e., they signify that no error has occurred in transmission). The other six points, shown in gray, represent errors in the received bit sequence.

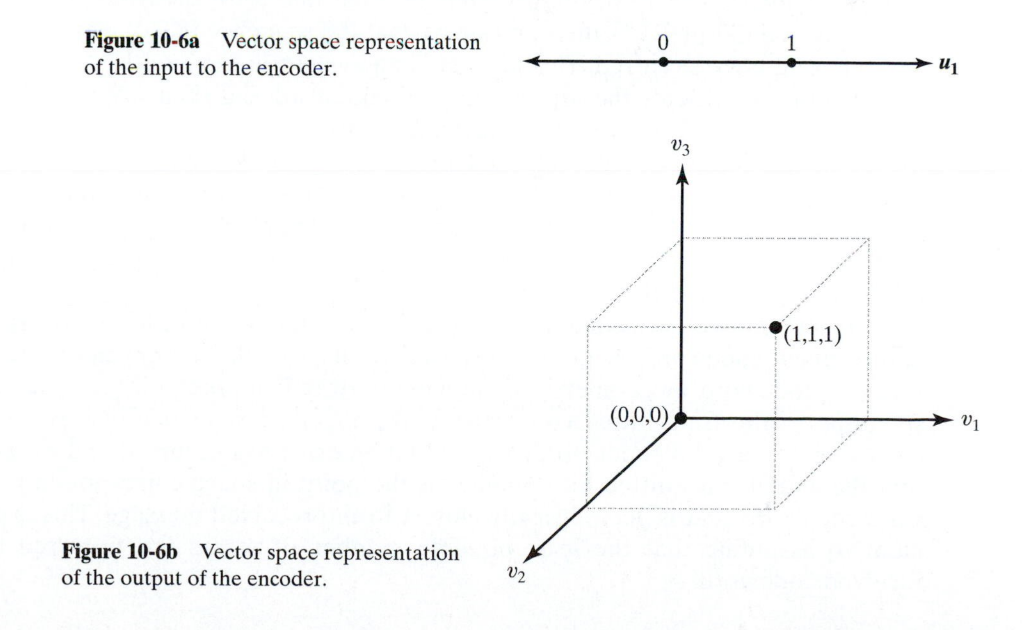

Figure 10-6a Vector space representation of the input to the encoder.

Figure 10-6b Vector space representation of the output of the encoder.

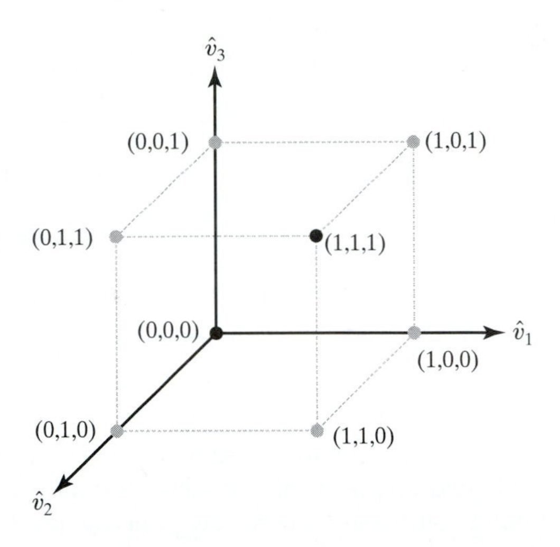

Figure 10-7 Vector space representation of the received bit sequence.

As described in Example 10.2, with majority voting we know that the received bit sequence is a valid codeword if all three received copies of the information bits are the same; in other words, if we receive the sequence (0,0,0) or the sequence (1,1,1). These two sequences correspond to the two black points in Figure 10-7. If we receive (1,1,1), we assume that the information bit from the source is 1, and if we receive (0,0,0), we assume that the information bit from the source is 0.

Suppose, however, that we receive a sequence in which the three bits are not the same. There are six possibilities: (0,0,1), (0,1,0), (0,1,1), (1,0,0), (1,0,1) and (1,1,0). These six sequences represent invalid codewords, signifying that an error has occurred, and correspond to the six gray points in Figure 10-7. As described in Example 10.2, if we receive an invalid codeword, we examine the three bits: If two are "1"s, then we assume that the sequence (1,1,1) was transmitted and that one error occurred; if two of the three received bits are "0"s, then we assume that the sequence (0,0,0) was transmitted and that one error occurred. This process is equivalent in Figure 10-7 to finding the gray dot that represents the invalid received codeword and then selecting the valid codeword (black dot) that is *geometrically closest* to it.

Examine Figure 10-7 carefully and make sure you understand this concept. As an example, consider the received bit sequence (0,1,0). This corresponds to the lowest left gray dot in Figure 10-7. Of the two valid codewords (the black dots), (0,0,0) is geometrically closer and hence we assume that (0,0,0) was transmitted, corresponding to an information bit 0 from the source.

To summarize our geometric interpretation so far, we consider the transmitted and received codewords as points in space. We observe that errors can corrupt the received codeword by geometrically moving it away from the point in space corresponding to the transmitted codeword. If the received sequence of bits does not correspond to a valid codeword, we note that an error has occurred and we assume that the actual transmitted bit sequence is the point in space corresponding to the valid codeword that is geometrically closest to the received message. This is equivalent to assuming that the least possible number of errors has occurred in the received codeword.

We can use our geometric interpretation to determine the error-detecting and error-correcting power of a code. In considering our majority voting example and in viewing Figure 10-7, we can observe the following:

1. All one-bit errors result in points corresponding to invalid codewords (gray dots), so all one-bit errors can be detected. Furthermore, we note that the point in Figure 10-7 corresponding to any one-bit error is closer to the correct valid codeword than it is to any incorrect valid codeword—for example, the point corresponding to (1,1,0) is geometrically closer to (1,1,1) than it is to (0,0,0)—so all one-bit errors can be successfully corrected.
2. All two-bit errors also result in points corresponding to invalid codewords (gray dots), so all two-bit errors can be detected. Note, however, that the point corresponding to a two-bit error is closer to an *incorrect* valid codeword than to the correct valid codeword. Two-bit errors can thus be detected but cannot be corrected. For example, if (1,1,1) is transmitted and the first two bits are received in error, then (0,0,1) is received. Since (0,0,1) is closer to (0,0,0) than it is to (1,1,1), if we try to correct the errors we will end up assuming that (0,0,0) was transmitted, which is incorrect.
3. All three-bit errors result in received sequences representing valid codewords, so three-bit errors cannot even be detected.

There is nothing in the work we have done so far that limits our geometric interpretation to a particular type of error-detecting or error-correcting code or to a particular block size. Generalizing our geometric interpretation, we can say that for an (n,k) block code, 2^k points are mapped from k-dimensional vector space into n-dimensional vector space. Note that there is a total of 2^n points in the n-dimensional vector space that can represent received bit sequences; the $2^n - 2^k$ unmapped points represent errors (in our majority voting case, these are the six gray dots in Figure 10-7). The error-detecting power of a code is limited to errors that move the received bit sequence away from a valid codeword by a geometric distance shorter than the distance to the closest incorrect valid codeword (in our majority voting case, a distance shorter than that produced by a three-bit error). The error-correcting power of the code is limited to errors that move the received message away from the correct valid codeword less than half the distance towards the closest incorrect valid codeword. Error-detecting power is generally greater than error-correcting power. You can use the power in a code either to detect the greater number of errors or to detect and correct the lesser number of errors, *but you cannot use it to do both*. For instance, in our majority voting example, we can either detect up to two errors per block or detect and correct up to one error per block, but we cannot do both.

Consider a code with 2^k valid codewords. Suppose we examine the distances between each pair of valid codewords and let $distance_{\text{min}}$ be the minimum distance between any two valid codewords in the entire code (i.e., $distance_{\text{min}}$ is the distance between the two codewords that are closest together). Our code can detect all errors that cause a received codeword to deviate from a valid codeword by a distance less than $distance_{\text{min}}$, and our code can detect and correct all errors that cause a received codeword to deviate from a valid codeword by a distance less than $distance_{\min}/2$. Again, remember that we can use the power in a code either to detect the greater number of errors or to detect and correct the lesser number of errors, but we cannot use it to do both.

Our geometric interpretation is good for both analysis and design of error-detecting and error-correcting codes. From a design standpoint, we want to develop a code that places the points corresponding to valid codewords as geometrically far apart as possible. From an analysis standpoint, once we know where all the valid codewords are located within the n-dimensional space, we can determine the code's error-detection and error-correction power by computing the distances between the valid points.

Calculating the distance between any two points in three-dimensional space—say, (x_1, y_1, z_1) and (x_2, y_2, z_2)— is straightforward; we just take the Pythagorean sum

$$\text{distance}_{R3} = \sqrt{(x_2 - x_1)^2 + (y_2 - y_1)^2 + (z_2 - z_1)^2} \tag{10.30}$$

This type of distance, based on Euclidean (planar) geometry, is known as the *Euclidean distance*. We can extend this concept to higher-dimensional space, defining the Euclidean distance between two points $\boldsymbol{A}$ and $\boldsymbol{B}$ in n-dimensional space, $\boldsymbol{A} = (a_1, a_2, a_3, \ldots a_n)$ and $\boldsymbol{B} = (b_1, b_2, b_3, \ldots b_n)$, as

$$\text{distance}_{Rn} = \sqrt{(b_1 - a_1)^2 + (b_2 - a_2)^2 + (b_3 - a_3)^2 + \cdots + (b_n - a_n)^2} \tag{10.31}$$

Since we are using $GF(2)$, each coordinate of each point is either 0 or 1, which further simplifies calculation of Euclidean distance. Consider our two points $(a_1, a_2, a_3, \ldots a_n)$ and $(b_1, b_2, b_3, \ldots b_n)$, where each a coordinate and each b coordinate is either 0 or 1. If $a_i = b_i$, then $(b_i - a_i)^2 = 0$, whereas if $a_i \neq b_i$, then $(b_i - a_i)^2 = 1$. The Euclidean distance between the two points $\boldsymbol{A}$ and $\boldsymbol{B}$ is just

$$\text{distance}_{Rn} = \sqrt{d} \tag{10.32}$$

where d is the number of dimensions in which the a_i and b_i coordinates disagree (i.e., the coordinate for one of the points is 1 and the same coordinate for the other point is 0). Since the points $\boldsymbol{A} = (a_1, a_2, a_3, \ldots a_n)$ and $\boldsymbol{B} = (b_1, b_2, b_3, \ldots b_n)$ represent codewords, we can also define d as the number of bits in which two codewords differ. We call d the *Hamming distance*.

Example 10.8 Determining the distance between two codewords

Find the Euclidean and Hamming distances between the two codewords 100110 and 001100.

Solution

The two codewords differ in their first, third, and fifth bits, so the Hamming distance between the two codewords is 3 and the Euclidean distance is $\sqrt{3}$.

Example 10.9 Determining the distance between codewords in a majority voting code

Find the Hamming and Euclidean distances between the two valid codewords in the majority voting code illustrated in Figure 10 6b.

Solution

The two codewords are 000 and 111. They differ in all three bits, so the Hamming distance between the two codewords is 3 and Euclidean distance is $\sqrt{3}$.

We now know that the error-detecting and error-correcting power of a code depends on $distance_{min}$, which we defined as the Euclidean distance between the two closest valid codewords in the entire code. We also know that the Euclidean distance between any two codewords is

$$distance_{Rn} = \sqrt{d} \tag{10.32R}$$

where d represents the number of bits in which the two codewords differ (the Hamming distance). We therefore know that

$$distance_{min} = \sqrt{d_{min}} \tag{10.33}$$

where d_{min} is the least number of bits in which two valid codewords differ. Equation (10.33) tells us that we can find the two closest valid codewords in a code by examining the entire set of valid codewords and then selecting the two that differ in the fewest bits. We call d_{min} the *minimum Hamming distance.*

We can also use d_{min} instead of the minimum Euclidean distance to directly determine the error-detecting and error-correcting power of the code, as illustrated in the following two theorems.

Theorem 10.1

An error control code can detect all errors of $d_{min} - 1$ bits or less.

Proof Consider a received codeword with n errors. The point representing the received codeword is a Euclidean distance of $\sqrt{n}$ from the correct valid codeword. We can guarantee that the code can detect the errors as long as

$$\sqrt{n} < distance_{min} \tag{10.34}$$

Combining Equations (10.33) and (10.34), we can guarantee that the code can detect the errors as long as

$$\sqrt{n} < \sqrt{d_{min}} \tag{10.35}$$

Since both n and d_{min} are positive, we can square both sides and retain the inequality, meaning that the code can detect the errors as long as

$$n < d_{min} \tag{10.36}$$

Since n and d_{min} are integers, Equation (10.36) is equivalent to stating that the code can guarantee detection of any $d_{min} - 1$ or fewer errors.

Theorem 10.2

An error control code can detect and correct all errors of $\left\lfloor \frac{d_{min} - 1}{2} \right\rfloor$ bits or less, where the bracketing $\lfloor x \rfloor$ means "the greatest integer less than or equal to x."

Proof Consider a received codeword with n errors. The point representing the received codeword is a Hamming distance of n from the correct valid codeword. We can guarantee that the code can detect and correct the errors as long as

$$n < \frac{d_{\min}}{2} \tag{10.37}$$

This is true because as long as Equation (10.37) is satisfied, the received codeword is closer to the correct valid codeword than it is to any other valid codeword. Since n and $d_{\min}$ are integers, Equation (10.37) can be restated as

$$t = \left\lfloor \frac{d_{\min} - 1}{2} \right\rfloor \tag{10.38}$$

where the code can guarantee detection and correction of all errors less than or equal to t bits.

Theorems 10.1 and 10.2 show us that error-detection and error-correction power can be easily measured once we know $d_{\min}$, the number of bits in which the two closest valid codewords differ.

Example 10.10 Determining error-detecting and error-correcting power of a code

Determine the error-detecting power and the error-correcting power of the following (5,2) code:

Data Bits	Codeword
00	00000
01	01101
10	11010
11	10111

Solution

We can determine the error-detecting and error-correcting power of the code if we can find $d_{\min}$, the minimum Hamming distance between two valid codewords. The Hamming distances between each pair of valid codewords are as follows:

Pair of Codewords	Hamming Distance
00000 and 01101	3
00000 and 11010	3
00000 and 10111	4
01101 and 11010	4
01101 and 10111	3
11010 and 10111	3

We see that $d_{\min} = 3$, so the code can detect up to two errors *or* detect and correct one error.

The concept of minimum Hamming distance is very useful for understanding the error-detecting and error-correcting power of a code, but finding d_{min} can be very time-consuming, especially if the code has a large number of valid codewords. Consider, for instance, our (7,4) Hamming code. This code has 16 valid codewords, which means that there are 120 different possible pairs of codewords. Calculating the Hamming distance between each pair of codewords will take quite a while. Let's see if we can determine an easier way to find d_{min}. To do this, we'll first need to define three terms: *weight, minimum weight*, and *linear block code*.

The *weight* of a codeword, represented by w, is the number of "1"s in that codeword. Weight is easy to determine by inspection; for instance, the weight of the fourth codeword in Example 10.10, 10111, is four. The *minimum weight of a code*, w_{min}, is defined as the minimum weight of any valid codeword in the code, other than the codeword represented by all "0"s. The minimum weight of the code in Example 10.10 is three.

A block code is called a *linear block code* if the modulo-2 sum of any two valid codewords is also a valid codeword. This additional restriction provides more structure, and therefore more error-detecting and error-correcting power for the code (the restriction ensures that all of the 2^k valid codewords lie within a k-dimensional subspace of the n-dimensional space). Most practical block codes, including Hamming codes, are linear block codes. Note that the codeword represented by all "0"s will always be a valid codeword in a linear block code, since any codeword modulo-2 added to itself will produce all "0"s.

We are now ready to determine a very simple way to find d_{min}.

Theorem 10.3

For any linear block code, $d_{\text{min}} = w_{\text{min}}$.

Proof Consider a linear block code with codewords $\mathbf{C}_1$, $\mathbf{C}_2$, $\mathbf{C}_3$, etc., and let Φ represent the all-zeroes codeword. The code is a linear block code, so for any two codewords $\mathbf{C}_i$ and $\mathbf{C}_j$, $\mathbf{C}_i \oplus \mathbf{C}_j$ is also a valid codeword. Let the notation $d(\mathbf{C}_i, \mathbf{C}_j)$ represent the distance between $\mathbf{C}_i$ and $\mathbf{C}_j$, and let the notation $w(\mathbf{C}_i)$ represent the weight of $\mathbf{C}_i$. Note that

$$w(\mathbf{C}_i) = d(\mathbf{C}_i, \Phi) \tag{10.39}$$

and

$$d(\mathbf{C}_i, \mathbf{C}_j) = d(\mathbf{C}_i \oplus \mathbf{C}_j, \mathbf{C}_j \oplus \mathbf{C}_j) = d(\mathbf{C}_i \oplus \mathbf{C}_j, \Phi) = w(\mathbf{C}_i \oplus \mathbf{C}_j) \tag{10.40}$$

Now suppose that the two codewords with the minimum Hamming distance are $\mathbf{C}_y$ and $\mathbf{C}_z$.

$$d_{\text{min}} = d(\mathbf{C}_y, \mathbf{C}_z) = w(\mathbf{C}_y \oplus \mathbf{C}_z) \tag{10.41}$$

Since the code is a linear block code, $\mathbf{C}_y \oplus \mathbf{C}_z$ is a valid codeword. So the codeword $\mathbf{C}_y \oplus \mathbf{C}_z$ has a weight equal to d_{min}. Can there be another nonzero codeword—say, $\mathbf{C}_x$—with a weight less than $w(\mathbf{C}_y \oplus \mathbf{C}_z)$? Well,

$$w(\mathbf{C}_x) = d(\mathbf{C}_x, \Phi) \tag{10.42}$$

which, by definition, must be greater than or equal to d_{min}.

So

$$w(\boldsymbol{C}_x) \geq d_{\text{min}} \tag{10.43}$$

But from Equation (10.41), $d_{\text{min}} = w(\boldsymbol{C}_y \oplus \boldsymbol{C}_z)$, so

$$w(\boldsymbol{C}_x) \geq w(\boldsymbol{C}_y \oplus \boldsymbol{C}_z) \tag{10.44}$$

Equation (10.44) tells us that there are no nonzero codewords with a weight smaller than the weight of $\boldsymbol{C}_y \oplus \boldsymbol{C}_z$, so from Equation (10.41),

$$d_{\text{min}} = w(\boldsymbol{C}_y \oplus \boldsymbol{C}_z) = w_{\text{min}} \tag{10.45}$$

Theorem 10.3 gives us a very simple way to determine the error-detecting and error-correcting power of a linear block code. We just examine all the valid codewords, determine which nonzero codeword has the minimum number of “1”s, and count the number of “1”s in that codeword. This value, the minimum weight of the code, is equal to the minimum Hamming distance d_{min}.

Example 10.11 Error-detecting and error-correcting power of a (7,4) Hamming code

Hamming codes are linear block codes. Verify that the (7,4) Hamming code described by

$$[\boldsymbol{G}] = \begin{bmatrix} 1 & 1 & 0 & 1 & 0 & 0 & 0 \\ 0 & 1 & 1 & 0 & 1 & 0 & 0 \\ 1 & 1 & 1 & 0 & 0 & 1 & 0 \\ 1 & 0 & 1 & 0 & 0 & 0 & 1 \end{bmatrix}$$

has the capability to correct all single-bit errors or to detect all two-bit errors within a block.

Solution

As with Example 10.10, we can determine the error-detecting and error-correcting power of the code if we can find d_{min}, the minimum Hamming distance between two valid codewords. We do not want to calculate the Hamming distances between each pair of codewords; as mentioned earlier, the 16 valid codewords can be grouped into 120 different pairs. Instead, let's find the weight of the code. The 16 valid seven-bit codewords are generated as follows:

$$[0\ 0\ 0\ 0]\bullet[\boldsymbol{G}] = [0\ 0\ 0\ 0\ 0\ 0\ 0]$$
$$[0\ 0\ 0\ 1]\bullet[\boldsymbol{G}] = [1\ 0\ 1\ 0\ 0\ 0\ 1]$$
$$[0\ 0\ 1\ 0]\bullet[\boldsymbol{G}] = [1\ 1\ 1\ 0\ 0\ 1\ 0]$$
$$[0\ 0\ 1\ 1]\bullet[\boldsymbol{G}] = [0\ 1\ 0\ 0\ 0\ 1\ 1]$$
$$[0\ 1\ 0\ 0]\bullet[\boldsymbol{G}] = [0\ 1\ 1\ 0\ 1\ 0\ 0]$$
$$[0\ 1\ 0\ 1]\bullet[\boldsymbol{G}] = [1\ 1\ 0\ 0\ 1\ 0\ 1]$$
$$[0\ 1\ 1\ 0]\bullet[\boldsymbol{G}] = [1\ 0\ 0\ 0\ 1\ 1\ 0]$$
$$[0\ 1\ 1\ 1]\bullet[\boldsymbol{G}] = [0\ 0\ 1\ 0\ 1\ 1\ 1]$$
$$[1\ 0\ 0\ 0]\bullet[\boldsymbol{G}] = [1\ 1\ 0\ 1\ 0\ 0\ 0]$$

$$[1\ 0\ 0\ 1]\bullet[G] = [0\ 1\ 1\ 1\ 0\ 0\ 1]$$
$$[1\ 0\ 1\ 0]\bullet[G] = [0\ 0\ 1\ 1\ 0\ 1\ 0]$$
$$[1\ 0\ 1\ 1]\bullet[G] = [1\ 0\ 0\ 1\ 0\ 1\ 1]$$
$$[1\ 1\ 0\ 0]\bullet[G] = [1\ 0\ 1\ 1\ 1\ 0\ 0]$$
$$[1\ 1\ 0\ 1]\bullet[G] = [0\ 0\ 0\ 1\ 1\ 0\ 1]$$
$$[1\ 1\ 1\ 0]\bullet[G] = [0\ 1\ 0\ 1\ 1\ 1\ 0]$$
$$[1\ 1\ 1\ 1]\bullet[G] = [1\ 1\ 1\ 1\ 1\ 1\ 1]$$

In examining the 16 codewords, we see that the minimum weight of any of the nonzero codewords is three. The code is a linear block code, $d_{\min} = w_{\min} = 3$, so the code can detect up to two errors per block or correct one error per block.

10.5 Cyclic Codes

In the previous section we developed Hamming codes, which have a minimum distance of three. These codes are single-error correcting *or* double-error detecting. Many applications, however, require more powerful error detection or error correction due to limited transmitter power or due to a particularly noisy channel. Also, in many applications such as cellular telephones and CDs, multiple errors may occur within a small group of consecutive bits. Such errors may be caused by an external disturbance (for instance, lightning or a power surge), by fading (if radio transmitters and receivers are used), or by dust particles or other imperfections in recording media such as floppy discs or CDs. Because these errors occur in clusters or bursts, they are called *burst errors*. The length of the burst is the number of bits between the first and last errors, inclusive.

Additional error-detecting or error-correcting power can be achieved by breaking up the data bits into smaller and smaller blocks and using a Hamming code on each block, but this approach requires a relatively large number of parity bits. Let's now develop another, more powerful, family of linear block codes called *cyclic codes*. These codes produce additional error-detecting and error-correcting power with low or moderate overhead by introducing additional mathematical structure.

An (n,k) linear block code is called a *cyclic code* if every cyclic shift of a codeword produces another codeword. Consider, for example, a codeword

$$[\mathbf{v}] = [v_1\ v_2\ v_3\ v_4\ \ldots\ v_{n-1}\ v_n]$$

Let's define a *cyclic shift of one place to the right* as

$$[\mathbf{v}]^{(1)} = [v_n\ v_1\ v_2\ v_3\ v_4\ \ldots\ v_{n-1}]$$

and in general, a *cyclic shift of* i *places to the right* as

$$[\mathbf{v}]^{(i)} = [v_{(n-i)+1}\ v_{(n-i)+2}\ v_{(n-i)+3}\ \ldots\ v_n\ v_1\ v_2\ \ldots\ v_{n-i}]$$

We will soon see how this additional structure provides additional error control power.

So far, we've developed two different mathematical representations for a code: a matrix representation and a geometric representation. We've seen that each of these representations is useful in analyzing and designing codes. Let's now develop a third representation, a *polynomial representation*. The polynomial representation, when used with polynomial mathematics, will be a good conceptual tool for creating, manipulating, and analyzing cyclic codes. Consider a codeword $v_1\ v_2\ v_3\ v_4 \ldots v_{n-1}\ v_n$. We can represent this codeword as a series of bits, as a $1 \times n$ vector (matrix representation), or as a point $\boldsymbol{V}$ in n-dimensional space (geometric representation). We can also represent the codeword as a polynomial $\boldsymbol{V}(x)$, written as

$$\boldsymbol{V}(x) = v_1 + v_2x + v_3x^2 + v_4x^3 + \cdots + v_nx^{n-1} \tag{10.46}$$

where x is just a dummy variable and $v_1, v_2, v_3, \ldots v_n$ are coefficients whose values are either 1 or 0. The highest value of i that has a nonzero v_i coefficient determines the *degree* of the polynomial, which will be $i - 1$.

We can generate codes and manipulate codewords using polynomial algebra (remembering to use $\oplus$ and $\bullet$ instead of conventional addition and multiplication, respectively). To generate a cyclic code, we develop an appropriate generator polynomial $\boldsymbol{G}(x)$ and then represent the encoding process as a product of an unencoded block of information multiplied by $\boldsymbol{G}(x)$. For example, the generator polynomial $\boldsymbol{G}(x) = 1 + x + x^3$ can be used to generate a (7,4) cyclic code. If four bits of unencoded information u_1, u_2, u_3, u_4 are represented in polynomial form as $\boldsymbol{U}(x) = u_1 + u_2x + u_3x^2 + u_4x^3$, a polynomial $\boldsymbol{V}(x)$ representing the seven-bit encoded block can be generated as

$$\boldsymbol{V}(x) = \boldsymbol{U}(x)\bullet\boldsymbol{G}(x) \tag{10.47}$$

Polynomial multiplication in $GF(2)$ is the same as conventional polynomial multiplication except that we use modulo-2 multiplication and addition to determine the values of the coefficients instead of conventional multiplication and addition. The process of using a generator polynomial is shown in Example 10.12.

Example 10.12 Using a generator polynomial

Use the generator polynomial $\boldsymbol{G}(x) = 1 + x + x^3$ to generate the seven-bit codeword corresponding to the unencoded four-bit block 1101.

Solution

The unencoded four-bit block 1101 can be represented in polynomial form as

$$\boldsymbol{U}(x) = 1 + 1x + 0x^2 + 1x^3$$

Multiplying by the generator polynomial yields

$$\begin{aligned}\boldsymbol{V}(x) &= \boldsymbol{U}(x)\bullet\boldsymbol{G}(x)\\ &= (1 + 1x + 0x^2 + 1x^3)\bullet(1 + x + x^3)\\ &= 1 + x + x^3 + x + x^2 + x^4 + 0x^2 + 0x^3 + 0x^5 + x^3 + x^4 + x^6\end{aligned}$$

Combining terms of the same order (and remembering to use modulo-2 addition),

$$\begin{aligned} V(x) &= 1 + x + x^3 + x + x^2 + x^4 + 0x^2 + 0x^3 + 0x^5 + x^3 + x^4 + x^6 \\ &= 1 + (1 \oplus 1)x + (1 \oplus 0)x^2 + (1 \oplus 0 \oplus 1)x^3 + (1 \oplus 1)x^4 + 0x^5 + x^6 \\ &= 1 + 0x + 1x^2 + 0x^3 + 0x^4 + 0x^5 + x^6 \end{aligned}$$

$V(x)$ corresponds to the encoded bit sequence 1010001.

Table 10-3, in Section 10.5.2, lists some typical generator polynomials. The theory behind the development of appropriate generator polynomials is quite interesting, but it is outside the scope of this book. If you're interested in finding out more, see Lin and Costello [10.3], Wicker [10.4], or another book devoted exclusively to error control coding.

We've seen how polynomial mathematics can be used to generate codewords. Polynomial mathematics is also useful for manipulating codewords. For example, consider a codeword represented by the polynomial $V(x)$. Let n be the number of bits in the codeword. A cyclic shift of i places to the right can be accomplished algebraically by first modulo-2 multiplying $V(x)$ by x^i, then dividing the result by $x^n + 1$, and then discarding the quotient and keeping the remainder. This process requires two practices that are not yet familiar to us: polynomial division using modulo-2 mathematics, and emphasizing the remainder rather than the quotient. Polynomial division using modulo-2 mathematics is similar to conventional polynomial division, except that all coefficients are either 0 or 1 and we must use modulo-2 subtraction. This is illustrated in Example 10.13. Be sure you understand modulo-2 subtraction and each step in the example.

Example 10.13 Polynomial manipulation to produce a cyclic shift

Use polynomial manipulation to cyclically shift the codeword 1001011 two places to the right.

Solution

First we express the codeword 1001011 as a polynomial

$$1001011 \Rightarrow 1 + 0x + 0x^2 + 1x^3 + 0x^4 + 1x^5 + 1x^6 = x^6 + x^5 + x^3 + 1$$

This last step lists the polynomial with the highest order of x first and with the zero coefficients eliminated. Our next step is to multiply the polynomial by x^2

$$x^2(x^6 + x^5 + x^3 + 1) = x^8 + x^7 + x^5 + x^2$$

Finally, we need to divide the resulting polynomial by $x^7 + 1$. As mentioned earlier, polynomial division using modulo-2 mathematics is the same as conventional polynomial division except that all coefficients are either 0 or 1 and we must employ modulo-2 subtraction. Let's start our division. Note that the zero coefficients have been inserted as "place holders."

$$x^7 + 1 \overline{\smash{)}\, x^8 + x^7 + 0x^6 + x^5 + 0x^4 + 0x^3 + x^2 + 0x + 0}$$

We know that multiplying the divisor $x^7 + 1$ by x produces a polynomial of the same order as the dividend $x^8 + x^7 + x^5 + x^2$, while multiplying the divisor by x^2 produces a polynomial of greater order than the dividend. Thus, x is the highest order in the quotient.

$$\begin{array}{r l} & x \\ x^7 + 1 & \overline{\smash{)}\, x^8 + x^7 + 0x^6 + x^5 + 0x^4 + 0x^3 + x^2 + 0x + 0} \end{array}$$

Multiplying the divisor by x yields $x^8 + x$, which we must modulo-2 subtract from the dividend.

$$\begin{array}{r l} & x \\ x^7 + 1 & \overline{\smash{)}\, x^8 + x^7 + 0x^6 + x^5 + 0x^4 + 0x^3 + x^2 + 0x + 0} \\ & \underline{x^8 + \qquad\qquad\qquad\qquad\qquad\qquad\qquad\qquad x \qquad} \\ & ???? \end{array}$$

How do we perform modulo-2 subtraction? Going back to the basics, we know that given any two variables A and B, subtraction is defined as

$$A - B = A + \overline{B} \tag{10.48}$$

where $\overline{B}$ is the additive inverse of B (i.e., $B + \overline{B} = 0$). Carrying this convention to modulo-2 mathematics and $GF(2)$,

$$(A - B)_{mod\,2} = (A + \overline{B})_{mod\,2} = A \oplus \overline{B} \tag{10.49}$$

Furthermore, we established in Section 10.2.1 that for modulo-2 addition and $GF(2)$, the additive inverse of any element is itself

$$\overline{B} = B \tag{10.7R}$$

Therefore,

$$(A - B)_{mod\,2} = A \oplus \overline{B} = A \oplus B \tag{10.50}$$

So, in $GF(2)$, modulo-2 subtraction is the same as modulo-2 addition. Performing this operation on our polynomial,

$$\begin{array}{r l} & x \\ x^7 + 1 & \overline{\smash{)}\, x^8 + x^7 + 0x^6 + x^5 + 0x^4 + 0x^3 + x^2 + 0x + 0} \\ & \underline{x^8 + \qquad\qquad\qquad\qquad\qquad\qquad\qquad\qquad x \qquad} \\ & 0x^8 + x^7 + 0x^6 + x^5 + 0x^4 + 0x^3 + x^2 + x + 0 \end{array}$$

We know that multiplying the divisor $x^7 + 1$ by 1 produces a polynomial of the same order as $x^7 + x^5 + x^2 + x$, while multiplying the divisor by x produces a polynomial of greater order than $x^7 + x^5 + x^2 + x$. Therefore, 1 is the next highest order in the quotient.

$$\begin{array}{r l} & x + 1 \\ x^7 + 1 & \overline{\smash{)}\, x^8 + x^7 + 0x^6 + x^5 + 0x^4 + 0x^3 + x^2 + 0x + 0} \\ & \underline{x^8 + \qquad\qquad\qquad\qquad\qquad\qquad\qquad\qquad x \qquad} \\ & 0x^8 + x^7 + 0x^6 + x^5 + 0x^4 + 0x^3 + x^2 + x + 0 \\ & \qquad\; \underline{x^7 + \qquad\qquad\qquad\qquad\qquad\qquad\qquad 1} \\ & \qquad\qquad\quad 0x^6 + x^5 + 0x^4 + 0x^3 + x^2 + x + 1 \end{array}$$

The resultant polynomial, $0x^6 + x^5 + 0x^4 + 0x^3 + x^2 + x + 1$, is the remainder because its order is lower than the order of the divisor $x^7 + 1$. This remainder corresponds to the codeword 1110010 (remember to read the polynomial's lowest order first), which is the original codeword 1001011 cyclically shifted two places to the right. The practice of performing division and then focusing on the remainder instead of the quotient is something new (kind of like peeling a banana, then throwing away the fruit and eating the skin). We will be using the remainder quite often in the manipulation of cyclic codes.

Cyclic codes are both useful and powerful. They are easy to generate and manipulate using hardware or software, since the $\oplus$ and $\bullet$ operations in $GF(2)$ can be performed by conventional logic gates or Boolean algebra commands. The power in a cyclic code lies in the extra structure and algebraic properties imposed by requiring all cyclic shifts of a codeword to also be codewords. Let's now investigate two types of cyclic codes: *cyclic redundancy check* (CRC) codes, which are useful for error detection (ARQ applications), and *Bose Chaudhuri Hocquenghem* (BCH) codes, which are useful for error detection and error correction.

10.5.1 Cyclic Redundancy Check Codes

Cyclic redundancy check (CRC) codes have very powerful error-detection capability and require relatively few redundant bits (Peterson and Brown [10.5]). The code is generated as follows:

> Consider a block of k unencoded information bits. This block is converted to polynomial form, a generator polynomial of degree j is selected, and the polynomial representing the unencoded data is then divided by the generator polynomial. This division produces a j-bit remainder. The remainder is converted from polynomial form back into a string of j bits, which we call the *CRC bits*. The CRC bits are then appended to the end of the block of unencoded information. We now have an encoded block of $k + j$ bits, with the first k bits representing the original information and the last j bits representing the cyclic redundancy check. This encoded block is then transmitted.

The receiver checks for errors in the received block of $k + j$ bits by taking the first k received bits and calculating the expected CRC bits. The expected CRC bits are then compared to the received CRC bits. If the two are equal, then it is assumed that no error occurred. If the two are not equal, then one or more errors have occurred.

The process of using a CRC code is illustrated in Example 10.14.

Example 10.14 Encoding using CRC

Consider the eight-bit block of information 11010011. Using the generator polynomial $x^3 + 1$,

a. Determine the three-bit cyclic redundancy check and the 11-bit transmitted block.

b. Explain the process used at the receiver to determine if an error occurred.

Solution

a. To generate the cyclic redundancy check code,

Step 1: Convert the block of information into polynomial form.

$$11010011 \Rightarrow 1 + 1x + 0x^2 + 1x^3 + 0x^4 + 0x^5 + 1x^6 + 1x^7$$
$$= x^7 + x^6 + x^3 + x + 1$$

Step 2: Divide the information polynomial by the CRC generator polynomial.

$$
\begin{array}{r|l}
 & x^4 + x^3 + x \\
\hline
x^3 + 1 & x^7 + x^6 + 0x^5 + 0x^4 + x^3 + 0x^2 + x + 1 \\
 & x^7 + x^4 \\
\hline
 & x^6 + 0x^5 + x^4 + x^3 + 0x^2 + x + 1 \\
 & x^6 + x^3 \\
\hline
 & x^4 + 0x^3 + 0x^2 + x + 1 \\
 & x^4 + x \\
\hline
 & 0x^2 + 0x + 1
\end{array}
$$

Step 3: Convert the remainder from polynomial form into a string of bits. This string is the CRC bits. Note that the degree of the remainder will be one less than the degree of the generator polynomial.

$$0x^2 + 0x + 1 \Rightarrow 100$$

Step 4: Append the CRC bits to the end of the original bit stream and transmit the entire sequence

11010011100

b. At the receiver,

Step 5: Separate the received message into its information bits and its CRC bits.

Step 6: Repeat transmitter Steps 2–4 with the received information bits to get the expected CRC bits.

Step 7: If the expected CRC bits match the received CRC bits, then no errors occurred. Otherwise, error(s) occurred.

CRC codes, in particular those employing 16 or 32 CRC bits, are used in many communications protocols because they offer powerful error detection with low overhead. For example, facsimile (CCITT Group IV) uses a 16-bit CRC based on the generator polynomial $x^{16} + x^{12} + x^5 + 1$ (ITU [10.6]). This code requires 16 redundant bits and can detect all single-bit, two-bit, and three-bit errors, any odd number of errors, any burst of errors shorter than 17 bits, 99.92% of all 17-bit burst errors, and 99.85% of all bursts of errors longer than 17 bits (Peterson and Brown [10.5]). In many applications, a 16- or 32-bit CRC can be used to effectively detect errors in information blocks that are thousands of bits long.

10.5.2 Bose Chaudhuri Hocquenghem Codes

Bose Chaudhuri Hocquenghem (BCH) codes are a family of cyclic codes with powerful error-correction capability (Hocquenghem [10.7], Bose and Ray-Chaudhuri [10.8]). For any positive integer $m \geq 3$, there exists an (n,k) BCH code which can detect and correct up to t errors and which has the following parameters and properties:

1. The encoded words are length $n = 2^m - 1$.
2. The number of parity bits (i.e., overhead) is $n - k \leq mt$.
3. Minimum distance of the code $d_{\min} \geq 2t + 1$.

Table 10-3 shows generator polynomials $\boldsymbol{G}(x)$ for the possible combinations of n, k, and t resulting from $m = 3, 4$, and 5. Codewords are generated by converting a block of k information bits into polynomial form and then modulo-2 multiplying by $\boldsymbol{G}(x)$, as shown in Equation (10.47). The theory behind the development of these generator polynomials is quite interesting, but it is outside the scope of this book. To find out more about this topic, see Lin and Costello [10.3], Wicker [10.4], or another book devoted exclusively to error control coding. There are, of course, generator polynomials for combinations of n, k, and t resulting from larger values of m. Lin and Costello [10.3], for instance, provides a table showing generator polynomials for all combinations of n, k, and t resulting from values of $m \leq 10$.

Table 10-3 Generator Polynomials for BCH Codes with $m \leq 5$

n	k	t	Generator Polynomials $\boldsymbol{G}(x)$
7	4	1	$x^3 + x + 1$
15	11	1	$x^4 + x + 1$
15	7	2	$x^8 + x^7 + x^6 + x^4 + 1$
15	5	3	$x^{10} + x^8 + x^5 + x^4 + x^2 + x^1 + 1$
31	26	1	$x^5 + x^2 + 1$
31	21	2	$x^{10} + x^9 + x^8 + x^6 + x^5 + x^3 + 1$
31	16	3	$x^{15} + x^{11} + x^{10} + x^9 + x^8 + x^7 + x^5 + x^3 + x^2 + x + 1$
31	11	5	$x^{20} + x^{18} + x^{17} + x^{13} + x^{10} + x^9 + x^7 + x^6 + x^4 + x^2 + 1$
31	6	7	$x^{25} + x^{24} + x^{21} + x^{19} + x^{18} + x^{16} + x^{15} + x^{14} + x^{13} + x^{11} + x^9 + x^5 + x^2 + x^1 + 1$
	etc. . . .		

10.6 Hybrid FEC/ARQ Codes

We've seen that CRC codes provide very powerful error detection capability with few parity bits and are therefore suitable for most ARQ applications. Suppose that the CRC encodes a k-bit block into an n-bit block and that the probability of an error in the received n-bit block is P. The overhead associated with the CRC itself is relatively small (k/n is close to 1 since the CRC requires very few parity bits), but the overhead associated with retransmission may be much larger, especially if errors occur relatively frequently. For practical values of P, the average number of redundant bits needed to transmit a k-bit block is $(n - k) + Pn + P^2n + P^3n + \ldots \approx (n - k) + Pn$, so if we express overhead as the average number of parity bits needed to ensure that one bit of information is accurately received,

$$Overhead_{ARQ} \approx \frac{(n - k) + Pn}{k} \text{ redundant bits per bit of information} \quad (10.51)$$

Equation (10.51) and common sense tell us that ARQ may not be an efficient strategy if errors occur relatively often.

We know that FEC codes require more parity bits per block than ARQ codes, but FEC codes can correct errors without requiring retransmission. The overhead associated with an (n,k) FEC code is

$$Overhead_{FEC} = \frac{(n - k)}{k} = \frac{n}{k} - 1 = \frac{1}{r} - 1 \text{ redundant bits per bit of information} \quad (10.52)$$

where r represents the code rate, as defined in Equation (10.20). FEC is an efficient strategy if single-bit errors occur relatively often, but suppose we have an environment where single-bit errors occur relatively frequently and we also have occasional multiple-bit errors. Making the FEC code sufficiently powerful to handle the occasional multiple-bit errors will require a large number of parity bits (and hence a large overhead), but these parity bits will rarely be needed. Using a weaker FEC requires less overhead, but the weaker FEC will "correct" the multiple-bit errors with wrong information (the code will "correct" the erroneous block to a wrong but valid codeword). This is very undesirable, since the user will be given wrong information, but—because FEC is used—the user will be confident that the information is correct.

Let's consider a hybrid FEC/ARQ code for environments with relatively frequent single-bit errors and occasional multiple-bit errors. Suppose we employ a low-overhead FEC code capable of correcting the single-bit errors (for example, a (15,11) Hamming code), and suppose we then group a number of FEC blocks together and protect all of these blocks with one CRC code. Consider, for instance, 5200 information bits. We could subdivide these bits into 26-bit unencoded blocks, and encode each block using a (31,26) BCH code. We can then group 20 encoded blocks together (producing a string of 620 bits) and use a single, 16-bit CRC to protect all 620 encoded bits. To transmit the 5200 information bits, we will need to send 6360 encoded bits (ten 636-bit groups). Upon reception, any single-bit errors in each 31-bit block will be detected and corrected. If a rare multiple-bit error occurs within a 20-block group, the CRC will detect the error and request retransmission of that group.

10.7 Correcting Burst Errors

As we briefly mentioned in earlier sections, burst errors occur in many applications, such as cellular telephones and CDs. These errors may be caused by an external disturbance (for instance, lightning or a power surge), by fading (if radio transmitters and receivers are used), or by dust particles or other imperfections in recording media. CRCs are often used to detect burst errors, but, as we know, these codes do not have error-correction capabilities. Powerful BCH codes can be used to detect and correct burst errors, but if the bursts occur infrequently, then block size must be quite large or else the overhead required by the strong BCH code will be excessive. Hybrid FEC/ARQ codes will not be efficient, since the errors will almost always occur in bursts. Let's examine other techniques for combatting burst errors.

10.7.1 Interleaving

Burst error protection can be provided by *interleaving*—a method that spreads out the effect of a burst error among several blocks. Consider a stream of encoded data as shown in Figure 10-8a. The encoded blocks are labeled *a, b, c*, and *d*, and a (7,4) Hamming code is used to provide the capability to correct all single-bit errors within each block. Now consider a burst error, as shown in Figure 10-8b, that destroys three bits of the second block. This three-bit burst error cannot be corrected by the Hamming code. Note, however, that if the stream of encoded data is rearranged prior to transmission, as shown in Figure 10-8c, then the three-bit burst error causes single-bit errors in three of the blocks. These errors, as shown in Figure 10-8d, can all be corrected by the (7,4) Hamming code.

This technique of rearranging the order of transmitted bits to distribute the effects of a burst error is called *interleaving with degree* x, where *x* is the number of bits in the transmitted data stream between consecutive bits from the same block. Figure 10-8c shows interleaving with degree 4. The larger the degree of interleaving (often called *depth*), the longer the burst of errors that can be corrected. Note, however, that a larger degree of interleaving requires a larger buffer and also produces a larger delay before

a_1	a_2	a_3	a_4	a_5	a_6	a_7	b_1	b_2	b_3	b_4	b_5	b_6	b_7	c_1	c_2	c_3	c_4	c_5	c_6	c_7	d_1	d_2	d_3	d_4	d_5	d_6	d_7

Figure 10-8a Four blocks of encoded bits, not interleaved.

a_1	a_2	a_3	a_4	a_5	a_6	a_7	b_1	b_2	b_3	b_4	b_5	b_6	b_7	c_1	c_2	c_3	c_4	c_5	c_6	c_7	d_1	d_2	d_3	d_4	d_5	d_6	d_7

Figure 10-8b A three-bit burst error cannot be corrected.

a_1	b_1	c_1	d_1	a_2	b_2	c_2	d_2	a_3	b_3	c_3	d_3	a_4	b_4	c_4	d_4	a_5	b_5	c_5	d_5	a_6	b_6	c_6	d_6	a_7	b_7	c_7	d_7

Figure 10-8c Four blocks of encoded bits, interleaved.

a_1	b_1	c_1	d_1	a_2	b_2	c_2	d_2	a_3	b_3	c_3	d_3	a_4	b_4	c_4	d_4	a_5	b_5	c_5	d_5	a_6	b_6	c_6	d_6	a_7	b_7	c_7	d_7

Figure 10-8d A three-bit burst error can now be corrected.

a complete block of data is received. Some applications, such as telephone and videoconferencing, may not tolerate long delays.

10.7.2 Reed-Solomon Codes

Reed-Solomon codes are an extension of the BCH codes, developed for nonbinary Galois fields. As we established in Section 10.2, $GF(2)$ has two elements, 0 and 1, and two operators, modulo-2 addition and modulo-2 multiplication. Let's consider a Galois field that has 2^m elements (where m is an integer) and that has the operations modulo-m addition and modulo-m multiplication. Since this field has 2^m elements, each element can be used to represent m bits in the same way that we can use a single symbol in M-ary modulation techniques to represent multiple bits (see Section 4.5 and Section 5.8). We can develop generator polynomials using the 2^m elements as coefficients, and we can extend the properties of BCH coding to develop codes that can now correct up to t erroneous *symbols* in a block. Such codes are very useful for burst error correction, since bursts will destroy multiple bits within the same symbol. Detailed development of Reed-Solomon coding is outside the scope of this book. To find out more about this topic, see Lin and Costello [10.3], Wicker [10.4], or another book devoted exclusively to error control coding. Reed-Solomon codes are extensively used on CDs to correct for burst errors caused by scratches and fingerprints.

10.8 Convolutional Codes and Viterbi Decoding

Like block codes, convolutional codes divide the bit stream from the source into k-bit blocks. Each k-bit block is then encoded into an n-bit block, but, unlike block codes, the values of the n bits depend not only on the values of the k bits in the corresponding source block *but also on the values of bits in previous* k-*bit source blocks*. Convolutional codes are more powerful than block codes because they have an extra layer of redundancy (they exploit past history), but they are also more computationally complex.

10.8.1 Convolutional Encoding

Convolutional codes are described using three parameters: n, k, and m. As with block codes, k represents the number of information bits in each block and n represents the number of encoded bits per block. m represents the number of previous k-bit blocks that influence the encoding of the present block. As an example, consider a (3,2,1) convolutional code. This code has two information bits per block and three encoded bits per block, and the coding of the present block involves information bits from one previous block as well as from the present block.

To mathematically describe a convolutional code, we will need to develop some new notation. Consider the following:

Let $u_1^i, u_2^i, \ldots, u_k^i$ represent the k information bits in the ith block from the source. Therefore $u_1^{i-1}, u_2^{i-1}, \ldots, u_k^{i-1}$ represents the k information bits in the $i-1$st block from the source; $u_1^{i-2}, u_2^{i-2}, \ldots, u_k^{i-2}$ represents the k information bits in the $i-2$nd block from the source; etc.

Let $v_1^i, v_2^i, \ldots v_n^i$ represent the n encoded bits for the ith block.

Example 10.15 A convolutional code

Consider a (3,2,1) convolutional code with the following encoding process

$$v_1^i = u_1^{i-1} \oplus u_1^i$$
$$v_2^i = u_2^{i-1} \oplus u_2^i$$
$$v_3^i = u_1^i \oplus u_2^i$$

If the two previous bits output by the source are 01 and the two present bits output by the source are 10, determine the encoded bits corresponding to the present information.

Solution

$u_1^{i-1} = 0, u_2^{i-1} = 1, u_1^i = 1$, and $u_2^i = 0$, so

$$v_1^i = 0 \oplus 1 = 1$$
$$v_2^i = 1 \oplus 0 = 1$$
$$v_3^i = 1 \oplus 0 = 1$$

The encoded bits corresponding to the present information are 111.

As engineers, we've seen this type of process before, where the output of a digital circuit depends not only on the present input to the circuit but also on past input(s) to the circuit. In fact, we spent considerable time studying this process in our digital logic course when we analyzed sequential networks. Let's think of the encoder as a sequential network, as shown in Figure 10-9.[4] Many of the tools used to analyze sequential networks will also be useful for analyzing convolutional codes, most notably the state table and state diagram. We will review these tools briefly as we use them; for a more detailed review, you can refer to a digital logic book such as Roth [10.9] or Katz [10.10].

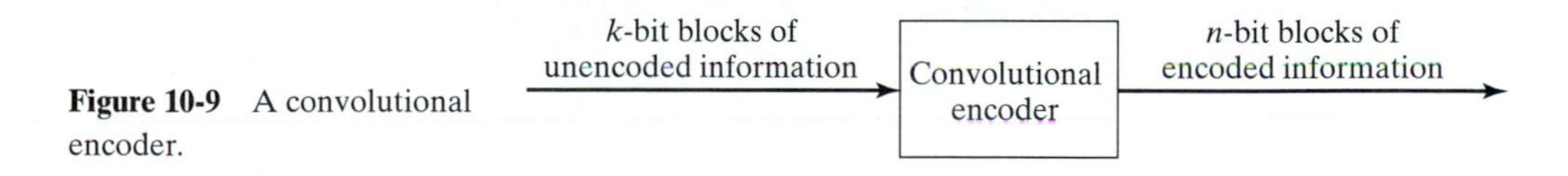

Figure 10-9 A convolutional encoder.

States are used to represent the history of previous input blocks that we must remember in order to correctly encode the present input block. For instance, the convolutional code in Example 10.15 requires that we remember one previous two-bit block. For this code we will need four states, one to represent the previous block being 00, one to represent the previous block being 01, a third to represent the previous block being 10, and a fourth to represent the previous block being 11. A *state table* lists all possible combinations of states and values of the bits in the present input block. For each combination, the table gives the codeword representing the encoded block.

[4]Convolutional encoders are often implemented using shift registers and other circuitry employed in sequential networks.

Example 10.16 State table for a convolutional code—Part 1

Develop a state table for the (3,2,1) convolutional code given in Example 10.15.

Solution

As just discussed, we need four states to represent each possible combination of the previous input block. Let's name the states as follows:

Let S_0 represent the state $u_1^{i-1} = 0, u_2^{i-1} = 0$ (i.e., the previous input block was 00).
Let S_1 represent the state $u_1^{i-1} = 0, u_2^{i-1} = 1$ (i.e., the previous input block was 01).
Let S_2 represent the state $u_1^{i-1} = 1, u_2^{i-1} = 0$ (i.e., the previous input block was 10).
Let S_3 represent the state $u_1^{i-1} = 1, u_2^{i-1} = 1$ (i.e., the previous input block was 11).

For each state there are four possible combinations of bits representing the present input block. We therefore need a table with 16 rows.

Table 10-4 fully describes the convolutional code given in Example 10.15. To better understand the table, let's examine the seventh row. In Example 10.15, we determined that if the two previous bits output by the source were 01 (which we now represent as S_1) and if the two present bits output by the source are 10, the encoded block is 111. This is shown in the seventh row of the table. We can determine the encoded block for each of the 15 other rows in the table in a similar way.

Table 10-4 State Table for Convolutional Code in Example 10.15

Previous State (Dependent on Previous Information)	Present Information Block u_1^i, u_2^i	3-Bit Encoded Block v_1^i, v_2^i, v_3^i
$S_0\ (u_1^{i-1}, u_2^{i-1} = 00)$	00	000
$S_0\ (u_1^{i-1}, u_2^{i-1} = 00)$	01	011
$S_0\ (u_1^{i-1}, u_2^{i-1} = 00)$	10	101
$S_0\ (u_1^{i-1}, u_2^{i-1} = 00)$	11	110
$S_1\ (u_1^{i-1}, u_2^{i-1} = 01)$	00	010
$S_1\ (u_1^{i-1}, u_2^{i-1} = 01)$	01	001
$S_1\ (u_1^{i-1}, u_2^{i-1} = 01)$	10	111
$S_1\ (u_1^{i-1}, u_2^{i-1} = 01)$	11	100
$S_2\ (u_1^{i-1}, u_2^{i-1} = 10)$	00	100
$S_2\ (u_1^{i-1}, u_2^{i-1} = 10)$	01	111
$S_2\ (u_1^{i-1}, u_2^{i-1} = 10)$	10	001
$S_2\ (u_1^{i-1}, u_2^{i-1} = 10)$	11	010
$S_3\ (u_1^{i-1}, u_2^{i-1} = 11)$	00	110
$S_3\ (u_1^{i-1}, u_2^{i-1} = 11)$	01	101
$S_3\ (u_1^{i-1}, u_2^{i-1} = 11)$	10	011
$S_3\ (u_1^{i-1}, u_2^{i-1} = 11)$	11	000

As we observed in Example 10.16, we use states in the first column of the state table to represent the history of previous information blocks. The encoder sits in a particular state, based on past inputs, waiting for the present information block from the source. As the encoder receives the present information block $(u_1^i, u_2^i, \ldots, u_k^i)$, the encoder outputs an encoded block $(v_1^i, v_2^i, \ldots v_n^i)$ and then transitions to a new state and awaits the next information block. Thinking of the encoder in this way, we can express the second column of the state table using the new state rather than the inputs. This is illustrated in Example 10.17.

Example 10.17 State table for a convolutional code—Part 2

Express the state table for the (3,2,1) convolutional code representing the present block of input information as a transition to a new state.

Solution

Table 10-5 provides the representation we need. The state table is now in the form we're used to seeing when we analyze sequential networks.

Table 10-5 State Table for Convolutional Code in Example 10.15, with Present Input Information Represented as a State

Previous State	Present State	3-Bit Encoded Block v_1^i, v_2^i, v_3^i
$S_0\ (u_1^{i-1}, u_2^{i-1} = 00)$	$S_0\ (u_1^i, u_2^i = 00)$	000
$S_0\ (u_1^{i-1}, u_2^{i-1} = 00)$	$S_1\ (u_1^i, u_2^i = 01)$	011
$S_0\ (u_1^{i-1}, u_2^{i-1} = 00)$	$S_2\ (u_1^i, u_2^i = 10)$	101
$S_0\ (u_1^{i-1}, u_2^{i-1} = 00)$	$S_3\ (u_1^i, u_2^i = 11)$	110
$S_1\ (u_1^{i-1}, u_2^{i-1} = 01)$	$S_0\ (u_1^i, u_2^i = 00)$	010
$S_1\ (u_1^{i-1}, u_2^{i-1} = 01)$	$S_1\ (u_1^i, u_2^i = 01)$	001
$S_1\ (u_1^{i-1}, u_2^{i-1} = 01)$	$S_2\ (u_1^i, u_2^i = 10)$	111
$S_1\ (u_1^{i-1}, u_2^{i-1} = 01)$	$S_3\ (u_1^i, u_2^i = 11)$	100
$S_2\ (u_1^{i-1}, u_2^{i-1} = 10)$	$S_0\ (u_1^i, u_2^i = 00)$	100
$S_2\ (u_1^{i-1}, u_2^{i-1} = 10)$	$S_1\ (u_1^i, u_2^i = 01)$	111
$S_2\ (u_1^{i-1}, u_2^{i-1} = 10)$	$S_2\ (u_1^i, u_2^i = 10)$	001
$S_2\ (u_1^{i-1}, u_2^{i-1} = 10)$	$S_3\ (u_1^i, u_2^i = 11)$	010
$S_3\ (u_1^{i-1}, u_2^{i-1} = 11)$	$S_0\ (u_1^i, u_2^i = 00)$	110
$S_3\ (u_1^{i-1}, u_2^{i-1} = 11)$	$S_1\ (u_1^i, u_2^i = 01)$	101
$S_3\ (u_1^{i-1}, u_2^{i-1} = 11)$	$S_2\ (u_1^i, u_2^i = 10)$	011
$S_3\ (u_1^{i-1}, u_2^{i-1} = 11)$	$S_3\ (u_1^i, u_2^i = 11)$	000

Now that we've developed the state table representation for a convolutional encoder, we can think of the encoder as a sequential network that is placed into a particular state based on previous input blocks. The encoder then receives the present input block, outputs an encoded block, and transitions to a new state to await the next information block. Each time a k-bit information block is input from the source, the encoder transitions from one state to another and outputs one n-bit codeword. For instance, using our (3,2,1) convolutional code and the second row of Table 10-5, if previous input blocks have placed the encoder in S_0 and if the present input block is 01, the encoder outputs 011 and transitions to S_1. We show this transition graphically in Figure 10-10. Circles are used to represent the states, a line and arrow represent the transition, and the bit sequence above the line and arrow represents the encoded output (i.e., the n-bit codeword).

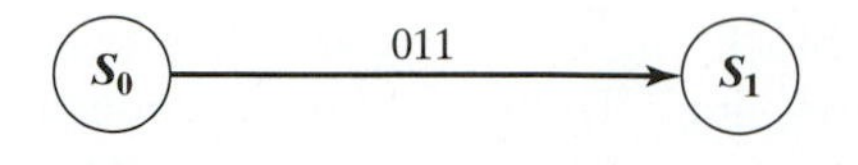

Figure 10-10 Representing the transition from S_0 to S_1 for the (3,2,1) convolutional code.

There are three more possible transitions from S_0, and these are shown in Figure 10-11. Note how this information is obtained from Table 10-5. Similarly, we can add the four possible transitions from S_1, as shown in Figure 10-12.

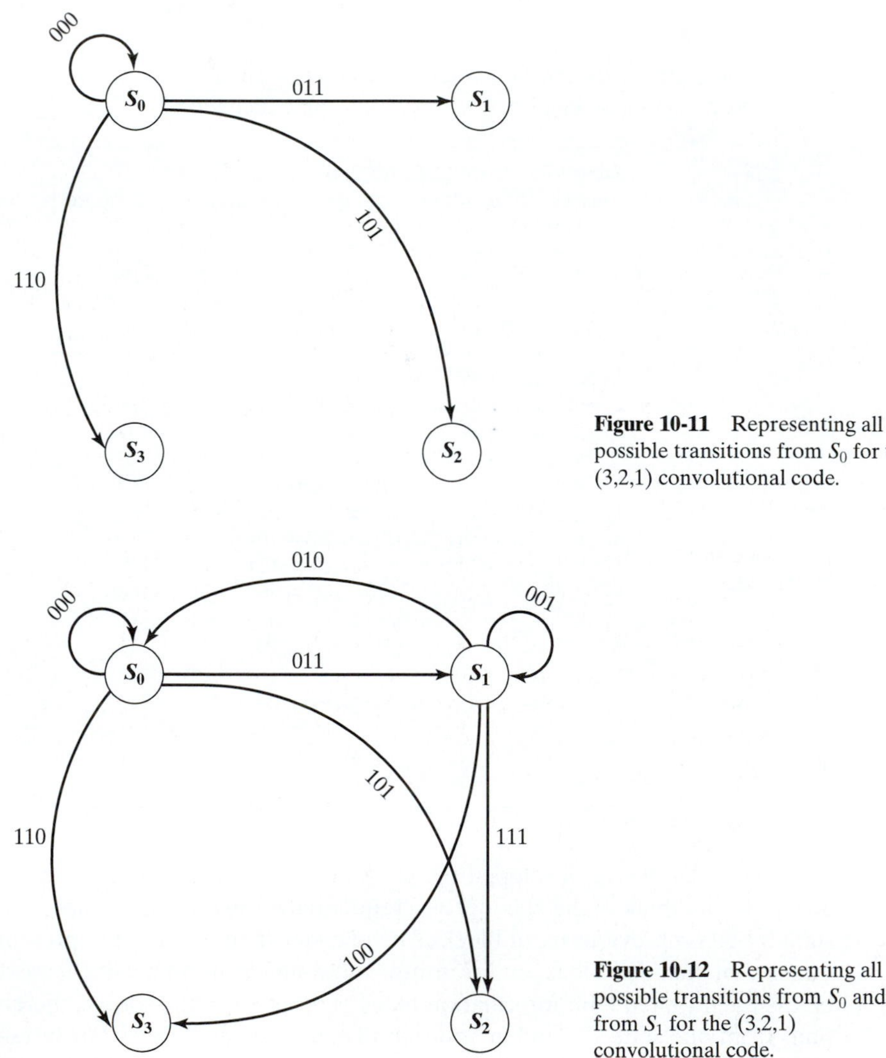

Figure 10-11 Representing all possible transitions from S_0 for the (3,2,1) convolutional code.

Figure 10-12 Representing all possible transitions from S_0 and from S_1 for the (3,2,1) convolutional code.

If we draw all possible transitions from each of the possible states, then we have completely described the encoding process. This graphical description is called a *state diagram*.

Example 10.18 State diagram for a convolutional code

Draw a state diagram for the (3,2,1) convolutional code given in Example 10.15.

Solution

Figure 10-12 shows all of the transitions from S_0 and from S_1, so all we have to do is add the four transitions from S_2 and the four transitions from S_3. The complete state diagram is shown in Figure 10-13. Be sure you understand how this state diagram can be formed from the state table (Table 10-5).

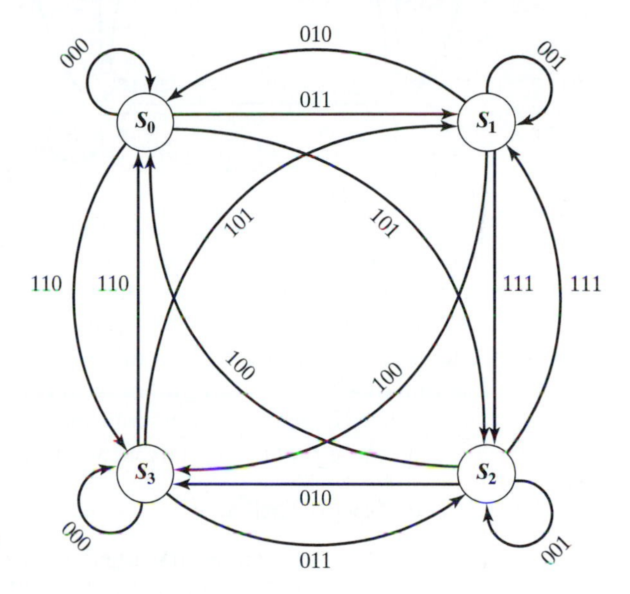

Figure 10-13 State diagram for the (3,2,1) convolutional code.

We can use a state diagram to encode a series of information bits. For example, suppose we use the (3,2,1) convolutional code in Example 10.18 and suppose the source outputs the six-bit sequence 011110. Assuming that the encoder is initialized to state S_0, we can trace the following path through the state diagram in Figure 10-13 and determine the appropriate sequence of output codewords:

1. The encoder is initialized to state S_0. The first information block (i.e., the first two bits from the source) is 01, causing the encoder to transition to state S_1 and to output the codeword 011. This process is shown by the line from S_0 to S_1 (with long dashes) in Figure 10-14.
2. The encoder is now in state S_1. The next information block (i.e., the third and fourth bits from the source) is 11, causing the encoder to transition to state S_3 and to output the codeword 100. This process is shown by the line from S_1 to S_3 (short dashes) in Figure 10-14.
3. The encoder is now in state S_3. The next information block (i.e., the fifth and sixth bits from the source) is 10, causing the encoder to transition to state S_2 and to output the codeword 011. This process is shown by the line from S_3 to S_2 (alternating dots and dashes) in Figure 10-14.

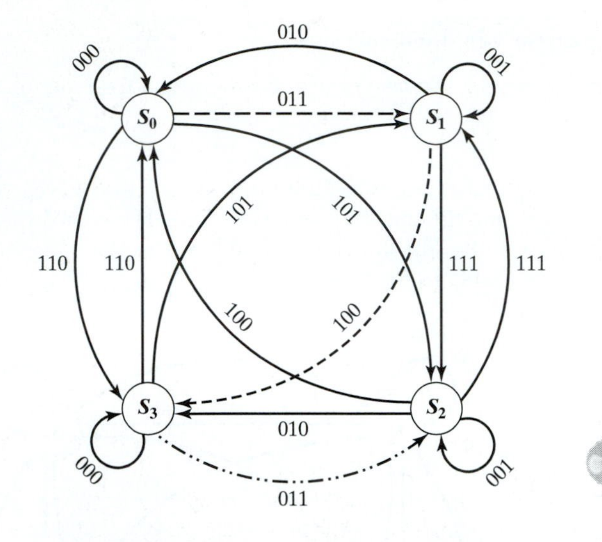

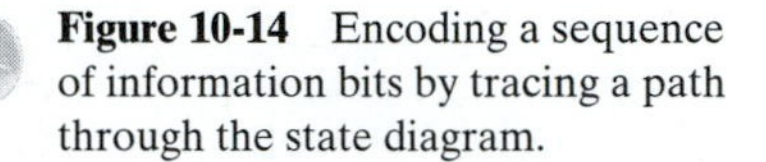
Figure 10-14 Encoding a sequence of information bits by tracing a path through the state diagram.

The encoded sequence is thus 011100011 and the encoder is in state S_2 awaiting the next two information bits from the source.

Tracing a path through the state diagram gives us the convolutionally encoded bit sequence for a series of input bits, but this process is unwieldy for long input sequences. The state diagram quickly becomes cluttered, and after a state is visited twice it becomes difficult to keep track of the order of the transitions. Let's see if we can find a way to spread out the state diagram so that it can more clearly track sequences of bits through time.

10.8.2 Creating a Trellis

Suppose we create a new diagram called a *trellis*. Let's stack all of the states into a column, and let's repeat the column for each block of input bits we receive from the source. Starting with the left column, each time a pair of input bits is received we draw an arrow to indicate a transition to the appropriate state in the next column to the right and we indicate the appropriate output bit sequence. Figure 10-15 shows the state

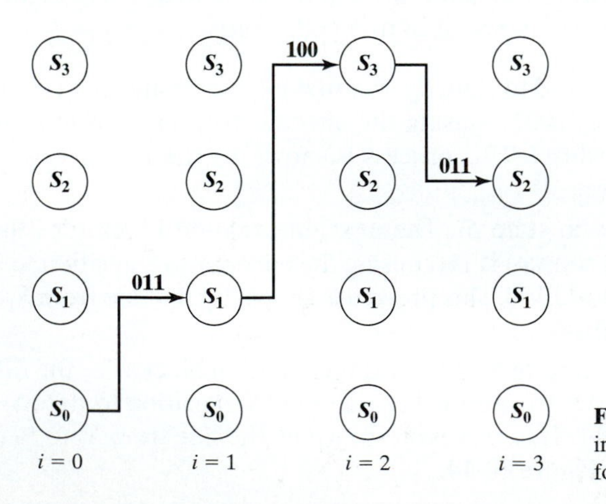

Figure 10-15 A trellis showing inputs, state transitions, and outputs for the (3,2,1) convolutional code.

transitions and outputs for our (3,2,1) convolutional code, given the same input sequence we used in Figure 10-14 (011110) and assuming that the encoder is initialized in state S_0. The output sequence is 011100011. Be sure you understand how the trellis is drawn, how a path is traced through the trellis, and how the output sequence is determined.

Example 10.19 Trellis for a convolutional code

A source outputs the bit sequence 01111010110000. Given the (3,2,1) convolutional code in Example 10.15, draw a trellis showing the state transitions and the encoded output sequence. Assume that the encoder is initialized in state S_0.

Solution

Dividing the input into two-bit blocks, we see that the sequence 01111010110000 corresponds to the states S_1, S_3, S_2, S_2, S_3, S_0, S_0. We can draw this path through the trellis, as shown in Figure 10-16. Now we need to determine the three-bit output sequence for each state transition. We can do this using either the state diagram in Figure 10-13 or the state table (Table 10-5). For example, transitioning from S_1 to S_3 produces the output sequence 100. The outputs for each state transition are added to Figure 10-16, completing the trellis. Reading the trellis, we see that the output sequence is 011100011001010110000.

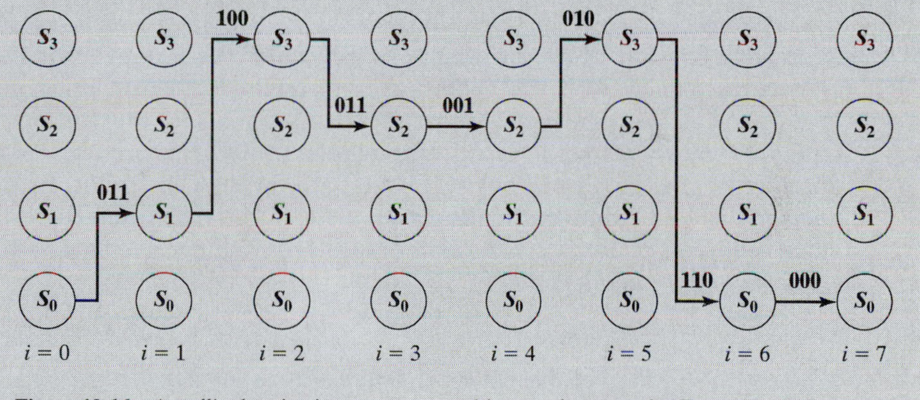

Figure 10-16 A trellis showing inputs, state transitions, and outputs for Example 10.19.

10.8.3 Decoding and the Viterbi Algorithm

We can also use a state diagram and trellis for decoding the received bits and for detecting and correcting errors. Suppose, for example, a communication system is using the (3,2,1) convolutional code in Example 10.15 and the system receives the encoded sequence 011100011001010110000. Assuming that the encoder and decoder are initialized in state S_0, the received sequence can be decoded as follows:

1. The first three bits of the received sequence are 011. We know that the encoder is initialized in state S_0, and using Table 10-5 and the state diagram in Figure 10-13, we see that the only transition from S_0 that produces the output 011 is the

transition to S_1. The first three received bits, 011, therefore tell us that the encoder has transitioned from S_0 to S_1. A transition from S_0 to S_1 occurs only when the present input block from the source is 01, so the first two bits from the source were 01.

2. The next three bits of the received sequence are 100. At the end of Step 1 the encoder was in state S_1, and we know from Table 10-5 and the state diagram in Figure 10-13 that the only transition from S_1 that produces the output 100 is the transition to S_3. A transition to S_3 occurs only when the present input block from the source is 11, so the next two bits from the source were 11.
3. The next three bits of the received sequence are 011. At the end of Step 2 the encoder was in state S_3, and the only transition from S_3 that produces the output 011 is the transition to S_2. A transition to S_2 occurs only when the present input block from the source is 10, so the next two bits from the source were 10.
4. The next three bits of the received sequence are 001. The only transition from S_2 that produces the output 001 is the transition to S_2. A transition to S_2 occurs only when the present input block from the source is 10, so the next two bits from the source were 10.
5. Continue in this manner for the remaining three-bit blocks of the received sequence.

The process we've just described works well when the received sequence has no errors, but what do we do when errors occur? The key to error detection and error correction is to think about paths through the trellis. Each possible path through the trellis corresponds to an encoded sequence of bits, but not all received sequences will correspond to paths through the trellis. If errors occur in the received bit sequence, the result will not be a valid path through the trellis. Error correction is accomplished by finding the valid path through the trellis that corresponds to the fewest number of total errors in the received sequence. This concept is called *maximum likelihood decoding*.

Example 10.20 Our first attempt at decoding a convolutional code with errors

A received bit sequence is 011101011. Given the (3,2,1) convolutional code in Example 10.15, determine whether or not errors have occurred. If so, correct the errors.

Solution

Assuming that the encoder and decoder are initialized to S_0, the first three bits, 011, correspond to a transition from S_0 to S_1. So far, so good. The next three bits, however, are 101, which is not a valid output for any transition from S_1. We therefore know that an error has occurred, and we need to look at all the paths through the trellis to see which one produces the fewest total number of errors. The first part of our path involves moving from the first column of the trellis ($i = 0$) to the second column ($i = 1$). Since the encoder is initialized to S_0, there are four possible paths to the second column: S_0 to S_0, S_0 to S_1, S_0 to S_2, and S_0 to S_3. Figure 10-17a shows these four paths. Correct bits are shown in bold black, errors are shown in outlined type.

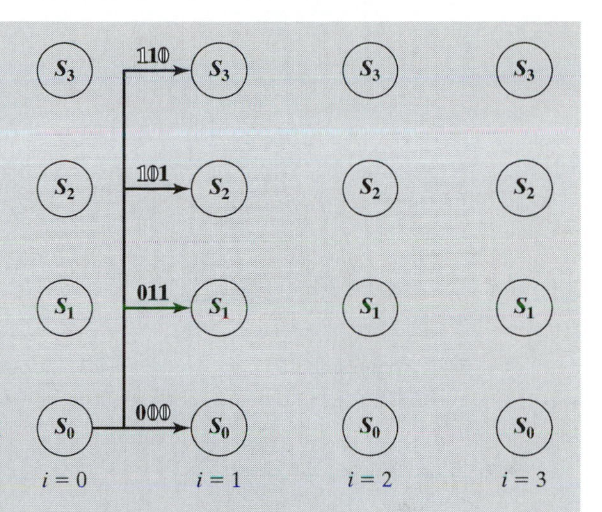

Figure 10-17a Valid paths from the first column of the trellis to the second. Correct bits in the received sequence are shown in bold black; errors are shown in outlined type.

In examining Figure 10-17a, we might be tempted to assume that the transition from S_0 to S_1 is correct since it involves no errors while the other three paths each involve two errors. We need to remember, however, that we are interested in the path through the trellis that produces the fewest *total* number of errors in going from the $i = 0$ column all the way to $i = 3$. This path may or may not pass through S_1 in the $i = 1$ column—we don't know yet.

Our next step is to investigate the errors involved in all possible paths from the $i = 1$ column to the $i = 2$ column. There will be 16 such paths, four originating from each of the states in the $i = 1$ column. Figure 10-17b shows the four transitions from S_0 in the $i = 1$ column; Figure 10-17c shows the four transitions from S_1 in the $i = 1$ column; Figure 10-17d shows the four transitions from S_2 in the $i = 1$ column; and Figure 10-17e shows the four transitions from S_3 in the $i = 1$ column. Using Figures 10-17b–e we can determine the total number of errors involved in any of the 16 possible paths from $i = 0$ to $i = 2$. For example, the path S_0, S_2, S_1 involves a total of three errors.

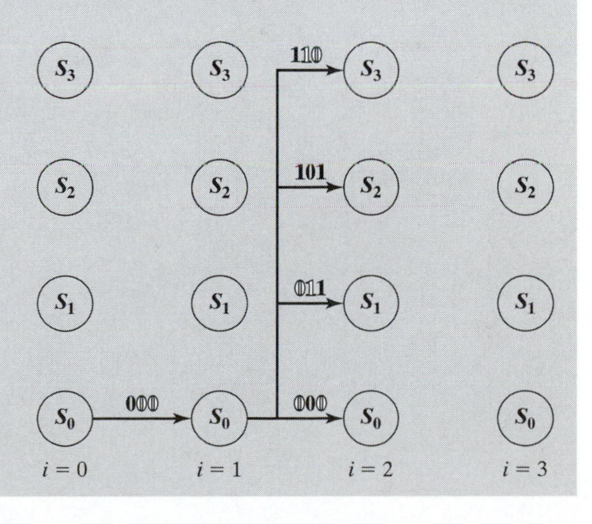

Figure 10-17b Paths from S_0 in the second column to each of the states in the third column.

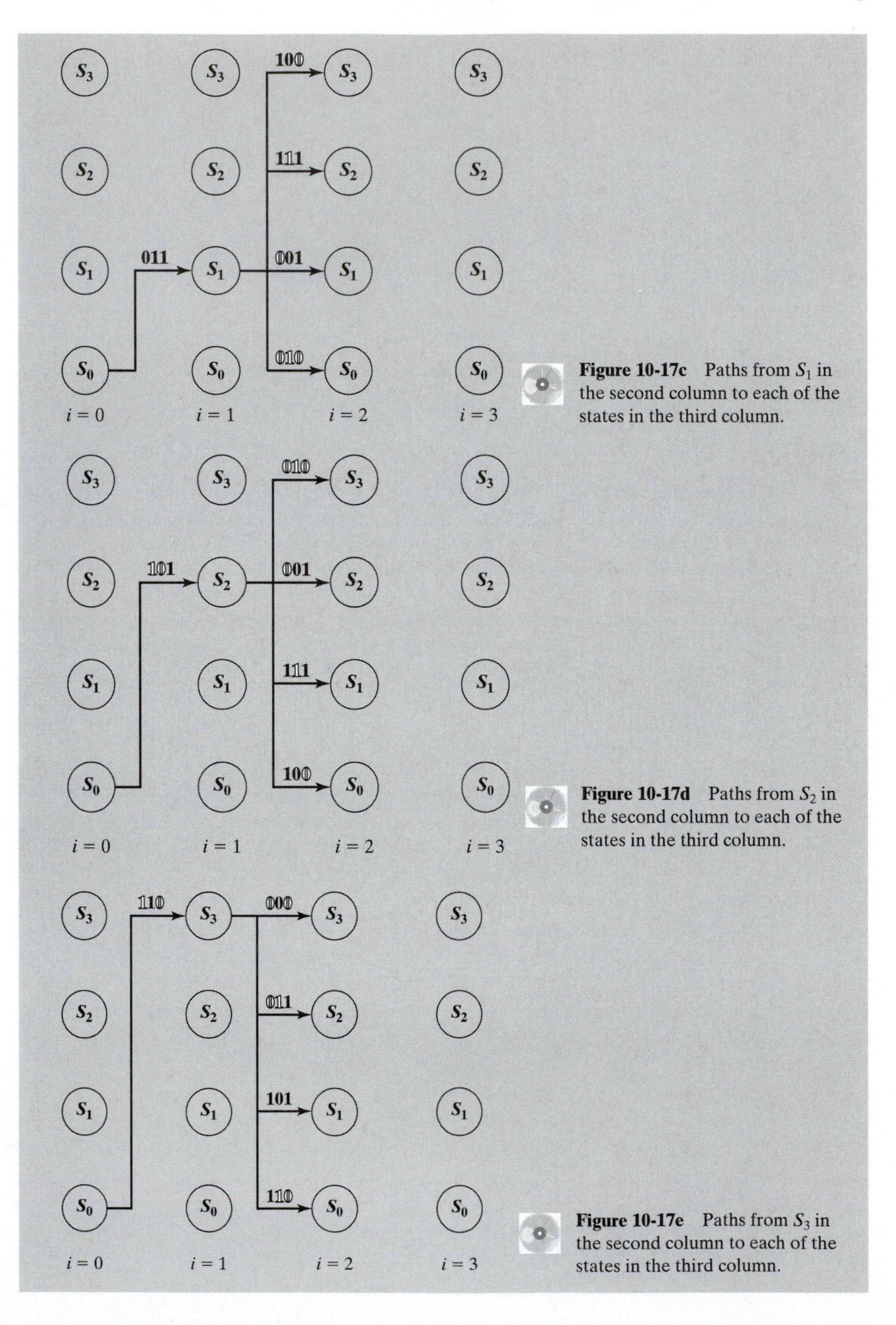

Figure 10-17c Paths from S_1 in the second column to each of the states in the third column.

Figure 10-17d Paths from S_2 in the second column to each of the states in the third column.

Figure 10-17e Paths from S_3 in the second column to each of the states in the third column.

To complete our trellis, we now need to extend our paths to the $i = 3$ column. The decoding process is quickly becoming unwieldy—when we're done there will be 64 possible paths through the trellis. If we were to print all of these paths in this book, we would easily see that the path through the trellis that assumes the least number of errors is S_0, S_1, S_3, S_2. This path, shown in Figure 10-17f, corresponds to the output sequence 011100011 and assumes that there was one error in the received bit sequence. We can directly decode the corrected received information bits from the corrected path; S_0, S_1, S_3, S_2 corresponds to the information 011110.

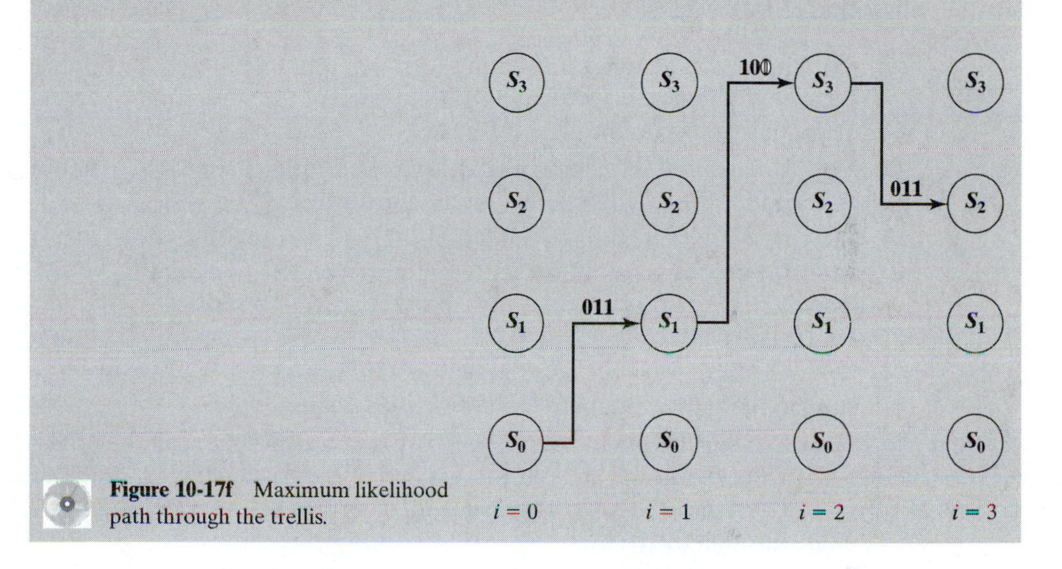

Figure 10-17f Maximum likelihood path through the trellis.

We've just seen that errors in the received bit sequence produce an invalid path through the trellis and that error correction is accomplished by finding the valid path through the trellis that corresponds to the fewest number of total errors. We've also seen, however, that there is a large number of possible valid paths through a trellis—the trellis in Example 10.20 produces 64 possible valid paths. Longer received bit sequences will produce an even larger number of valid paths, and comparing each valid path to the received bit sequence can very quickly become a long, tiresome task. Is there a simple, methodical, fast way to perform maximum likelihood decoding (i.e., to find the path through the trellis that produces the least number of errors)? The answer is yes, using a technique known as the *Viterbi algorithm.*

Let's define the *cost* of a path through the trellis as the number of errors that path assumes. For instance, in Example 10.20 the cost of the path S_0, S_1, S_3, S_3 is three errors. The *least cost path* through the trellis in Example 10.20 is S_0, S_1, S_3, S_2, which assumes only one error. The Viterbi algorithm is based on a simple principle. Moving through a trellis from left to right, suppose we want to determine the least cost path from point a to point c that also passes through point b. The least cost path from a to c which passes through b is the concatenation of the least cost path from a to b plus the least cost path from b to c. Furthermore, the cost of this path is the sum of the least cost of the path from a to b plus the least cost of the path from b to c. This doesn't sound earthshaking, but its implications are very important: Once we have determined the least cost path from one point to another through a portion of the trellis, that path must be part of the least cost path through any longer portion of trellis that includes these two points. This allows us to work through a trellis fairly quickly, keeping only the most promising paths and eliminating the others before they become long and unmanageable. Cost calculations are also simplified.

Given a convolutional code and a received bit sequence, we can apply the Viterbi algorithm as follows:

1. Start in the first column at the initial state (usually S_0) and determine the cost (i.e., the number of errors) of the path to each of the states in the second column. Trace these paths and record the costs.
2. Now for each state S_j in the third column, determine the cost of a path from each state in the second column to S_j in the third column.
3. In examining the results of Steps 1 and 2, you can easily see the least cost paths from the initial state in the first column to each state S_j in the third column. Keep these least cost paths and record their costs, but discard all other paths in the trellis.[5]
4. For each state S_j in the fourth column, determine the cost of a path from each state in the third column to S_j in the fourth column.
5. In examining the results of Steps 3 and 4, you can now easily see the least cost paths from the initial state in the first column to each state S_j in the fourth column. Keep these least cost paths and record their costs, but discard all other paths through the trellis.[5]
6. Repeat Steps 4 and 5 for the fifth, sixth, seventh columns, etc., until you reach the last column in the trellis.
7. You now have the least cost paths from the initial state in the first column to each state in the last column. Determine which of these parts has the lowest cost and, using the states and state transitions along this path, determine the sequence of decoded, corrected information bits.

Example 10.21 Using the Viterbi algorithm

Suppose we use the (3,2,1) convolutional code in Example 10.15 and receive the sequence 011101011. Correct the received sequence using the Viterbi algorithm and determine the information transmitted by the source.

Solution

Step 1: We know the system starts in S_0. The first three received bits are 011. Figure 10-18a shows the cost of a transition from S_0 to each state in Column 2.

[5]If there are two paths that both produce the least cost to a state S_j, then we can arbitrarily choose one of the paths. We should, however, mark this path because if, after Step 7, the lowest cost path through the entire trellis contains segments where arbitrary choices were made, then we have exceeded the error-correcting power of the code.

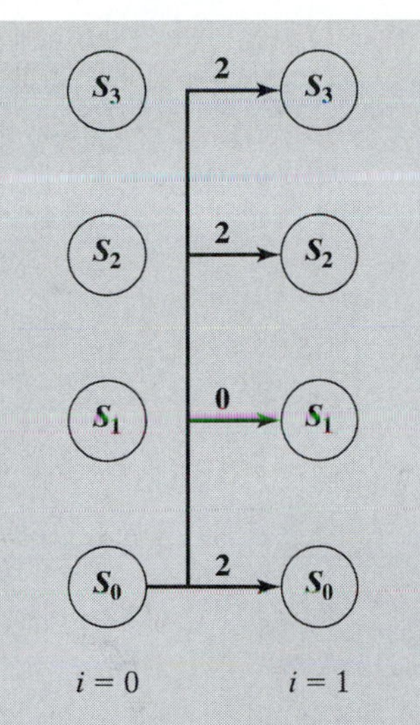

Figure 10-18a Initial transitions in the trellis and their costs.

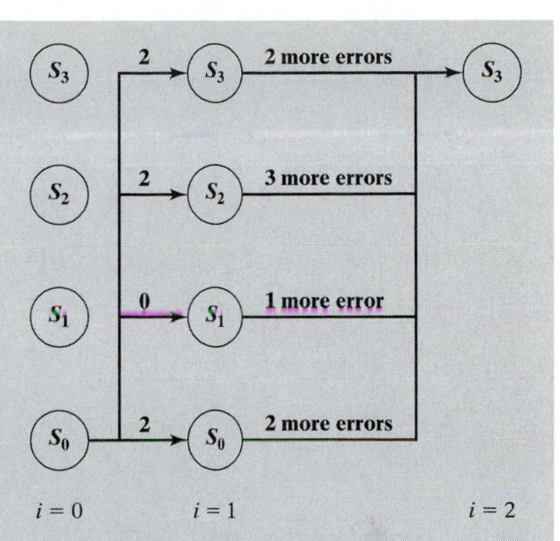

Figure 10-18b Costs of all possible transitions from second column to S_3 in third column.

Steps 2 and 3: We now look at the third column and determine the minimum total cost of transitioning to each state in the third column. We know that the second encoded block of received bits is 101. Let's begin by examining the cost of transitioning to S_3 in the third column. Transitioning from S_3 in the second column to S_3 in the third column assumes two additional errors. The path S_0, S_3, S_3 therefore assumes a total of $2 + 2 = 4$ errors. Transitioning from S_2 in the second column to S_3 in the third column assumes three additional errors, so the path S_0, S_2, S_3 assumes $2 + 3 = 5$ total errors. Similarly, we can calculate the costs of transitioning from S_1 in the second column to S_3 in the third column and from S_0 in the second column to S_3 in the third column. These costs are shown in Figure 10-18b.

In examining Figure 10-18b, we see that the least total cost of transitioning from S_0 in the first column to S_3 in the third is the path S_0, S_1, S_3, which assumes a total of $0 + 1 = 1$ error. We can eliminate the three more expensive paths shown in Figure 10-18b, leaving only the path shown in Figure 10-18c.

Next, we need to examine the costs of transitioning to S_2 in the third column. As we can see from Figure 10-18d, the least cost path from S_0 in the first column to S_2 in the third column is S_0, S_1, S_2, which assumes a total of one error. We can now eliminate the three more expensive paths shown in Figure 10-18d, leaving only the path shown in Figure 10-18e.

We can examine S_1 in the third column in a similar way to determine the least cost path from S_0 in the first column to S_1 in the third column; and we can do the same with the transition from S_0 in the first column to S_0 in the third column. Figure 10-18f shows the least total cost of transitioning from S_0 in the first column to each of the four states in the third column. The bold black numbers indicate the total number of errors assumed along the path. Note that the transition from S_0 in the first column to S_0 in the third column is the same (three errors) whether we go through S_2 in the second column or S_1 in the second column. We can arbitrarily eliminate one of these two paths (but remember the footnote on page 502).

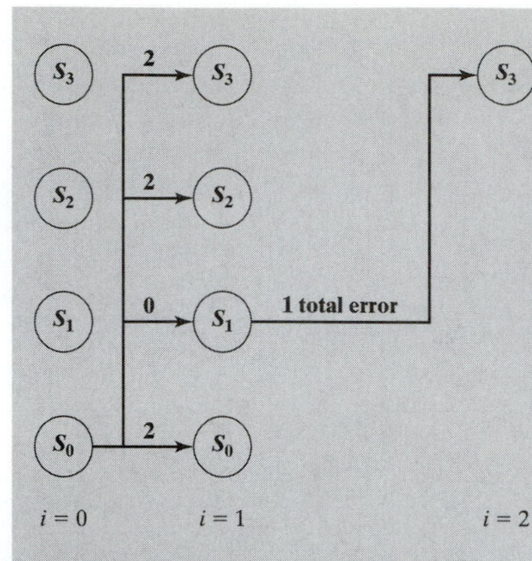

Figure 10-18c Least cost path from S_0 in first column to S_3 in third column.

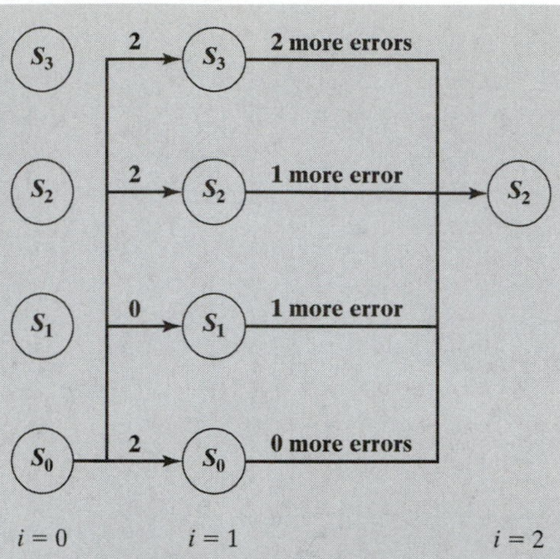

Figure 10-18d Costs of all possible transitions from second column to S_2 in third column.

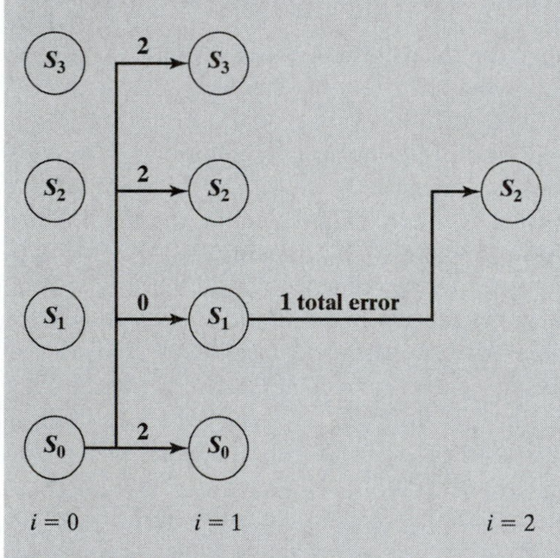

Figure 10-18e Least cost path from S_0 in first column to S_2 in third column.

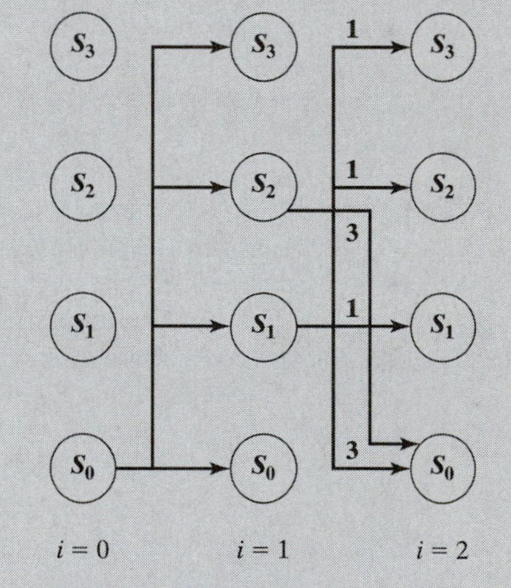

Figure 10-18f Least cost paths from S_0 in first column to each state in third column.

Steps 4 and 5: Just as we did in Steps 2 and 3, we can calculate the minimum additional cost in moving from the third column to each state in the fourth column. The minimum additional costs are shown in Figure 10-18g. As just mentioned, since there are two paths from S_0 in the first column to S_0 in the third column that have the same cost, one can be arbitrarily eliminated (S_0, S_2, S_0 was eliminated and the chosen path, S_0, S_1, S_0, was marked with an asterisk). We can also eliminate the path from S_0 in the first column to S_3 in the second column, since that path is not part of any of the least cost paths to a state in the third column. We can eliminate the path from S_0 in the first column to S_0 in the second column for the same reason.

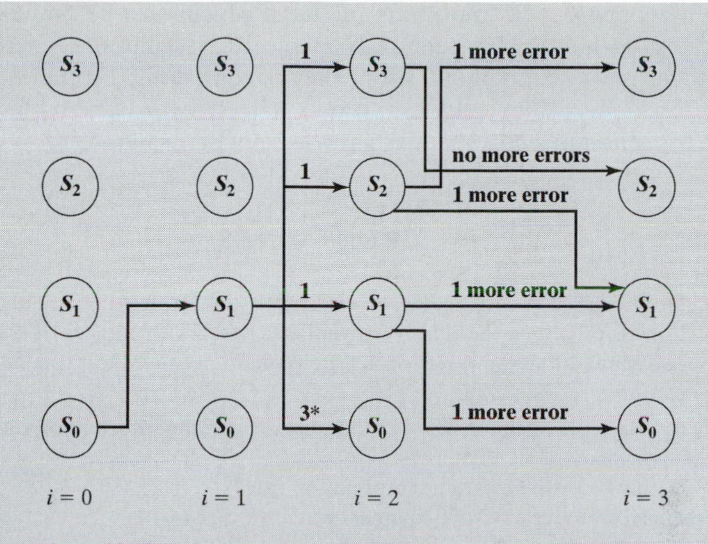

Figure 10-18g Additional costs for transitioning to fourth column.

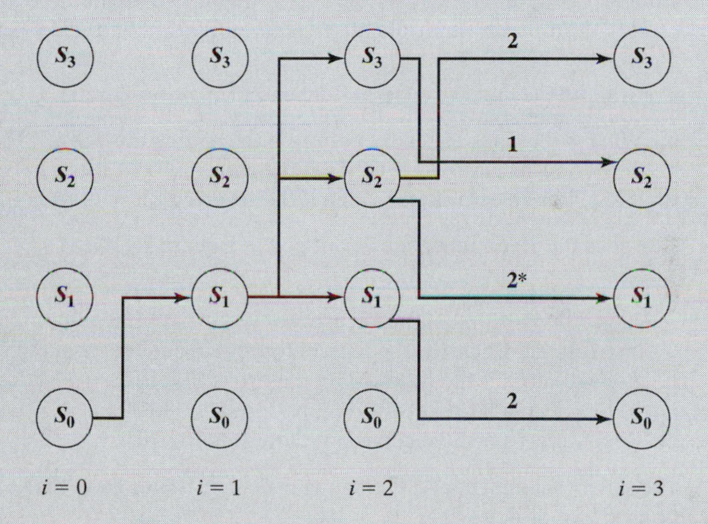

Figure 10-18h Least cost paths from S_0 in first column to each state in final column.

We can now easily determine the total costs from S_0 in the first column to each state in the fourth column. These total costs are shown in Figure 10-18h. In examining Figure 10-18h, we see that the least cost path is S_0, S_1, S_3, S_2, which corresponds to the information 011110.

Without the Viterbi algorithm, the number of paths that must be evaluated through the trellis grows exponentially. For example, given a convolutional code with M states and n blocks of input data, there are M^n possible paths through the trellis, each involving n state transitions and all of which must be evaluated to determine the least cost path. As we've seen in Example 10.21, using the Viterbi algorithm, we can eliminate most of the paths before they become long. Furthermore, the cost calculations for the remaining paths are simple, since they involve adding the cost of a single state transition to a cost that has already been determined.

PROBLEMS

10.1 Define and compare automatic retransmission request (ARQ) and forward error correction (FEC). State the relative advantages and disadvantages of these two types of coding and name one example of each type (you do not need to explain the details of how your example codes work).

10.2
- **a.** State the Shannon-Hartley theorem concerning information rate and channel capacity. Explain its significance.
- **b.** How do we determine information rate?
- **c.** How do we measure channel capacity?

10.3 Why is mathematical structure important in developing error control codes?

10.4 Consider a mathematical system consisting of two elements, 0 and 1, and using conventional addition and multiplication as its operators. Is this system a field? Justify your answer.

10.5 Consider the Hamming code developed in Equation (10.15).

- **a.** What is the transmitted bit pattern if the source bits are 1011?
- **b.** Show how an error is corrected if the received bit pattern is 1001001.
- **c.** Show how an error is corrected if the received bit pattern is 1000011.

10.6 Consider the Hamming code developed in Equation (10.15).

- **a.** What is the transmitted bit pattern if the received bit pattern is 1000110? Did the received bit pattern contain an error or did it not? If it contained an error, which bit was in error, and what was the correct four-bit sequence from the source?
- **b.** Repeat Part a if the received bit pattern is 1111010.
- **c.** Repeat Part a if the received bit pattern is 1000100.
- **d.** Repeat Part a if the received bit pattern is 1011010.

10.7 The parity submatrix of a Hamming code consists of rows that contain unique combinations of two or more "1"s.

- **a.** Why must each row contain at least two "1"s?
- **b.** Why must each row be unique?
- **c.** If each row is unique and contains a combination of two or more "1"s, why is the code produced by the generator matrix capable of detecting and correcting all single-bit errors?

10.8 Consider the following channel encoder:

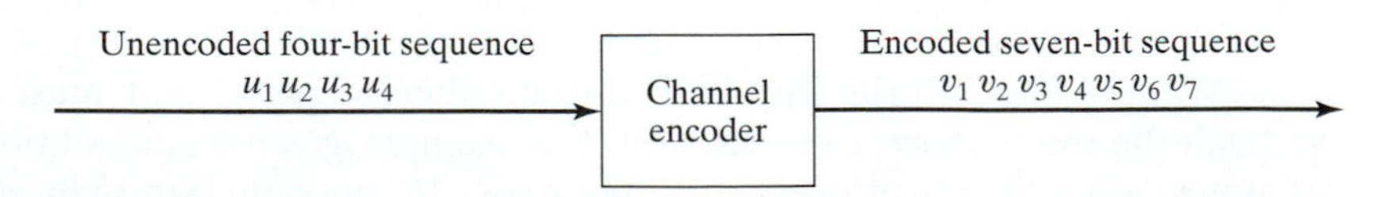

Figure P10-8 (7,4) channel encoder.

where

$$\begin{aligned} v_1 &= u_1 \oplus u_2 \oplus u_4 \\ v_2 &= u_1 \oplus u_2 \oplus u_3 \\ v_3 &= u_1 \oplus u_3 \oplus u_4 \\ v_4 &= u_1 \\ v_5 &= u_2 \\ v_6 &= u_3 \\ v_7 &= u_4 \end{aligned}$$

a. Does the channel encoder produce a Hamming code? Justify your answer.
b. What is the rate of the code?
c. Determine the generator matrix for the code.
d. Show how the generator matrix is used to encode the four-bit information sequence 1011. What is the seven-bit encoded sequence corresponding to 1011?
e. Determine the matrix $[\boldsymbol{H}]^T$ used to decode the received codewords.
f. Show how $[\boldsymbol{H}]^T$ decodes the seven-bit encoded sequence you produced in Part d.
g. Suppose an error occurred in the fifth bit of the encoded sequence you produced in Part d. Determine the syndrome of the received, erroneous codeword.

10.9 Suppose a Hamming code is constructed using six parity bits. What is the maximum size block of information bits for which this code can provide single-bit error-detection and error-correction capability?

10.10 Equation (10.22) provides a generator matrix for a (15,11) Hamming code. Create a different generator matrix for a (15,11) Hamming code.

10.11 Construct a Hamming error correction code that uses 11 data bits and four check bits and that can correct any received message that has an error in one of its 15 received bits. Give an example of how the code detects and corrects an error in, say, the sixth bit.

10.12 Explain the geometric interpretation of error control coding. How can this interpretation help us analyze and design linear block error control codes?

10.13 Explain the relationship between Hamming distance and Euclidean distance. Why is Hamming distance a useful parameter in measuring the error-detecting and error-correcting power of a linear block code?

10.14 Suppose we create a linear block code with a minimum weight of 3. Such a code is capable of detecting and correcting a single-bit error or detecting but not correcting a two-bit error. Why can't this code detect and correct a two-bit error? What happens if we try to detect and correct a two-bit error?

10.15 Use polynomial manipulation to cyclically shift the codeword 111010 three places to the right.

10.16 Suppose a communication system uses a CRC code for error detection with a generator polynomial of $x^5 + x + 1$. If the data sequence from the source encoder is 1011011100011, what is the transmitted bit sequence?

10.17 Using Table 10-3, find an appropriate generator polynomial for a (15,7) BCH code and determine the codeword corresponding to the data sequence 1110010. Describe the error-detecting and error-correcting power of this BCH code.

10.18 Consider a convolutional code with the following encoding process:

$$\begin{aligned} v_1^i &= u_1^i \oplus u_2^{i-1} \\ v_2^i &= u_2^i \oplus u_1^{i-1} \\ v_3^i &= u_1^i \oplus u_2^i \end{aligned}$$

a. What is the size of each block of unencoded bits?
b. What is the size of each block of encoded bits?
c. What is the constraint length of the code?
d. Determine the state table for the convolutional code.
e. Draw the state diagram for the convolutional code.
f. Draw the trellis diagram if the input information is 100111111011. Assume that the encoder starts in the 00 state.

10.19 Consider the convolutional encoder described in Problem (10.18). Suppose the received bit sequence is 101001010010010. The encoder starts in the 00 state. Using Viterbi decoding, determine the correct transmitted sequence and the corresponding information from the source. Be sure to show all your steps in the Viterbi decoding sequence.

10.20 How many different states are needed to represent the state diagram or trellis diagram of a (5,3,2) convolutional code?

References

CHAPTER 1

[1.1] Shannon, C. E. 1948. A Mathematical Theory of Communication. *Bell Sys. Tech. J.* 27:379–423, 623–56.

CHAPTER 2

[2.1] Sklar, B. 2001. *Digital Communications: Fundamentals and Applications*. 2nd ed. Englewood Cliffs, N.J.: Prentice Hall.

CHAPTER 3

[3.1] Schwartz, M. 1990. *Information Transmission, Modulation, and Noise*. 4th ed. New York: McGraw-Hill.

CHAPTER 4

[4.1] Papoulis, A. 2002. *Probability, Random Variables, and Stochastic Processes*. 4th ed. New York: McGraw-Hill.

[4.2] Gardner, F. M., and W. C. Lindsey, eds. 1980. *IEEE Trans. Commun*. COM-28. (Special issue on synchronization.)

[4.3] Franks, L. E. 1980. Carrier and Bit Synchronization in Data Communication—A Tutorial Review. *IEEE Trans. Commun*. COM-28:1107–129.

[4.4] Scholtz, R. A. 1980. Frame Synchronization Techniques. *IEEE Trans. Commun.* COM-28:1204–213.

[4.5] Carter, C. R. 1980. Survey of Synchronization Techniques for a TDMA Satellite—Switched System. *IEEE Trans. Commun.* COM-28:1291–301.

[4.6] Razavi, B., ed. 1996. *Monolithic Phase-Locked Loops and Clock Recovery Circuits: Theory and Design*. New York: IEEE Press.

[4.7] Roza, E. 1996. Analysis of Phase-Locked Timing Extraction Circuits. In *Monolithic Phase-Locked Loops and Clock Recovery Circuits: Theory and Design*, edited by B. Razavi. New York: IEEE Press.

[4.8] Messerschmitt, D. 1996. Frequency Detection for PLL Acquisition in Timing and Carrier Recovery. In *Monolithic Phase-Locked Loops and Clock Recovery Circuits: Theory and Design*, edited by B. Razavi. New York: IEEE Press.

[4.9] Stiffler, J. J. 1971. *Theory of Synchronous Communications*. Englewood Cliffs, N.J.: Prentice Hall.

[4.10] Lindsey, W. C., and M. K. Simon. 1991. *Telecommunication Systems Engineering*. Chapter 9. New York: Dover Publications.

[4.11] Lee, E., and D. Messerschmitt. 1993. *Digital Communication*. 2nd ed. Boston/London: Kluwer Academic Publishers.

[4.12] Mehrotra, A. 1994. *Cellular Radio Performance Engineering*. Norwood, Mass.: Archtech House Publishers.

[4.13] Lucky, R. W., and H. R. Rudin. 1967. An Automatic Equalizer for General Purpose Communication Channels. *Bell Sys. Tech. J.* 46:2179–194.

[4.14] Salz, J. 1973. Optimum Mean-Square Decision Feedback Equalization. *Bell Sys. Tech. J.* 52(8):1341–373.

[4.15] Forney, G. D. 1972. Maximum-Likelihood Sequence Estimation of Digital Sequences in the Presence of Intersymbol Interference. *Trans. Inform. Theory*. IT-18:363–78.

[4.16] Falconer, D. D., and L. Ljung. 1978. Application of Fast Kalman Estimation to Adaptive Equalization. *IEEE Trans. Inform. Theory.* COM-26:1439–446.

[4.17] Lucky, R. W., J. Salz, and E. J. Weldon. 1968. *Principles of Data Communications*. New York: McGraw-Hill.

[4.18] Lee, E., and D. Messerschmitt. 1993. *Digital Communication*. 2nd ed. Boston/London: Kluwer Academic Publishers.

[4.19] Proakis, J. G. 2001. *Digital Communications*. 4th ed. New York: McGraw-Hill.

[4.20] Ziemer, R., and R. Peterson. 1985. *Digital Communications and Spread Spectrum Systems*. New York: Macmillan.

CHAPTER 5

[5.1] Papoulis, A. 2002. *Probability, Random Variables, and Stochastic Processes*. 4th ed. New York: McGraw-Hill.

[5.2] Shanmugam, K. S. 1983. *Digital and Analog Communication Systems.* New York: J. Wiley & Sons.

[5.3] Lindsey, W. C., and M. K. Simon. 1991. *Telecommunication Systems Engineering*. New York: Dover Publications.

[5.4] Simon, M. K., S. M. Hinedi, and W. C. Lindsey. 1995. *Digital Communication Techniques: Signal Design and Detection.* Englewood Cliffs, N.J.: Prentice Hall.

[5.5] Carlson, A. B., P. B. Crilly, and J. C. Rutledge. 2002. *Communication Systems*. 4th ed. New York: McGraw-Hill.

[5.6] Proakis, J. G. 2001. *Digital Communications*. 4th ed. New York: McGraw-Hill

[5.7] Taub, H., and D. L. Schilling. 1986. *Principles of Communication Systems*. 2nd ed. New York: McGraw-Hill.

CHAPTER 6

[6.1] Shanmugam, K. S. 1983. *Digital and Analog Communication Systems.* New York: J. Wiley & Sons.

[6.2] Taub, H., and D. L. Schilling. 1986. *Principles of Communication Systems*. 2nd ed. New York: McGraw-Hill.

[6.3] Carson, J. R., and T. C. Fry. 1937. Variable Frequency Electric Circuit Theory with Application to the Theory of Frequency Modulation. *Bell Sys. Tech. J.* 16:513–40.

[6.4] Downing, J. J. 1964. *Modulation Systems and Noise*. Englewood Cliffs, N.J.: Prentice Hall.

CHAPTER 7

[7.1] Bellamy, J. 2002. *Digital Telephony*. 3rd ed. New York: J. Wiley & Sons.

[7.2] Pickholtz, R. L., L. B. Milstein, and D. L. Schilling. 1991. Spread Spectrum for Mobile Communications. *IEEE Trans. Vehicular Tech.* 40:313–22.

[7.3] Lee, W. C. Y. 1991. Overview of Cellular CDMA. *IEEE Trans. Vehicular Tech.* 40:291–302.

[7.4] Simon, M. K., J. K. Omura, R. A. Scholtz, and B. K. Levitt. 2001. *Spread Spectrum Communications Handbook*. New York: McGraw-Hill.

CHAPTER 8

[8.1] Lyons, R. G. 1997. *Understanding Digital Signal Processing*. Reading, Mass.: Addison-Wesley.

[8.2] Sklar, B. 2001. *Digital Communications: Fundamentals and Applications.* 2nd ed. Englewood Cliffs, N.J.: Prentice Hall. ISBN 0-13-084788-7.

[8.3] Kondoz, A. 1999. *Digital Speech: Coding for Low Bit Rate Communications Systems*. New York: J. Wiley & Sons.

[8.4] Deller, J. R., J. H. L. Hansen, and J. G. Proakis. 2000. *Discrete-Time Processing of Speech Signals*. Piscataway, N.J.: IEEE Press.

[8.5] Mitra, S. 2001. *Digital Signal Processing: A Computer-Based Approach*. 2nd ed. New York: McGraw-Hill.

CHAPTER 9

[9.1] Huffman, D. A. 1952. A Method for the Construction of Minimum Redundancy Codes. *Proc. IRE*. 40:1098–101.

[9.2] Blahut, R. E. 1987. *Principles and Practice of Information Theory*. Reading, Mass.: Addison-Wesley.

[9.3] Shanmugam, K. S. 1983. *Digital and Analog Communication Systems.* New York: J. Wiley & Sons.

[9.4] Shannon, C. E. 1951. Prediction and Entropy of Printed English. *Bell Sys. Tech. J.* 30:50–64.

[9.5] Storer, J. A. 1998. *Data Compression Methods and Theory*. Rockville, Md.: Computer Science Press.

[9.6] Ziv, J., and A. Lempel. 1978. Compression of Individual Sequences via Variable-Rate Coding. *IEEE Trans. Inform. Theory*. 24(5):530–36.

[9.7] Welch, T. 1984. A Technique for High-Performance Data Compression. *IEEE Comput.* 17:8–19.

[9.8] International Telecommunications Union. 1999. *Standardization of Group 3 Facsimile Terminals for Document Transmission*. Recommendation T.4. Geneva. Amended Feb. 2000 and Nov. 2000.

[9.9] Sayood, K. 2000. *Introduction to Data Compression*. 2nd ed. San Francisco: Morgan Kaufman.

[9.10] International Telecommunications Union. 1988. *Facsimile Coding Schemes and Coding Control Functions for Group 4 Facsimile Apparatus*. Recommendation T.6. Geneva.

[9.11] Gonzalez, R. C., and R. E. Woods. 1992. *Digital Image Processing*. Reading, Mass.: Addison-Wesley.

CHAPTER 10

[10.1] Wilson, S. G. 1996. *Digital Modulation and Coding*. Englewood Cliffs, N.J.: Prentice-Hall.

[10.2] Shannon, C. E. 1948. A Mathematical Theory of Communication. *Bell Sys. Tech. J.* 27:379–423, 623–56.

[10.3] Lin, S., and D. J. Costello. In press. *Error Control Coding: Fundamentals and Applications*. 2nd ed. Englewood Cliffs, N.J.: Prentice Hall.

[10.4] Wicker, S. B. 1995. *Error Control Systems for Digital Communication and Storage*. Englewood Cliffs, N.J.: Prentice Hall.

[10.5] Peterson, W. W., and D. T. Brown. 1961. Cyclic Codes for Error Detection. *Proc. IRE*. 49:228–35.

[10.6] International Telecommunications Union. 1996. *Signalling Data Link* (CCITT Signalling System No. 7). Recommendation Q.703. Geneva.

[10.7] Hocquenghem, A. 1959. Codes Correcteurs d'Erreurs. *Chiffres*. 2:147–56.

[10.8] Bose, R. C., and D. K. Ray-Chaudhuri. 1960. On a Class of Error Correcting Binary Group Codes. *Inf. Control*. 3:68–79.

[10.9] Roth, C. H., Jr. 2003. *Fundamentals of Logic Design*. 5th ed. Boston: PWS (Brooks/Colc).

[10.10] Katz, R. H. 2003. *Contemporary Logic Design*. 2nd ed. Redwood City, Calif.: Addison-Wesley Benjamin/Cummings Publishing Co.

Answers to Selected Problems

CHAPTER 2

2.07 $s(t) = 0.857 - 0.670\cos\left(\frac{2\pi t}{7}\right) + 0.323\sin\left(\frac{2\pi t}{7}\right) - 0.366\cos\left(\frac{4\pi t}{7}\right)$

$$+ 0.459\sin\left(\frac{4\pi t}{7}\right) + 0.0943\cos\left(\frac{6\pi t}{7}\right) - 0.413\sin\left(\frac{6\pi t}{7}\right) - 0.0707\cos\left(\frac{8\pi t}{7}\right)$$

$$- 0.310\sin\left(\frac{8\pi t}{7}\right) + 0.146\cos\left(\frac{10\pi t}{7}\right) + 0.183\sin\left(\frac{10\pi t}{7}\right) + \cdots$$

2.10 $s(t) = 0.857 + 0.744\cos\left(\frac{2\pi t}{7} - 0.857\pi\right) + 0.587\cos\left(\frac{4\pi t}{7} - 0.714\pi\right)$

$$+ 0.424\cos\left(\frac{6\pi t}{7} \times 0.429\pi\right) + 0.318\cos\left(\frac{8\pi t}{7} + 0.571\pi\right)$$

$$+ 0.235\cos\left(\frac{10\pi t}{7} - 0.857\pi\right) + \ldots$$

2.22 Bandwidth = 400 Hz; 90.3% in-band power

CHAPTER 3

3.3 d. (1) 0.5v (2) 0.32v (3) 0.134v (4) 0.042v (5) 6th bit is a "1" but received signal < 0v

3.9 $\alpha_{max} = 0.7$

3.14 a. data pattern = "011010011"

3.19 02E0 = −0.0910; 02E1 = −0.1765; 02E2 = −0.2605; ...

CHAPTER 4

4.8 $A \geq 1.23$v

4.13 $A \geq 1.56$v

4.32 a. Bandwidth = 30 kHz; (b) +8.22v, +2.74v, −2.74v, −8.22v

CHAPTER 5

5.4 Waveform B is PSK with f_c = 200 kHz, data = "110100", Bandwidth = 200 kHz, f_L = 100 kHz, f_H = 300 kHz

5.23 b. 4.5 volts^2 c. P_b = 0.00361

5.31 c. For QPSK, P_s = 0.0456

CHAPTER 6

6.5 P_{trans} = 1.27 volts^2

6.9 a. 310 kHz

6.26 b. (ii) 11.6% (v) 96.0%

CHAPTER 7

7.3 a. 15 kbit/sec

b. The second user is assigned six slots per frame; the third user is assigned ten slots per frame.

7.12 b. The frequency bands for the four signals are: 1.57−1.77 MHz, 1.79−1.99 MHz, 2.01−2.21 MHz, and 2.23−2.43 MHz.

7.21 G_p = 8

CHAPTER 8

8.7 maximum quantization error = 0.3125 volts

8.10 a. dynamic range = 6 volts

CHAPTER 9

9.3 (b) 3.32 bits (e) 0.152 bits

9.7 (b) 2.26 bits per message

9.14 (a) 1953 binary digits

(c) 111111,111111,111111,001011,010001,100011,001011,010001,100011

CHAPTER 10

10.6 (b) The fourth bit $\hat{v}_4$ was recieved in error—the correct received sequence is 1110010 and the information bits are 0010.

10.8 (c) $[G] = \begin{bmatrix} 1 & 1 & 1 & 1 & 0 & 0 & 0 \\ 1 & 1 & 0 & 0 & 1 & 0 & 0 \\ 0 & 1 & 1 & 0 & 0 & 1 & 0 \\ 1 & 0 & 1 & 0 & 0 & 0 & 1 \end{bmatrix}$

10.16 101101110001111011

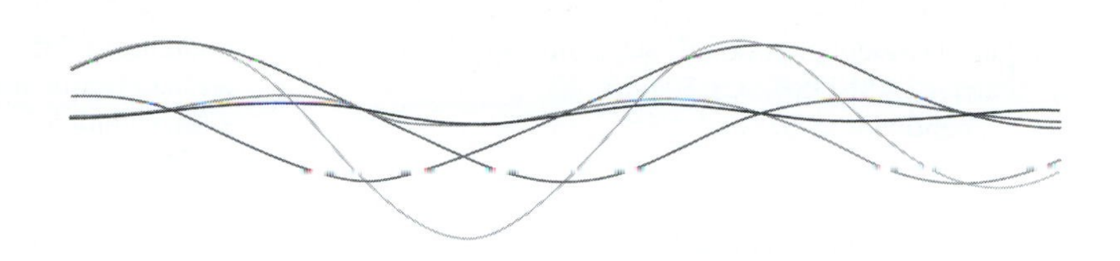

Index

A

A-law companding. *See* companding
accuracy
as a parameter in performance-vs.-cost tradeoffs, 4
See also fidelity (for analog systems); probability of bit error and probability of symbol error (for digital systems); signal-to-noise ratio; specific error control codes; specific modulation techniques
adaptive differential pulse coded modulation (ADPCM), 411
additive Gaussian white noise (AWGN). *See* noise, additive white Gaussian
ADPCM. *See* adaptive differential pulse coded modulation
aliasing, 392, 394, 395–400
AM. *See* amplitude modulation
AM-DSB-C. *See* amplitude modulation-double sideband-carrier
AM-DSB-SC. *See* amplitude modulation-double sideband-suppressed carrier
AM-SSB. *See* amplitude modulation-single sideband
AM-VSB. *See* amplitude modulation-vestigial sideband
amplitude modulation (AM). *See* specific amplitude modulation techniques
amplitude modulation-double sideband-carrier (AM-DSB-C), 315–26
bandwidth (same as AM-DSB-SC), 319
comparison with other AM, FM, and PM techniques, 355, 356
envelope detector, 315, 316, 318, 319
magnitude spectrum, 319
modulation index, 322
noncoherent receiver, 315–19
power, 319–22
power efficiency (η), 321
signal-to-noise ratio, 322–26
threshold effect, 326
transmission, 320

amplitude modulation-double sideband-suppressed carrier (AM-DSB-SC), 306–15
bandwidth, 307
coherent receiver, 309–15
comparison with other AM, FM, and PM techniques, 355, 356
magnitude spectrum, 306, 307
sidebands (defined), 307
signal-to-noise ratio, 313, 315
transmission, 309
amplitude modulation-single sideband (AM-SSB), 326–29
bandwidth, 326
coherent receiver, 327–29
comparison with other AM, FM, and PM techniques, 355, 356
magnitude spectrum, 326–29, 328, 329
signal-to-noise ratio, 327
transmission, 327
amplitude modulation-vestigial sideband (AM-VSB), 327, 329–33
bandwidth, 329, 330
coherent receiver, 330–33
comparison with other AM, FM, and PM techniques, 355, 356
filter transfer function (transmitter), 329, 330, 333
magnitude spectrum, 329–33
signal-to-noise ratio, 333
transmission, 329–32
amplitude shift keying (ASK)
average normalized power spectrum, 213–17
bandwidth, 218, 219
bit error rate (coherent demodulation), 236
bit error rate (noncoherent demodulation), 256–65
coherent receiver, 230-32
comparison with other digital bandpass modulation techniques, 271–73
decomposing into a baseband signal multiplied by a carrier, 213–15, 220
defined, 212
noncoherent receiver, 254, 255
amplitude spectrum (defined), 35
analog signal (defined), 390
analog-to-digital conversion, 388–417
See also adaptive differential pulse coded modulation; continuously variable slope delta modulation; delta modulation; differential pulse coded modulation; quantization; sampling
angle modulation, 335, 338
See also frequency modulation; phase modulation
anti-alias filter, 398–400
ARQ. *See* automatic retransmission request
ASK. *See* amplitude shift keying
associative property, 462, 463
autocorrelation
defined, 242
ensemble average (same as conventional definition), 242
properties if random process is wide sense stationary, 244
relationship to average normalized power spectral density in a wide sense stationary random process. *See* Wiener-Khintchine theorem
time average, 253. *See also* ergodicity
automatic retransmission request (ARQ)
contrasted to forward error correction codes, 488
defined, 459
overhead generated by, 488
See also specific error control technique
average codeword length, 425, 426, 446
See also entropy
average information rate (of a source), 424
See also Shannon's first theorem; Shannon-Hartley theorem
average normalized power
defined for periodic signal, 38
defined for series of pulses, 83
spectrum, 41–44
See also specific modulation techniques
AWGN. *See* noise, additive white Gaussian

B

B-frame (for MPEG compression), 454
balance
in tree diagram for a prefix code, 430, 431, 436, 437
property for a pseudo-random (PN) sequence, 375–77
bandpass
channel (defined), 77
sampling. *See* sampling, bandpass signal
signaling. *See* specific modulation techniques
bandwidth
as a parameter in performance-vs.-cost tradeoffs, 4

defined (formally), 48
effect on distortion, 45–48, 85–88
related to channel capacity, 460
See also specific modulation techniques
baseband
channel (defined) 76, 77
sampling. *See* sampling, baseband signal
signaling. *See* pulse amplitude modulation
BCH code, 485, 487, 490
BER (bit error rate). *See* probability of bit error
Bessel function, 260, 339, 340
binary digit (contrasted to bit), 422
bipolar signaling, 168, 169, 320
bit error rate (BER). *See* probability of bit error
bit rate. *See* transmission speed
block code
contrasted with convolutional code, 490
cyclic (defined), 481
defined, 468
linear (defined), 479
See also specific error control code
blocking capacitor, 318, 319, 324, 326, 345, 346
burst error
correction, 489, 490
defined, 481
detection, 487
bus (topology), 363, 364

C

capacity. *See* channel, capacity
carrier recovery, 172, 189
carrier signal (defined), 212
See also specific bandpass modulation techniques
Carson's rule, 342, 343
CDM. *See* code division multiplexing
CDMA. *See* code division multiplexing
channel
attenuation (and effect on accuracy), 98, 99, 123–27
bandpass (defined), 77
baseband (defined), 76, 77
capacity (defined), 460
defined, 2, 76
distortion, 8, 45, 76, 77, 190. *See also* distortion
encoding. *See* error control coding
equalization. *See* equalization
ideal, 44, 77, 84–87
check bits. *See* parity bits
chip rate. *See* code division multiplexing, chip rate
clock recovery, 181–85
closure property, 462, 463
code division multiplexing (CDM)
chip rate, 371, 372
defined, 363, 370–71, 374, 379
direct sequence spread spectrum (DSSS), 371–83
frequency-hopped spread spectrum (FHSS), 381–85
processing gain, 373, 375, 380–82
signal-to-noise ratio, 380, 381
spreading and dispreading. *See* spreading code
code rate (for error control codes), 469
coding. *See* analog-to-digital conversion; compression; error control coding
coherent receiver
defined, 228
See also specific modulation techniques
communication (defined), 1
communication system (block diagram), 1, 2, 75, 76
commutative group, 462
commutative property, 462, 463
companding, 355, 407
complex exponential form of Fourier series. *See* Fourier series
compression, 421–54
See also specific compression techniques
conditional probability. *See* random variable
constellation
8-ary PSK, 287
16-ary PSK, 299
defined, 275
QAM, 299
QPSK, 275
continuously variable slope delta modulation (CVSD), 411, 415–17
convolutional code
defined, 490
encoding with sequential network, 491, 493
Viterbi decoding, 490, 497, 501–06
correlation (defined), 156
correlation receiver
derivation as optimal coherent receiver, 149–56
See also coherent receiver for specific modulation techniques
counters (used in PAM modulator), 113, 114
CRC. *See* cyclic redundancy check code

crystal oscillator. *See* oscillator
CVSD. *See* continuously variable slope delta modulation
cyclic code
  defined, 481
  polynomial manipulation, 482–86
  See also specific cyclic codes
cyclic redundancy check code (CRC), 485–88
cyclic shift, 375, 481, 483, 485

D

damped sinc. *See* raised cosine
damping factor. *See* raised cosine
data compression. *See* compression
dB. *See* decibel
DCT. *See* discrete cosine transform
decibel (dB) (defined), 156
decoding matrix, 471
de-emphasis. *See* frequency modulation
degree, of polynomial (defined), 482
delay distortion, 190–91
delta function. *See* impulse function
delta modulation (DM), 411–16
demodulation
  defined, 4
  See also specific modulation techniques
despreading. *See* spreading code
deterministic mathematics (defined and contrasted to stochastic mathematics), 127
dictionary encoding (compression)
  defined, 439
  dynamic dictionary, 441–44
  LZW, 442–44
  static dictionary, 439–41
dictionary guaranteed progress, 441, 442
differential phase shift keying (DPSK)
  average normalized power spectrum (same as phase shift keying), 222–24
  bit error rate, 270, 271
  defined, 267, 268
  DQPSK. *See* differential quaternary phase shift keying
  M-ary. *See* phase shift keying, M-ary
  noncoherent receiver, 268, 269
differential pulse coded modulation (DPCM), 407–11, 414, 415, 453
differential quaternary phase shift keying (DQPSK)
  accuracy compared to QPSK, 286
  decomposition into quadrature components, 285, 286
  defined, 286, 287
  demodulation (noncoherent), 285, 286
  symbol error rate, 286
digital-to-analog conversion. *See* continuously variable slope delta modulation; delta modulation
direct sequence spread spectrum. *See* code division multiplexing
discrete cosine transform (DCT), 447–51, 453
discrete signal (defined), 391
discriminator, 345–348, 352, 353
distortion
  amplitude (defined), 190
  effects of channel. *See* bandwidth; channel
  effects of quantization. *See* quantization
  phase (defined), 190
distributive property, 463
DM. *See* delta modulation
DPCM. *See* differential pulse coded modulation
DPSK. *See* differential phase shift keying
DQPSK. *See* differential quaternary phase shift keying
duality property (of Fourier transform), 67, 90
dynamic dictionary. *See* dictionary encoding (compression)
dynamic range, 400–04, 409–11

E

encoding. *See* analog-to-digital conversion; compression; error control coding
encoding rate (defined), 391
energy. *See* normalized energy
ensemble (defined), 238
  See also random process
entropy
  defined, 423
  relationship to average codeword length, 424–26
envelope detector, 254, 255, 315–320
EPROM (in PAM modulator circuit), 113, 114, 115
equalization, 190–99
  least mean squared (LMS), 198–99
  transversal, 196–98
  zero-forcing (ZF), 198

equipment complexity
as a parameter in performance-vs.-cost tradeoffs, 4
subdivided into hardware complexity, software complexity, and required development effort, 4
See also specific modulation techniques
equiprobable (defined), 125
ergodicity
defined, 252, 253
implications, 253
time-averaged parameters. *See* autocorrelation, time average; mean, time average
error control coding, 458–506
ARQ vs. FEC, 459, 488
defined, 458
overhead, 488
See also specific error control codes
error correcting power. *See* Hamming distance, minimum; specific error control codes
error detecting power. *See* Hamming distance, minimum; specific error control codes
error propagation, 446, 453, 459
error term (DPCM and DM), 409–13
error vector, 472
Euclidean distance
compared to Hamming distance, 476, 477
defined, 476
Euler's identity, 23–25
exclusive-OR gate
for generating pseudo-random sequences, 377
for modulo-2 addition, 461
for spreading, 371, 372, 378, 381
expected value (defined), 130
eye diagram, 194, 195

F

facsimile, 444–47, 487
FDM. *See* frequency division multiplexing
FEC. *See* forward error correction
fidelity, 312
See also signal-to-noise ratio; specific analog modulation techniques
field (defined), 463
FM. *See* frequency modulation
folded frequency response, 198
forward error correction (FEC)
contrasted to ARQ codes, 488
defined, 459
overhead generated by, 488
See also specific error control techniques
Fourier series
comparing the different forms, 30, 31
complex exponential (or two-sided) form, 23–30
harmonics (defined), 14
one-sided form, 19–22
trigonometric form, 13–19
two-sided form (complex exponential form), 23–30
Fourier transform
derivation from Fourier series, 52–55
properties, 66, 67
relationship to Fourier series, 55, 59
frame
for moving images, 451–54
for time division multiplexing, 365
free running frequency. *See* voltage controlled oscillator
frequency. *See* instantaneous frequency
frequency deviation. *See* maximum frequency deviation
frequency division multiplexing (FDM), 368–70
frequency domain analysis
contrasted with time domain analysis, 6
defined, 6
justifications for using, 6–13
See also Fourier series; Fourier transform
frequency hopping. *See* code division multiplexing
frequency modulation (FM), 334–56
bandwidth, 343
comparison with other analog modulation techniques, 355–56
defined, 334
magnitude spectrum, 338–43
maximum frequency deviation. *See* maximum frequency deviation
pre-emphasis/de-emphasis, 354–55
receiver, 345–46
signal-to-noise ratio, 351–54
transmitter, 343–45

frequency scaling property (of Fourier transform), 67
frequency shift keying (FSK)
average normalized power spectrum, 223, 225
bandwidth, 225
bit error rate (coherent demodulation), 233
bit error rate (noncoherent demodulation), 265, 266
coherent receiver, 232–34
comparison with other digital bandpass modulation techniques, 271–73
decomposing into two ASK signals, 224
defined, 219, 221
M-ary (coherently demodulated), 292–97
M-ary (noncoherently demodulated), 297, 298
noncoherent receiver, 255, 256
frequency shift property (of Fourier transform), 66
frequency synthesizer, 173, 178–81
FSK. *See* frequency shift keying

G

Galois field, 461,463, 476, 482, 484, 485, 490
Gaussian probability density
application to predicting effects of thermal noise, 133
calculating probability distribution using Q-function and Q-table, 137–44
defined, 130
See also noise, additive white Gaussian
generator matrix, 467–69
generator polynomial, 482, 483, 485–87, 490
geometric interpretation of error control coding, 472, 482
See also Euclidean distance; Hamming distance; weight
GF(2). *See* Galois field
graphical interpretation
error control coding. *See* geometric interpretation
probability of bit error, 156–61
gray code, 205, 206, 287, 291
gray scale, 447
group (defined), 462
guardbands
for frequency division multiplexing, 370
for sampling, 397, 398, 400, 401, 404

H

Hamming code, 464–72
Hamming distance
defined, 476
relationship to Euclidean distance, 476
relationship of minimum Hamming distance to error correcting power, 478–80
relationship of minimum Hamming distance to error detecting power, 477–80
relationship to minimum Hamming distance to minimum weight, 479, 480
Huffman coding, 431–39, 441, 446, 450
hybrid FEC/ARQ techniques, 488, 489

I

identity element (defined), 462
I-frame (for MPEG compression), 453, 454
image compression. *See* compression; DCT; MPEG; JPEG; facsimile
impulse function
defined, 68
representing discrete components in a spectral density, 68, 69
used in sampling, 392, 393
in-band power. *See* bandwidth, effect on distortion
information
qualitative definition, 2, 75
quantitative definition, 422
See also entropy; Shannon's first theorem
instantaneous frequency (defined), 335
instantaneous phase (defined), 334
integrate and dump, 156
See also correlation receiver
intensity matrix, 448, 449
interleaving, 489
International Telecommunications Union (ITU), 446, 447
intersymbol interference (ISI)
caused by channel, 191, 193–96
caused by pulse shape, 89, 90, 92–94, 96, 101, 102, 111, 126, 147
defined, 89
used in combining error control coding and modulation, 458
inverse Fourier transform, 54, 55
inverse (defined), 462
ISI. *See* intersymbol interference

ITU. *See* International Telecommunications Union

J

jitter. *See* timing jitter
Joint Photographic Experts Group. *See* JPEG
joint probability density function. *See* random variable
joint probability distribution function. *See* random variable
JPEG, 447, 450, 453

L

least cost path, 501, 502, 506
See also Viterbi decoding
L'Hopital's rule, 34, 35
limiter (for FM and PM), 346, 349, 351
linear block code. *See* block code, linear
linearity (defined), 7, 8
look-up table. *See* EPROM
LZW compression, 442, 443

M

M-ary modulation and demodulation. *See* specific modulation techniques
magnitude spectrum
defined, 10
See also specific modulation techniques
matched filter, 149, 154, 155, 156, 159, 160, 207
See also correlation receiver
matrix representation (of error control coding), 482
See also decoding matrix; error vector; generator matrix; syndrome
maximal length PN (pseudo-random) code, 377
maximum frequency deviation
defined, 336
practical limitations, 337
relationship to bandwidth in PM and FM. *See* Carson's rule
maximum phase deviation
defined, 336
practical limitations, 336
maximum likelihood decoding, 498, 501
mean
defined for random process, 240
defined for random variable, 130, 237
ensemble average (same as conventional definition for random process), 240
time average, 252. *See also* ergodicity
See also random process; random variable; wide sense stationary
maximal length PN sequence. *See* pseudo-random sequence
media topology. *See* topology
modified Huffman, 446
modified READ, 446, 447
modulation
defined, 3
See also specific modulation techniques
modulation index (for AM), 322
modulation property (of Fourier transform), 66
modulo-2 mathematics, 461–63
motion compensation, 453
Motion Pictures Experts Group (MPEG). *See* MPEG
MPEG, 453, 454
multiplexing
defined, 362
See also code division multiplexing; frequency division multiplexing; time division multiplexing

N

"negative frequency" components
existence in two-sided spectra, 28
physical interpretation, 29, 30, 52
noise
additive white Gaussian, 134, 136, 144, 146, 208, 229, 239, 241, 242, 264, 266
filtered, 125, 134–36, 145
narrowband, 258, 311, 314, 322, 324, 326, 347, 350, 352
quadrature representation of narrowband noise, 258
power spectral density, 134
processing, 158
thermal, 128–30, 132–37
See also Gaussian probability density; signal-to-noise ratio; specific modulation techniques
noise margin (defined), 98
noncoherent receiver
defined, 236
See also specific modulation techniques

normal distribution. *See* Gaussian probability density
normalized energy
  defined, 60, 61
  spectral density, 62–65
Nyquist
  bandwidth, 196
  criterion, 194, 196
  pulse shaping, 105
  rate (sampling), 395

O

one-sided form of Fourier series. *See* Fourier series
one-sided spectrum
  relationship to two-sided spectrum, 28–30, 52
  See also Fourier series; Fourier transform
OOSK (on-off shift keying). *See* amplitude shift keying
orthogonality
  Fourier series components, 14, 40, 41, 61
  in bandpass signaling, 233, 234, 279, 282, 285, 286, 293–295, 297, 298, 300
  in multiplexing, 362, 363
  spreading codes, 370, 378
oscillator, 172, 173, 179, 180
  See also voltage controlled oscillator
overhead (of error control codes), 460, 461, 466, 469, 481, 487, 488, 489

P

packet
  components, 367, 368
  defined, 367
PAM. *See* pulse amplitude modulation
parity bits, 465, 466, 468–71, 481, 487, 488
Parseval's theorem
  energy, 61
  power, 41, 49
P-frame (for MPEG compression), 453, 454
perceptual quality (of color images), 450
phase. *See* instantaneous phase
phase deviation. *See* maximum phase deviation
phase locked loop (PLL), 172–89
  acquisition range, 188
  capture range, 188
  lock range, 177, 178
  loop bandwidth, 180, 181
  loop filter, 173, 175–78, 180
  theory of operation, 172–89
phase modulation (PM), 334–56
  bandwidth, 343
  comparison with other analog modulation techniques, 355–56
  defined, 334
  magnitude spectrum, 338–43
  maximum phase deviation. *See* maximum phase deviation
  receiver, 345–46
  signal-to-noise ratio, 349–51
  transmitter, 343–45
phase shift keying (PSK)
  average normalized power spectrum, 222–24
  bandwidth, 222
  bit error rate (coherent demodulation), 227–30
  bit error rate (noncoherent demodulation). *See* differential phase shift keying
  coherent receiver, 227–29
  comparison with other digital bandpass modulation techniques, 271–73
  decomposing into a baseband signal multiplied by a carrier, 222, 223
  defined, 221, 222
  DPSK. *See* differential phase shift keying
  DQPSK. *See* differential quaternary phase shift keying
  M-ary (coherently demodulated), 287–92
  M-ary (noncoherently demodulated), 292
  noncoherent receiver. *See* differential phase shift keying
  QPSK. *See* quaternary phase shift keying
phase spectrum (defined), 11. *See also* specific modulation techniques
phasor representation
  of Fourier series components, 19, 20
  of noise, 323–26, 347–49
pilot tone, 181, 182
pixel, 444–452
PLL. *See* phase locked loop
PM. *See* phase modulation
PN sequence. *See* pseudo-random sequence

polynomial representation (of error control coding), 482
See also generator polynomial
postdetection filter (FM, PM), 346
power
as a parameter in performance-vs.-cost tradeoffs, 4
See also average normalized power
pre-emphasis. *See* frequency modulation
prefix code, 427, 431, 432, 436
primary colors, 447
probabilistic mathematics. *See* stochastic mathematics
probability density function. *See* random variable
probability distribution function. *See* random variable
probability of bit error
compared to probability of symbol error, 204–06, 291, 292, 295
defined, 124
See also specific modulation techniques
probability of symbol error
compared to probability of bit error, 204–06, 291, 292, 295
defined, 201
See also specific modulation techniques
processing gain. *See* code division multiplexing, processing gain
pseudo-random (PN) sequence, 374–78
PSK. *See* phase shift keying
pulse amplitude modulation (PAM)
average normalized power spectrum. *See* specific pulse shape
bandwidth. *See* specific pulse shape
bit error rate (simple receiver), 145
bit error rate (correlation receiver), 156
correlation receiver, 149–72
defined, 88
M-ary, 200–06
pulse shaping. *See* raised cosine pulse; rectangular pulse; sinc pulse
raised cosine pulse. *See* raised cosine pulse
rectangular pulse. *See* rectangular pulse
simple receiver, 123–27, 144–49
sinc pulse. *See* sinc pulse
susceptibility to timing jitter, 96–101, 108–111
transmitter, 111–15

Q

Q function, 138–44
See also Gaussian probability density
Q table, 141
See also Gaussian probability density
QAM. *See* quadrature amplitude modulation
QPSK. *See* quaternary phase shift keying
quadrature amplitude modulation (QAM)
defined, 298
receiver, 299–301
symbol error rate, 301
quadrature representation
narrowband noise, 258
QAM signal, 300
QPSK signal. *See* quadrature phase shift keying
quadrature phase shift keying. *See* quaternary phase shift keying
quantization
companding, 355, 407
error, 402–04
noise, 403
nonuniform, 404–07
uniform, 400–04
quaternary phase shift keying (QPSK), 274–84
accuracy compared to DQPSK, 286
average normalized power spectral density, 276–78
bandwidth (relative to binary PSK), 276–78
bit error rate, 281–83
coherent demodulation, 279–84
defined, 274
decomposition into quadrature components, 275–79
DQPSK. *See* differential quaternary phase shift keying
noncoherent demodulation. *See* differential phase shift keying
symbol error rate, 283, 284

R

raised cosine pulse (for pulse amplitude modulation)
bandwidth, 104
comparison with other pulse shapes for PAM, 111
damped sinc in time domain, 101, 103–06

raised cosine pulse (for pulse amplitude modulation)(*continued*)
 damping factor, 101, 102, 103
 defined, 104
 magnitude spectrum, 104
 pulse shaping (for PAM), 101–11
 rolloff factor, 104
random process
 autocorrelation. *See* autocorrelation
 defined, 238
 ergodicity. *See* ergodicity
 mean. *See* mean
random variable
 conditional probability (defined), 132
 defined, 127
 expected value. *See* expected value
 mean. *See* mean
 joint probability density function, 132
 joint probability distribution function, 132
 probability density function, 129
 probability distribution, 127
 standard deviation. *See* standard deviation
 variance. *See* variance
Rayleigh probability density (or distribution), 258, 259, 262, 265, 266, 325, 326, 347
receiver
 coherent vs. noncoherent, 236
 defined, 2, 76
 See also specific modulation techniques
reconstruction (of sampled signal), 395
rectangular pulse (for pulse amplitude modulation)
 average normalized power spectral density, 83
 bandwidth, 85–88
 comparison with other pulse shapes for PAM, 111
 normalized energy spectral density, 79–82
 optimum pulse width, 86
Reed-Solomon code, 490
relative element address designate. *See* modified READ
Rician probability density (or distribution), 260, 261, 262, 265, 266
ring (topology), 363
rolloff factor. *See* raised cosine
run length coding, 445, 446, 447, 448, 450
run property (for a pseudo-random (PN) sequence), 375–77

S

sampling, 390–400
 aliasing, 392, 394, 395–400
 anti-alias filter, 398–400
 bandpass signal, 399, 400
 baseband signal, 392–99
 Nyquist rate, 395
sample function (defined), 238
 See also random process
Schwarz's inequality, 152, 166
sequential network (for convolutional encoding), 491, 493
Shannon, Claude, 1, 422, 424, 460
Shannon's first theorem, 424, 425, 439, 461
Shannon-Hartley theorem, 460, 461
shift register
 in convolutional encoder, 491
 in PAM modulator, 113, 114
 in PN generator, 377, 378
signal-to-noise ratio (SNR), 151
 AM systems, 313, 315, 322–27, 333
 defined, 145
 FM systems, 351–54
 PM systems, 349–51
 relationship to bit error rate and symbol error rate. *See* specific digital modulation technique
sinc function (defined), 32
sinc pulse (for pulse amplitude modulation)
 bandwidth, 90
 comparison with other pulse shapes for PAM, 111
 defined, 32
 intersymbol interference (avoided), 92
 magnitude spectrum, 90
 optimum pulse width, 93, 94
 pulse shaping (for PAM), 89–101
SNR. *See* signal-to-noise ratio
source
 defined, 2, 75
 encoding. *See* compression
spatial dependency
 defined, 451
 exploited for image compression. *See* discrete cosine transform; facsimile; JPEG; MPEG; run length coding
speed. *See* transmission speed
spread spectrum. *See* code division multiplexing
spreading code

despreading, 372–76
function, 371–76
generation, 377, 378
properties, 375–77
See also code division multiplexing; pseudo-random sequences
standard deviation, 137
star (topology), 363
state diagram (for representation of convolutional code), 491, 495–98
state table (for representation of convolutional code), 491–93, 495, 497
statistical TDM. *See* time division multiplexing
stochastic mathematics
defined, 127
See also random process; random variable
suppressed carrier. *See* amplitude modulation-double sideband-suppressed carrier
symbol
defined, 200
period, 201
See also specific M-ary modulation technique
symbol error rate. *See* probability of symbol error
synchronization, 172–89
synchronous TDM. *See* time division multiplexing
syndrome, 470, 471, 472
systematic code (defined), 468

T

T-1 system, 365–66
tapped-delay line, 197, 198
TDM. *See* time division multiplexing
temporal dependency
defined, 451
exploited for moving image compression, 451–54
terminal (defined), 179
thermal noise. *See* noise, thermal
threshold detector, 156, 300
threshold effect (in AM systems), 326
time-averaged parameters. *See* autocorrelation, time average; ergodicity; mean, time average
time division multiplexing (TDM)
defined, 362, 364
statistical TDM, 367, 368
synchronous TDM, 367
time slots, 364–67
time domain analysis
contrasted with frequency domain analysis, 6
defined, 6
time scaling property (of Fourier transform), 67
time shift property (of Fourier transform), 66, 154
time slots. *See* time division multiplexing
timing jitter
defined, 96
susceptibility of various PAM pulse shapes, 96–8, 101, 102, 104, 105, 111
timing recovery. *See* clock recovery
topology, 363
tradeoffs (performance vs. cost)
general discussion of, and parameters in, performance-vs.-cost tradeoffs, 4
See also specific compression techniques; specific error control coding techniques; specific modulation techniques
transform pair, 65
transmission speed
as a parameter involved in performance-vs.-cost tradeoffs, 4
See also bandwidth; specific modulation techniques
transmitter
defined, 2, 76
functions, 2, 76, 459
See also specific modulation techniques
tree
diagram (for data compression), 426–28, 430–32, 434, 436, 437, 441–43
topology, 364
trellis, 496–503, 505
trigonometric form of Fourier series. *See* Fourier series
two-sided form of Fourier series. *See* Fourier series
two-sided spectrum
relationship to one-sided spectrum, 28–30, 52
See also Fourier series; Fourier transform

U

uniform quantization. *See* quantization, uniform
unipolar signaling, 168, 169, 232, 254–56, 320
user (defined), 2, 76

V

variance
defined, 130, 131, 237
See also random process; random variable

VCO. *See* voltage controlled oscillator
vestigial sideband. *See* amplitude modulation-vestigial sideband
Viterbi decoding, 490, 497, 501–06
voltage controlled oscillator (VCO), 173–76, 179, 181, 186–89
  free running frequency, 188, 189
  theory of operation, 173
  use as FM or PM modulator, 343–45

W

weight (of a code)
  defined, 479
  relationship of minimum weight to error detecting and error correcting power, 479–81
white noise. *See* noise, additive white Gaussian
wide sense stationary (WSS)
  defined, 244
  properties, 244
  See also Wiener-Khintchine theorem
Wiener-Khintchine theorem
  implications, 245, 250–52
  proven, 245–50
  stated, 245
WSS. *See* wide sense stationary

Z

ZIP command, 442

The Fourier Transform and Its Properties

1. Fourier transform $$S(f) = \int_{-\infty}^{\infty} s(t)e^{-j2\pi ft}\,dt$$

2. Inverse Fourier transform $$s(t) = \int_{-\infty}^{\infty} S(f)e^{j2\pi ft}\,df$$

3. Convolution $$s_1(t)*s_2(t) \leftrightarrow S_1(f)S_2(f)$$

4. Time shift $$s(t - t_o) \leftrightarrow S(f)e^{-j2\pi ft_o}$$

5. Frequency shift $$S(f + f_o) \leftrightarrow s(t)e^{-j2\pi f_o t}$$

6. Linearity $$as_1(t) + bs_2(t) \leftrightarrow aS_1(f) + bS_2(f)$$

7. Modulation $$s(t)\cos(2\pi f_o t) \leftrightarrow \tfrac{1}{2}[S(f - f_o) + S(f + f_o)]$$

8. Time scaling $$s(at) \leftrightarrow \frac{1}{|a|}S\left(\frac{f}{a}\right)$$

9. Frequency scaling $$S(af) \leftrightarrow \frac{1}{|a|}s\left(\frac{t}{a}\right)$$

10. Differentiation $$\frac{d^n s(t)}{dt^n} \leftrightarrow (j2\pi f)^n S(f)$$

11. Integration $$\int_{-\infty}^{t} s(\lambda)d\lambda = (j2\pi f)^{-1}S(f) + \tfrac{1}{2}S(0)\delta(f)$$

12. Multiplication $$s_1(t)s_2(t) \leftrightarrow S_1(f)*S_2(f)$$

13. Duality $$S(t) \leftrightarrow s\{(-f)\}$$